Electronic Devices and Amplifier Circuits

with MATLAB® Computing

Second Edition

Steven T. Karris

Orchard Publications
www.orchardpublications.com

Electronic Devices and Amplifier Circuits with MATLAB® Computing, Second Edition

Direct all inquiries to Orchard Publications, info@orchardpublications.com

Library of Congress Cataloging-in-Publication Data

Library of Congress Control Number (LCCN) 2008934432

TXu1–254-969

ISBN-13: 978–1–934404–13–3

ISBN-10: 1–934404–13–6

Disclaimer

Preface

This text is an undergraduate level textbook presenting a thorough discussion of state–of–the art electronic devices. It is self–contained; it begins with an introduction to solid state semiconductor devices. The prerequisites for this text are first year calculus and physics, and a two–semester course in circuit analysis including the fundamental theorems and the Laplace transformation. No previous knowledge of MATLAB® is required; the material in Appendix A and the inexpensive MATLAB Student Version is all the reader needs to get going. Our discussions are based on a PC with Windows XP platforms but if you have another platform such as Macintosh, please refer to the appropriate sections of the MATLAB's User Guide which also contains instructions for installation. Additional information including purchasing may be obtained from The MathWorks, Inc., 3 Apple Hill Drive, Natick, MA 01760-2098. Phone: 508 647–7000, Fax: 508 647–7001, e–mail: info@mathwork.com and web site http://www.mathworks.com.This text can also be used without MATLAB.

This is our fourth electrical and computer engineering-based text with MATLAB applications. My associates, contributors, and I have a mission to produce substance and yet inexpensive texts for the average reader. Our first three texts* are very popular with students and working professionals seeking to enhance their knowledge and prepare for the professional engineering examination.

The author and contributors make no claim to originality of content or of treatment, but have taken care to present definitions, statements of physical laws, theorems, and problems.

Chapter 1 is an introduction to the nature of small signals used in electronic devices, amplifiers, definitions of decibels, bandwidth, poles and zeros, stability, transfer functions, and Bode plots. Chapter 2 is an introduction to solid state electronics beginning with simple explanations of electron and hole movement. This chapter provides a thorough discussion on the junction diode and its volt-ampere characteristics. In most cases, the non-linear characteristics are plotted with simple MATLAB scripts. The discussion concludes with diode applications, the Zener, Schottky, tunnel, and varactor diodes, and optoelectronics devices. Chapters 3 and 4 are devoted to bipolar junction transistors and FETs respectively, and many examples with detailed solutions are provided. Chapter 5 is a long chapter on op amps. Many op amp circuits are presented and their applications are well illustrated.

The highlight of this text is Chapter 6 on integrated devices used in logic circuits. The internal construction and operation of the TTL, NMOS, PMOS, CMOS, ECL, and the biCMOS families

* *These are Circuit Analysis I, ISBN 978–0–9709511–2–0, Circuit Analysis II, ISBN 978–0–9709511–5–1, and Signals and Systems, 978–1–934404–11–9.*

of those devices are fully discussed. Moreover, the interpretation of the most important parameters listed in the manufacturers data sheets are explained in detail. Chapter 7 is an introduction to pulse circuits and waveform generators. There, we discuss the 555 Timer, the astable, monostable, and bistable multivibrators, and the Schmitt trigger.

Chapter 8 discusses to the frequency characteristic of single-stage and cascade amplifiers, and Chapter 9 is devoted to tuned amplifiers. Sinusoidal oscillators are introduced in Chapter 10.

This is the second edition of this title, and includes several Simulink models. Also, two new appendices have been added. Appendix A, is an introduction to MATLAB. Appendix B is an introduction to Simulink, Appendix C is an introduction to Proportional-Integral-Derivative (PID) controllers, Appendix D describes uncompensated and compensated networks, and Appendix D discusses the substitution, reduction, and Miller's theorems.

The author wishes to express his gratitude to the staff of The MathWorks™, the developers of MATLAB® and Simulink® for the encouragement and unlimited support they have provided me with during the production of this text.

A companion to this text, *Digital Circuit Analysis and Design with Simulink® Modeling and Introduction to CPLDs and FPGAs*, ISBN 978–1–934404–05–8 is recommended as a companion. This text is devoted strictly on Boolean logic, combinational and sequential circuits as interconnected logic gates and flip–flops, an introduction to static and dynamic memory devices. and other related topics.

Although every effort was made to correct possible typographical errors and erroneous references to figures and tables, some may have been overlooked. Our experience is that the best proofreader is the reader. Accordingly, the author will appreciate it very much if any such errors are brought to his attention so that corrections can be made for the next edition. We will be grateful to readers who direct these to our attention at info@orchardpublications.com. Thank you.

Orchard Publications
Fremont, California 94538–4741
United States of America
www.orchardpublications.com
info@orchardpublications.com

Table of Contents

1 *Basic Electronic Concepts and Signals*

2 *Introduction to Semiconductor Electronics – Diodes*

MATLAB Computing
Pages 3–13, 3–39, 3–113

4 ***Field Effect Transistors and PNPN Devices***

MATLAB Computing
Pages 4–5, 4–10

5 *Operational Amplifiers*

6 *Integrated Circuits*

8 *Frequency Characteristics of Single–Stage and Cascaded Amplifiers*

9 *Tuned Amplifiers*

10 *Sinusoidal Oscillators*

Chapter 1

Basic Electronic Concepts and Signals

Electronics may be defined as the science and technology of electronic devices and systems. Electronic devices are primarily non–linear devices such as diodes and transistors and in general integrated circuits (ICs) in which small signals (voltages and currents) are applied to them. Of course, electronic systems may include resistors, capacitors and inductors as well. Because resistors, capacitors and inductors existed long ago before the advent of semiconductor diodes and transistors, these devices are thought of as electrical devices and the systems that consist of these devices are generally said to be electrical rather than electronic systems. As we know, with today's technology, ICs are becoming smaller and smaller and thus the modern IC technology is referred to as *microelectronics*.

1.1 Signals and Signal Classifications

A *signal* is any waveform that serves as a means of communication. It represents a fluctuating electric quantity, such as voltage, current, electric or magnetic field strength, sound, image, or any message transmitted or received in telegraphy, telephony, radio, television, or radar. Figure 1.1 shows a typical signal $f(t)$ that varies with time where $f(t)$ can be any physical quantity such as voltage, current, temperature, pressure, and so on.

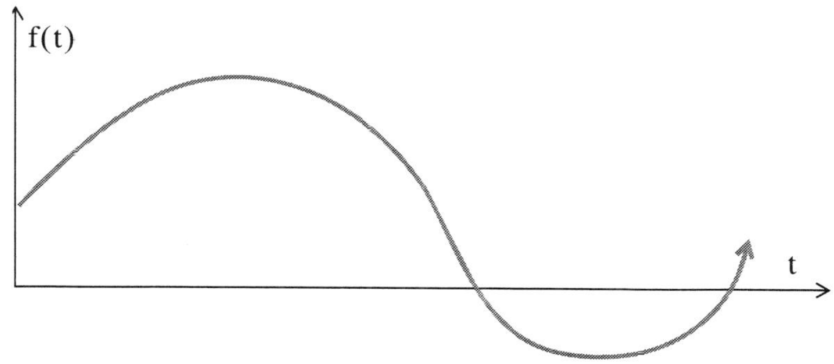

Figure 1.1. Typical waveform of a signal

We will now define the average value of a waveform.

Consider the waveform shown in Figure 1.2. The *average value* of $f(t)$ in the interval $a \le t \le b$ is

$$f(t)_{ave}\Big|_a^b = \frac{\text{Area}}{\text{Period}} = \frac{\int_a^b f(t)dt}{b-a} \qquad (1.1)$$

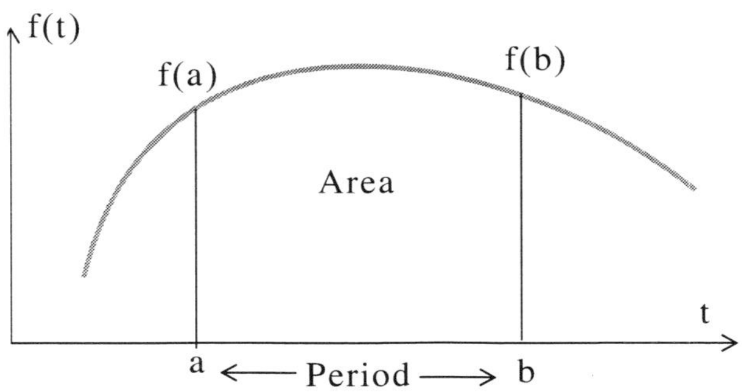

Figure 1.2. Defining the average value of a typical waveform

A *periodic time function* satisfies the expression

$$f(t) = f(t + nT) \tag{1.2}$$

for all time t and for all integers n. The constant T is the *period* and it is the smallest value of time which separates recurring values of the waveform.

An *alternating waveform* is any periodic time function whose average value over a period is zero. Of course, all sinusoids are alternating waveforms. Others are shown in Figure 1.3.

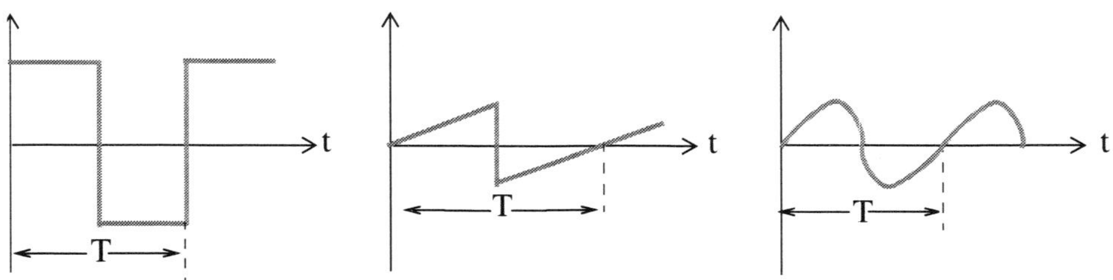

Figure 1.3. Examples of alternating waveforms

The effective (or RMS) value of a periodic current waveform $i(t)$ denoted as I_{eff} is the current that produces heat in a given resistor R at the same average rate as a direct (constant) current I_{dc}, that is,

$$\text{Average Power} = P_{ave} = RI_{eff}^2 = RI_{dc}^2 \tag{1.3}$$

Also, in a periodic current waveform $i(t)$ the instantaneous power $p(t)$ is

$$p(t) = Ri^2(t) \tag{1.4}$$

and

$$P_{ave} = \frac{1}{T}\int_0^T p(t)dt = \frac{1}{T}\int_0^T Ri^2 dt = \frac{R}{T}\int_0^T i^2 dt \tag{1.5}$$

Equating (1.3) with (1.5) we obtain

$$RI_{eff}^2 = \frac{R}{T}\int_0^T i^2 dt$$

or

$$I_{eff}^2 = \frac{1}{T}\int_0^T i^2 dt \qquad (1.6)$$

or

$$I_{RMS} = I_{eff} = \sqrt{\frac{1}{T}\int_0^T i^2 dt} = \sqrt{Ave(i^2)} \qquad (1.7)$$

where RMS stands for Root Mean Squared, that is, the effective value I_{eff} or I_{RMS} value of a current is computed as the *square root of the mean (average) of the square of the current.*

Warning 1: In general, $Ave(i^2) \neq (i_{ave})^2$. $Ave(i^2)$ implies that the current i must first be squared and the average of the squared value is to be computed. On the other hand, $(i_{ave})^2$ implies that the average value of the current must first be found and then the average must be squared.

Warning 2: In general, $P_{ave} \neq V_{ave} \cdot I_{ave}$. If $v(t) = V_p \cos\omega t$ and $i(t) = I_p \cos(\omega t + \theta)$ for example, $V_{ave} = 0$ and $I_{ave} = 0$, it follows that $P_{ave} = 0$ also. However,

$$P_{ave} = \frac{1}{T}\int_0^T pdt = \frac{1}{T}\int_0^T vidt \neq 0$$

In introductory electrical engineering books it is shown* that if the peak (maximum) value of a current of a sinusoidal waveform is I_p, then

$$I_{RMS} = I_p/\sqrt{2} = 0.707I_p \qquad (1.8)$$

and we must remember that (1.8) applies to sinusoidal values only.

1.2 Amplifiers

An *amplifier* is an electronic circuit which increases the magnitude of the input signal. The symbol of a typical amplifier is a triangle as shown in Figure 1.4.

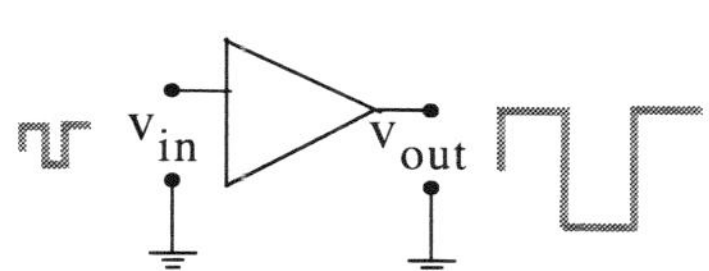

Figure 1.4. Symbol for electronic amplifier

* *Please refer to Circuit Analysis I with MATLAB Applications, ISBN 978–0–9709511–2–0.*

An electronic (or electric) circuit which produces an output that is smaller than the input is called an *attenuator*. A resistive voltage divider* is a typical attenuator.

An amplifier can be classified as a voltage, current or power amplifier. The *gain* of an amplifier is the ratio of the output to the input. Thus, for a voltage amplifier

$$\text{Voltage Gain} = \frac{\text{Output Voltage}}{\text{Input Voltage}}$$

or

$$G_v = V_{out}/V_{in}$$

The *current gain* G_i and *power gain* G_p are defined similarly.

1.3 Decibels

The ratio of any two values of the same quantity (power, voltage or current) can be expressed in **decibels (dB)**. For instance, we say that an amplifier has $10\ dB$ power gain, or a transmission line has a power loss of $7\ dB$ (or gain $-7\ dB$). If the gain (or loss) is $0\ dB$, the output is equal to the input. We should remember that a negative voltage or current gain G_v or G_i indicates that there is a $180°$ phase difference between the input and the output waveforms. For instance, if an op amp has a gain of -100 (dimensionless number), it means that the output is $180°$ out–of–phase with the input. For this reason we use absolute values of power, voltage and current when these are expressed in dB terms to avoid misinterpretation of gain or loss.

By definition,

$$dB = 10\log\left|\frac{P_{out}}{P_{in}}\right| \tag{1.9}$$

Therefore,

$10\ dB$ represents a power ratio of 10

$10n\ dB$ represents a power ratio of 10^n

It is useful to remember that

$20\ dB$ represents a power ratio of 100

$30\ dB$ represents a power ratio of $1,000$

$60\ dB$ represents a power ratio of $1,000,000$

* *Please refer to Circuit Analysis I with MATLAB Applications, ISBN 978–0–9709511–2–0.*

Also,

1 dB represents a power ratio of approximately 1.25

3 dB represents a power ratio of approximately 2

7 dB represents a power ratio of approximately 5

From these, we can estimate other values. For instance, $4\ dB = 3\ dB + 1\ dB$ which is equivalent to a power ratio of approximately $2 \times 1.25 = 2.5$. Likewise, $27\ dB = 20\ dB + 7\ dB$ and this is equivalent to a power ratio of approximately $100 \times 5 = 500$.

Since $y = \log x^2 = 2\log x$ and $P = V^2/Z = I^2 \cdot Z$, if we let $Z = 1$ the dB values for voltage and current ratios become

$$dB_v = 10\log\left|\frac{V_{out}}{V_{in}}\right|^2 = 20\log\left|\frac{V_{out}}{V_{in}}\right| \tag{1.10}$$

and

$$dB_i = 10\log\left|\frac{I_{out}}{I_{in}}\right|^2 = 20\log\left|\frac{I_{out}}{I_{in}}\right| \tag{1.11}$$

1.4 Bandwidth and Frequency Response

Like electric filters, amplifiers exhibit a band of frequencies over which the output remains nearly constant. Consider, for example, the magnitude of the output voltage $|V_{out}|$ of an electric or electronic circuit as a function of radian frequency ω as shown in Figure 1.5.

As shown in Figure 1.5, the *bandwidth* is $BW = \omega_2 - \omega_1$ where ω_1 and ω_2 are the *cutoff frequencies*. At these frequencies, $|V_{out}| = \sqrt{2}/2 = 0.707$ and these two points are known as the *3–dB down* or *half–power points*. They derive their name from the fact that since power $p = v^2/R = i^2 \cdot R$, for $R = 1$ and for v or $i = \sqrt{2}/2 = 0.707$ the power is $1/2$, that is, it is "halved".

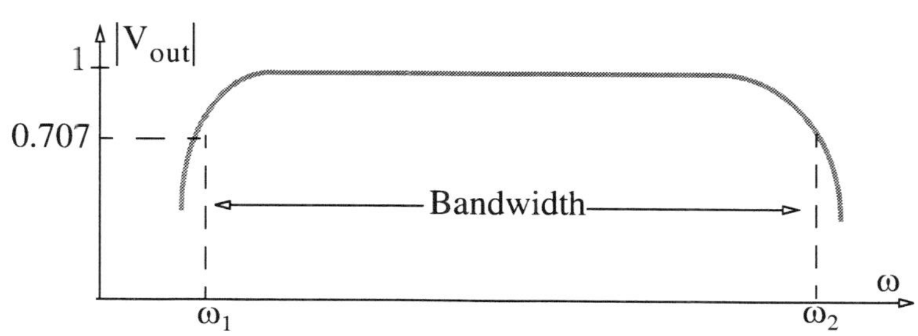

Figure 1.5. Definition of bandwidth

Alternately, we can define the bandwidth as the frequency band between half–power points. We recall from the characteristics of electric filters, the low–pass and high–pass filters have only one cutoff frequency whereas band–pass and band–elimination (band–stop) filters have two. We may think that low–pass and high–pass filters have also two cutoff frequencies where in the case of the low–pass filter the second cutoff frequency is at $\omega = 0$ while in a high–pass filter it is at $\omega = \infty$.

We also recall also that the output of circuit is dependent upon the frequency when the input is a sinusoidal voltage. In general form, the output voltage is expressed as

$$V_{out}(\omega) = |V_{out}(\omega)|e^{j\varphi(\omega)} \tag{1.12}$$

where $|V_{out}(\omega)|$ is known as the *magnitude response* and $e^{j\varphi(\omega)}$ is known as the *phase response*. These two responses together constitute the *frequency response* of a circuit.

Example 1.1

Derive and sketch the magnitude and phase responses of the RC low–pass filter shown in Figure 1.6.

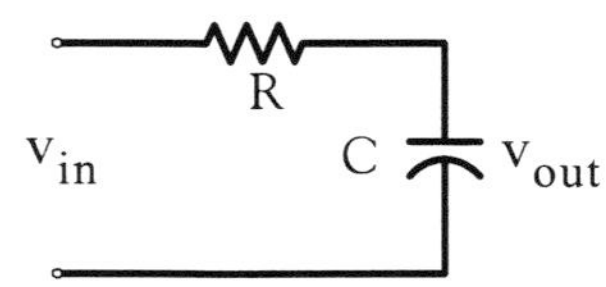

Figure 1.6. RC low–pass filter

Solution:

By application of the voltage division expression we obtain

$$V_{out} = \frac{1/j\omega C}{R + 1/j\omega C}V_{in}$$

$$\frac{V_{out}}{V_{in}} = \frac{1}{1 + j\omega RC} \tag{1.13}$$

or

$$\frac{V_{out}}{V_{in}} = \frac{1}{\sqrt{1+\omega^2R^2C^2}\angle\tan^{-1}(\omega RC)} = \frac{1}{\sqrt{1+\omega^2R^2C^2}}\angle-\tan^{-1}(\omega RC) \tag{1.14}$$

and thus the magnitude is

$$\left|\frac{V_{out}}{V_{in}}\right| = \frac{1}{\sqrt{1+\omega^2R^2C^2}} \tag{1.15}$$

and the phase angle (sometimes called argument and abbreviated as arg) is

$$\varphi = \arg\left(\frac{V_{out}}{V_{in}}\right) = -\tan^{-1}(\omega RC) \qquad (1.16)$$

To sketch the magnitude, we let ω assume the values 0, $1/RC$, and ∞. Then,

as $\omega \to 0$, $|V_{out}/V_{in}| \cong 1$

for $\omega = 1/RC$, $|V_{out}/V_{in}| = 1/(\sqrt{2}) = 0.707$

and as $\omega \to \infty$, $|V_{out}/V_{in}| \cong 0$

To sketch the phase response, we use (1.16). Then,

as $\omega \to -\infty$, $\phi \cong -\tan^{-1}(-\infty) \cong 90^{\circ}$

as $\omega \to 0$, $\phi \cong -\tan^{-1}0 \cong 0$

for $\omega = 1/RC$, $\phi \cong -\tan^{-1}1 \cong -45^{\circ}$

as $\omega \to \infty$, $\phi \cong -\tan^{-1}(\infty) \cong -90^{\circ}$

The magnitude and phase responses of the RC low–pass filter are shown in Figure 1.7.

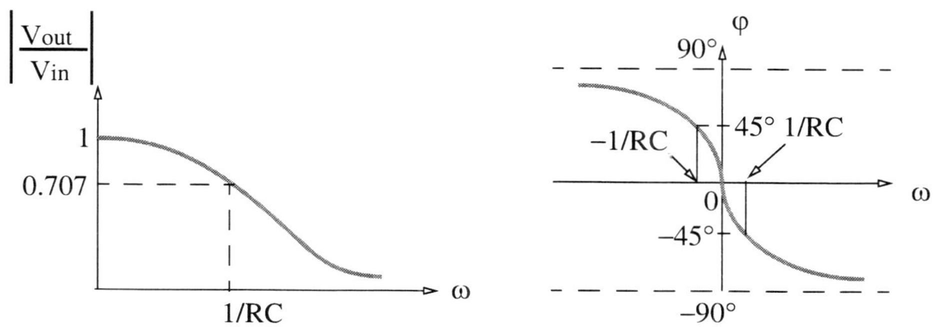

Figure 1.7. Magnitude and phase responses for the low–pass filter of Figure 1.6

We can use MATLAB* to plot the magnitude and phase angle for this low-pass filter using relation (1.13) expressed in MATLAB script below and plot the magnitude and phase angle in semilog scale.

```
w=1:10:100000; RC=1; xf=1./(1+j.*w.*RC);...
mag=20.*log10(abs(xf)); phase=angle(xf).*180./pi;...
subplot(121); semilogx(w,mag); subplot(122); semilogx(w,phase)
```

* *For an introduction to MATLAB, please refer to Appendix A*

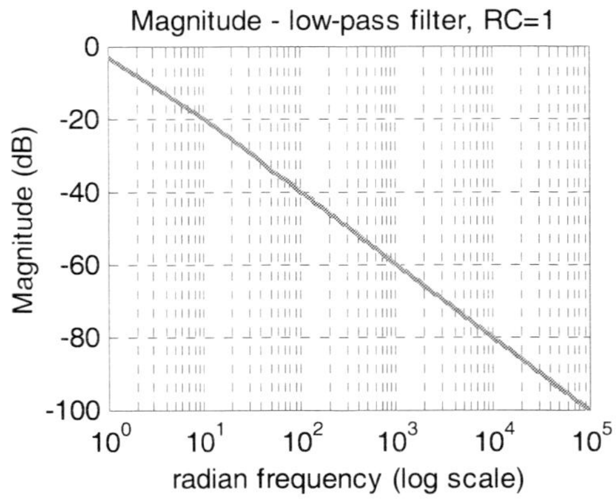

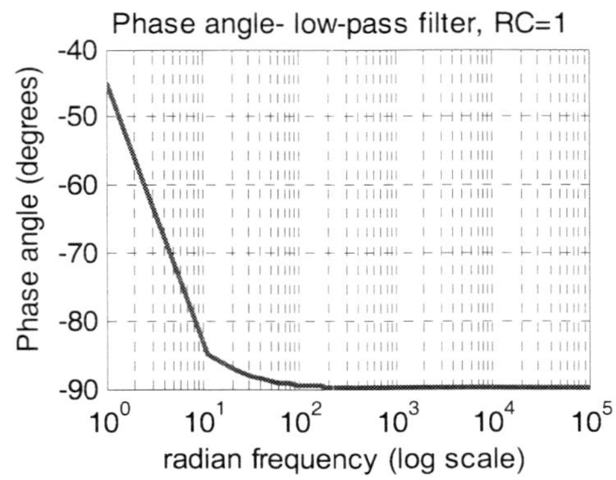

1.5 Bode Plots

The magnitude and phase responses of a circuit are often shown with asymptotic lines as approximations. Consider two frequency intervals expressed as

$$u_2 - u_1 = \log_{10}\omega_2 - \log_{10}\omega_1 = \log\left(\frac{\omega_2}{\omega_1}\right) \tag{1.17}$$

then two common frequency intervals are (1) the *octave* for which $\omega_2 = 2\omega_1$ and (2) the *decade* for which $\omega_2 = 10\omega_1$.

Now, let us consider a circuit whose gain is given as

$$G(\omega)_v = C/\omega^k \tag{1.18}$$

where C is a constant and k is a non–zero positive integer. Taking the common log of (1.18) and multiplying by 20 we obtain

$$20\log_{10}\{G(\omega)_v\} = 20\log_{10}C - 20k\log_{10}\omega$$

$$\{G(\omega)_v\}_{dB} = 20\log_{10}C - 20k\log_{10}\omega \tag{1.19}$$

We observe that (1.19) represents an equation of a straight line with abscissa $\log_{10}\omega$, slope of $-20k$, and $\{G(\omega)_v\}$ intercept at $20\log_{10}10C = \text{constant}$. We can choose the slope to be either $-20k\ dB/decade$ or $-6k\ dB/octave$. Thus, if $k = 1$, the slope becomes $-20\ dB/decade$ as illustrated in the plot of Figure 1.8.

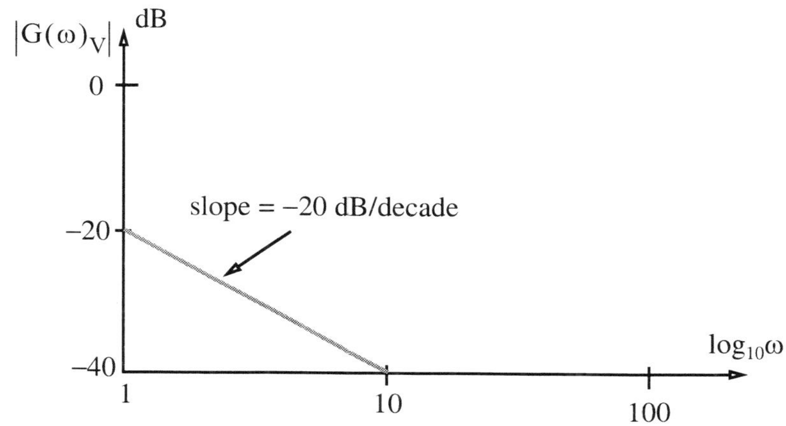

Figure 1.8. Plot of relation (1.19) for $k = 1$

Then, any line parallel to this slope will represent a *drop* of 20 dB/decade. We observe also that if the exponent k in (1.18) is changed to 2, the slope will be −40 dB/decade.

We can now approximate the magnitude and phase responses of the low–pass filter of Example 1.1 with asymptotic lines as shown in Figure 1.9.

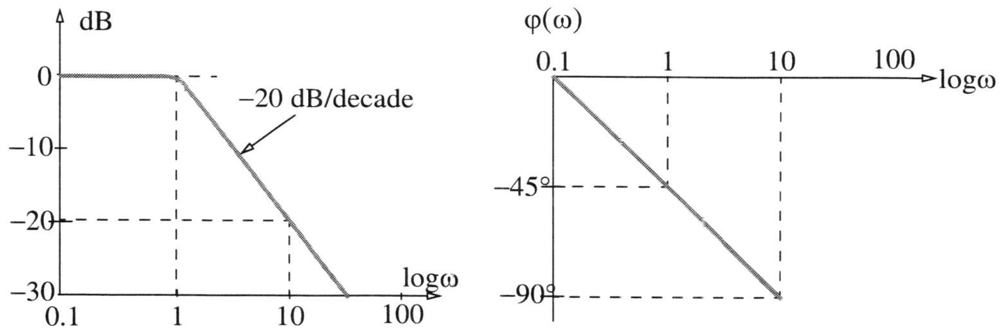

Figure 1.9. Magnitude and phase responses for the low–pass filter of Figure 1.6.

1.6 Transfer Function

Let us consider the continuous–time,* linear,† and time–invariant‡ system of Figure 1.10.

* *A continuous–time signal is a function that is defined over a continuous range of time.*

† *A linear system consists of linear devices and may include independent and dependent voltage and current sources. For details, refer to Circuit Analysis I with MATLAB Applications, ISBN 978–0–9709511–2–0.*

‡ *A time–invariant system is a linear system in which the parameters do not vary with time.*

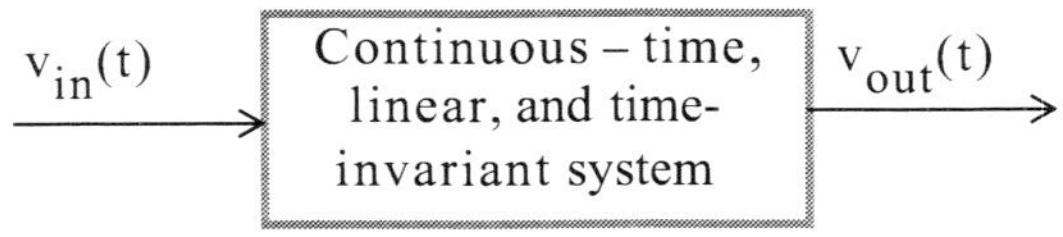

Figure 1.10. Input–output block diagram for linear, time–invariant continuous–time system

We will assume that initially no energy is stored in the system. The input–output relationship can be described by the differential equation of

$$b_m \frac{d^m}{dt^m} v_{out}(t) + b_{m-1} \frac{d^{m-1}}{dt^{m-1}} v_{out}(t) + b_{m-2} \frac{d^{m-2}}{dt^{m-2}} v_{out}(t) + \ldots + b_0 v_{out}(t) =$$
$$a_n \frac{d^n}{dt^n} v_{in}(t) + a_{n-1} \frac{d^{n-1}}{dt^{n-1}} v_{in}(t) + a_{n-2} \frac{d^{n-2}}{dt^{n-2}} v_{in}(t) + \ldots + a_0 v_{in}(t) \tag{1.20}$$

For practically all electric networks, $m \geq n$ and the integer m denotes the order of the system.

Taking the Laplace transform* of both sides of (1.20) we obtain

$$(b_m s^m + b_{m-1} s^{m-1} + b_{m-2} s^{m-2} + \ldots + b_0) V_{out}(s) =$$
$$(a_n s^n + a_{n-1} s^{n-1} + a_{n-2} s^{n-2} + \ldots + a_0) V_{in}(s)$$

Solving for $V_{out}(s)$ we obtain

$$V_{out}(s) = \frac{(a_n s^n + a_{n-1} s^{n-1} + a_{n-2} s^{n-2} + \ldots + a_0)}{(b_m s^m + b_{m-1} s^{m-1} + b_{m-2} s^{m-2} + \ldots + b_0)} V_{in}(s) = \frac{N(s)}{D(s)} V_{in}(s)$$

where $N(s)$ and $D(s)$ are the numerator and denominator polynomials respectively.

The *transfer function* $G(s)$ is defined as

$$G(s) = \frac{V_{out}(s)}{V_{in}(s)} = \frac{N(s)}{D(s)} \tag{1.21}$$

Example 1.2

Derive the transfer function $G(s)$ of the network in Figure 1.11.

* *The Laplace transform and its applications to electric circuit analysis is discussed in detail in Circuit Analysis II, ISBN 978–0–9709511–5–1.*

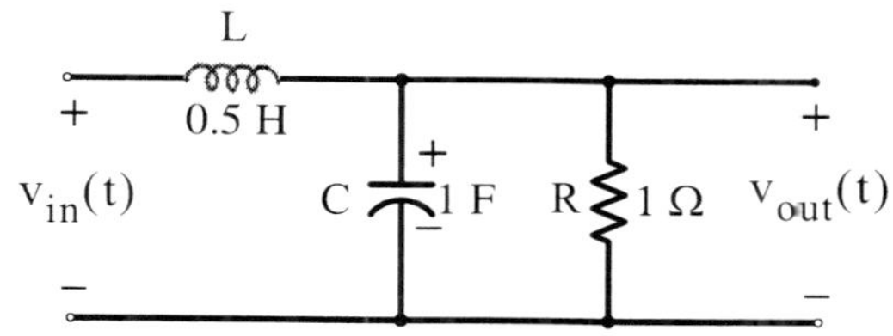

Figure 1.11. Network for Excmple 1.2

Solution:

The given circuit is in the t – domain.* The transfer function $G(s)$ exists only in the s – domain† and thus we redraw the circuit in the s – domain as shown in Figure 1.12.

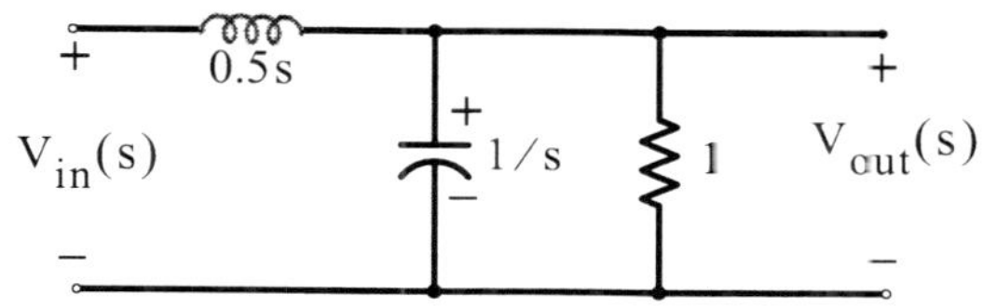

Figure 1.12. Circuit of Example 1.2 in the s – domain

For relatively simple circuits such as that of Figure 1.12, we can readily obtain the transfer function with application of the voltage division expression. Thus, parallel combination of the capacitor and resistor yields

$$\frac{1/s \times 1}{1/s + 1} = \frac{1}{s+1}$$

and by application of the voltage division expression

$$V_{out}(s) = \frac{1/(s+1)}{0.5s + 1/(s+1)} V_{in}(s)$$

or

$$G(s) = \frac{V_{out}(s)}{V_{in}(s)} = \frac{2}{s^2 + s + 2}$$

1.7 Poles and Zeros

Let

$$F(s) = \frac{N(s)}{D(s)} \tag{1.22}$$

* *For brevity, we will denote the time domain as t – domain*

† *Henceforth, the complex frequency, i.e.,* $s = \sigma + j\omega$*, will be referred to as the* s – domain.

where $N(s)$ and $D(s)$ are polynomials and thus (1.22) can be expressed as

$$F(s) = \frac{N(s)}{D(s)} = \frac{b_m s^m + b_{m-1} s^{m-1} + b_{m-2} s^{m-2} + \ldots + b_1 s + b_0}{a_n s^n + a_{n-1} s^{n-1} + a_{n-2} s^{n-2} + \ldots + a_1 s + a_0} \qquad (1.23)$$

The coefficients a_k and b_k for $k = 0, 1, 2, \ldots, n$ are real numbers and, for the present discussion, we have assumed that the highest power of $N(s)$ is less than the highest power of $D(s)$, i.e., $m < n$. In this case, $F(s)$ is a *proper rational function*. If $m \ge n$, $F(s)$ is an *improper rational function.*

It is very convenient to make the coefficient a_n of s^n in (12.2) unity; to do this, we rewrite it as

$$F(s) = \frac{N(s)}{D(s)} = \frac{\frac{1}{a_n}(b_m s^m + b_{m-1} s^{m-1} + b_{m-2} s^{m-2} + \ldots + b_1 s + b_0)}{s^n + \frac{a_{n-1}}{a_n} s^{n-1} + \frac{a_{n-2}}{a_n} s^{n-2} + \ldots + \frac{a_1}{a_n} s + \frac{a_0}{a_n}} \qquad (1.24)$$

The roots of the numerator are called the *zeros* of $F(s)$, and are found by letting $N(s) = 0$ in (1.24). The roots of the denominator are called the *poles** of $F(s)$ and are found by letting $D(s) = 0$. However, in most engineering applications we are interested in the nature of the poles.

1.8 Stability

In general, a system is said to be *stable* if a finite input produces a finite output. We can predict the stability of a system from its impulse response† $h(t)$. In terms of the impulse response,

1. A system is stable if the impulse response $h(t)$ goes to zero after some time as shown in Figure 1.13.
2. A system is *marginally stable* if the impulse response $h(t)$ reaches a certain non–zero value but never goes to zero as shown in Figure 1.14.
3. A system is *unstable* if the impulse response $h(t)$ reaches infinity after a certain time as shown in Figure 1.15.

* *The zeros and poles can be distinct (different from one another), complex conjugates, repeated, of a combination of these. For details refer to Circuit Analysis II with MATLAB Applications, ISBN 978–0–9709511–5–1.*

† *For a detailed discussion on the impulse response, please refer to Signals and Systems with MATLAB Computing and Simulink Modeling, Fourth Edition, ISBN 978–1–934404–11–9.*

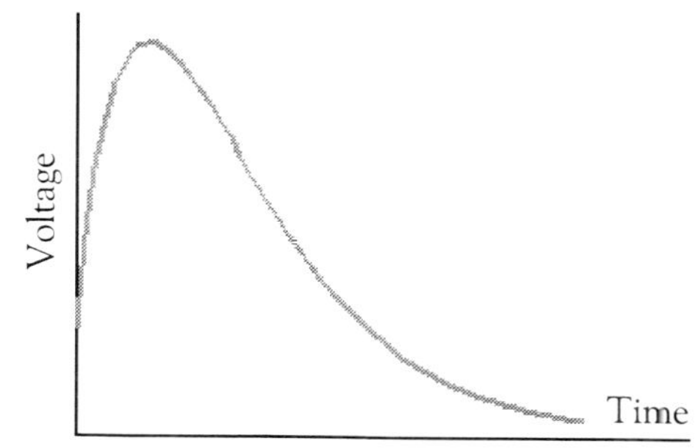

Figure 1.13. Characteristics of a stable system

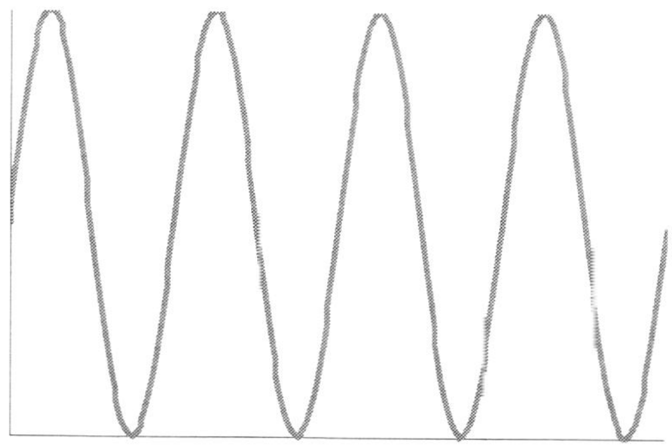

Figure 1.14. Characteristics of a marginally stable system

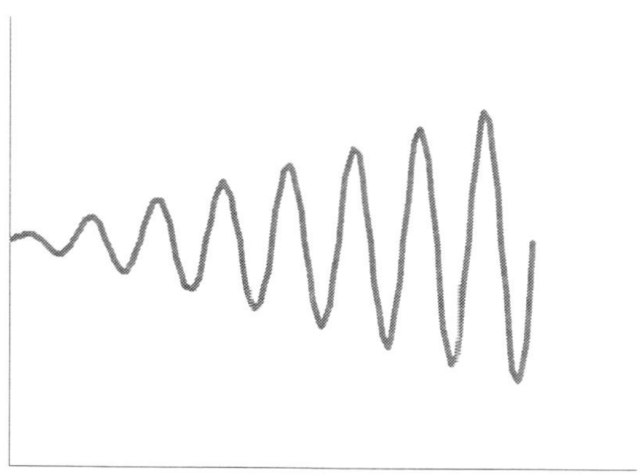

Figure 1.15. Characteristics of an unstable system

We can plot the poles and zeros of a transfer function $G(s)$ on the complex frequency plane of the complex variable $s = \sigma + j\omega$. A system is stable only when all poles lie on the left–hand half–plane. It is marginally stable when one or more poles lie on the $j\omega$ axis, and unstable when one or more poles lie on the right–hand half–plane. However, the location of the zeros in the $s-\text{plane}$ is immaterial, that is, the nature of the zeros do not determine the stability of the system.

We can use the MATLAB* function **bode(sys)** to draw the Bode plot of a Linear Time Invariant (LTI) System where **sys = tf(num,den)** creates a continuous–time transfer function **sys** with numerator **num** and denominator **den**, and **tf** creates a transfer function. With this function, the frequency range and number of points are chosen automatically. The function **bode(sys,{wmin,wmax})** draws the Bode plot for frequencies between **wmin** and **wmax** (in radians/second) and the function **bode(sys,w)** uses the user–supplied vector **w** of frequencies, in radians/second, at which the Bode response is to be evaluated. To generate logarithmically spaced frequency vectors, we use the command **logspace(first_exponent,last_exponent, number_of_values)**. For example, to generate plots for 100 logarithmically evenly spaced points for the frequency interval $10^{-1} \le \omega \le 10^{2}$ r/s, we use the statement logspace(–1,2,100).

The **bode(sys,w)** function displays both magnitude and phase. If we want to display the magnitude only, we can use the **bodemag(sys,w)** function.

MATLAB requires that we express the numerator and denominator of $G(s)$ as polynomials of s in descending powers.

Example 1.3

The transfer function of a system is

$$G(s) = \frac{3(s-1)(s^2+2s+5)}{(s+2)(s^2+6s+25)}$$

a. is this system stable?

b. use the MATLAB **bode(sys,w)** function to plot the magnitude of this transfer function.

Solution:

a. Let us use the MATLAB solve('eqn1','eqn2',...,'eqnN') function to find the roots of the quadratic factors.

```
syms s; equ1=solve('s^2+2*s+5–0'), equ2=solve('s^2+6*s+25–0')

equ1 =
[-1+2*i]   [-1-2*i]

equ2 =
[-3+4*i]   [-3-4*i]
```

The zeros and poles of $G(s)$ are shown in Figure 1.16.

* *An introduction to MATLAB is included as Appendix A.*

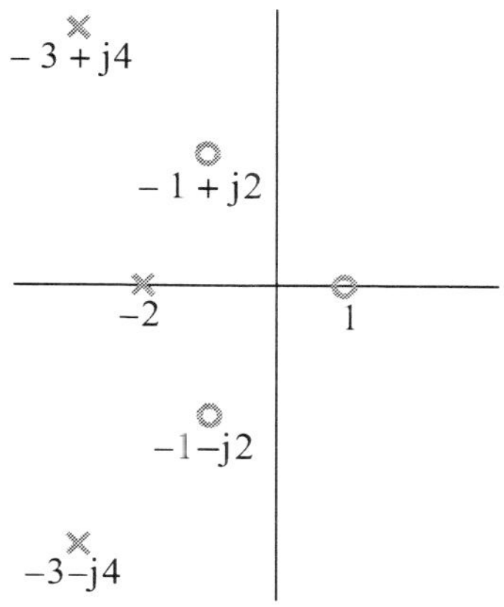

Figure 1.16. Poles and zeros of the transfer function of Example 1.3

From Figure 1.16 we observe that all poles, denoted as ×, lie on the left–hand half–plane and thus the system is stable. The location of the zeros, denoted as o, is immaterial.

b. We use the MATLAB **expand(s)** symbolic function to express the numerator and denominator of $G(s)$ in polynomial form

```
syms s; n=expand((s-1)*(s^2+2*s+5)), d=expand((s+2)*(s^2+6*s+25))
n =
s^3+s^2+3*s-5

d =
s^3+8*s^2+37*s+50
```

and thus

$$G(s) = \frac{3(s^3+s^2+3s-5)}{(s^3+8s^2+37s+50)}$$

For this example we are interested in the magnitude only so we will use the script

```
num=3*[1 1 3 -5]; den=[1 8 37 50]; sys=tf(num,den);...
w=logspace(0,2,100); bodemag(sys,w); grid
```

The magnitude plot is shown in Figure 1.17.

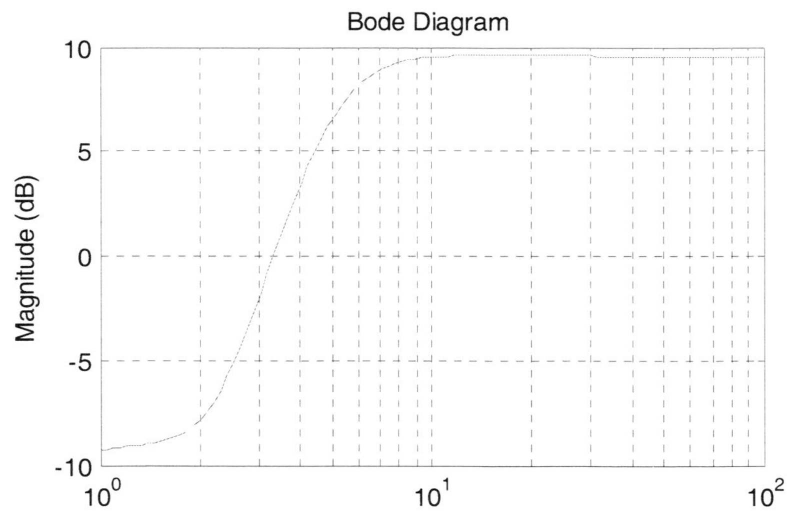

Figure 1.17. Bode plot for Example 1.3

Example 1.4

It is known that a voltage amplifier has a frequency response of a low–pass filter, a DC gain of 80 dB, attenuation of –20 dB per decade, and the 3 dB cutoff frequency occurs at 10 KHz. Determine the gain (in dB) at the frequencies 1 KHz, 10 KHz, 100KHz, 1 MHz, 10 MHz, and 100 MHz.

Solution:

Using the given data we construct the asymptotic magnitude response shown in Figure 1.18 from which we obtain the data shown on the table below.

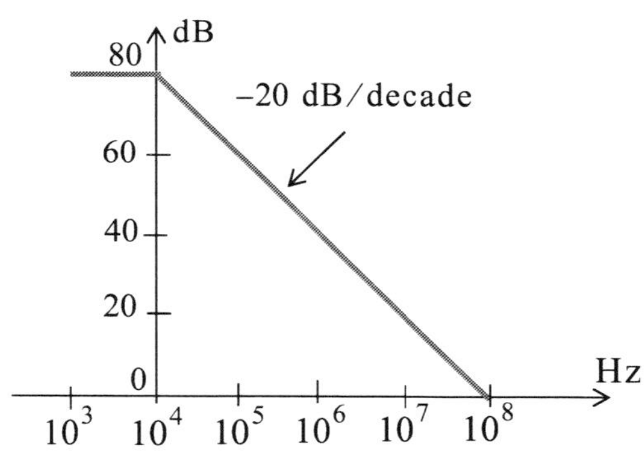

Figure 1.18. Asymptotic magnitude response for Example 1.4

Frequency	1 KHz	10 KHz	100 KHz	1 MHz	10 MHz	100 MHz
Gain (dB)	80	77	60	40	20	0

1.9 Voltage Amplifier Equivalent Circuit

Amplifiers are often represented by equivalent circuits* also known as *circuit models*. The equivalent circuit of a voltage amplifier is shown in Figure 1.19.

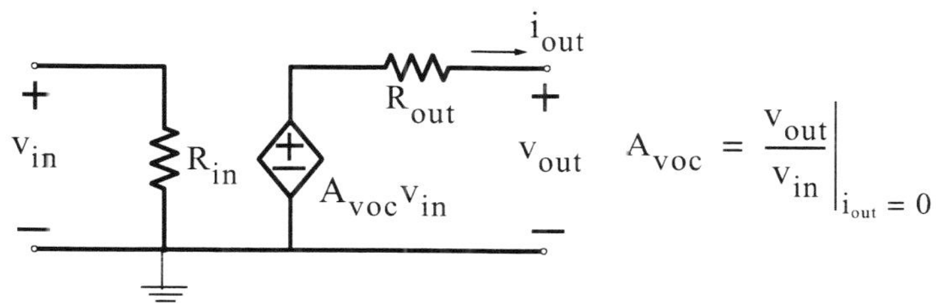

Figure 1.19. Circuit model for voltage amplifier where A_{voc} *denotes the open circuit voltage gain*

The ideal characteristics for the circuit of Figure 1.19 are $R_{in} \to \infty$ and $R_{out} \to 0$.

Example 1.5

For the voltage amplifier of Figure 1.20, find the overall voltage gain $A_v = v_{load}/v_S$. Then, use MATLAB to plot the magnitude of A_v for the range $10^3 \le \omega \le 10^8$. From the plot, estimate the 3 dB cutoff frequency.

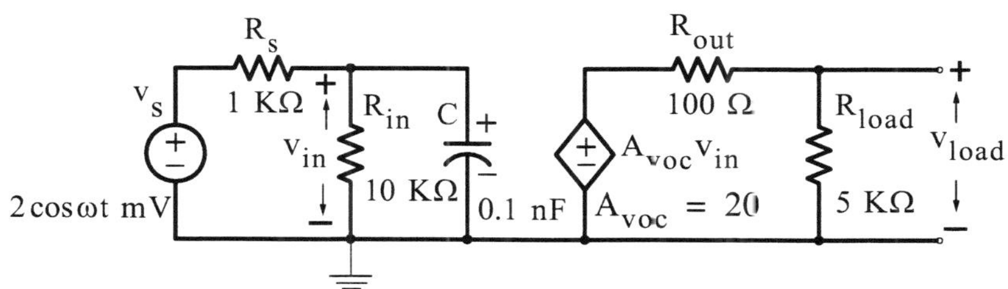

Figure 1.20. Amplifier circuit for Example 1.5

Solution:

The $s-\text{domain}$ equivalent circuit is shown in Figure 1.21.

* *Readers who have a copy of Circuit Analysis I, ISBN 978–0–9709511–2–0, are encouraged to review Chapter 4 on equivalent circuits of operational amplifiers. Refer also to Chapter 5, this text.*

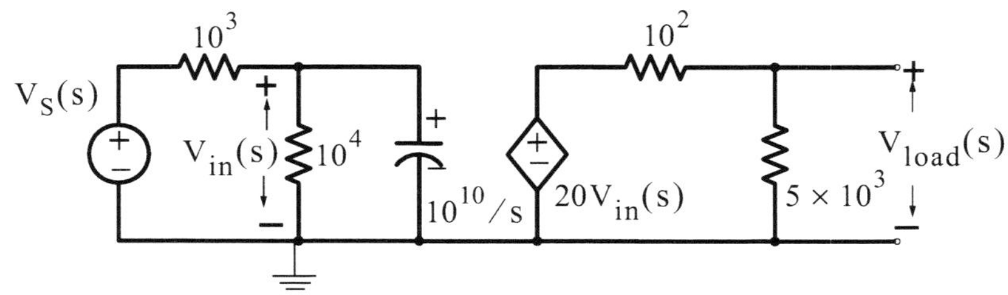

Figure 1.21. The s – domain *circuit of Figure 1.20*

The parallel combination of the 10^4 resistor and $10^{10}/s$ capacitor yields

$$Z(s) = 10^4 \parallel 10^{10}/s = \frac{10^{14}/s}{10^4 + 10^{10}/s} = \frac{10^{14}}{10^4 s + 10^{10}}$$

and by the voltage division expression

$$V_{in}(s) = \frac{10^{14}/(10^4 s + 10^{10})}{10^3 + 10^{14}/(10^4 s + 10^{10})} V_S(s) = \frac{10^{14}}{10^7 s + 1.1 \times 10^{14}} V_S(s) \tag{1.25}$$

Also,

$$V_{load}(s) = \frac{5 \times 10^3}{10^2 + 5 \times 10^3} 20V_{in}(s) = \frac{10^5}{5.1 \times 10^3} V_{in}(s) = 19.61 V_{in}(s) \tag{1.26}$$

By substitution of (1.25) into (1.26) we obtain

$$V_{load}(s) = \frac{19.61 \times 10^{14}}{10^7 s + 1.1 \times 10^{14}} V_S(s)$$

$$G_v(s) = \frac{V_{load}(s)}{V_S(s)} = \frac{19.61 \times 10^{14}}{10^7 s + 1.1 \times 10^{14}} \tag{1.27}$$

and with MATLAB

```
num=[0  19.61*10^14]; den=[10^7  1.1*10^14]; sys=tf(num,den);...
w=logspace(3,8,1000); bodemag(sys,w); grid
```

The plot is shown in Figure 1.22 and we observe that the cutoff frequency occurs at about 22 dB where

$$f_C \approx 10^7/2\pi \approx 1.59 \text{ MHz}$$

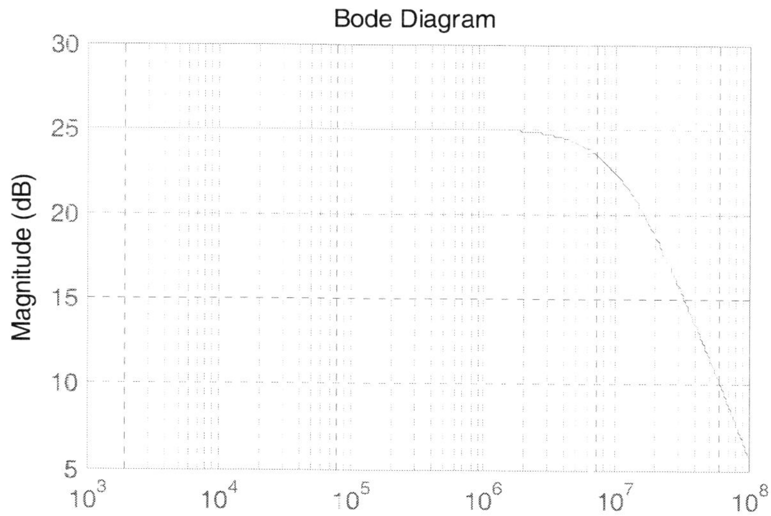

Figure 1.22. Bode plot for the voltage amplifier of Example 1.5

1.10 Current Amplifier Equivalent Circuit

The equivalent circuit of a current amplifier is shown in Figure 1.23.

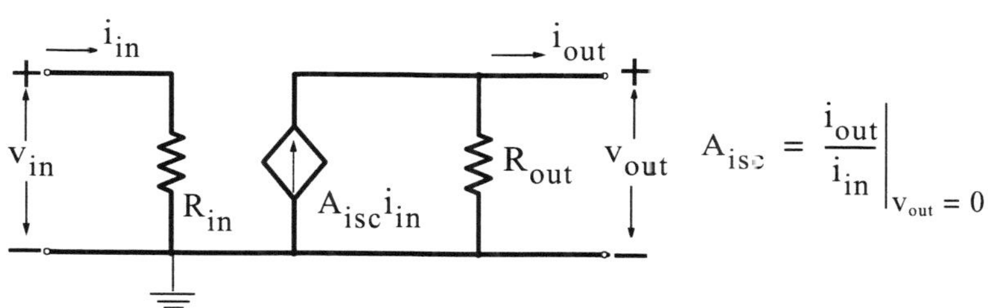

Figure 1.23. Circuit model for current amplifier where A_{isc} denotes the short circuit current gain

The ideal characteristics for the circuit of Figure 1.23 are $R_{in} \to 0$ and $R_{out} \to \infty$.

Example 1.6

For the current amplifier of Figure 1.24, derive an expression for the overall current gain $A_i = i_{load}/i_s$.

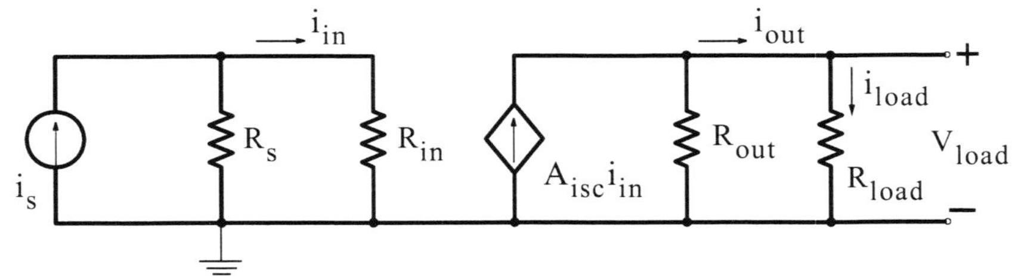

Figure 1.24. Current amplifier for Example 1.6

Solution:

Using the current division expression we obtain

$$i_{in} = \frac{R_s}{R_s + R_{in}} i_s \tag{1.28}$$

Also,

$$i_{load} = \frac{R_{out}}{R_{out} + R_{load}} A_{isc} i_{in} \tag{1.29}$$

Substitution of (1.28) into (1.29) yields

$$i_{load} = \frac{R_{out}}{R_{out} + R_{load}} A_{isc} \frac{R_S}{R_S + R_{in}} i_s \tag{1.30}$$

or

$$A_i = \frac{i_{load}}{i_s} = \frac{R_{out}}{R_{out} + R_{load}} \frac{R_S}{R_S + R_{in}} A_{isc}$$

In Sections 1.9 and 1.10 we presented the voltage and current amplifier equivalent circuits also known as circuit models. Two more circuit models are the *transresistance* and *transconductance* equivalent circuits and there are introduced in Exercises 1.4 and 1.5 respectively.

1.11 Summary

- A signal is any waveform that serves as a means of communication. It represents a fluctuating electric quantity, such as voltage, current, electric or magnetic field strength, sound, image, or any message transmitted or received in telegraphy, telephony, radio, television, or radar.
- The average value of a waveform $f(t)$ in the interval $a \le t \le b$ is defined as

$$f(t)_{ave}\Big|_a^b = \frac{\text{Area}}{\text{Period}} = \frac{\int_a^b f(t)dt}{b-a}$$

- A periodic time function satisfies the expression

$$f(t) = f(t+nT)$$

for all time t and for all integers n. The constant T is the period and it is the smallest value of time which separates recurring values of the waveform.

- An alternating waveform is any periodic time function whose average value over a period is zero.
- The effective (or RMS) value of a periodic current waveform $i(t)$ denoted as I_{eff} is the current that produces heat in a given resistor R at the same average rate as a direct (constant) current I_{dc} and it is found from the expression

$$I_{RMS} = I_{eff} = \sqrt{\frac{1}{T}\int_0^T i^2 dt} = \sqrt{Ave(i^2)}$$

where RMS stands for Root Mean Squared, that is, the effective value I_{eff} or I_{RMS} value of a current is computed as the square root of the mean (average) of the square of the current.

- If the peak (maximum) value of a current of a sinusoidal waveform is I_p, then

$$I_{RMS} = I_p/(\sqrt{2}) = 0.707 I_p$$

- An amplifier is an electronic circuit which increases the magnitude of the input signal.
- An electronic (or electric) circuit which produces an output that is smaller than the input is called an attenuator. A resistive voltage divider is a typical attenuator.
- An amplifier can be classified as a voltage, current or power amplifier. The gain of an amplifier is the ratio of the output to the input. Thus, for a voltage amplifier

$$\text{Voltage Gain} = \frac{\text{Output Voltage}}{\text{Input Voltage}}$$

or

$$G_v = V_{out}/V_{in}$$

The current gain G_i and power gain G_p are defined similarly.

- The ratio of any two values of the same quantity (power, voltage or current) can be expressed in decibels (dB). By definition,

$$dB = 10\log|P_{out}/P_{in}|$$

The dB values for voltage and current ratios are

$$dB_v = 20\log|V_{out}/V_{in}|$$

$$dB_i = 20\log|I_{out}/I_{in}|$$

- The bandwidth is $BW = \omega_2 - \omega_1$ where ω_1 and ω_2 are the cutoff frequencies. At these frequencies, $|V_{out}| = \sqrt{2}/2 = 0.707$ and these two points are known as the 3–dB down or half–power points.
- The low–pass and high–pass filters have only one cutoff frequency whereas band–pass and band–stop filters have two. We may think that low–pass and high–pass filters have also two cutoff frequencies where in the case of the low–pass filter the second cutoff frequency is at $\omega = 0$ while in a high–pass filter it is at $\omega = \infty$.
- We also recall also that the output of circuit is dependent upon the frequency when the input is a sinusoidal voltage. In general form, the output voltage is expressed as

$$V_{out}(\omega) = |V_{out}(\omega)|e^{j\varphi(\omega)}$$

where $|V_{out}(\omega)|$ is referred to as the *magnitude response* and $e^{j\varphi(\omega)}$ is referred to as the *phase response*. These two responses together constitute the frequency response of a circuit.

- The magnitude and phase responses of a circuit are often shown with asymptotic lines as approximations and these are referred to as Bode plots.
- Two frequencies ω_1 and ω_2 are said to be separated by an octave if $\omega_2 = 2\omega_1$ and separated by a decade if $\omega_2 = 10\omega_1$.
- The transfer function of a system is defined as

$$G(s) = \frac{V_{out}(s)}{V_{in}(s)} = \frac{N(s)}{D(s)}$$

where the numerator $N(s)$ and denominator $D(s)$ are as shown in the expression

$$V_{out}(s) = \frac{(a_n s^n + a_{n-1} s^{n-1} + a_{n-2} s^{n-2} + \ldots + a_0)}{(b_m s^m + b_{m-1} s^{m-1} + b_{m-2} s^{m-2} + \ldots + b_0)} V_{in}(s) = \frac{N(s)}{D(s)} V_{in}(s)$$

- In the expression

$$F(s) = \frac{N(s)}{D(s)} = \frac{\frac{1}{a_n}(b_m s^m + b_{m-1} s^{m-1} + b_{m-2} s^{m-2} + \ldots + b_1 s + b_0)}{s^n + \frac{a_{n-1}}{a_n} s^{n-1} + \frac{a_{n-2}}{a_n} s^{n-2} + \ldots + \frac{a_1}{a_n} s + \frac{a_0}{a_n}}$$

where $m < n$, the roots of the numerator are called the zeros of $F(s)$, and are found by letting $N(s) = 0$. The roots of the denominator are called the poles of $F(s)$ and are found by letting $D(s) = 0$.

- The zeros and poles can be real and distinct, or repeated, or complex conjugates, or combinations of real and complex conjugates. However, in most engineering applications we are interested in the nature of the poles.

- A system is said to be stable if a finite input produces a finite output. We can predict the stability of a system from its impulse response $h(t)$.

- Stability can easily be determined from the transfer function $G(s)$ on the complex frequency plane of the complex variable $s = \sigma + j\omega$. A system is stable only when all poles lie on the left–hand half–plane. It is marginally stable when one or more poles lie on the $j\omega$ axis, and unstable when one or more poles lie on the right–hand half–plane. However, the location of the zeros in the $s-$plane is immaterial.

- We can use the MATLAB function **bode(sys)** to draw the Bode plot of a system where **sys = tf(num,den)** creates a continuous–time transfer function **sys** with numerator **num** and denominator **den**, and **tf** creates a transfer function. With this function, the frequency range and number of points are chosen automatically. The function **bode(sys,{wmin,wmax})** draws the Bode plot for frequencies between **wmin** and **wmax** (in radians/second) and the function **bode(sys,w)** uses the user–supplied vector **w** of frequencies, in radians/second, at which the Bode response is to be evaluated. To generate logarithmically spaced frequency vectors, we use the command **logspace(first_exponent,last_exponent, number_of_values)**.

  The **bode(sys,w)** function displays both magnitude and phase. If we want to display the magnitude only, we can use the **bodemag(sys,w)** function.

- Amplifiers are often represented by equivalent circuits also known as circuit models. The common types are the voltage amplifier, the current amplifier, the transresistance amplifier, and the transconductance amplifier.

1.12 Exercises

1. Following the procedure of Example 1.1, derive and sketch the magnitude and phase responses for an RC high–pass filter.

2. Derive the transfer function $G(s)$ for the network shown below.

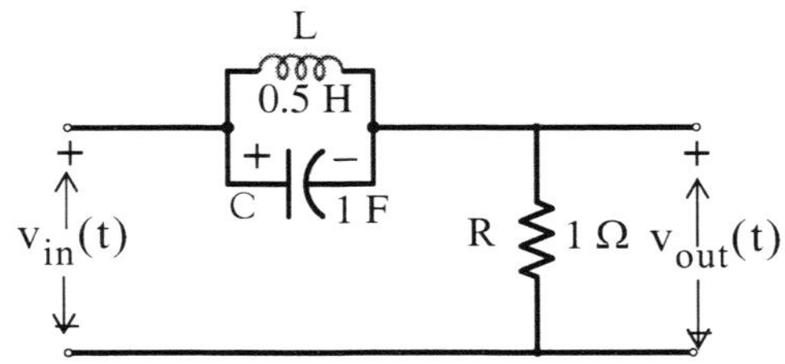

3. A system has poles at -4, $-2+j, -2-j$, and zeros at -1, $-3+j2$, and $-3-j2$. Derive the transfer function of this system given that $G(\infty) = 10$.

4. The circuit model shown below is known as a *transresistance amplifier* and the ideal characteristics for this amplifier are $R_{in} \to 0$ and $R_{out} \to 0$.

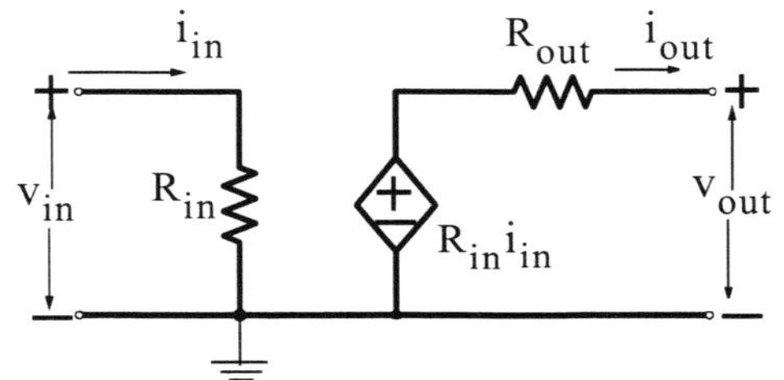

$$R_{in} = \left.\frac{v_{out}}{i_{in}}\right|_{i_{out}=0}$$

With a voltage source v_s in series with resistance R_s connected on the input side and a load resistance R_{load} connected to the output, the circuit is as shown below.

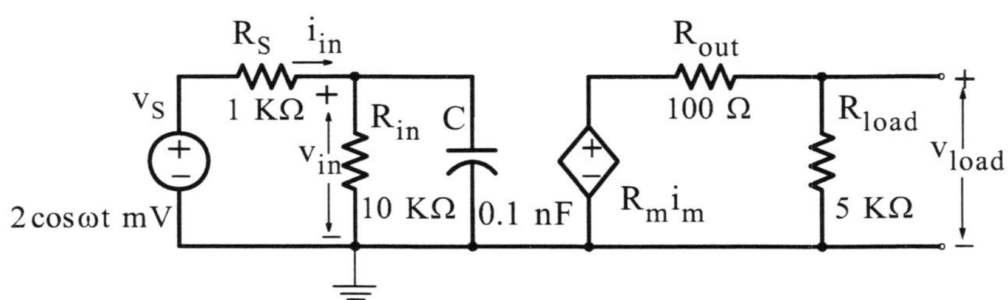

Find the overall voltage gain $A_v = v_{load}/v_s$ if $R_m = 100\ \Omega$. Then, use MATLAB to plot the magnitude of A_v for the range $10^3 \le \omega \le 10^8$. From the plot, estimate the 3 dB cutoff frequency.

5. The circuit model shown below is known as a *transconductance amplifier* and the ideal characteristics for this amplifier are $R_{in} \to \infty$ and $R_{out} \to \infty$.

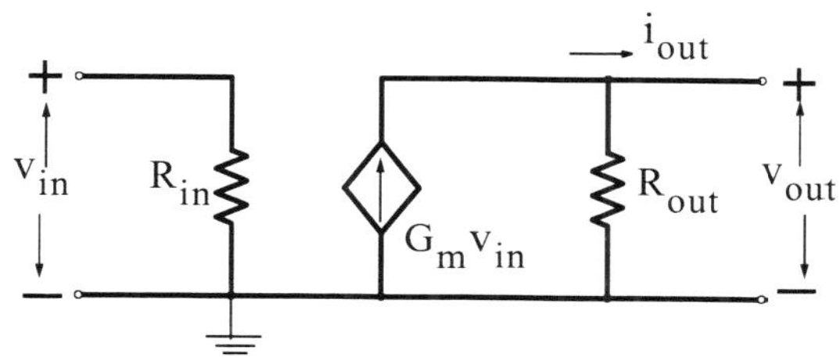

$$G_m = \left.\frac{i_{out}}{v_{in}}\right|_{v_{out} = 0}$$

With a voltage source v_s in series with resistance R_s connected on the input side and a load resistance R_{load} connected to the output, the circuit is as shown below.

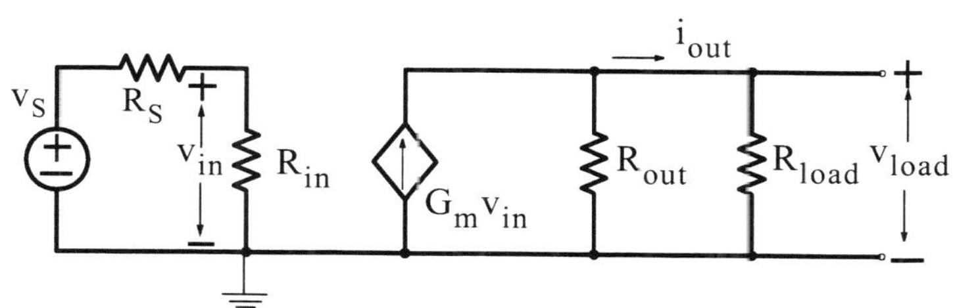

Derive an expression for the overall voltage gain $A_v = v_{load}/v_s$

6. The circuit shown below is an R-C high-pass filter.

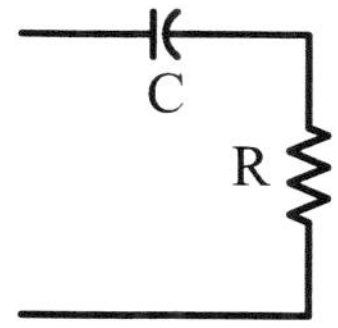

a. Derive the transfer function

b. Use MATLAB to plot the magnitude and phase angle for this high-pass filter. Hint: Use the procedure in Example 1.

7. For the low-pass filter circuit in Example 1 show that the output across the capacitor at high frequencies, i.e., for $\omega \gg 1/RC$, approximates an integrator, that is, the voltage across the capacitor is the integral of the input voltage.

8. For the high-pass filter circuit in Exercise 6 above 1 show that the output across the resistor at low frequencies, i.e., for $\omega \ll 1/RC$, approximates a differentiator, that is, the voltage across the resistor is the derivative of the input voltage.

1.13 Solutions to End–of–Chapter Exercises

Dear Reader:

The remaining pages on this chapter contain solutions to all end–of–chapter exercises.

You must, for your benefit, make an honest effort to solve these exercises without first looking at the solutions that follow. It is recommended that first you go through and solve those you feel that you know. For your solutions that you are uncertain, look over your procedures for inconsistencies and computational errors, review the chapter, and try again. Refer to the solutions as a last resort and rework those problems at a later date.

You should follow this practice with all end–of–chapter exercises in this book.

1.

$$V_{out} = \frac{R}{R + 1/j\omega C}V_{in}$$

or

$$G(j\omega) = \frac{V_{out}}{V_{in}} = \frac{j\omega RC}{1 + j\omega RC} = \frac{j\omega RC + \omega^2 R^2 C^2}{1 + \omega^2 R^2 C^2} = \frac{\omega RC(j + \omega RC)}{1 + \omega^2 R^2 C^2}$$

$$= \frac{\omega RC\sqrt{1 + \omega^2 R^2 C^2}\angle \text{atan}(1/(\omega RC))}{1 + \omega^2 R^2 C^2} = \frac{1}{\sqrt{1 + 1/(\omega^2 R^2 C^2)}}\angle \text{atan}(1/(\omega RC)) \qquad (1)$$

The magnitude of (1) is

$$|G(j\omega)| = \frac{1}{\sqrt{1 + 1/(\omega^2 R^2 C^2)}} \qquad (2)$$

and the phase angle or argument, is

$$\theta = \arg\{G(j\omega)\} = \text{atan}(1/\omega RC) \qquad (3)$$

We can obtain a quick sketch for the magnitude $|G(j\omega)|$ versus ω by evaluating (2) at $\omega = 0$, $\omega = 1/RC$, and $\omega \to \infty$. Thus,

As $\omega \to 0$,

$$|G(j\omega)| \cong 0$$

For $\omega = 1/RC$,

$$|G(j\omega)| = 1/\sqrt{2} = 0.707$$

and as $\omega \to \infty$,

$$|G(j\omega)| \cong 1$$

We will use the MATLAB script below to plot $|G(j\omega)|$ versus radian frequency ω. This is shown on the plot below where, for convenience, we let $RC = 1$.

```
w=0:0.02:100; RC=1; magGs=1./sqrt(1+1./(w.*RC).^2); semilogx(w,magGs); grid
```

We can also obtain a quick sketch for the phase angle, i.e., $\theta = \arg\{G(j\omega)\}$ versus ω, by evaluating (3) at $\omega = 0$, $\omega = 1/RC$, $\omega = -1/RC$, $\omega \to -\infty$, and $\omega \to \infty$. Thus,

as $\omega \to 0$,

$$\theta \cong -\text{atan}0 \cong 0°$$

For $\omega = 1/RC$,

$$\theta = -\text{atan}1 = -45°$$

For $\omega = -1/RC$,

$$\theta = -\text{atan}(-1) = 45°$$

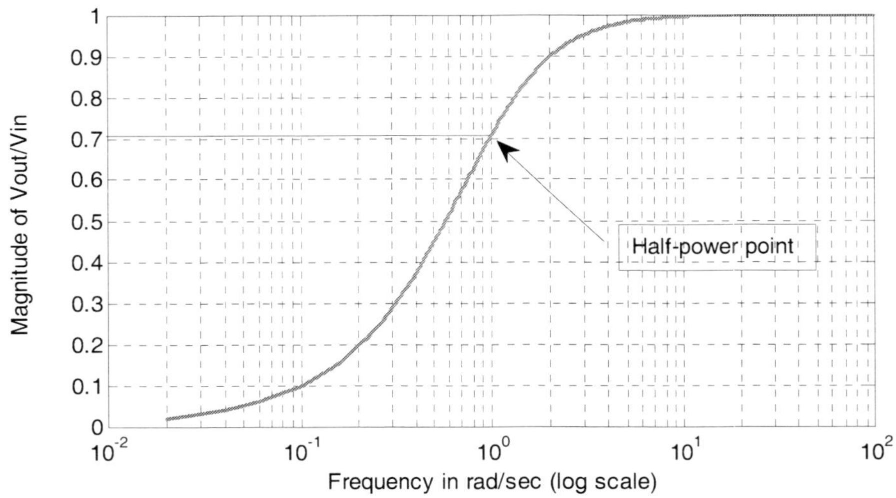

As $\omega \rightarrow -\infty$,

$$\theta = -\text{atan}(-\infty) = 90°$$

and as $\omega \rightarrow \infty$,

$$\theta = -\text{atan}(\infty) = -90°$$

We will use the MATLAB script below to plot the phase angle θ versus radian frequency ω. This is shown on the plot below where, for convenience, we let $RC = 1$.

```
w=-8:0.02:8; RC=1; argGs=atan(1./(w.*RC)).*180./pi; plot(w,argGs); grid
```

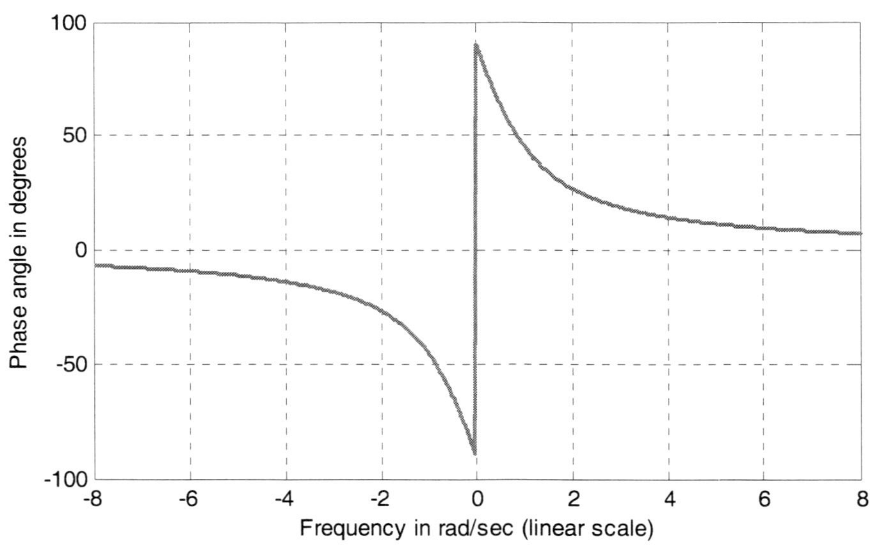

2. We draw the s – domain equivalent shown below.

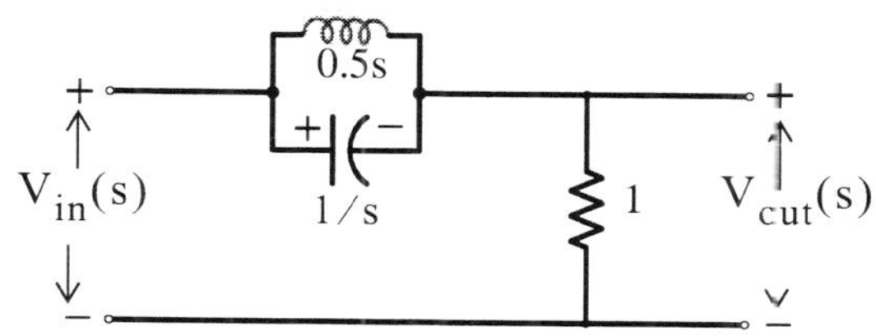

Parallel combination of the inductor and capacitor yields

$$\frac{s/2 \cdot 1/s}{s/2 + 1/s} = \frac{s}{s^2+2}$$

and by application of the voltage division expression we obtain

$$V_{out}(s) = \frac{1}{s/(s^2+2)+1}V_{in}(s)$$

$$G(s) = \frac{V_{out}(s)}{V_{in}(s)} = \frac{s^2+2}{s^2+s+2}$$

3. The transfer function has the form

$$G(s) = \frac{K[s-(-1)][s-(-3+j2)][s-(-3-j2)]}{[s-(-4)][s-(-2+j)][s-(-2-j)]}$$

$$= \frac{K(s+1)(s^2+6s+13)}{(s+4)(s^2+4s+5)} = \frac{K(s^3+7s^2+19s+13)}{s^3+8s^2+21s+20}$$

To determine the value of the constant K we divide all terms of G(s) by s^3 and we obtain

$$G(s) = \frac{K(1+7/s+19/s^2+13/s^3)}{1+8/s+21/s^2+20/s^3}$$

and as $s \to \infty$, $G(s) \approx K$. It is given that $G(\infty) = 10$, then $K = 10$ and the final form of the transfer function is

$$G(s) = \frac{10(s^3+7s^2+19s+13)}{s^3+8s^2+21s+20} = \frac{10(s+1)(s^2+6s+13)}{(s+4)(s^2+4s+5)}$$

```
num=10*[1 7 19 13]; den=[1 8 21 20]; w=logspace(0,2,100); bode(num,den,w);grid
```

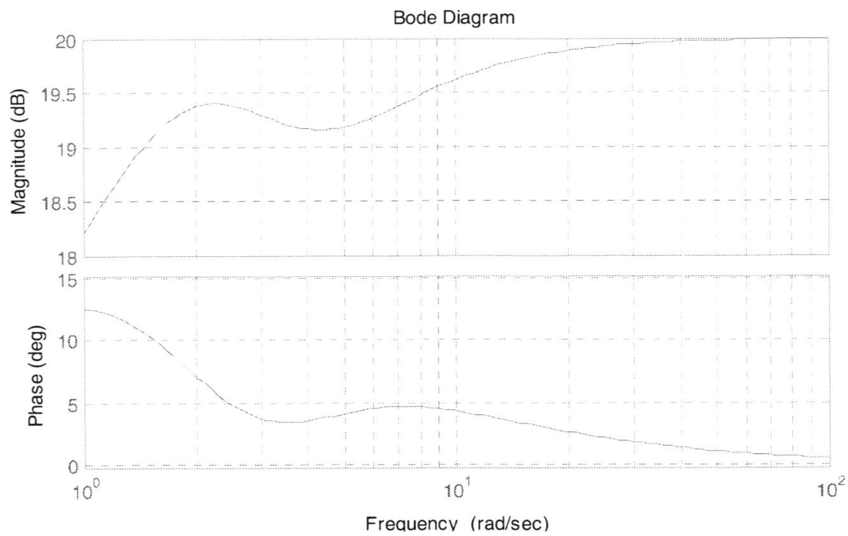

4.

The s – domain equivalent circuit is shown below.

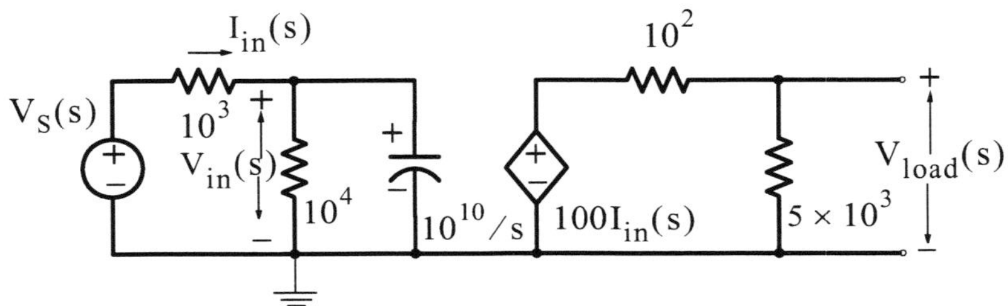

The parallel combination of the 10^4 resistor and $10^{10}/s$ capacitor yields

$$Z(s) = 10^4 \parallel 10^{10}/s = \frac{10^{14}/s}{10^4 + 10^{10}/s} = \frac{10^{14}}{10^4 s + 10^{10}}$$

and by the voltage division expression

$$V_{in}(s) = \frac{10^{14}/(10^4 s + 10^{10})}{10^3 + 10^{14}/(10^4 s + 10^{10})} V_S(s) = \frac{10^{14}}{10^7 s + 1.1 \times 10^{14}} V_S(s)$$

Also,

$$V_{load}(s) = \frac{5 \times 10^3}{10^2 + 5 \times 10^3} 100 I_{in}(s) = \frac{5 \times 10^5}{5.1 \times 10^3} I_{in}(s) = 98 I_{in}(s) \quad (1)$$

where

$$I_{in}(s) = \frac{V_{in}(s)}{Z(s)} = \frac{10^{14}V_S(s)/(10^7 s + 1.1\times 10^{14})}{10^{14}/(10^4 s + 10^{10})} = \frac{V_S(s)}{(10^4 s + 10^{10})(10^7 s + 1.1\times 10^{14})}$$

and by substitution into (1) we obtain

$$V_{load}(s) = \frac{98}{(10^4 s + 10^{10})(10^7 s + 1.1\times 10^{14})}V_S(s)$$

Then,

$$G_v(s) = \frac{V_{load}(s)}{V_S(s)} = \frac{98}{(10^4 s + 10^{10})(10^7 s + 1.1\times 10^{14})}$$

or

$$G_v(s) = \frac{V_{load}(s)}{V_S(s)} = \frac{98}{10^{11}s^2 + 1.2\times 10^{18}s + 1.1\times 10^{24}}$$

and with the MATLAB script below,

```
num=[0  0  98]; den=[10^11  1.2*10^18  1.1*10^24]; sys=tf(num,den);...
w=logspace(5,11,1000); bodemag(sys,w); grid
```

we obtain the Bode plot below.

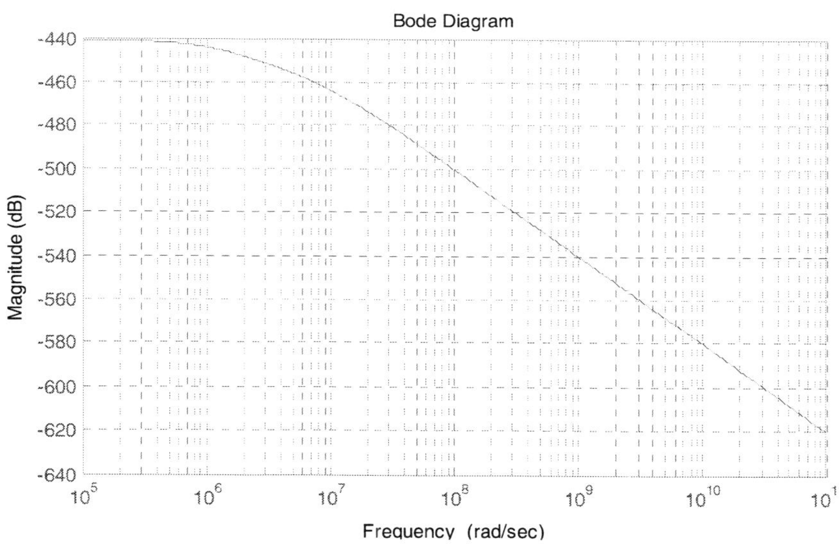

This plot shows a high attenuation of the source voltage v_S and thus the transresistance circuit model should not be used as a voltage amplifier.

5.

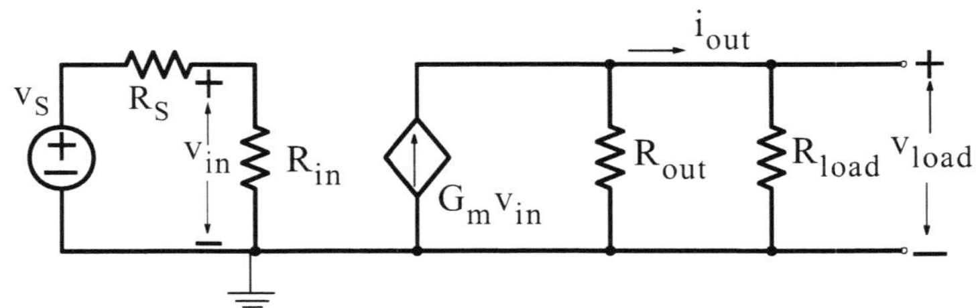

By the voltage division expression

$$v_{in} = \frac{R_{in}}{R_S + R_{in}} v_S \quad (1)$$

Also,

$$v_{load} = \frac{R_{out} R_{load}}{R_{out} + R_{load}} G_m v_{in} \quad (2)$$

Substitution of (1) into (2) yields

$$v_{load} = \frac{R_{out} R_{load}}{R_{out} + R_{load}} G_m \frac{R_{in}}{R_S + R_{in}} v_S$$

$$A_v = \frac{v_{load}}{v_S} = \left(\frac{R_{in}}{R_S + R_{in}}\right)\left(\frac{R_{out} R_{load}}{R_{out} + R_{load}}\right) G_m$$

6.

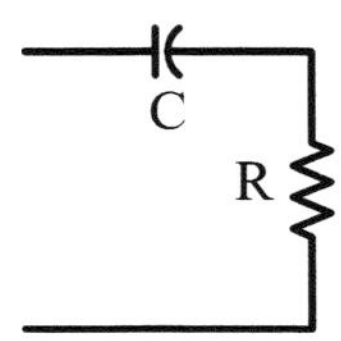

Application of the voltage division expression yields

$$V_{out} = \frac{R}{R + 1/j\omega C} V_{in}$$

or

$$G(j\omega) = \frac{V_{out}}{V_{in}} = \frac{j\omega RC}{1 + j\omega RC}$$

```
w=1:10:100000; RC=1; ; xf=(j.*w.*RC)./(1+j.*w.*RC);...
mag=20.*log10(abs(xf)); phase=angle(xf).*180./pi;...
subplot(121); semilogx(w,mag); subplot(122); semilogx(w,phase)
```

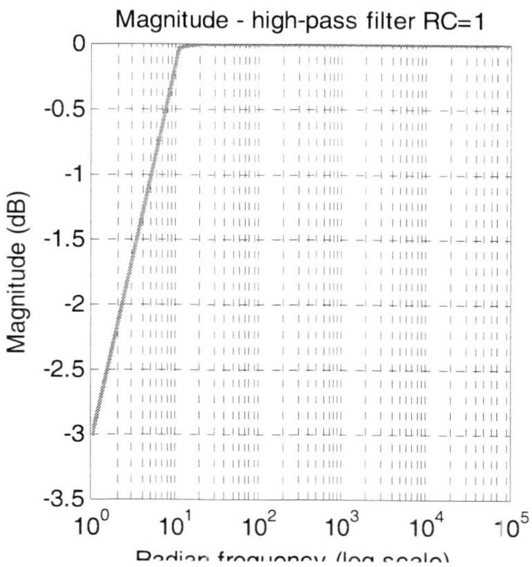

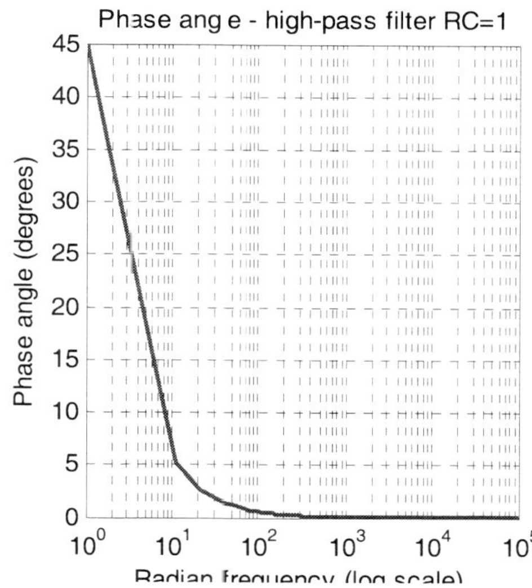

7.

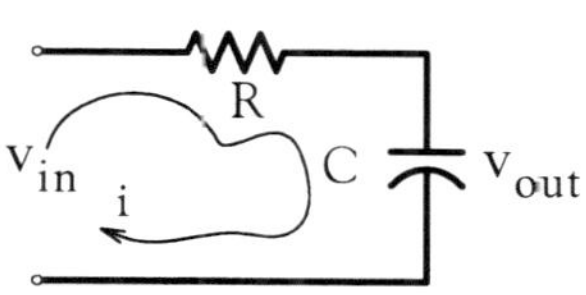

$$i = \frac{v_{in}}{R + 1/j\omega C}$$

and for $\omega \gg 1/RC$ or $R \gg 1/\omega C$

$$i \approx \frac{v_{in}}{R}$$

Also,

$$v_{out} = v_C = \frac{1}{C}\int_0^t i\,dt$$

Thus,

$$v_{out} \approx \frac{1}{RC}\int_0^t v_{in}\,dt$$

We can illustrate this with a Simulink* model as follows:

The transfer function of the low-pass filter above is readily derived from the voltage division expression, i.e.,

$$V(s)_{out} = \frac{1/sC}{R + 1/sC}V(s)_{in}$$

from which

* *For an introduction to Simulink, please refer to Appendix B*

$$G(s) = \frac{V(s)_{out}}{V(s)_{in}} = \frac{1/sC}{R + 1/sC} = \frac{1}{RCs + 1}$$

and for $RC = 1$,

$$G(s) = \frac{1}{s+1}$$

Next, we create the Simulink model shown below.

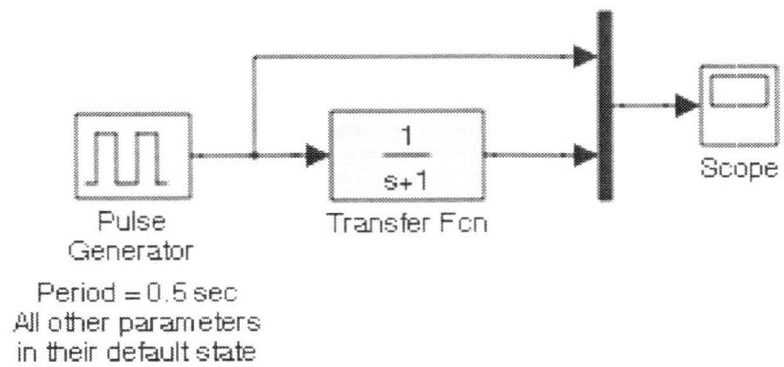

After the execution command is issued, the **Scope** block displays the input and output waveforms below.

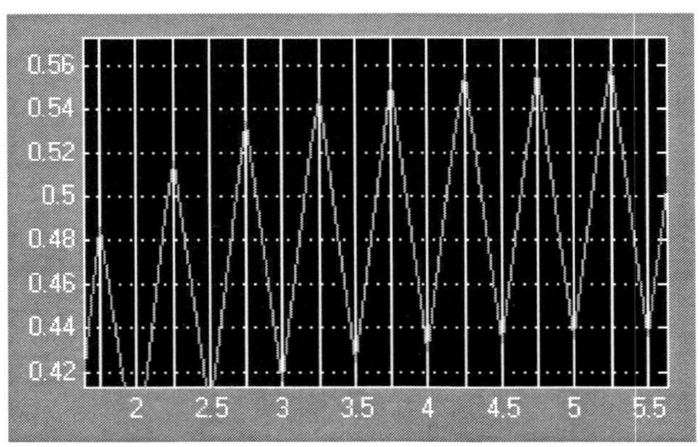

8.

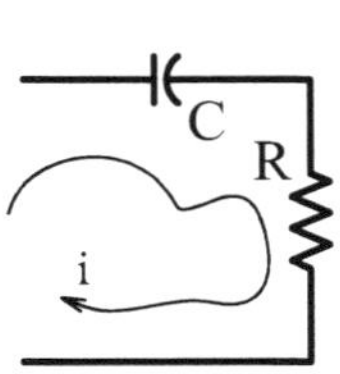

$$i = \frac{v_{in}}{R + 1/j\omega C}$$

and for $\omega \ll 1/RC$ or $R \ll 1/\omega C$

$$i \approx \frac{v_{in}}{1/j\omega C}$$

or

$$v_{in} \approx \frac{i}{1/j\omega C} \approx v_C$$

Also,

$$v_{out} = iR = C\frac{dv_C}{dt}R = RC\frac{dv_C}{dt}$$

Thus,

$$v_{out} \approx RC\frac{dv_{in}}{dt}$$

We can illustrate this with a Simulink model as follows:

The transfer function of the high-pass filter above is readily derived from the voltage division expression, i.e.,

$$V(s)_{out} = \frac{R}{R + 1/sC}V(s)_{in}$$

from which

$$G(s) = \frac{V(s)_{out}}{V(s)_{in}} = \frac{RCs}{RCs + 1} = \frac{1}{RCs + 1}$$

and for $RC = 1$,

$$G(s) = \frac{s}{s + 1}$$

Next, we create the Simulink model shown below.

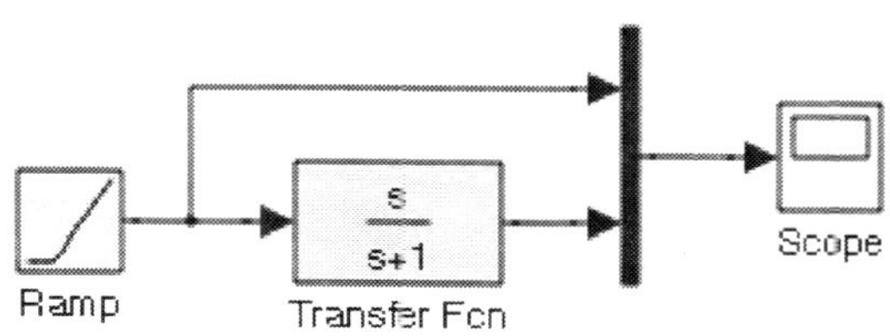

After the execution command is issued, the **Scope** block displays the input and output waveforms below.

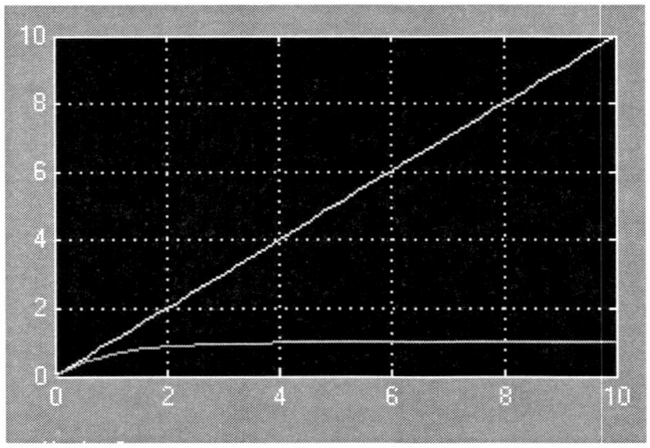
10
8
6
4
2
0
0
2
4
6
8
10

Chapter 2

Introduction to Semiconductor Electronics - Diodes

This chapter begins with an introduction to semiconductor electronics. The electron and hole movement is explained and illustrated in simple terms. The N–type and P–type semiconductors are discussed and majority and minority carriers are defined. The junction diode, its characteristics, and applications are discussed. The chapter concludes with the introduction of other types of diodes, i.e., Zener diodes, tunnel diodes, and others.

2.1 Electrons and Holes

We recall from the Periodic Table of Elements that silicon is classified as a semiconductor and it is widely used in the fabrication of modern technology electronic devices. Silicon has four valence electrons* and Figure 2.1 shows a partial silicon crystal structure in a two–dimensional plane where we observe that atoms combine to form an octet of valence electrons by sharing electrons; this combination is referred to as *covalent bonding*.

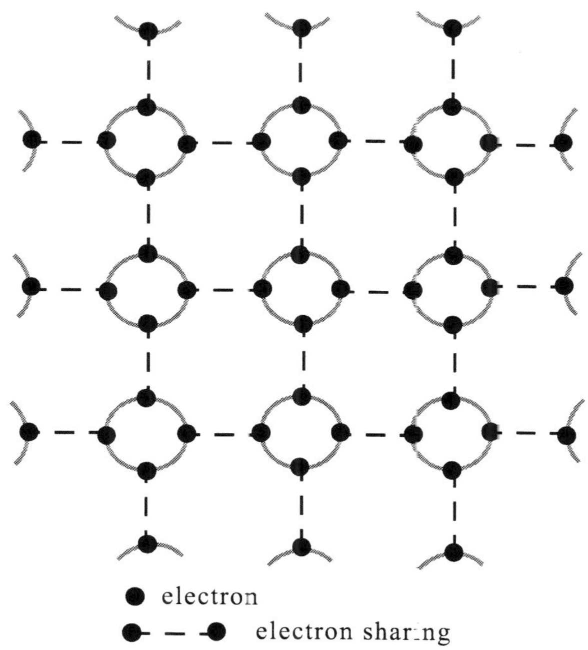

Figure 2.1. Partial silicon crystal structure

* *Valence electrons are those on the outer orbit.*

Thermal (heat) energy can dislodge (remove) an electron from the valence orbit of the silicon atom and when this occurs, the dislodged electron becomes a *free electron* and thus a vacancy (empty) space is created and it is referred to as a *hole*. The other electrons which stay in the valence orbit are called *bound electrons*. Figure 2.2 shows a free electron that has escaped from the valence orbit and the hole that has been created. Therefore, in a crystal of pure silicon that has been thermally agitated there is an equal number of free electrons and holes.

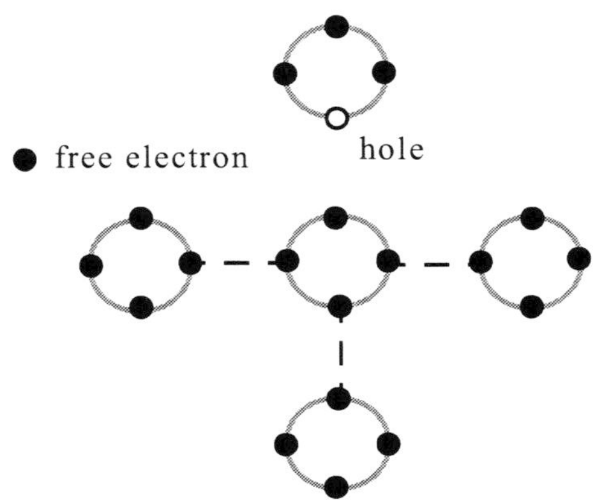

Figure 2.2. Free electron and the created hole in a partial silicon crystal

When a free electron approaches a hole it is attracted and "captured" by that hole. Then, that free electron becomes once again a bound electron and this action is called *recombination*. Accordingly, in a silicon crystal that has been thermally agitated we have two types of current movement; the *free electron movement* and the *hole movement*. The movement of holes can be best illustrated with the arrangement in Figure 2.3.

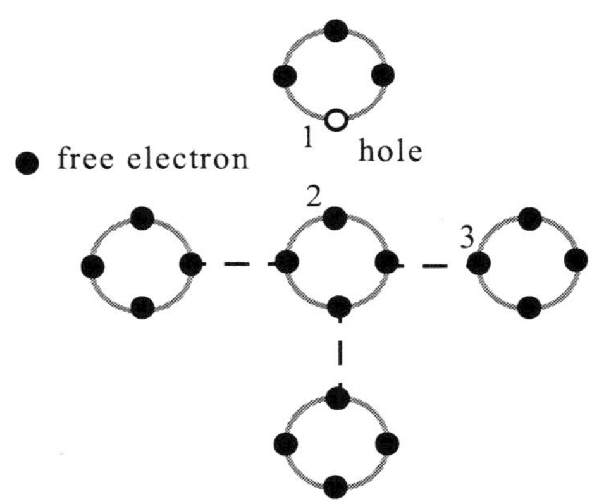

Figure 2.3. Free electron and hole movement at random

Figure 2.3 shows that a hole exists in position 1. Let us now suppose that the bound electron in position 2 is attracted by the hole in position 1. A new hole has now been created in position 2 and thus we say that the hole has moved from position 1 to position 2. Next, the hole in position

2 may attract the bound electron from position 3 and the hole now appears in position 3. This continued process is called hole movement and it is opposite to the free electron movement.
The free electron and hole movement is a random process. However, if we connect a voltage source as shown in Figure 2.4, the hole and free electron movement takes place in an orderly fashion.

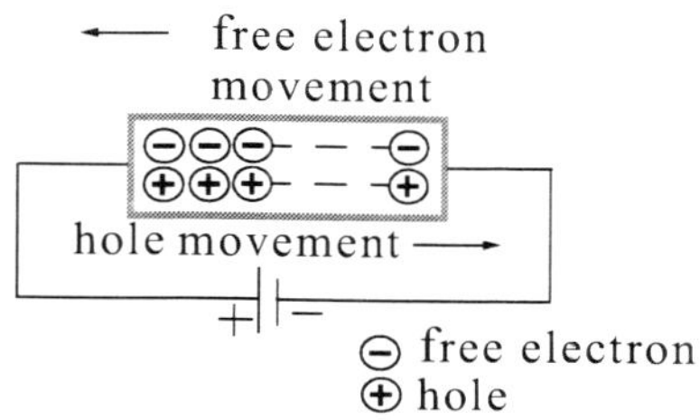

Figure 2.4. Free electron and hole movement when an external voltage is applied

We should keep in mind that holes are just vacancies and not positive charges although they move the same way as positive charges. We should also remember that in both N–type and P–type materials, current flow in the external circuit consists of electrons moving out of the negative terminal of the battery and into the positive terminal of the battery. Hole flow, on the other hand, only exists within the material itself.

Doping is a process where *impurity atoms** which are atoms with five valence electrons such as phosphorous, arsenic, and antimony, or atoms with three valence electrons such as boron, aluminum, and gallium, are added to *melted silicon*. The silicon is first melted to break down its original crystal structure and then impurity atoms are added. The newly formed compound then can be either an *N–type semiconductor* or a *P–type semiconductor* depending on the impurity atoms that were added as shown in Figure 2.5.

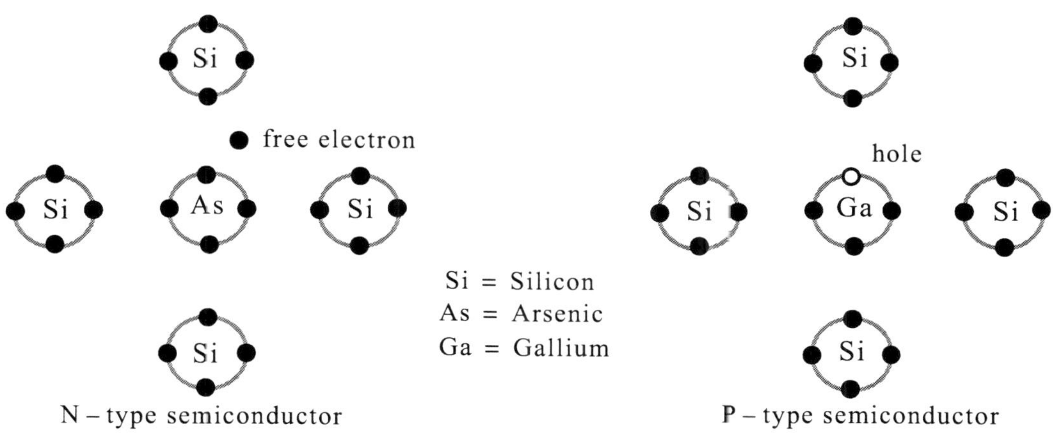

Figure 2.5. N–type and P–type semiconductors

* *Atoms with five valence electrons are often referred to as pentavalent atoms and atoms with three valence electrons are referred to as trivalent atoms.*

An N–type semiconductor has more free electrons than holes and for this reason the free electrons are considered to be the *majority carriers* and the holes the *minority carriers*. Conversely, a P–type semiconductor has more holes than free electrons and thus the holes are the majority carriers and the free electrons are the minority carriers.

We should remember that although the N–type material has an excess of free electrons, it is still electrically neutral. This is because the donor atoms in the N material were left with positive charges (the protons outnumbered the electrons) after the free electrons became available by covalent bonding. Therefore, for every free electron in the N material there is a corresponding positively charged atom to balance it and the N material has a net charge of zero.

By the same reasoning, the P–type material is also electrically neutral because the excess of holes is exactly balanced by the number of free electrons.

2.2 The Junction Diode

A junction diode is formed when a piece of P–type material and a piece of N–type material are joined together as shown in Figure 2.6 where the area between the P–type and N–type materials is referred to as the *depletion region*. The depletion region is shown in more detail in Figure 2.7.

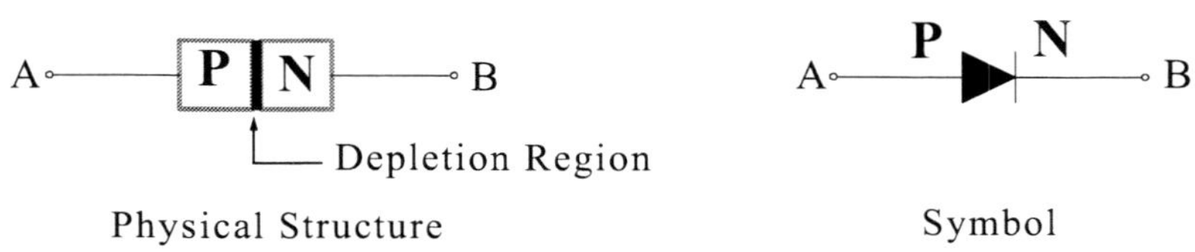

Figure 2.6. Formation of a junction diode and its symbol

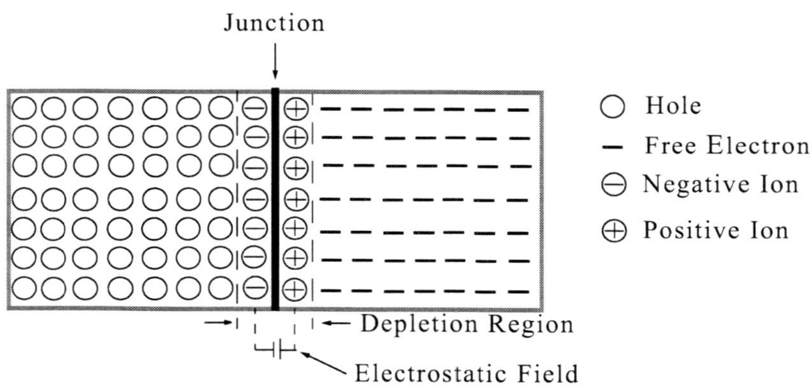

Figure 2.7. The PN junction barrier formation

We would think that if we join the N and P materials together by one of the processes mentioned earlier, all the holes and electrons would pair up. This does not happen. Instead the electrons in the N material diffuse (move or spread out) across the junction into the P material and fill some of the holes. At the same time, the holes in the P material diffuse across the junction into the N material and are filled by N material electrons. This process, called *junction recombination*, reduces the number of free electrons and holes in the vicinity of the junction. Because there is a depletion, or lack of free electrons and holes in this area, it is known as the depletion region.

The loss of an electron from the N–type material created a positive ion in the N material, while the loss of a hole from the P material created a negative ion in that material. These ions are fixed in place in the crystal lattice structure and cannot move. Thus, they make up a layer of fixed charges on the two sides of the junction as shown in Figure 2–7. On the N side of the junction, there is a layer of positively charged ions; on the P side of the junction, there is a layer of negatively charged ions. An electrostatic field, represented by a small battery in Figure 2–7, is established across the junction between the oppositely charged ions. The diffusion of electrons and holes across the junction will continue until the magnitude of the electrostatic field is increased to the point where the electrons and holes no longer have enough energy to overcome it, and are repelled by the negative and positive ions respectively. At this point equilibrium is established and, for all practical purposes, the movement of carriers across the junction ceases. For this reason, the electrostatic field created by the positive and negative ions in the depletion region is called a *barrier*.

The action just described occurs almost instantly when the junction is formed. Only the carriers in the immediate vicinity of the junction are affected. The carriers throughout the remainder of the N and P material are relatively undisturbed and remain in a balanced condition.

If we attach a voltage source to a junction diode with the plus (+) side of the voltage source connected to the P–type material and the minus (–) side to the N–type as shown in Figure 2.8, a *forward–biased* PN junction is formed.

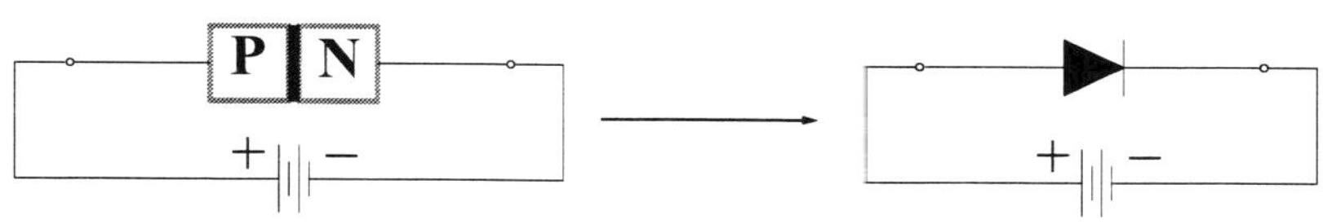

Figure 2.8. Forward–biased junction diode

When a junction diode is forward–biased, conventional current will flow in the direction of the arrow on the diode symbol.

If we reverse the voltage source terminals as shown in Figure 2.9, a *reverse–biased* PN junction is formed.

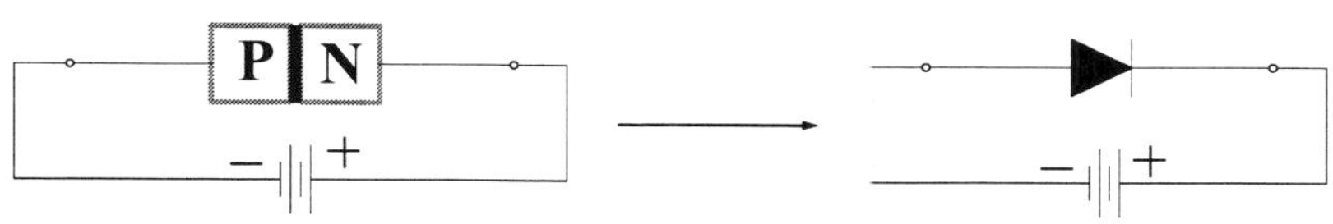

Figure 2.9. Reverse–biased junction diode

When a junction diode is reverse–biased, ideally no current will flows through the diode.

The P–type side of the junction diode is also referred to as the *anode* and the N–type side as the *cathode.* These designations and the notations for the voltage V_D across the diode and the current I_D through the diode are shown in Figure 2.10 where the direction of the current I_D through the diode is the direction of the conventional* current flow.

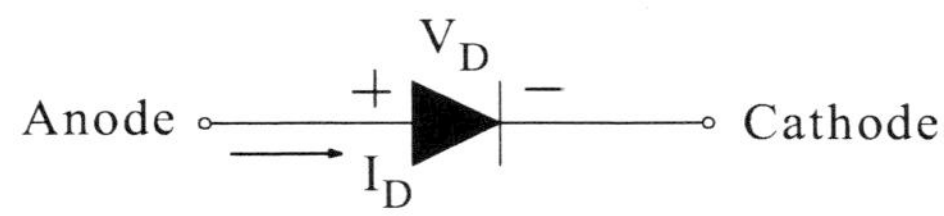

Figure 2.10. Voltage and current designations for a junction diode

Figure 2.11 shows the *ideal* $i_D - v_D$ *characteristics* of a junction diode.

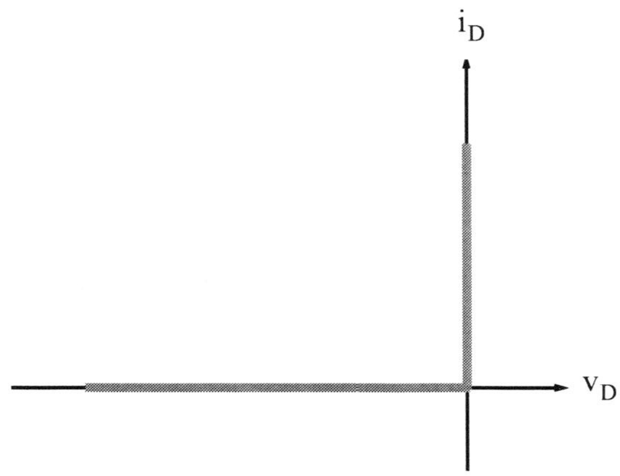

Figure 2.11. Ideal $i_D - v_D$ characteristics of a junction diode

With reference to Figure 2.11 we see that when $v_D > 0$, ideally $i_D \to \infty$, and when $v_D < 0$, ideally $i_D \to 0$. However, the actual $i_D - v_D$ relationship in a forward–biased junction diode is the nonlinear relation

$$i_D = I_r[e^{(qv_D/nkT)} - 1] \tag{2.1}$$

where i_D and v_D are as shown in Figure 2.10, I_r is the reverse current, that is, the current which would flow through the diode if the polarity of v_D is reversed, q is charge of an electron, that is, $q = 1.6 \times 10^{-19}$ coulomb, the coefficient n varies from 1 to 2 depending on the current level and the nature or the recombination near the junction, k = *Boltzmann's constant*, that is,

* *It is immaterial whether we use the electron current flow or the conventional current flow. The equations for the voltage–current relationships are the same as proved in Circuit Analysis I with MATLAB Applications, ISBN 978–0–9709511–2–0.*

$k = 1.38 \times 10^{-23}$ joule/Kelvin, and T is the absolute temperature in degrees Kelvin, that is, $T = 273 + \text{temperature in } ^{\circ}C$. It is convenient to combine q, k and T in (2.1) into one variable V_T known as *thermal voltage* where

$$V_T = kT/q \qquad (2.2)$$

and by substitution into (1),

$$i_D = I_r[\epsilon^{(v_D/nV_T)} - 1] \qquad (2.3)$$

Thus, for $T = 300\ ^{\circ}K$ we obtain

$$V_T\big|_{300\ ^{\circ}K} = kT/q = 1.38 \times 10^{-23} \times 300/1.6 \times 10^{-19} \approx 26mV \qquad (2.4)$$

We will use the MATLAB script below to plot the instantaneous current i_D versus the instantaneous voltage v_D for the interval $0 \le v_D \le 10$ v, with $n = 1$, and temperature at 27 °C.

```
vD=0: 0.001: 1; iR=10^(-15); n=1; VT=26*10^(-3);...
iD=iR.*(exp(vD./(n.*VT))-1); plot(vD,iD); axis([0 1 0 0.005]);...
xlabel('Diode voltage vD, volts'); ylabel('Diode current iD, amps');...
title('iD-vD characteristics for a forward-biased junction diode'); grid
```

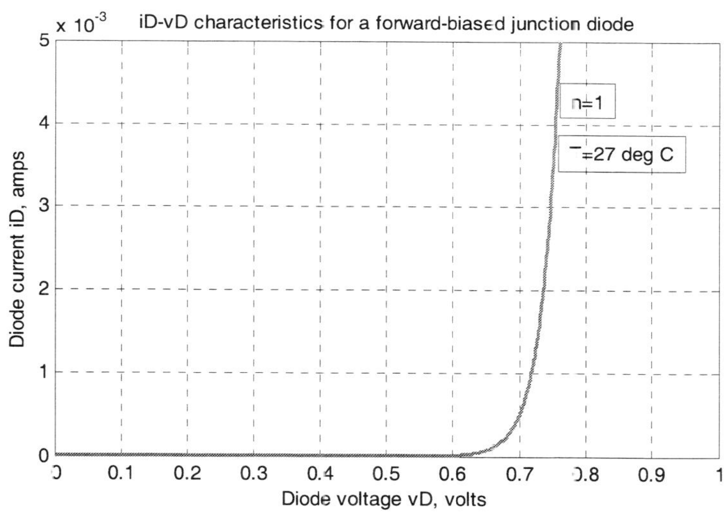

Figure 2.12. Voltage–current characteristics of a forward–biased junction diode.

The curve of Figure 2.12 shows that in a junction diode made with silicon and an impurity, conventional current will flow in the direction of the arrow of the diode as long as the voltage drop v_D across the diode is about 0.65 volt or greater. We also observe that at $v_D = 0.7$ V, the current through the diode is $i_D \approx 0.5$ mA.

When a junction diode is reverse–biased, as shown in Figure 2.13, a very small current will flow and if the applied voltage exceeds a certain value the diode will reach its *avalanche* or *Zener region*. The voltage–current characteristics of a reverse biased junction diode are shown in Figure 2.13 where V_Z is referred to as the Zener diode voltage. Zener diodes are discussed on the next section.

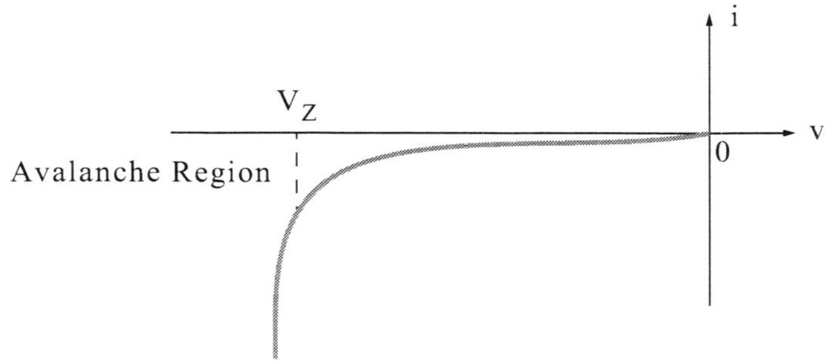

Figure 2.13. The reverse biased region of a junction diode

Commercially available diodes are provided with a given rating (volts, watts) by the manufacturer, and if these ratings are exceeded, the diode will burn–out in either the forward–biased or the reverse–biased direction.

The maximum amount of average current that can be permitted to flow in the forward direction is referred to as the *maximum average forward current* and it is specified at a special temperature, usually 25 °C. If this rating is exceeded, structure breakdown can occur.

The maximum peak current that can be permitted to flow in the forward direction in the form of recurring pulses is referred to as the *peak forward current.*

The maximum current permitted to flow in the forward direction in the form of nonrecurring pulses is referred to as the *maximum surge current*. Current should not equal this value for more than a few milliseconds.

The maximum reverse–bias voltage that may be applied to a diode without causing junction breakdown is referred to as the *Peak Reverse Voltage* (PRV) and it is the most important rating.

All of the above ratings are subject to change with temperature variations. If, for example, the operating temperature is above that stated for the ratings, the ratings must be decreased.

There are many types of diodes varying in size from the size of a pinhead used in subminiature circuitry, to large 250–ampere diodes used in high–power circuits. A typical diode is identified as XNYYYY where X denotes the number of semiconductor junctions (1 for diodes, 2 for transistors, and 3 a tetrode which has three junctions), N identifies the device as a semiconductor, and YYYY is an identification number. For instance, 1N4148 is a semiconductor diode and 2N3904 is a transistor. We will discuss transistors in Chapter 3.

The N side of a typical junction diode has a black band as shown in Figure 2.14.

Figure 2.14. Diode symbol and orientation

Diodes are used in various applications where it is desired to have electric current flow in one direction but to be blocked in the opposite direction as shown in Figure 2.15.

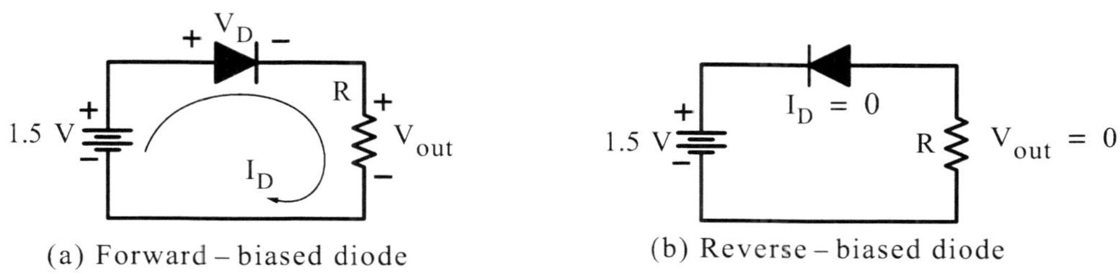

Figure 2.15. Diodes in DC Circuits

In the circuit of Figure 2.15(a) the diode is forward–biased, so current flows, and thus $V_{out} = 1.5 - V_D = 1.5 - 0.7 = 0.8\ V$. In the circuit of Figure 2.15(b) the diode is reverse–biased, no current flows, and thus $V_{out} = 0$.

2.3 Graphical Analysis of Circuits with Non–Linear Devices

As we've seen the junction diode $i - v$ characteristics are non–linear and thus we cannot derive the voltage–current relationships with Ohm's law. However, we will see later that for small signals (voltages or currents) these circuits can be represented by linear equivalent circuit models. If a circuit contains only one non–linear device, such as a diode, and all the other devices are linear, we can apply Thevenin's theorem to reduce the circuit to a Thevenin equivalent in series with the non–linear element. Then, we can analyze the circuit using a graphical solution. The procedure is illustrated with the following example.

Example 2.1

For the circuit of Figure 2.16, the $i - v$ characteristics of the diode D are shown in Figure 2.17 where V_{TH} and R_{TH} represent the Thevenin* equivalent voltage and resistance respectively of another circuit that has been reduced to its Thevenin equivalent. We wish to find the voltage v_D across the diode and the current i_D through this diode using a graphical solution.

* *For a thorough discussion on Thevenin's equivalent circuits, refer to Circuit Analysis I with MATLAB Applications, ISBN 978–0–9709511–2–0.*

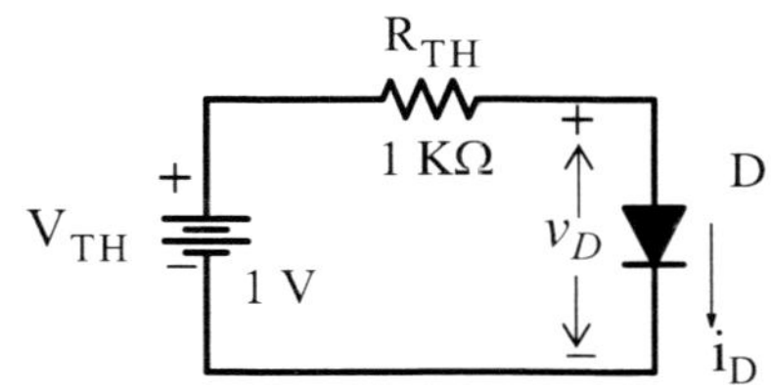

Figure 2.16. Circuit for Example 2.1

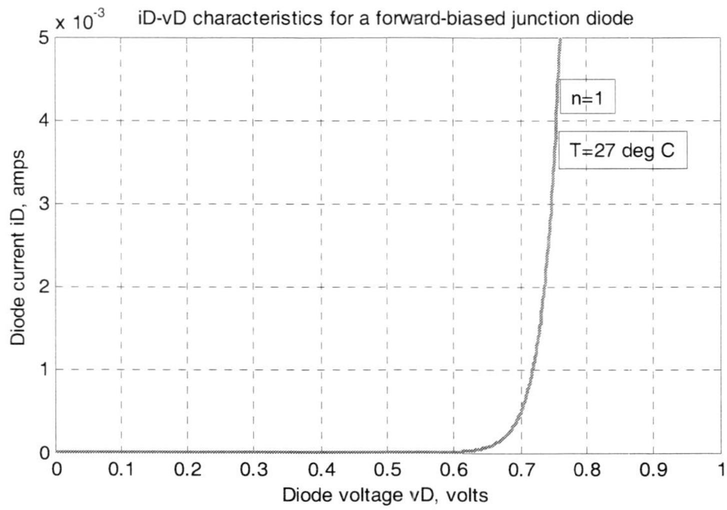

Figure 2.17. Voltage–current characteristics of the diode of Example 2.1

Solution:

The current i_D through the diode is also the current through the resistor. Then, by KVL

$$v_R + v_D = 1 \text{ V}$$

$$Ri_D = -v_D + 1$$

$$i_D = -\frac{1}{R}v_D + \frac{1}{R} \qquad (2.5)$$

We observe that (2.5) is an equation of a straight line and two points of this straight line can be obtained by first letting $v_D = 0$, then $i_D = 0$. We obtain the straight line shown in Figure 2.18 which is plotted on the same graph as the given diode $i-v$ characteristics. This line is referred to as a *load line*.

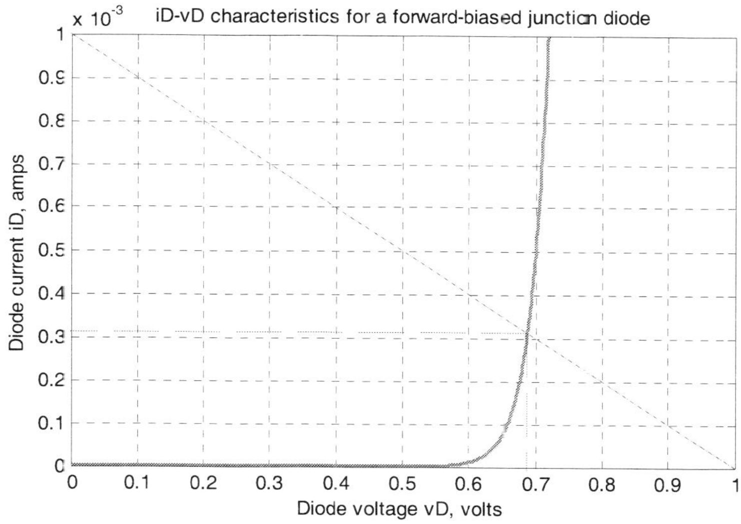

Figure 2.18. Curves for determining voltage and current in the diode of Example 2.1

The intersection of the non–linear curve and the load line yields the voltage and the current of the diode where we find that $v_D \approx 0.68V$ and $i_D \approx 0.32\ mA$.

Check:

Since this is a series circuit, $i_R = 0.32\ mA$ also. Therefore, the voltage drop v_R across the resistor is $v_R = 1\ K\Omega \times 0.32\ mA = 0.32\ V$. Then, by KVL

$$v_R + v_D = 0.32 + 0.68 = 1\ V$$

The relation of (2.3) gives us the current when the voltage is known. Quite often, we want to find the voltage when the current is known. To do that we rewrite (2.3) as

$$i_D + I_r = I_r e^{(v_D/nV_T)}$$

and since $i_D \gg I_r$, the above relation reduces to

$$i_D = I_r e^{(v_D/nV_T)} \qquad (2.6)$$

or

$$\frac{i_D}{I_r} = e^{(v_D/nV_T)}$$

Taking the natural logarithm of both sides we obtain

$$v_D/nV_T = \ln\frac{i_D}{I_r}$$

$$v_D = nV_T \ln(i_D/I_r) \tag{2.7}$$

Recalling that

$$\log_a x = \frac{\log_b x}{\log_b a}$$

we obtain

$$\ln(i_D/I_r) = \log_e(i_D/I_r) = \frac{\log_{10}(i_D/I_r)}{\log_{10} e} = \frac{\log_{10}(i_D/I_r)}{0.4343} = 2.3\log_{10}(i_D/I_r)$$

and thus (2.7) may also be written as

$$v_D = 2.3nV_T \log_{10}(i_D/I_r) \tag{2.8}$$

Example 2.2

Derive an expression for the voltage change $\Delta v = V_2 - V_1$ corresponding to a current change $\Delta i = I_2 - I_1$.

Solution:

From (2.3)

$$i_D = I_r[e^{(v_D/nV_T)} - 1]$$

$$i_D + I_r = I_r e^{(v_D/nV_T)}$$

and since $i_D \gg I_r$

$$i_D \approx I_r e^{(v_D/nV_T)}$$

Let

$$I_{D1} \approx I_r e^{V_{D1}/nV_T}$$

and

$$I_{D2} \approx I_r e^{V_{D2}/nV_T}$$

By division we obtain

$$\frac{I_{D2}}{I_{D1}} \approx \frac{I_r e^{V_{D2}/nV_T}}{I_r e^{V_{D1}/nV_T}} = e^{(V_{D2} - V_{D1})/nV_T}$$

Taking the natural log of both sides we obtain

$$\ln\frac{I_{D2}}{I_{D1}} = \ln(e^{(V_{D2}-V_{D1})/nV_T}) = \frac{1}{nV_T}(V_{D2}-V_{D1})$$

$$\Delta v = V_{D2}-V_{D1} = nV_T\ln(I_{D2}/I_{D1}) = 2.3nV_T\log_{10}(I_{D2}/I_{D1}) \qquad (2.9)$$

Example 2.3

Experiments have shown that the reverse current I_r increases by about 15% per $1\ °C$ rise, and it is known that for a certain diode $I_r = 10^{-14}$ A at $27\ °C$. Compute I_r at $52\ °C$.

Solution:

$$I_r\big|_{52\ °C} = 10^{-14}(1+0.15)^{(52-27)°C} = 10^{-14}(1.15)^{25} \approx 3.3\times10^{-13}\ A$$

This represents about 97% increase in reverse current when the temperature rises from $27\ °C$ to $52\ °C$.

2.4 Piecewise Linear Approximations

The analysis of electronic circuits that contain diodes is greatly simplified with the use of diode models where we approximate the diode forward–biased characteristics with two straight lines as shown in Figure 2.19.

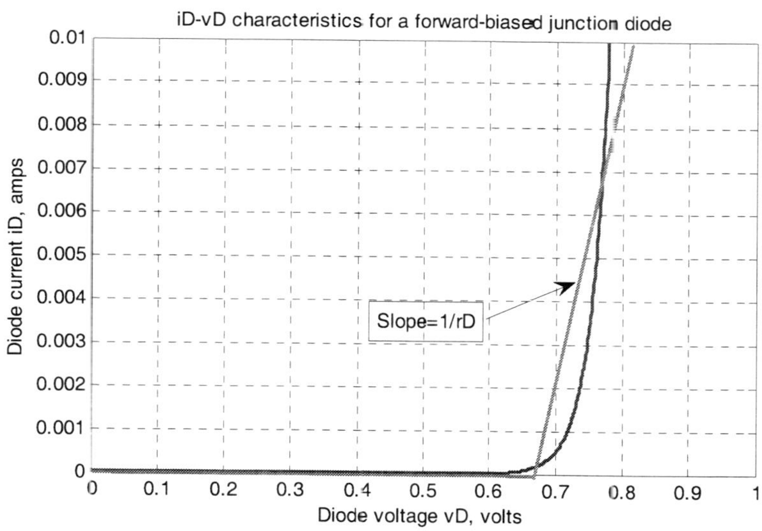

Figure 2.19. Straight lines for forward–biased diode characteristics approximations

Using the approximation with the straight lines shown in Figure 2.19, we can now represent a typical junction diode with the equivalent circuit shown in Figure 2.20.

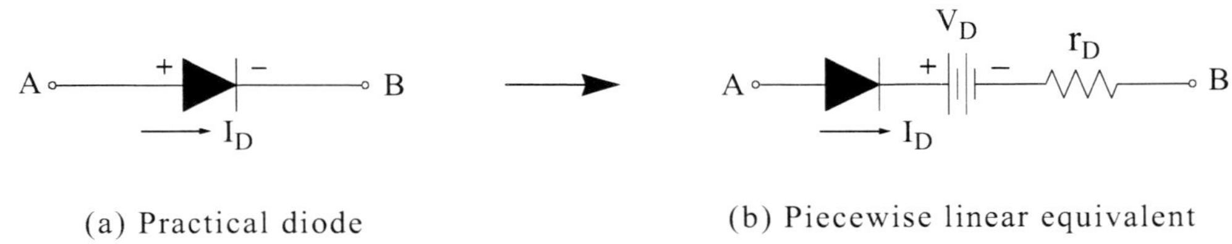

Figure 2.20. Representation of a practical diode by its piecewise linear equivalent

In Figure 2.20(b), the diode represents an ideal diode whose $i-v$ characteristics are shown in Figure 2.21, the horizontal solid line in Figure 2.19 represents the small voltage source V_D in Figure 2.20(b), and the reciprocal of the slope of the line in Figure 2.19 is represented by the resistance r_D shown in Figure 2.20(b). For convenience, these representations are also illustrated in Figure 2.21.

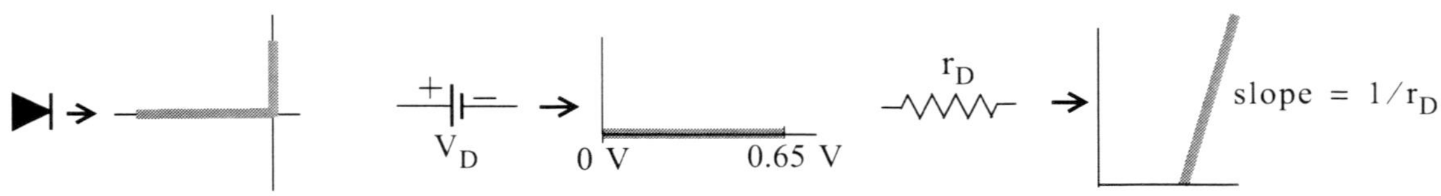

Figure 2.21. The components of a practical junction diode

Example 2.4

In the circuit of Figure 2.22(a) the diodes are identical and the piecewise linear $i-v$ characteristics are shown in Figure 2.22(b). Find the voltage V_{out}.

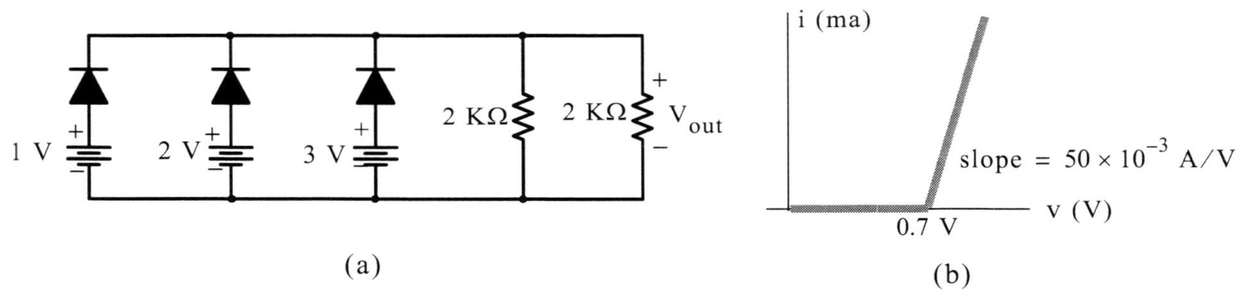

Figure 2.22. Circuit and piecewise linear $i-v$ characteristics for Example 2.4

Solution:

In Figure 2.23 we have replaced the diodes by their piecewise linear equivalents and have combined the two parallel resistors. Also, for each branch we have combined the diode voltage $V_D = 0.7\ V$ with the applied voltages, and $r_D = 1/slope = 50 \times 10^{-3}\ A/V = 20\ \Omega$.

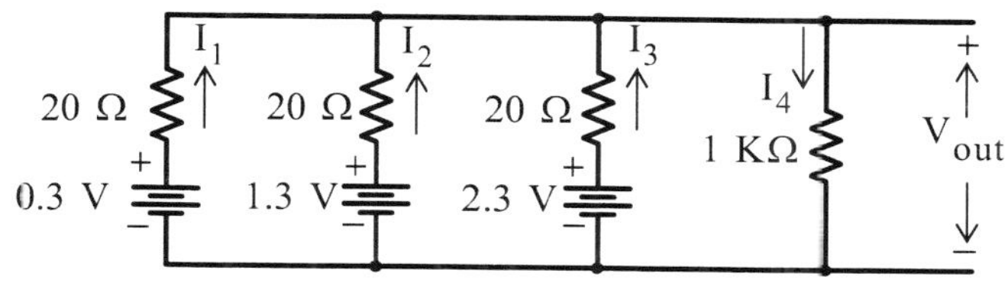

Figure 2.23. Piecewise linear equivalent circuit for Example 2.4

Let us follow the procedure below to find out if we can arrive to a valid answer. By Kirchoff's Current Law (KCL)

$$I_1 + I_2 + I_3 = I_4$$

$$\frac{0.3 - V_{out}}{20} + \frac{1.3 - V_{out}}{20} + \frac{2.3 - V_{out}}{20} = \frac{V_{out}}{1000}$$

$$\frac{15 - 50V_{out} + 65 - 50V_{out} + 115 - 50V_{out} - V_{out}}{1000} = 0$$

$$151V_{out} = 195$$

$$V_{out} = 195/151 = 1.29 \approx 1.3\ V$$

Check:

$$I_1 \approx (0.3 - 1.3)/20 \approx -50\ mA$$

$$I_2 \approx (1.3 - 1.3)/20 \approx 0$$

$$I_3 = (2.3 - 1.3)/20 \approx 50\ mA$$

$$I_4 = 1.3/1000 = 1.3\ mA$$

$$I_1 + I_2 + I_3 \neq I_4$$

We see that the current I_1 cannot be negative, that is, it cannot flow on the opposite direction of the one shown. Also, the current I_2 is zero. Therefore, we must conclude that only the diode on the right side conducts and by the voltage division expression

$$V_{out} = \frac{1000}{20 + 1000} \times 1.3 \approx 1.3\ V$$

2.5 Low Frequency AC Circuits with Junction Diodes

When used with AC circuits of low frequencies, diodes, usually with $1.8 \le n \le 2.0$ are biased to operate at some point in the neighborhood of the relatively linear region of the $i - v$ characteristics where $0.65 \le v_D \le 0.8$ V. A bias point denoted as Q whose coordinates are $Q(V_D, I_D)$ is shown in Figure 2.24 for a junction diode with $n = 2$.

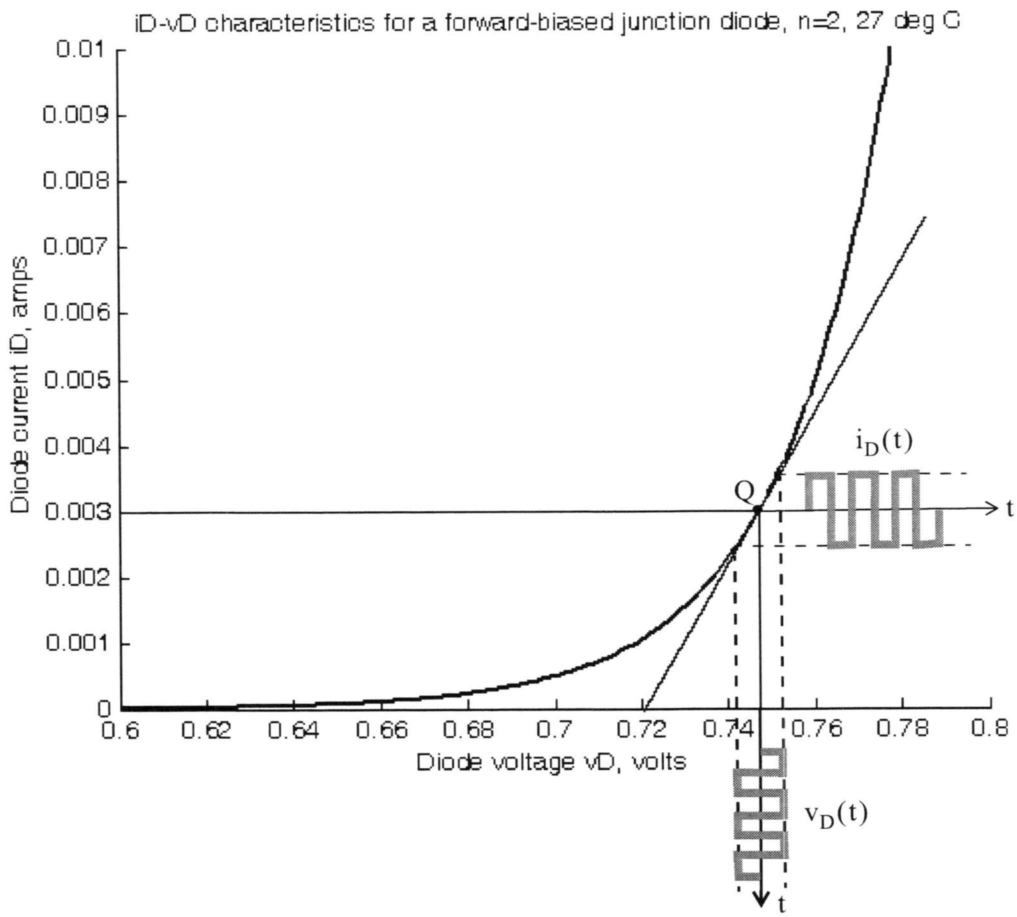

Figure 2.24. Junction diode biased at point Q and changes in i_D *corresponding to changes in* v_D

Figure 2.24 shows how changes in $v_D(t)$ result in changes in $i_D(t)$.

We can derive an expression that relates $v_D(t)$ and $i_D(t)$ in a junction diode. The current I_D produced by the bias DC voltage V_D is

$$I_D = I_r e^{V_D/nV_T} \tag{2.10}$$

and with an AC voltage $v_D(t)$ superimposed the sum $v_T(t)$ of the DC and AC voltages is

$$v_T(t) = V_D + v_D(t) \tag{2.11}$$

The total diode current $i_T(t) = I_D + i_D(t)$ corresponding to the total voltage in (2.11) is

$$i_T(t) = I_D + i_D(t) = I_r e^{(V_D + v_D)/nV_T} = I_r e^{V_D/nV_T} e^{v_D/nV_T}$$

and in analogy with (2.10)

$$i_D(t) = I_D e^{v_D/nV_T} \quad (2.12)$$

If $v_D/nV_T \le 0.1$ we can use Maclaurin's series* expansion on (2.12) using the relation

$$f(x) = f(0) + f'(0)x + \frac{f''(0)}{2!}x^2 + \ldots + \frac{f^{(n)}(0)}{n!}x^n \quad (2.13)$$

and the first two terms of the series yield

$$i_D(t) \approx I_D + \frac{I_D}{nV_T}v_D(t) \quad (2.14)$$

We must remember that the approximation in (2.14) is a small–signal approximation and should be used only when $v_D/nV_T \le 0.1$.

The ratio I_D/nV_T in (2.14) is denoted as g_D and is referred to as *incremental conductance*. Its reciprocal nV_T/I_D is denoted as r_D is called *incremental resistance*.

Example 2.5

In the circuit of Figure 2.25, the current through the diode is $I_D = 1 \text{ mA}$ when $V_D = 0.7 \text{ V}$, and it is known that $n = 2$. Find the DC voltage V_{out} and the AC voltage v_{out} at 27 °C where $V_T = 26 \text{ mV}$.

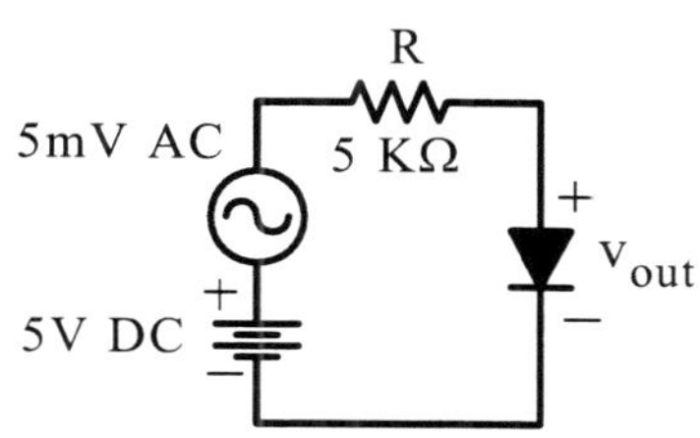

Figure 2.25. Circuit for Example 2.5

Solution:

Replacing the diode with the piecewise linear equivalent we obtain the circuit of Figure 2.26.

* *For a detailed discussion on Taylor and Maclaurin's series refer to Numerical Analysis Using MATLAB and Excel, ISBN 978–1–934404–03–4.*

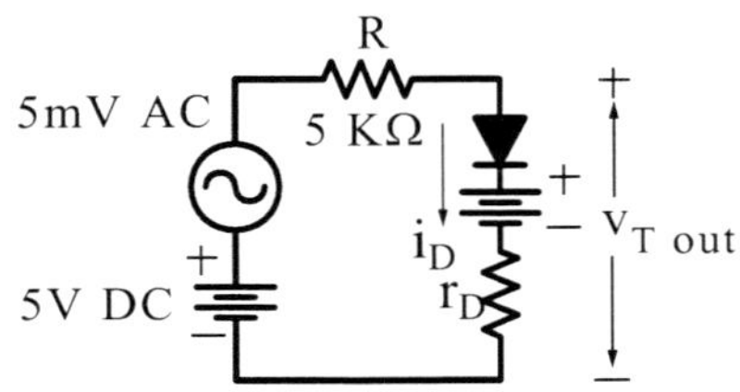

Figure 2.26. The piecewise linear equivalent of the circuit in Example 2.5

We apply the superposition principle* for this example. We fist consider the DC voltage source acting alone, by suppressing (shorting out) the AC voltage source to find V_{out}. Then, we consider the AC voltage source acting alone by suppressing (shorting out) the DC voltage source to find v_{out}. The total output voltage will be the sum of these two, that is, $v_{T\ out} = V_{out} + v_{out}$.

With the DC voltage source acting alone and with the assumption that $r_D \ll 5\ K\Omega$, the current I_D is

$$I_D = \frac{5 - 0.7}{5 \times 10^3} = 0.86mA$$

and since we are told that $I_D\big|_{v\,=\,0.7V} = 1mA$, by linear interpolation,

$$V_{D\ actual} = 0.86 \times 0.7 = 0.6\ V$$

Therefore,

$$V_{out} = 0.6\ VDC$$

With the AC voltage source acting alone the value of the incremental resistance r_D is the reciprocal of (2.14) and thus

$$r_D = \frac{nV_T}{I_D} = \frac{2 \times 26 \times 10^{-3}}{0.86 \times 10^{-3}} = 60.5\ \Omega$$

Then,

$$v_{D\ peak} = \frac{r_D}{R + r_D} \cdot 5mV = \frac{60.5 \times 5 \times 10^{-3}}{5000 + 60.5} = 60\ \mu VAC$$

Therefore

$$v_{T\ out} = V_{out} + v_{out} = 0.6\ VDC + 60\ \mu VAC$$

* *Generally, the superposition principle applies only to linear circuits. However, it is also applicable for this example since it is applied to a piecewise linear equivalent circuit.*

2.6 Junction Diode Applications in AC Circuits

Diodes are also used in AC circuits where it is desired to convert AC voltages to DC voltages. The circuit of Figure 2.27 is a *half–wave rectifier.*

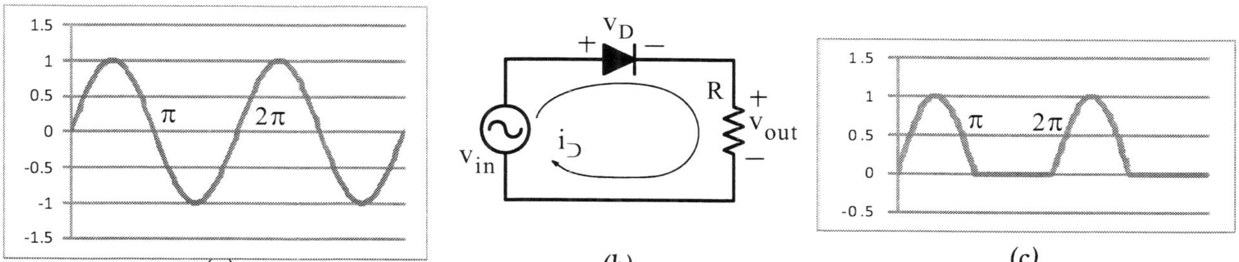

(a) (b) (c)

Figure 2.27. Half–wave rectifier circuit

In the half–wave rectifier circuit of Figure 2.27(b), the diode is forward–biased during the positive half–cycle from 0 to π of the input voltage v_{in} and so current flows through the diode and resistor where it develops an output voltage drop $v_{out} = v_{in} - v_D$. For instance, if the maximum value of v_{in} is 10 V volts, the maximum value of v_{out} will be $v_{out} = 10 - 0.7 = 9.3\ V$. The diode is reverse–biased during the negative half–cycle from π to 2π so no current flows, and thus $v_{out} = 0\ V$.

The waveform in Figure 2.27(c) can be simulated* with the **Saturation** block in the model in Figure 2.28.

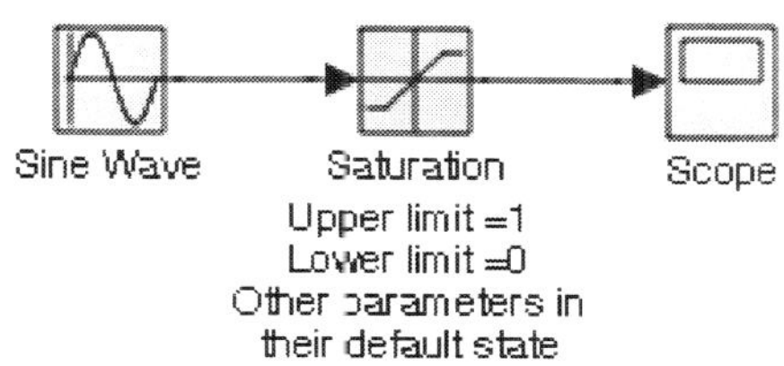

Figure 2.28. Simulink model for creating half–rectification waveform with a saturation block

The **Scope** block in Figure 2.28 displays the half–rectification waveform shown in Figure 2.29.

* *This and other simple Simulink models presented in this text could be parts of a large model. For a brief introduction to Simulink please refer to Appendix B in this text, and for a complete coverage refer to* ***Introduction to Simulink with Engineering Applications****, Second Edition, ISBN 978–1–934404–09–6.*

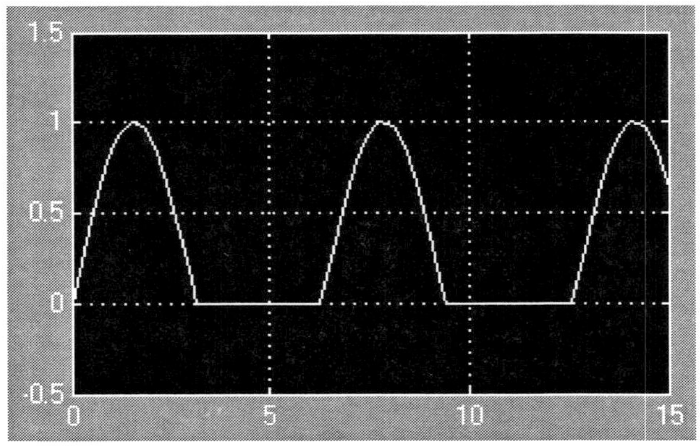

Figure 2.29. Half–rectification waveform displayed by the Scope block in Figure 2.28

Example 2.6

Design a DC voltmeter* that will have a 10 volt full–scale using a milliammeter with 1 milliampere full–scale and internal resistance 20 Ω, a junction diode, and an external resistor R whose value must be found. The input is an AC voltage with a value of 63 volts peak–to–peak. Assume that the diode is ideal.

Solution:

Typically, a voltmeter is a modified milliammeter where an external resistor R_V is connected in series with the milliammeter as shown in Figure 2.30 where

$$I = \text{current through circuit}$$

$$R_M = \text{internal resistance of milliameter}$$

$$R_V = \text{external resistor in series with } R_M$$

$$V_M = \text{voltmeter full scale reading}$$

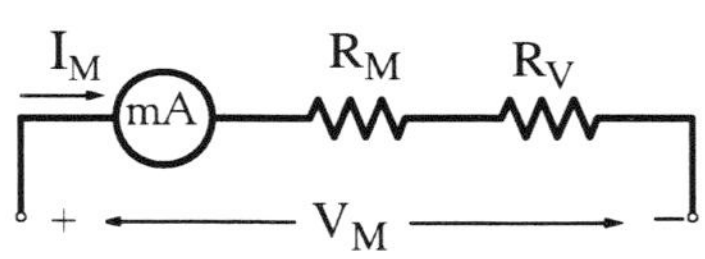

R_V = Voltmeter internal resistance
V_M= Voltmeter range

Figure 2.30. Typical voltmeter circuit

Because the available input voltage is AC and we want to read DC (average) values, we insert a junction diode is series as shown in Figure 2.31, and we need to find the value of R_V so that the

* *For a detailed discussion on electronic instruments refer to Circuit Analysis I with MATLAB Applications, ISBN 978–0–9709511–2–0.*

meter will read 1 mA full scale but it is now being labeled as 10 V DC.

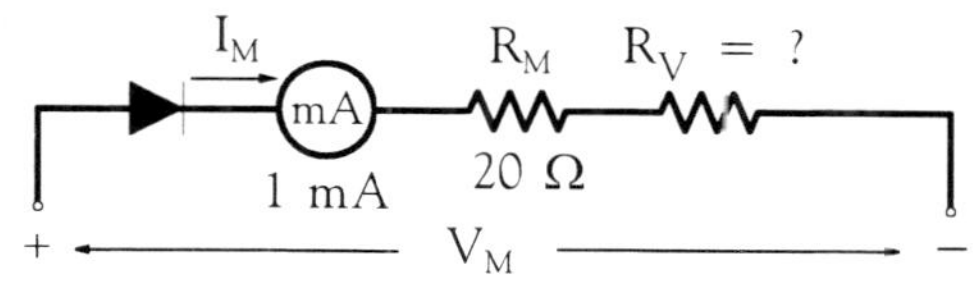

Figure 2.31. Voltmeter for Example 2.6

Since the diode does not conduct during the negative half cycle, it follows that

$$I_{peak} = I_p = \frac{V_{p-p}/2}{R_M + R_V} = \frac{31.5}{20 + R_V}$$

and

$$I_{ave} = \frac{\text{Area}}{\text{Period}} = \frac{\int_0^{2\pi} I_p \sin t dt}{2\pi} = \frac{\frac{31.5}{20 + R_V}\int_0^{\pi} \sin t dt}{2\pi} = \frac{31.5(-\cos t)|_0^{\pi}}{(20 + R_V)2\pi} = \frac{31.5}{(20 + R_V)\pi}$$

For full–scale reading we want $I_{ave} = 1$ mA. Therefore,

$$\frac{31.5}{(20 + R_V)\pi} = 10^{-3}$$

$$20 + R_V = \frac{31.5}{\pi} \times 10^3$$

$$R_V = 10\ \text{K}\Omega$$

Of course, we can label our voltmeter for any value such as 50 V, 100 V, and so on and can use any AC waveform with different peak–to–peak values.

Figure 2.32(a) shows a half–wave rectifier circuit consisting of a transformer,* a junction diode, and a load resistor, and Figure 2.32(b) shown the waveforms of the input voltage v_{in} and the load voltage v_{load}.

* *For a detailed discussion on transformers, refer to Circuit Analysis II with MATLAB Applications, ISBN 978–0–9709511–5–1.*

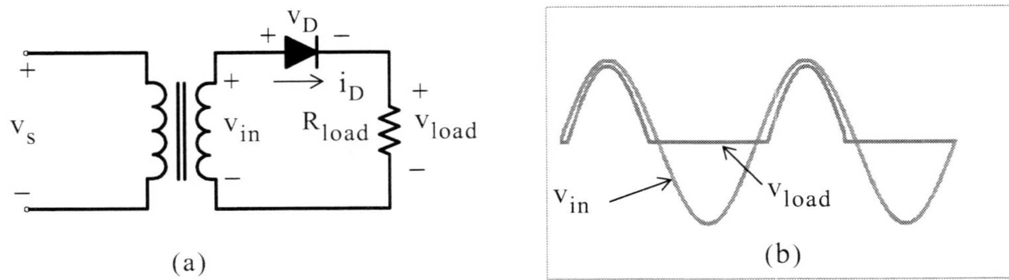

Figure 2.32. Half–wave rectifier circuit and input and output waveforms

When using diodes in rectifier circuits we must calculate:

a. the maximum current that the diode will allow without being damaged, and

b. The *Peak Inverse Voltage* (PIV) that the diode can withstand without reaching the reverse–biased breakdown region.

If the applied voltage is $v_S = V_p \sin\omega t$, then $PIV = V_p$ but in practice we must use diodes whose reverse–biased breakdown voltages 70% or greater than the value of V_p. As shown in Figure 2.32(b), the diode begins conducting sometime after the input voltage shown by the dotted curve rises to about $0.7\ V$. We derive the angle by which the solid curve lags the dotted curve with reference to Figure 2.33* as follows.

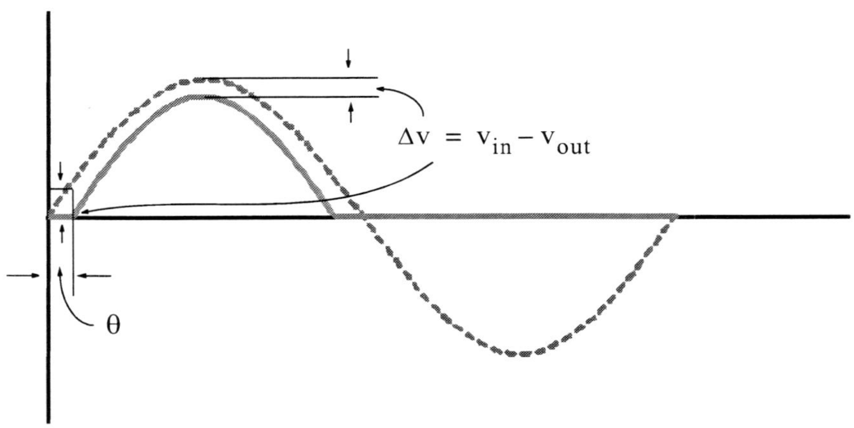

Figure 2.33. Waveforms for the derivation of conduction angle θ for a half–wave rectifier

* *These waveforms can be created in a Simulink model with a Sine Wave block, an Abs block, a Dead Zone block, and displayed in a Scope block.*

From Figure 2.33 we observe that conduction begins at $\Delta v = v_{in} - v_{out}$ corresponding to angle θ. The input waveform is a sinusoid of the form $v_{in} = V_p \sin\omega t$ or, for simplicity, $v_{in} = V_p \sin x$, and at $x = \theta$, $v_{in} = \Delta v$ and thus $\Delta v = V_p \sin\theta$ or

$$\theta = \sin^{-1}\Delta v / V_p \tag{2.15}$$

We also observe that the conduction terminates at the angle $(\pi - \theta)$ and therefore the entire conduction angle is

$$(\pi - \theta) - \theta = (\pi - 2\theta) \tag{2.16}$$

We can also find the average value of the waveform of v_{out}. We begin with the definition of the average value, that is,

$$V_{out}(ave) = \frac{Area}{Period} = \frac{1}{2\pi}\int_{\theta}^{(\pi-\theta)} (V_p \sin\phi - \Delta v) d\phi = \frac{1}{2\pi}(-V_p \cos\phi - \Delta v\phi)\Big|_{\phi=\theta}^{\pi-\theta}$$

$$V_{out}(ave) = \frac{1}{2\pi}[-V_p \cos(\pi-\theta) - \Delta v(\pi-\theta) + V_p\cos\theta + \Delta v\theta] = \frac{1}{2\pi}[2V_p\cos\theta - (\pi - 2\theta)\Delta v]$$

Generally, the angle θ is small and thus $\cos\theta \approx 1$ and $(\pi - 2\theta) \approx \pi$. Therefore, the last relation above reduces to

$$V_{out}(ave) \approx \frac{V_p}{\pi} - \frac{\Delta v}{2} \tag{2.17}$$

Figure 2.34 shows a *full–wave bridge rectifier* circuit with input the sinusoid $v_{in}(t) = A\sin\omega t$ as shown in Figure 2.35, and the output of that circuit is $v_{out}(t) = |A\sin\omega t|$ as shown in Figure 2.36.

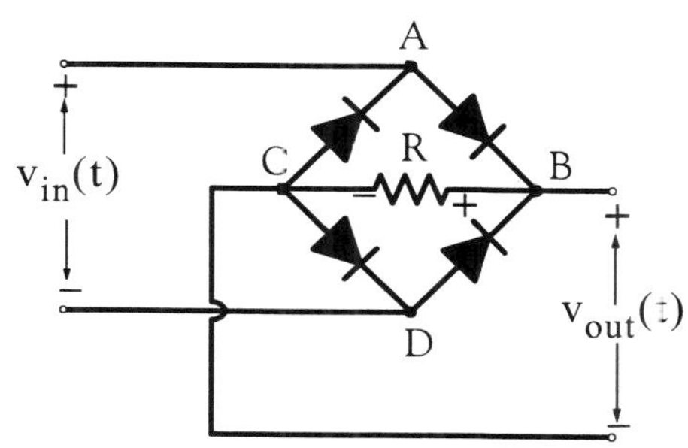

Figure 2.34. Full–wave rectifier circuit

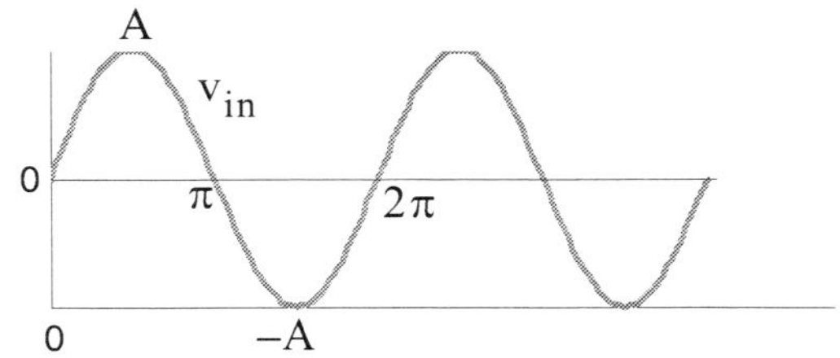

Figure 2.35. Input waveform for the circuit in Figure 2.34

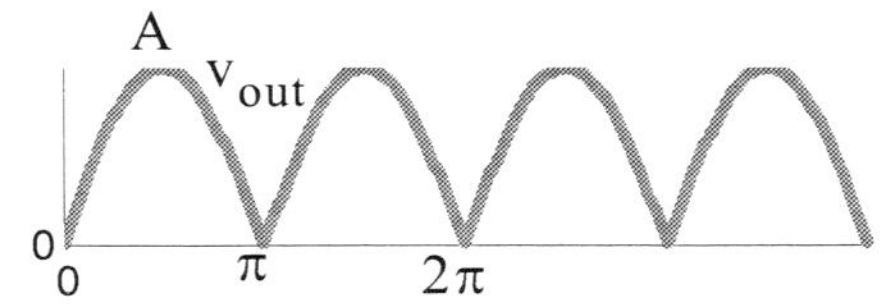

Figure 2.36. Output waveform for the circuit of Figure 2.34

From Figure 2.34 we see that during the positive half cycle conventional current flows from the voltage source to Point A, then to Point B, it goes through the resistor from Point B to Point C, and through Point D returns to the negative terminal of the voltage source. During the negative half cycle the lower terminal of the voltage source becomes the positive terminal, current flows from Point D to Point B, it goes through the resistor from Point B to Point C, and through Point A returns to the upper (now negative) terminal of the voltage source. We observe that during both the positive and negative half–cycles the current enters the right terminal of the resistor, and thus v_{out} is the same for both half–cycles as shown in the output waveform of Figure 2.36.

Figure 2.37 shows the input and output waveforms of the full–wave bridge rectifier on the same graph. It is to be noted that the difference in amplitude between v_{in} and v_{out} is denoted as $2\Delta v$ because in a full–wave bridge rectifier circuit there are two diodes in the conduction path instead of one as shown in Figure 2.33 for the half–wave rectifier. Accordingly, v_{out} lags v_{in} by the angle $\theta = \sin^{-1}(2\Delta v/v_{in})$. Also, the output is zero for an angle $2\theta = 2\sin^{-1}(2\Delta v/v_{in})$ centered around the zero crossing points.

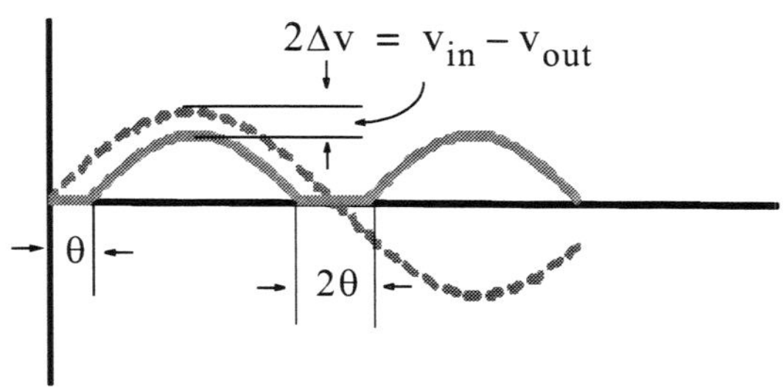

Figure 2.37. Input and output waveforms for the full–wave bridge rectifier of Figure 2.34

Figure 2.38(a) shows a full–wave rectifier with a center–tapped transformer secondary winding and Figure 2.38(b) shows the input and output waveforms.

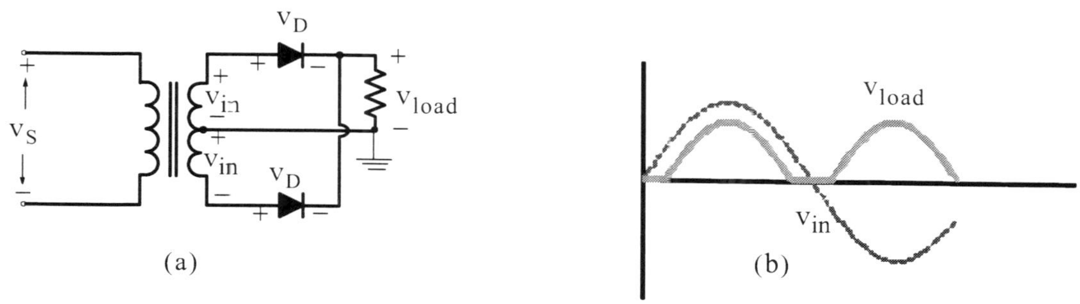

Figure 2.38. Full–wave rectifier with centered tapped secondary winding

The waveform in Figure 2.36 can be simulated with the **Saturation** block in the model in Figure 2.39.

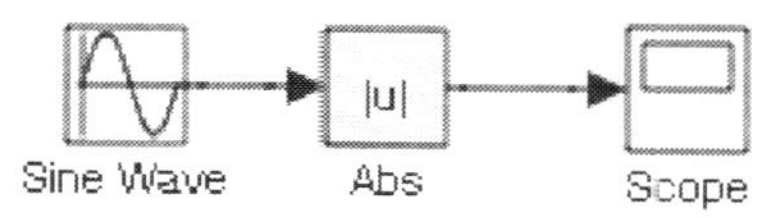

Figure 2.39. Simulink model for creating full–rectification waveform with an Abs block

The **Scope** block in Figure 2.39 displays the half–rectification waveform shown in Figure 2.40.

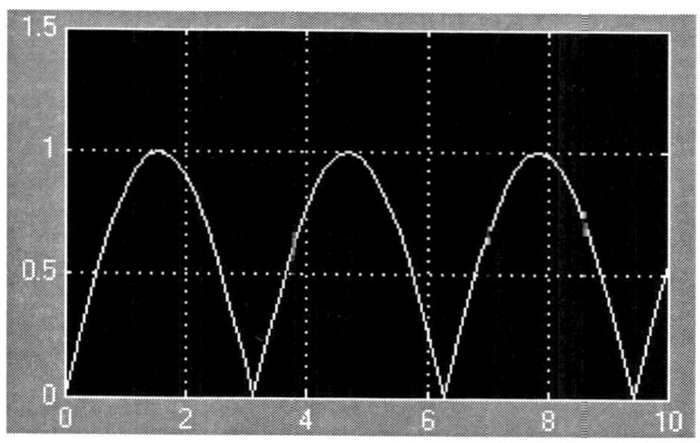

Figure 2.40. Full–rectification waveform displayed by the Scope block in Figure 2.39

The output voltages v_{out} from the half and full wave rectifiers are often called *pulsating DC voltages.* These voltages can be smoothed–out with the use of electric filters as illustrated with the following example.

Example 2.7

It is shown* in Fourier Analysis textbooks that the trigonometric Fourier series for the waveform of a full–wave rectifier with even symmetry is given by

$$v_R(t) = \frac{2A}{\pi} - \frac{4A}{\pi}\left\{\frac{\cos 2\omega t}{3} + \frac{\cos 4\omega t}{15} + \ldots\right\} \tag{2.18}$$

where A is the amplitude and this waveform appears across the resistor R of the full–wave rectifier in Figure 2.41 where the inductor and capacitor form a filter to smooth out the pulsating DC.

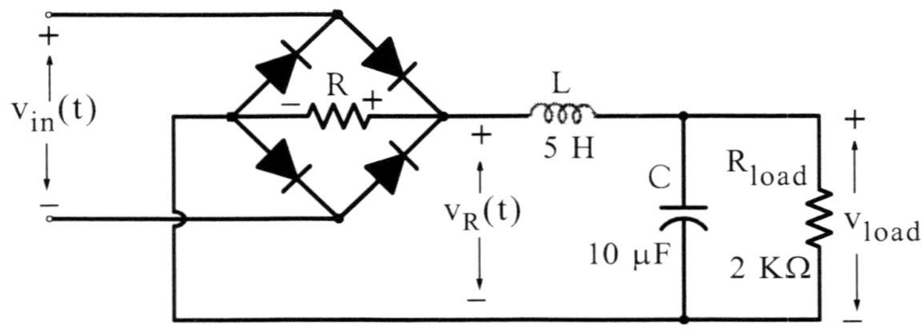

Figure 2.41. Circuit for Example 2.7

Compute and sketch the voltage v_{load} assuming that $v_{in}(t) = 120\ V\ RMS$ operating at the fundamental frequency $f = 60\ Hz$.

Solution:

We replace the given circuit by its phasor equivalent as shown in Figure 2.42 where the inductive reactance is $X_L = j\omega_n L$ and the capacitive reactance is $X_C = 1/j\omega_n C$.

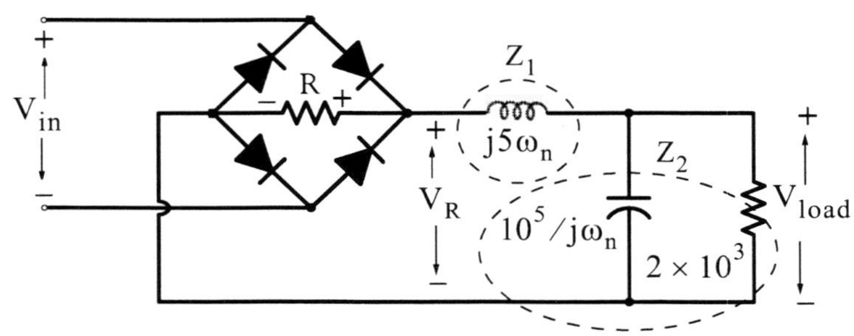

Figure 2.42. Phasor equivalent circuit for Example 2.7

For simplicity, we let

$$Z_1 = j5\omega_n$$

* *Please refer to Chapter 7 of Signals and Systems with MATLAB Computing and Simulink Modeling, ISBN 978–1–934404–11–9*

and

$$Z_2 = \frac{2\times 10^3 \times 10^5/j\omega_n}{2\times 10^3 + 10^5/j\omega_n} = \frac{2\times 10^8}{10^5 + 2\times 10^3 \times j\omega_n}$$

By the voltage division expression

$$V_{load} = \frac{Z_2}{Z_1+Z_2}V_R = \frac{(2\times 10^8)/(10^5+2\times 10^3\times j\omega_n)}{j5\omega_n + (2\times 10^8)/(10^5+2\times 10^3\times j\omega_n)}V_R$$
$$= \frac{(2\times 10^8)}{(j5\omega_n)(10^5+2\times 10^3\times j\omega_n)+(2\times 10^8)}V_R \qquad (2.19)$$

We will now compute the components of V_{load} for $n = 0, 2,$ and 4 and using superposition we will add these three terms.

We were given that

$$v_R(t) = \frac{2A}{\pi} - \frac{4A}{\pi}\left\{\frac{\cos 2\omega t}{3} + \frac{\cos 4\omega t}{15} + \dots\right\} \qquad (2.20)$$

and so for $n = 0$,

$$v_R(t)\big|_{n=0} = \frac{2A}{\pi}$$

and since for this DC case the inductor is just a short and the capacitor an open,

$$V_{load}\big|_{n=0} = V_R = \frac{2A}{\pi} \qquad (2.21)$$

For $n = 2$, $2\omega = 2\times 2\pi f = 4\times\pi\times 60 = 754\ r/s$ and from (2.20)

$$v_R(t)\big|_{n=2} = -\frac{4A}{3\pi}\cos 754t$$

and in the phasor $(j\omega)$ domain

$$V_R\big|_{n=2} = -\frac{4A}{3\pi}\angle 0^\circ$$

Then, from (2.19)

$$V_{load}\big|_{n=2} = \frac{Z_2}{Z_1+Z_2}V_R\Bigg|_{n=2} = \left[\frac{(2\times 10^8)}{(j5\times 754)(10^5+2\times 10^3\times j754)+(2\times 10^8)}\right]V_R$$

We will use MATLAB to find the magnitude and phase angle of the bracketed expression above.

```
bracket1=(2*10^8)/((j*5*754)*(10^5+2*10^3*j*754)+2*10^8);...
abs(bracket1), angle(bracket1)*180/pi
```

```
ans =
    0.0364
ans =
 -176.0682
```

Therefore,

$$V_{load}\Big|_{n=2} = \frac{Z_2}{Z_1 + Z_2}V_R\Big|_{n=2} = (0.0364\angle -176.1°) \times \left(-\frac{4A}{3\pi}\angle 0°\right) = -0.0154A\angle -176.1° \qquad (2.22)$$

For $n = 4$, $4\omega = 4 \times 2\pi f = 8 \times \pi \times 60 = 1,508\ r/s$ and from (2.20)

$$v_R(t)\Big|_{n=4} = -\frac{4A}{15\pi}\cos 1508t$$

and in the phasor $(j\omega)$ domain

$$V_R\Big|_{n=4} = -\frac{4A}{15\pi}\angle 0°$$

Then, from (2.19)

$$V_{load}\Big|_{n=4} = \frac{Z_2}{Z_1 + Z_2}V_R\Big|_{n=4} = \left[\frac{(2 \times 10^8)}{(j5 \times 1508)(10^5 + 2 \times 10^3 \times j1508) + (2 \times 10^8)}\right]V_R$$

We will again use MATLAB to find the magnitude and phase angle of the bracketed expression above.

```
bracket2=(2*10^8)/((j*5*1508)*(10^5+2*10^3*j*1508)+2*10^8);...
abs(bracket2), angle(bracket2)*180/pi
```

```
ans =
    0.0089
ans =
 -178.0841
```

Therefore,

$$V_{load}\Big|_{n=4} = \frac{Z_2}{Z_1 + Z_2}V_R\Big|_{n=4} = (0.089\angle -178.1°) \times \left(-\frac{4A}{15\pi}\angle 0°\right) = -0.00076\angle -178.1° \qquad (2.23)$$

Combining (2.21), (2.22), and (2.23) we obtain

$$\begin{aligned} V_{load} &= V_{load}\Big|_{n=0} + V_{load}\Big|_{n=2} + V_{load}\Big|_{n=4} \\ &= A(2/\pi - 0.0154\angle -176.1° - 0.00076A\angle -178.1°) \end{aligned} \qquad (2.24)$$

and in the time domain

$$v_{load}(t) = A(2/\pi - 0.0154\cos(2\omega t - 176.1°) - 0.00076\cos(4\omega t - 178.1)) \quad (2.25)$$

This is the filtered output voltage across the $2\ K\Omega$ resistor.

Let us plot (2.25) using MATLAB with $A = 20$ and $\omega = 377$.

```
t=0: 1: 2000; vL=20.*(2./pi–0.0154.*cos(2.*377.*t–176.1.*pi./180)...
–0.00076.*cos(4.*377.*t–178.1.*pi./180)); plot(t,vL); axis([0 2000 0 20]); hold on;...
plot(t,(40/pi–0.0154)); axis([0 2000 0 20]); xlabel('Time (sec)');...
ylabel('Load voltage (vL)'); title('Filtered output voltage for Example 2.7'); grid
```

The plot is shown in Figure 2.43 and we observe that the ripple is approximately ±0.5 V about the average (DC) value.

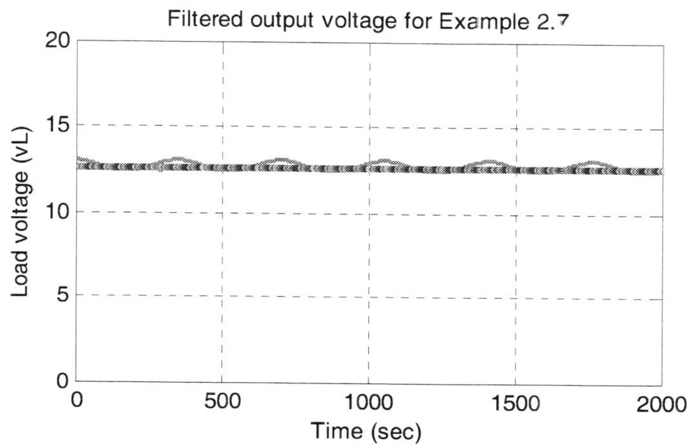

Figure 2.43. Voltage across the $2\ K\Omega$ resistor in Figure 2.42

The waveform in Figure 2.43 can be simulated with the Simulink model in Figure 2.44. The value $A = 20$ must be entered at the MATLAB command prompt.

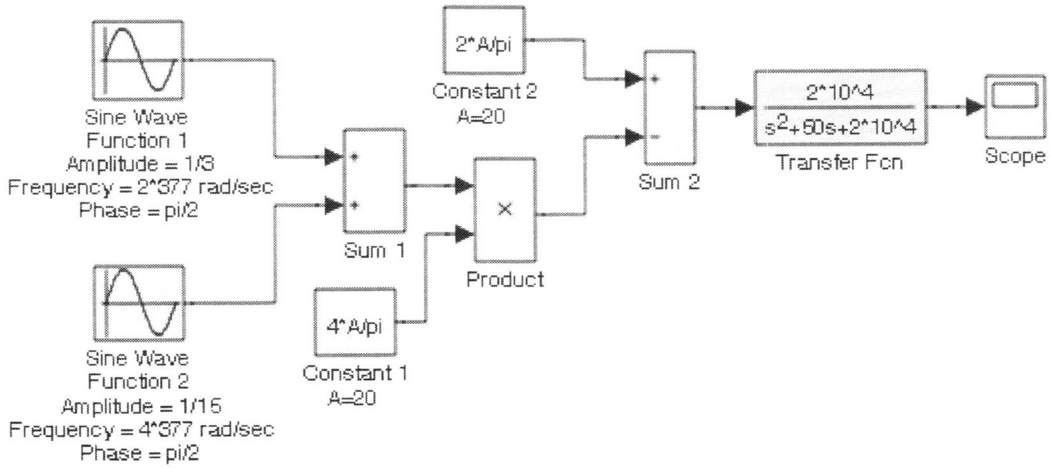

Figure 2.44. Simulink model for the circuit in Figure 2.42

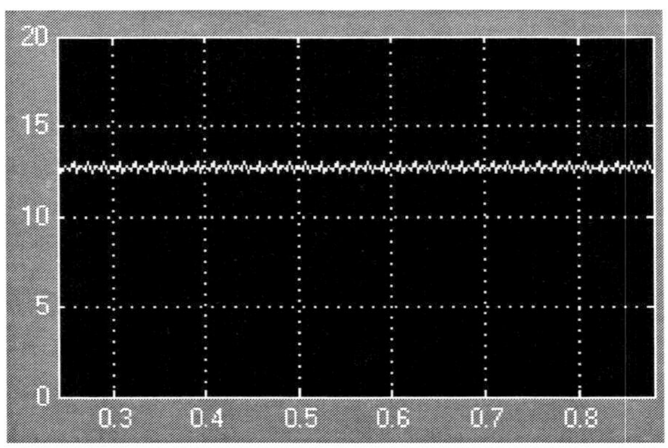

Figure 2.45. Full–rectification waveform displayed by the Scope block in Figure 2.44

2.7 Peak Rectifier Circuits

The circuit of Figure 2.46(a) is referred to as *peak rectifier*. Figure 2.46(b) shows the input v_{in} and output v_{out} waveforms and we observe that the value of v_{out} is approximately equal to the peak of the input sinewave v_{in}. We have assumed that the diode is ideal and thus as v_{in} is applied and reaches its positive peak value, the voltage v_{out} across the capacitor assumes the same value. However, when v_{in} starts decreasing, the diode becomes reverse–biased and the voltage across the capacitor remains constant since there is no path to discharge.

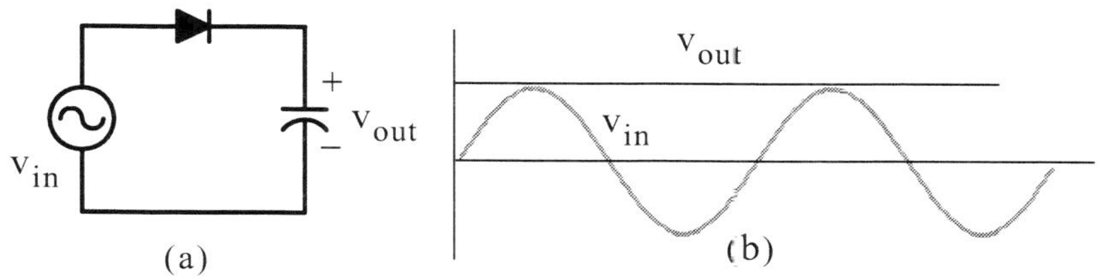

Figure 2.46. Peak rectifier circuit

The peak rectifier circuit shown in Figure 2.46(a) will behave like an *AC to DC converter* if we add a resistor across the capacitor as shown in Figure 2.47(a). Then, assuming that the diode is ideal and that the time constant $\tau = RC$ is much greater than the discharge interval, the voltage across the resistor–capacitor combination will be as shown in Figure 2.47(b).

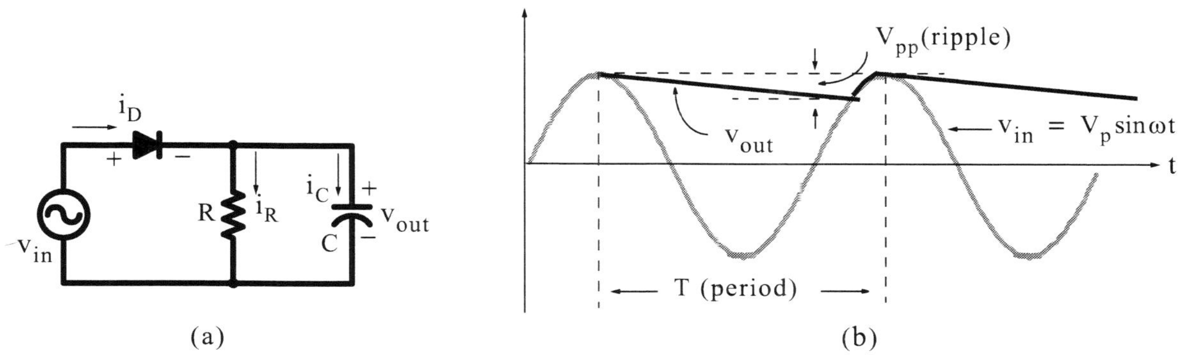

Figure 2.47. Peak rectifier used as an AC to DC converter

From the waveforms of Figure 2.47(b) se see that peak–to–peak ripple voltage $V_{pp}(ripple)$ can be made sufficiently small by choosing the time constant $\tau = RC$ much larger that the period T, that is, $RC \gg T$. Also, from Figure 2.47(b) we see that the average output voltage $V_{out}(ave)$ when the diode conducts is

$$V_{out}(ave) = V_p - \frac{1}{2}V_{pp}(ripple) \tag{2.26}$$

Next, we need to find an expression for $V_{pp}(ripple)$ when the diode does not conduct.

We recall from basic circuit theory* that in an simple RC circuit the capacitor voltage v_C discharges as the decaying exponential

$$v_C = v_{out} = V_p e^{-(t/RC)} \tag{2.27}$$

or

$$v_{out} = V_p - V_{pp}(ripple) = V_p e^{-(t/RC)} \tag{2.28}$$

* *Please refer to Circuit Analysis I with MATLAB Applications, ISBN 978–0–9709511–2–0.*

and since we want $RC \gg T$, we can simplify (2.28) by recalling that

$$e^{-x} = 1 - x + \frac{x^2}{2!} - \frac{x^3}{3!} + \ldots$$

and for small x,

$$e^{-x} \approx 1 - x$$

Therefore, we can express (2.28) as

$$V_{pp}(\text{ripple}) = V_p - V_p e^{-(t/RC)} = V_p(1 - e^{-(t/RC)}) = V_p\left(1 - 1 - \left(-\frac{T}{RC}\right)\right)$$

or

$$V_{pp}(\text{ripple}) = V_p\frac{T}{RC} \tag{2.29}$$

and from (2.29) we observe that for $V_{pp}(\text{ripple})$ to be small, we must make $RC \gg T$

2.8 Clipper Circuits

Clipper (or *limiter*) circuits consist of diodes, resistors, and sometimes DC sources to clip or limit the output to a certain level. Clipper circuits are used in applications where it is necessary to limit the input to another circuit so that the latter would not be damaged.

The input–output characteristics of clipper circuits are typically those of the forward–biased and reverse–biased diode characteristics except that the output is clipped to a a certain level. Figure 2.48 shows a clipper circuit and its input–output characteristics where the diode does not conduct for $v_{in} < 0.6\ V$ and so $v_{out} = v_{in}$. The diode conducts for $v_{in} \geq 0.7\ V$ and thus $v_{out} \approx 0.7\ V$.

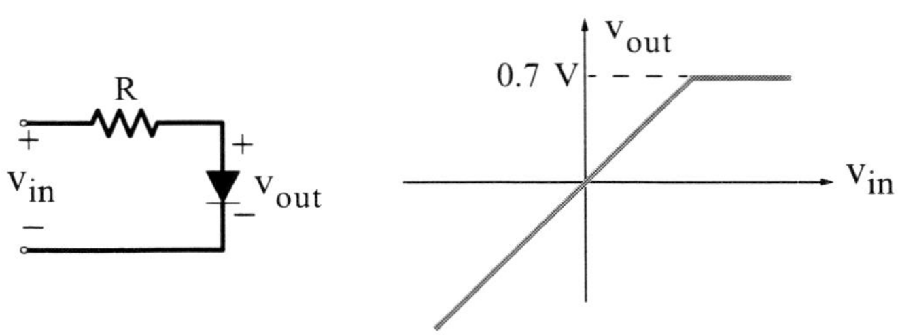

Figure 2.48. Circuit where $v_{out} = v_{in}$ *when diode does not conduct and* $v_{out} \approx 0.7\ V$ *when* $v_{in} \geq 0.7\ V$

Figure 2.49 shows a clipper circuit and its input–output characteristics where the diode does not conduct for $v_{in} > -0.6\ V$ but it conducts for $v_{in} \leq 0.7\ V$ and thus $v_{out} \approx -0.7\ V$.

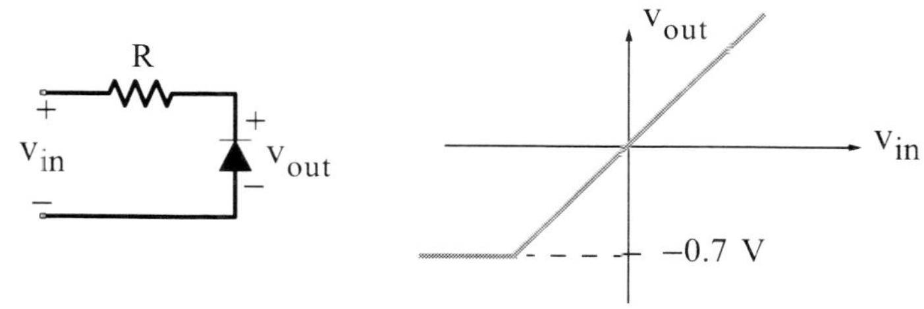

Figure 2.49. Circuit where $v_{out} = v_{in}$ *if diode does not conduct and* $v_{out} \approx -0.7\ V$ *if* $v_{in} \le -0.7\ V$

Figure 2.50 shows a clipper circuit with two diodes in parallel with opposite polarities. This circuit is effectively a combination of the clipper circuits shown in Figures 2.48 and 2.49 and both diodes are not conducting when $-0.7 \le v_{in} \le 0.7$ and for this interval $v_{out} = v_{in}$. Outside this interval, one or the other diode conducts. This circuit is often referred to as a *hard limiter.*

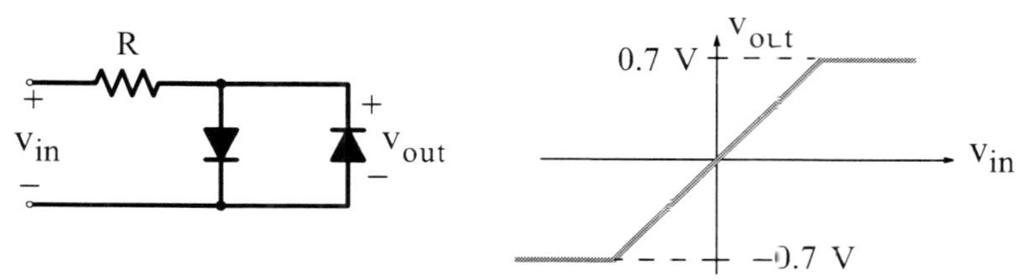

Figure 2.50. Circuit where $v_{out} = v_{in}$ *when* $-0.7 \le v_{in} \le 0.7$ *and* $v_{out} \approx \pm 0.7\ V$ *outside this interval*

Occasionally, it is desirable to raise the limit level to a value other than $v_{out} \approx \pm 0.7\ V$. This can be accomplished by placing a DC source in series with the diode as shown in Figure 2.51.

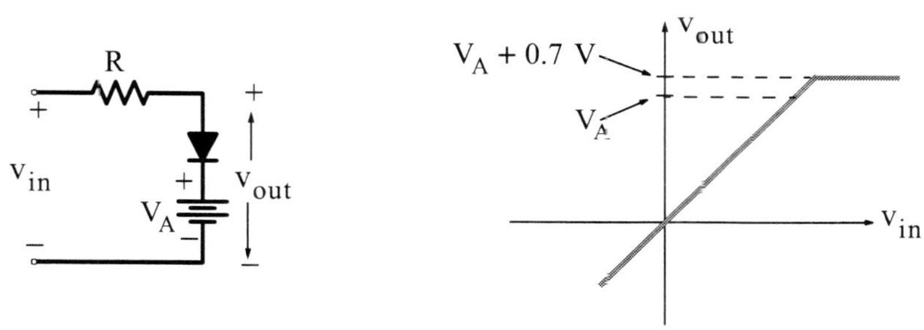

Figure 2.51. Circuit where $v_{out} = v_{in}$ *if diode does not conduct and* $v_{out} \approx V_A + 0.7\ V$ *if it conducts*

Example 2.8

In the circuit of Figure 2.52, all three resistors have the same value, $V_A = V_B$ with the polarities shown, the diodes are identical with $v_D = 0.7$, and $r_D \ll R$. Derive expressions for and sketch the $v_{out} - v_{in}$ characteristics.

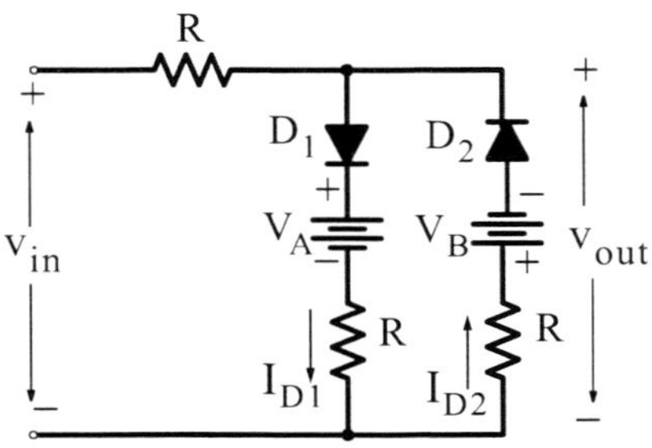

Figure 2.52. Circuit for Example 2.8

Solution:

Diode D_1 conducts when $v_{in} > V_A$ and under this condition diode D_2 does not conduct. Then,

$$I_{D1} = \frac{v_{in} - (V_A + v_D)}{R + R} = \frac{v_{in} - V_A - 0.7}{2R} \tag{2.30}$$

and

$$v_{out1} = 0.7 + V_A + RI_{D1} = 0.7 + V_A + R\frac{(v_{in} - V_A - 0.7)}{2R} = 0.5(v_{in} + V_A + 0.7) \tag{2.31}$$

Diode D_2 conducts when $v_{in} < -V_B$ and under this condition diode D_1 does not conduct. Then, in accordance with the passive sign convention* we obtain

$$I_{D2} = -\frac{v_{in} - (V_B - v_D)}{R + R} = -\frac{v_{in} - V_B + 0.7}{2R} \tag{2.32}$$

and

$$v_{out2} = -0.7 - V_B - RI_{D1} = -0.7 - V_B - R\left(-\frac{v_{in} - V_B + 0.7}{2R}\right) = 0.5(v_{in} - V_B - 0.7) \tag{2.33}$$

For range $-V_B < v_{in} < V_A$, $v_{out} = v_{in}$, and the $v_{out} - v_{in}$ characteristics are shown in Figure 2.53.

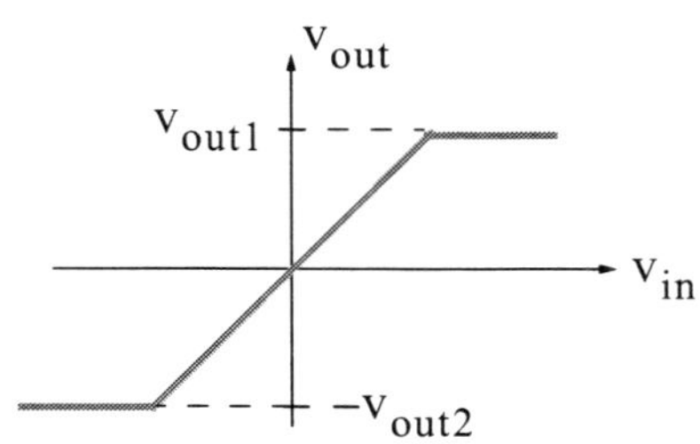

Figure 2.53. The $v_{out} - v_{in}$ characteristics for the circuit of Example 2.8

* *Please refer to Circuit Analysis I with MATLAB Applications, ISBN 978–0–9709511–2–0.*

2.9 DC Restorer Circuits

A DC restorer (or clamped capacitor) is a circuit that can restore a DC voltage to a desired DC level. The circuit of Figure 2.54 is a DC restorer and its operation can be best illustrated with the following example.

Example 2.9

In the circuit of Figure 2.54, the input v_{in} is as shown in Figure 2.55. Compute and sketch the waveform for the output v_{out}.

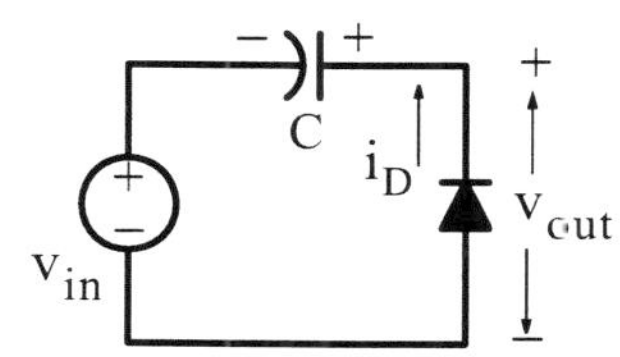

Figure 2.54. DC Restorer circuit

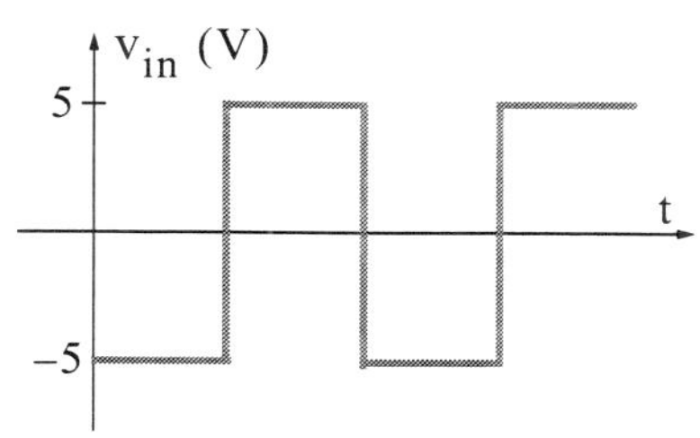

Figure 2.55. Input waveform for the circuit of Example 2.9

Solution:

From the circuit of Figure 2.54 we observe that the diode conducts only when $v_{in} < 0$ and thus the capacitor charges to the negative peak of the input which for this example is $-5\ V$, and the capacitor voltage is $v_C = 5\ V$ with polarity as shown in Figure 2.54. We observe that there is no path for the capacitor to discharge and therefore $v_C = 5\ V$ always. Since $v_{out} = v_{in} + v_C$, when $v_{in} = -5\ V$, $v_{out} = -5 + 5 = 0$ and the output has shifted upwards to zero volts as shown in Figure 2.56. When the input voltage is positive, $v_{in} = 5\ V$, and $v_{out} = 5 + 5 = 10\ V$. The waveform for the output voltage is as shown in Figure 2.56.

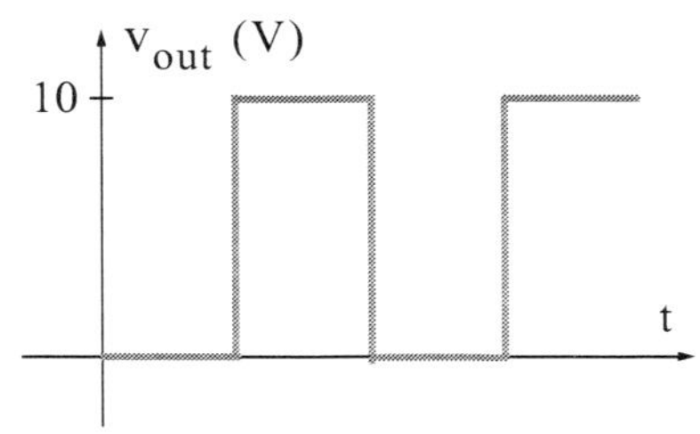

Figure 2.56. Output waveform for the circuit of Example 2.9

2.10 Voltage Doubler Circuits

The circuit of Figure 2.57 is referred to as a *voltage doubler.*

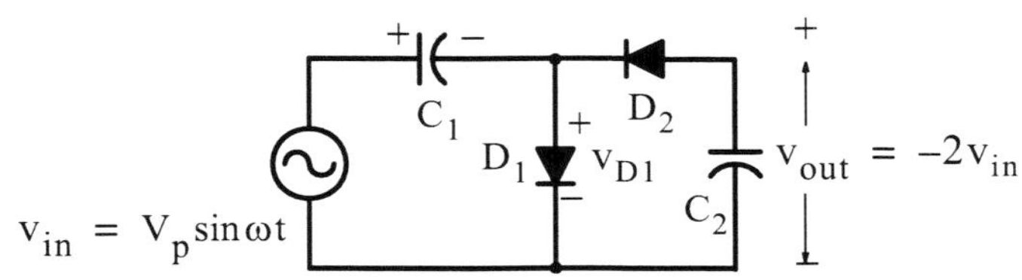

Figure 2.57. A voltage doubler circuit

With reference to the circuits of Figures 2.46 and 2.54, we observe that the circuit of Figure 2.57 is a combination of a DC restorer circuit consisting of the capacitor C_1 and diode D_1 and a peak rectifier consisting of C_2 and diode D_2. During the positive half–cycle of the input waveform diode D_1 conducts, the peak value V_p of the input v_{in} appears across the capacitor C_1 of and thus $v_{D1} = 0$. During the negative half–cycle of the input waveform diode D_2 conducts, capacitor C_2 is being charged to a voltage which is the sum of v_{in} and v_{C1} and thus the output voltage is $v_{out} = v_{in} + v_{C1} = -V_p \sin\omega t - V_p \sin\omega t = -2V_p \sin\omega t = -2v_{in}$ as shown in Figure 2.58.

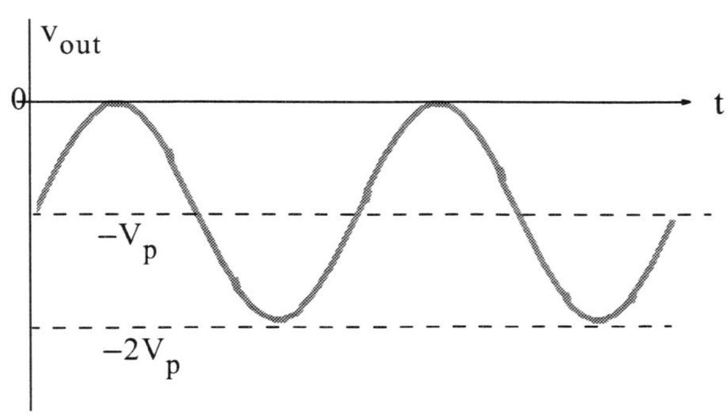

Figure 2.58. Output waveform for the voltage doubler circuit of Figure 2.57

Voltage triplers and *voltage quadruplers* can also be formed by adding more diodes and capacitor to a voltage doubler.

2.11 Diode Applications in Amplitude Modulation (AM) Detection Circuits

Junction diodes are also used in AM radio signals to remove the lower envelope of a modulated signal. A typical AM Envelope Detector circuit is shown in Figure 2.59. It is beyond the scope of this text to describe the operation of this circuit; it is described in electronic communications systems books. Our intent is to show the application of the junction diode in such circuits.

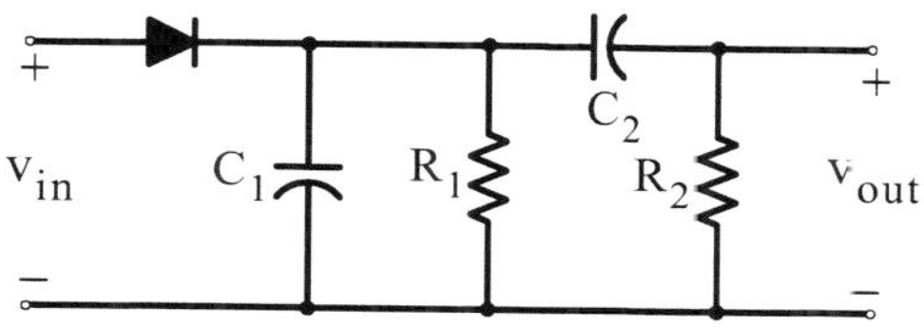

Figure 2.59. Typical AM Envelope Detector circuit

The operation of an AM audio frequency signal is shown in Figure 2.60 where the magnitude of $v_m(t)$ is always less than V_C; hence the envelope of the modulated wave never drops to zero. The waveform of the envelope of the modulated signal is the same as the waveform of the modulating signal $v_m(t)$. Thus if the information in the modulated carrier is to be preserved, the waveform of the envelope must be preserved. The information can be recovered by recovering the waveform of the envelope; the peak rectifier circuit is used for this purpose. The peak rectifier is also known as *diode detector*, and in this application it is frequently called an *envelope detector*. v_m. The diode removes the lower envelope and the upper envelope. In the circuit of Figure 2.59 capacitor C_2 and resistor R_2 form a high–pass filter that removes the carrier frequency.

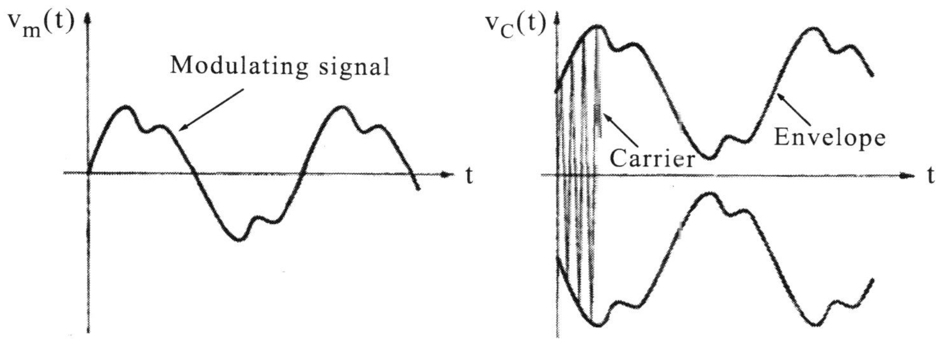

Figure 2.60. Input and output signals for the circuit of Figure 2.59

2.12 Diode Applications in Frequency Modulation (FM) Detection Circuits

Figure 2.61 shows how junction diodes can be used in FM circuits. This circuit is known as the

Foster–Seely discriminator and it is used to remove the carrier, which varies in accordance with the input audio frequency, and produce the audio frequency at its output.

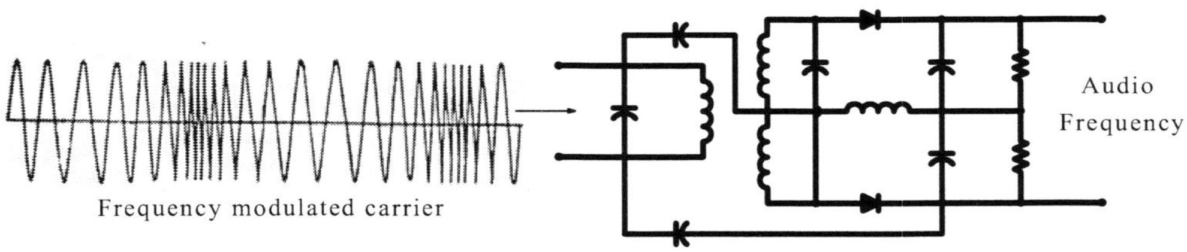

Figure 2.61. Foster–Seely discriminator circuit

2.13 Zener Diodes

By special manufacturing processes the reverse voltage breakdown of a diode can be made almost vertical as shown on the $i - v$ characteristics curve of Figure 2.62.

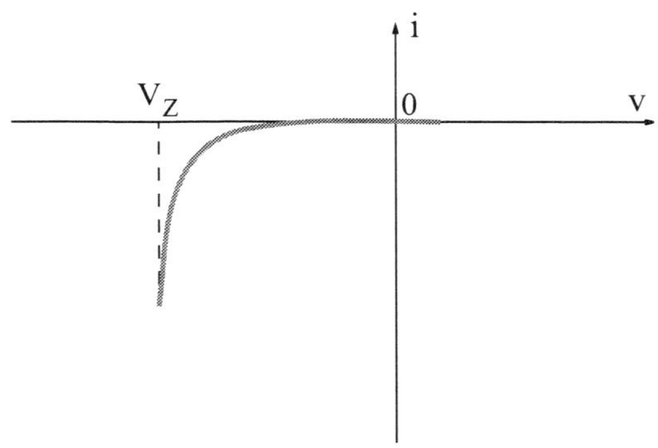

Figure 2.62. Zener diode operating region

These diodes are called *Zener diodes* and they are designed to operate in the breakdown region. The Zener diode is always connected as a reverse–biased diode and its voltage rating V_Z and the maximum power it can absorb are given by the manufacturer. The Zener diode symbol is shown in Figure 2.63.

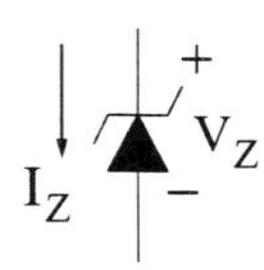

Figure 2.63. Symbol for a Zener diode

Zener diodes from small voltage ratings of 5 volts to large voltage ratings of 250 volts are commercially available and we can calculate the maximum current that a Zener diode can withstand from its voltage and power ratings. For instance, a 2 watt, 25 volt Zener diode can withstand a maximum current of 80 milliamperes.

One important application of the Zener diode is in voltage regulation. A voltage regulator is a circuit or device that holds an output voltage constant during output load variations. Example 2.10 below illustrates how a Zener diode accomplishes this.

Example 2.10

For the circuits (a) and (b) in Figure 2.64 it is required that the output voltage v_{out} remains constant at 25 volts regardless of the value variations of the load resistor R_L. Calculate the output voltage v_{out} for both circuits and then show how a 25 volt Zener diode can be used with these circuits to keep the output voltage constant at 25 volts.

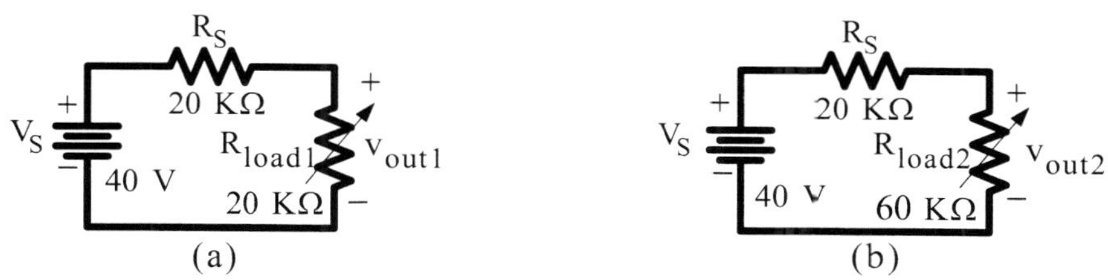

Figure 2.64. Circuits for Example 2.10

Solution:

For the circuit of Figure 2.64(a) the output voltage v_{out1} is found by application of the voltage division expression. Thus,

$$v_{out1} = \frac{R_{load1}}{R_S + R_{load1}} V_S = \frac{20\ K\Omega}{20\ K\Omega + 20\ K\Omega} 40\ V = 20\ V$$

Likewise, for the circuit of Figure 2.64(b) the output voltage v_{out2} is

$$v_{out2} = \frac{R_L}{R_S + R_{load2}} V_S = \frac{60\ K\Omega}{20\ K\Omega + 60\ K\Omega} 40\ V = 30\ V$$

These calculations show that as the load R_{load} varies from $20\ K\Omega$ to $60\ K\Omega$ the output voltage varies from 20 volts to 30 volts and this variation indicates a poor voltage regulation. To improve the regulation we connect a 25 volt Zener diode in parallel with the load as shown in Figure 2.65 and the output voltages v_{out1} and v_{out2} will be the same, that is, 25 volts.

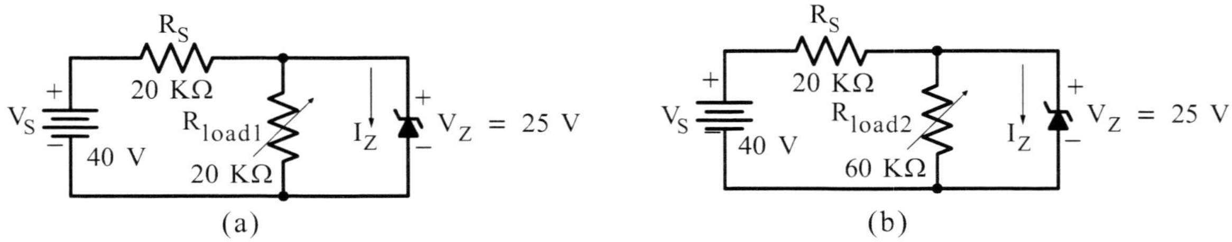

Figure 2.65. Circuits for Example 2.10 with Zener diodes to improve regulation

In Example 2.10 above we did not take into consideration the manufacturer's specifications. Example 2.11 below is a more realistic example and shows that a Zener diode may or may not conduct depending on the value of the load resistor.

The Zener diode model is very similar to the junction diode model and it is shown in Figure 2.66(a). Of course, the Zener diode is operating at a different region where V_{Z0} and r_Z are as shown in Figure 2.66(b).

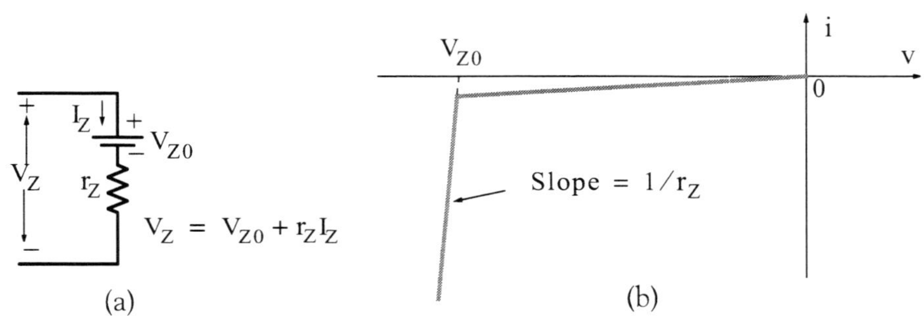

Figure 2.66. The Zener diode model and the definitions of V_{Z0} and r_Z

Thus, V_{Z0} is v–axis intercept of the straight line with slope $1/r_Z$ and $r_Z = \Delta v/\Delta i$. Like the incremental resistance r_D of a junction diode, the incremental resistance r_Z of a Zener diode is typically about 20 Ω and it is specified by the manufacturer.

Example 2.11

For the circuit of Figure 2.67, the manufacturer's datasheet for the Zener diode shows that $V_Z = 10$ V when $I_Z = 6$ mA. The datasheet provides also the $i-v$ characteristics but the value of V_{Z0} is not given. The current I_{ZBD} in Figure 2.68 represents the current where the Zener diode begins entering the breakdown region.

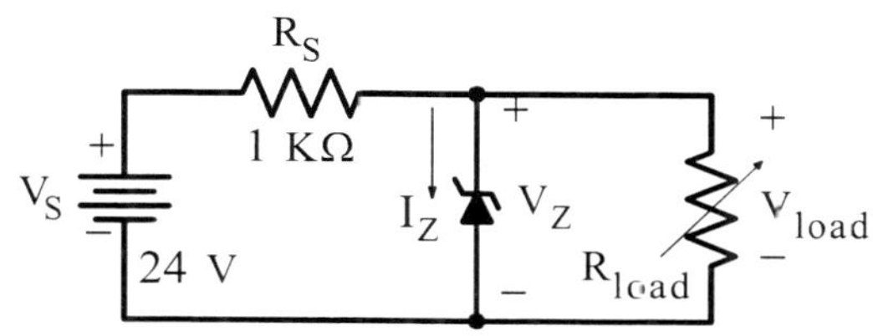

Figure 2.67. Circuit for Example 2.11

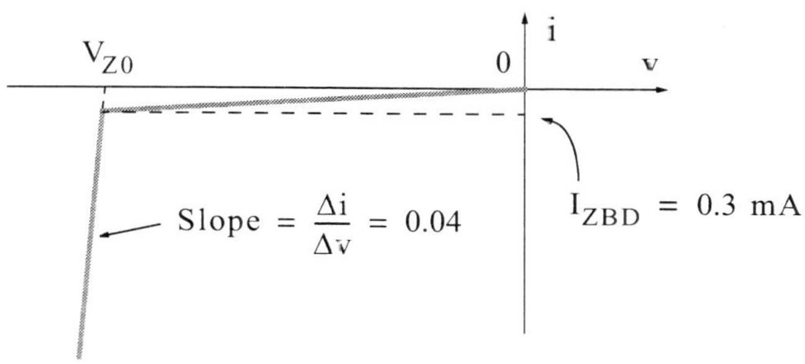

Figure 2.68. Graph for Example 2.11

Compute:

a. The output voltage V_{load} if the load resistor R_{load} is disconnected.

b. The output voltage V_{load} if the load resistor R_{load} is connected and adjusted to $1\ K\Omega$.

c. The output voltage V_{load} if the load resistor R_{load} is connected and adjusted to $200\ \Omega$.

d. The minimum value of the load resistor R_{load} for which the Zener diode will be conducting in the breakdown region.

Solution:

a.

From Figure 2.66(a),

$$V_Z = V_{Z0} + r_Z I_Z \tag{2.34}$$

and replacing the Zener diode by its equivalent we obtain the circuit shown in Figure 2.69 where the slope is

$$1/r_Z = \Delta i/\Delta v = 0.04$$

$$r_Z = \Delta v/\Delta i = 1/0.04 = 25\ \Omega$$

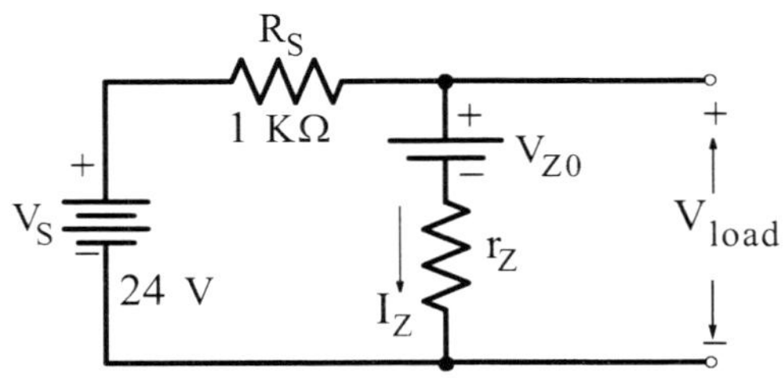

Figure 2.69. Zener diode replaced by its equivalent circuit and the load is disconnected

From (2.34),

$$V_{Z0} = V_Z - r_Z I_Z = 10 - 25 \times 6 \times 10^{-3} = 8.5 \text{ V}$$

and

$$I_Z = \frac{V_S - V_{Z0}}{R_S + r_Z} = \frac{24 - 8.5}{1000 + 25} = 15.2 \text{ mA}$$

Thus

$$v_{out} = V_{Z0} + r_Z I_Z = 8.5 + 25 \times 15.2 \times 10^{-3} = 8.8 \text{ V}$$

b.

When the load resistor is connected and adjusted to $1 \text{ K}\Omega$ the circuit is as shown in Figure 2.70.

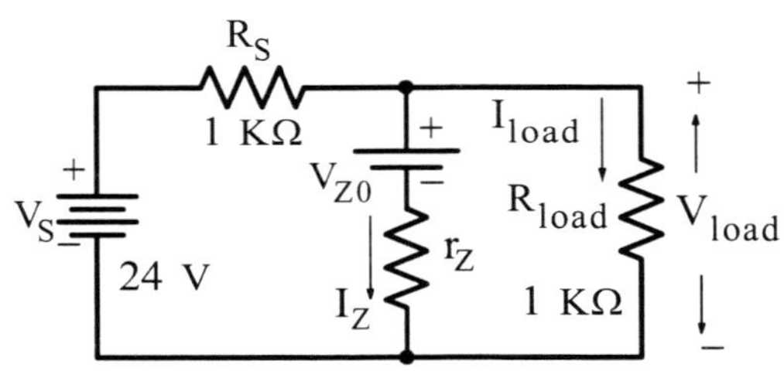

Figure 2.70. Circuit of Example 2.11 with load resistor connected and adjusted to $1 \text{ K}\Omega$

We will assume that the Zener diode still operates well within the breakdown region so that the addition of the load resistor does not change the output voltage significantly. Then,

$$I_{load} \approx \frac{V_{load}}{R_{load}} = \frac{8.8}{10^3} = 8.8 \text{ mA}$$

This means that the Zener diode current I_Z will be reduced to

$$I'_Z = 15.2 - 8.8 = 6.4 \text{mA}$$

and thus the output voltage will be

$$V_{load} = V_{Z0} + r_Z I'_Z = 8.5 + 25 \times 6.4 \times 10^{-3} = 8.5 \text{ V}$$

The percent change in output voltage is

$$\% \text{ change} = \frac{8.5 - 8.8}{8.5} \times 100 = -3.53$$

c.

When the load resistor is connected and adjusted to $200\ \Omega$ the circuit is as shown in Figure 2.71.

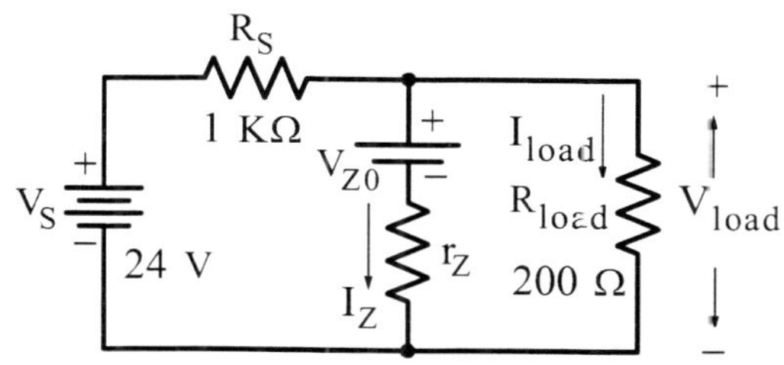

Figure 2.71. Circuit of Example 2.11 with load resistor connected and adjusted to $200\ \Omega$

We will again assume that the Zener diode still operates well within the breakdown region so that the addition of the load resistor does not change the output voltage significantly. Then,

$$I_{load} \approx \frac{V_{load}}{R_{load}} = \frac{8.8}{200} = 44 \text{ mA}$$

But this is almost three times higher than the current of $I_Z = 15.2 \text{ mA}$ found in (a). Therefore, we conclude that the Zener diode does not conduct and, with the Zener diode branch being an open circuit, by the voltage division expression we obtain

$$V_{load} = \frac{R_{load}}{R_S + R_{load}} V_S = \frac{200}{1000 + 200} \times 24 = 4 \text{ V}$$

and since this voltage is less than 10 V the Zener diode does not conduct.

d.

From the graph of Figure 2.68 it is reasonable to assume that when $I_{ZBD} = 0.3 \text{ mA}$, then $V_{ZBD} \approx 9.7 \text{ V}$. Then, the current I_{RS} through the resistor which would allow the Zener diode to conduct will be

$$I_{RS} = \frac{V_S - V_{ZBD}}{R_S} = \frac{24 - 9.7}{10^3} = 14.3 \text{ mA}$$

This is the total current supplied by the V_S source and since $I_{ZBD} = 0.3 \text{ mA}$, the current through the load resistor, whose value is to be determined, will be

$$I_{load} = I_{RS} - I_{ZBD} = 14.3 - 0.3 = 14 \text{ mA}$$

Therefore,

$$R_{load} = \frac{V_{ZBD}}{I_{load}} = \frac{9.7}{14 \times 10^{-3}} = 693\ \Omega$$

Zener diodes can also be used as limiters. Figure 2.72 shows how a single Zener diode can be used to limit one side of a sinusoidal waveform to V_Z while limiting the other side to approximately zero.

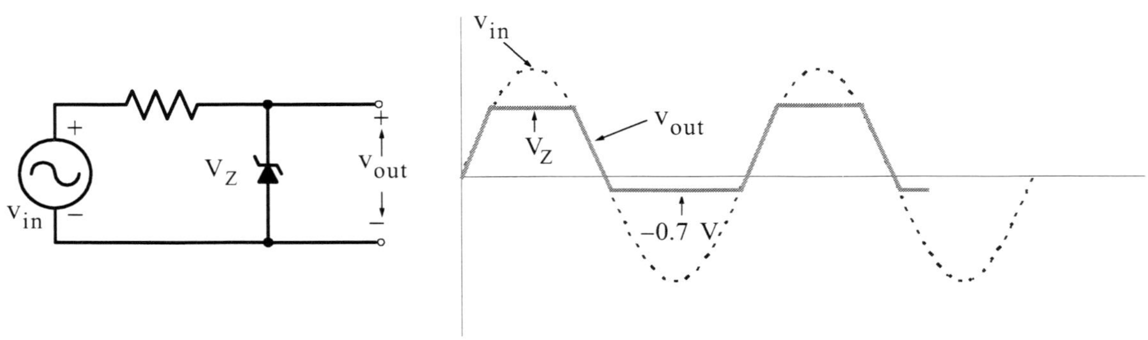

Figure 2.72. Zener limiter with one Zener diode

Figure 2.73 shows a circuit with two opposing Zener diodes limiting the input waveform to V_Z on both polarities.

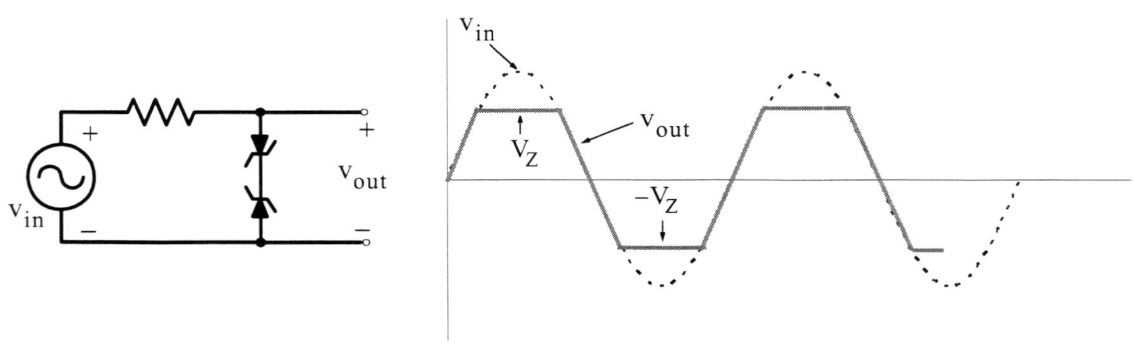

Figure 2.73. Zener limiter with two opposing Zener diodes

2.14 The Schottky Diode

A *Schottky diode* is a junction of a lightly doped n –type semiconductor with a metal electrode as shown in Figure 2.74(a). Figure 2.74(b) shows the Schottky diode symbol.

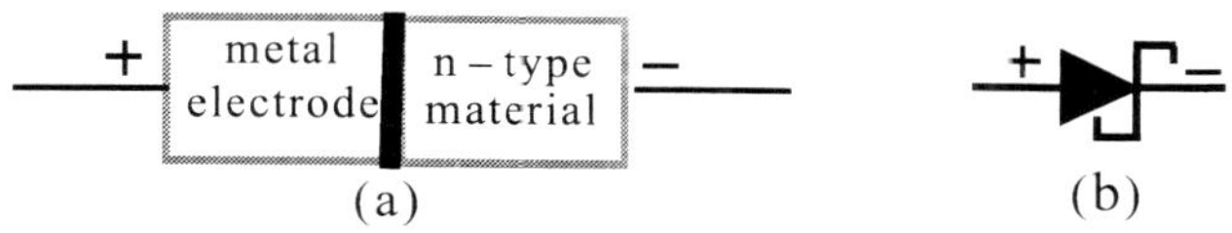

Figure 2.74. The components of a Schottky diode and its symbol

The junction of a doped semiconductor – usually n –type – with a special metal electrode can produce a very fast switching diode which is mainly used in high (up to 5 MHz) frequency circuits or high speed digital circuits. When the diode is forward–biased, the electrons move from the n –type material to the metal and give up their energy quickly. Since there are no minority carriers (holes) the conduction stops quickly and changes to reverse–bias.

Schottky diodes find wide application as rectifiers for high frequency signals and also are used in the design of galliun arsenide circuits. The forward voltage drop of a conducting Schottky diode is typically 0.3 to 0.5 volt compared to the 0.6 to 0.8 found in silicon junction diodes.

2.15 The Tunnel Diode

The conventional junction diode uses semiconductor materials that are lightly doped with one impurity atom for ten–million semiconductor atoms. This low doping level results in a relatively wide depletion region. Conduction occurs in the normal junction diode only if the voltage applied to it is large enough to overcome the potential barrier of the junction. In 1958, Leo Esaki, a Japanese scientist, discovered that if a semiconductor junction diode is heavily doped with impurities, it will have a region of negative resistance. This type of diode is known as *tunnel diode*.

In a tunnel diode the semiconductor materials used in forming a junction are doped to the extent of one–thousand impurity atoms for ten–million semiconductor atoms. This heavy doping produces an extremely narrow depletion zone similar to that in the Zener diode. Also because of the heavy doping, a tunnel diode exhibits an unusual current–voltage characteristic curve as compared with that of an ordinary junction diode. The characteristic curve for a tunnel diode and its symbol are shown in Figure 2.75.

From Figure 2.75 we see that the current increases to a peak value with a small applied forward bias voltage (point 1 to 2), then the current decreases with an increasing forward bias to a minimum current (point 2 to 3), and it starts increasing again with further increases in the bias voltage point 3 to 4.)

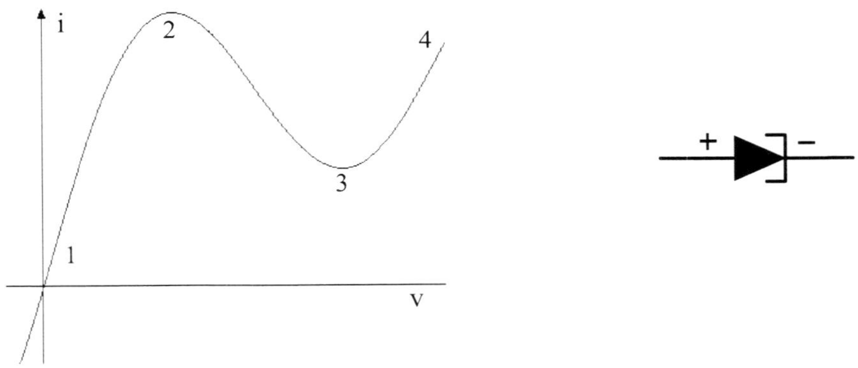

Figure 2.75. Tunnel diode characteristics and symbol

The portion of the characteristic curve between points 2 and 3 is the region of *negative resistance* and an explanation of why a tunnel diode has a region of negative resistance is best understood by using energy levels discussed in semiconductor physics texts.

The negative resistance characteristic makes the tunnel diode useful in oscillators and microwave amplifiers. The *unijunction transistor*, discussed in Chapter 4, has similar oscillator characteristics. A typical tunnel diode oscillator is shown in Figure 2.76.

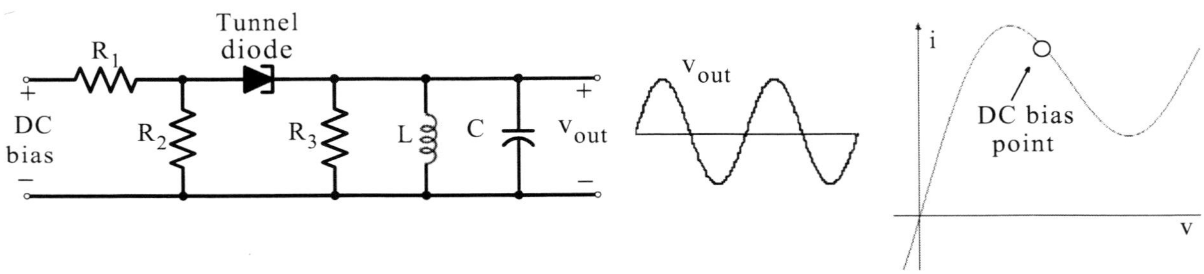

Figure 2.76. Tunnel diode oscillator circuit

Because the negative resistance region of the tunnel diode is its most important characteristic, it is desirable that the so–called *peak–to–valley ratio* I_P/V_P must be quite high. Typical values of I_P/V_P are 3.5 for silicon, 6 for germanium, and 15 for gallium arsenide (GaAs), and the corresponding values of V_P are 65, 55, and 150 mV respectively. For this reason, silicon is not used in the manufacturing of tunnel diodes. The voltage at which the current begins to rise again is denoted as V_V and typical values for silicon, germanium, and GaAs are 420, 350, and 500 mV respectively.

A variation of the tunnel diode is the *backward diode* that is used as a rectifier in which "forward" (p side negative) direction occurs without the usual voltage offset of a conventional junction diode. The "reverse" direction corresponds to the conventional forward direction so that the "reverse" breakdown voltage is typically 0.65 V. The backward diode is also used in circuits with small amplitude waveforms.

Example 2.12

A DC voltage source, a tunnel diode whose $i-v$ characteristics are as shown, and a resistor are connected in series as shown in Figure 2.77. Find the current i

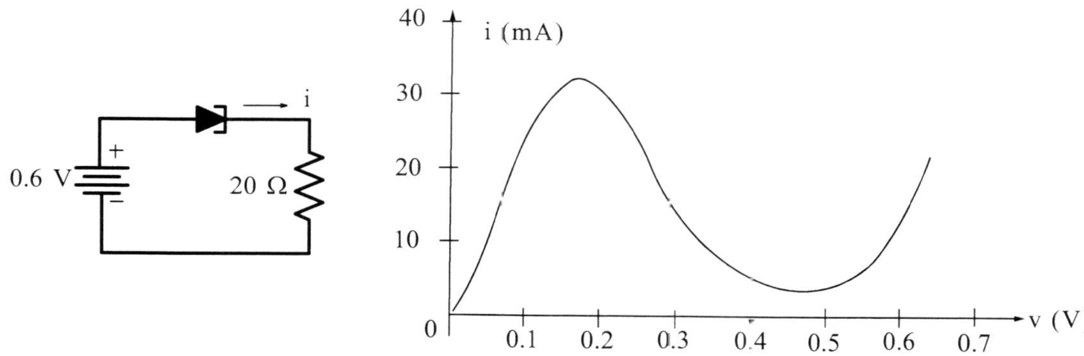

Figure 2.77. Circuit and tunnel diode $i-v$ characteristics for Example 2.12

Solution:

Let the voltage across the tunnel diode be V_{TD} and the voltage across the resistor be V_R. Then,

$$V_{TD} - V_R = 0.6\ V$$

or

$$V_{TD} + 20i = 0.6\ V \tag{2.35}$$

When $i = 0$, (2.35) reduces to $V_{TD} = 0.6$ and when $V_{TD} = 0$, $i = 0.6/20 = 30\ mA$, and with these values we draw the load line shown in Figure 2.78. We observe that when $V_{TD} \approx 0.11\ V$, $i \approx 25\ mA$, when $V_{TD} \approx 0.29\ V$, $i \approx 16\ mA$, and when $V_{TD} \approx 0.52\ V$, $i \approx 4\ mA$.

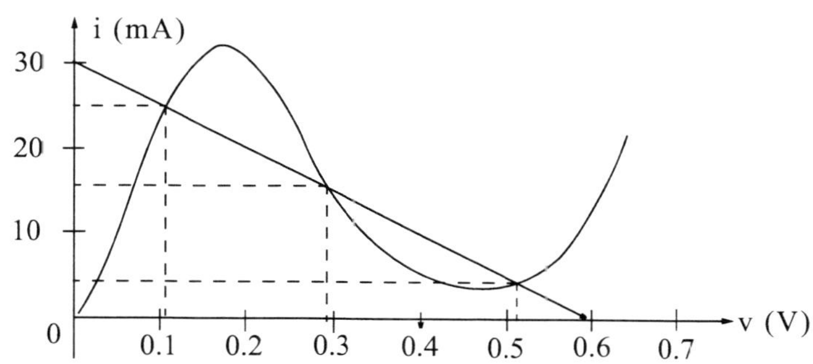

Figure 2.78. Graphical solution for the determination of the current i for the circuit of Figure 2.77

2.16 The Varactor

The *varactor*, or *varicap*, is a diode that behaves like a variable capacitor, with the PN junction functioning like the dielectric and plates of a common capacitor. For this reason, the symbol for a varactor is as shown in Figure 2.79.

Figure 2.79. Symbol for varactor

A varactor diode uses a PN junction in reverse bias and has a structure such that the capacitance of the diode varies with the reverse voltage. A voltage controlled capacitance is useful in tuning applications. Typical capacitance values are small, in the order of picofarads.

Presently, varactors are replacing the old variable capacitor tuning circuits as in television tuners. Figure 2.80 shows a typical varactor tuning circuit where the DC control voltage V_C changes the capacitance of the varactor to form a resonant circuit.

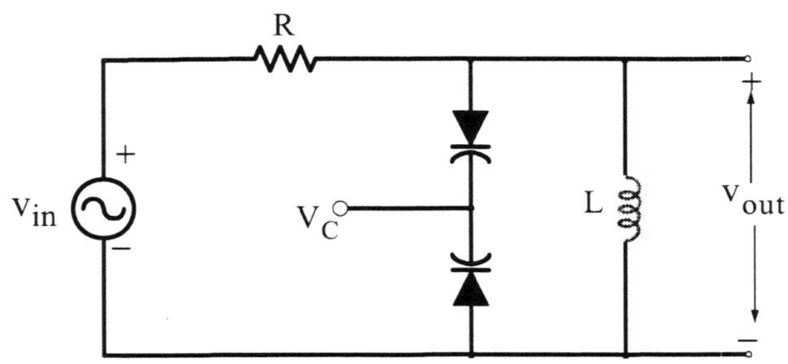

Figure 2.80. A varactor tuner circuit

2.17 Optoelectronic Devices

Optoelectronic devices either produce light or use light in their operation. The first of these, the *light–emitting diode* (LED), was developed to replace the fragile, short–life incandescent light bulbs used to indicate on/off conditions on instrument panels. A light–emitting diode, when forward biased, produces visible light. The light may be red, green, or amber, depending upon the material used to make the diode.

The circuit symbols for all optoelectronic devices have arrows pointing either toward them, if they use light, or away from them, if they produce light. The LED is designated by a standard diode symbol with two arrows pointing away from the cathode as shown in Figure 2.81 where the arrows indicate light leaving the diode.

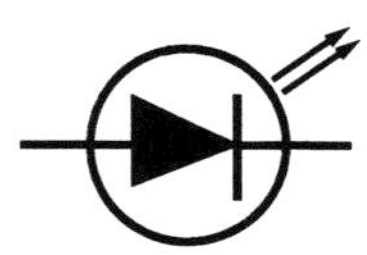

Figure 2.81. Symbol for LED

The LED operating voltage is small, about 1.6 volts forward bias and generally about 10 milliamperes. The life expectancy of the LED is very long, over 100,000 hours of operation. For this reason, LEDs are used widely as "power on" indicators of digital voltmeters, frequency counters, etc. and as displays for pocket calculators where they form seven–segment displays, as shown in Figure 2.82.

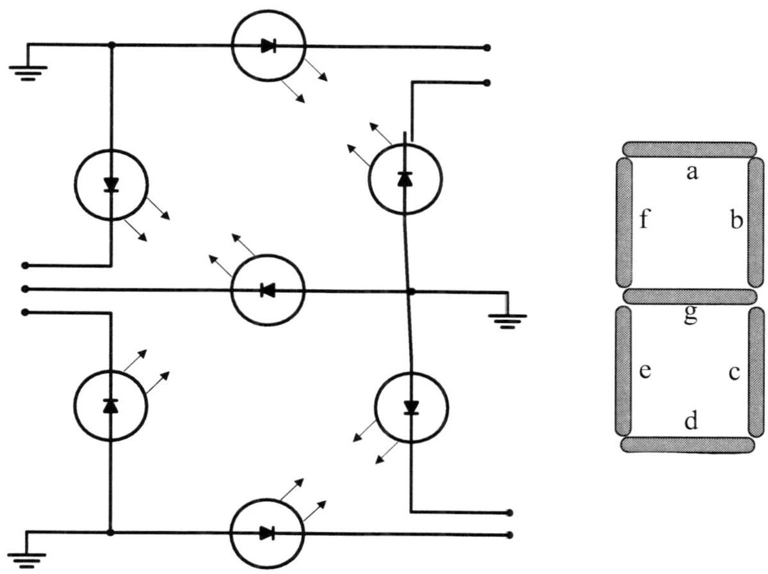

Figure 2.82. LEDs arranged for seven segment display

In Figure 2.82 the anodes are internally connected. When a negative voltage is applied to the proper cathodes, a number is formed. For example, if negative voltage is applied to all cathodes the number 8 is produced, and if a negative voltage is changed and applied to all cathodes except LED b and e the number 5 is displayed.

Seven–segment displays are also available in common–cathode form, in which all cathodes are at the same potential. When replacing LED displays, one must ensure the replacement display is the same type as the faulty display. Since both types look alike, the manufacturer's number should be checked.

Laser diodes are LEDs specifically designed to produce coherent light with a narrow bandwidth and are suitable for CD players and optical communications.

Another optoelectronic device in common use today is the *photodiode*. Unlike the LED, which produces light, the photodiode receives light and converts it to electrical signals. Basically, the photodiode is a light–controlled variable resistor. In total darkness, it has a relatively high resistance and therefore conducts little current. However, when the PN junction is exposed to an external light source, internal resistance decreases and current flow increases. The photodiode is operated with reverse–bias and conducts current in direct proportion to the intensity of the light source. The symbol for a photodiode is shown in Figure 2.83 where the arrows pointing toward the diode indicate that light is required for operation of the device.

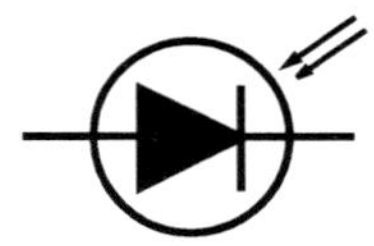

Figure 2.83. Symbol for photodiode

Switching the light source on or off changes the conduction level of the photodiode and varying the light intensity controls the amount of conduction. Photodiodes respond quickly to changes in light intensity, and for this reason are extremely useful in digital applications such as photographic light meters and optical scanning equipment.

A *phototransistor* is another optoelectronic device that conducts current when exposed to light. We will discuss phototransistor in the next chapter.

An older device that uses light in a way similar to the photodiode is the photoconductive cell, or *photocell*, shown with its schematic symbol in Figure 2.84. Like the photodiode, the photocell is a light–controlled variable resistor. However, a typical light–to–dark resistance ratio for a photocell is 1: 1000. This means that its resistance could range from 1000 ohms in the light to 1000 kilohms in the dark, or from 2000 ohms in the light to 2000 kilohms in the dark, and so forth. Of course, other ratios are also available. Photocells are used in various types of control and timing circuits as, for example, the automatic street light controllers in most cities. The symbol for a photocell is shown in Figure 2.84.

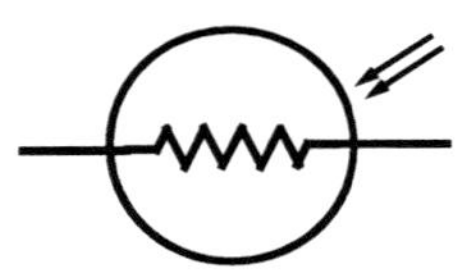

Figure 2.84. Symbol for photocell

A *solar cell* is another device that converts light energy into electrical energy. A solar cell acts much like a battery when exposed to light and produces about 0.45 V volt across its terminals, with current capacity determined by its size. As with batteries, solar cells may be connected in series or parallel to produce higher voltages and currents. The device is finding widespread application in communications satellites and solar–powered homes. The symbol for a solar cell is shown in Figure 2.85.

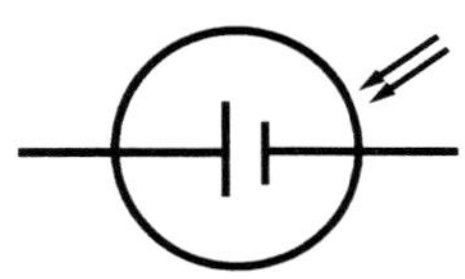

Figure 2.85. Symbol for solar cell

Figure 2.86 shows an *optical coupler*, also referred to as an *optoisolator*. The latter name is derived from the fact that it provides isolation between the input and output. It consists of an LED and a photodiode and each of these devices are isolated from each other.

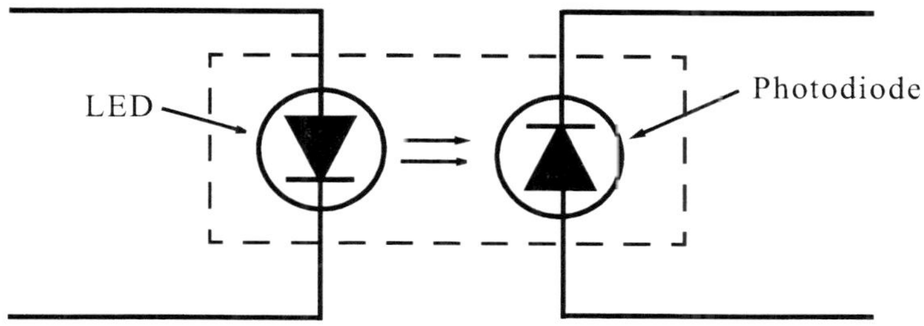

Figure 2.86. An optical coupler

Isolation between the input and output is desirable because it reduces electromagnetic interference. Their most important application is in fiber optic communications links.

2.18 Summary

- In a crystal of pure silicon that has been thermally agitated there is an equal number of free electrons and holes. A free electron is one which has escaped from its valence orbit and the vacancy it has created is called a hole.
- Doping is a process where impurity atoms which are atoms with five valence electrons such as phosphorous, arsenic, and antimony, or atoms with three valence electrons such as boron, aluminum, and gallium, are added to melted silicon.
- An N–type semiconductor has more free electrons than holes and for this reason the free electrons are considered to be the majority carriers and the holes the minority carriers.
- A P–type semiconductor has more holes than free electrons and thus the holes are the majority carriers and the free electrons are the minority carriers.
- A junction diode is formed when a piece of P–type material and a piece of N–type material are joined together. The area between the P–type and N–type materials is referred to as the depletion region.
- The electrostatic field created by the positive and negative ions in the depletion region is called a barrier.
- A forward–biased PN junction is formed if we attach a voltage source to a junction diode with the plus (+) side of the voltage source connected to the P–type material and the minus (–) side to the N–type. When a junction diode is forward–biased, conventional current will flow in the direction of the arrow on the diode symbol.
- A reverse–biased PN junction is formed if we attach a voltage source to a junction diode with the plus (+) side of the voltage source connected to the N–type material and the minus (–) side to the P–type.
- The P–type side of the junction diode is also referred to as the anode and the N–type side as the cathode.
- The $i_D - v_D$ relationship in a forward–biased junction diode is the nonlinear relation

$$i_D = I_r[e^{(qv_D/nkT)} - 1]$$

where i_D is the current through the diode, v_D is the voltage drop across the diode, I_r is the reverse current, that is, the current which would flow through the diode if the polarity of v_D is reversed, q is charge of an electron, that is, $q = 1.6 \times 10^{-19}$ coulomb, the coefficient n varies from 1 to 2 depending on the current level and the nature or the recombination near the junction, $k = Boltzmann's\ constant$, that is, $k = 1.38 \times 10^{-23}$ joule/Kelvin, and T is the absolute temperature in degrees Kelvin, that is, $T = 273 + \text{temperature in } {}^oC$. It is convenient to combine q, k, and T into one variable V_T known as thermal voltage where $V_T = kT/q$ and thus

$$i_D = I_r[e^{(v_D/nV_T)} - 1]$$

and at $T = 300\ °K$, $V_T \approx 26mV$

- In a junction diode made with silicon and an impurity, conventional current will flow in the direction of the arrow of the diode as long as the voltage drop v_D across the diode is about 0.65 volt or greater. In a typical junction diode at $v_D = 0.7\ V$, the current through the diode is $i_D \approx 1\ mA$.
- When a junction diode is reverse–biased, a very small current will flow and if the applied voltage exceeds a certain value the diode will reach its avalanche or Zener region. Commercially available diodes are provided with a given rating (volts, watts) by the manufacturer, and if these ratings are exceeded, the diode will burn–out in either the forward–biased or the reverse–biased direction.
- The maximum reverse–bias voltage that may be applied to a diode without causing junction breakdown is referred to as the Peak Reverse Voltage (PRV) and it is the most important rating.
- The analysis of electronic circuits containing diodes can be performed by graphical methods but it is greatly simplified with the use of diode models where we approximate the diode forward–biased characteristics with two straight lines representing the piecewise linear equivalent circuit.
- When used with AC circuits of low frequencies, diodes, usually with $1.8 \le n \le 2.0$ are biased to operate at some point in the neighborhood of the relatively linear region of the $i - v$ characteristics where $0.65 \le v_D \le 0.8\ V$. A bias point, denoted as Q, is established at the intersection of a load line and the linear region. The current $i_D(t)$ can be found from the relation

$$i_D(t) \approx I_D + \frac{I_D}{nV_T}v_D(t)$$

provided that $v_D/nV_T \le 0.1$.

- The ratio I_D/nV_T, denoted as g_D, is referred to as incremental conductance and its reciprocal nV_T/I_D, denoted as r_D, is called incremental resistance.
- Junction diodes are extensively used in half–wave and full–wave rectifiers to convert AC voltages to pulsating DC voltages. A filter is used at the output to obtain a relatively constant output with a small ripple.

- A peak rectifier circuit is a simple series circuit with a capacitor, a diode, and an AC voltage source where the output voltage across the capacitor is approximately equal to the peak value of the input AC voltage. If a resistor R is placed in parallel with the capacitor C and their values are chosen such that the time constant $\tau = RC$ is much greater than the discharge interval, the peak rectifier circuit will behave like an AC to DC converter.
- Clipper (or limiter) circuits consist of diodes, resistors, and sometimes DC sources to clip or limit the output to a certain level. Clipper circuits are used in applications where it is necessary to limit the input to another circuit so that the latter would not be damaged.
- A DC restorer (or clamped capacitor) is a circuit that can restore a DC voltage to a desired DC level. In its basic form is a series circuit with a voltage source, a capacitor, and a diode, and the output is the voltage across the diode.
- A voltage doubler circuit is a combination of a DC restorer and a peak rectifier. Voltage triplers and voltage quadruplers are also possible with additional diodes and capacitors.
- Zener diodes are designed to operate in the reverse voltage breakdown (avalanche) region. A Zener diode is always connected as a reverse–biased diode and its voltage rating V_Z and the maximum power it can absorb are given by the manufacturer. Their most important application is in voltage regulation.
- Zener diodes can also be used as limiters and limiting can occur during the positive half–cycle of the input voltage, during the negative half–cycle, or during both the positive and negative half–cycles of the input voltage.
- A Schottky diode is a junction of a lightly doped n–type semiconductor with a metal electrode. The junction of a doped semiconductor – usually n–type – with a special metal electrode can produce a very fast switching diode which is mainly used in high (up to 5 MHz) frequency circuits or high speed digital circuits. Schottky diodes find wide application as rectifiers for high frequency signals and also are used in the design of galliun arsenide (GaAs) circuits. The forward voltage drop of a conducting Schottky diode is typically 0.3 to 0.5 volt compared to the 0.6 to 0.8 found in silicon junction diodes.
- A tunnel diode is one which is heavily doped with impurities, it will have a region of negative resistance. The negative resistance characteristic makes the tunnel diode useful in oscillators and microwave amplifiers.
- The backward diode is a variation of the tunnel diode. It is used as a rectifier and in circuits with small amplitude waveforms.
- A *varactor*, or *varicap*, is a diode that behaves like a variable capacitor, with the PN junction functioning like the dielectric and plates of a common capacitor. A varactor diode uses a PN junction in reverse bias and has a structure such that the capacitance of the diode varies with the reverse voltage. A voltage controlled capacitance is useful in tuning applications. Typical

capacitance values are small, in the order of picofarads. Varactors have now replaced the old variable capacitor tuning circuits as in television tuners.

- Optoelectronic devices either produce light or use light in their operation.
- A Light–Emitting Diode (LED) is an optoelectronic device that, when forward biased, produces visible light. The light may be red, green, or amber, depending upon the material used to make the diode. The life expectancy of the LED is very long, over 100,000 hours of operation. For this reason, LEDs are used widely as "power on" indicators of digital voltmeters, frequency counters, etc. and as displays for pocket calculators where they form seven–segment displays.
- Laser diodes are LEDs specifically designed to produce coherent light with a narrow bandwidth and are suitable for CD players and optical communications.
- A photodiode is another optoelectronic device which receives light and converts it to electrical signals. Photodiodes respond quickly to changes in light intensity, and for this reason are extremely useful in digital applications such as photographic light meters and optical scanning equipment.
- A photocell is an older device that uses light in a way similar to the photodiode. Photocells are used in various types of control and timing circuits as, for example, the automatic street light controllers in most cities.
- A solar cell is another device that converts light energy into electrical energy. A solar cell acts much like a battery when exposed to light and as with batteries, solar cells may be connected in series or parallel to produce higher voltages and currents. Solar cells find widespread application in communications satellites and solar–powered homes.
- An optical coupler, sometimes referred to as an optoisolator, consists of an LED and a photodiode and each of these devices are isolated from each other. Isolation between the input and output is desirable because it reduces electromagnetic interference. Their most important application is in fiber optic communications links.

2.19 Exercises

1. Plot the $i - v$ characteristics of a forward–biased junction diode with $n = 2$ at $27\ °C$.

2. Show that for a decade (factor 10) change in current i_D of a forward–biased junction diode the voltage v_D changes by a factor of $2.3nV_T$. Hint: Start with the approximate relation $i_D \approx I_r e^{v_D/nV_T}$, form the ratio V_{D2}/V_{D1} corresponding to the ratio I_{D2}/I_{D1}, and plot in semi-log scale.

3. Suggest an experiment that will enable one to compute the numerical value of n.

4. It is known that for a certain diode $I_r = 10^{-13}\ A$ at $50\ °C$. and the reverse current I_r increases by about 15% per $1\ °C$ rise. At what temperature will I_r double in value? What conclusion can we draw from the result?

5. When a junction diode operates at the reverse–bias region $v < 0$ and before avalanche occurs, the current I_r is almost constant. Assuming that $|v| \gg V_T$ we can show that $i_D \approx -I_r$ and for this reason I_r is often referred to as the *saturation current* and whereas in the forward–biased region I_r has a typical value of $10^{-14}\ A$ to $10^{-15}\ A$ at $27\ °C$, a typical value in the reverse–biased region is $10^{-9}\ A$. Also, the reverse current doubles for every $10\ °C$ rise in temperature. In the circuit below, the applied voltage V_S is not sufficient to drive the diode into the avalanche region, and it is known that $V_R = 0.5\ V$ at $20\ °C$.

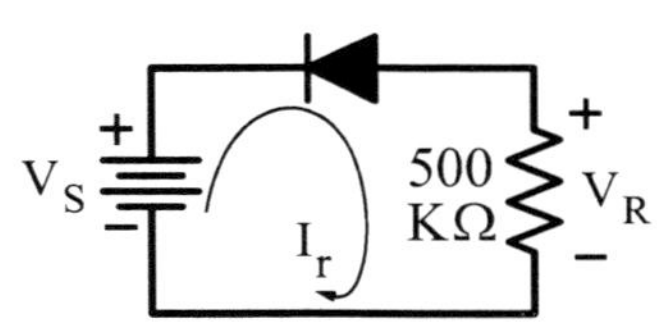

Find:

a. V_R at $0\ °C$

b. V_R at $40\ °C$

6. For the circuit below, find the value of V_{out} assuming that all three diodes are ideal. Compare your answer with that of Example 2.4. Make any reasonable assumptions.

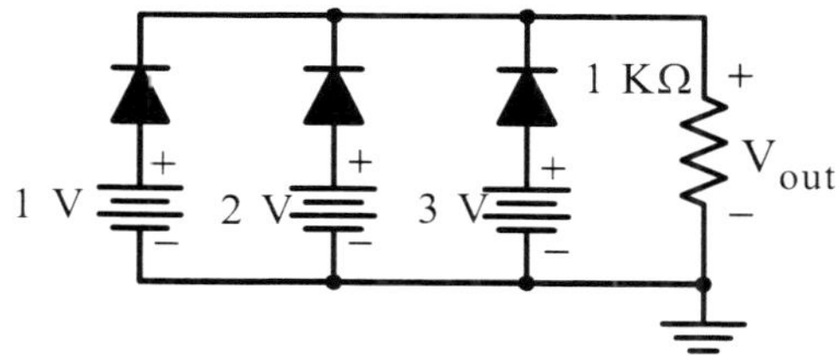

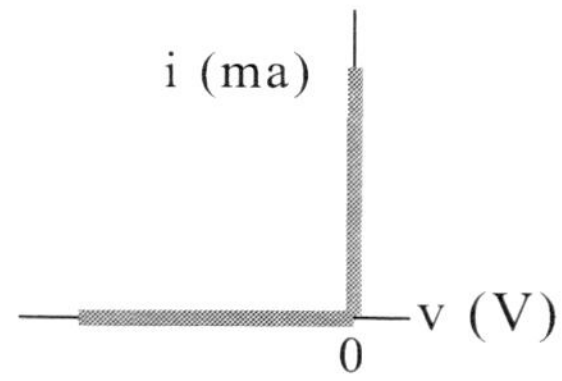

7. For the circuit shown below, the diodes are identical and it is known that at $V_D = 0.65$, $I_D = 0.5\ mA$. It is also known that the voltage across each diode changes by $0.1\ V$ per decade change of current. Compute the value of R so that $V_{out} = 3\ V$.

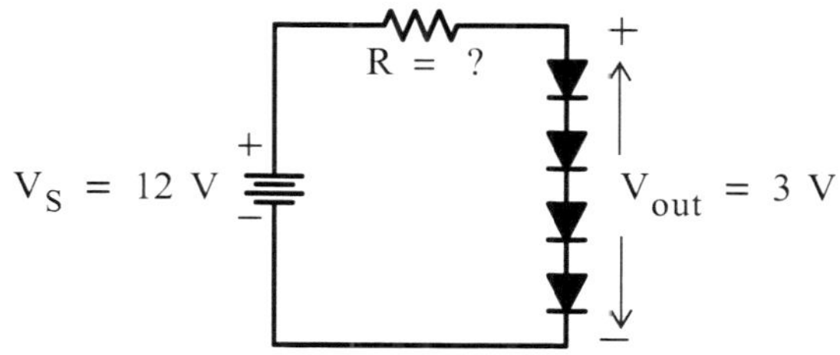

8. For the circuit shown below, the diodes are identical and when $V_S = 12\ V$ exactly, $V_{out} = 2.8\ V$. Assuming that $n = 1.5$ and $T = 27\ °C$ find:

a. The percent change in V_{out} when V_S changes by $\pm 10\ \%$.

b. The percent change in V_{out} when a load resistor $V_{load} = 2\ K\Omega$ is connected across the four diodes.

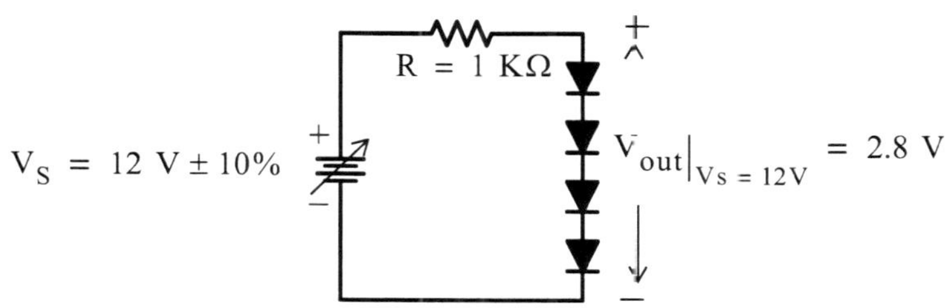

9. It is known that a junction diode with $n = 1.92$ allows a current $I_D = 1\ mA$ when $V_D = 0.7\ V$ at $T = 27\ °C$. Derive the equation of the straight line tangent at $I_D = 1\ mA$ and the point where this straight line crosses the v_D axis.

10. For the half–wave rectifier circuit below, the transformer is a step–down 10:1 transformer and the primary voltage $v_s = 120\ RMS$. The diode voltage $v_D = 0.7\ V$ and the diode resistance $r_D \ll R_{load}$. Find:

a. the diode peak current I_{peak}

b. the angle θ by which v_{load} lags the transformer secondary voltage v_{in}

c. the conduction angle

d. the average value of v_{load}

e. the minimum theoretical Peak Inverse Voltage (PIV)

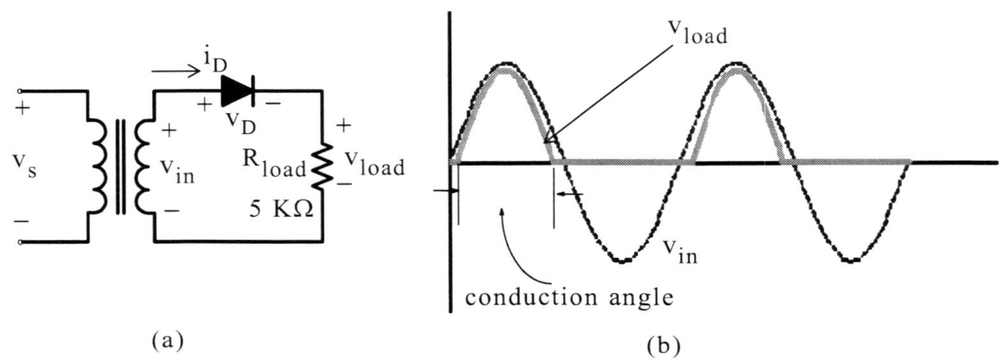

11. For the full–wave bridge rectifier shown below it is given that $v_{in}\ (peak) = 17\ V$, $v_D = 0.7\ V$, $R = 200\ \Omega$, and the diode resistance $r_D \ll R$. Find:

a. the diode peak current I_{peak}

b. the average value of v_{load}

c. the minimum theoretical Peak Inverse Voltage (PIV)

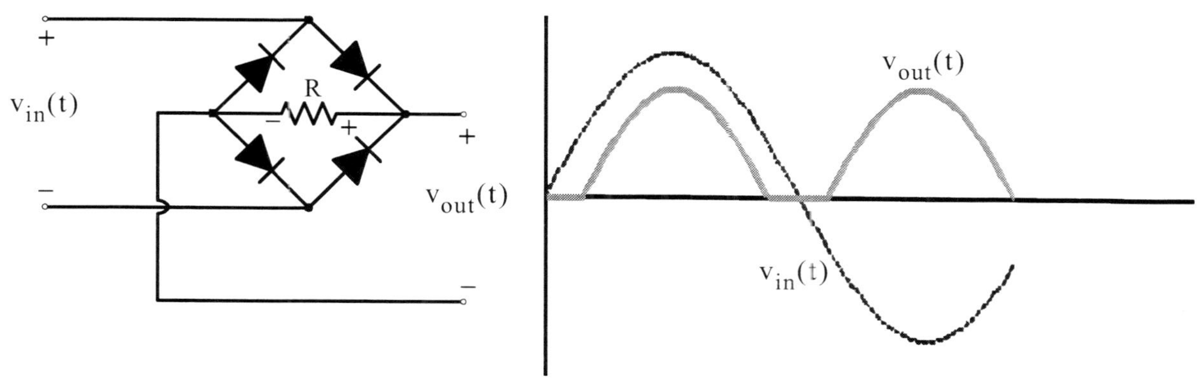

12. For the peak rectifier shown below, find the value of the capacitor so that the peak–to–peak ripple will be 1 volt.

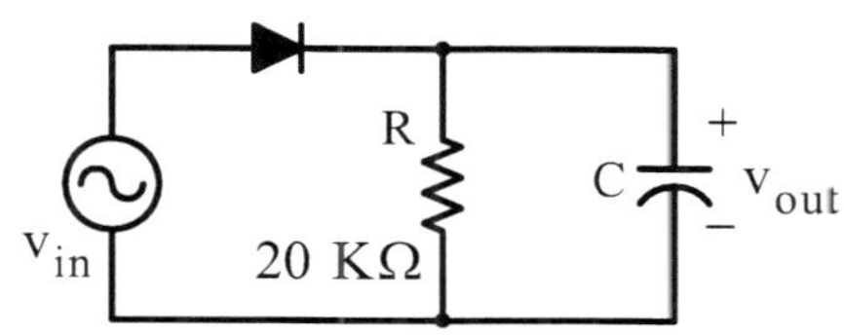

$v_{in} = 177 \sin 377t$

13. A circuit and its input waveform are shown below. Compute and sketch the waveform for the output v_{out}.

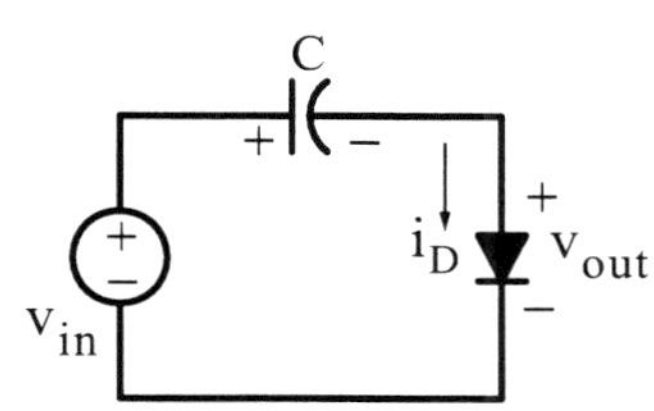

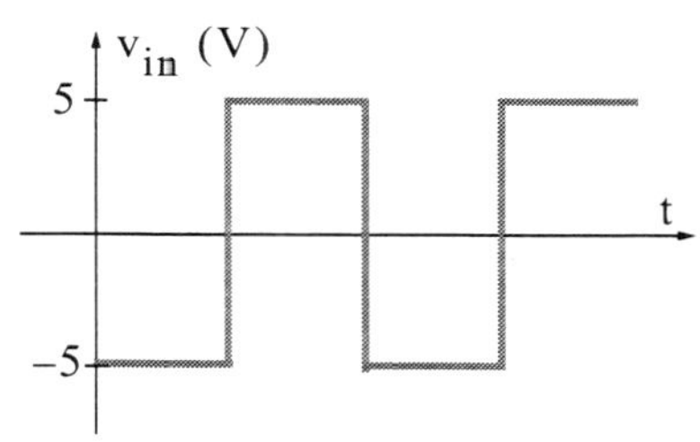

14. The nominal value of a Zener diode is $V_Z = 12\ V$ at $I_Z = 10\ mA$, and $r_Z = 30\ \Omega$.

   a. Find V_Z if $I_Z = 5\ mA$

   b. Find V_Z if $I_Z = 20\ mA$

   c. Find V_{Z0} of the Zener diode model

15. A tunnel diode has the idealized piecewise linear characteristics shown below where the valley current, the peak current, the peak voltage, the valley voltage, and the peak forward voltage have the values $I_V = 1\ mA$, $I_P = 10\ mA$, $V_P = 50\ mV$, $V_V = 350\ mV$, and $V_F = 500\ mV$ respectively.

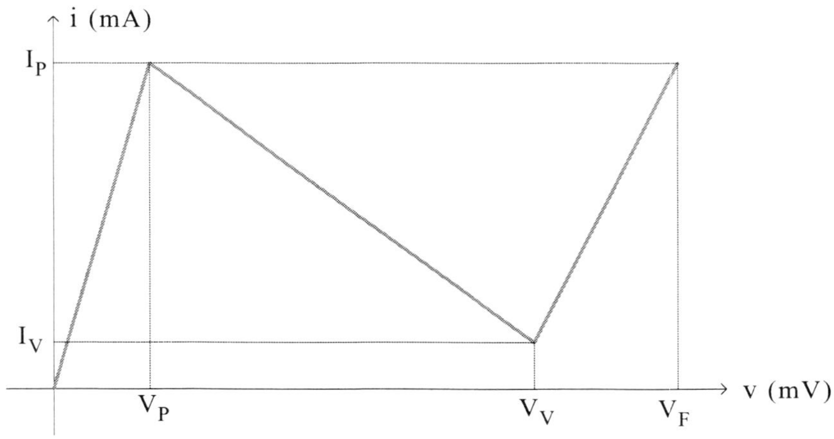

Shown below is a resistor R, referred to as a *tunnel resistor*, placed in parallel with the tunnel diode. Determine the value of this resistor so that this parallel circuit will exhibit no negative resistance region.

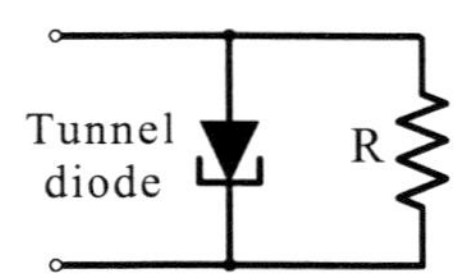

16. A typical solar cell for converting the energy of sunlight into electrical energy in the form of heat is shown in the Figure (a) below. The terminal voltage characteristics are shown in Figure (b) below.

   a. Determine the approximate operating point that yields the maximum power. What is the value of the maximum power output?

   b. What should the value of the resistor be to absorb maximum power from the solar cell?

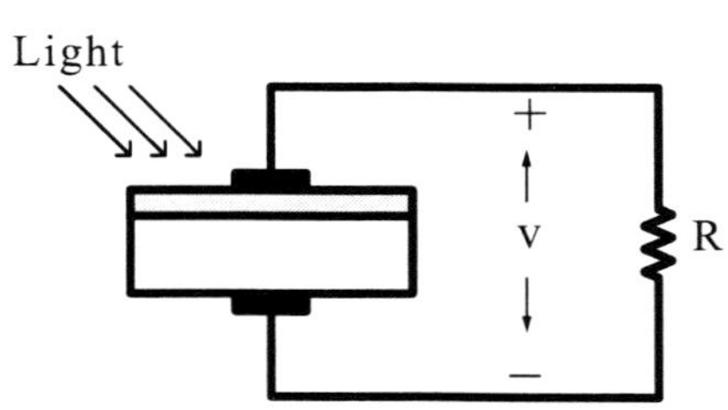

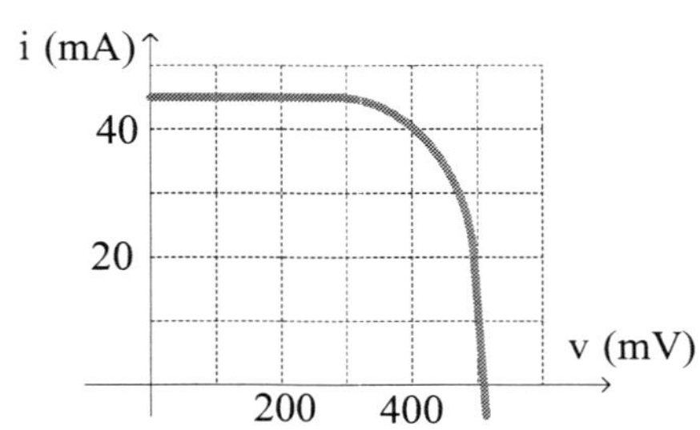

2.20 Solutions to End–of–Chapter Exercises

1.
```
vD=0: 0.01: 1; iR=10^(–15); n=2; VT=26*10^(–3);...
iD=iR.*(exp(vD./(n.*VT))–1); plot(vD,iD);...
xlabel('Diode voltage vD, volts'); ylabel('Diode current iD, amps');...
title('iD–vD characteristics for a forward–biased junction diode'); grid
```

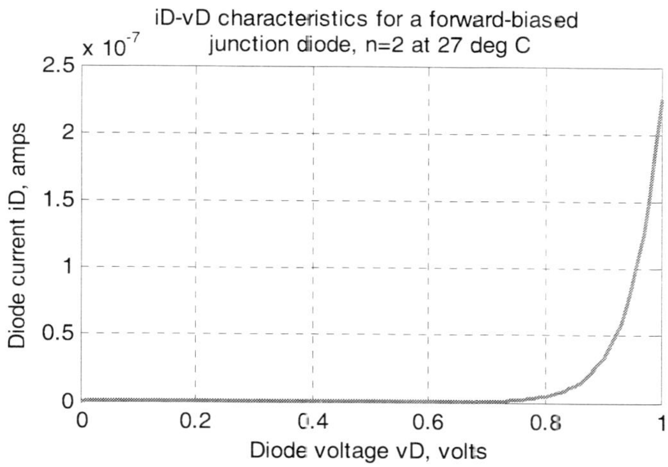

We observe that when $n = 2$, the diode begins to conduct at approximately $7.8\ V$.

2. Let $I_{D1} = I_r e^{V_{D1}/nV_T}$ and $I_{D2} = I_r e^{V_{D2}/nV_T}$. Then,

$$I_{D2}/I_{D1} = e^{V_{D2}/nV_T}/e^{V_{D1}/nV_T} = e^{(V_{D2}-V_{D1})/nV_T}$$

or

$$V_{D2} - V_{D1} = nV_T \ln(I_{D2}/I_{D1}) = 2.3nV_T \log_{10}(I_{D2}/I_{D1})$$

For convenience, we let $V_{D2} - V_{D1} = y$ and $I_{D2}/I_{D1} = x$ and we use the following MATLAB script to plot $y = 2.3nV_T \log_{10}x$ on semilog scale.

```
x=1: 10: 10^6; y=2.3.*1.*26.*10.^(–3).*log10(x); semilogx(x,y);...
xlabel('x=ID2/ID1'); ylabel('y=VD2–VD1'); title('Plot for Exercise 2'); grid
```

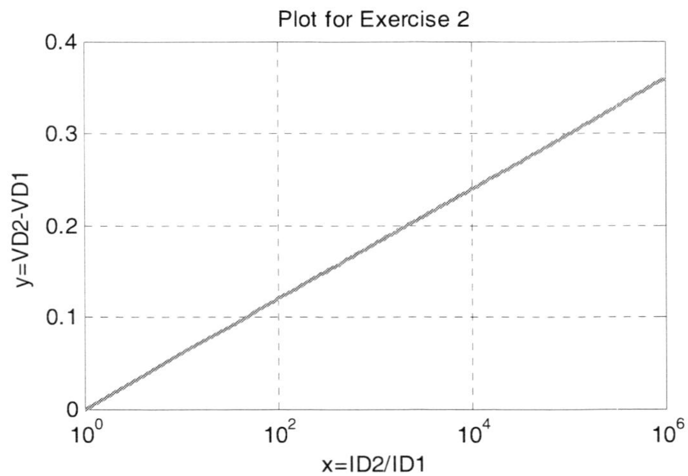

3.

This can be done with a variable power supply V_S, $1\ K\Omega$ resistor, a DC voltmeter V, a DC ammeter A and the diode D whose value of n is to be found, connected as shown below. We assume that the voltmeter internal resistance is very high and thus all current produced by V_s flows through the diode.

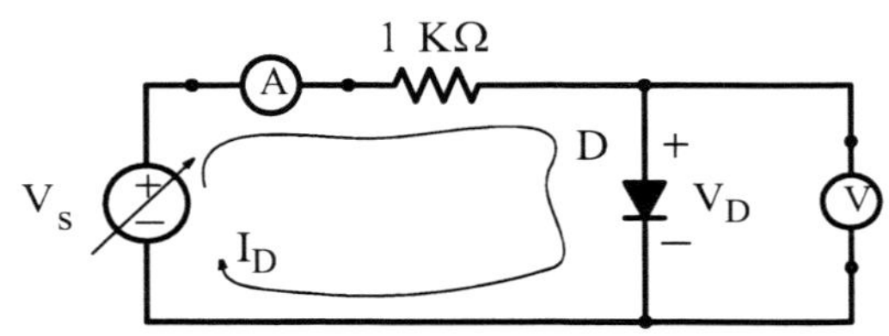

We can now adjust the variable power supply V_S to obtain two pairs $i-v$, say I_{D1}, V_{D1} and I_{D2}, V_{D2} and use the relation $V_{D2} - V_{D1} = 2.3nV_T \log_{10}(I_{D2}/I_{D1})$ to find n. However, for better accuracy, we can adjust V_S to obtain the values of several pairs, plot these on semilog paper and find the best straight line that fits these values.

4.

$$10^{-13}(1+0.15)^{(x-50)} = 2\times 10^{-13}$$

$$1.15^{(x-50)} = 2$$

$$\frac{(1.15)^x}{(1.15)^{50}} = 2$$

$$(1.15)^x = 2\times(1.15)^{50} = 2.167\times 10^3$$

$$\log_{10}(1.15)^x = x\log_{10}(1.15) = \log_{10}(2.167\times 10^3)$$

$$x\log_{10}(1.15) = \log_{10}(2.167) + 3\log_{10}(10) = 0.336 + 3 = 3.336$$

$$x = 3.336/\log_{10}1.15 = 54.96\ ^\circ C \approx 55\ ^\circ C$$

Therefore, we can conclude that the reverse current I_r doubles for every 5 °C rise in temperature.

5.

At 20 °C,

$$I_r\big|_{20\ ^\circ C} = \frac{0.5\ V}{500\ K\Omega} = 1\ \mu A$$

and since the reverse current doubles for every 10 °C rise in temperature, we have:

a.

$$I_r\big|_{40\ ^\circ C} = 2\times 2\times 1\ \mu A = 4\ \mu A$$

and

$$V_R\big|_{40\ ^\circ C} = 4\times 10^{-6}\times 500\times 10^3 = 2V$$

b.

$$I_r\big|_{0\ ^\circ C} = 0.5\times 0.5\times 1\ \mu A = 0.25\ \mu A$$

and

$$V_R\big|_{0\ ^\circ C} = 0.25\times 10^{-6}\times 500\times 10^3 = 0.125\ V$$

6. We assume that diode D_3 and the resistor are connected first. Then, $V_{out} = 3\ V$ and stays at this value because diodes D_1 and D_2, after being connected to the circuit, are reverse–biased.

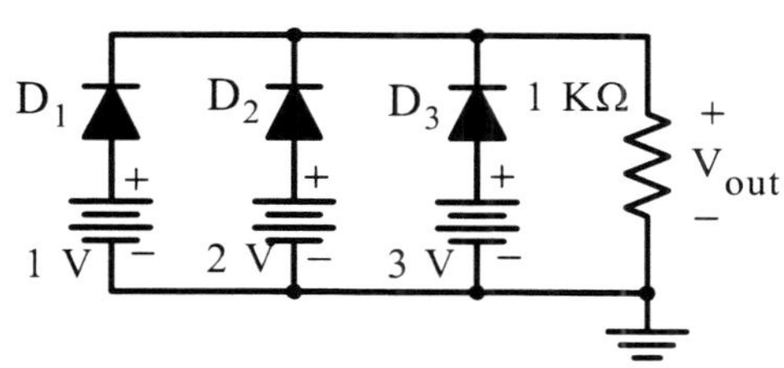

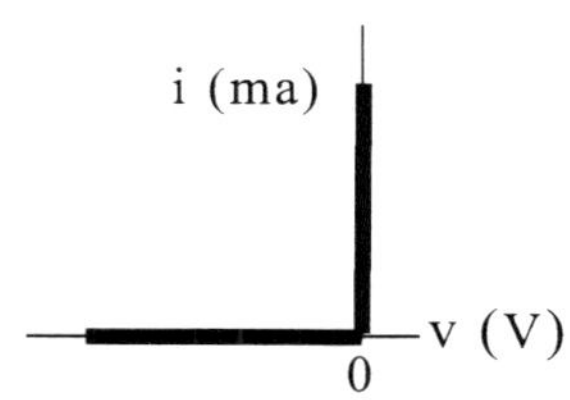

The value of $V_{out} = 3\ V$ is unrealistic when compared with the realistic value which we found in Example 2.4. This is because we've assumed ideal diodes, a physical impossibility.

7.

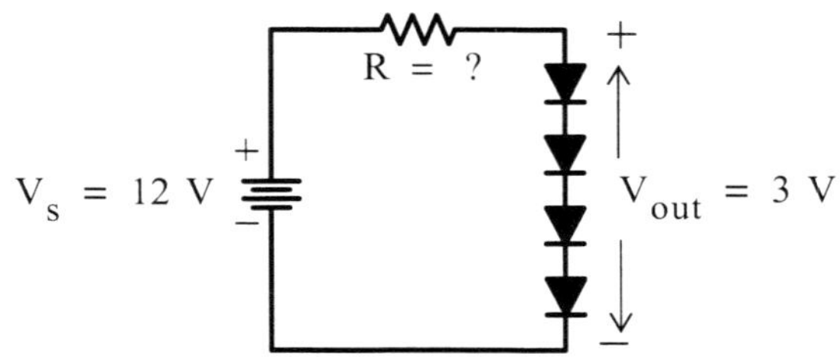

Since the diodes are identical, for the output voltage to be $V_{out} = 3\ V$, the voltage drop across each diode must be $3\ V/4 = 0.75\ V$, and since the change is $\Delta V_D = 0.75 - 0.65 = 0.1\ V$, the current is $0.5 \times 10 = 5\ mA$. Then,

$$R = \frac{V_S - V_{out}}{I_D} = \frac{12-3}{5 \times 10^{-3}} = 1.8\ K\Omega$$

8.

a.

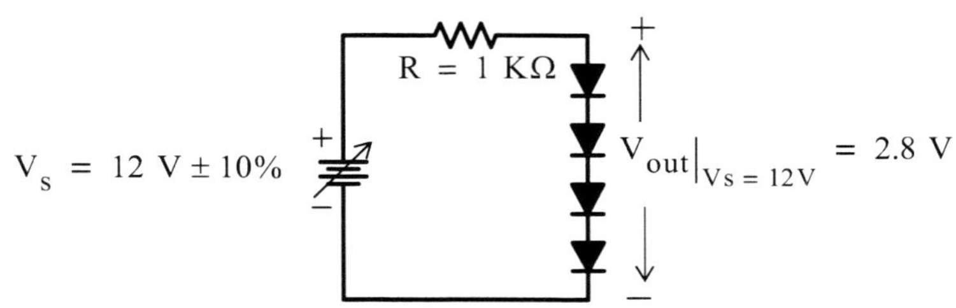

$$I_D = \frac{V_S - V_{out}}{R} = \frac{12-2.8}{10^3} = 9.2\ mA$$

The incremental resistance r_D per diode is found from $r_D = nV_T/I_D$ with $n = 1.5$. Thus, for all four diodes

$$4r_D = \frac{4 \times 1.5 \times 26 \times 10^{-3}}{9.2 \times 10^{-3}} = 17\ \Omega$$

and for $\pm 10\ \%$ or $\pm 1.2\ V$ change we find that

$$\Delta V_{out} = \frac{17\ \Omega}{1000 + 17} = 16.7 mV$$

Therefore,

$$V'_{out} = 2.8 \pm 0.0167\ V$$

and

$$\%\Delta V_{out} = \frac{2.8167 - 2.8}{2.8} \times 100 = \pm 59.6\ \%$$

b.

The $2\ K\Omega$ load when connected across the four diodes will have a voltage of $2.8\ V$ and thus the current through this load will be

$$I_{load} = \frac{2.8\ V}{2\ K\Omega} = 1.4\ mA$$

Therefore, the current through the four diodes decreases by $1.4\ mA$ and the decrease in voltage across all four diodes will be

$$\Delta V_{out} = -1.4\ mA \times 17\ \Omega = -23.8\ mV$$

9.

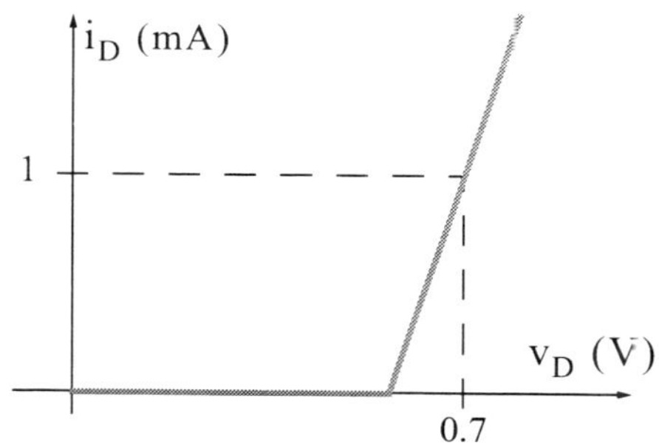

From analytic geometry, the equation of a straight line in an $x-y$ plane is $y = mx + b$ where m is the slope and b is the y intercept. For this exercise the equation of the straight line is

$$i_D = mv_D + b \quad (1)$$

where

$$m = \frac{1}{r_D} = \frac{I_D}{nV_T} = \frac{10^{-3}}{1.92 \times 26 \times 10^{-3}} = 0.02$$

and by substitution into (1) above

$$i_D = 0.02v_D + b \quad (2)$$

Next, we find the i_D intercept using the given data that when $v_D = 0.7\ V$, $i_D = 1\ mA$. Then, by substitution into (2) we obtain

$$10^{-3} = 0.02 \times 0.7 + b$$

from which $b = -0.013$ and thus the equation of the straight line becomes

$$i_D = 0.02v_D - 0.013 \quad (3)$$

Now, from (3) above we find that the v_D axis intercept is $0 = 0.02v_D - 0.013$ or $v_D = 0.65\ V$.

10.

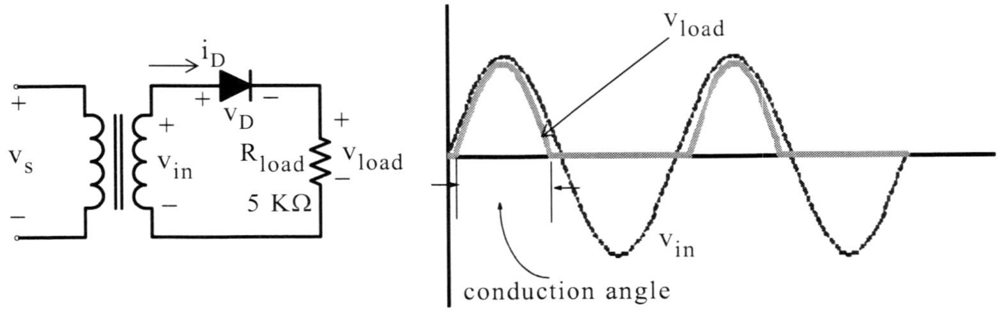

(a) (b)

a.

It is given that the transformer is a step down type with ratio *10:1* ; therefore, the secondary voltage is 12 V RMS or $v_{in} = 12\sqrt{2}\text{ peak} = 17\text{ V peak}$. Then,

$$I_{peak} = \frac{v_{in} - v_D}{r_D + R_{load}}$$

and since $r_D \ll R_{load}$

$$I_{peak} = \frac{v_{in} - v_D}{R_{load}} = \frac{17 - 0.7}{5 \times 10^3} = 3.26\text{ mA}$$

b.

From (2.15), $\theta = \sin^{-1}\Delta v / V_p$ where $\Delta v = V_p - V_{out} = v_D = 0.7\text{ V}$. Then,

$$\theta = \sin^{-1}0.7/17 = 2.36°$$

c. The conduction angle is

$$\pi - 2\theta = 180° - 2 \times 2.36° = 175.28°$$

d.

From (2.17)

$$V_{load}(ave) \approx \frac{V_p}{\pi} - \frac{\Delta v}{2} = \frac{17}{\pi} - \frac{0.7}{2} = 5.06\text{ V}$$

e.

The minimum theoretical Peak Inverse Voltage (PIV) is

$$PIV = V_p = 17\text{ V}$$

but for practical purposes we should choose the value of

$$1.7 \times PIV = 1.7 \times 17 \approx 30\text{ V}$$

11.

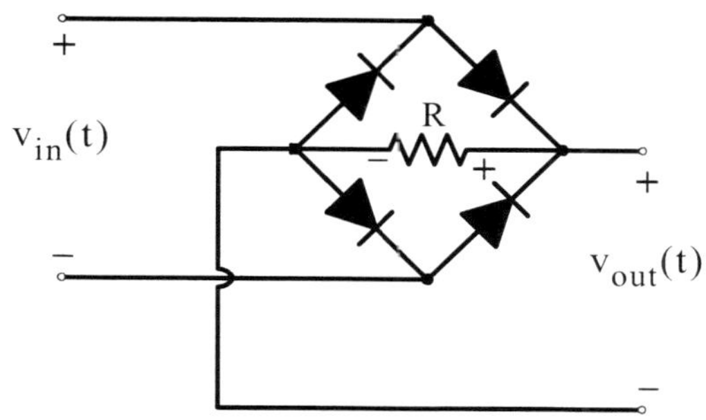

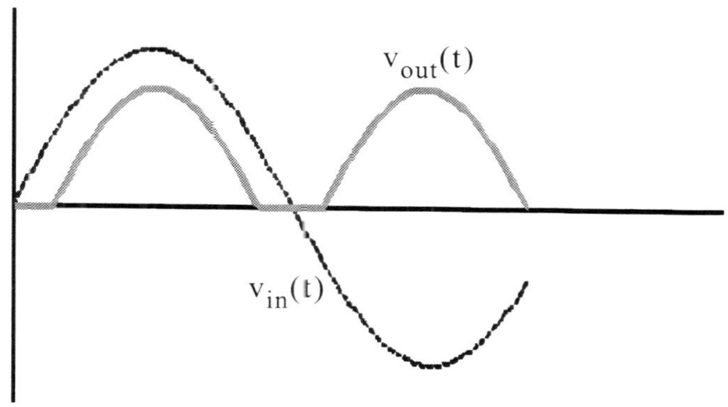

a.

The diode peak current I_{peak} is

$$I_{peak} = \frac{v_{in} - 2v_D}{r_D + R}$$

and since $r_D \ll R$

$$I_{peak} \approx \frac{v_{in} - 2v_D}{R} = \frac{17 - 2 \times 0.7}{200} = 78 \text{ mA}$$

b.

We start with the definition of the average value, that is,

$$V_{out}(ave) = \frac{Area}{Period} = \frac{1}{\pi}\int_{\theta}^{(\pi-\theta)} (V_p \sin\phi - 2\Delta v)d\phi = \frac{1}{\pi}(-V_p \cos\phi - 2\Delta v\phi)\Big|_{\phi=\theta}^{\pi-\theta}$$

or

$$V_{out}(ave) = \frac{1}{\pi}[-V_p\cos(\pi-\theta) - 2\Delta v(\pi-\theta) + V_p\cos\theta + 2\Delta v\theta]$$

$$= \frac{1}{\pi}[2V_p\cos\theta - 2\Delta v(\pi - 2\theta)]$$

Generally, the angle θ is small and thus $\cos\theta \approx 1$ and $(\pi - 2\theta) \approx \pi$. Therefore, the last relation above reduces to

$$V_{out}(ave) \approx \frac{2V_p}{\pi} - 2\Delta v = \frac{2 \times 17}{\pi} - 2 \times 0.7 = 9.42 \text{ V}$$

c.

The minimum theoretical Peak Inverse Voltage (PIV) is

$$PIV = V_p - v_D = 17 - 0.7 = 16.3 \text{ V}$$

but for practical purposes we should choose the value of about 30 V.

12.

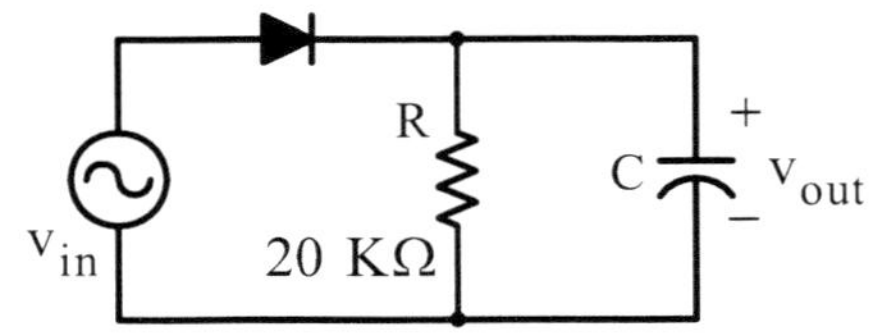

Here, $\omega = 377\ r/s$ or $f = \omega/2\pi = 377/2\pi = 60\ Hz$, and $T = 1/f = 1/60$. Rearranging (2.29), we obtain

$$C = V_p \cdot \frac{T}{V_{pp}(\text{ripple}) \cdot R \cdot 60} = 177 \cdot \frac{1}{1 \times 20 \times 10^3 \times 60} = 148\ \mu F$$

13.

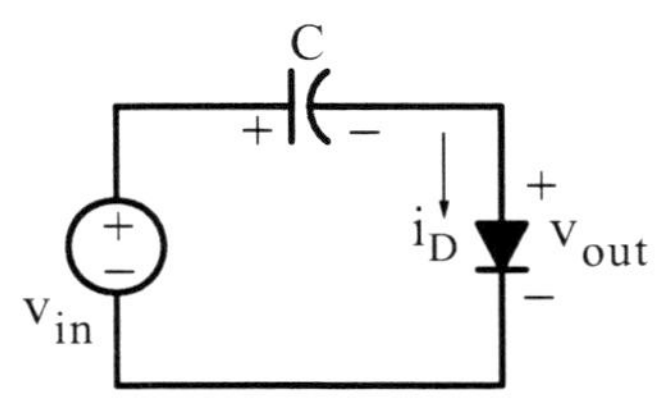

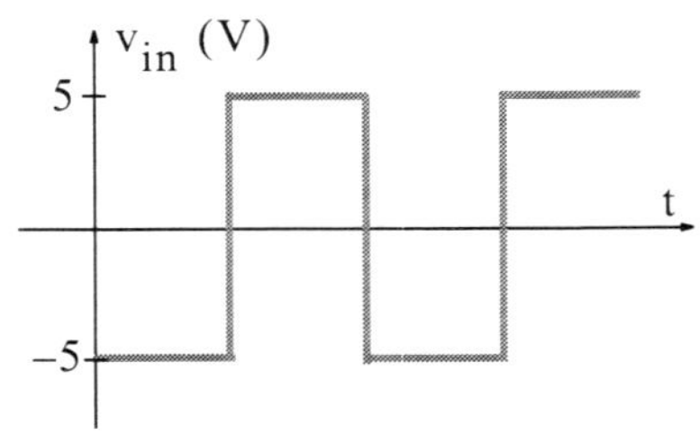

From the circuit above we observe that the diode conducts only when $v_{in} > 0$ and thus the capacitor charges to the positive peak of the input which for this example is $5\ V$, and the capacitor voltage is $v_C = 5\ V$ with polarity shown. Since $v_{out} = v_{in} - v_C$, when $v_{in} = 5\ V$, $v_{out} = 5 - 5 = 0$ and the output has shifted upwards to zero volts as shown below. When the input voltage is negative, $v_{in} = -5\ V$, and $v_{out} = -5 - 5 = -10\ V$. The waveform for the output voltage is as shown below.

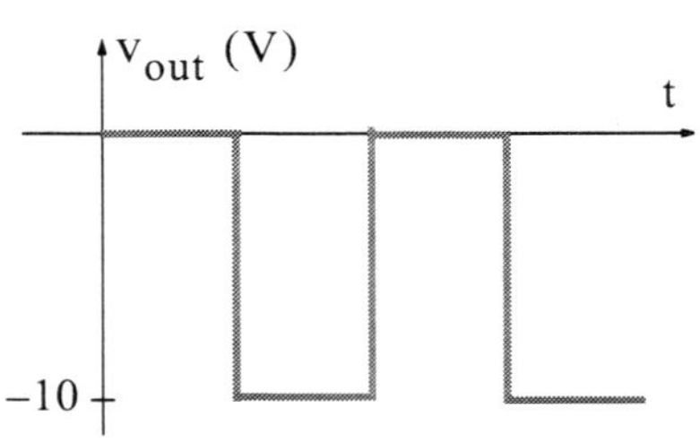

14.

a. By definition $r_Z = \Delta V_Z/\Delta I_Z$ or $\Delta V_Z = r_Z \Delta I_Z$. Then,

$$V_Z\big|_{I_Z = 5mA} = V_Z\big|_{I_Z = 10mA} - r_Z \Delta I_Z = 12 - 30 \times (10 - 5) \times 10^{-3} = 11.85\ V$$

b.

$$V_Z\big|_{I_Z = 20mA} = V_Z\big|_{I_Z = 10mA} - r_Z \Delta I_Z = 12 - 30 \times (10 - 20) \times 10^{-3} = 12.3 \text{ V}$$

c. From (2.34) $V_Z = V_{Z0} + r_Z I_Z$ where V_Z and I_Z are the nominal values. Then

$$V_{Z0} = V_Z - r_Z I_Z = 12 - 30 \times 10 \times 10^{-3} = 11.7 \text{ V}$$

15.

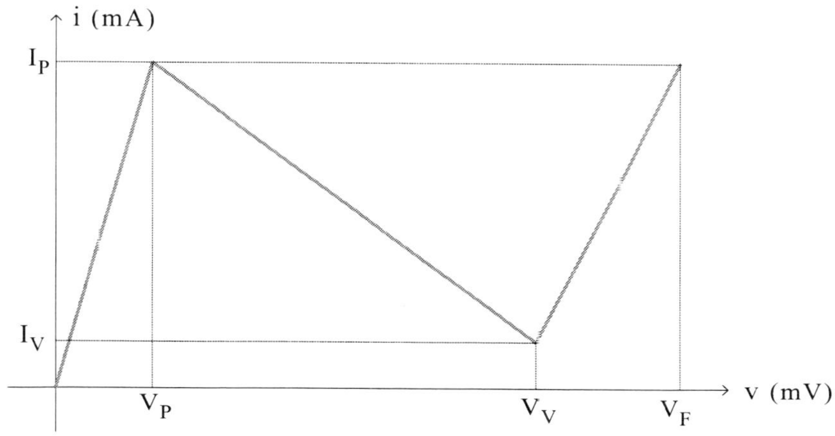

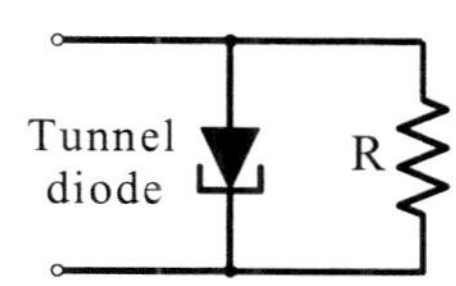

For the region of negative resistance the slope m is

$$m = \frac{\Delta i}{\Delta v} = \frac{1 - 10}{350 - 50} = -0.03$$

and thus

$$\frac{di_D}{dv} = -0.03 \ \Omega^{-1} \quad (1)$$

where i_D and v are as shown in the circuit below.

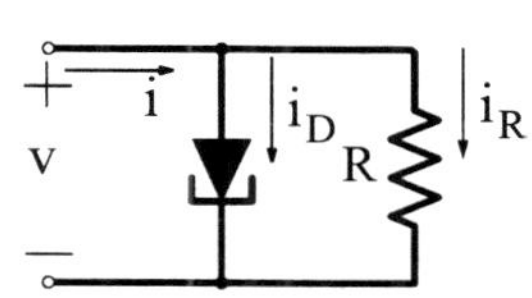

To eliminate the negative resistance region, we must have

$$\frac{di}{dv} \ge 0 \quad (2)$$

for all values of v. Since

$$i = i_D + i_R = i_D + v/R$$

it follows that

$$\frac{di}{dv} = \frac{di_D}{dv} + \frac{1}{R}$$

and (2) above will be satisfied if

$$\frac{1}{R} \ge -\frac{di_D}{dv} \quad (3)$$

From (1) and (3)

$$\frac{1}{R} \ge -(-0.03)$$

and thus the maximum value of the resistor should be $R \approx 33.33\ \Omega$. With this value, the slope of the total current i versus the voltage v across the parallel circuit will be zero. For a positive slope greater than zero we should choose a resistor with a smaller value. Let us plot the total current i versus the voltage v with the values $R_1 \approx 33.33\ \Omega$ and $R_2 = 25\ \Omega$.

We will use the following MATLAB script for the plots.

```
v=51:0.1:349; m=–0.03; b=11.5; id=m*v+b;...
R1=33.33; i1=id+v/R1; R2=25; i2=id+v/R2; plot(v,id,v,i1,v,i2); grid
```

and upon execution of this script we obtain the plot shown below.

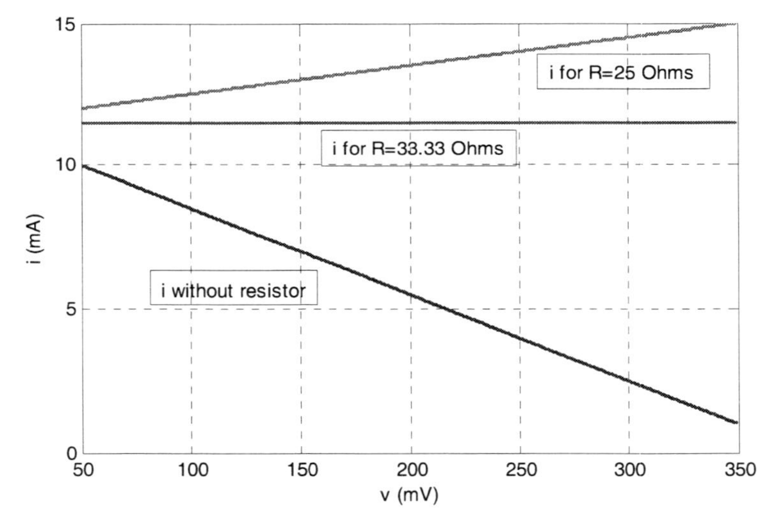

16.

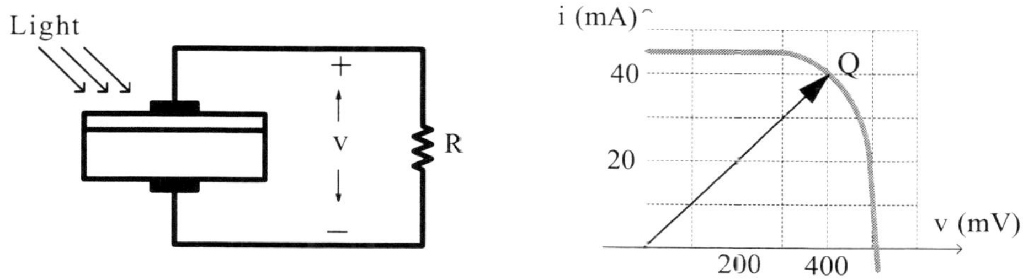

a. By inspection, the operating point that will yield maximum power is denoted as Q where

$$P_{max} = 400 \times 40 = 16 \text{ mw}$$

b. The value of R that will receive maximum power is

$$R = \left(\frac{v}{i}\right)_{max} = \frac{400}{40} = 10\ \Omega$$

NOTES:

Chapter 3

Bipolar Junction Transistors

This is a long chapter devoted to bipolar junction transistors. The NPN and PNP transistors are defined and their application as amplifiers is well illustrated with numerous examples. The small and large signal equivalent circuits along with the h−parameter and T−equivalent circuits are presented, and the Ebers−Moll model is discussed in detail.

3.1 Introduction

Transistors are three terminal devices that can be formed with the combination of two separate PN junction materials into one block as shown in Figure 3.1.

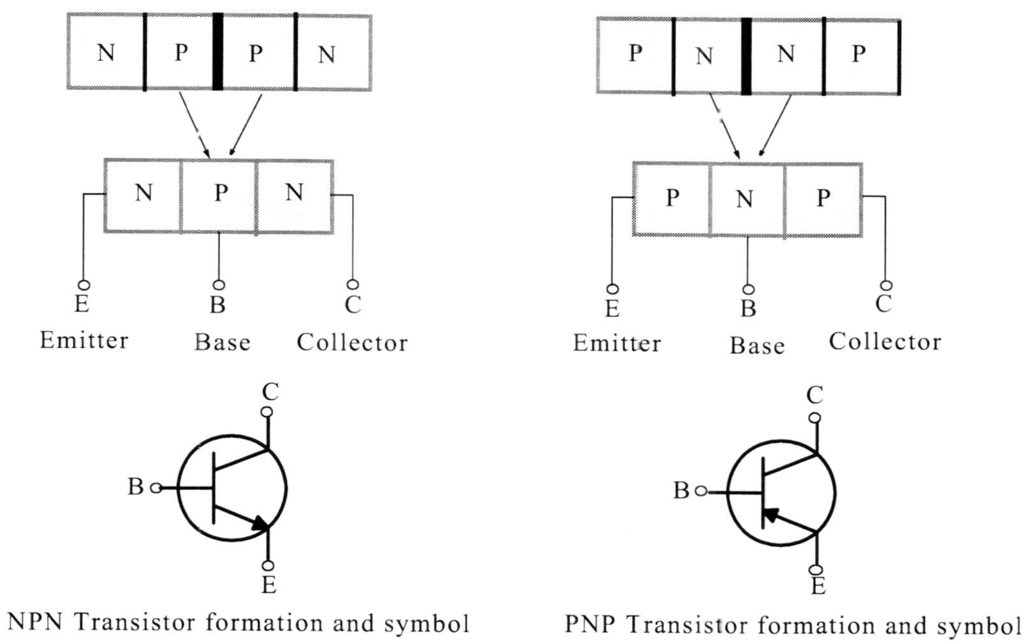

Figure 3.1. NPN and PNP transistor construction and symbols

As shown in Figure 3.1, an NPN transistor is formed with two PN junctions with the P−type material at the center, whereas a PNP transistor is formed with two PN junctions with the N−type material at the center. The three terminals of a transistor, whether it is an NPN or PNP transistor, are identified as the *emitter*, the *base*, and the *collector*. Can a transistor be used just as a diode? The answer is yes, and Figure 3.2 shows several possible configurations and most integrated circuits employ transistors to operate as diodes.

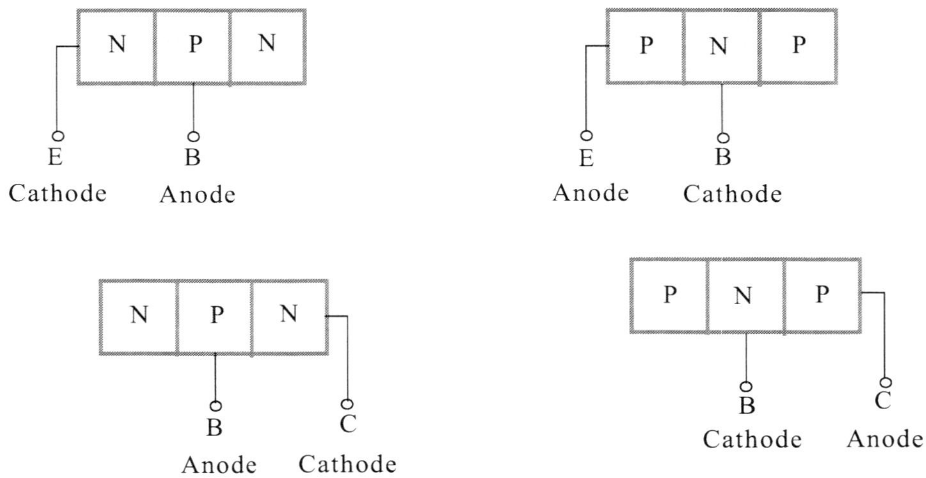

Figure 3.2. Transistors configured as diodes

Transistors are used either as amplifiers or more commonly as electronic switches. We will discuss these topics on the next section. Briefly, a typical NPN transistor will act as a closed switch when the voltage V_{BE} between its base and emitter terminals is greater than $0.7\ V$ but no greater than $5\ V$ to avoid possible damage. The transistor will act as an open switch when the voltage V_{BE} is less than $0.6\ V$. Figure 3.3 shows an NPN transistor used as an electronic switch to perform the operation of inversion, that is, the transistor inverts (changes) an input of $5\ V$ to an output of $0\ V$ when it behaves like a closed switch as in Figure 3.3, and it inverts an input of $0\ V$ to an output of $5\ V$ when it behaves like an open switch as in Figure 3.4.

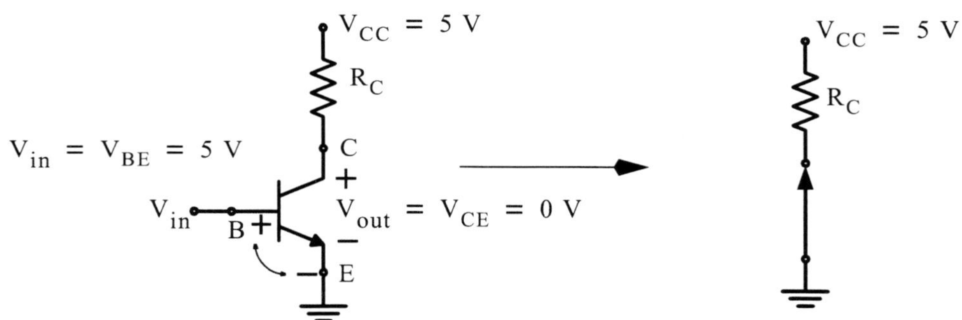

Figure 3.3. NPN transistor as electronic closed switch – inverts $5\ V$ to $0\ V$

Like junction diodes, most transistors are made of silicon. Gallium Arsenide (GaAs) technology has been under development for several years and its advantage over silicon is its speed, about six times faster than silicon, and lower power consumption. The disadvantages of GaAs over silicon is that arsenic, being a deadly poison, requires very special manufacturing processes and, in addition, it requires special handling since it is extremely brittle. For these reasons, GaAs is much more expensive than silicon and it is usually used only in superfast computers.

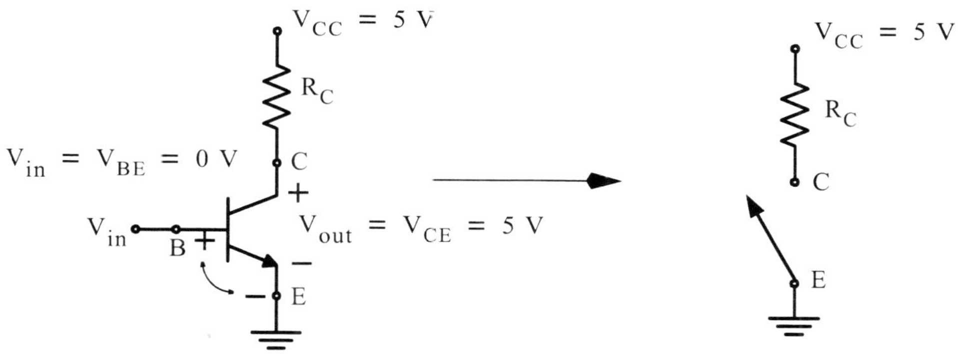

Figure 3.4. NPN transistor as electronic open switch – inverts 0 V to 5 V

3.2 NPN Transistor Operation

For proper operation, the NPN and PNP transistors must be biased as shown in Figure 3.5.

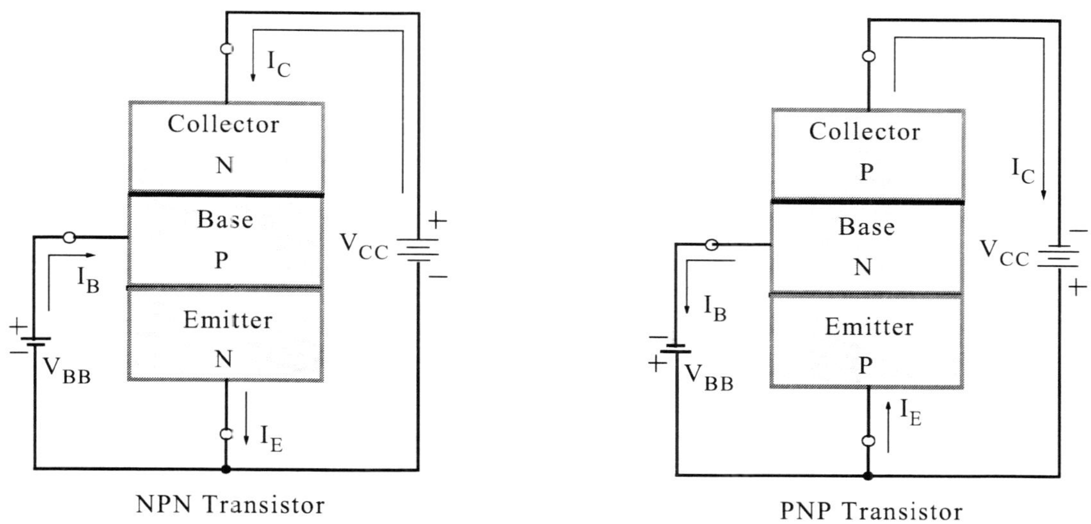

Figure 3.5. Biased NPN and PNP Transistors for proper operation

The bias voltage sources are V_{BB} for the base voltage and V_{CC} for the collector voltage. Typical values for V_{BB} are about 1 V or less, and for V_{CC} about 10 V to 12 V. The difference in these bias voltages is necessary to cause current flow from the collector to the emitter in an NPN transistor and from the emitter to collector in a PNP transistor.

3.3 The Bipolar Junction Transistor as an Amplifier

When a transistor is used as an amplifier, it is said to be operating in the active mode. Since a transistor is a 3–terminal device, there are three currents, the base current, denoted as i_B,* the collector current, denoted as i_C, and the emitter current, denoted as i_E. They are shown in Figures 3.5 and 3.6.

Figure 3.6. The base, collector, and emitter currents in a transistor

For any transistor, NPN or PNP, the three currents are related as

$$i_B + i_C = i_E \tag{3.1}$$

We recall from Chapter 2, equation (2.3), that $i_D = I_r[e^{(v_D/nV_T)} - 1]$. In a transistor, $n \approx 1$, and the collector current is

$$i_C = I_r e^{v_{BE}/V_T} \tag{3.2}$$

where I_r is the reverse (saturation) current, typically 10^{-12} A to 10^{-15} A as in junction diodes, v_{BE} is the base–to–emitter voltage, and $V_T \approx 26$ mV at $T = 300$ °K.

A very useful parameter in transistors is the *common–emitter gain* β, a constant whose value typically ranges from 75 to 300. Its value is specified by the manufacturer. Please refer to Section 3.19. The base current i_B is much smaller than the collector current i_C and these two currents are related in terms of the constant β as

$$i_B = i_C/\beta \tag{3.3}$$

and with (3.2) we obtain

$$i_B = (I_r/\beta)e^{v_{BE}/V_T} \tag{3.4}$$

From (3.1) and (3.3) we obtain

$$i_E = i_B + i_C = i_C/\beta + i_C = \frac{\beta+1}{\beta} i_C \tag{3.5}$$

and with (3.2)

$$i_E = \frac{\beta+1}{\beta} I_r e^{v_{BE}/V_T} \tag{3.6}$$

Another important parameter in transistors is the *common–base current gain* denoted as α and it

* *It is customary to denote instantaneous voltages and currents with lower case letters and the bias voltages and currents with upper case letters.*

is related to β as

$$\alpha = \frac{\beta}{\beta + 1} \tag{3.7}$$

From(3.7) it is obvious that $\alpha < 1$ and from (3.5) and (3.7)

$$i_C = \alpha i_E \tag{3.8}$$

Also, from (3.6) and (3.7)

$$i_E = \frac{1}{\alpha} I_r e^{v_{BE}/V_T} \tag{3.9}$$

and we can express β in terms of α by rearranging (3.7). Then,

$$\beta = \frac{\alpha}{1 - \alpha} \tag{3.10}$$

Another lesser known ratio is the *common–collector current gain* ratio denoted as γ and it is defined as the ratio of the change in the emitter current to the change in the base current. Thus,

$$\alpha = \frac{di_C}{di_E} \tag{3.11}$$

$$\beta = \frac{di_C}{di_B} \tag{3.12}$$

$$\gamma = \frac{di_E}{di_B} \tag{3.13}$$

and their relationships are

$$\alpha = \frac{\beta}{\beta + 1} \qquad \beta = \frac{\alpha}{1 - \alpha} \qquad \gamma = \beta + 1 \tag{3.14}$$

Example 3.1

A transistor manufacturer produces transistors whose α values vary from 0.992 to 0.995. Find the β range corresponding to this α range.

Solution:

For $\alpha = 0.992$ (3.10) yields

$$\beta = \frac{\alpha}{1 - \alpha} = \frac{0.992}{1 - 0.992} = 124$$

and for $\alpha = 0.995$,

$$\beta = \frac{\alpha}{1-\alpha} = \frac{0.995}{1-0.995} = 199$$

Therefore, for the range

$$0.992 \le \alpha \le 0.995$$

the corresponding β range is

$$124 \le \beta \le 199$$

3.3.1 Equivalent Circuit Models – NPN Transistors

We can draw various equivalent circuit models with dependent voltage and current sources* for NPN transistors using the relations (3.1) through (3.10). To illustrate, let us draw an equivalent circuit using relations (3.4), (3.3), and (3.1) which are repeated here for convenience.

$$i_B = (I_r/\beta)e^{v_{BE}/V_T} \tag{3.15}$$

$$i_C = \beta i_B \tag{3.16}$$

$$i_E = i_B + i_C = (I_r/\beta)e^{v_{BE}/V_T} + \beta i_B \tag{3.17}$$

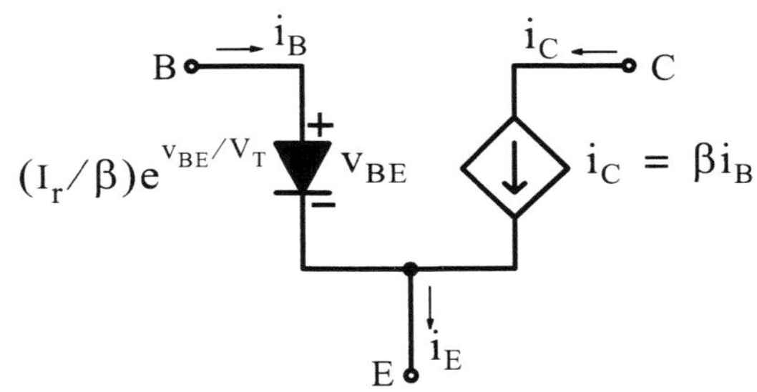

Figure 3.7. NPN transistor equivalent circuit model for relations (3.15), (3.16), and (3.17)

If we know I_r, β, i_B, and the operating temperature, we can find the other parameters for the circuit model in Figure 3.7.

Example 3.2

For a given NPN transistor, $I_r = 2 \times 10^{-14}$, $\beta = 200$, $i_B = 5\ \mu A$, and $T = 27\ °C$. Find the numerical values of the parameters shown in Figure 3.7. The collector bias voltage V_{CC} (not

* *Dependent sources are discussed in detail in Chapters 1 through 4, Circuit Analysis I with MATLAB Applications, ISBN 978–0–9709511–2–0.*

shown) is $10V$.

Solution:

$$\frac{I_r}{\beta} = \frac{2 \times 10^{-14}}{200} = 10^{-16}$$

We find v_{BE} from (3.15), i.e.,

$$i_B = (I_r/\beta)e^{v_{BE}/V_T}$$

Then,

$$5 \times 10^{-6} = 10^{-16}e^{v_{BE}/V_T}$$

$$e^{v_{BE}/V_T} = \frac{5 \times 10^{-6}}{10^{-16}} = 5 \times 10^{10}$$

$$v_{BE}/V_T = \ln(5 \times 10^{10})$$

$$v_{BE} = V_T(\ln 5 + 10\ln 10) = 26 \times 10^{-3}(1.61 + 23.03) \approx 0.64\ V$$

The collector bias voltage V_{CC} is used for proper transistor operation and its value is not required for the above calculations.

3.3.2 Equivalent Circuit Models – PNP Transistors

Relations (3.15), (3.16), and (3.17) apply also to PNP transistor equivalent circuits except that v_{BE} needs to be replaced by v_{EB} as shown in Figure 3.8.

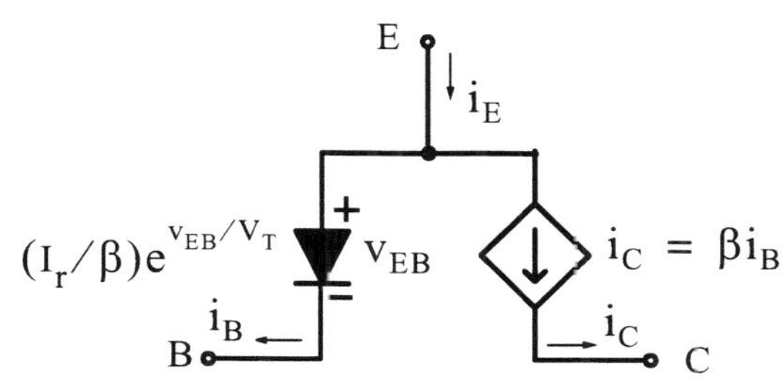

Figure 3.8. PNP transistor equivalent circuit model for relations (3.15), (3.16), and (3.17)

For easy reference we summarize the current–voltage relationships for both NPN and PNP transistors in the active mode in Table 3.1.

TABLE 3.1 NPN and PNP transistor current–voltage characteristics

NPN Transistor			PNP Transistor		
$\alpha = \frac{\beta}{\beta+1} = \frac{i_C}{i_E}$	$\beta = \frac{\alpha}{1-\alpha} = \frac{i_C}{i_B}$	$i_E = i_B + i_C$	$\alpha = \frac{\beta}{\beta+1} = \frac{i_C}{i_E}$	$\beta = \frac{\alpha}{1-\alpha} = \frac{i_C}{i_B}$	$i_E = i_B + i_C$
$i_B = \left(\frac{I_r}{\beta}\right)e^{\frac{v_{BE}}{V_T}}$	$i_C = I_r e^{\frac{v_{BE}}{V_T}}$	$i_E = \left(\frac{I_r}{\alpha}\right)e^{\frac{v_{BE}}{V_T}}$	$i_B = \left(\frac{I_r}{\beta}\right)e^{\frac{v_{EB}}{V_T}}$	$i_C = I_r e^{\frac{v_{EB}}{V_T}}$	$i_E = \left(\frac{I_r}{\alpha}\right)e^{\frac{v_{EB}}{V_T}}$
$i_B = i_C/\beta$	$i_C = \beta i_B$	$i_B = i_C/\alpha$	$i_B = i_C/\beta$	$i_C = \beta i_B$	$i_B = i_C/\alpha$
$i_B = i_E/(\beta+1)$	$i_C = \alpha i_E$	$i_E = (\beta+1)i_B$	$i_B = i_E/(\beta+1)$	$i_C = \alpha i_E$	$i_E = (\beta+1)i_B$
$i_B = (1-\alpha)\,i_E$	$V_T = 26\text{ mV at } T = 27\text{ °C}$		$i_B = (1-\alpha)\,i_E$	$V_T = 26\text{ mV at } T = 27\text{ °C}$	
$v_{BE} = V_T[\ln(\beta) - \ln(I_r) + \ln(i_B)]$			$v_{BE} = V_T[\ln(\beta) - \ln(I_r) + \ln(i_B)]$		

The relations in Table 3.1 are very useful in establishing voltage and current levels at various points on an NPN or PNP transistor.

Example 3.3

An NPN transistor with $\beta = 150$ is to operate in the common (grounded) base configuration. A DC power supply at $V_S = \pm 12\text{ V}$ is available and with two external resistors, one connected between the collector and V_{CC} and the other between the emitter and V_{EE}, we want to keep the collector current I_C* at 1.6 mA and the collector voltage V_C at 4 V. Find the values of the resistors given that when $V_{BE} = 0.7\text{ V}$, $I_C = 1.2\text{ mA}$. The circuit operates at $T = 27\text{ °C}$.

Solution:

Since the transistor is to operate at the common base configuration, after connecting the resistors and the bias voltages, our circuit is as shown in Figure 3.9.

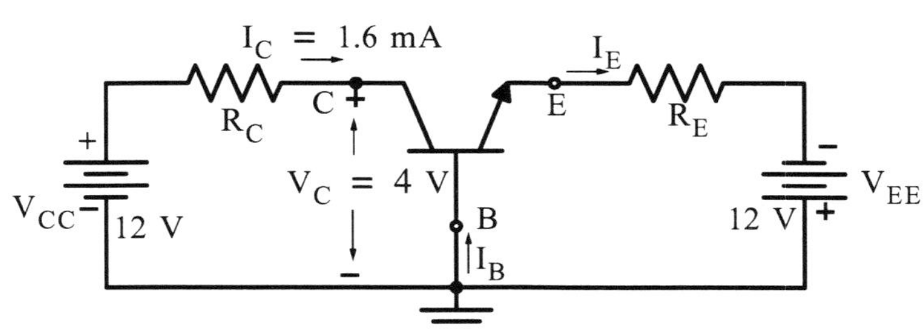

Figure 3.9. Transistor circuit for Example 3.3 – Computations for R_C

* As stated earlier, we use upper case letters for DC (constant) values, and lower case letters for instantaneous values.

Application of Kirchoff's Voltage Law (KVL) on the collector side of the circuit with $I_C = 1.6\ mA$ and $V_C = 4\ V$ yields

$$R_C I_C + V_C = V_{CC}$$

$$R_C = \frac{V_{CC} - V_C}{I_C} = \frac{12 - 4}{1.6 \times 10^{-3}} = 5\ K\Omega$$

We are given that for $V_{BE} = 0.7\ V$, $I_C = 1.2\ mA$ and we need to find V_{BE} at $I_C = 1.6\ mA$. We find V_{BE} from the ratio

$$\frac{1.6\ mA}{1.2\ mA} = \frac{I_r e^{V_{BE}/V_T}}{I_r e^{0.7/V_T}} = e^{(V_{BE} - 0.7)/V_T}$$

$$\ln\left(\frac{1.6}{1.2}\right) = (V_{BE} - 0.7)/V_T$$

$$V_{BE} = 0.7 + 26 \times 10^{-3} \ln\left(\frac{1.6}{1.2}\right) = 0.708$$

Next,

$$V_{BE} = V_B - V_E$$

and since the base is grounded, $V_B = 0$, and $V_E = -V_{BE} = -0.708$ as shown in Figure 3.10.

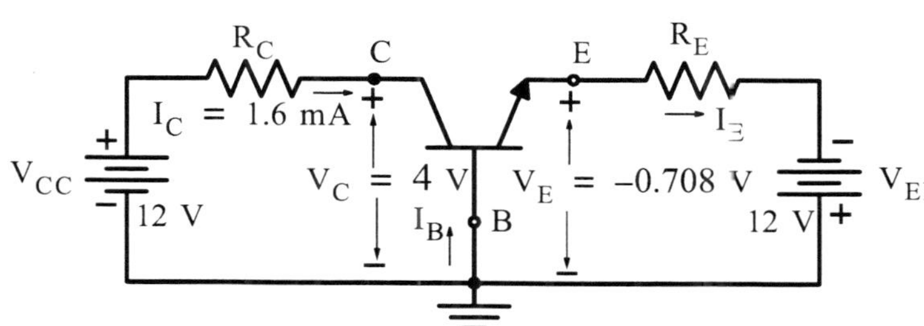

Figure 3.10. Transistor circuit for Example 3.3 – Computations for R_E

From Table 3.1

$$I_C = \alpha I_E$$

or

$$I_E = I_C/\alpha$$

and since $\beta = 150$,

$$\alpha = \beta/(\beta + 1) = 150/151 = 0.993$$

and

$$I_E = I_C/\alpha = (1.6\ mA)/0.993 = 1.61\ mA$$

Then, by KVL

$$-V_E + R_E I_E = V_{EE}$$

$$R_E = \frac{V_{EE} + V_E}{I_E} = \frac{-(-12) - 0.708}{1.61 \times 10^{-3}} = 7\ K\Omega$$

3.3.3 Effect of Temperature on the $i_C - v_{BE}$ Characteristics

As with diodes, the base–emitter voltage v_{BE} decreases approximately 2 mV for each 1 °C rise in temperature when the emitter current i_E remains constant.

Example 3.4

For the PNP transistor circuit in Figure 3.11, at $T = 27\ °C$ the emitter to base voltage $v_{BE} = 0.7\ V$ and the emitter current I_E is held constant at all temperature changes. Find the changes in emitter voltage V_E and collector voltage V_C if the temperature rises to $T = 50\ °C$.

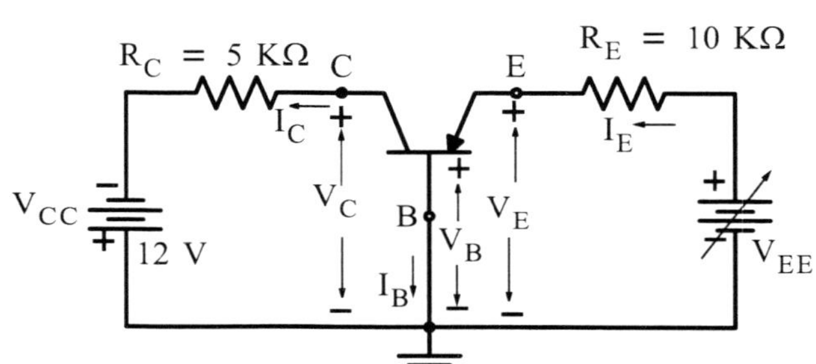

Figure 3.11. Circuit for Example 3.4

Solution:

Since the base is grounded, $v_{EB} = V_E$ and since this voltage decreases by 2 mV for each 1 °C temperature rise, the change in V_E is

$$\Delta V_E = (50 - 27)\ °C \times (-2\ mV/1\ °C) = -46\ mV$$

and at $T = 50\ °C$,

$$V_E = 0.7 - 0.046 = 0.654\ V$$

There is no change in the collector voltage V_C because the emitter current I_E is held constant, and since $I_C = \alpha I_E$, the voltage V_C remains unchanged.

3.3.4 Collector Output Resistance – Early Voltage

From basic circuit analysis theory, we recall that a current source has a parallel resistance attached to it and it is referred to as output resistance. Ideally, this resistance should be infinite and we can connect any passive load to the current source. However, most integrated circuits use transistors as loads instead of resistors R_C, and when active loads (transistors) are used, we should consider the finite output resistance that is in parallel with the collector. This resistance is in the order of $100\ K\Omega$ or greater. This output resistance looking into the collector is defined as

$$r_{out} = \left.\frac{\partial v_{CE}}{\partial i_C}\right|_{V_{BE} = \text{constant}} \tag{3.18}$$

or

$$r_{out} = \frac{V_A}{I_C} \tag{3.19}$$

where V_A is the *Early voltage* supplied by the manufacturer, and I_C is the DC collector current.

Before we consider the next example, let us illustrate a transistor in the common–emitter mode with the resistive circuit shown in Figure 3.12.

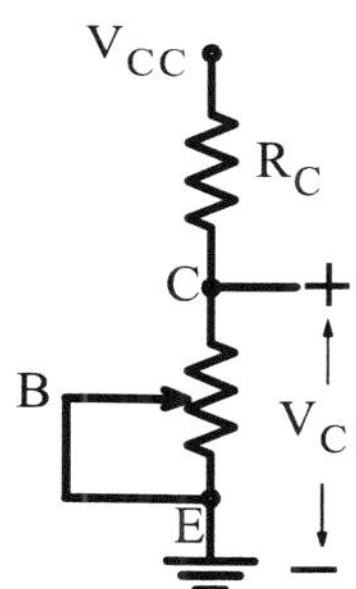

Figure 3.12. Representing a transistor as a resistive circuit with a potentiometer

When the potentiometer resistance is decreased (the wiper moves upwards) the current through the collector resistor R_C increases and the voltage drop across R_C increases. This voltage drop subtracts from the supply voltage V_{CC} and the larger the voltage drop, the smaller the voltage V_C at the collector. Conversely, when the potentiometer resistance is increased (the wiper moves downwards) the current through the collector resistor R_C decreases and the voltage drop across R_C decreases. This voltage drop subtracts from the supply voltage V_{CC} and the smaller the voltage drop, the larger the voltage V_C at the collector.

We recall also that a resistor serves as a current limiter and it develops a voltage drop when current flows through it. A transistor is a current–in, current–out device. We supply current to the base of the transistor and current appears at its collector. The current into the base of the transistor is in the order of a few microamps while the current at the collector is in the order of a few milliamps. The transistor circuit and the waveforms shown in Figure 3.13 will help us understand the transistor operation in the common–emitter mode.

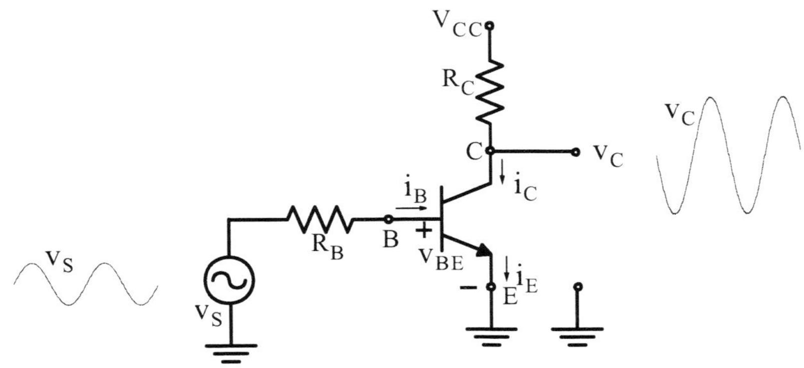

Figure 3.13. Transistor operation in the common–emitter mode

For the circuit of Figure 3.13, for the base–to–emitter voltage interval $0 \le v_{BE} \le 0.65$, no current flows into the base. But when the base–to–emitter voltage is $v_{BE} \approx 0.7\ V$, a small current flows into the base. A further increase in the supply voltage v_S has no effect on v_{BE} which remains fairly constant at $0.7\ V$, but the base current continues to increase causing an increase in the collector current i_C. This, in turn causes the voltage drop across the resistor R_C to increase and thus the collector voltage v_C decreases. and for this reason the output voltage v_C appears as $180°$ out–of–phase with the input (supply) voltage v_S. As v_S decreases, less current flows into the base and the collector current decreases also causing the voltage drop across the resistor R_C to decrease, and consequently the collector voltage v_C increases.

Example 3.5

The datasheet for the NPN transistor in Figure 3.14 indicates that the Early voltage is $V_A = 80$. The base to emitter voltage V_{BE} is held constant at $0.7\ V$ and when V_{CC} is adjusted so that $V_{CE} = 1\ V$, the collector current $I_C = 0.8\ mA$. Find the values of I_C as V_{CE} varies from $1\ V$ to $10\ V$ and plot I_C versus V_{CE}. Do you expect a linear relationship between I_C and V_{CE}?

Solution:

Relations (3.18) and (3.19) are applicable here. Thus,

$$r_{out} = \frac{V_A}{I_C} = \frac{80}{0.8} = 100\ K\Omega$$

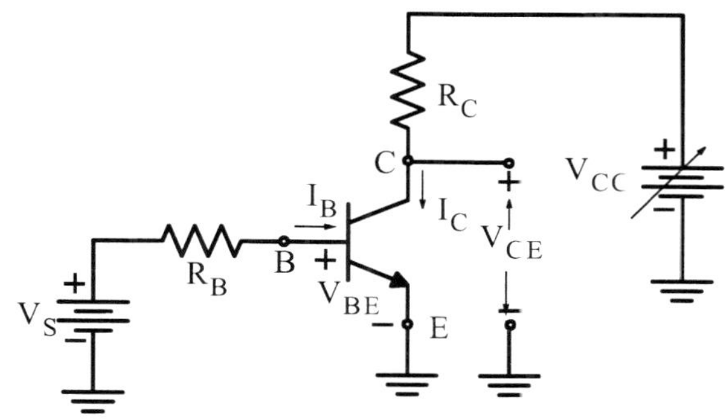

Figure 3.14. Circuit for Example 3.5

and

$$\Delta i_C = \frac{\Delta v_{CE}}{r_{out}} = 10^{-5}\Delta v_{CE}$$

This expression shows that there is a linear relationship between I_C and v_{CE}. The MATLAB script below performs all computations and plots I_C versus v_{CE}. The plot is shown in Figure 3.15.

```
vCE=1: 10; vA=80; iC1=0.8*10^(−3); r0=vA/iC1; iC=zeros(10,3) ...
iC(:,1)=vCE'; iC(:,2)=(vCE./r0)'; iC(:,3)=(iC1+vCE./r0)'; fprintf(' \r');...
disp('vCE   delta iC    new iC');...
disp('-------------------------');...
fprintf('%2.0f \t %2.2e \t %2.2e\n',iC');...
plot(vCE,iC(:,3)); xlabel('vCE (V)'); ylabel('iC (A)'); grid;...
title('iC vs vCE for Example 3.5')
```

```
vCE    delta iC     new iC
-------------------------
 1   1.00e-005   8.10e-004
 2   2.00e-005   8.20e-004
 3   3.00e-005   8.30e-004
 4   4.00e-005   8.40e-004
 5   5.00e-005   8.50e-004
 6   6.00e-005   8.60e-004
 7   7.00e-005   8.70e-004
 8   8.00e-005   8.80e-004
 9   9.00e-005   8.90e-004
10   1.00e-004   9.00e-004
```

Generally, the i_C versus v_{CE} relation is non–linear. It is almost linear when a transistor operates in the active region, and non–linear when it operates in the cutoff and saturation regions. Table 3.2 shows the three modes of operation in a bipolar transistor and the forward or reverse–biasing of the emitter–base and collector–base junctions.

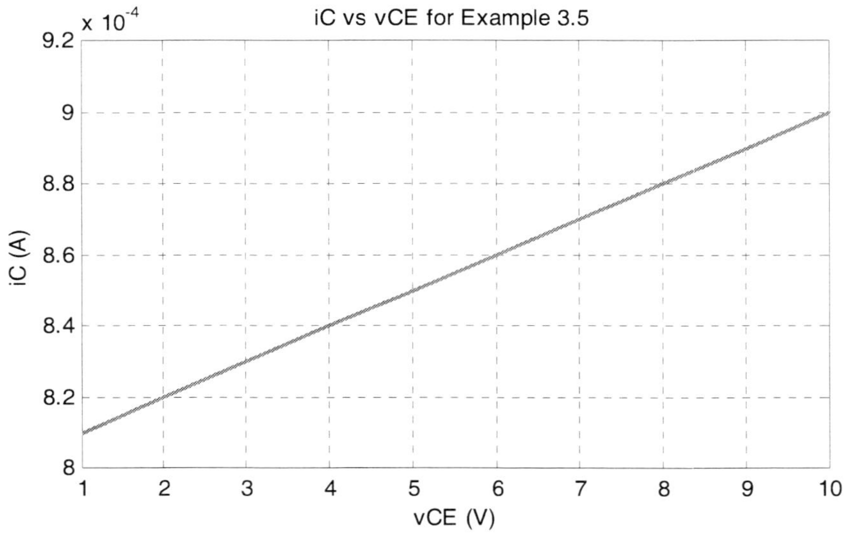

Figure 3.15. Plot for Example 3.5

TABLE 3.2 Region of operation for bipolar transistors

Region of Operation	Emitter–Base junction	Collector–base junction
Active	Forward	Reverse
Saturation	Forward	Forward
Cutoff	Reverse	Reverse

Example 3.6

For the circuit in Figure 3.16, $\beta = 120$ and $V_{BE} = 0.7$. Find V_E, I_E, I_C, V_C, and determine whether this circuit with the indicated values operates in the active, saturation, or cutoff mode.

Solution:

From the given circuit, by observation

$$V_E = V_B - V_{BE} = 5 - 0.7 = 4.3\ \text{V}$$

and

$$I_E = \frac{V_E}{R_E} = \frac{4.3}{2.5 \times 10^3} = 1.72\ \text{mA}$$

It is given that $\beta = 120$. Then,

$$\alpha = \frac{\beta}{\beta + 1} = \frac{120}{121} = 0.992$$

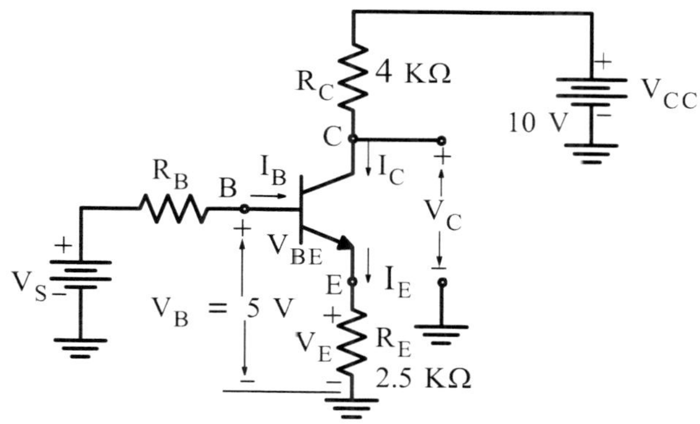

Figure 3.16. Circuit for Example 3.6

and

$$I_C = \alpha I_E = 0.992 \times 1.72 = 1.71\ \text{mA}$$

Then,

$$I_B = 1.72\ \text{mA} - 1.71\ \text{mA} = 0.01\ \text{mA} = 10\ \mu\text{A}$$

Finally, we find the collector voltage V_C as

$$V_C = V_{CC} - I_C R_C = 10 - 1.71 \times 10^{-3} \times 4 \times 10^3 = 3.16\ \text{V}$$

We observe that the transistor is an NPN type and for the active mode operation the base–collector junction PN must be reverse–biased. It is not because $V_C < V_B$ and thus we conclude that with the given values the transistor is in saturation mode.

Example 3.7

For a PNP transistor circuit with the base grounded, it is given that $V_{EE} = 12\ \text{V}$, $V_{CC} = -12\ \text{V}$, $R_E = 5\ \text{K}\Omega$, $R_C = 3\ \text{K}\Omega$,and $\beta = 150$. Find V_E, I_E, I_C, V_C, and I_B. Is the circuit operating in the active mode?

Solution:

The circuit is as shown in Figure 3.17. With $V_{EE} = 12\ \text{V}$ it is reasonable to assume that the emitter–base junction is forward–biased and since the base is grounded, we have

$$V_E = V_{EB} = 0.7\ \text{V}$$

and

$$I_E = \frac{V_{EE} - V_E}{R_E} = \frac{12 - 0.7}{5 \times 10^3} = 2.3\ \text{mA}$$

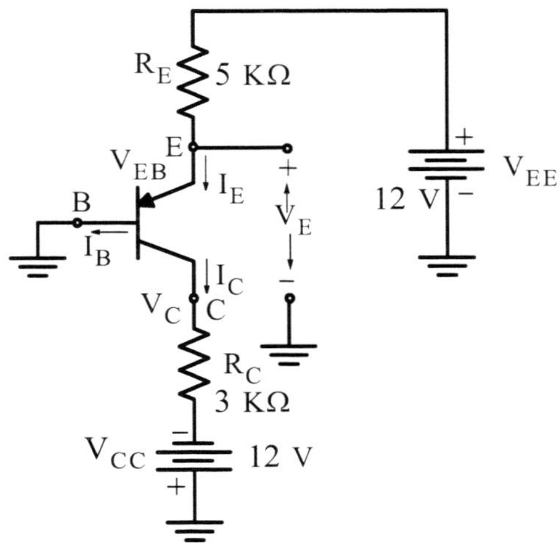

Figure 3.17. PNP transistor for Example 3.7

With $\beta = 150$

$$\alpha = \frac{\beta}{\beta + 1} = \frac{150}{151} = 0.993$$

Then,

$$I_C = \alpha I_E = 0.993 \times 2.3 \times 10^{-3} = 2.28 \text{ mA}$$

and

$$V_C = R_C I_C - V_{CC} = 3 \times 10^3 \times 2.28 \times 10^{-3} - 12 = -5.16 \text{ V}$$

With this value of V_C, the collector–base junction is reverse–biased and the PNP transistor is in the active mode. The base current is

$$I_B = I_E - I_C = 2.3 \times 10^{-3} - 2.28 \times 10^{-3} = 20 \ \mu\text{A}$$

Example 3.8

For the circuit in Figure 3.18, it is known that $\beta = 120$. Find I_B, I_E, I_C, V_B, V_C, and V_E. Is the transistor operating in the active mode?

Solution:

To simplify the part of the circuit to the left of the base, we apply Thevenin's theorem at points x and y as shown in Figure 3.19, and denoting the Thevenin equivalent voltage and resistance as V_{TH} and R_{TH} respectively, we find that

$$V_{TH} = V_{xy} = \frac{R_2}{R_1 + R_2} V_S = \frac{30}{60 + 30} 12 = 4 \text{ V}$$

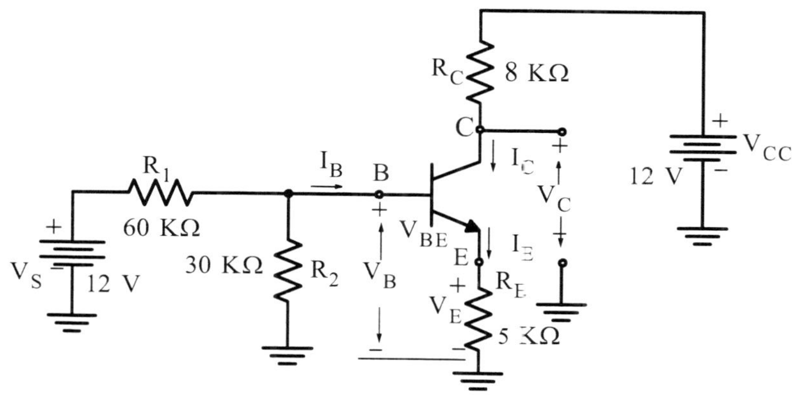

Figure 3.18. Circuit for Example 3.8

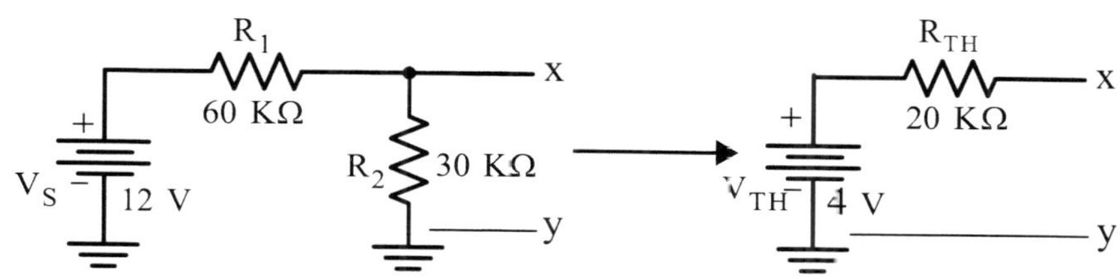

Figure 3.19. Application of Thevenin's theorem to the circuit of Example 3.8

and

$$R_{TH} = \frac{R_1 \times R_2}{R_1 + R_2} = \frac{60 \times 30}{60 + 30} = 20\ K\Omega$$

The circuit in Figure 3.18 is thus reduced to that in Figure 3.20.

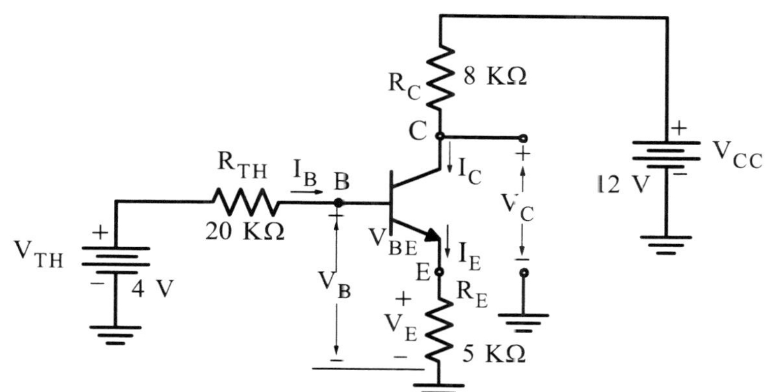

Figure 3.20. The circuit in Figure 3.18 after application of Thevenin's theorem

Application of KVL around the left part of the circuit of Figure 3.20 yields

$$R_{TH}I_B + V_{BE} + R_EI_E = V_{TH}$$

$$(20 \times 10^3)I_B + 0.7 + (5 \times 10^3)I_E = 4$$

$$4I_B + I_E = \frac{3.3}{5 \times 10^3} = 0.66 \times 10^{-3}$$

From Table 3.1, $I_E = (\beta + 1)I_B$. Then,

$$4I_B + (\beta + 1)I_B = 0.66 \times 10^{-3}$$

$$125I_B = 0.66 \times 10^{-3}$$

$$I_B = 5.28 \times 10^{-6}\ A = 5.28\ \mu A$$

and

$$I_E = (\beta + 1)I_B = 121 \times 5.28 \times 10^{-6} = 0.639\ mA$$

Next,

$$V_E = R_E I_E = 5 \times 10^3 \times 0.639 \times 10^{-3} = 3.2\ V$$

and

$$V_B = V_{BE} + V_E = 0.7 + 3.2 = 3.9\ V$$

Also,

$$I_C = I_E - I_B = 0.639 \times 10^{-3} - 5.28 \times 10^{-6} = 0.634\ mA$$

and

$$V_C = V_{CC} - R_C I_C = 12 - 8 \times 10^3 \times 0.634 \times 10^{-3} = 6.93\ V$$

Since this is an NPN transistor and $V_C > V_B$, the base–collector PN junction is reverse–biased and thus the transistor is in active mode.

3.4 Transistor Amplifier Circuit Biasing

In our previous discussion, for convenience, a separate voltage source V_{BE} has been used to provide the necessary forward–bias voltage and another voltage source V_{CC} to establish a suitable collector voltage V_C where $V_C = V_{CC} - R_C I_C$. However, it is not practical to use a separate emitter–base bias voltage V_{BE}. This is because conventional batteries are not available for $0.7\ V$. For this reason we use resistors in the order of kilohms to form voltage dividers with desired values. In addition to eliminating the battery, some of these biasing methods compensate for slight variations in transistor characteristics and changes in transistor conduction resulting from temperature irregularities.

Figure 3.21 shows the basic NPN transistor amplifier where resistor R_B provides the necessary forward bias for the emitter–base junction. Conventional current flows from V_{CC} through R_B to the base then to the grounded emitter. Since the current in the base circuit is very small (a few hundred microamperes) and the forward resistance of the transistor is low, only a few tenths of a volt of positive bias will be felt on the base of the transistor. However, this is enough voltage on the base, along with ground on the emitter and the large positive voltage on the collector to properly bias the transistor.

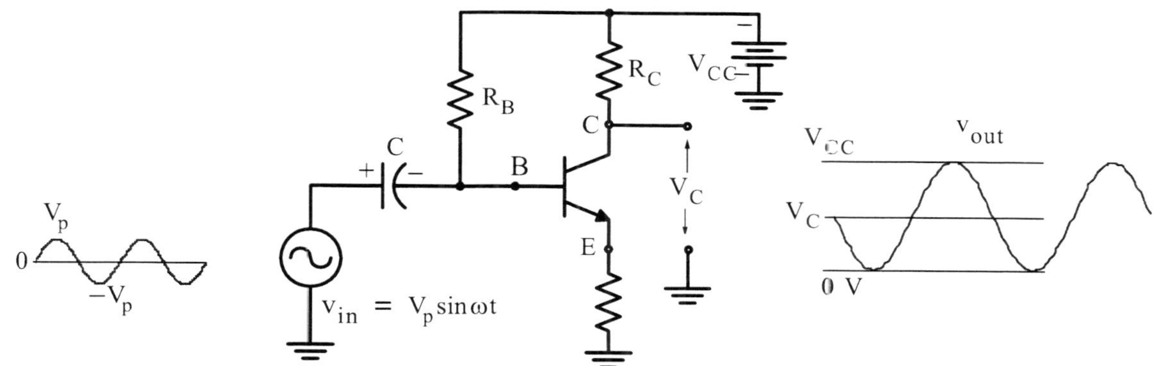

Figure 3.21. The basic NPN transistor amplifier biased with a resistive network

With the transistor properly biased, direct current flows continuously, with or without an input signal, throughout the entire circuit. The direct current flowing through the circuit develops more than just base bias; it also develops the collector voltage V_C as it flows from V_{CC} through resistor R_C and, as we can see on the output graph, the output signal starts at the V_C level and either increases or decreases. These DC voltages and currents that exist in the circuit before the application of a signal are known as *quiescent* voltages and currents (the quiescent state of the circuit). The DC quiescent point is the DC bias point Q with coordinates (V_{BE}, I_B). We will discuss the Q point in detail in a later section.

The collector resistor R_C is placed in the circuit to keep the full effect of the collector supply voltage off the collector. This permits the collector voltage V_C to change with an input signal, which in turn allows the transistor to amplify voltage. Without R_C in the circuit, the voltage on the collector would always be equal to V_{CC}.

The coupling capacitor C is used to pass the ac input signal and block the dc voltage from the preceding circuit. This prevents DC in the circuitry on the left of the coupling capacitor from affecting the bias on the transistor. The coupling capacitor also blocks the bias of the transistor from reaching the input signal source.

The input to the amplifier is a sine wave that varies a few millivolts above and below zero. It is introduced into the circuit by the coupling capacitor and is applied between the base and emitter. As the input signal goes positive, the voltage across the base–emitter junction becomes more positive. This in effect increases forward bias, which causes base current to increase at the same rate as that of the input sine wave. Collector and emitter currents also increase but much more than the base current. With an increase in collector current, more voltage is developed across R_C. Since the voltage across R_C and the voltage across the transistor (collector to emitter) must add up to V_{CC}, an increase in voltage across R_C results in an equal decrease in voltage across the transistor. Therefore, the output voltage from the amplifier, taken at the collector of the transistor with respect to the emitter, is a negative alternation of voltage that is larger than the input, but has the same sine wave characteristics.

During the negative alternation of the input, the input signal opposes the forward bias. This action decreases base current, which results in a decrease in both collector and emitter currents. The decrease in current through R_C decreases its voltage drop and causes the voltage across the transistor to rise along with the output voltage. Therefore, the output for the negative alternation of the input is a positive alternation of voltage that is larger than the input but has the same sine wave characteristics.

By examining both input and output signals for one complete alternation of the input, we can see that the output of the amplifier is an exact reproduction of the input except for the reversal in polarity and the increased amplitude (a few millivolts as compared to a few volts).

Figure 3.22 shows the basic PNP transistor amplifier. As we already know, the primary difference between the NPN and PNP amplifier is the polarity of the source voltage V_{CC}. With a negative V_{CC}, the PNP base voltage is slightly negative with respect to ground, which provides the necessary forward bias condition between the emitter and base.

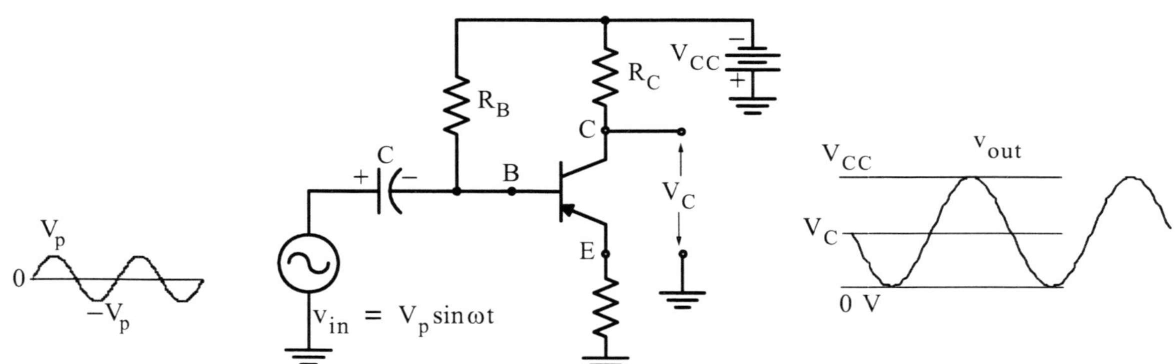

Figure 3.22. The basic PNP transistor amplifier

When the PNP input signal goes positive, it opposes the forward bias of the transistor. This action cancels some of the negative voltage across the emitter–base junction, which reduces the current through the transistor. Therefore, the voltage across R_C decreases, and the voltage across the transistor increases. Since V_{CC} is negative, the voltage on the collector V_C goes in a negative direction toward V_{CC}. Thus, the output is a negative alternation of voltage that varies at the same rate as the sine wave input, but it is opposite in polarity and has a much larger amplitude.

During the negative alternation of the input signal, the transistor current increases because the input voltage aids the forward bias. Therefore, the voltage across R_C increases, and consequently, the voltage across the transistor decreases or goes in a positive direction. This action results in a positive output voltage, which has the same characteristics as the input except that it has been amplified and the polarity is reversed.

In summary, the input signals in the preceding circuits were amplified because the small change in base current caused a large change in collector current. And, by placing resistor R_C in series with the collector, voltage amplification was achieved.

3.5 Fixed Bias

The biasing method used in the transistor circuits in Figures 3.21 and 3.22 is known as *fixed bias*. Even though with modern technology transistors are components (parts) of integrated circuits, or ICs, some are used as single devices. To bias a transistor properly, one must establish a constant DC current in the emitter so that it will not be very sensitive to temperature variations and large variations in the value of β among transistors of the same type. Also, the Q point must be chosen so that it will allow maximum signal swing from positive to negative values.

Figure 3.23 shows another configuration of an NPN transistor with fixed bias.

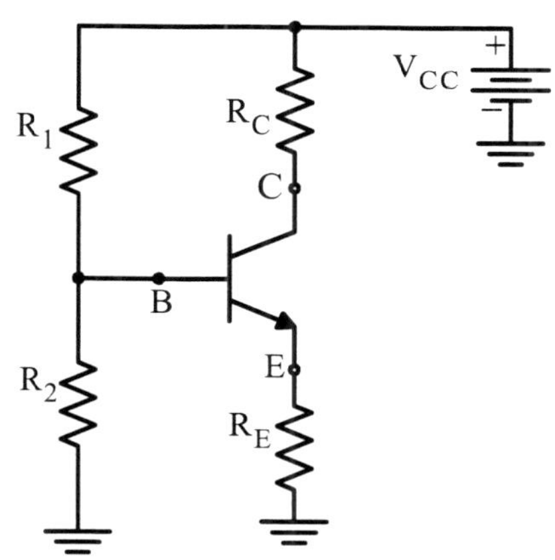

Figure 3.23. NPN Transistor with fixed bias

Following the procedure of Example 3.8 we can simplify the circuit in Figure 3.23 with the use of Thevenin's theorem to the circuit in Figure 3.24.

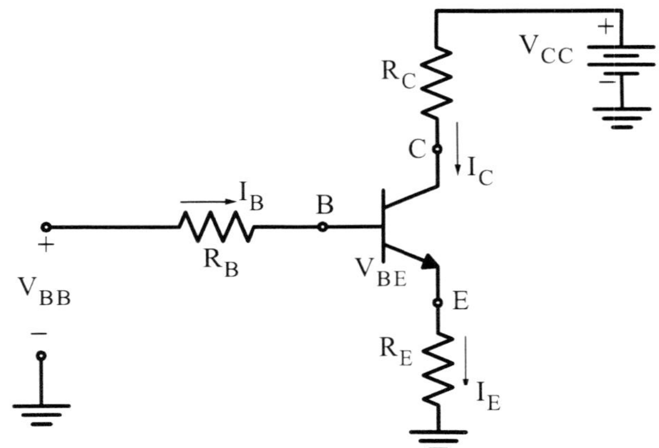

Figure 3.24. The circuit in Figure 3.23 after application of Thevenin's theorem

With reference to Figures 3.23 and 3.24 we obtain the following relations:

$$V_{BB} = \frac{R_2}{R_1 + R_2} V_{CC} \tag{3.20}$$

$$R_B = \frac{R_1 R_2}{R_1 + R_2} \tag{3.21}$$

It was stated earlier that it is imperative to keep variations in temperature and changes in β values to a minimum and this can be achieved by making the emitter current I_E fairly constant. Therefore, let us now derive an expression for I_E.

From Figure 3.24 with application of KVL we obtain

$$R_B I_B + V_{BE} + R_E I_E = V_{BB}$$

and from Table 3.1

$$I_B = \left(\frac{1}{\beta + 1}\right) I_E$$

Then,

$$R_B\left(\frac{1}{\beta + 1}\right) I_E + V_{BE} + R_E I_E = V_{BB}$$

or

$$I_E = \frac{V_{BB} - V_{BE}}{R_E + R_B/(\beta + 1)} = \frac{V_{BB} - 0.7}{R_E + R_B/(\beta + 1)} \tag{3.22}$$

From expression (3.22) we see that the emitter current I_E will be fairly constant if $V_{BB} \gg 0.7$ and $R_E \gg R_B/(\beta + 1)$. Accordingly, R_B must be small and since $R_B = R_1 R_2/(R_1 + R_2)$, we should use small resistance values for R_1 and R_2. Designers recommend that the sum of R_1 and R_2 is such that the current through them (assuming that the base current is zero) is about half of the emitter current I_E. It is also recommended that V_{BB}, V_{CB} or V_{CB}, and the product $R_C I_C$ each be close to one–third of the value of V_{CC}.

It is often said that the emitter resistor R_E provides a negative feedback action which stabilizes the bias current. To understand how this is done, let us assume that the emitter current I_E increases. In this case, the voltages $R_E I_E$ and V_E will also increase. But the base voltage being determined by the voltage division provided by resistors R_1 and R_2 will remain relatively constant and according to KVL, the base–to–emitter voltage V_{BE} will decrease. And since $I_C = I_r e^{V_{BE}/V_T}$, both the collector current I_C and emitter current I_E will decrease. Obviously, this is a contradiction to our original assumption (increase in I_E) and thus we say that the emitter resistor R_E provides a negative feedback action.

Let us consider the transistor circuit in Figure 3.23 which is repeated as Figure 3.25 for convenience.

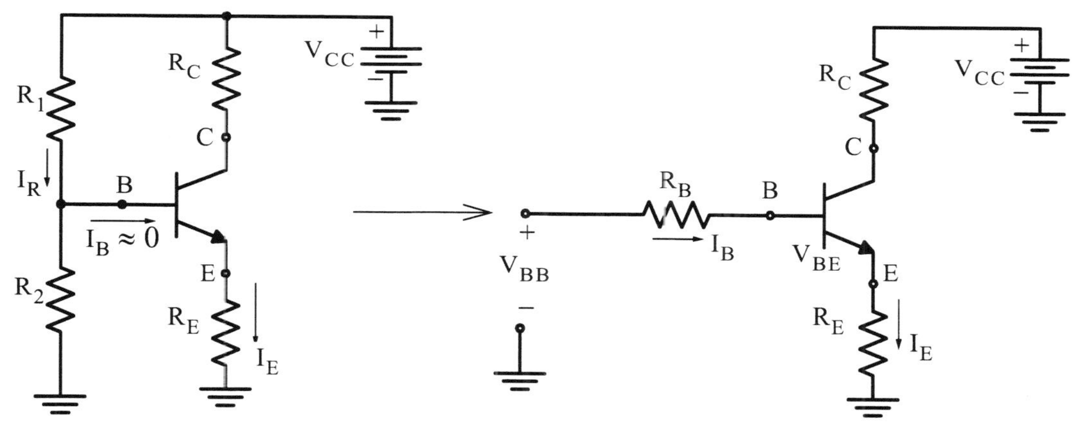

Figure 3.25. NPN transistor with fixed bias

Circuit designers maintain that if we specify the V_{CC} voltage, the I_E current, and the value of β, we can determine the appropriate values of the four resistors for proper biasing by letting the values of V_{BB}, V_{CB} or V_{CB}, and the product $R_C I_C$ each be close to one–third of the value of V_{CC}, and assuming that the base current is negligible, the current I_R through resistors R_1 and R_2 is $I_R = 0.5 I_E$. For convenience, we will denote the sum of R_1 and R_2 as R_{eq}.

Example 3.9

For the transistor circuit of Figure 3.26 $\beta = 120$. Find the values of the four resistors for appropriate fixed biasing.

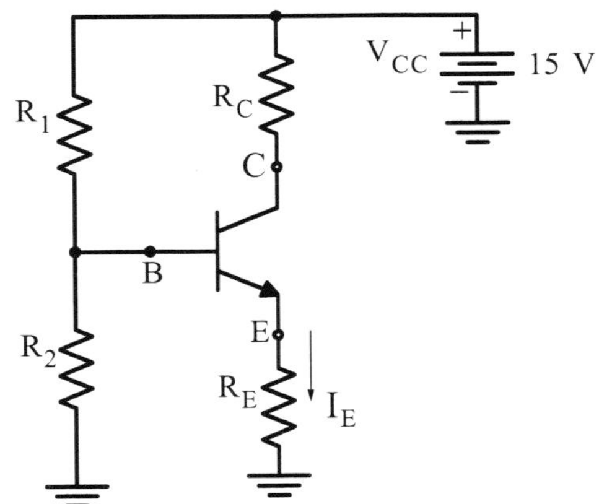

Figure 3.26. Transistor circuit for Example 3.9

Solution:

The given circuit and its Thevenin equivalent are shown in Figure 3.27. The Thevenin equivalent is shown just to indicate the value of V_{BB} which will be used in the calculations.

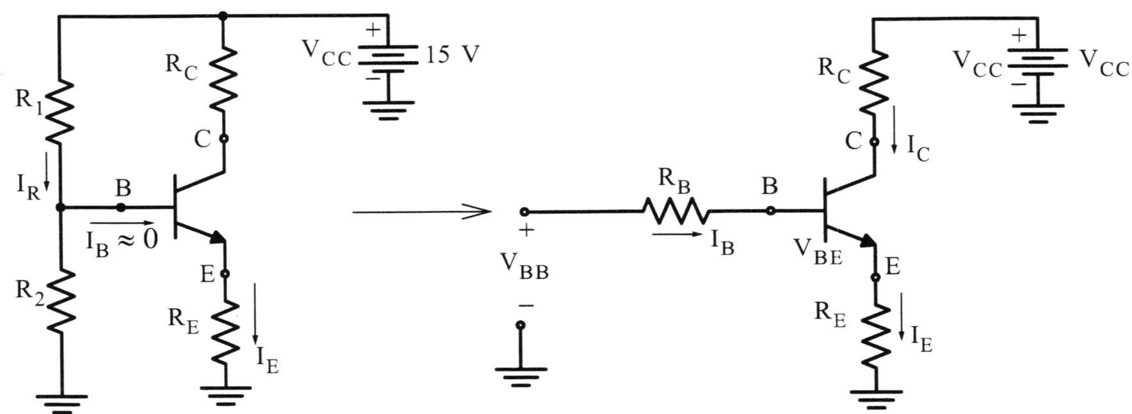

Figure 3.27. The transistor circuit in Example 3.9 and its Thevenin equivalent

As stated above, it is suggested that the values of V_{BB}, V_{CB} or V_{CB}, and the product $R_C I_C$ each be close to one–third of the value of V_{CC}, and assuming that the base current is negligible, the current I_R through resistors R_1 and R_2 is $I_R = 0.5 I_E$. Then,

$$V_{BB} = \frac{1}{3}V_{CC} = \frac{1}{3}15 = 5 \text{ V}$$

and with reference to the Thevenin equivalent circuit above, since $I_B \approx 0$ by KVL

$$V_{BE} + R_E I_E = V_{BB}$$

$$R_E = \frac{V_{BB} - V_{BE}}{I_E} = \frac{5 - 0.7}{1 \times 10^{-3}} = 4.3\ K\Omega$$

Also,

$$R_{eq} I_R = (R_1 + R_2) I_R = V_{CC}$$

$$R_{eq} = (R_1 + R_2) = \frac{V_{CC}}{I_R} = \frac{V_{CC}}{0.5 I_E} = \frac{15}{0.5 \times 10^{-3}} = 300\ K\Omega$$

To find $R_{eq} = (R_1 + R_2)$ we use the Thevenin equivalent voltage expression, that is,

$$V_{TH} = V_{EB} = 5\ V = \frac{R_2}{R_1 + R_2} V_{CC} = \frac{R_2}{R_1 + R_2} 15\ V = \frac{R_2}{300\ K\Omega} 15\ V$$

$$\frac{R_2}{300\ K\Omega} = \frac{5\ V}{15\ V} = \frac{1}{3}$$

and thus

$$R_2 = 100\ K\Omega$$

and

$$R_1 = 300\ K\Omega - R_2 = 300\ K\Omega - 100\ K\Omega = 200\ K\Omega$$

We will find the value of R_C from the relation

$$R_C I_C = \frac{1}{3} V_{CC}$$

or

$$R_C = \frac{(1/3)V_{CC}}{I_C} = \frac{(1/3)V_{CC}}{\alpha I_E} = \frac{(1/3)V_{CC}}{(\beta/(\beta+1))I_E} = \frac{5}{(120/121) \times 1 \times 10^{-3}} = 4.96\ K\Omega$$

3.6 Self-Bias

The fixed bias arrangement discussed in the previous section is thermally unstable. If the temperature of the transistor rises for any reason (due to a rise in ambient temperature or due to current flow through it), the collector current will increase. This increase in current also causes the DC quiescent point to move away from its desired position (level). This reaction to temperature is undesirable because it affects amplifier gain (the number of times of amplification) and could result in distortion, as we will see later in this chapter. A better method of biasing, known as *self–bias* is obtained by inserting the bias resistor directly between the base and collector, as shown in Figure 3.28.

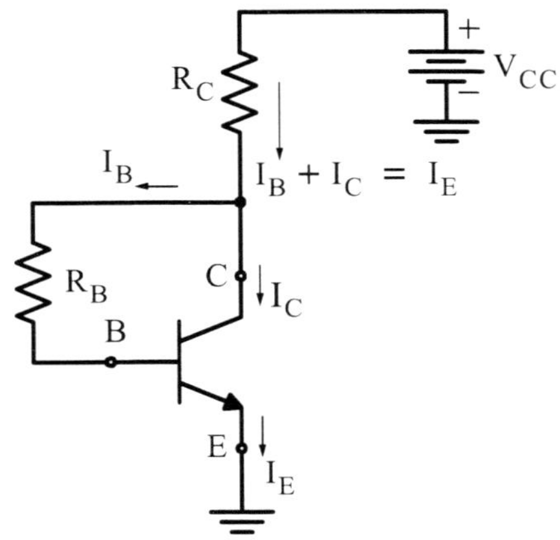

Figure 3.28. NPN transistor amplifier with self–bias

By tying the collector to the base in this manner, feedback voltage can be fed from the collector to the base to develop forward bias. Now, if an increase of temperature causes an increase in collector current, the collector voltage V_C will fall because of the increase of voltage produced across the collector resistor R_L. This drop in V_C will be fed back to the base and will result in a decrease in the base current. The decrease in base current will oppose the original increase in collector current and tend to stabilize it. The exact opposite effect is produced when the collector current decreases.

From Figure 3.28,

$$R_C I_E + R_B I_B + V_{BE} = V_{CC}$$

$$R_C I_E + R_B \frac{1}{\beta + 1} I_E + V_{BE} = V_{CC}$$

$$I_E = \frac{V_{CC} - V_{BE}}{R_C + R_B/(\beta + 1)} \tag{3.23}$$

From (3.23) we see that to maintain the emitter current I_E fairly constant, we should choose the collector and base resistors such that $R_B/(\beta + 1) \ll R_C$.

Example 3.10

For the transistor circuit in Figure 3.29 $\beta = 120$ and we want the collector voltage to vary in accordance with $V_C = 3\sin\omega t \text{ V} + V_{BE}$ and $I_E = 1 \text{ mA}$. Find the values of R_C and R_B to meet these specifications.

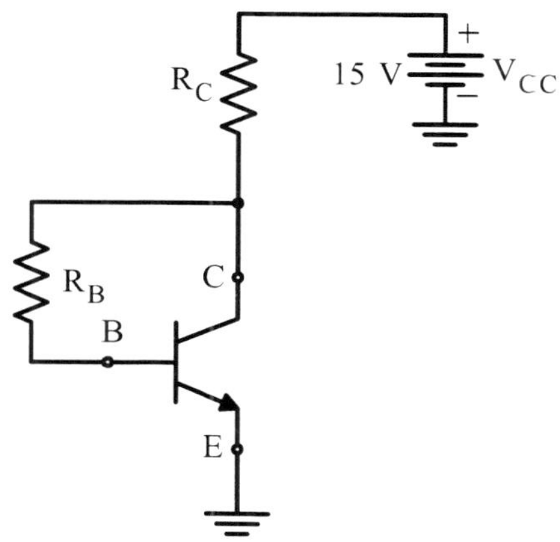

Figure 3.29. Transistor circuit for Example 3.10

Solution:

We assign current directions as shown in Figure 3.30.

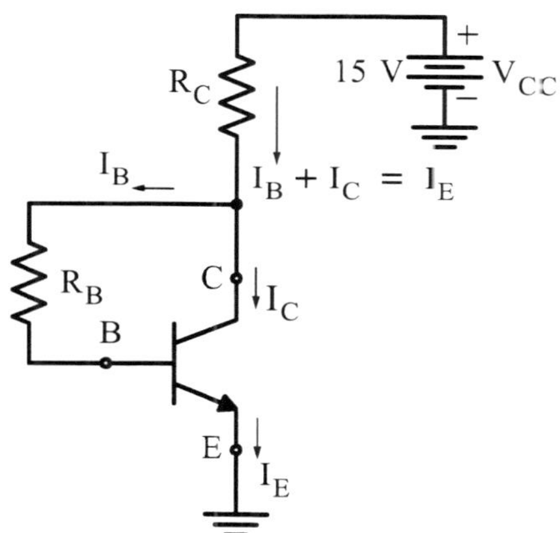

Figure 3.30. Circuit for Example 3.10 with assigned current directions

By inspection,

$$V_C = R_B I_B + V_{BE} = R_B\left(\frac{1}{\beta+1}\right) I_E + 0.7 = |3\sin\omega t| + 0.7 = 3 + 0.7$$

or

$$R_B = \frac{3}{1 \times 10^{-3} \times (1/121)} = 363\ \text{K}\Omega$$

Also, by inspection

$$V_C = V_{CC} - R_C I_E$$

$$R_C = \frac{V_{CC} - V_C}{I_E} = \frac{15 - 3.7}{1 \times 10^{-3}} = 11.3\ \text{K}\Omega$$

3.7 Amplifier Classes and Operation

In the previous discussions we assumed that for every portion of the input signal there was an output from the amplifier. This is not always the case with all types of amplifiers. It may be desirable to have the transistor conducting for only a portion of the input signal. The portion of the input for which there is an output determines the class of operation of the amplifier. There are four classes of amplifier operations. They are Class A, Class B, Class AB, and Class C.

Before discussing the different classes of amplifiers, we should remember that every amplifier has some unavoidable limitations on its performance. The most important that we need to be concerned about when choosing and using them are:

- **Limited bandwidth.** For each amplifier there is an upper frequency beyond which it finds it impossible to amplify signals.
- **Noise.** All electronic devices tend to add some random noise to the signals passing through them, hence degrading the SNR (signal to noise ratio). This, in turn, limits the accuracy of any measurement or communication.
- **Limited output voltage, current, and power levels.** A given amplifier cannot output signals above a particular level; there is always a finite limit to the output signal size.
- **Distortion.** The actual signal pattern will be altered due non–linearities in the amplifier. This also reduces the accuracy of measurements and communications.
- **Finite gain.** A given amplifier may have a high gain, but this gain cannot normally be infinite so may not be large enough for a given purpose. This is why we often use multiple amplifiers or stages to achieve a desired overall gain.

Let us first discuss the limits to signal size. Figure 3.31(a) shows a simple amplifier being used to drive output signals into a resistive load. The power supply voltages are $+V_{CC}$ and $-V_{EE}$ and thus the output voltage v_{out} will be limited to the range

$$-V_{EE} < v_{out} < +V_{CC}$$

From our earlier discussion we can think of the transistor as a variable resistor between the collector resistor R_C and emitter resistor R_E. We denote this variable resistor, i.e., the transistor, as R_{tr} and its value depends on the input voltage v_{in}. The two extreme values of the variable resistor are $R_{tr} = \infty$ (open circuit) and $R_{tr} = 0$ (short circuit). These conditions are shown in Figures 3.31(b) and 3.31(c).

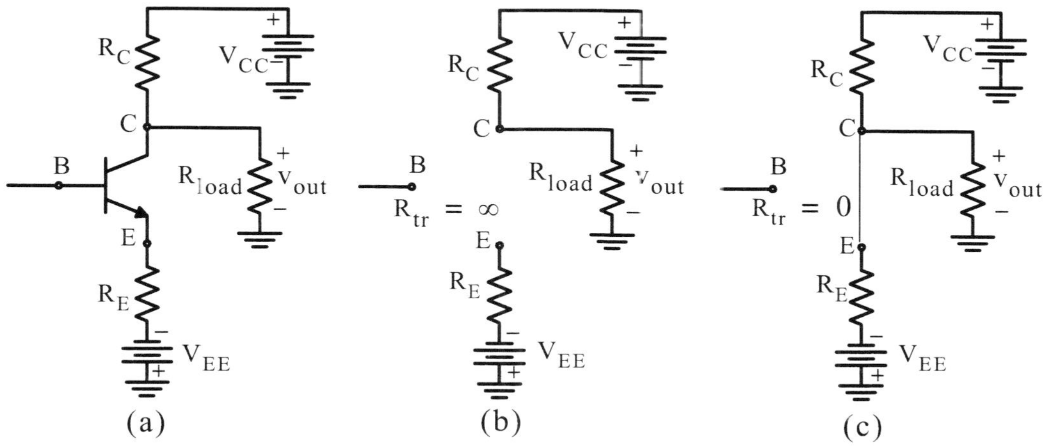

Figure 3.31. Simple amplifier with resistive load

Application of the voltage division expression for the circuit in Figure 3.31(b) yields the maximum voltage across the load resistor, that is,

$$v_{out}(max) = \frac{R_{load}}{R_C + R_{load}} \cdot V_{CC} \tag{3.24}$$

and the corresponding maximum current is

$$i_{out}(max) = \frac{V_{CC}}{R_C + R_{load}} \tag{3.25}$$

Obviously, we want the power delivered to the load resistor to be maximum; therefore, we must make the collector resistor R_C much smaller than the load resistor R_{load}, that is, $R_C \ll R_{load}$ to maximize the current through the load resistor. For convenience, we choose R_C to be one–tenth of the value of R_{load}, that is,

$$R_C = 0.1R_{load} \tag{3.26}$$

Next, by application of Thevenin's theorem for the circuit of Figure 3.31(c) where the load resistor is disconnected, we redraw the circuit as shown in Figure 3.32.

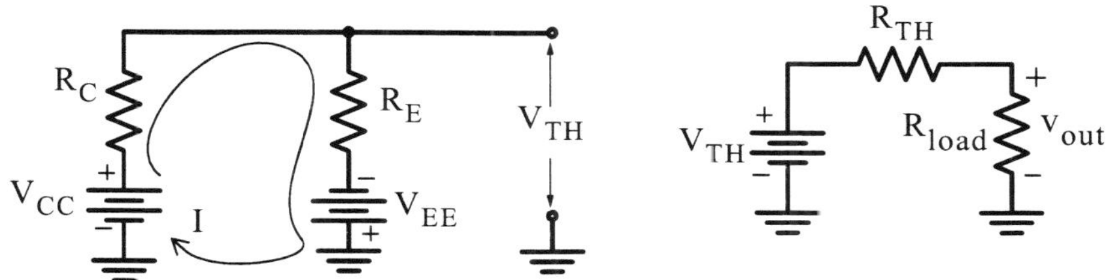

Figure 3.32. Thevenin equivalent for the circuit in Figure 3.31(c).

From Figure 3.32,

$$I = \frac{V_{CC} + V_{EE}}{R_C + R_E}$$

$$V_{TH} = V_{CC} - R_C I = \frac{R_E V_{CC} - R_C V_{EE}}{R_C + R_E} \tag{3.27}$$

$$R_{TH} = \frac{R_C R_E}{R_C + R_E} \tag{3.28}$$

and thus

$$v_{out}(min) = \frac{R_{load}}{R_{TH} + R_{load}} V_{TH} \tag{3.29}$$

From (3.29) we see that for $v_{out}(min)$ to be close to $-V_{EE}$, the Thevenin resistance R_{TH} shown in (3.28) and (3.29) should be minimized. We already have minimized R_C to be $R_C = 0.1 R_{load}$, so let us make R_E one–tenth of R_C, that is, $R_E = 0.1 R_C$, or

$$R_E = 0.01 R_{load} \tag{3.30}$$

Example 3.11

Let us suppose that the circuit of Figure 3.33 is to be used as the output stage of an audio system whose load resistance is $R_{load} = 8\ \Omega$ and the supply voltages are $V_{CC} = +24V$ and $V_{EE} = -24V$. Taken into account the recommendations discussed above for sizing the emitter and collector resistors, find the power absorbed by the combination of the collector resistor, the transistor, and the emitter resistor when $v_{out} = 0\ V$ and the amplifier is on.

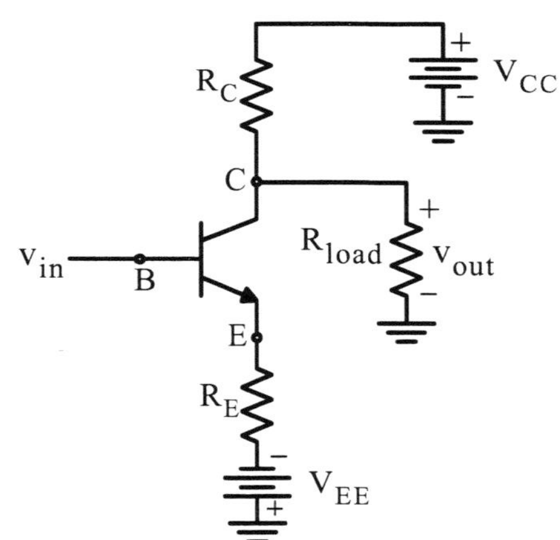

Figure 3.33. Circuit for Example 3.11

Solution:

With $R_C = 0.1R_{load} = 0.1 \times 8 = 0.8\ \Omega$ and $v_{out} = 0\ V$, there is no current through the load resistor and since $R_C I_C + v_{out} = 24\ V$, $I_C = 24/0.8 = 30\ A$, that is, the amplifier will be drawing $30\ A$ from the positive power supply. Since there is no current through the load, this current will flow through the transistor and the emitter resistor and into the negative power supply. Thus, the power absorbed by the combination of the collector resistor R_C, the transistor, and the emitter resistor R_E will be $30 \times (24 + 24) = 1440\ w = 1.44\ Kw$. This is indeed a very large amount of power and thus this amplifier is obviously very inefficient.

For a comparison of output signals for the different amplifier classes of operation, please refer to Figure 3.34 during the following discussion.

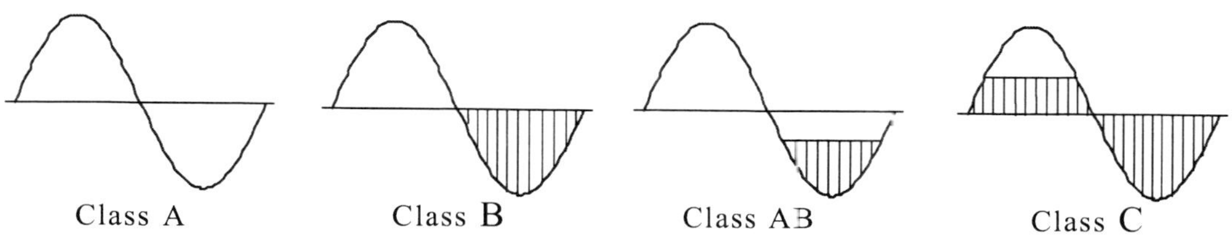

Figure 3.34. Output signals for Class A, Class B, Class AB, and Class C amplifiers

We should remember that the circuits presented in our subsequent discussion are only the output stages of an amplifier to provide the necessary drive to the load.

3.7.1 Class A Amplifier Operation

Class A amplifiers are biased so that variations in input signal polarities occur within the limits of cutoff and saturation. In a PNP transistor, for example, if the base becomes positive with respect to the emitter, holes will be repelled at the PN junction and no current can flow in the collector circuit. This condition is known as cutoff. Saturation occurs when the base becomes so negative with respect to the emitter that changes in the signal are not reflected in collector–current flow.

Biasing an amplifier in this manner places the DC operating point between cutoff and saturation and allows collector current to flow during the complete cycle (360 degrees) of the input signal, thus providing an output which is a replica of the input. Figure 3.35 is an example of a Class A amplifier. Although the output from this amplifier is 180 degrees out–of–phase with the input, the output current still flows for the complete duration of the input.

Class A amplifiers are used as amplifiers in radio, radar, and sound systems.

The output stages of Class A amplifiers carry a fairly large current. This current is referred to a *quiescent current* and it is defined as the current in the amplifier when the output voltage is zero.

In Example 3.11 we learned that the arrangement of the circuit in Figure 3.33 is very inefficient. However, a Class A amplifier can be made more efficient if we employ a *push–pull** arrangement as shown in Figure 3.36.

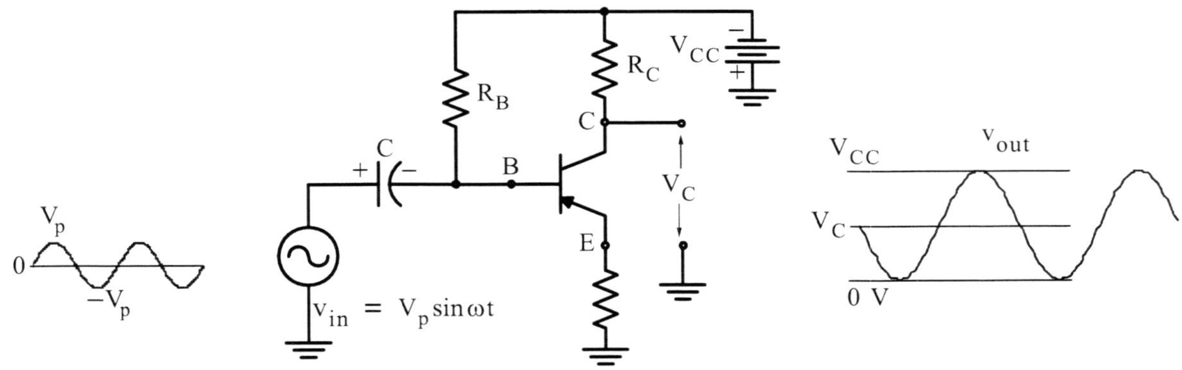

Figure 3.35. Typical Class A amplifier

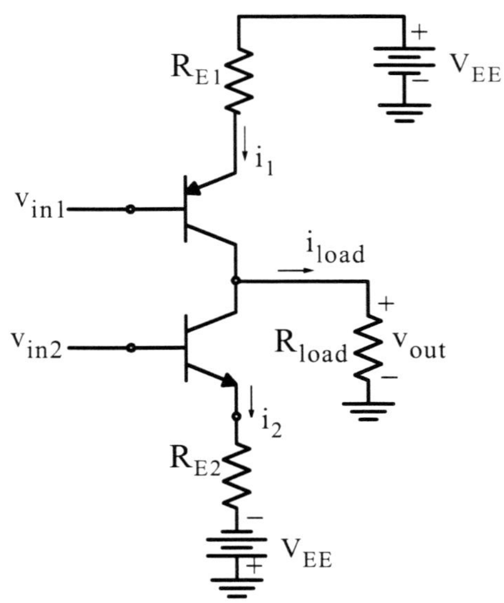

Figure 3.36. Push–Pull Output stage for a Class A amplifier

From the circuit in Figure 3.36, it is evident that we can control the currents i_1 and i_2 by varying the input voltages v_{in1} and v_{in2}. It is convenient to set the quiescent current, denoted as I_Q, to one–half the maximum current drawn by the load. Then, we can adjust the currents i_1 and i_2 to be equal and opposite. Then,

* *The expression push–pull (or double ended) derives its name from the fact that one of the two transistors pushes (sources) current into the load during the positive cycle while the other pulls (sinks) current from the load during the negative cycle.*

$$i_1 = I_Q + \frac{i_{load}}{2} \tag{3.31}$$

and

$$i_2 = I_Q - \frac{i_{load}}{2} \tag{3.32}$$

The currents i_1 and i_2 in each transistor vary from 0 to $2I_Q$; therefore, the load current is within the range $-2I_Q \le i_{load} \le 2I_Q$.

Example 3.12

In the circuit in Figure 3.37, $+V_{EE} = 24\ V$, $-V_{EE} = -24\ V$, and $R_{load} = 8\ \Omega$. Compute the power absorbed by the circuit if we want to apply up to $24\ V$ to the $8\ \Omega$ load.

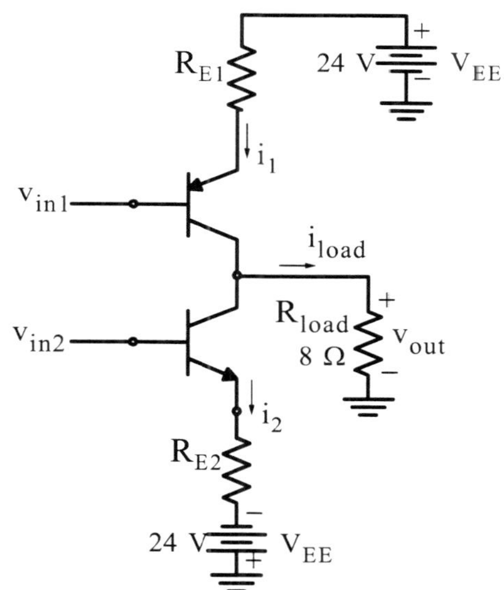

Figure 3.37. Circuit for Example 3.12

Solution:

The maximum load current is

$$i_{load}(max) = \frac{24\ V}{8\ \Omega} = 3\ A$$

and the quiescent current is

$$I_Q = \frac{3\ A}{2} = 1.5\ A$$

Then, each transistor will absorb

$$P = 24 \times 1.5 = 36\ w$$

and the total power absorbed by the circuit is

$$P_{total} = 2 \times 36 = 72 \text{ w}$$

This is not very efficient, but it is much more efficient than the circuit in the previous example.

3.7.2 Class B Amplifier Operation

The circuit in Figure 3.38 shows the output stage of a Class B amplifier. This consists also of a push–pull arrangement but the bases (inputs) are tied by two diodes. The current in the diodes is supplied by two current sources denoted as I_{bias}.

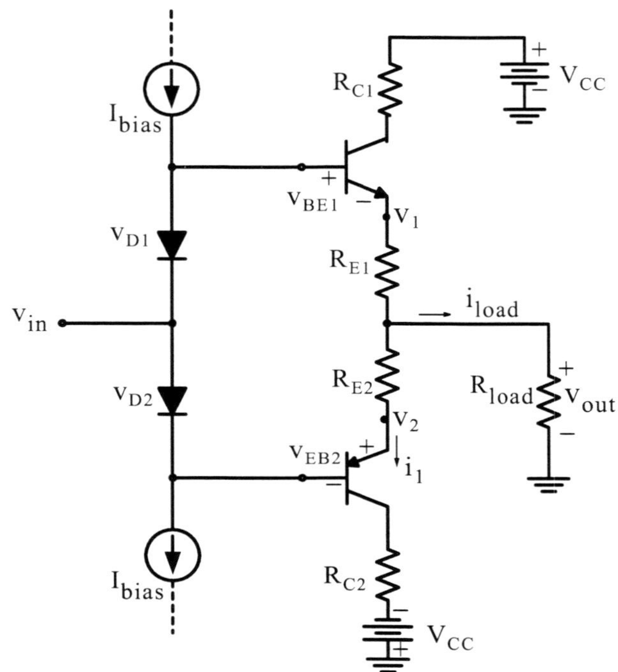

Figure 3.38. Output stage in a typical Class B amplifier

When v_{in} goes positive, the upper transistor conducts and the lower transistor is cutoff. Then, the input to the base of the upper transistor is $v_{in} + v_{D1}$ and the voltage at the emitter terminal of the upper transistor is $v_1 = v_{in} + v_{D1} - v_{BE1}$, and since $v_{BE1} = v_{D1}$, we find that

$$v_1 = v_{in} \text{ for } v_{in} > 0 \tag{3.33}$$

When v_{in} goes negative, the upper transistor is cutoff and the lower transistor conducts. then the input to the base of the lower transistor is $v_{in} - v_{D2}$ and the voltage at the emitter terminal of the lower transistor is $v_2 = v_{EB2} + v_{in} - v_{D2}$, and since $v_{EB2} = v_{D2}$, we find that

$$v_2 = v_{in} \text{ for } v_{in} < 0 \qquad (3.34)$$

From (3.33) and (3.34) we conclude that

$$v_1 = v_2 = v_{in} \qquad (3.35)$$

Therefore, when $v_{in} = 0$, $v_1 = 0$, and $v_2 = 0$ also, and the current and the output of both transistors including the quiescent currents I_{bias} will be zero also, and the power absorbed by the circuit will be zero. Accordingly, it appears that this arrangement has perfect efficiency. However, this ideal condition is never achieved because no two diodes or two transistors are exactly identical. There exists a range where both transistors are cutoff when the input signal changes polarity and this results in *crossover distortion* as shown in Figure 3.39. This distortion is due to the non–linearities in transistor devices where the output does not vary linearly with the input.

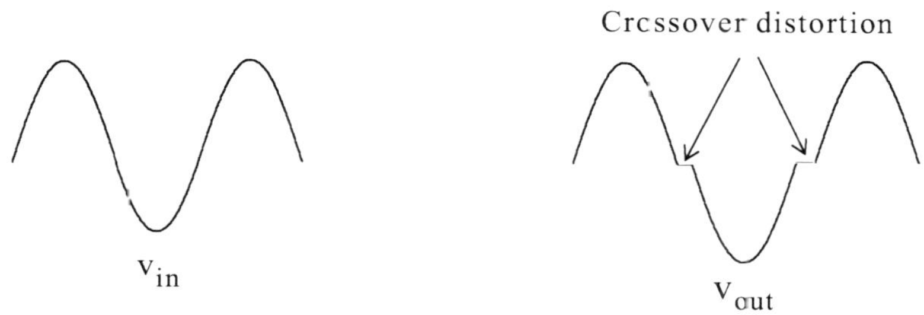

Figure 3.39. Crossover distortion in Class B amplifier

The efficiency of a typical Class B amplifier varies between 65 to 75 percent.

3.7.3 Class AB Amplifier Operation

We've seen that Class A amplifiers are very inefficient, and Class B amplifiers although are efficient, they produce crossover distortion. Class AB amplifiers combine the advantages of Class A and Class B amplifiers while they minimize the problems associated with them. Two possible arrangements for the output stage of a typical Class AB amplifier are shown in Figure 3.40.

In Figure 3.40(a) the voltage v_1 at the upper transistor is $v_1 = v_{in} + v_{D1} + v_{D2} - v_{BE1}$, and since $v_{BE1} = v_{D1} = v_{D2}$, we find that $v_1 = v_{in} + v_D$ and similarly for the lower transistor $v_2 = v_{in} - v_D$. Then, when $v_{in} = 0$, we have

$$v_1 - v_2 = 2v_D \qquad (3.36)$$

and the quiescent current, assuming that $R_{E1} = R_{E2}$, is

$$I_{bias} = \frac{v_1 - v_2}{R_{E1} + R_{E2}} = \frac{2v_D}{2R_E} = \frac{v_D}{R_E} \qquad (3.37)$$

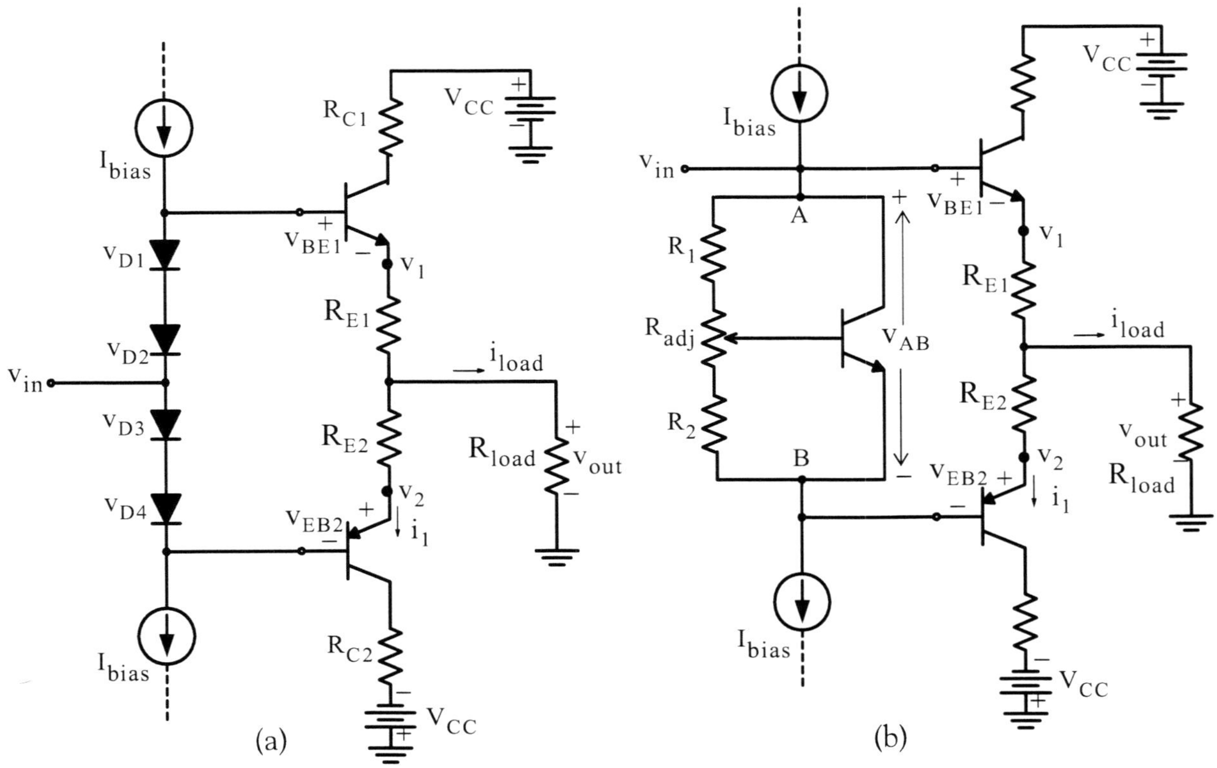

Figure 3.40. Two possible arrangements for the output stage in a typical Class AB amplifier

For small output signals that require currents in the range $-2I_{bias} < i_{load} < 2I_{bias}$ both transistors will conduct and will behave as Class A amplifier. But for larger signals, one transistor will conduct and supply the current required by the load, while the other will be cutoff. In other words, for large signals the circuit of Figure 3.40(a) acts like a Class B amplifier and hence the name Class AB amplifier.

For sinusoidal signals into the load R_{load}, the RMS voltage will be

$$v_{out}(RMS) = \frac{\sqrt{2}v_D}{R_E} \cdot R_{load} \tag{3.38}$$

and the output power will be

$$P_{load} = \frac{2v_D^2}{R_E^2} \cdot R_{load} \tag{3.39}$$

For maximum power, we should make R_E as small as possible, so let $R_E = 0.5\ \Omega$, and let $R_{load} = 8\ \Omega$ and $v_D = 0.5\ V$. Then,

$$P_{load} = \frac{2v_D^2}{R_E^2} \cdot R_{load} = \frac{0.5}{0.25} \times 8 = 20 \text{ w}$$

If we let $V_{CC} = 24 \text{ V}$ and $I_{bias} = 1 \text{ A}$, the total power absorbed will be

$$P_{total} = 24 \times 1 + 20 = 44 \text{ w}$$

For the circuit in Figure 3.40(b), the voltage v_{AB} can be adjusted to any desired value by selecting appropriate values for resistors R_1 and R_2. The adjustable resistor R_{adj} is set to a position to yield the desired value of the quiescent current I_{bias}.

From the above discussion, we have seen that the Class AB amplifier maintains current flow at all times so that the output devices can begin operation nearly instantly without the crossover distortion in Class B amplifiers. However, complete current is not allowed to flow at any one time thus avoiding much of the inefficiency of the Class A amplifier. Class AB designs are about 50 percent efficient (half of the power supply is power is turned into output to drive speakers) compared to Class A designs at 20 percent efficiency.

Class AB amplifiers are the most commonly used amplifier designs due to their attractive combination of good efficiency and high–quality output (low distortion and high linearity close to but not equal to Class A amplifiers).

3.7.4 Class C Amplifier Operation

Class C amplifiers have high efficiency and find wide applications in continuous wave (CW) laser and radar applications, in frequency modulation (FM), and phase and pulse amplification. However, Class C amplifiers cannot be used with amplitude modulation (AM) because of the high distortion. A typical Class C output stage is shown in Figure 3.41.

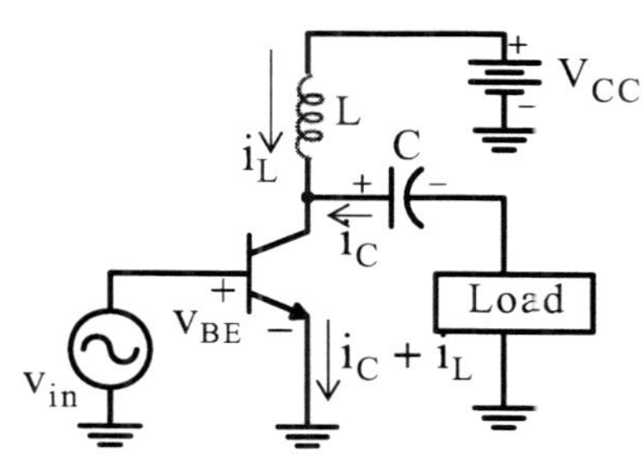

Figure 3.41. Output stage in a typical Class C amplifier

In Class C amplifier operation, the transistor in Figure 3.41 behaves as a open–closed switch. The load can be thought of as an antenna. During the positive half–cycle the transistor behaves like a closed switch, current i_L flows through the inductor and creates a magnetic field, and at the same time the capacitor discharges and thus the two currents $i_C + i_L$ flow through the emitter to the ground. During the negative half–cycle the transistor behaves like an open switch, the magnetic field in the inductor collapses and the current i_L will flow through the capacitor and the load.

Other classes of amplifiers such as Class D, Class E, and others have been developed by some manufacturers. These are for special applications and will not be discussed in this text. For more information on these, the interested reader may find information on the Internet.

3.8 Graphical Analysis

The operation of a simple transistor circuit can also be described graphically. We will use the circuit in Figure 3.42 for our graphical analysis.

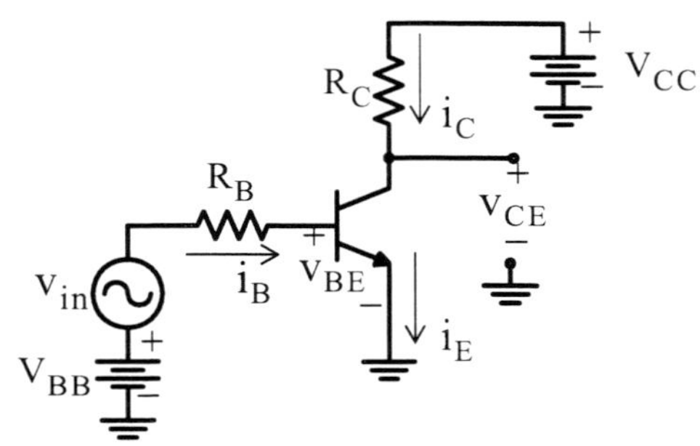

Figure 3.42. Circuit showing the variables used in the graphs in Figures 3.43 and 3.44

We start with a plot of i_B versus v_{BE} to determine the point where the curve

$$i_B = (I_r/\beta)e^{v_{BE}/nV_T} \tag{3.40}$$

and the equation of the straight line intersect

$$i_B = \frac{V_{BB} - v_{BE}}{R_B} \tag{3.41}$$

The equation of (3.41) was obtained with the AC source v_{in} shorted out. This equation can be expressed as

$$i_B = -\frac{1}{R_B}v_{BE} + \frac{V_{BB}}{R_B} \tag{3.42}$$

We recognize (3.42) as the equation of a straight line of the form $y = mx + b$ with slope $-1/R_B$. This equation and the curve of equation (3.40) are shown in Figure 3.43.

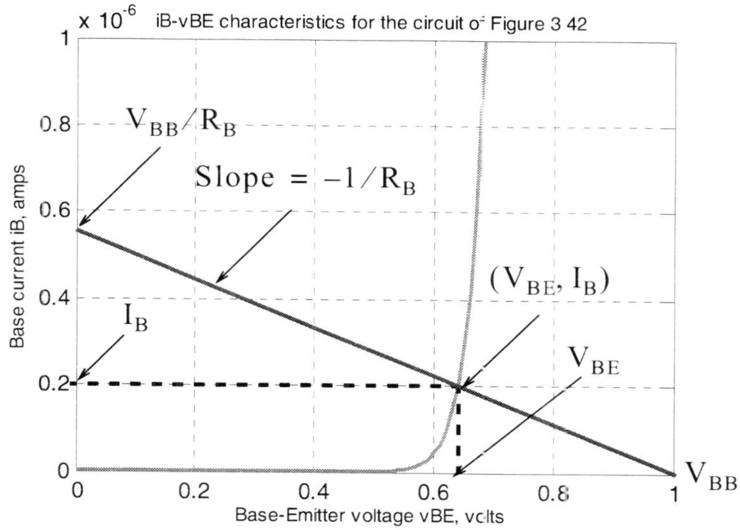

Figure 3.43. Plot showing the intersection of Equations (3.40) and (3.42)

We used the following MATLAB script to plot the curve of equation (3.40).

```
vBE=0: 0.01: 1; iR=10^(−15); beta = 100; n=1; VT=27*10^(−3);...
iB=(iR./beta).*exp(vBE./(n.*VT)); plot(vBE,iB); axis([0 1 0 10^(−6)]);...
xlabel('Base−Emitter voltage vBE, volts'); ylabel('Base current iB, amps');...
title('iB−vBE characteristics for the circuit of Figure 3.42'); grid
```

From Figure 3.43 we obtain the values of V_{BE} and I_B on the v_{BE} and i_B axes respectively. Next, we refer to the family of curves of the collector current i_C versus collector–emitter voltage v_{CE} for different values of i_B as shown in Figure 3.44. where the straight line with slope $-1/R_C$ is derived from the relation

$$v_{CE} = V_{CC} - R_C i_C \tag{3.43}$$

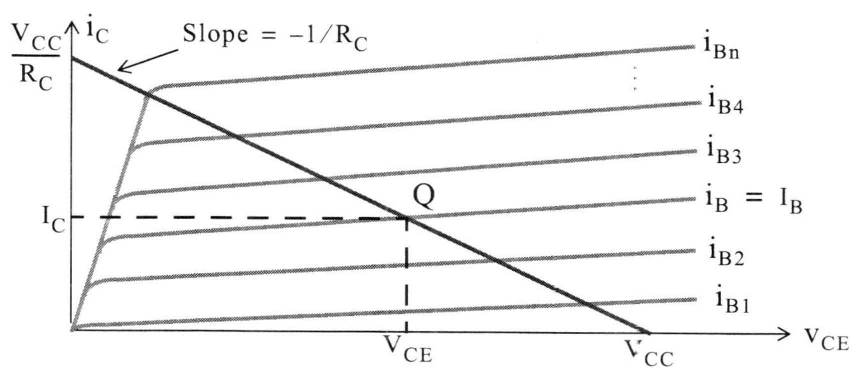

Figure 3.44. Family of curves for different values of i_B

Solving (3.43) for i_C we obtain

$$i_C = -\frac{1}{R_C}v_{CE} + \frac{V_{CC}}{R_C}^{*} \tag{3.44}$$

As shown in Figure 3.44, this straight line, commonly known as *load line*, and the curve $i_B = I_B$ intersect at point Q whose coordinates are the DC bias values V_{CE} and I_C. Obviously, the value of the collector resistor R_C must be chosen such that the load line is neither a nearly horizontal nor a nearly vertical line.

Example 3.13

For the circuit in Figure 3.45, the input voltage v_i is a sinusoidal waveform. Using the i_B versus v_{BE} and i_C versus v_{CE} curves shown in Figure 3.46, sketch the waveforms for v_{BE}, i_B, v_{CE}, and i_C.

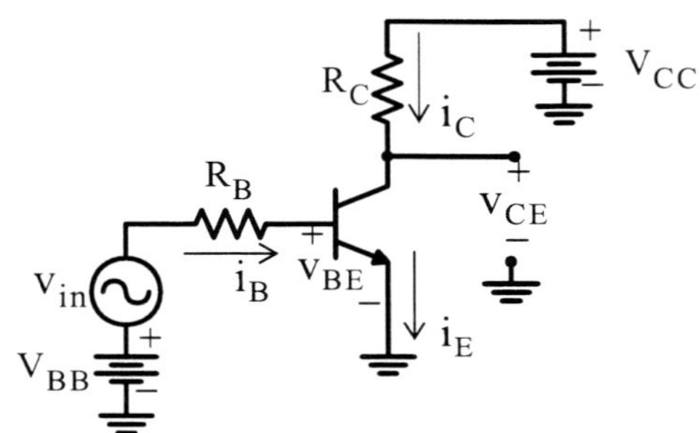

Figure 3.45. Circuit for Example 3.13

Solution:

Let $v_{in} = V_p \sin\omega t$. We draw three parallel lines with slope $-1/R_B$, one corresponding to input $v_{in} = 0$, the second at $v_{in} = V_p$, and the third at $v_{in} = -V_p$ as shown in Figure 3.47. The input voltage v_{in} is superimposed on the DC bias voltage V_{BB}.

* *We observe that (3.40) describes an equation of a straight line of the form* $y = mx + b$ *where the slope is* $m = -1/R_C$ *and the ordinate axis intercept is* $b = V_{CC}/R_C$.

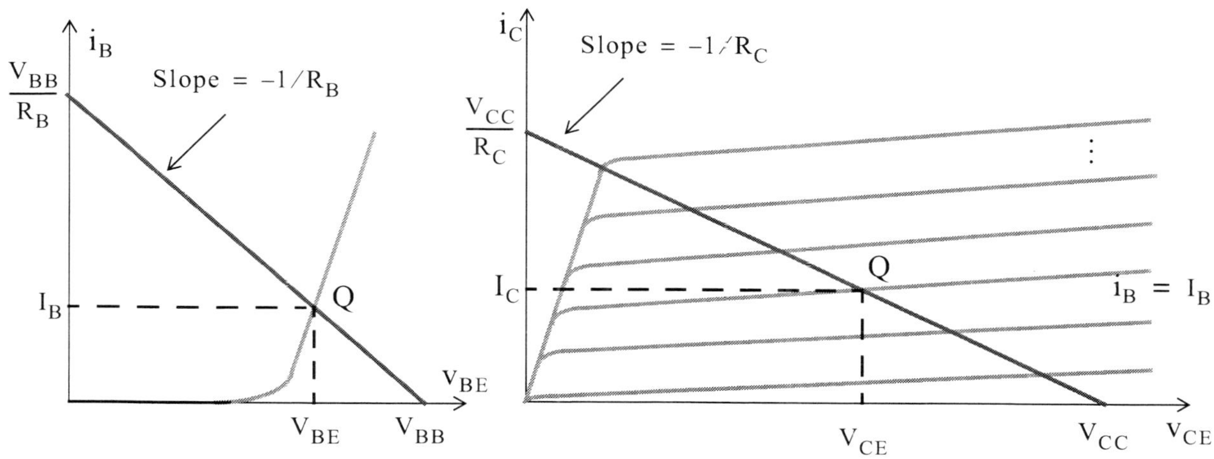

Figure 3.46. i_B *versus* v_{BE} *and* i_C *versus* v_{CE} *curves for Example 3.13*

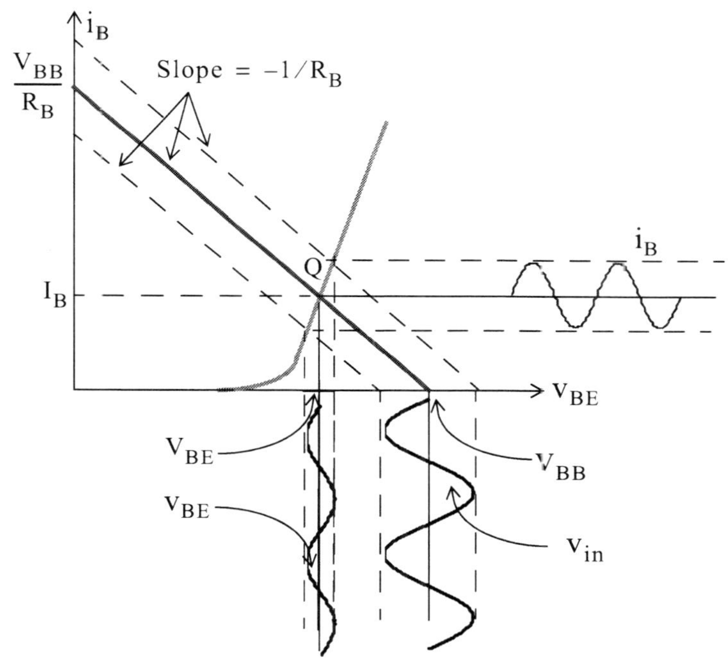

Figure 3.47. Graphical representation of v_{BE} *and* i_B *when the input voltage* v_{in} *is a sinusoid*

Figure 3.48 is the graphical representation of v_{CE} and i_C when the input voltage v_{in} is a sinusoid.

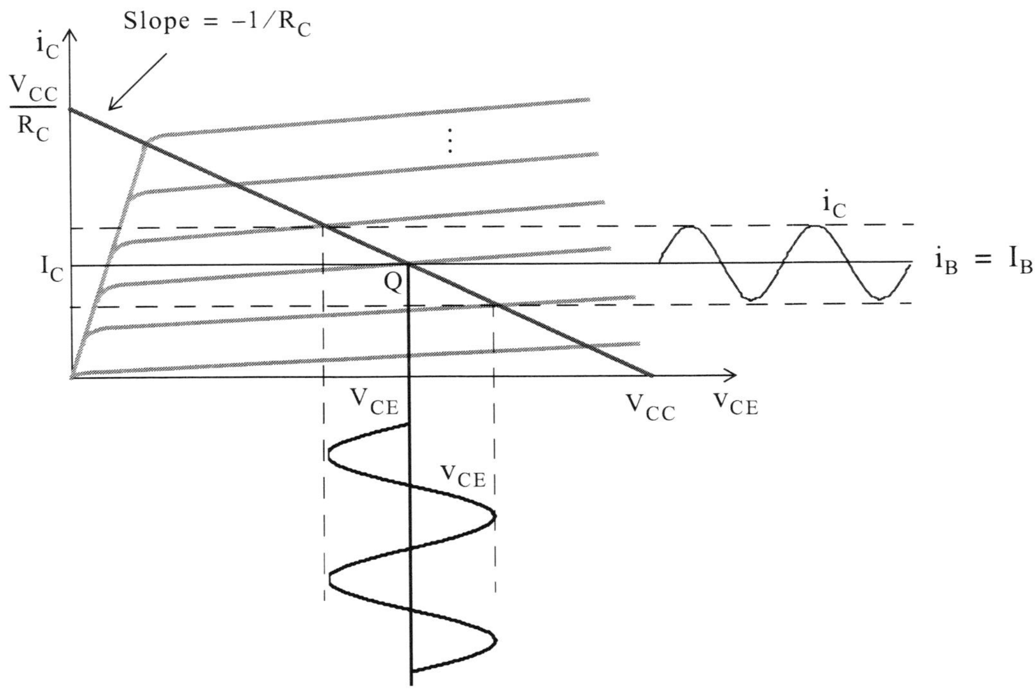

Figure 3.48. Graphical representation of v_{CE} *and* i_C *when the input voltage* v_{in} *is a sinusoid*

3.9 Power Relations in the Basic Transistor Amplifier

In our subsequent discussion we will denote time–varying quantities with lower case letters and lower case subscripts. We will represent average (DC) values with upper case letters and upper case subscripts. We will use lower case letters with upper case subscripts for the sum of the instantaneous and average values.

Let us consider the circuit of Figure 3.49. The collector current and the collector–to–emitter voltage can be expressed as

$$i_C = I_C + i_c \tag{3.45}$$

and

$$v_{CE} = V_{CE} + v_{ce} \tag{3.46}$$

where I_C and V_{CE} are the average values i_c and v_{ce} are time–varying components whose average value is zero.

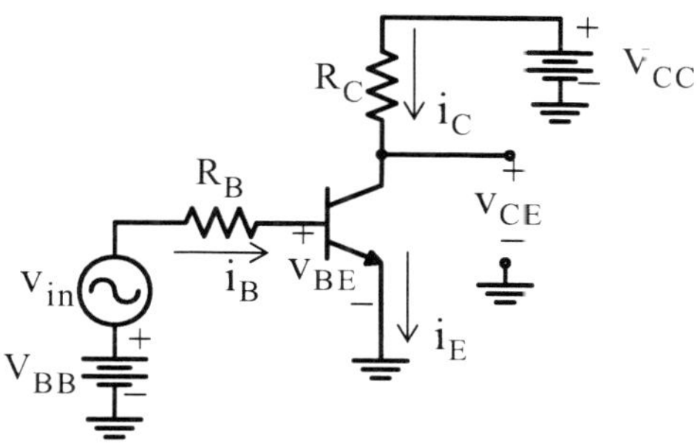

Figure 3.49. Circuit for the derivation of power relations

The power drawn from the collector supply at each instant is

$$p_{CC} = V_{CC} i_C = V_{CC} I_C + V_{CC} i_c \tag{3.47}$$

Since V_{CC} is constant and i_c has zero average value, the average value of the term $V_{CC} i_c$ is zero.* Therefore, the average power drawn from the collector supply is

$$P_{CC} = V_{CC} I_C \tag{3.48}$$

and if there is negligible or no distortion, the current I_C and power P_{CC} are both independent of the signal amplitude.

The power absorbed by the load at each instant is

$$P_{LOAD} = R_{LOAD} i_C^2 = R_{LOAD}(I_C + i_c)^2 = R_{LOAD} I_C^2 + 2R_{LOAD} I_C i_c + R_{LOAD} i_c^2 \tag{3.49}$$

The term $2R_{LOAD} I_C i_c$ is zero since $2R_{LOAD} I_C$ is constant and the average of i_c is zero; hence the average power absorbed by the load is

$$P_{LOAD} = R_{LOAD} I_C^2 + R_{LOAD}(i_c^2)_{ave} = R_{LOAD} I_C^2 + R_{LOAD}(i_{c\ RMS})^2 \tag{3.50}$$

The power absorbed by the transistor at each instant is

$$p_C = v_{CE} i_C = (V_{CC} - R_{LOAD} i_C) i_C = V_{CC} i_C - R_{LOAD} i_C^2 = p_{CC} - p_{LOAD} \tag{3.51}$$

Therefore, the average power absorbed by the transistor is

$$P_C = P_{CC} - P_{LOAD} \tag{3.52}$$

* *By definition* Average = Area/Period = $\left(\int_0^T i dt\right)/T$ *and thus if the value of the integral is zero, the average is also zero.*

3.10 Piecewise-Linear Analysis of the Transistor Amplifier

The circuit shown in Figure 3.50 is a model to represent the transistor where the two ideal diodes are included to remind us of the two PN junctions in the transistor.

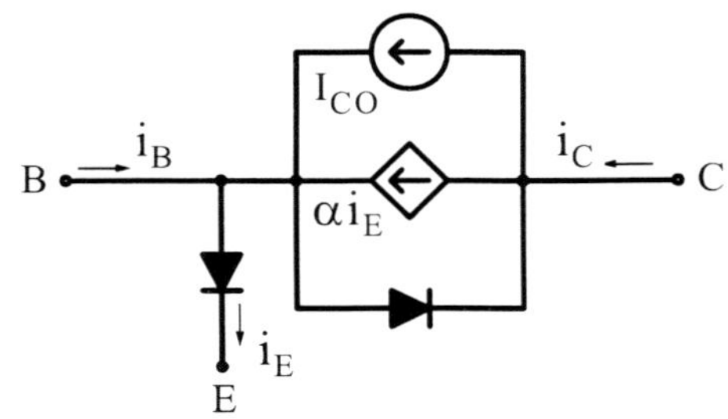

Figure 3.50. A transistor model

In Figure 3.50, I_{CO} is the current into the reverse-biased collector with $i_E = 0$. By KCL,

$$i_C = \alpha i_E + I_{CO} \tag{3.53}$$

and since $i_E = i_C + i_B$

$$i_C = \alpha(i_C + i_B) + I_{CO} \tag{3.54}$$

$$i_C - \alpha i_C = \alpha i_B + I_{CO}$$

$$(1-\alpha)i_C = \alpha i_B + I_{CO}$$

or

$$i_C = \frac{\alpha}{(1-\alpha)}i_B + \frac{1}{(1-\alpha)}I_{CO} \tag{3.55}$$

From Table 3.1

$$\frac{\alpha}{1-\alpha} = \beta$$

Also,

$$\frac{\alpha}{1-\alpha} + 1 = \beta + 1$$

$$\frac{1}{1-\alpha} = \beta + 1$$

and by substitution into (3.55)

$$i_C = \beta i_B + (\beta + 1)I_{CO} \tag{3.56}$$

Thus, the equivalent circuit in Figure 3.50 may be redrawn as shown in Figure 3.51.

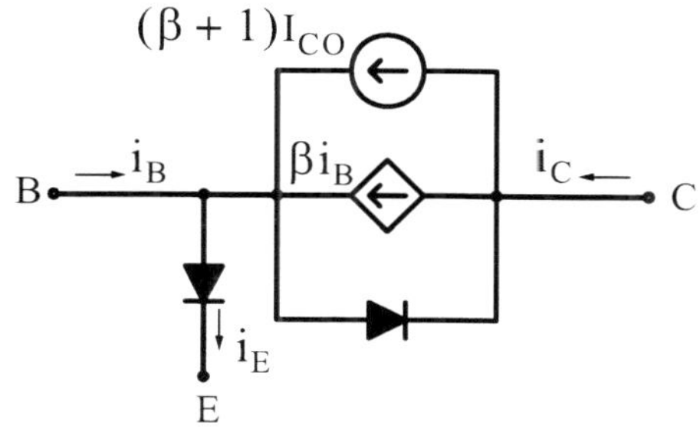

Figure 3.51. An alternative transistor model.

The transistor models shown in Figures 3.50 and 3.51 are essentially ideal models. An improved transistor model is shown in Figure 3.52 where for silicon type of transistors $V_D \approx 0.7\ V$ and r_b, referred to as the *base spreading resistance*, is included to account for the small voltage drop in the base of the transistor.

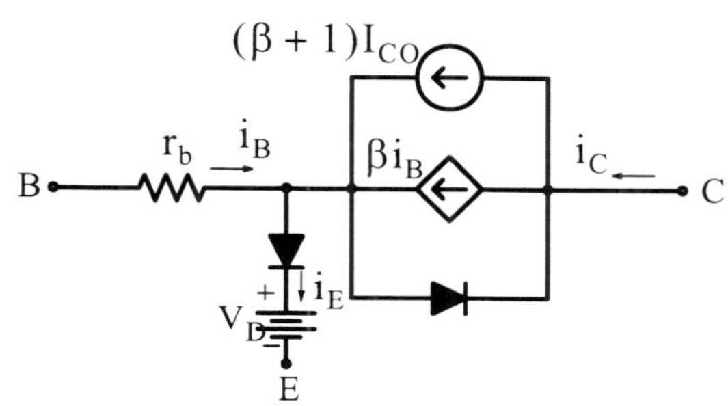

Figure 3.52. A more accurate model for the transistor

Analogous to the base resistance r_b are the *emitter diffusion resistance* defined as

$$r_e = \left.\frac{\partial v_E}{\partial i_E}\right|_{i_C = \text{constant}} \tag{3.57}$$

and the collector resistance

$$r_c = \left.\frac{\partial v_C}{\partial i_C}\right|_{i_E = \text{constant}} \tag{3.58}$$

Typical values for r_b are between $50\ \Omega$ to $250\ \Omega$, for r_e are between $10\ \Omega$ to $25\ \Omega$, and r_c is very high, in excess of $1\ M\Omega$. The model in Figure 3.52 is for an NPN transistors. It applies also to PNP transistors provided that the diodes, voltage polarities, and current directions are reversed.

The transistor amplifier in Figure 3.53 can be analyzed by piece–wise linear methods with the aid of the transistor model in Figure 3.52.

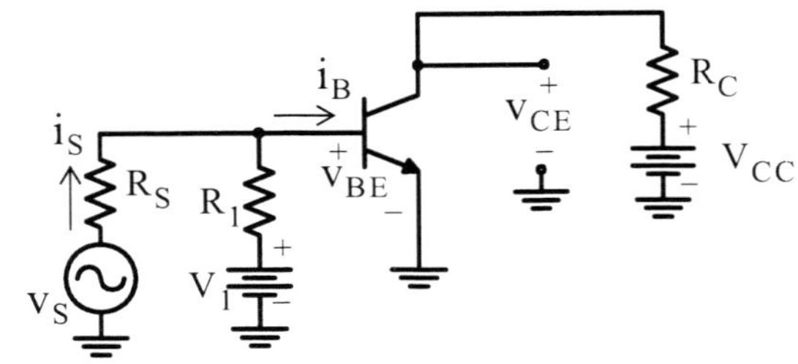

Figure 3.53. Basic transistor amplifier to be analyzed by piece–wise linear methods

For convenience, we neglect the small effects of r_b, V_D, and $(\beta + 1)I_{CO}$ in Figure 3.52 and thus the model is now as shown in Figure 3.54.

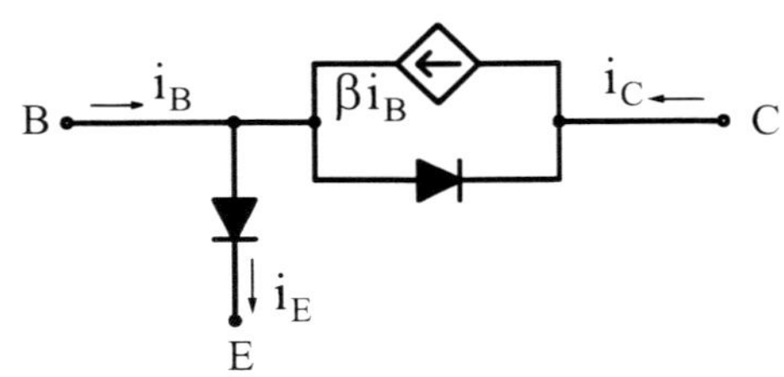

Figure 3.54. The piece–wise linear model in Figure 3.52 where r_b, V_D, and $(\beta + 1)I_{CO}$ are neglected

When the transistor in the amplifier in Figure 3.53 is replaced by the piece–wise linear model in Figure 3.54, we obtain the circuit shown in Figure 3.55(a).

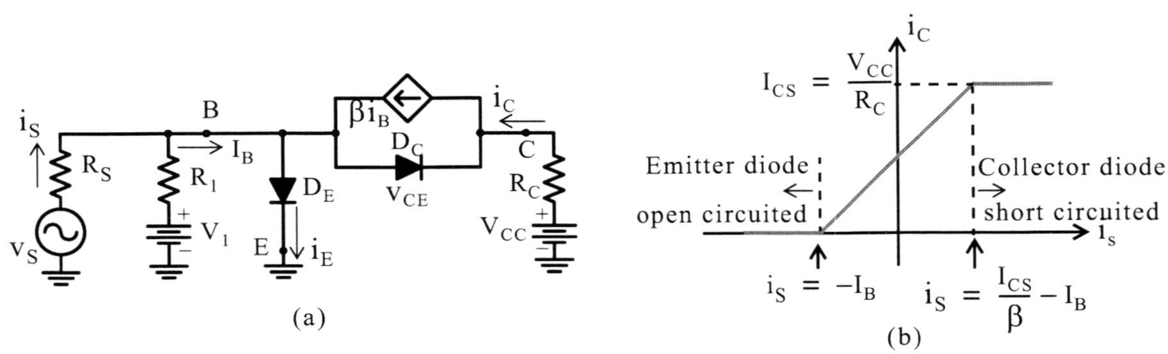

Figure 3.55. Piecewise linear model for the circuit in figure 3.53 and its current transfer characteristics

The current transfer characteristics are constructed by determining the points at which the diodes change from the conducting to the non–conducting state. The current I_{CS} represents the collector saturation current.

Let us now suppose that the source current i_s reaches a value that causes the reverse voltage across the collector diode D_C to become zero and the emitter diode D_E is conducting and allows a current i_E to flow through it. When this occurs, there is no voltage drop across either diode, all three terminals of the transistor are at ground potential, and the current in the collector diode D_C is zero. This condition establishes the right–hand break on the transfer characteristics and its value is determined as follows:

The collector current under the conditions stated above is

$$i_C = \frac{V_{CC}}{R_C} = \beta i_B = I_{CS}$$

or

$$i_B = \frac{V_{CC}}{\beta R_C} = \frac{I_{CS}}{\beta}$$

and since the base is grounded,

$$i_S = i_B - \frac{V_1}{R_1} = \frac{I_{CS}}{\beta} - I_B \qquad (3.59)$$

Next, let us now suppose that the source current i_S becomes negative and reaches a value that causes the collector diode D_C to become reverse–biased and the current in the emitter diode D_E to reach a zero value. When this occurs, there is no current in either diode, and the base terminal is at ground potential. This condition establishes the left–hand break on the transfer characteristics and its value is determined as follows:

$$i_B = -i_C = -\beta i_B \qquad (3.60)$$

But this equation is true only if $\beta = -1$. For $\beta \neq -1$, this equation is satisfied only if $i_B = 0$. Therefore, with the base terminal at ground potential and $i_B = 0$, (3.59) reduces to

$$i_S = -I_B \qquad (3.61)$$

Example 3.14

A DC power supply with a transistor regulator is shown in Figure 3.56. The resistor R_1 provides a suitable current to sustain the breakdown condition of the Zener diode. Any change in supply voltage causes a compensating change in the voltage drop across the transistor from collector to emitter and the load voltage v_{load} is thereby held constant in spite of changes in the input voltage or load resistance R_{load}. The transistor parameters are $\beta = 100$, $r_b = 75\ \Omega$, $V_{BE} = 0.7\ V$ and the current I_{CO} into the reverse–biased collector is negligible.

Find the values of the load voltage v_{load}, the collect–to–emitter voltage v_{CE}, and the power P_C absorbed by the transistor.

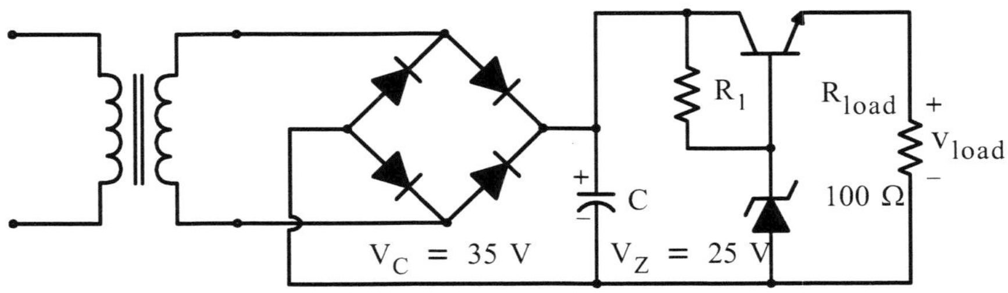

Figure 3.56. DC power supply for Example 3.14

Solution:

We replace the transistor with its piece–wise linear model as shown in Figure 3.57.

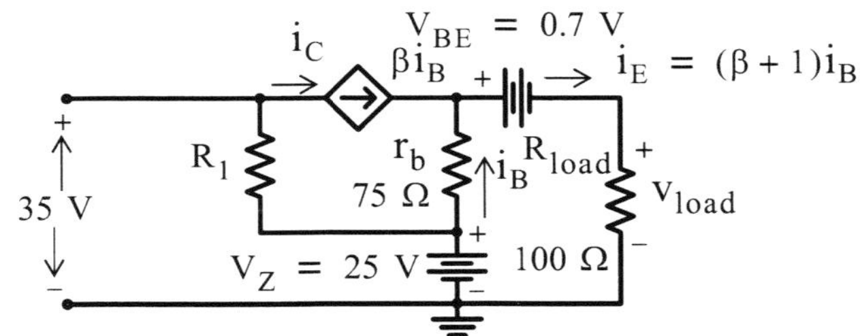

Figure 3.57. The piece–wise linear model for the transistor in Figure 3.55

From Figure 3.57,

$$r_b i_B + V_{BE} + R_{load} i_E = V_Z$$

$$V_Z - V_{BE} = r_b i_B + (\beta + 1) R_{load} i_B$$

$$i_B = \frac{V_Z - V_{BE}}{r_b + (\beta + 1) R_{load}} = \frac{25 - 0.7}{75 + 101 \times 100} = \frac{24.3}{10175} = 2.4\text{mA}$$

$$v_{load} = (\beta + 1) R_{load} i_B = 101 \times 100 \times 2.4 \times 10^{-3} = 24.24\text{V}$$

We observe that the load voltage is independent of the supply voltage of 35 V. This occurs because the collector can be represented as an ideal current source βi_B. However, if the supply voltage falls below the Zener voltage of 25 V, the collector–base junction will no longer be reverse–biased, and the voltage regulation action of the transistor will fail. Also, when this occurs, the breakdown state will not be sustained in the Zener diode.

The collector current is

$$i_C = \beta i_B = 100 \times 2.4 = 240\ \text{mA}$$

and the collector–to–emitter voltage is

$$v_{CE} = \text{Supply Voltage} - \text{Load Voltage} = 35 - 24.24 = 10.76\ V$$

The power absorbed by the transistor is

$$P_C = v_{CE} i_C = 10.76 \times 0.24 = 2.58\ w$$

3.11 Incremental Linear Models

In our discussion on piece–wise linear models on the preceding section, the small voltage drop between base and emitter is small in comparison with the bias voltage and thus can be neglected. However, the base–to–emitter voltage cannot be neglected when only the increments of voltage and currents is considered. Also, when calculating increments of current and voltage, it is often necessary to account for the small effects of variations in collector voltage on both the input and output circuits. For these reasons the incremental model for the transistor provides a better approximation than the piece–wise linear approximation.

The base–to–emitter voltage v_{BE} and the collector current i_C are functions of the base current i_B and collector–to–emitter voltage v_{CE}. In other words,

$$v_{BE} = f(i_B, v_{CE}) \tag{3.62}$$

and

$$i_C = f(i_B, v_{CE}) \tag{3.63}$$

If i_B and v_{CE} are changed in small increments, the resulting increment in v_{BE} can be expressed as

$$\Delta v_{BE} \approx dv_{BE} = \frac{\partial v_{BE}}{\partial i_B} di_B + \frac{\partial v_{BE}}{\partial v_{CE}} dv_{CE} \tag{3.64}$$

and the increment in i_C can be written as

$$\Delta i_C \approx di_C = \frac{\partial i_C}{\partial i_B} di_B + \frac{\partial i_C}{\partial v_{CE}} dv_{CE} \tag{3.65}$$

The partial derivative in the first term of (3.64) has the dimensions of resistance and it is denoted as r_n, and that in the second term is a dimensionless voltage ratio denoted as μ. It is also convenient to denote these derivatives in lower case letters with lower case subscripts. Then, (3.64) is expressed as

$$v_{be} = r_n i_b + \mu v_{ce} \tag{3.66}$$

Likewise, the partial derivative in the first term of (3.65) is a dimensionless current ratio denoted as β, and that in the second term has the dimensions of conductance and it is denoted as g_o. Then, (3.65) is expressed as

$$i_c = \beta i_b + g_o v_{ce} \qquad (3.67)$$

The relations of (3.66) and (3.67) along with $i_e = i_b + i_c$ suggest the circuit shown in Figure 3.58 known as the *hybrid incremental network model* for the transistor. It is referred to as hybrid model because of the mixed set of voltages and currents as indicated by the expressions of (3.66) and (3.67).

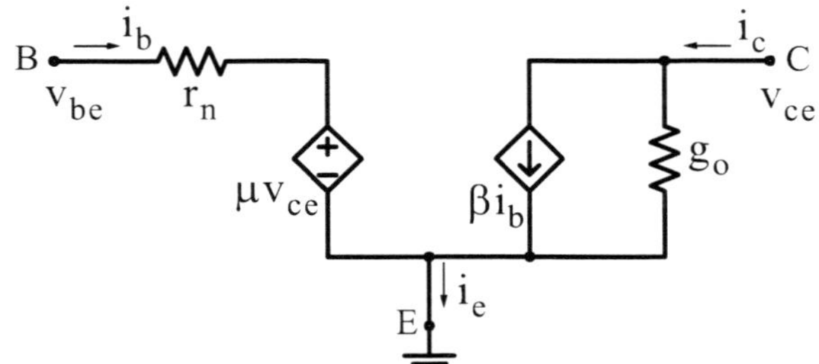

Figure 3.58. The hybrid incremental model for a transistor in the common–emitter configuration

The input resistance r_n is the slope of the input voltage and current characteristics and it accounts for the voltage drop across the base–emitter junction. Likewise, the output conductance g_o is the slope of the output current and voltage characteristics.

The voltage amplification factor μ is related to the input characteristics caused by a change in v_{CE}, and the current amplification factor β is related to the output characteristics caused by a change in i_B.

Typical values for the parameters of relations (3.66) and (3.67) are $r_n = 2\ K\Omega$, $\mu = 5 \times 10^{-4}$, $\beta = 100$, and $g_o = 2 \times 10^{-5}\ \Omega^{-1}$, and since the value of μ is a very small number, the voltage source μv_{ce} in Figure 3.58 can be replaced by a short circuit, and thus the model reduces to that shown in Figure 3.59.

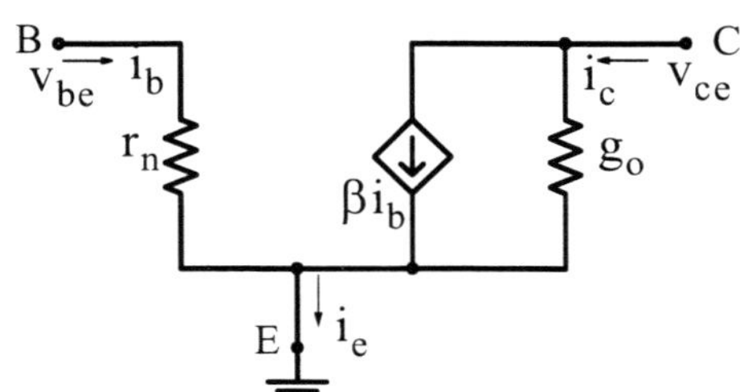

Figure 3.59. The hybrid incremental model for the transistor with $\mu v_{ce} = 0$

The transistor hybrid parameters* provide us with a means to evaluate voltages, currents, and power in devices that are connected externally to the transistor. Let us, for example, consider the circuit of Figure 3.60 which is an incremental model for the transistor amplifier in Figure 3.53.

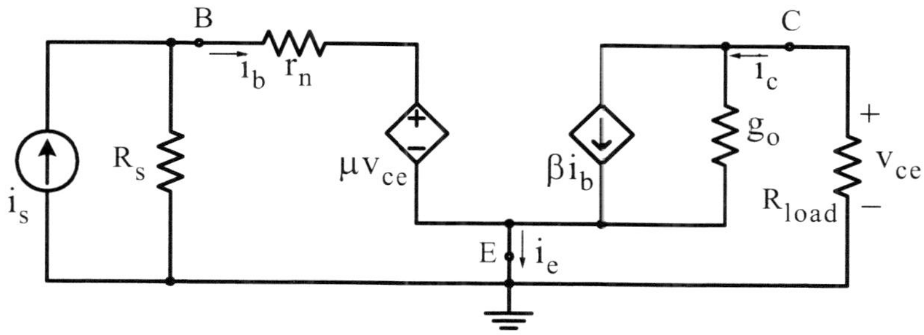

Figure 3.60. Transistor incremental model with external devices

Now, we let R_{eq} represent the parallel combination $r_o = 1/g_o$ and R_{load}. Then,

$$v_{ce} = -R_{eq}\beta i_b \tag{3.68}$$

and

$$\mu v_{ce} = -\mu R_{eq}\beta i_b \tag{3.69}$$

Hence, the voltage of the μv_{ce} is proportional to the current flowing through the source, and from this fact we can replace the voltage source with a resistance $-\mu\beta R_{eq}$, and thus the transistor incremental model in Figure 3.60 can be redrawn as shown in Figure 3.61.

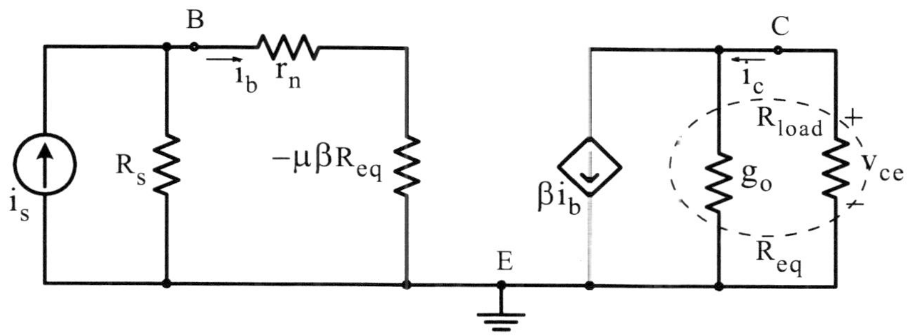

Figure 3.61. The circuit in Figure 3.60 with the voltage source μv_{ce} replaced by the resistance $-\mu\beta R_{eq}$

The negative resistance $-\mu\beta R_{eq}$ is always much smaller than r_n and thus the net input resistance to the transistor is always positive. Therefore, the negative resistance $-\mu\beta R_{eq}$ can be replaced by a short circuit, and assuming that the base current i_b is unaffected by this assumption, the voltages and currents in the collector side of the circuit are not affected.

* *The h–equivalent transistor circuits are discussed in Section 3.15.*

The current amplification A_c is defined as

$$A_c = i_c / i_s \tag{3.70}$$

where

$$i_c = g_o v_{ce} + \beta i_b = \frac{1}{r_o}(-R_{load}\, i_c) + \beta \frac{R_1}{R_1 + r_n} i_s$$

$$i_c + \frac{R_{load}}{r_o} i_c = \beta \frac{R_1}{R_1 + r_n} i_s$$

$$\left(1 + \frac{R_{load}}{r_o}\right) i_c = \beta \frac{R_1}{R_1 + r_n} i_s$$

$$\left(\frac{r_o + R_{load}}{r_o}\right) i_c = \beta \frac{R_1}{R_1 + r_n} i_s$$

$$\frac{i_c}{i_s} = \frac{\beta(R_1/(R_1 + r_n))}{(r_o + R_{load})/r_o} = \frac{\beta R_1 r_o}{(R_1 + r_n)(r_o + R_{load})} = \frac{R_1}{(R_1 + r_n)} \beta \frac{r_o}{(r_o + R_{load})}$$

and thus the current amplification A_c is

$$A_c = \frac{i_c}{i_s} = \frac{R_1}{(R_1 + r_n)} \beta \frac{r_o}{(r_o + R_{load})} \tag{3.71}$$

The parameters r_n, μ, β, and g_o are normally denoted by the h (hybrid) parameters* as $r_n = h_{11} = h_{ie}$, $\mu = h_{12} = h_{re}$, $\beta = h_{21} = h_{fe}$, and $g_o = h_{22} = h_{oe}$. These designations along with the additional notations $v_{be} = v_1$, $i_b = i_1$, $v_{ce} = v_2$, and $i_c = i_2$, provide a symmetrical form for the relations of (3.66)and (3.67) as follows:

$$\begin{aligned} v_1 &= h_{11} i_1 + h_{12} v_2 \\ i_2 &= h_{21} i_1 + h_{22} v_2 \end{aligned} \tag{3.72}$$

or

$$\begin{aligned} v_1 &= h_{ie} i_1 + h_{re} v_2 \\ i_2 &= h_{fe} i_1 + h_{oe} v_2 \end{aligned} \tag{3.73}$$

In (3.73) the subscript i denotes the input impedance with the output short–circuited, the sub-

* *For a detailed discussion of the* z, y, h, *and* g *parameters refer to Circuit Analysis II with MATLAB Applications, ISBN 978–0–9709511–5-1.*

script r denotes the reverse transfer voltage ratio with the input terminals open–circuited, the subscript f denotes the forward transfer current ratio with the output short circuited, and the subscript o denotes the output admittance with the input terminals open–circuited. The second subscript e indicates that the parameters apply for the transistor operating in the common–emitter mode. A similar set of symbols with the subscript b replacing the subscript e denotes the hybrid parameters for a transistor operating in the common–base mode, and a set with the subscript c replacing the letter e denotes the hybrid parameters for a transistor operating in the common–collector mode.

Values for the hybrid parameters at a typical quiescent operating point for the common–emitter mode are provided by the transistor manufacturers. Please refer to the last section in this chapter. Table 3.3 lists the h–parameter equations for the three bipolar transistor configurations.

TABLE 3.3 h–parameter equations for transistors

Parameter	*Common–Base*	*Common–Emitter*	*Common–Collector*
h_{11}	h_{ib}	$h_{ie} \approx h_{11}/(1+h_{21})$	$h_{ic} \approx h_{11}/(1+h_{21})$
h_{12}	h_{rb}	$h_{re} \approx h_{11}h_{22}/(1+h_{21}) - h_{12}$	$h_{rc} \approx 1$
h_{21}	h_{fb}	$h_{fe} \approx -h_{21}/(1+h_{21})$	$h_{fc} \approx -1/(1+h_{21})$
h_{22}	h_{ob}	$h_{oe} \approx h_{22}/(1+h_{21})$	$h_{oc} \approx h_{22}/(1+h_{21})$

Example 3.15

For the amplifier circuit in Figure 3.62 it is known that $r_n = h_{11} = 2\ K\Omega$, $\beta = h_{21} = 100$, $\mu = h_{12} = 5 \times 10^{-4}$, and $g_o = h_{22} = 2 \times 10^{-5}\ \Omega^{-1}$. Find the small signal current amplification $A_c = i_{load}/i_s$.

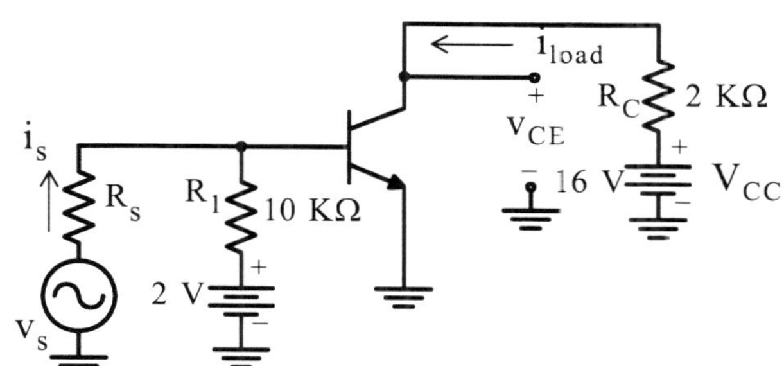

Figure 3.62. Transistor amplifier for Example 3.15

Solution:

The incremental model of this transistor amplifier is shown in Figure 3.63 where

$$R_{eq} = \frac{(1/g_o)R_{load}}{1/g_o + R_{load}} = \frac{50 \times 10^3 \times 2 \times 10^3}{52 \times 10^3} = 1.923\ K\Omega$$

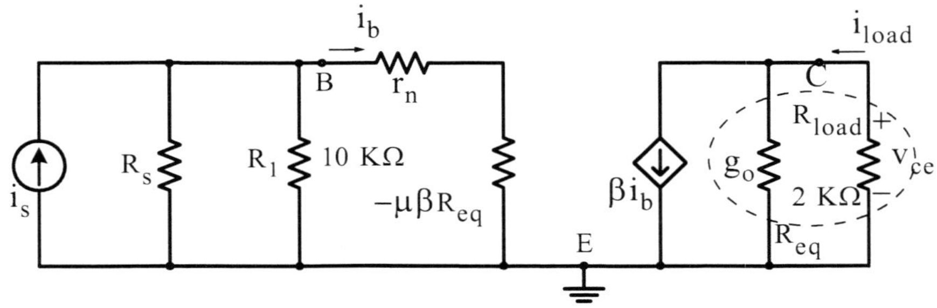

Figure 3.63. The incremental model for the transistor circuit in Figure 3.62

and thus the magnitude of the equivalent resistance reflected into the input part of the circuit is

$$|\mu\beta R_{eq}| = 5 \times 10^{-4} \times 100 \times 1.293 \times 10^3 = 64.65\ \Omega = 0.06465\ K\Omega$$

and since this is much smaller than $r_n = h_{11} = 2\ K\Omega$, it can be neglected.

The current gain A_c can be found from the relation (3.71). Then,

$$A_c = \frac{i_{load}}{i_s} = \frac{R_1}{(R_1 + r_n)}\beta\frac{r_o}{(r_o + R_{load})} = \frac{10}{12} \times 100 \times \frac{50}{52} = 80$$

3.12 Transconductance

Another useful parameter used in small signal analysis at high frequencies is the *transconductance,* denoted as g_m, and defined as

$$g_m = \left.\frac{di_C}{dv_{BE}}\right|_{i_C = I_C} \tag{3.74}$$

and as a reminder, we denote time–varying quantities with lower case letters and lower case subscripts. Thus, the transconductance g_m is the slope at point Q on the i_C versus v_{BE} characteristics at $i_C = I_C$ as shown in Figure 3.64.

An approximate value for the transconductance at room temperature is $g_m \approx 40I_C$. This relation is derived as follows:

From (3.2)

$$i_C = I_r e^{v_{BE}/V_T} \tag{3.75}$$

and with (3.74)

$$g_m = \left.\frac{di_C}{dv_{BE}}\right|_{i_C = I_C} = I_r \frac{1}{V_T} e^{v_{EE}/V_T} \tag{3.76}$$

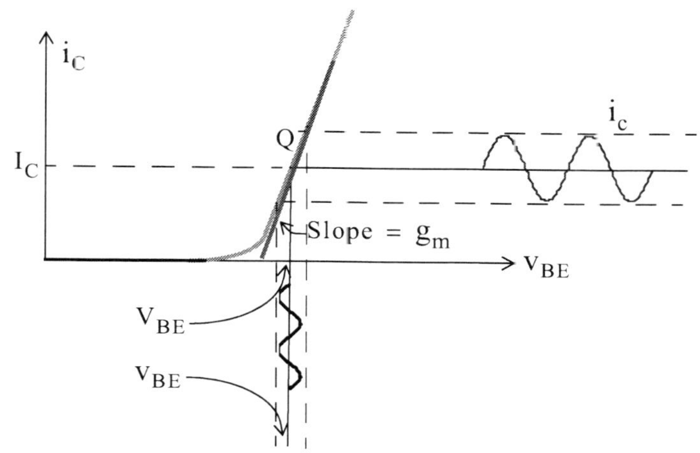

Figure 3.64. The transconductance g_m defined

By substitution of (3.75) into (3.76) we obtain

$$g_m = i_C/V_T \tag{3.77}$$

and with $V_T = 26 \times 10^{-3}$ V

$$g_m \approx 40 i_C \tag{3.78}$$

and thus we see that the transconductance is proportional to the collector current i_C. Therefore, a transistor can be viewed as an amplifier with a transconductance of 40 millimhos for each milliampere of collector current.

3.13 High–Frequency Models for Transistors

The incremental models presented in the previous section do not take into consideration the high–frequency effects in the transistor. Figures 3.65(a) and 3.65(b) are alternate forms for representing a transistor.

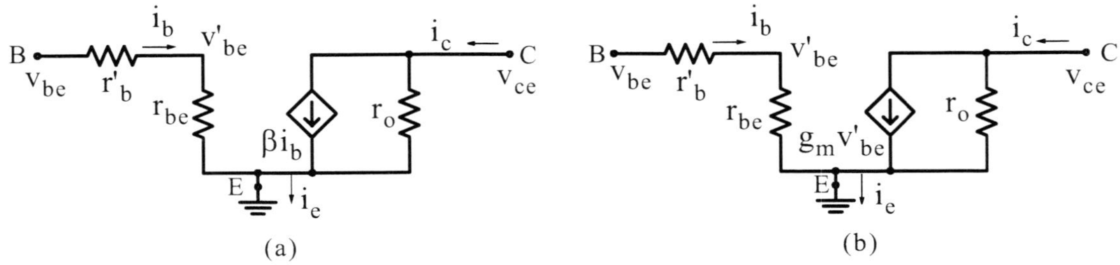

Figure 3.65. Alternate forms for the transistor model

In Figure 3.65(a) the input impedance r_n is separated into two parts; one part, denoted as r'_b, accounts for the ohmic base spreading resistance; the other part, denoted as r_{be}, is a nonlinear resistance and accounts for the voltage drop v'_{be} across the emitter junction. Thus,

$$r_n = r'_b + r_{be} \tag{3.79}$$

From Figure 3.65(a)

$$i_b = \frac{v'_{be}}{r_{be}} \tag{3.80}$$

and thus

$$\beta i_b = \frac{\beta}{r_{be}} v'_{be} = g_m v'_{be}{}^{*} \tag{3.81}$$

and thus the model in Figure 3.65(a) can also be represented as that in Figure 3.65(b).

From (3.76)

$$g_m = \beta / r_{be} \tag{3.82}$$

or

$$r_{be} = \frac{\beta}{g_m} = \frac{\beta}{40 i_C} \tag{3.83}$$

Thus, when $i_C = 1$ ma, $r_{be} = 25\beta\ \Omega$, and in general, when i_C is in milliamps,

$$r_{be} = (25\beta)/i_C \tag{3.84}$$

The incremental models we have discussed thus far are valid only for frequencies of about 2 MHz or less. For higher frequencies, the effects of junction capacitances must be taken into account. Figure 3.66 shows a model for the transistor at high frequencies referred to as *hybrid–π* model.

* *Since* $i_c = \beta i_b = \beta \frac{v'_{be}}{r_{be}}$, *it follows that* $\left.\frac{i_c}{v'_{be}}\right|_{v_{ce}=0} = \left.\frac{di_c}{dv'_{be}}\right|_{v_{CE} = \text{constant}} = g_m$

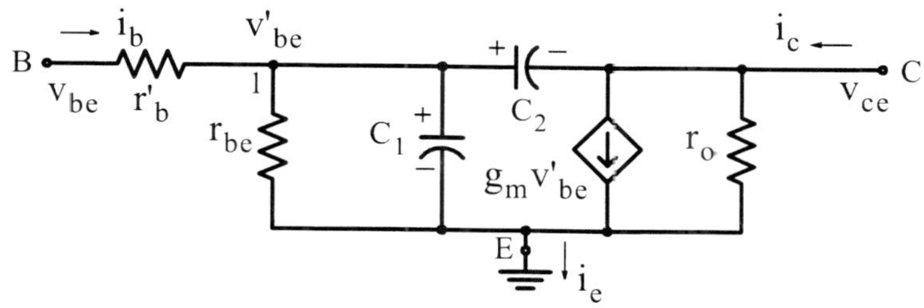

Figure 3.66. The hybrid–π model for the transistor at high frequencies

The capacitor C_1 represents the capacitance that exists across the forward–biased emitter junction while the capacitor C_2 represents a much smaller capacitance that exists across the reverse–biased collector junction.

At low frequencies the capacitors act as open circuits and thus do not affect the transistor performance. At high frequencies, however, the capacitors present a relatively low impedance and thereby reduce the amplitude of the signal voltage v'_{be}. This reduction in v'_{be} causes in turn a reduction in the strength of the controlled source $g_m v'_{be}$ and a reduction in the collector current i_c. We can derive some useful relations by determining the short–circuit collector current i_c when a sinusoidal input current is applied between the base and emitter terminals.

With the output short–circuited, capacitor C_2 is in parallel with C_1 between Node 1 and ground, and the equivalent impedance is

$$Z_{be} = r_{be} \| 1/j\omega(C_1 + C_2) = j\omega(C_1 + C_2) = \frac{r_{be} \times 1/j\omega(C_1 + C_2)}{r_{be} + 1/j\omega(C_1 + C_2)} = \frac{r_{be}}{r_{be}j\omega(C_1 + C_2) + 1}$$

However, C_1 is typically 100 times as large as C_2 and it can be neglected. Then,

$$Z_{be} = \frac{r_{be}}{r_{be}j\omega C_1 + 1} \qquad (3.85)$$

and denoting the phasor* quantities with bolded capical letters, we obtain

$$\mathbf{V'}_{be} = Z_{be}\mathbf{I}_b = \frac{r_{be}}{r_{be}j\omega C_1 + 1}\mathbf{I}_b \qquad (3.86)$$

and

$$\mathbf{I}_c = g_m\mathbf{V'}_{be} = \frac{g_m r_{be}}{r_{be}j\omega C_1 + 1}\mathbf{I}_b$$

* *Phasors are rotating vectors and are used to represent voltages and currents in complex form. Impedances and admittances are complex quantities but not phasors. For a detailed discussion, refer to Circuit Analysis I, ISBN 978-0-9709511-2-0.*

Using the relation of (3.82) we obtain

$$\mathbf{I}_c = \frac{\beta}{r_{be}j\omega C_1 + 1}\mathbf{I}_b \tag{3.87}$$

Thus, if the amplitude of the input current is held constant as the frequency is increased, the amplitude of the collector current decreases and approaches zero at very high frequencies.

The coefficient of $\mathbf{I}_b$ in (3.86) must be a dimensionless constant; therefore the quantity $1/r_{be}C_1$ has the dimensions of frequency. Also, it is customary to represent C_1 as C_e, and it is helpful to define a new symbol ω_β as

$$\omega_\beta \equiv \frac{1}{r_{be}C_e} \tag{3.88}$$

By substitution of (3.88) into (3.87) we obtain

$$\mathbf{I}_c = \frac{\beta}{j\omega/\omega_\beta + 1}\mathbf{I}_b \tag{3.89}$$

and the magnitute of the collector current is

$$|\mathbf{I}_c| = \frac{\beta}{\sqrt{(\omega/\omega_\beta)^2 + 1}}|\mathbf{I}_b| \tag{3.90}$$

Therefore, at the frequency $\omega = \omega_\beta$, the magnitude of the collector current $\mathbf{I}_c$ is reduced to $1/\sqrt{2}$ times its low–frequency value. Thus, ω_β serves as a useful measure of the band of frequencies over which the short–circuit current amplification remains reasonably constant and nearly equal to its low–frequency value. For this reason, ω_β is referred to as the β *cutoff frequency*.

The frequency at which the current amplification is unity, that is, the frequency at which $|\mathbf{I}_c| = |\mathbf{I}_b|$, is found from (3.90) as

$$\beta = \sqrt{(\omega/\omega_\beta)^2 + 1} \tag{3.91}$$

As we know, a typical value for β is 100; therefore, in (3.91) the term $(\omega/\omega_\beta)^2$ is much larger than unity and $\omega \gg \omega_\beta$. Letting $\omega = \omega_T$, we obtain

$$\omega_T \approx \beta\omega_\beta \qquad f_T \approx \beta f_\beta \tag{3.92}$$

The relation of (3.92) is referred to as the current gain–bandwidth product for the transistor and it is an important *figure of merit** for a transistor. Also, from (3.82), (3.88), and (3.92)

$$\omega_T = \frac{g_m}{C_e} \qquad (3.93)$$

or

$$f_T = \frac{g_m}{2\pi C_e} \qquad (3.94)$$

Example 3.16

The specifications for a certain transistor state that $\beta = 100$, the current gain–bandwidth product is $f_T = 200\ MHz$, and the collector current is $i_C = 1\ mA$. Using the hybrid–π model of Figure 3.66, find:

a. the transconductance g_m

b. the base–emitter capacitance C_e

c. the base–emitter resistance r_{be}

d. the β cutoff frequency

Solution:

a. From (3.78)

$$g_m \approx 40 i_C = 40\ \text{millimhos}$$

b. From (3.94)

$$C_e = \frac{g_m}{2\pi f_T} = \frac{40 \times 10^{-3}}{2\pi \times 2 \times 10^8} \approx 32\ pF$$

c. From (3.83)

$$r_{be} = \frac{\beta}{g_m} = \frac{100}{40 \times 10^{-3}} = 2.5\ K\Omega$$

d. From (3.92)

$$f_\beta = \frac{f_T}{\beta} = \frac{200 \times 10^6}{100} = 2\ MHz$$

and this indicates that at this frequency the current amplification is still large.

* *The figure of merit is useful in comparing different devices for their over-all performance.*

3.14 The Darlington Connection

Figure 3.67 shows two transistors in a common collector configuration known as the *Darlington connection*. The circuit has high input impedance and low output impedance.

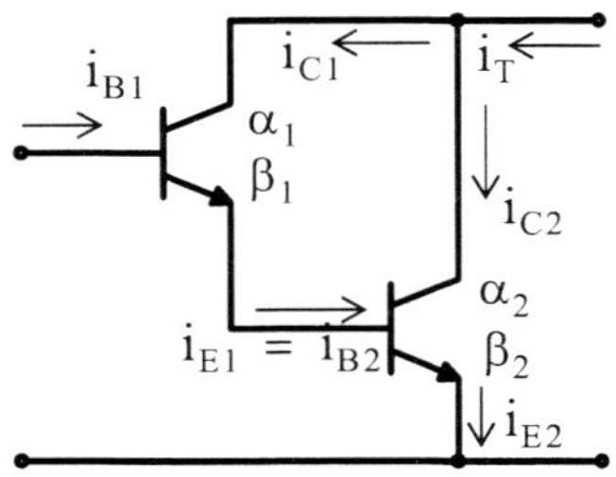

Figure 3.67. The Darlington connection

The combined α_T and β_T are evaluated as follows:

$$i_{E1} = i_{B2} = i_{C1} + i_{B1} \tag{3.95}$$

$$i_{C1} = \alpha_1 i_{E1} = \alpha_1 i_{B2} \tag{3.96}$$

$$i_{B2} = \alpha_1 i_{B2} + i_{B1} \tag{3.97}$$

$$i_{B2} = \frac{i_{B1}}{1-\alpha_1} \tag{3.98}$$

Also,

$$i_{E2} = i_{C2} + i_{B2} = \alpha_2 i_{E2} + i_{B2} \tag{3.99}$$

$$i_{E2} = \frac{i_{B2}}{1-\alpha_2} \tag{3.100}$$

Then,

$$i_T = i_{C1} + i_{C2} = \alpha_1 i_{E1} + \alpha_2 i_{E2} = \alpha_1 i_{B2} + \alpha_2 i_{E2} = \frac{\alpha_1 i_{B1}}{1-\alpha_1} + \frac{\alpha_2 i_{B2}}{1-\alpha_2} \tag{3.101}$$

From (3.98)

$$i_{B1} = (1-\alpha_1) i_{B2} \tag{3.102}$$

and by substitution into (3.101)

$$i_T = \frac{\alpha_1(1-\alpha_1) i_{B2}}{1-\alpha_1} + \frac{\alpha_2 i_{B2}}{1-\alpha_2} = \alpha_1 i_{B2} + \frac{\alpha_2 i_{B2}}{1-\alpha_2} \tag{3.103}$$

Also, from (3.100)

$$i_{B2} = (1-\alpha_2) i_{E2} \tag{3.104}$$

and by substitution into (3.103)

$$i_T = \frac{\alpha_1(1-\alpha_1)i_{B2}}{1-\alpha_1} + \frac{\alpha_2 i_{B2}}{1-\alpha_2} = \alpha_1(1-\alpha_2)i_{E2} + \frac{\alpha_2(1-\alpha_2)i_{E2}}{1-\alpha_2}$$
$$= (\alpha_1(1-\alpha_2)i_{E2} + \alpha_2 i_{E2}) \qquad (3.105)$$

Let the overall α value be denoted as α_T; then, $\alpha_T = i_T/i_{E2}$ and dividing (3.105) by i_{E2} we obtain

$$\alpha_T = \frac{i_T}{i_{E2}} = \alpha_1(1-\alpha_2) + \alpha_2 = \alpha_1 + \alpha_2 - \alpha_1\alpha_2 \qquad (3.106)$$

From Table 3.1,

$$\beta = \frac{\alpha}{1-\alpha} = \frac{i_C}{i_B} \qquad (3.107)$$

and with (3.102), (3.104), (3.105), and (3.107)

$$\beta_T = \frac{i_T}{i_{B1}} = \frac{\alpha_1(1-\alpha_2)i_{E2} + \alpha_2 i_{E2}}{(1-\alpha_1)(1-\alpha_2)i_{E2}} = \frac{\alpha_1(1-\alpha_2)+\alpha_2}{(1-\alpha_1)(1-\alpha_2)} = \frac{\alpha_1}{(1-\alpha_1)} + \frac{\alpha_2}{(1-\alpha_1)(1-\alpha_2)}$$
$$= \frac{\alpha_1}{(1-\alpha_1)} + \frac{1}{(1-\alpha_1)}\frac{\alpha_2}{(1-\alpha_2)} = \frac{\alpha_1}{(1-\alpha_1)} + \frac{\alpha_1}{(1-\alpha_1)/\alpha_1}\frac{\alpha_2}{(1-\alpha_2)} = \beta_1 + (\beta_1\beta_2)/\alpha_1$$

and since $\alpha \approx 1$

$$\beta_T \approx \beta_1 + \beta_1\beta_2 \qquad (3.108)$$

Also, since $\beta_1 \ll \beta_1\beta_2$

$$\beta_T \approx \beta_1\beta_2 \qquad (3.109)$$

3.15 Transistor Networks

In this section we will represent the common–base, common–emitter, and common collector transistor circuits by their h–equivalent and T–equivalent circuits, and we will derive the equations for input resistance, output resistance, voltage gain, and current gain.

3.15.1 The h-Equivalent Circuit for the Common-Base Transistor

Figure 3.68 shows the common–base configuration of an NPN transistor.

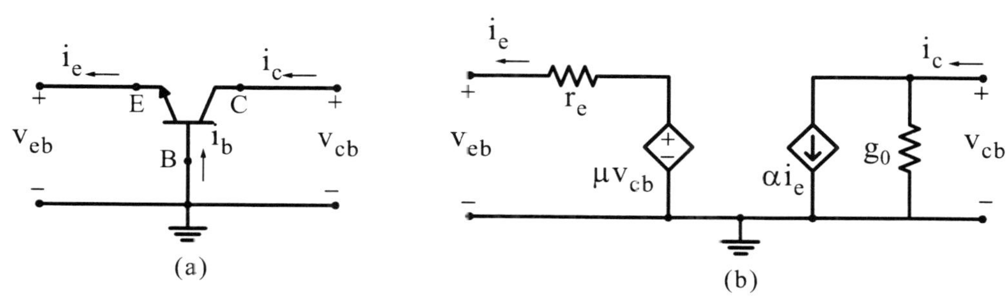

Figure 3.68. Common–base transistor circuit and its equivalent

From the equivalent circuit in Figure 3.68(b) we can draw the *h–parameter equivalent circuit* shown in Figure 3.69.

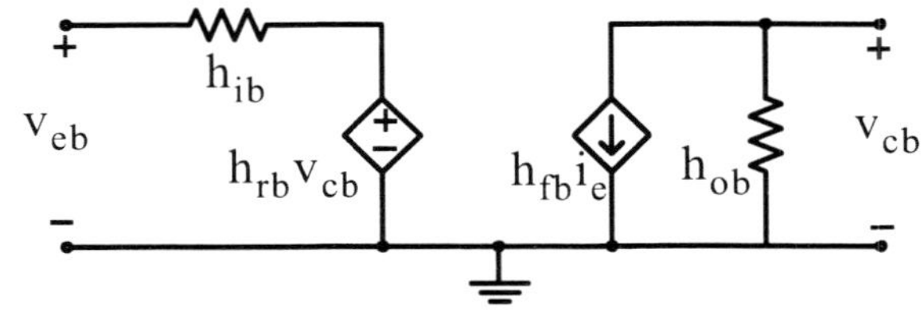

Figure 3.69. The h–parameter equivalent circuit for common–base transistor

From Figures 3.68 and (3.69), se observe that

$$h_{ib} = r_e \qquad h_{ob} = g_o \qquad h_{fb} = \alpha \qquad h_{rb} = \mu$$

Typical values for the h–parameter equivalent circuit for the common–base transistor are:

$$h_{ib} = 30\ \Omega \qquad h_{rb} = 5 \times 10^{-4} \qquad h_{fb} = 0.98 \qquad h_{ob} = 6 \times 10^{-6}\ \Omega^{-1}$$

As indicated above, the input resistance $h_{ib} = r_e$ has a small value, typically in the 25 to 35 Ω. This can be shown as follows:

From (3.2)

$$i_C = I_r e^{v_{BE}/V_T} \tag{3.110}$$

or

$$i_C = I_C + i_c = I_r e^{(v_{BE} + v_{be})/V_T} = I_r e^{v_{BE}/V_T} \cdot e^{v_{be}/V_T} \tag{3.111}$$

and since

$$I_C = I_r e^{v_{BE}/V_T} \tag{3.112}$$

by substitution of (3.112) into (3.111) we obtain

$$i_C = I_C e^{v_{be}/V_T} \tag{3.113}$$

Normally, $v_{be} \ll V_T$ and we recall that

$$e^x = 1 + x + \frac{x^2}{2!} + \frac{x^3}{3!} + \frac{x^4}{4!} + \ldots$$

Retaining only the first two terms of this series, by substitution into (3.113) we obtain

$$i_C = I_C\left(1 + \frac{v_{be}}{V_T}\right) = I_C + \frac{I_C}{V_T}v_{be} \tag{3.114}$$

and the signal component of i_C is

$$i_c = \frac{I_C}{V_T} v_{be} \tag{3.115}$$

Also,

$$i_e = \frac{i_c}{\alpha} = \frac{I_C}{\alpha V_T} v_{be} = \frac{I_E}{V_T} v_{be}$$

and thus

$$r_e = \frac{v_{be}}{i_e} = \frac{V_T}{I_E} \tag{3.116}$$

For $V_T = 26\ mV$ and $I_E = 1\ mA$,

$$r_{in} = r_e = 26\ \Omega \tag{3.117}$$

To find the overall voltage gain, we connect a voltage source v_s with its internal resistance R_s at the input, and a load resistor R_C at the output as shown in Figure 3.70.

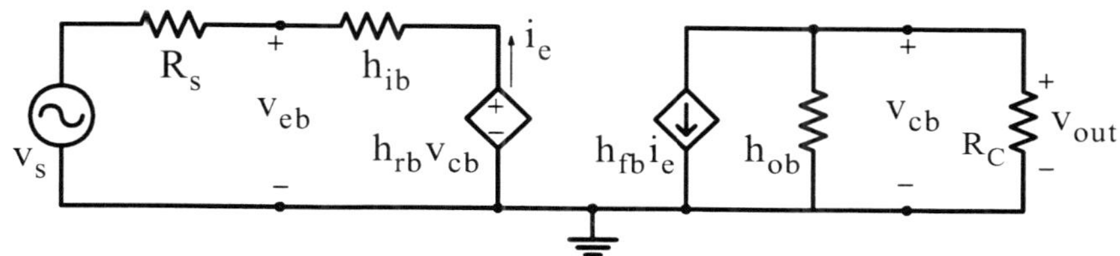

Figure 3.70. Circuit for the computation of voltage gain in a common base–transistor amplifier

The conductance $h_{ob} = g_o$ is in the order of $0.5 \times 10^{-6}\ \Omega^{-1}$ or $r_o = 2\ M\Omega$, and since it is much larger than R_C, it can be neglected (open circuit). Also, $h_{rb} = \mu = 5 \times 10^{-4}$ and the voltage source $h_{rb}v_{cb}$ can also be neglected (short circuit). With these observations, the circuit of Figure 3.70 is simplified to that in Figure 3.71.

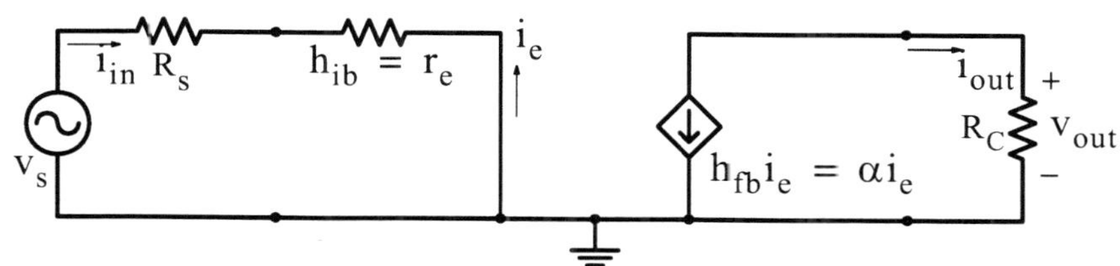

Figure 3.71. Simplified circuit for the computation of voltage gain in a common base transistor

From Figure 3.71,

$$v_{out} = -R_C h_{fb} i_e = -R_C \alpha i_e \tag{3.118}$$

and

$$i_e = \frac{-v_s}{R_s + h_{ib}} = \frac{-v_s}{R_s + r_e} \tag{3.119}$$

Substitution of (3.119) into (3.118) and division by v_s yields the overall voltage gain A_v as

$$A_v = \frac{v_{out}}{v_s} = \frac{\alpha R_C}{R_s + r_e} \tag{3.120}$$

and since $\alpha \approx 1$ and $r_e \approx 26\Omega$, the voltage gain depends on the values of R_s and R_C.

The current gain A_i is

$$A_i = \frac{i_{out}}{i_{in}} = \frac{-\alpha i_e}{-i_e} = \alpha \tag{3.121}$$

and the output resistance R_{out} is

$$R_{out} = R_C \tag{3.122}$$

Example 3.17

An NPN transistor is connected in a common–base configuration with $V_T = 26\ mV$, $i_E = 1\ mA$, $\beta = 120$, $R_S = 2\ K\Omega$, and $R_C = 5\ K\Omega$. Find the voltage and current gains, and input and output resistances.

Solution:

$$\alpha = \frac{\beta}{\beta + 1} = \frac{120}{121} = 0.992$$

$$h_{ib} = r_e = V_T / i_E = 26\ mV/1\ mA = 26\ \Omega$$

$$A_V = \frac{\alpha R_C}{R_s + r_e} = \frac{0.992 \times 5 \times 10^3}{(2 + 0.026) \times 10^3} = 2.45$$

$$A_i = \alpha = 0.992$$

$$R_i = r_e = 26\ \Omega$$

$$R_o = R_C = 5\ K\Omega$$

3.15.2 The T–Equivalent Circuit for the Common–Base Transistor

Transistor equivalent circuits can also be expressed as T–equivalent circuits. Figure 3.72 shows the common–base T–equivalent circuit.

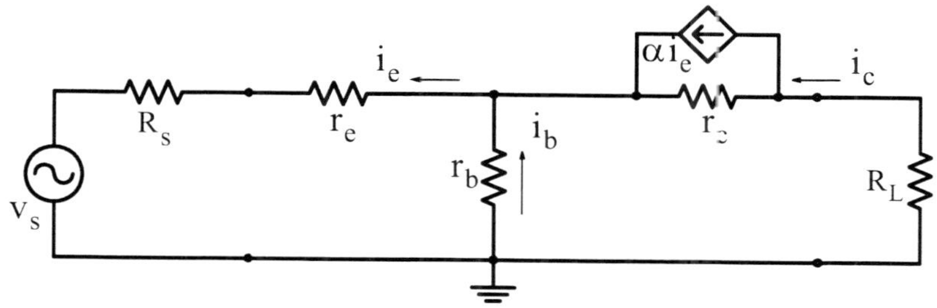

Figure 3.72. T–equivalent model for the common–base transistor

Typical numerical values for the 2N525 transistor are as follows:

$$\alpha = 0.978$$

$$r_c = 1.67\ \mathrm{M}\Omega$$

$$r_e = 12.5\ \Omega$$

$$r_b = 840\ \Omega$$

and since

$$r_e \ll r_c$$

$$r_b \ll r_c$$

$$R_L < r_c$$

and with these assumptions the input and output resistances and voltage and current gains for the T–equivalent model for the common–base transistor are:

$$\text{Input resistance} = r_{in} = r_e + r_b(1-\alpha) \tag{3.123}$$

$$\text{Output resistance} = r_{out} = r_e \cdot \frac{r_e + r_b(1-\alpha) + R_S}{r_e + r_b + R_S} \tag{3.124}$$

$$\text{Voltage gain} = A_v = \frac{\alpha R_L}{r_e + r_b(1-\alpha)} \tag{3.125}$$

$$\text{Current gain} = A_i = \alpha \tag{3.126}$$

3.15.3 The h–Equivalent Circuit for the Common–Emitter Transistor

Figure 3.73 shows the common–emitter configuration of an NPN transistor.

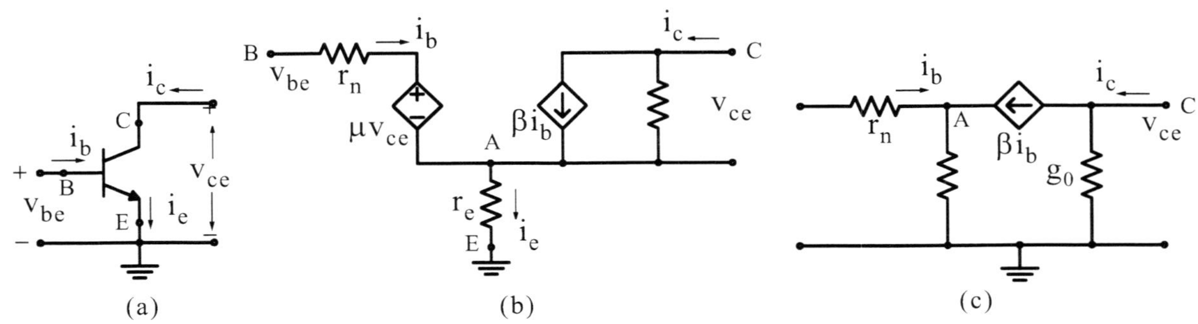

Figure 3.73. Common–emitter transistor circuit and its equivalent

The equivalent circuit in Figure 3.73(b), is same as in Figure 3.58 with the addition of the emitter resistor r_e. The conductance g_0 is in the order of $27 \times 10^{-6}\ \Omega^{-1}$ or $r_0 = 40\ K\Omega$, and since it is much larger than a typical R_C resistance, it can be neglected. Also, $h_{re} = \mu = 3.4 \times 10^{-4}$ and the dependent voltage source μv_{ce} can also be neglected. Therefore, for simplicity and compactness, we can represent the circuit of Figure 3.73(b) as that in Figure 3.73(c).

The *h–parameter equivalent circuit* in Figure 3.73(b) is shown in Figure 3.74 where

$$h_{ie} = r_n \qquad h_{oe} = g_0 \qquad h_{fe} = \beta \qquad h_{re} = \mu$$

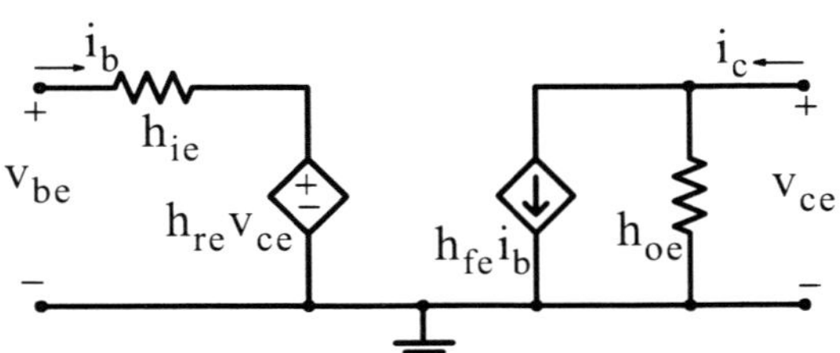

Figure 3.74. The h–parameter equivalent circuit for common–emitter transistor

Typical values for the h–parameter equivalent circuit for the common–emitter transistor are:

$$h_{ie} = 1.5\ K\Omega \qquad h_{re} = 3.4 \times 10^{-4} \qquad h_{fe} = 80 \qquad h_{oe} = 27 \times 10^{-6}\ \Omega^{-1}$$

Figure 3.75 shows the h–parameter equivalent circuit in Figure 3.74 with a signal source v_S and its internal resistance R_S connected on the left side, and a load resistor R_L connected on the right side. Using the circuit in Figure 3.75 we can compute the exact voltage and current gains, and input and output resistances.

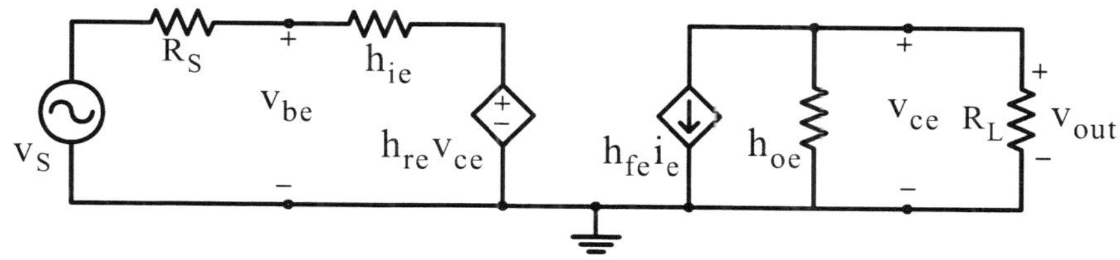

Figure 3.75. Circuit for exact computation in voltage and current gains and input and output resistances

As stated above, the conductance $h_{oe} = g_o$ is in the order of $27 \times 10^{-6}\ \Omega^{-1}$ or $r_o = 40\ K\Omega$, and since it is much larger than R_C it can be neglected. Also, $h_{re} = \mu = 3.4 \times 10^{-4}$ and the voltage source $h_{re}v_{ce}$ can also be neglected. Therefore, to compute the input and output resistances and overall voltage and current gains to a fairly accurate values, we can use the simplified circuit in Figure 3.73(c), we connect a voltage source v_S with its internal resistance R_S at the input, and a load resistor R_C at the output as shown in Figure 3.76 where an additional resistor R_E is connected in series with the emitter resistor r_e to increase the input resistance, and R'_S represents the series combination of R_S and r_n.*

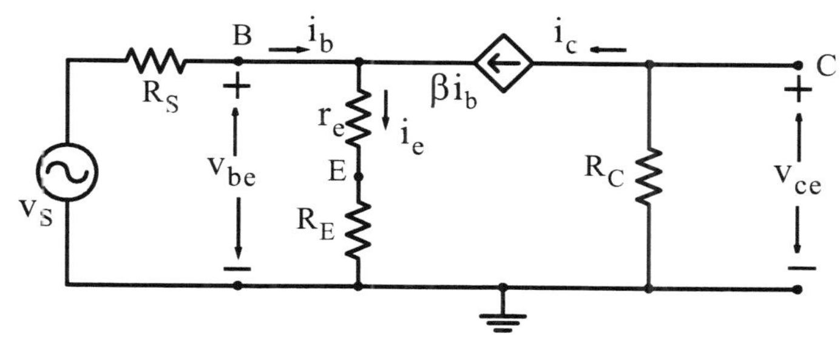

Figure 3.76. The simplified common–emitter transistor equivalent circuit

From the equivalent circuit in Figure 3.76,

$$r_{in} = \frac{v_{be}}{i_b} = \frac{(r_e + R_E)i_e}{i_b} = \frac{(r_e + R_E)(i_b + i_c)}{i_b} = \frac{(r_e + R_E)(i_b + \beta i_b)}{i_b} = (r_e + R_E)(\beta + 1)$$

or

$$r_{in} = (r_e + R_E)(\beta + 1) \tag{3.127}$$

and since $\beta \gg 1$

$$r_{in} \approx \beta(r_e + R_E) \tag{3.128}$$

The resistor R_E is referred to as the *emitter degeneration resistance* because it causes negative feedback. That is, if the collector current i_c increases, the emitter current i_e will also increase since

* *Typically,* $r_n \ll r_e + R_E$.

$i_c = \alpha i_e$ and any increase in the base current i_b will be negligible. Then, the voltage drop across the resistor R_E will rise and the voltage drop across the resistor r_e will decrease to maintain the voltage v_{be} relatively constant. But a decrease in the voltage drop across the resistor r_e means a decrease in the emitter current i_e and consequently a decrease in the collector current i_c.

Relation (3.128) indicates that the input resistance r_{in} can be controlled by choosing an appropriate value for the external resistor R_E.

From Figure 3.76 we observe that the output resistance is the collector resistance R_C, that is,

$$r_{out} = R_C \tag{3.129}$$

The overall voltage gain is

$$A_V = \frac{v_{out}}{v_S} = \frac{v_{out}}{v_b} \cdot \frac{v_b}{v_S} \tag{3.130}$$

From Figure 3.76

$$v_{out} = v_{ce} = -\beta i_b R_C \tag{3.131}$$

and

$$v_b = (r_e + R_E)i_e = (r_e + R_E)(i_b + i_c) = (r_e + R_E)(i_b + \beta i_b) = (r_e + R_E)(1 + \beta)i_b \tag{3.132}$$

From (3.131) and (3.132)

$$\frac{v_{out}}{v_b} = \frac{-\beta i_b R_C}{(r_e + R_E)(1 + \beta)i_b} = \frac{-\beta R_C}{(r_e + R_E)(\beta + 1)} \tag{3.133}$$

Also,

$$v_b = \frac{r_{in}}{r_{in} + R_s} v_S$$

and with (3.127)

$$\frac{v_b}{v_S} = \frac{r_{in}}{r_{in} + R_S} = \frac{(r_e + R_E)(\beta + 1)}{(r_e + R_E)(\beta + 1) + R_S} \tag{3.134}$$

Substitution of (3.133) and (3.134) into (3.130) yields

$$A_V = \frac{v_{out}}{v_S} = \frac{v_{out}}{v_b} \cdot \frac{v_b}{v_S} = \frac{-\beta R_C}{(r_e + R_E)(\beta + 1)} \cdot \frac{(r_e + R_E)(\beta + 1)}{(r_e + R_E)(\beta + 1) + R_S}$$

$$A_V = \frac{-(\beta + 1)R_C}{(\beta + 1)(r_e + R_E) + R_S} \approx \frac{-\beta R_C}{\beta(r_e + R_E) + R_S} \tag{3.135}$$

and the minus (–) sign indicates that the output is 180° out–of–phase with the input.

From (3.134) we observe that the introduction of the external resistor R_E results in reduction of the overall gain.

The current gain is

$$A_i = \frac{i_{out}}{i_b} = \frac{-\beta i_b}{i_b} = -\beta \quad (3.136)$$

Example 3.18

An NPN transistor is connected in a common–emitter configuration with $V_T = 26\ mV$, $i_E = 1\ mA$, $\beta = 120$, $R_S = 2\ K\Omega$, and $R_C = 5\ K\Omega$.

a. Find the voltage and current gains, and input and output resistances if $R_E = 0$.

b. Find the voltage and current gains, and input and output resistances if $R_E = 200\Omega$.

c. Find the maximum value of the applied signal v_S so that v_{be} or v_b under the conditions of (a) and (b) will not exceed $5\ mV$.

Solution:

a.

From (3.116)

$$h_{ie} = r_e = V_T/i_E = 26\ mV/1\ mA = 26\ \Omega$$

From (3.128)

$$r_{in} \approx \beta(r_e + R_E) = 120(26 + 0) = 3.12\ K\Omega$$

and from (3.129)

$$r_{out} = R_C = 5\ K\Omega$$

The voltage and current gains are found from (3.135) and (3.136). Thus, with $R_E = 0$, the voltage gain is

$$A_V \approx \frac{-\beta R_C}{\beta r_e + R_s} = \frac{-120 \times 5 \times 10^3}{120 \times 26 + 2 \times 10^3} = -117.2$$

and the current gain is

$$A_i = -\beta = -120$$

b.

$$r_{in} \approx \beta(r_e + R_E) = 120(26 + 200) = 27.12\ K\Omega$$

$$r_{out} = R_C = 5\ K\Omega$$

$$A_V \approx \frac{-\beta R_C}{\beta(r_e + R_E) + R_s} = \frac{-120 \times 5 \times 10^3}{120(26 + 200) + 2 \times 10^3} = -20.6$$

$$A_i = -\beta = -120$$

We observe that whereas the addition of R_E has increased the input resistance by 88.5%, the voltage gain has decreased by 82.4%.

c.

Without R_E,

$$r_{in} \approx \beta r_e = 120 \times 26 = 3.12\ K\Omega$$

and

$$v_{be} = \frac{r_{in}}{r_{in} + R_s} v_S$$

$$v_{S(max)} = \frac{r_{in} + R_s}{r_{in}} v_{be} = \frac{3.12 + 2}{3.12} \times 5\ mV = 8.2\ mV$$

With R_E,

$$r_{in} \approx \beta(r_e + R_E) = 120(26 + 200) = 27.12\ K\Omega$$

Then,

$$v_{S(max)} = \frac{r_{in} + R_s}{r_{in}} v_{be} = \frac{27.12 + 2}{27.12} \times 5\ mV = 5.4\ mV$$

We observe that for an increase of 88.5% in input resistance, the maximum value of the applied signal v_S decreases by 51.8%.

3.15.4 The T-Equivalent Circuit for the Common-Emitter Transistor

Figure 3.77 shows the common–emitter T–equivalent circuit.

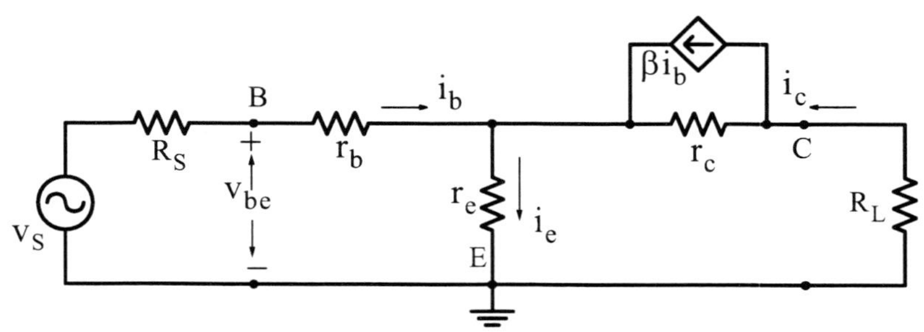

Figure 3.77. T–equivalent model for the common–emitter transistor

As with the T-equivalent model for the common-base transistor, we assume that

$$r_e \ll r_c$$

$$r_b \ll r_c$$

$$R_L < r_c$$

and with these assumptions the input and output resistances and voltage and current gains for the T-equivalent model for the common-emitter transistor are:

$$\text{Input resistance} = r_{in} = r_b + \frac{r_e}{1-\alpha} \tag{3.137}$$

$$\text{Output resistance} = r_{out} = r_c(1-\alpha) + r_e \cdot \frac{\alpha r_c + R_S}{r_e + r_b + R_S} \tag{3.138}$$

$$\text{Voltage gain} = A_v = \frac{-\alpha R_L}{r_e + r_b(1-\alpha)} \tag{3.139}$$

$$\text{Current gain} = A_i = -\beta \tag{3.140}$$

3.15.5 The h-Equivalent Circuit for the Common-Collector Transistor

Figure 3.78 shows the common-collector or emitter-follower configuration of an NPN transistor. The emitter-follower is useful in applications where a high-resistance source is to be connected to a low-resistance load.

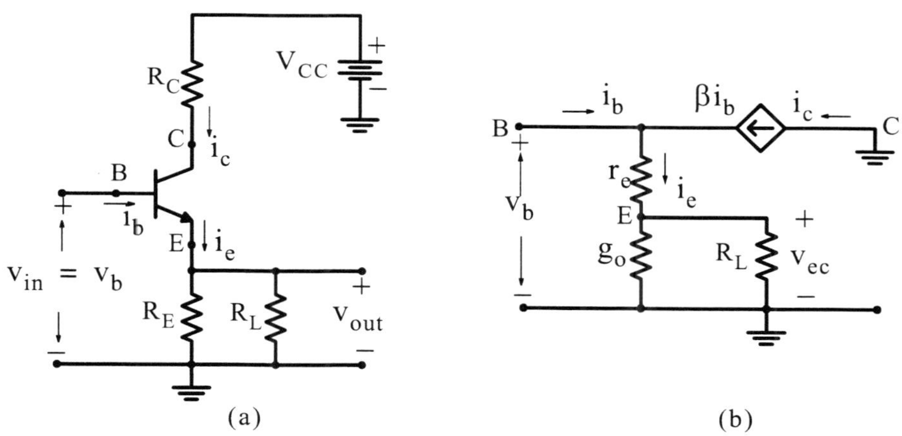

Figure 3.78. Common-collector or emitter-follower transistor circuit and its equivalent

From the equivalent circuit in Figure 3.78(b) we can draw the *h-parameter equivalent circuit* shown in Figure 3.79 where

$$h_{ic} = r_e \qquad h_{oc} = g_o \qquad h_{fc} = \beta \qquad h_{rc} = \mu$$

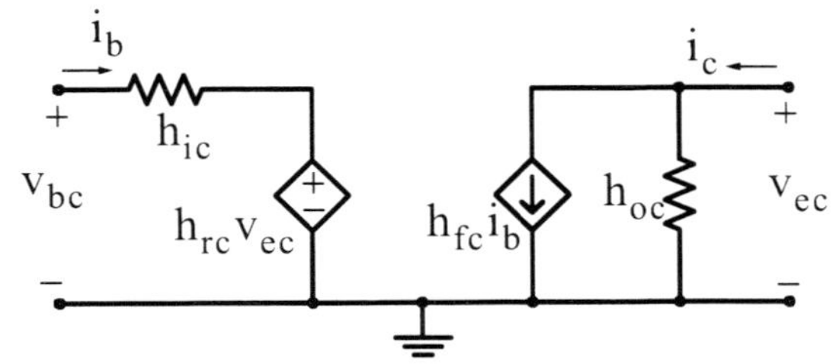

Figure 3.79. The h–parameter equivalent circuit for common–collector transistor

Typical values for the h–parameter equivalent circuit for the common–collector transistor are:

$$h_{ic} = 1.5\ K\Omega \qquad h_{rc} = 1 \qquad h_{fc} = -45 \qquad h_{oc} = 27 \times 10^{-6}\ \Omega^{-1}$$

Figure 3.80 shows the h–parameter equivalent circuit in Figure 3.79 with a signal source v_s and its internal resistance R_S connected on the left side, and a load resistor R_L connected on the right side.

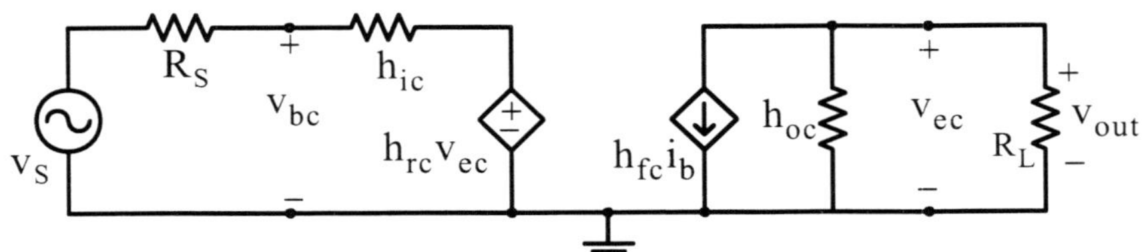

Figure 3.80. Circuit for the computation of voltage gain in a common–collector transistor amplifier

To find the input and output resistances and overall voltage and current gains, we denote the conductance $h_{oc} = g_o$ with its reciprocal r_o in the circuit in Figure 3.78(b), we connect a voltage source v_s with its internal resistance R_s at the input, and the circuit is as shown in Figure 3.81.

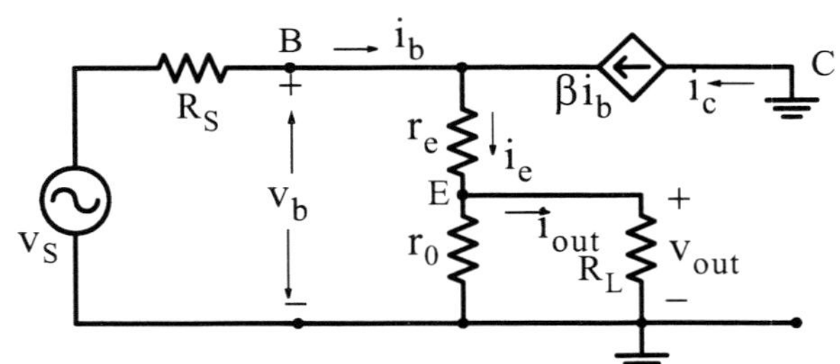

Figure 3.81. The simplified common–collector transistor amplifier equivalent circuit

From the equivalent circuit in Figure 3.81,

$$r_{in} = \frac{v_b}{i_b} = \frac{(r_e + r_o \| R_L)i_e}{i_b} = \frac{(r_e + r_o \| R_L)(i_b + i_c)}{i_b}$$

$$= \frac{(r_e + r_o \| R_L)(i_b + \beta i_b)}{i_b} = (r_e + r_o \| R_L)(\beta + 1) \tag{3.141}$$

and since $\beta \gg 1$

$$r_{in} \approx (r_e + r_o \| \beta R_L)$$

Also, since $r_o \gg R_L$ and $r_e \ll R_L$

$$r_{in} \approx \beta R_L \tag{3.142}$$

Relation (3.142) indicates that the input resistance r_{in} can be controlled by choosing an appropriate value for the load resistor R_L.

The output resistance r_{out} cannot be determined by inspection; therefore we will remove the load resistor R_L and the voltage source v_S, and we will connect a test voltage source v_x across the emitter to ground terminals, that is, across the resistor r_o and we will find the output resistance from the relation of the equivalent circuit derive it from the relation $r_{out} = v_x/i_x$ as shown in Figure 3.82 where the current i_b is shown as $(1-\alpha)i_e$.*

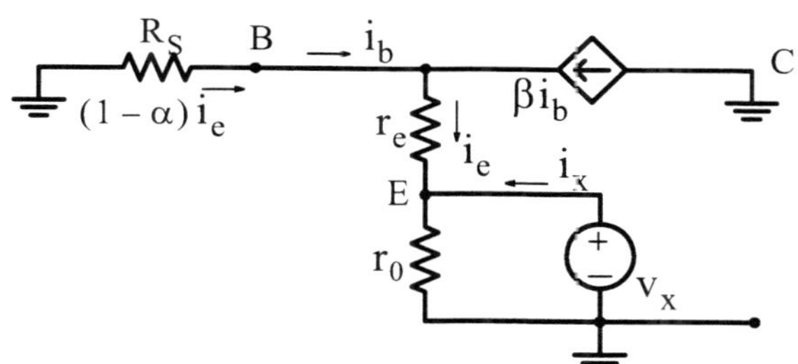

Figure 3.82. Equivalent circuit for the computation of the output resistance

From Figure 3.82 we obverse that the voltage drop across R_S is equal to the sum of the voltage drops across r_e and r_0, that is,

$$-R_S(1-\alpha)i_e = r_e i_e + v_x$$

$$v_x = -R_S(1-\alpha)i_e - r_e i_e$$

* *See Table 3.1*

$$-\frac{v_X}{i_e} = R_S(1-\alpha) + r_e$$

$$i_e = \frac{-v_X}{R_S(1-\alpha) + r_e} \tag{3.143}$$

Also,

$$r_0(i_e + i_X) = v_X$$

or

$$i_X = \frac{v_X}{r_0} - i_e \tag{3.144}$$

Substitution of (3.143) into (3.144) yields

$$i_X = \frac{v_X}{r_0} + \frac{v_X}{R_S(1-\alpha) + r_e}$$

and division of both sides by v_X yields

$$\frac{i_X}{v_X} = \frac{1}{r_{out}} = \frac{1}{r_o} + \frac{1}{R_S(1-\alpha) + r_e}$$

The last relation above reminds us of the formula for the combination of two parallel resistors. Then,

$$r_{out} = r_o \| R_S(1-\alpha) + r_e = r_o \| \frac{R_S}{\beta+1} + r_e \approx r_o \| \frac{R_S}{\beta} = \frac{r_o \cdot R_S/\beta}{r_o + R_S/\beta} \tag{3.145}$$

and obviously the output resistance is quite low. The voltage gain is

$$A_v = \frac{v_{out}}{v_S} = \frac{v_{out}}{v_b} \cdot \frac{v_b}{v_S}$$

where from Figure 3.81 and relation (3.141),

$$v_b = \frac{r_{in}}{R_S + r_{in}} v_S$$

$$\frac{v_b}{v_S} = \frac{r_{in}}{R_S + r_{in}} \approx \frac{\beta R_L}{R_S + \beta R_L} \tag{3.146}$$

$$v_{out} = \frac{r_o \| R_L}{r_e + r_o \| R_L} \cdot v_b$$

$$\frac{v_{out}}{v_b} = \frac{r_o \| R_L}{r_e + r_o \| R_L} \approx \frac{R_L}{r_e + R_L} \tag{3.147}$$

and thus

$$A_v = \frac{v_{out}}{v_S} = \frac{v_{out}}{v_b} \cdot \frac{v_b}{v_S} \approx \frac{R_L}{r_e + R_L} \cdot \frac{\beta R_L}{R_S + \beta R_L} = \frac{\beta R_L^2}{(r_e + R_L) \cdot (R_S + \beta R_L)} \tag{3.148}$$

Relation (3.148) reveals that the voltage gain of the emitter–follower is less than unity. The current gain is

$$A_i = \frac{i_{out}}{i_b}$$

where by the current division expression

$$i_{out} = \frac{r_o}{r_o + R_L} \cdot i_e = \frac{r_o}{r_o + R_L} \cdot (\beta + 1) i_b$$

and thus

$$A_i = \frac{i_{out}}{i_b} = \frac{(\beta + 1) r_o}{r_o + R_L} \approx \frac{\beta r_o}{r_o + R_L}$$

Since $r_o \gg R_L$,

$$A_i \approx \beta \tag{3.149}$$

Example 3.19

Figure 3.83 shows the equivalent circuit of a typical emitter–follower and it is given that $\beta = 80$. Find the input and output resistances and voltage and current gains.

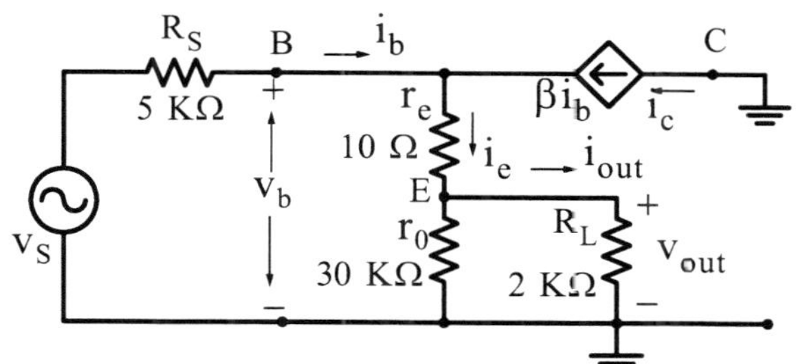

Figure 3.83. Emitter–follower transistor amplifier equivalent circuit for Example 3.19

Solution:

From (3.142)

$$r_{in} \approx \beta R_L = 80 \times 2 \times 10^3 = 160\ K\Omega$$

From (3.145)

$$r_{out} \approx \frac{r_o \cdot R_S/\beta}{r_o + R_S/\beta} = \frac{30 \times 10^3 \times 5 \times 10^3/80}{30 \times 10^3 + 5 \times 10^3/80} = 62.4\ \Omega$$

From (3.148)

$$A_v \approx \frac{\beta R_L^2}{(r_e + R_L) \cdot (R_S + \beta R_L)} = \frac{80 \times 4 \times 10^6}{(10 + 2 \times 10^3)(5 \times 10^3 + 80 \times 2 \times 10^3)} = 0.965$$

This is the voltage gain with a load resistor connected to the circuit. With this resistor disconnected, the input resistance as given by (3.141) where $\beta \gg 1$, is reduced to

$$r_{in} \approx \beta(r_e + r_o) \tag{3.150}$$

and (3.146) becomes

$$\frac{v_b}{v_S} = \frac{r_{in}}{R_S + r_{in}} \approx \frac{\beta(r_e + r_o)}{R_S + \beta(r_e + r_o)} \tag{3.151}$$

Also, (3.147) becomes

$$\frac{v_{out}}{v_b} = \frac{r_o}{r_e + r_o} \tag{3.152}$$

Then, the voltage gain with the load resistor disconnected is

$$A_v\Big|_{R_L \to \infty} = \frac{v_{out}}{v_S} = \frac{v_{out}}{v_b} \cdot \frac{v_b}{v_S} = \frac{r_o}{r_e + r_o} \cdot \frac{\beta(r_e + r_o)}{R_S + \beta(r_e + r_o)} \tag{3.153}$$

and since $r_e \ll r_o$ this relation reduces to

$$A_v\Big|_{R_L \to \infty} = \frac{\beta r_o}{R_S + \beta r_o} \tag{3.154}$$

With the given values

$$A_v\Big|_{R_L \to \infty} = \frac{80 \times 30 \times 10^3}{5 \times 10^3 + 80 \times 30 \times 10^3} = 0.998$$

and from (3.149)

$$A_i \approx \beta \approx 80$$

3.15.6 The T–Equivalent Circuit for the Common–Collector Transistor Amplifier

Figure 3.84 shows the common–collector T–equivalent circuit.

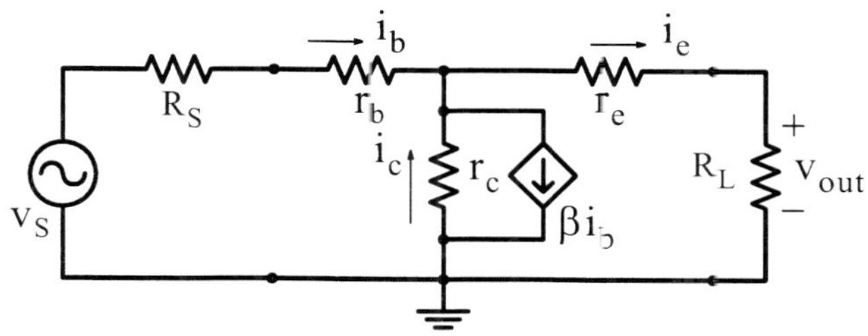

Figure 3.84. T–equivalent model for the common–collector transistor amplifier

As with the T-equivalent models for the common-base and common-emitter transistors, we assume that

$$r_e \ll r_c$$

$$r_b \ll r_c$$

$$R_L < r_c$$

and with these assumptions the input and output resistances and voltage and current gains for the T–equivalent model for the common–collector transistor amplifier are:

$$\text{Input resistance} = r_{in} = \frac{R_L}{1-\alpha} \tag{3.155}$$

$$\text{Output resistance} = r_{out} = r_e + (r_b + R_S)(1-\alpha) \tag{3.156}$$

$$\text{Voltage gain} = A_v \approx 1 \tag{3.157}$$

$$\text{Current gain} = A_i \approx \beta \tag{3.158}$$

Table 3.4 summarizes the three possible transistor amplifier configurations.

TABLE 3.4 Phase, input and output resistances, and voltage and current gains for transistor amplifiers

	Common–Base	Common–Emitter	Common Collector
Input/Output Phase	0°	180°	0°
Input Resistance r_{in}	Low	Moderate	High
Output Resistance r_{out}	Equal to R_C	High	Low
Voltage Gain A_v	Depends on ratio R_C/R_S	High	Close to unity
Current Gain A_i	Close to unity	High	Large

3.16 Transistor Cutoff and Saturation Regions

As mentioned earlier, a transistor can be in a cutoff, active, or saturation region. The conditions are shown in Table 3.2 and are repeated below for convenience.

Region of Operation	Emitter–Base junction	Collector–base junction
Active	Forward	Reverse
Saturation	Forward	Forward
Cutoff	Reverse	Reverse

Let us consider the transistor circuit in Figure 3.85. We will refer to it in our subsequent discussion to define the cutoff, active, and saturation regions.

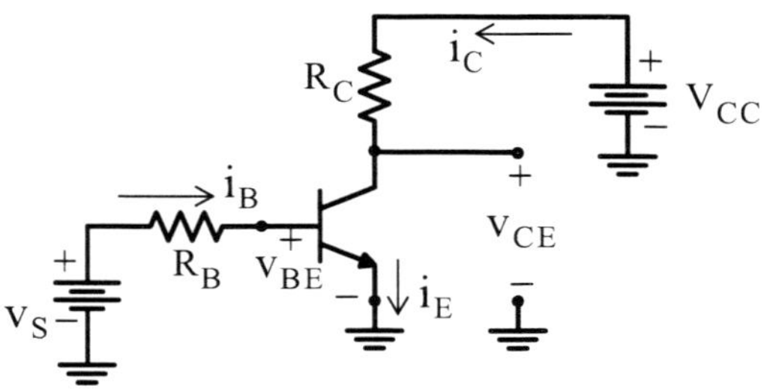

Figure 3.85. Transistor circuit for defining the regions of operation

3.16.1 Cutoff Region

If v_S is such that $v_{BE} < 0.6\ V$, the emitter–base junction will be reverse–biased and since V_{CC} is positive, the collector–base junction will also be reverse–biased, and the transistor will be in the cutoff mode. Then,

$$i_B = 0 \qquad i_E = 0 \qquad i_C = 0 \qquad v_{CE} = V_{CC}$$

3.16.2 Active Region

If v_S is such that $v_{BE} \ge 0.7\ V$, the emitter–base junction will be forward–biased and since V_{CC} is positive, the collector–base junction will be reverse–biased, and transistor will be in the active mode. Then, if $v_{BE} = 0.7\ V$, we will have

$$i_B = \frac{v_S - V_{BE}}{R_B} = \frac{v_S - 0.7}{R_B} \qquad i_C = \beta i_B \qquad v_C = V_{CC} - R_C i_C \qquad v_{CB} = v_C - v_{BE}$$

and if $v_{CB} \ge 0.7\ V$, the transistor will be in the active mode. However, if $v_{CB} < 0.7\ V$, the transistor will be in the saturation mode which is discussed in the next subsection.

3.16.3 Saturation Region

The saturation region is reached by supplying a base current larger than I_{CM}/β where I_{CM} is the maximum current the collector will deliver while in the active mode. Thus, we can find the maximum current the collector can have while in the active mode, and we can then determine whether the transistor is in the active mode or the saturation mode.

With reference to the circuit in Figure 3.85, the maximum current I_{CM} the collector can have while in the active mode can be found from the relation

$$R_C I_{CM} + v_{BE} = V_{CC}$$

or

$$I_{CM} = \frac{V_{CC} - v_{BE}}{R_C} = \frac{V_{CC} - 0.7}{R_C} \tag{3.159}$$

and if I_{BM} is the base current corresponding to the collector current I_{CM}, it follows that

$$I_{BM} = I_{CM}/\beta \tag{3.160}$$

We can also find the maximum value of the applied signal voltage v_S that will keep the transistor in the active mode from the relation

$$v_{S\ max} = R_B I_{BM} + 0.7 \tag{3.161}$$

If we increase the base current above I_{BM}, there will be a corresponding increase in I_{CM} and since $v_C = V_{CC} - R_C I_C$, v_C will decrease and if it falls below the value of v_{BE}, the collector–base junction will become forward–biased and if it reaches a value of $0.5\ V$, any further increase in the base current will result in a very small increase in the collector current, and whereas $\beta = di_C/di_B$ in the active mode, we can see that this relation does not hold when the transistor is in the saturation mode. In this case β is referred to as the *current gain at saturation* and it is denoted as β_{sat}.

When a transistor is deeply into saturation, the collector to emitter voltage is denoted as $v_{CE\ sat}$ and its value is approximately $0.2\ V$, that is,

$$v_{CE\ sat} \approx 0.2\ V \tag{3.162}$$

and the corresponding collector current $i_{C\ sat}$ is found from the relation

$$i_{C\ sat} = \frac{V_{CC} - v_{CE\ sat}}{R_C} \tag{3.163}$$

Example 3.20

For the transistor circuit in Figure 3.86 it is known that it is in the active mode and $\beta > 50$. Find v_E, i_E, v_C, i_C, and i_B.

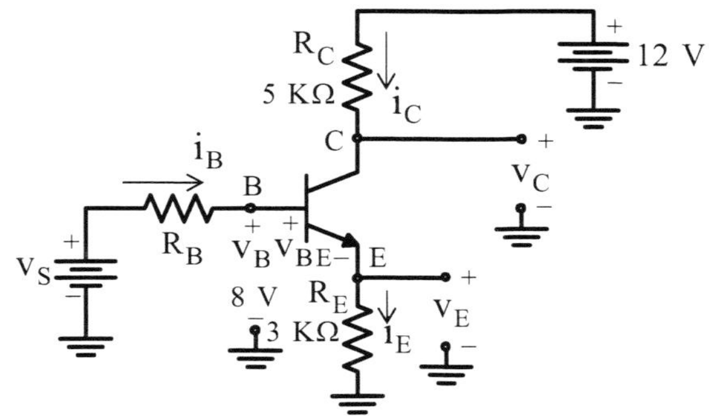

Figure 3.86. Circuit for Example 3.20

Solution:

We do not know whether the transistor operates in the cutoff, active, or saturation region, but since $v_B = 8\ V$, it is safe to assume that it operates either in the active or saturation region. Assuming active mode of operation, we obtain

$$v_E = v_B - v_{BE} = 8 - 0.7 = 7.3\ V$$

$$i_E = \frac{v_E}{R_E} = \frac{7.3}{3 \times 10^3} = 2.43\ mA$$

and since $i_C \approx \alpha i_E$, let us assume that

$$i_C \approx 2.3\ mA$$

Then, in active mode of operation

$$v_C = V_{CC} - R_C i_C = 12 - 5 \times 10^3 \times 2.3 \times 10^{-3} = 0.5\ V$$

and since this value is much less that $v_B = 8\ V$, we conclude that the transistor is deeply into saturation. Then, in saturation

$$v_C = v_{CE\ sat} + v_E = 0.2 + 7.3 = 7.5\ V$$

$$i_C = \frac{V_{CC} - v_C}{5 \times 10^3} = \frac{12 - 7.5}{5 \times 10^3} = 0.90\ mA$$

and

$$i_B = i_E - i_C = 2.43 - 0.90 = 1.53\ mA$$

and this is indeed a very large base current. We now can find the value of β at saturation.

$$\beta_{sat} = \frac{i_C}{i_B} = \frac{0.90}{1.53} = 0.59$$

This value also indicates that the transistor is deeply into saturation.

3.17 The Ebers–Moll Transistor Model

The Ebers–Moll model of the bipolar transistor provides a simple, closed–form expression for the collector, base, and emitter currents of a bipolar transistor in terms of the terminal voltages. It is valid in all regions of operation of the bipolar transistor, transitioning between them smoothly. The Ebers–Moll bipolar transistor model expresses each of the terminal currents in terms of a forward component I_F, which only depends on the base–to–emitter voltage, and a reverse component I_R, which only depends on the base–to–collector voltage.

Let α_F denote the current amplification in the normal operation (V_{BE} = forward – biased and V_{CB} = reverse – biased) and α_R denote the inverted operation (V_{BE} = reverse – biased and V_{CB} = forward – biased) common–base current gains of the bipolar transistor, and β_F and β_R denote the normal and inverted operations respectively of the common–emitter gains. Then, the terminal currents can be expressed as

$$I_C = I_F - \frac{I_R}{\alpha_R} \qquad I_E = \frac{I_F}{\alpha_R} - I_R \qquad I_E = \frac{I_F}{\beta_F} + \frac{I}{\beta} \tag{3.164}$$

These current gains are related to one another by

$$\alpha_{F,R} = \frac{\beta_{F,R}}{\beta_{F,R}+1} \qquad \beta_{F,R} = \frac{\alpha_{F,R}}{1-\alpha_{F,R}} \tag{3.165}$$

By substituting these relationships into (3.164), we can show that the Ebers–Moll model satisfies Kirchhoff's current law, that is,

$$I_E = I_C + I_B \tag{3.166}$$

If the emitter–base junction is forward biased and the collector–base junction is reverse biased, then $I_F \gg I_R$ and the bipolar transistor is said to be in the forward–active mode of operation. In this case, the terminal currents are given by

$$I_C = I_F \qquad I_E = \frac{I_F}{\alpha_F} \qquad I_B = \frac{I_F}{\beta_F} \tag{3.167}$$

Figure 3.87 shows how an NPN and a PNP transistors are biased.

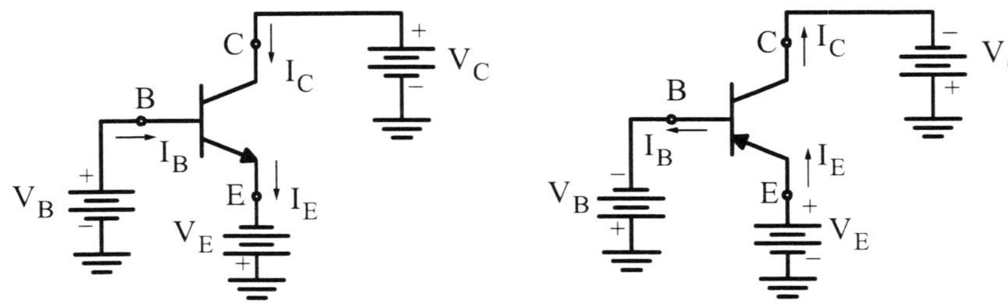

NPN Transistor PNP Transistor

Figure 3.87. Biased bipolar transistors showing the direction of conventional current flow

To bias an NPN bipolar transistor into the forward–active region, we should ensure that $V_{BE} > 4V_T$ and $V_C \ge V_B$. In this case, I_C, I_E, and I_B are positive and are flowing in the directions indicated. To bias a PNP bipolar transistor into the forward–active region, we should ensure that $V_{EB} > 4V_T$ and $V_C \le V_B$. In this case, I_C, I_E, and I_B are positive and are flowing in the directions indicated.

By using (3.166) and (3.167), we can express the terminal currents in terms of one another as

$$I_C = \beta_F I_B = \alpha_F I_E \qquad I_E = \frac{I_C}{\alpha_F} = (\beta_F + 1)I_B \qquad I_B = \frac{I_C}{\beta_F} = \frac{I_E}{\beta_F + 1} \tag{3.168}$$

If the collector–base junction is forward biased, and the emitter–base junction is reverse biased, then $I_R \gg I_F$ and the transistor is said to be in the reverse–active mode of operation in which the collector and emitters effectively reverse their roles. In this case the terminal currents are given by

$$I_C = -I_R/\alpha_R \qquad I_E = -I_R \qquad I_B = I_R/\beta_R \tag{3.169}$$

By using (3.166) and (3.169), we can express the terminal currents in terms of one another as

$$I_E = -\beta_R I_B = \alpha_R I_C \qquad I_C = -(\beta_R + 1)I_B = I_E/\alpha_R \tag{3.170}$$

If both junctions are forward biased simultaneously, the transistor is said to be in saturation.

Let I_S be the saturation current and V_T the thermal voltage. Then for the NPN bipolar transistor biased as shown in Figure 3.87, the forward current component, I_F, is given by

$$I_F = I_S(e^{V_{BE}/V_T} - 1) = I_S(e^{(V_B - V_E)/V_T} - 1) \tag{3.171}$$

Likewise, the reverse current component I_R is

$$I_R = I_S(e^{V_{BC}/V_T} - 1) = I_S(e^{(V_B - V_C)/V_T} - 1) \tag{3.172}$$

By substitution of (3.171) and (3.172) into (3.164) we express the terminal currents in terms of the terminal voltages as

$$
\begin{aligned}
I_C &= I_S(e^{(V_B - V_E)/V_T} - 1) - \frac{I_S}{\alpha_R}(e^{(V_B - V_C)/V_T} - 1) \\
I_E &= \frac{I_S}{\alpha_R}(e^{(V_B - V_E)/V_T} - 1) - I_S(e^{(V_B - V_C)/V_T} - 1) \\
I_B &= \frac{I_S}{\beta_F}(e^{(V_B - V_E)/V_T} - 1) - \frac{I_S}{\beta_R}(e^{(V_B - V_C)/V_T} - 1)
\end{aligned}
\qquad (3.173)
$$

If $V_{BE} > 4V_T$ and $V_C \ge V_B$, the transistor is biased deeply into the forward–active region and (3.172) reduces to

$$
I_C \approx I_S(e^{(V_B - V_E)/V_T}) \qquad I_E \approx \frac{I_S}{\alpha_F}(e^{(V_B - V_E)/V_T}) \qquad I_B \approx \frac{I_S}{\beta_F}(e^{(V_B - V_E)/V_T}) \qquad (3.174)
$$

If $V_{BC} > 4V_T$ and $V_E > V_B$, the transistor is biased deeply into the reverse–active region and (3.173) reduces to

$$
I_C \approx -\frac{I_S}{\alpha_R}(e^{(V_B - V_C)/V_T}) \qquad I_E \approx -I_S(e^{(V_B - V_C)/V_T}) \qquad I_B \approx \frac{I_S}{\beta_R}(e^{(V_B - V_C)/V_T}) \qquad (3.175)
$$

If $V_{BE} > 4V_T$ and $V_{BC} > 4V_T$, the transistor is deeply saturated and (3.173) can be expressed as

$$
\begin{aligned}
I_C &\approx I_S(e^{(V_B - V_E)/V_T}) - \frac{I_S}{\alpha_R}(e^{(V_B - V_C)/V_T}) \\
I_E &\approx \frac{I_S}{\alpha_F}(e^{(V_B - V_E)/V_T}) - I_S(e^{(V_B - V_C)/V_T}) \\
I_B &\approx \frac{I_S}{\beta_F}(e^{(V_B - V_E)/V_T}) + \frac{I_S}{\beta_R}(e^{(V_B - V_C)/V_T})
\end{aligned}
\qquad (3.176)
$$

If $V_{BE} < -4V_T$ and $V_{BC} < -4V_T$, the transistor is deeply into the cutoff region and (3.173) can be expressed as

$$
I_C \approx \frac{I_S}{\beta_R} \qquad I_E \approx \frac{I_S}{\beta_F} \qquad I_B \approx -\frac{I_S}{\beta_F} - \frac{I_S}{\beta_R} \qquad (3.177)
$$

For the PNP transistor shown in Figure 3.86, the forward current component, I_F, is given by

$$
I_F = I_S(e^{V_{EB}/V_T} - 1) = I_S(e^{(V_E - V_{EB})/V_T} - 1) \qquad (3.178)
$$

and the reverse current component, I_R, is given by

$$I_R = I_S(e^{V_{CB}/V_T} - 1) = I_S(e^{(V_C - V_B)/V_T} - 1) \tag{3.179}$$

From the Ebers–Moll model equations we can derive relationships for small signal parameters. First, we will find the incremental resistance seen looking into the base terminal of an NPN transistor with the emitter voltage held fixed. From (3.174), that is,

$$I_B \approx \frac{I_S}{\beta_F}(e^{(V_B - V_E)/V_T}) \tag{3.180}$$

we obtain this incremental base resistance by differentiating I_B with respect to V_B, which yields

$$r_b = \frac{\partial V_B}{\partial I_B} = \left(\frac{\partial I_B}{\partial V_B}\right)^{-1} = \left(\frac{1}{V_T}\underbrace{\frac{I_S}{\beta_F}(e^{(V_B - V_E)/V_T})}_{I_B}\right)^{-1}$$

or

$$r_b = \frac{V_T}{I_B} \tag{3.181}$$

Next, we shall compute the incremental resistance seen looking into the emitter terminal with the base voltage held constant. From (3.174), that is,

$$I_E \approx \frac{I_S}{\alpha_F}(e^{(V_B - V_E)/V_T}) \tag{3.182}$$

we obtain this incremental emitter resistance by differentiating $-I_E$ with respect to V_E, which yields

$$r_e = \frac{\partial V_E}{\partial(-I_E)} = -\left(\frac{\partial I_E}{\partial V_E}\right)^{-1} = -\left(\left(-\frac{1}{V_T}\right)\left(\underbrace{\frac{I_S}{\alpha_F}(e^{(V_B - V_E)/V_T})}_{I_E}\right)\right)^{-1}$$

or

$$r_e = V_T/I_E \tag{3.183}$$

Now, we shall define the incremental transconductance gain of the NPN transistor as the partial derivative of the collector current with respect to the base voltage. From (3.174)

$$I_C \approx I_S(e^{(V_B - V_E)/V_T}) \tag{3.184}$$

we can obtain this incremental transconductance gain as

$$g_m = \frac{\partial I_C}{\partial V_B} = \frac{1}{V_T} \cdot \underbrace{I_S(e^{(V_B - V_E)/V_T})}_{I_C} \tag{3.185}$$

or

$$g_m = \frac{I_C}{V_T} \tag{3.186}$$

Using the relations

$$I_C = \beta_F I_B = \alpha_F I_E \qquad I_E = \frac{I_C}{\alpha_F} = (\beta_F + 1)I_B \qquad I_B = \frac{I_C}{\beta_F} = \frac{I_E}{\beta_F + 1} \tag{3.187}$$

we can also express these small–signal parameters in terms of one another as

$$r_b = \frac{\beta_F}{g_m} = (\beta_F + 1)r_e \qquad r_e = \frac{\alpha_F}{g_m} = \frac{r_b}{\beta_F + 1} \qquad g_m = \frac{\beta_F}{r_b} = \frac{\alpha_F}{r_e} \tag{3.188}$$

3.18 Schottky Diode Clamp

In the saturation region, the collector–base diode is forward–biased. Due to the large diffusion capacitance, it takes a considerably long time to drive the transistor out of saturation. The *Schottky diode* alleviates this problem if connected between the base and the collector as shown in Figure 3.88.

The Schottky diode has the property that it turns on at a lower voltage than the PN junction. Therefore, when a transistor is in the saturation region, the current between the base and the collector is carried by the Schottky diode.

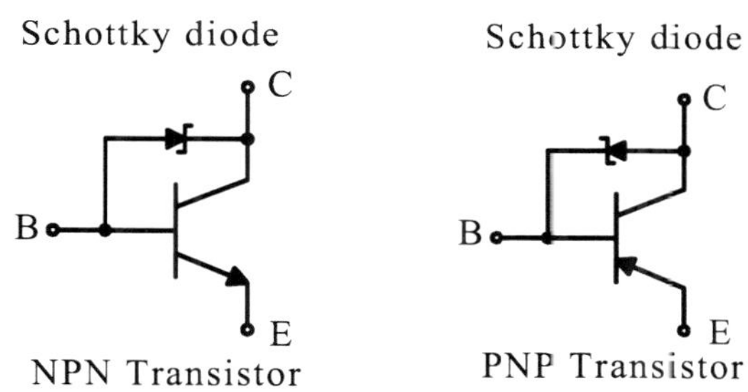

Figure 3.88. NPN and PNP transistors with Schottky diodes

3.19 Transistor Specifications

Transistors are available in a variety of shapes and sizes, each with its own unique characteristics. The specifications usually cover the items listed below, and the values given are typical.

1. Features, e.g., NPN Silicon Epitaxial Planar Transistor for switching and amplifier applications, and mechanical data, e.g., case, weight, and packaging options.

2. Maximum Ratings and Thermal Characteristics, e.g., collector–emitter, collector–base, and emitter base voltages, collector current, power dissipation at room temperature, thermal resistance, etc.

3. Electrical Characteristics, e.g., breakdown voltages, saturation voltages, cutoff currents, noise figure, delay, rise, fall, and storage times, etc. For example, some of the electrical characteristics for the 2N3904 NPN are listed as follows:

h_{FE} (DC current gain): min 100, typical 300 at $V_{CE} = 1\ V$, $I_C = 10\ mA$

h_{ie} (Input impedance): min $1\ K\Omega$, max $10\ K\Omega$ at $V_{CE} = 1\ V$, $I_C = 10\ mA$

h_{re} (Voltage feedback ratio): min 0.5×10^{-4}, max 8×10^{-4} at $V_{CE} = 10\ V$, $I_C = 1\ mA$, $f = 100\ MHz$

h_{fe} (Small signal current gain): min 100, max 400 at $V_{CE} = 10\ V$, $I_C = 1\ mA$, $f = 1\ KHz$

h_{oe} (Output admittance): min $1\ \mu\Omega^{-1}$, max $40\ \mu\Omega^{-1}$ at $V_{CE} = 10\ V$, $I_C = 1\ mA$, $f = 1\ KHz$

f_T (Gain–Bandwidth product): $300\ MHz$ at $V_{CE} = 20\ V$, $I_C = 10\ mA$, $f = 100\ MHz$

3.20 Summary

- Transistors are three terminal devices that can be formed with the combination of two separate PN junction materials into one block.
- An NPN transistor is formed with two PN junctions with the P–type material at the center, whereas a PNP transistor is formed with two PN junctions with the N–type material at the center.
- The three terminals of a transistor, whether it is an NPN or PNP transistor, are identified as the emitter, the base, and the collector.
- Transistors are used either as amplifiers or as electronic switches.
- Like junction diodes, most transistors are made of silicon. Gallium Arsenide (GaAs) technology has been under development for several years and its advantage over silicon is its speed, about six times faster than silicon, and lower power consumption. The disadvantages of GaAs over silicon is that arsenic, being a deadly poison, requires very special manufacturing processes.
- Since a transistor is a 3–terminal device, there are three currents, the base current, denoted as i_B, the collector current, denoted as i_C, and the emitter current, denoted as i_E.
- For any transistor, NPN or PNP, the three currents are related as

$$i_B + i_C = i_E$$

- In a transistor the collector current is defined as

$$i_C = I_r e^{v_{BE}/V_T}$$

where I_r is the reverse (saturation) current, typically 10^{-12} A to 10^{-15} A as in junction diodes, v_{BE} is the base–to–emitter voltage, and $V_T \approx 26\ mV$ at $T = 300\ °K$.

- A very useful parameter in transistors is the common–emitter gain β, a constant whose value ranges from 75 to 300. Its value is specified by the manufacturer. The base current i_B is much smaller than the collector current i_C and these two currents are related in terms of the constant β as

$$i_B = i_C/\beta$$

- Another important parameter in transistors is the *common–base current gain* denoted as α and relates the collector and emitter currents as

$$i_C = \alpha i_E$$

- The parameters α and β are related as

$$\alpha = \frac{\beta}{\beta + 1}$$

$$\beta = \frac{\alpha}{1 - \alpha}$$

- As with diodes, the base–emitter voltage v_{BE} decreases approximately 2 mV for each 1 °C rise in temperature when the emitter current i_E remains constant.
- In a transistor, the output resistance looking into the collector is defined as

$$r_{out} = \left.\frac{\partial v_{CE}}{\partial i_C}\right|_{v_{BE} = constant} = \frac{V_A}{I_C}$$

where V_A is the *Early voltage* supplied by the manufacturer, and I_C is the DC collector current.

- Two common methods of biasing a transistor are the fixed bias method and the self–bias method.
- There are four classes of amplifier operations: Class A, Class AB, Class B, and Class C.
- Class A amplifiers operate entirely in the active region and thus the output if a faithful reproduction of the input signal with some amplification. In a Class A amplifier the efficiency is very low. Class A amplifiers are used for audio and frequency amplification.
- Class AB amplifiers are biased so that the collector current is cutoff for a portion of one cycle of the input and so the efficiency is higher than that of a Class A amplifier. Class AB amplifiers are normally used as push–pull amplifiers to alleviate the crossover distortion of Class B amplifiers.
- Class B amplifiers are biased so that the collector current is cutoff during one–half of the input signal. Therefore, the efficiency in a Class B amplifier is higher than that of a Class AB amplifier. Class B amplifiers are used in amplifiers requiring high power output.
- Class C amplifiers are biased so that the collector current flows for less than one–half cycle of the input signal. Class C amplifiers have the highest efficiency and are used for radio frequency amplification in transmitters.
- The operation of a simple transistor circuit can also be described graphically using i_B versus v_{BE} and i_C versus v_{CE} curves.
- The average power drawn from the collector supply is

$$P_{CC} = V_{CC}I_C$$

- The average power absorbed by the load is

$$P_{LOAD} = R_{LOAD} I_C^2 + R_{LOAD} (i_{c\ RMS})^2$$

- The average power absorbed by the transistor is

$$P_C = P_{CC} - P_L$$

- The piecewise–linear analysis is a practical method of analyzing transistor amplifiers.
- The incremental model for the transistor provides a better approximation than the piece–wise linear approximation.
- In the incremental model for the transistor the input resistance r_n is the slope of the input voltage and current characteristics and it accounts for the voltage drop across the base–emitter junction. Likewise, the output conductance g_o is the slope of the output current and voltage characteristics. The voltage amplification factor μ is related to the input characteristics caused by a change in v_{CE}, and the current amplification factor β is related to the output characteristics caused by a change in i_B.
- Transistor hybrid parameters provide us with a means to evaluate voltages, currents, and power in devices that are connected externally to the transistor. The parameters r_n, μ, β, and g_o are normally denoted by the h (hybrid) parameters as $r_n = h_{11} = h_{ie}$, $\mu = h_{12} = h_{re}$, $\beta = h_{21} = h_{fe}$, and $g_o = h_{22} = h_{oe}$. These designations along with the additional notations $v_{be} = v_1$, $i_b = i_1$, $v_{ce} = v_2$, and $i_c = i_2$, provide a symmetrical form for the relations of (3.62)and (3.63) as follows:

$$v_1 = h_{11} i_1 + h_{12} v_2$$
$$i_2 = h_{21} i_1 + h_{22} v_2$$

- In the incremental model for the transistor the current amplification A_c is defined as

$$A_c = \frac{i_c}{i_s}$$

- The transconductance, denoted as g_m, and defined as

$$g_m = \left.\frac{di_C}{dv_{BE}}\right|_{i_C = I_C} = \frac{i_C}{V_T}$$

An approximate value for the transconductance at room temperature is $g_m \approx 40 I_C$.

- The high–frequency models for transistors take into consideration the nonlinear resistance r_{be} that accounts for the voltage drop across the emitter junction, and the capacitance C_e that exists across the forward–biased emitter junction. For higher frequencies, the effects of junction capacitances must be taken into account. Accordingly, the β cutoff frequency, denoted as ω_β, is defined as

$$\omega_\beta \equiv \frac{1}{r_{be}C_e}$$

- If i_C is in milliamps,

$$r_{be} = \frac{25\beta}{i_C}$$

- The current gain–bandwidth product, denoted as ω_T, is an important figure of merit for a transistor and it can be found from the relations

$$\omega_T \approx \beta\omega_\beta = \frac{g_m}{C_e}$$

- The Darlington connection consists of two transistors with a common collector point and exhibits a high input impedance and low output impedance. The combined α_T and β_T are

$$\alpha_T = \alpha_1 + \alpha_2 - \alpha_1\alpha_2$$

and

$$\beta_T \approx \beta_1 + \beta_1\beta_2$$

- The common–base, common–emitter, and common collector transistor circuits can also be represented by their h–equivalent and T–equivalent circuits.
- The Ebers–Moll model of the bipolar transistor provides a simple, closed–form expression for the collector, base, and emitter currents of a bipolar transistor in terms of the terminal voltages. It is valid in all regions of operation of the bipolar transistor, transitioning between them smoothly.
- The Schottky diode has the property that it turns on at a lower voltage than the PN junction. Therefore, when a transistor is in the saturation region, the current between the base and the collector is carried by the Schottky diode.

3.21 Exercises

1. It is known that in a NPN transistor, when the base–to–emitter voltage $v_{BE} = 0.7\ V$, the collector current $i_C = 1.2\ mA$ at temperature $T = 27\ °C$. Find the range of changes in v_{BE} at this temperature for the range $0.2 \le i_C \le 5\ mA$.

2. It is known that in a NPN transistor, when the base–to–emitter voltage $v_{BE} = 0.7\ V$, the emitter current $i_E = 1.2\ mA$, and $i_B = 12\ \mu A$ at temperature $T = 27\ °C$. Find β, α, and I_r.

3. For the NPN transistor circuit below, it is known that $V_E = -0.72\ V$ and $\beta = 115$. The circuit operates at $T = 27\ °C$. Find I_E, I_B, I_C, and V_C.

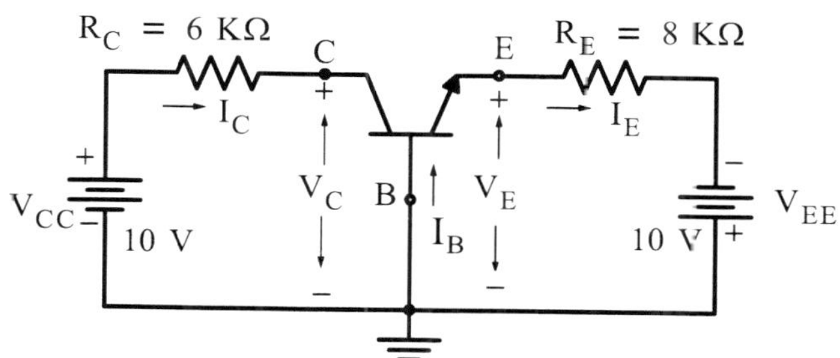

4. For the PNP transistor circuit below, it is known that $R_B = 130\ K\Omega$, $V_E = 2\ V$ and $V_{EB} = 0.7\ V$. The circuit operates at $T = 27\ °C$. Find α, β, and V_C.

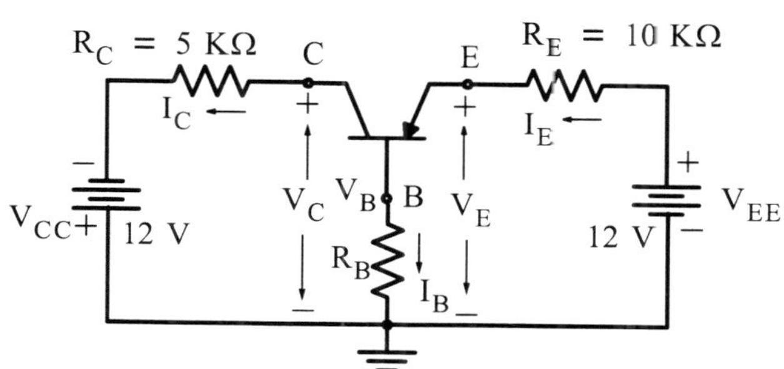

5. Find the output resistance r_{out} for a bipolar junction transistor whose Early voltage is $V_A = 75\ V$ and the DC collector current is

   a. $I_C = 0.1\ mA$

   b. $I_C = 1\ mA$

   c. $I_C = 5\ mA$

6. For the circuit below, $V_B = 3.7\ V$, $R_E = 3\ K\Omega$, and $\beta = 110$. Find V_E, I_E, I_C, V_C, I_B, and determine whether this circuit with the indicated values operates in the active, saturation, or cutoff mode.

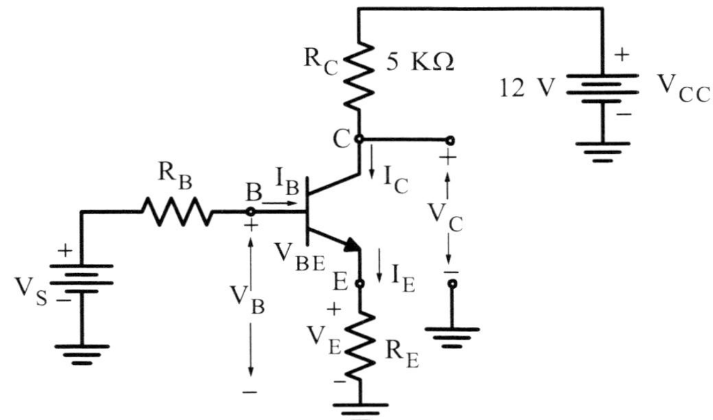

7. For the circuit below, $\beta = 150$. Find I_E, V_E, I_C, V_C, I_B, and determine whether this circuit with the indicated values operates in the active, saturation, or cutoff mode. Hint: The base–to–emitter junction of an NPN transistor is considered to be reverse–biased if $V_B = V_E = 0$

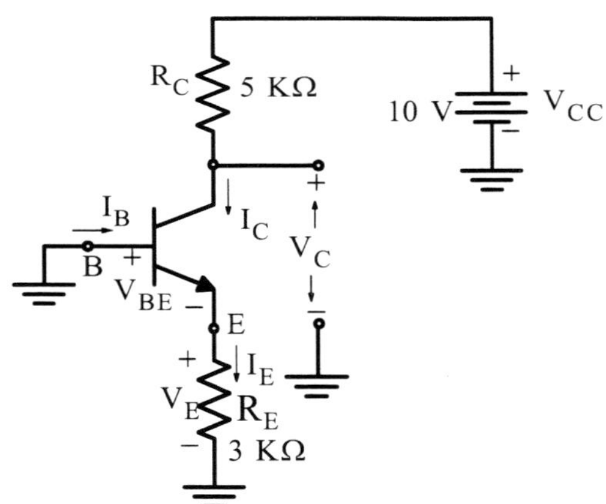

8. For the circuit below, $\beta = 110$. Find the highest voltage to which the base can be set so that the transistor will be in the active mode. Hint: The collector–base junction will still be reverse–biased if we make $V_B = V_C$.

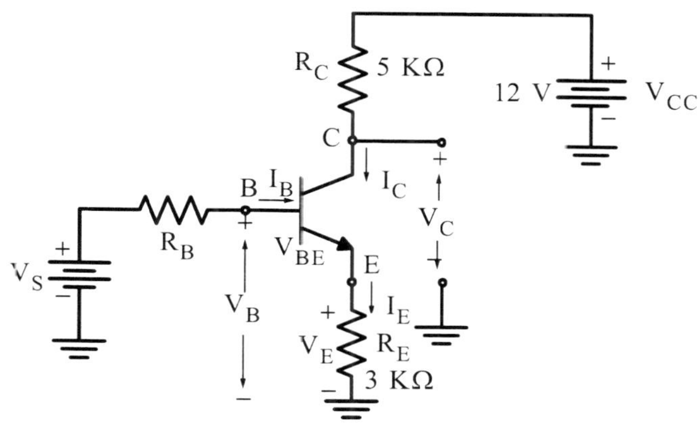

9. For an NPN transistor circuit $\beta = 100$, $V_{CC} = 11.3\ V$, $V_E = 3.5\ V$ and with the reverse–biased collector–base junction set at $V_{CB} = 1.8\ V$ we want the collector current to be $I_C = 0.8\ mA$. What should the values of R_C and R_E be to achieve this value?

10. For a PNP transistor circuit with $\beta = 120$, $V_{EE} = 12\ V$, $V_B = 0\ V$, $V_{CC} = -12\ V$, and $R_E = 3\ K\Omega$, what would the largest value of R_C be so that the transistor operates at the active mode?

11. For a PNP transistor circuit with $\beta = 150$, $V_{EE} = 12\ V$, $V_B = 0\ V$, $V_{CC} = -12\ V$, $I_E = 0.8\ mA$, and $V_{CB} = -3\ V$, what should the values of R_C and R_E be so that the transistor operates at the active mode?

12. For the circuit below, it is known that $\beta = 120$ for both transistors. Find all indicated voltages and currents. Are both the transistors operating in the active mode? What is the total power absorbed by this circuit?

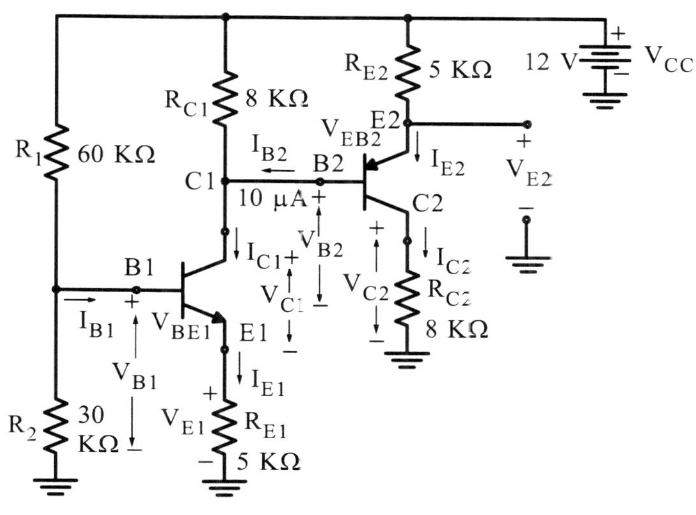

13. For the NPN transistor circuit below, $\beta = 100$ and the output volt–ampere characteristic curves are approximately horizontal lines.

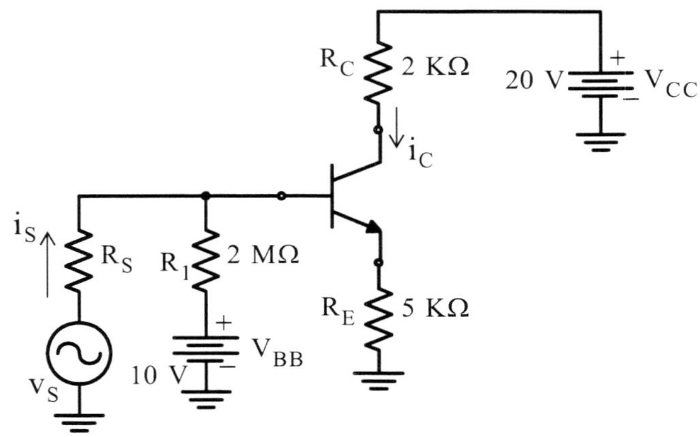

a. Sketch a family of these curves for $i_B = 2.5, 5.0, 7.5,$ and $10\ \mu A$. Construct a load line, and indicate the current and the voltage at which the load line intersects the axes.

b. Find the quiescent collector current I_C and collector–to–emitter voltage V_{CE}.

c. Determine graphically and plot i_C versus i_S for $-15 < i_S < 15\ \mu A$.

14. For the PNP transistor circuit below, $\beta = 70$ and the output volt–ampere characteristic curves are approximately horizontal lines.

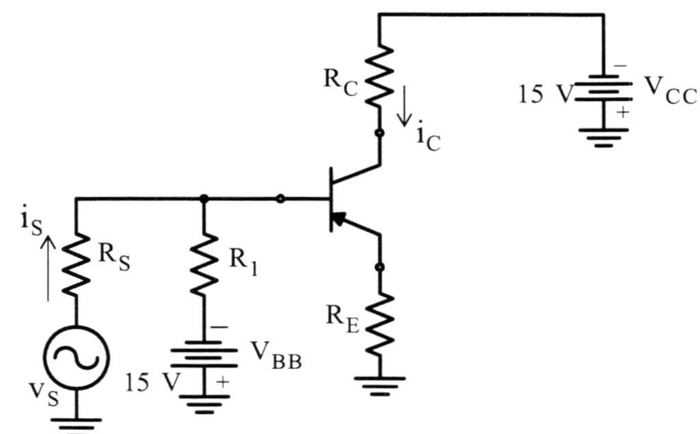

a. Find the values of R_1 and R_C required to locate the quiescent operating point Q at $i_C = 1$ mA and $v_{EC} = 5$ V.

b. Sketch the load line on the i_C versus v_{EC} coordinates. Show the current and voltage at which the load line intersects the axes, and indicate the quiescent point on this line. It is not necessary to draw the collector characteristics.

c. If the input signal is a sinusoidal current, approximately what is the greatest amplitude that the signal i_s can have without waveform distortion at the output?

15. For the transistor model shown below, it is known that $\beta = 80$, $i_c = 5\ mA$, $r'_b = 50\ \Omega$, and $r_o = 30\ K\Omega$.

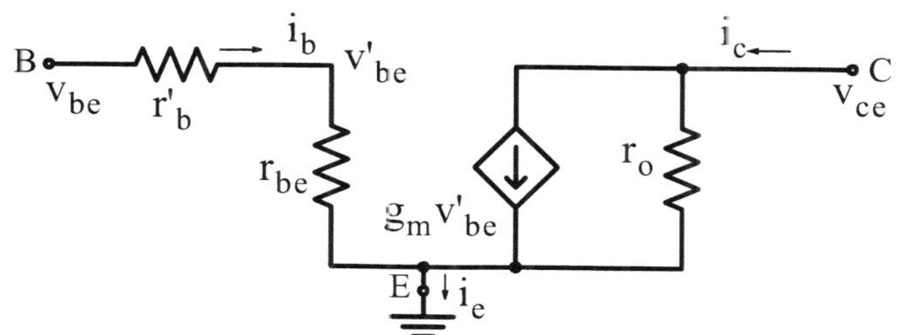

a. Find the transconductance g_m and the base–emitter resistance r_{be}.

b. Repeat part (a) for $\beta = 60$, $i_c = 1\ mA$, $r'_b = 50\ \Omega$, and $r_o = 30\ K\Omega$.

16. For the transistor circuit below, $i_C = 1\ mA$ and the hybrid parameters are $r_n = 2.5\ K\Omega$, $\mu = 2 \times 10^{-4}$, $\beta = 100$, and $r_o = 100\ K\Omega$.

a. Find the values of R_B and R_C required for a quiescent point Q at $i_C = 1\ mA$ and $v_{CE} = 5\ V$.

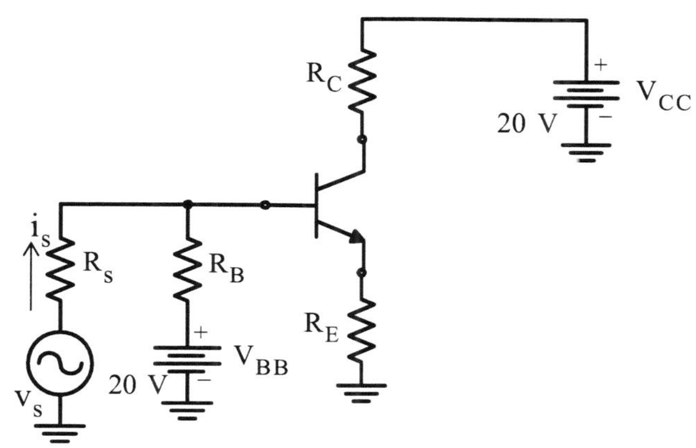

b. Using the hybrid representation for the transistor, draw an incremental model.

c. Find the current amplification A_c.

17. Determine whether the PNP transistor shown below is operating in the cutoff, active, or saturation region

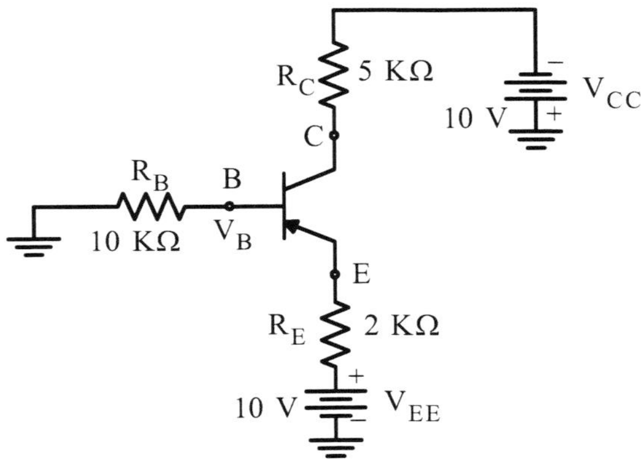

18. The equations describing the h parameters can be used to represent the network shown below. This network is a transistor equivalent circuit for the common–emitter configuration and the h parameters given are typical values for such a circuit. Compute the voltage and current gains for this network if a voltage source of $v_1 = \cos\omega t \text{ mV}$ in series with $800\ \Omega$ is connected at the input (left side), and a $5\ K\Omega$ load is connected at the output (right side).

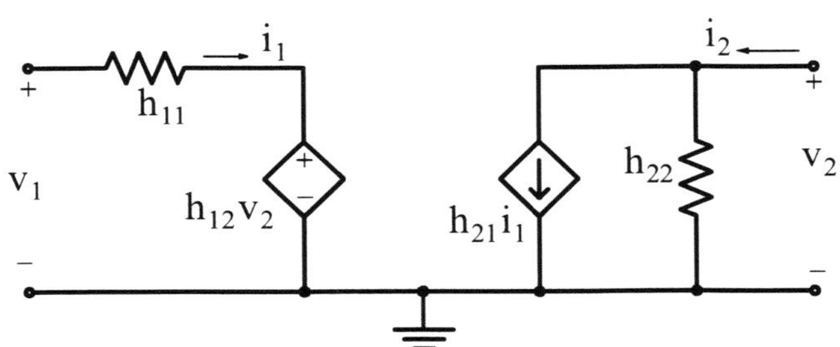

$$h_{11} = 1.2\ K\Omega$$

$$h_{12} = 2 \times 10^{-4}$$

$$h_{21} = 50$$

$$h_{22} = 50 \times 10^{-6}\ \Omega^{-1}$$

3.22 Solutions to End–of–Chapter Exercises

1. From (3.2)

$$i_C = I_r e^{v_{BE}/V_T} \quad (1)$$

and as we know from Chapter 2 at $T = 27\ °C$, $V_T = 26\ mV$. From the given data

$$1.2 \times 10^{-3} = I_r e^{0.7/V_T} \quad (2)$$

Division of (1) by (2) yields

$$\frac{i_C}{1.2 \times 10^{-3}} = \frac{e^{v_{BE}/V_T}}{e^{0.7/V_T}}$$

$$i_C = 1.2 \times 10^{-3} e^{(v_{BE} - 0.7)/V_T}$$

$$\ln(i_C) = \ln(1.2 \times 10^{-3} e^{(v_{BE} - 0.7)/V_T}) = \ln(1.2 \times 10^{-3}) + (v_{BE} - 0.7)/V_T$$

$$\ln(i_C) - \ln(1.2 \times 10^{-3}) = (v_{BE} - 0.7)/V_T$$

$$\ln\left(\frac{i_C}{1.2 \times 10^{-3}}\right) = (v_{BE} - 0.7)/V_T$$

$$V_T \ln\left(\frac{i_C}{1.2 \times 10^{-3}}\right) = v_{BE} - 0.7$$

$$v_{BE} = 0.7 + V_T \ln\left(\frac{i_C}{1.2 \times 10^{-3}}\right) = 0.7 + 26 \times 10^{-3} \ln\left(\frac{i_C}{1.2 \times 10^{-3}}\right) \quad (3)$$

With $i_C = 0.2\ mA$, from (3),

$$v_{BE} = 0.7 + 26 \times 10^{-3} \ln\left(\frac{0.2 \times 10^{-3}}{1.2 \times 10^{-3}}\right) = 0.7 + 26 \times 10^{-3} \times (-1.792) = 0.653\ V$$

and with $i_C = 5\ mA$,

$$v_{BE} = 0.7 + 26 \times 10^{-3} \ln\left(\frac{5 \times 10^{-3}}{1.2 \times 10^{-3}}\right) = 0.7 + 26 \times 10^{-3} \times 1.427 = 0.737\ V$$

Therefore, for $0.2 \le i_C \le 5\ mA$, the v_{BE} range is $0.653 \le v_{BE} \le 0.737\ V$

2. From (3.1)

$$i_B + i_C = i_E$$

$$i_C = i_E - i_B = 1.2 \times 10^{-3} - 0.012 \times 10^{-3} = 1.188 \times 10^{-3} = 1.188 \text{ mA}$$

From (3.3),

$$i_B = i_C / \beta$$

$$\beta = i_C / i_B = 1.188 \text{ mA} / 12 \text{ }\mu\text{A} = 99$$

From (3.7),

$$\alpha = \beta / (\beta + 1) = 99 / (99 + 1) = 0.99$$

From (3.2),

$$i_C = I_r e^{v_{BE}/V_T}$$

$$I_r = \frac{i_C}{e^{v_{BE}/V_T}} = \frac{1.188 \times 10^{-3}}{e^{0.7/(26 \times 10^{-3})}} = \frac{1.188 \times 10^{-3}}{e^{26.92}} = \frac{1.188 \times 10^{-3}}{4.911 \times 10^{11}} = 2.42 \times 10^{-15} \text{ A}$$

3.

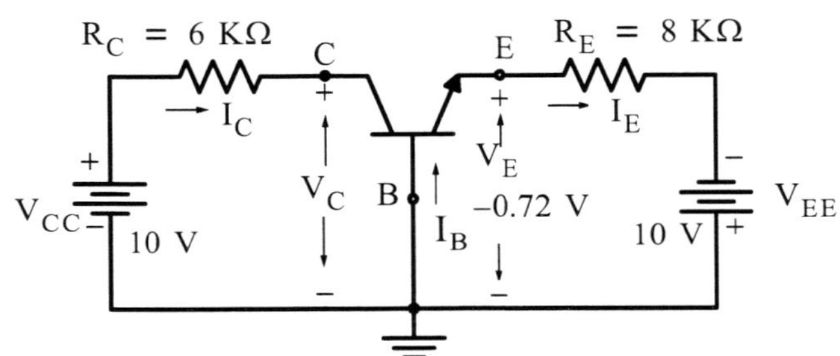

$$-V_E + R_E I_E = V_{EE}$$

$$I_E = \frac{V_{EE} + V_E}{R_E} = \frac{10 - 0.72}{8 \times 10^3} = 1.16 \text{ mA}$$

From Table 3.1, $I_B = I_E / (\beta + 1)$ and with $\beta = 115$ we find that

$$I_B = \frac{1.16 \text{ mA}}{115 + 1} = 10^{-5} \text{ A} = 10 \text{ }\mu\text{A}$$

$$I_C = I_E - I_B = 1.16 \times 10^{-3} - 10^{-5} = 1.15 \times 10^{-3} = 1.15 \text{ mA}$$

By application of KVL on the collector (left) side of the circuit above, we obtain

$$V_C = V_{CC} - R_C I_C = 10 - 6 \times 10^3 \times 1.15 \times 10^{-3} = 3.1 \text{ V}$$

4.

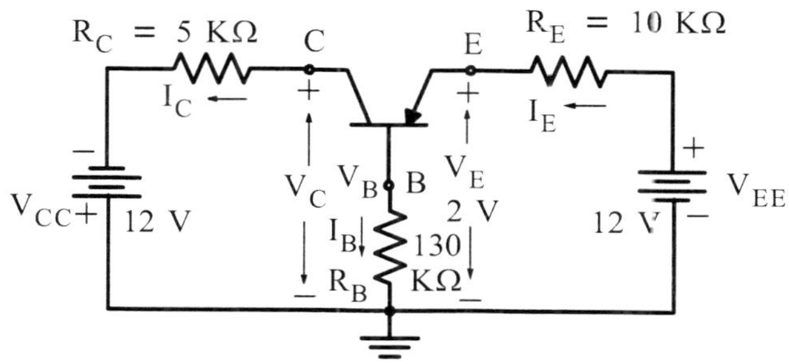

From Table 3.1, $\alpha = I_C/I_E$, $\beta = I_C/I_B$, and from the circuit above, $V_C = R_C I_C - 12$, and $V_B = V_E - V_{BE} = 2 - 0.7 = 1.3\ V$. Therefore, we need to find I_E, I_B, and I_C.

$$R_E I_E + V_E = 12\ V$$

$$I_E = \frac{12-2}{10\ K\Omega} = 1\ mA$$

$$I_B = \frac{V_B}{R_B} = \frac{1.3\ V}{130\ K\Omega} = 0.01\ \ mA$$

$$I_C = I_E - I_B = 1 - 0.01 = 0.99\ mA$$

$$\alpha = \frac{I_C}{I_E} = \frac{0.99}{1} = 0.99$$

$$\beta = \frac{I_C}{I_B} = \frac{0.99}{0.01} = 99$$

$$V_C = R_C I_C - 12 = 5\times 10^3 \times 0.99 \times 10^{-3} - 12 = -7.05\ V$$

5.

a.

$$r_{out} = \left.\frac{V_A}{I_C}\right|_{I_C = 0.1\ mA} = \frac{75}{10^{-4}} = 750\ K\Omega$$

b.

$$r_{out} = \left.\frac{V_A}{I_C}\right|_{I_C = 1\ mA} = \frac{75}{10^{-3}} = 75\ K\Omega$$

c.

$$r_{out} = \left.\frac{V_A}{I_C}\right|_{I_C = 5\ mA} = \frac{75}{5\times 10^{-3}} = 5\ K\Omega$$

6.

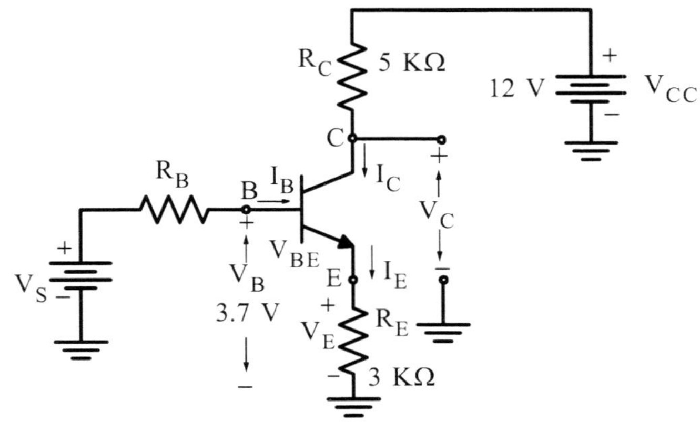

Since $V_B = 3.7\ V$ it is reasonable to assume that the base–emitter junction is forward–biased and thus $V_{BE} = 0.7\ V$. Then,

$$V_E = V_B - V_{BE} = 3.7 - 0.7 = 3\ V$$

and

$$I_E = \frac{V_E}{R_E} = \frac{3}{3 \times 10^{-3}} = 1\ mA$$

It is given that $\beta = 110$. Then,

$$\alpha = \frac{\beta}{\beta + 1} = \frac{110}{111} = 0.991$$

and

$$I_C = \alpha I_E = 0.991 \times 1 = 0.991\ mA$$

The collector voltage is

$$V_C = V_{CC} - R_C I_C = 12 - 5 \times 10^3 \times 0.991 \times 10^{-3} = 7.05V$$

This is an NPN transistor and the collector (N) to base (P) must be reverse–biased for the transistor to operate in the active mode. Since $V_C > V_B$, the transistor is indeed operating in the active mode and our assumption that the base–emitter junction is forward–biased, is correct.

Finally,

$$I_B = I_E - I_C = 1 - 0.991 = 9\ \mu A$$

7.

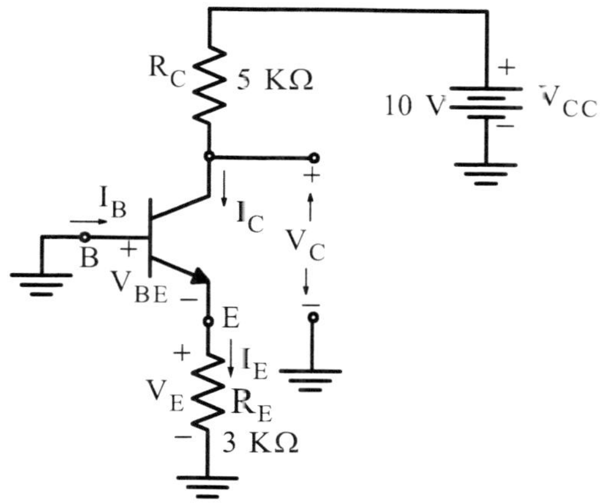

Since the base is grounded, $V_B = 0$, the emitter-base junction does not conduct, $V_{BE} = 0$, $I_B = 0$, $I_E = 0$, and $V_E = 0$. The collector current I_C is also zero because $I_C = I_E - I_B = 0$ and thus $V_C = V_{CC} - R_C I_C = 10V$. Under those conditions the transistor behaves like an open switch and thus it is operating in the cutoff mode.

8.

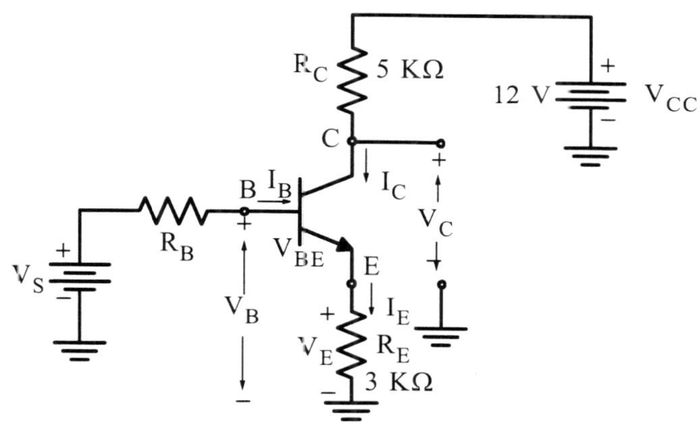

The transistor will be in the active mode for $V_B > 0.7V$ and for $V_B \le V_C$. Since $\beta = 110$,

$$\alpha = \frac{\beta}{\beta + 1} = \frac{110}{111} = 0.991$$

and

$$V_B = V_{BE} + V_E = V_{BE} + R_E I_E$$

$$I_E = \frac{V_B - V_{BE}}{R_E} = \frac{V_B - 0.7}{3 \times 10^3}$$

Then,

$$I_C = \alpha I_E = 0.991\frac{V_B - 0.7}{3 \times 10^3}$$

$$V_C = V_{CC} - R_C I_C = V_B = 5 \times 10^3 \times 0.991\frac{V_B - 0.7}{3 \times 10^3} = 5 \times 0.991\frac{V_B - 0.7}{3}$$

$$= 1.652(V_B - 0.7)$$

$$V_B = 12 - 5 \times 10^3 \times 0.991\frac{V_B - 0.7}{3 \times 10^3} = 12 - 5 \times 0.991\frac{V_B - 0.7}{3}$$

$$= (12 - 1.652(V_B - 0.7))$$

$$V_B + 1.652(V_B - 0.7) = 12$$

$$2.652 V_B = 10.84$$

or

$$V_B = \frac{10.84}{2.652} = 4.09 \text{ V}$$

9.

$$V_C = V_{CC} - R_C I_C = 11.3 - 0.8 \times 10^{-3} R_C \quad (1)$$

Also

$$V_C = V_{CB} + V_B = 1.8 + 3.5 = 5.3 \text{ V} \quad (2)$$

From (1) and (2)

$$R_C = \frac{11.3 - 5.3}{0.8 \times 10^{-3}} = 7.5 \text{ K}\Omega$$

With $\beta = 100$,

$$\alpha = \frac{\beta}{\beta + 1} = \frac{100}{101} = 0.99$$

then,

$$I_E = I_C / \alpha = 0.8/0.99 = 0.81 \text{ mA}$$

The emitter voltage is

$$V_E = V_B - V_{BE} = 3.5 - 0.7 = 2.8 \text{ V}$$

and thus

$$R_E = \frac{V_E}{I_E} = \frac{2.8}{0.81 \times 10^{-3}} = 3.5 \text{ K}\Omega$$

10.

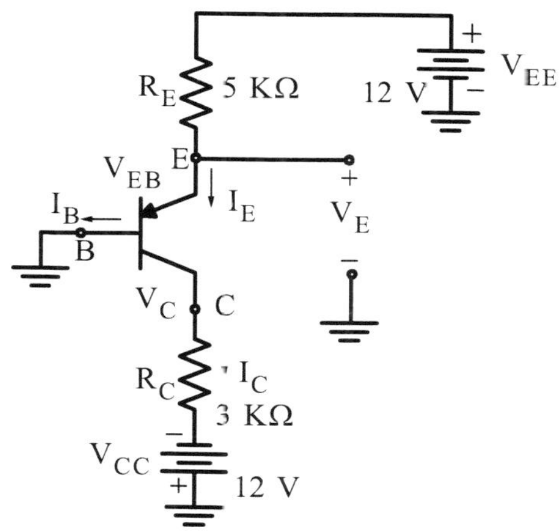

A PNP transistor will still be in the active mode if $V_C = V_B$. For this exercise the base is grounded so $V_B = 0$ and we can let $V_C \approx 0$ also. Then,

$$V_C = R_C I_C - V_{CC} = R_C \alpha I_E - 12 \approx 0$$

or

$$R_C = \frac{12}{\alpha I_E} \quad (1)$$

where

$$\alpha = \frac{\beta}{\beta + 1} = \frac{120}{121} = 0.992 \quad (2)$$

and from

$$V_E = V_{EB} = 0.7 = V_{EE} - R_E I_E = 12 - R_E I_E$$

we obtain

$$I_E = \frac{12 - 0.7}{3 \times 10^3} = 3.77\text{ mA} \quad (3)$$

By substitution of (2) and (3) into (1) we obtain

$$R_C = \frac{12}{\alpha I_E} = \frac{12}{0.992 \times 3.77 \times 10^{-3}} = 3.21\text{ K}\Omega$$

11.

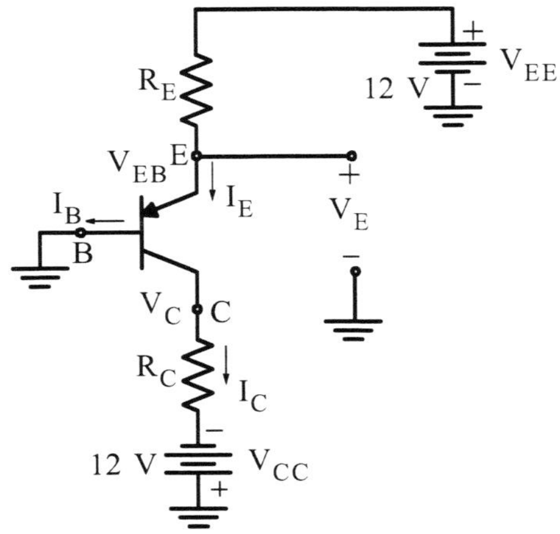

Since the base is grounded,

$$V_C = V_{CB} = -3\ V$$

and

$$V_E = V_{EB} = 0.7\ V$$

Then,

$$R_E I_E + V_E = V_{EE} = 12\ V$$

$$R_E = \frac{12 - 0.7}{0.8 \times 10^{-3}} = 14\ K\Omega$$

Also, since

$$I_C = \alpha I_E = \frac{\beta}{\beta + 1} I_E = 0.993 \times 0.8 = 0.795\ mA$$

then,

$$V_C + R_C I_C = -V_{CC}$$

or

$$R_C = \frac{-(-12) - 3}{0.795 \times 10^{-3}} = 11.3\ K\Omega$$

12.

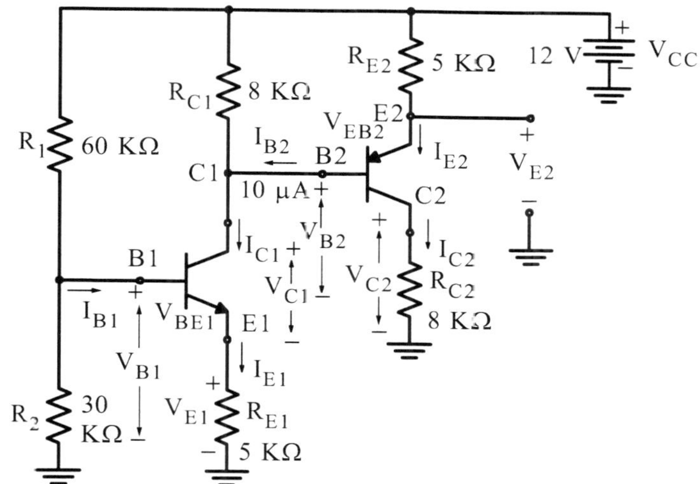

The left part in the circuit above is the same as that in Example 3.8 where we applied Thevenin's theorem. Therefore the given circuit is redrawn as shown below.

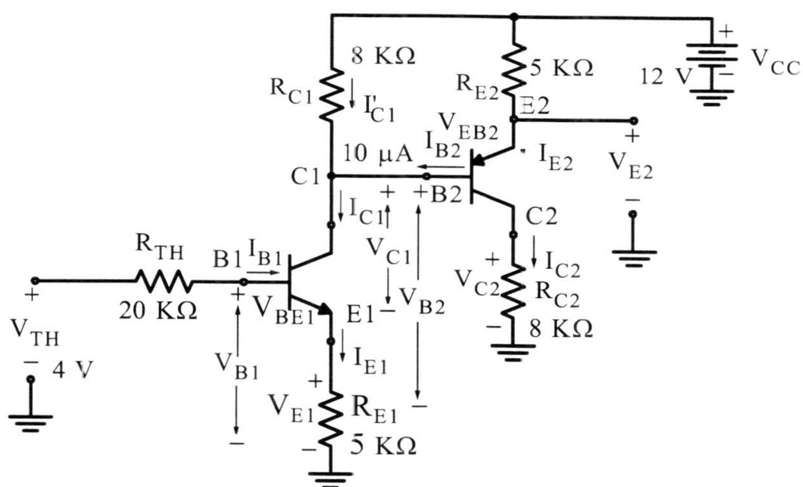

Application of KVL around the left part of the circuit yields

$$R_{TH}I_{B1} + V_{BE1} + R_{E1}I_{E1} = V_{TH}$$

$$(20 \times 10^3)I_{B1} + 0.7 + (5 \times 10^3)I_{E1} = 4$$

$$4I_{B1} + I_{E1} = \frac{3.3}{5 \times 10^3} = 0.66 \times 10^{-3}$$

From Table 3.1, $I_E = (\beta + 1)I_B$. Then,

$$4I_{B1} + (\beta + 1)I_{B1} = 0.66 \times 10^{-3}$$

$$125I_{B1} = 0.66 \times 10^{-3}$$

$$I_{B1} = 5.28 \times 10^{-6}\ A = 5.28\ \mu A$$

and

$$I_{E1} = (\beta + 1)I_{B1} = 121 \times 5.28 \times 10^{-6} = 0.639\ mA$$

Then,

$$V_{E1} = R_{E1}I_{E1} = 5 \times 10^{3} \times 0.639 \times 10^{-3} = 3.2\ V$$

and

$$V_{B1} = V_{BE1} + V_{E1} = 0.7 + 3.2 = 3.9\ V$$

Also,

$$I_{C1} = I_{E1} - I_{B1} = 0.639 \times 10^{-3} - 5.28 \times 10^{-6} = 0.634\ mA$$

Then,

$$I'_{C1} = I_{C1} - I_{B2} = 0.634\ mA - 10\ \mu A = 0.614\ mA$$

and

$$V_{C1} = V_{CC} - R_{C1}I'_{C1} = 12 - 8 \times 10^{3} \times 0.614 \times 10^{-3} = 7.09\ V$$

Since this is an NPN transistor and $V_{C1} > V_{B1}$, the base–collector PN junction is reverse–biased and thus the transistor is in active mode.

The emitter voltage V_{E2} of the PNP transistor is

$$V_{E2} = V_{EB2} + V_{C1} = 0.7 + 7.09 = 7.79\ V$$

and the emitter current I_{E2} is

$$I_{E2} = \frac{V_{CC} - V_{E2}}{R_{E2}} = \frac{12 - 7.79}{5 \times 10^{3}} = 0.842\ mA$$

It is given that $\beta = 120$. Then,

$$I_{C2} = \alpha I_{E2} = \frac{\beta}{\beta + 1}I_{E} = 0.992 \times 0.842 = 0.835\ mA$$

and

$$V_{C2} = R_{C2}I_{C2} = 8 \times 10^{3} \times 0.835 \times 10^{-3} = 6.68\ V$$

To find V_{B2} we observe that

$$V_{EC2} = V_{E2} - V_{C2} = 7.79 - 6.68 = 1.11\ V$$

Then,

$$V_{BC2} = V_{EC2} - V_{EB2} = 1.11 - 0.7 = 0.41 \text{ V}$$

and

$$V_{B2} = V_{BC2} + V_{C2} = 0.41 + 6.68 = 7.09 \text{ V}$$

As expected, this is the same value as that of V_{C1} as it can be seen from the circuit above.

This is an PNP transistor and since $V_{C2} < V_{B2}$ the base–collector junction is reverse–biased and the PNP transistor is also in the active mode. The total power absorbed by this circuit is

$$P = V_{CC}I_T = V_{CC}(I_{B1} + I'_{C1} + I_{E2}) = 12(5.28\ \mu\text{A} + 0.614\ \text{mA} + 0.842\ \text{mA}) = 17.5\ \text{mw}$$

13.

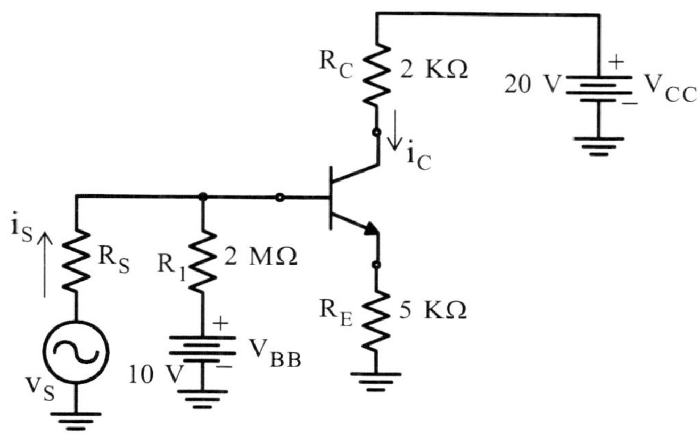

a. With $\beta = 100$, $i_C = \beta i_B = 100 i_B$ and for $i_B = 2.5, 5.0, 7.5,$ and $10\ \mu\text{A}$, we obtain

$$i_C = 0.25, 0.50, 0.75, \text{ and } 1.00 \text{ mA}$$

and these are shown on the plot below.

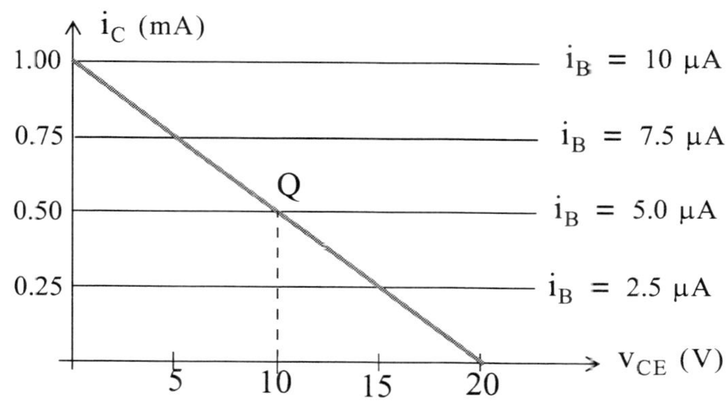

The load line is obtained from the relation

$$v_{CE} + R_C i_C = V_{CC} = 20 \text{ V}$$

When $i_C = 0$, $v_{CE} = 20\ V$, and when $v_{CE} = 0$, $i_C = V_{CC}/R_C = 20/20\ K\Omega = 1\ mA$. The load line then intercepts the v_{CE} axis at $20\ V$ and the i_C axis at $1\ mA$.

b. As shown above, the Q point is the intersection of the load line and the line $i_B = I_B$. Thus with reference to the circuit above and $i_s = 0$, $I_B = 10\ V/2\ M\Omega = 5\ \mu A$, and the Q point is as shown above and we observe that $I_C = 0.5\ mA$ and $V_{CE} = 10\ V$.

c.

$$i_B = I_B + i_s = \frac{V_{BB}}{R_1} + i_s = 5 + i_s$$

and

$$i_C = \beta i_B = \beta(5 + i_s)$$

Then, for the range $-15 < i_s < 15\ \mu A$, we obtain the corresponding collector current values as shown on the table below.

$i_s\ (\mu A)$	−15	−10	−5	0	5	10	15
$i_C\ (mA)$	0	0	0	5	10	10	10

The plot below shows the current transfer characteristics

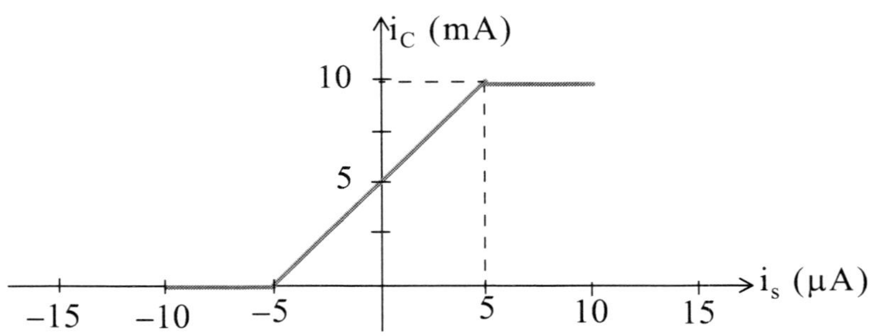

14.

a.

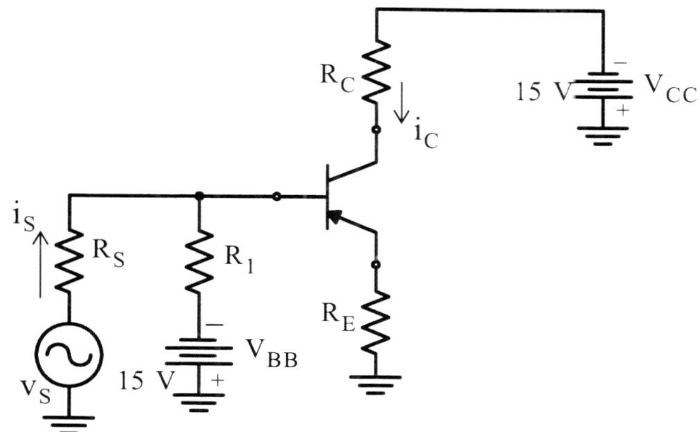

$$v_{EC} + R_C i_C = V_{CC}$$

$$5 + 10^{-3} R_C = 15$$

$$R_C = \frac{15 - 5}{10^{-3}} = 10\ K\Omega$$

and with $\beta = 70$

$$R_1 = \frac{V_{BB}}{I_B} = \frac{15}{I_C/70} = \frac{15 \times 70}{10^{-3}} = 1.05\ M\Omega$$

b.

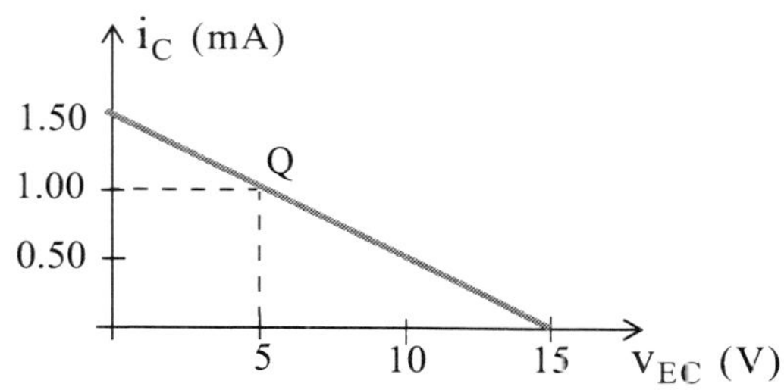

The slope of the load line is

$$m = \frac{i_{C2} - i_{C1}}{v_{EC2} - v_{EC1}} = \frac{0 - 1}{15 - 5} = -0.1$$

The i_C intercept, denoted as i_{CQ}, is found from the straight line equation

$$i_C = m v_{EC} + i_{CQ}$$

or

$$i_{CQ} = i_C - m v_{EC}$$

and with $i_C = 1 \text{ mA}$ and $v_{CE} = 5$

$$i_{CQ} = 1 - (-0.1) \times 5 = 1.5 \text{ mA}$$

c.

From the plot of part (b) we see that when $v_{EC} = 0$ (short circuit) the collector current is $i_C = 1.5 \text{ mA}$, and since $i_B = i_C/\beta = 1.5/70 = 21.4 \ \mu\text{A}$, and

$$i_B = I_B + i_s = \frac{V_{BB}}{R_1} = \frac{15}{1.05 \times 10^6} + i_s = 14.3 \ \mu\text{A} + i_s$$

we obtain

$$i_s = i_B - I_B = 21.4 - 14.3 = 7.1 \ \mu\text{A}$$

15.

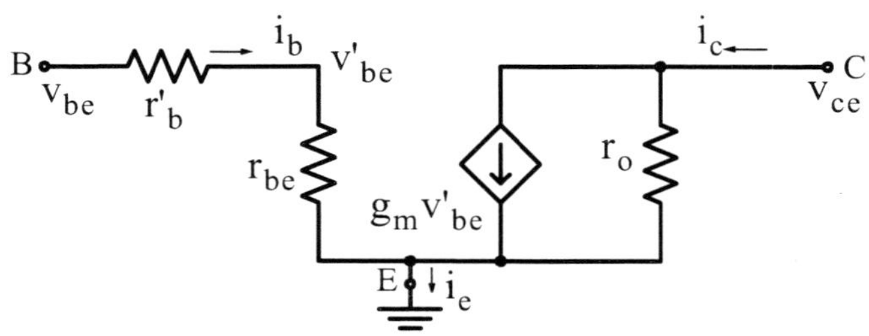

a. From (3.78),

$$g_m \approx 40 i_c = 40 \times 5 \times 10^{-3} = 200 \text{ millimhos}$$

and from (3.83)

$$r_{be} = \frac{\beta}{g_m} = \frac{80}{200} = 400 \ \Omega$$

b.

$$g_m \approx 40 i_c = 40 \times 10^{-3} = 40 \text{ millimhos}$$

and

$$r_{be} = \frac{\beta}{g_m} = \frac{60}{40} = 1.5 \text{ K}\Omega$$

16.

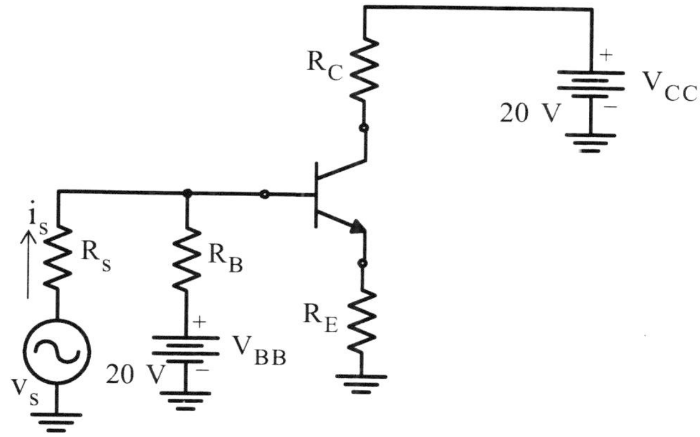

a.

$$R_C i_C + v_{CE} = V_{CC}$$

$$R_C = \frac{V_{CC} - v_{CE}}{i_C} = \frac{20 - 5}{10^{-3}} = 15\ \mathrm{K\Omega}$$

$$I_B = I_C / \beta = 10^{-3}/100 = 10\ \mu A$$

$$R_B = \frac{V_{BB}}{I_B} = \frac{20}{10^{-5}} = 2\ \mathrm{M\Omega}$$

b. The incremental model circuit in Figure 3.63 is applicable here and it is shown below with the given parameters and the values obtained in part (a).

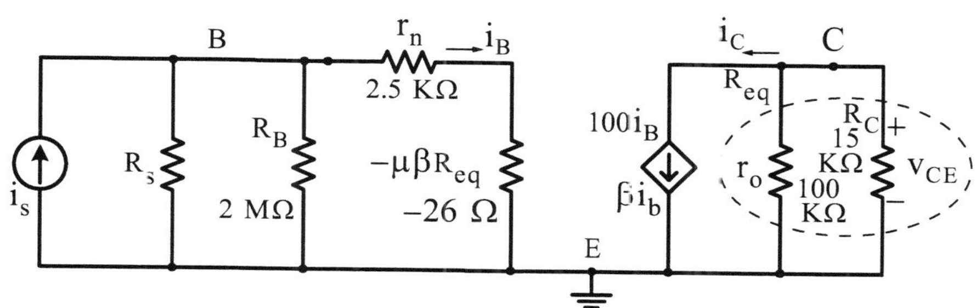

c. From (3.67)

$$A_c = \frac{i_c}{i_s} = \frac{R_B}{(R_B + r_n)}\beta\frac{r_o}{(r_o + R_C)}$$

or

$$A_c = \frac{2 \times 10^6}{2.0025 \times 10^6} \times 100 \times \frac{10^5}{1.15 \times 10^5} \approx 87$$

17.

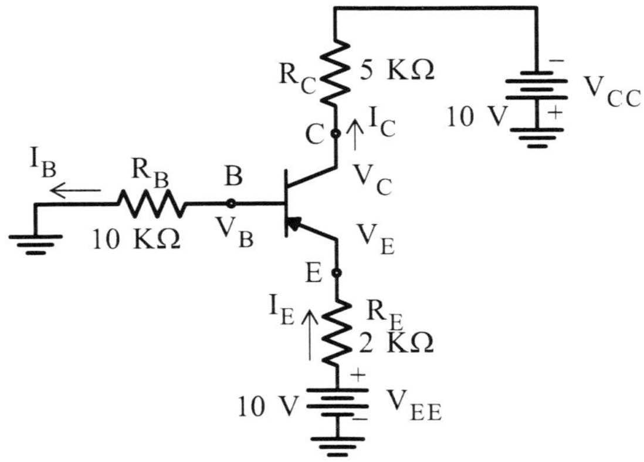

Let us assume that the transistor is in the active mode. Then,

$$V_{EB} = 0.7\ V$$

$$I_B = \frac{V_B}{R_B} = \frac{V_B}{10\ K\Omega} = 0.1V_B\ mA \quad (1)$$

$$V_E = V_{EB} + V_B = 0.7 + V_B$$

Also,

$$V_E = 10 - (2\ K\Omega)I_E$$

$$I_E = \frac{10 - V_E}{2\ K\Omega} = \frac{10 - (0.7 + V_B)}{2\ K\Omega} = 4.65 - 0.5V_B\ mA \quad (2)$$

where the numerical value of V_B in the last expression above is in mA, but since we've assumed that the transistor is in the active mode, its value should be just a few microamps. Therefore,

$$I_E \approx 4.65\ mA$$

and this value implies that the transistor operates in the saturation mode. In this case, $V_{CE\ sat} \approx 0.2\ V$, and thus

$$V_C = V_E - V_{CE\ sat} = 0.7 + V_B - 0.2 = V_B + 0.5\ V$$

The collector current is

$$I_{C\ sat} = \frac{V_C - (-10)}{5\ K\Omega} = \frac{V_B + 0.5 + 10}{5\ K\Omega} = 0.2V_B + 2.1\ mA \quad (3)$$

and with $I_E = I_B + I_C$ and (1), (2), and (3), we obtain

$$4.65 - V_B = 0.1V_B + 0.2V_B + 2.1$$

Solving for V_B,

$$V_B = \frac{2.55}{1.3} = 1.96 \text{ V}$$

Then

$$I_B = \frac{1.96}{10 \text{ K}\Omega} = 0.196 \text{ mA}$$

$$I_{C\ sat} = 0.2V_B + 2.1 \approx 3.03 \text{ mA}$$

Also,

$$\beta_{sat} = \frac{I_{C\ sat}}{I_B} = \frac{3.03}{0.196} \approx 15.5$$

18. We recall that

$$v_1 = h_{11}i_1 + h_{12}v_2 \quad (1)$$

$$i_2 = h_{21}i_1 + h_{22}v_2 \quad (2)$$

With the voltage source $v_1 = \cos\omega t$ mV in series with 800 Ω connected at the input and a 5 KΩ load connected at the output the network is as shown below.

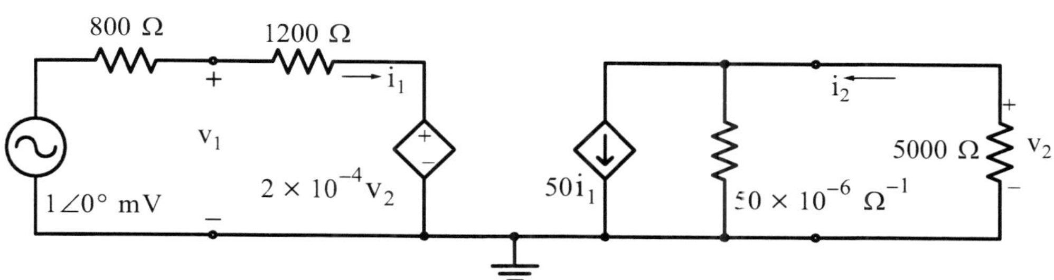

The network above is described by the equations

$$(800 + 1200)i_1 + 2 \times 10^{-4}v_2 = 10^{-3}$$

$$50i_1 + 50 \times 10^{-6}v_2 = i_2 = \frac{-v_2}{5000}$$

or

$$2 \times 10^3 i_1 + 2 \times 10^{-4}v_2 = 10^{-3}$$

$$50i_1 + 250 \times 10^{-6}v_2 = 0$$

We write the two equations above in matrix form and use MATLAB for the solution.

```
A=[2*10^3  2*10^(-4); 50  250*10^(-6)]; B=[10^(-3) 0]'; X=A\B;...
fprintf(' \n'); fprintf('i1 = %5.2e A \t',X(1)); fprintf('v2 = %5.2e V',X(2))
```

```
i1 = 5.10e-007 A      v2 = -1.02e-001 V
```

Therefore,

$$i_1 = 0.51\ \mu A \quad (3)$$

$$v_2 = -102\ mV \quad (4)$$

Next, we use (1) and (2) to find the new values of v_1 and i_2

$$= 1.2 \times 10^3 \times 0.51 \times 10^{-6} + 2 \times 10^{-4} \times (-102 \times 10^{-3}) = 0.592\ m$$

$$i_2 = 50 \times 0.51 \times 10^{-6} \times 50 \times 10^{-6} \times (-102 \times 10^{-3}) = 20.4\ \mu A$$

The voltage gain is

$$G_V = \frac{v_2}{v_1} = \frac{-102\ mV}{0.592\ mV} = -172.3$$

and the minus (–) sign indicates that the output voltage in 180° out–of–phase with the input.

The current gain is

$$G_I = \frac{i_2}{i_1} = \frac{20.4\ \mu A}{0.51\ \mu A} = 40$$

and the output current is in phase with the input current.

Chapter 4

Field Effect Transistors and PNPN Devices

This chapter begins with a discussion of Field Effect Transistors (FETs), characteristics, and applications. Other PNPN devices, the four–layer diode, the silicon controlled rectifier (SCR), the silicon controlled switch (SCS), and the triac are introduced with some of their applications. The chapter includes also a brief discussion on unijunction transistors, and diacs.

4.1 The Junction Field Effect Transistor (JFET)

The *Field–Effect Transistor* (FET) is another semiconductor device. The Junction FET (JFET) is the earlier type and the *Metal Oxide Semiconductor* FET (MOSFET) is now the most popular type. In this section we will discuss the JFET and we will discuss the MOSFET in the next section.

Figure 4.1(a) shows the basic JFET amplifier configuration and the output volt–ampere characteristics are shown in Fig. 4.1(b). These characteristics are similar to those for the junction transistor except that the parameter for this family is the input voltage rather than the input current. Like the old vacuum triode, the FET is a voltage–controlled device.

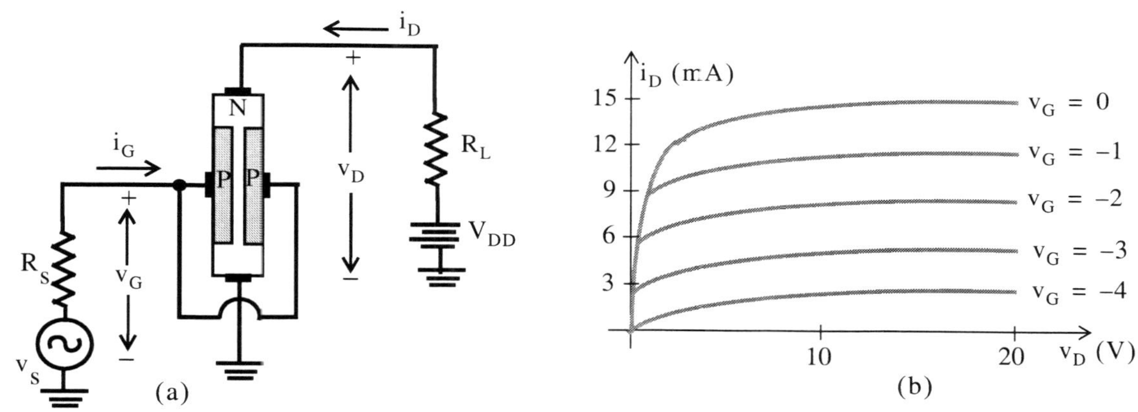

Figure 4.1. Pictorial representation and output volt–ampere characteristics for a typical JFET

The lower terminal in the N material is called the *source*, and the upper terminal is called the *drain*; the two regions of P material, which are usually connected together externally, are called *gates*. P–N junctions exist between the P and N materials, and in normal operation the voltage applied to the gates biases these junctions in the reverse direction. A potential barrier exists across the junctions, and the electrons carrying the current i_D in the N material are forced to flow through the channel between the two gates. If the voltage applied to the gates is changed, the width of the transition region at the junction changes; thus the width of the channel changes, resulting in a change in the resistance between source and drain. In this way the current in the output circuit is controlled by the gate voltage. A small potential applied to the gates, 5 to 10 volts, is sufficient to reduce the channel width to zero and to cut off the flow of current in the output circuit.

One of the most attractive features of the JFET is the fact that the input resistance, measured between gate and source, can be made very large, from 1 to 100 megohms. This high input resistance results from the fact that the input voltage v_G biases the junctions in the reverse direction and consequently is required to deliver only the small leakage current across the junctions. There are many circuits in which a high input resistance is required, and the ordinary junction transistor is not suitable for such applications.

JFETs have been made with input resistances exceeding 1 megohm, with transconductances in the range of 1 to 5 millimhos, and with the ability to provide a voltage amplification of about 10 at frequencies up to 10 MHz. The current in a JFET is carried only by majority carriers and for this reason the FET is also known as *unipolar transistor*, in contrast to the ordinary transistor in which both majority and minority carriers participate in the conduction process, and it is known as *bipolar transistor*.

To better understand the JFET operation, let us review the depletion region which we discussed in Chapter 2. Figure 4.2 shows a PN junction and the depletion region when no bias is applied.

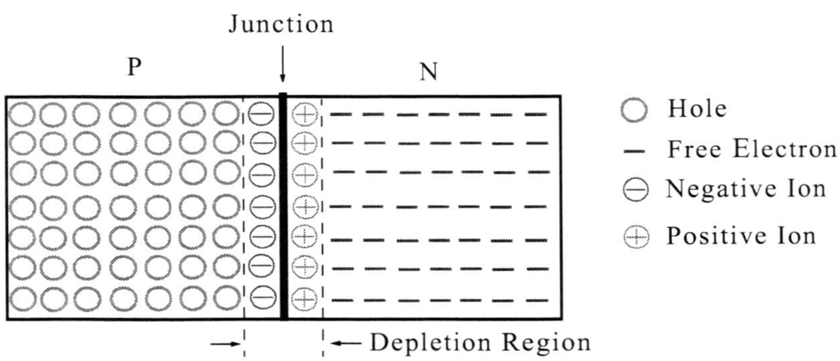

Figure 4.2. PN junction and depletion region of a typical PN junction

The width of the depletion region depends on the applied bias. Forward–biasing, where a voltage source is connected with the plus (+) on the P side and the minus (–) on the N side, makes the depletion region narrower by repelling the holes and free electrons toward the junction, and if the bias is about 0.65 V, the free electrons will cross the junction and join with the holes. Thus, the forward biasing causes the depletion region to decrease resulting in a low resistance at the junction and a relatively large current across it.

If a reverse–biasing is applied, where a voltage source is connected with the plus (+) on the N side and the minus (–) on the P side, the free electrons move towards the right side while the holes move towards the left side causing the depletion region to widen. A high resistance between the terminals is developed, and only a small current flows between the terminals.

The cross–sectional area of a typical JFET channel can be controlled by variations in the voltage applied to the gate. This is illustrated in Figure 4.3.

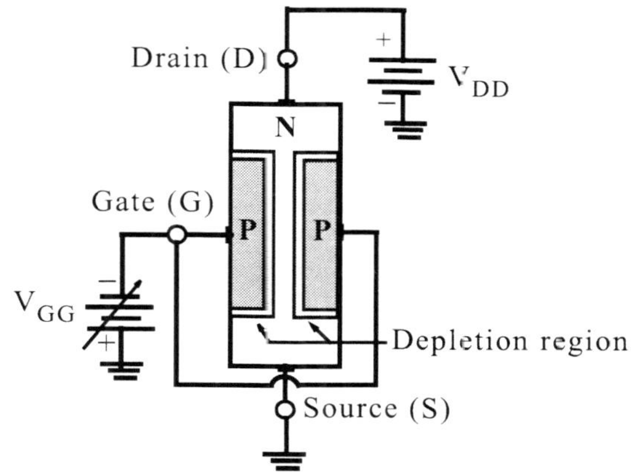

Figure 4.3. JFET operation with adjustable gate bias

In Figure 4.3, let us adjust the voltage source at the gate so that $V_{GG} = 0$, that is, the gate is grounded, and set the voltage at the drain to $V_{DD} = 10\ V$. Let us assume that under those conditions the drain–to–source current through the channel is $I_{DS} = 10\ mA$. Therefore, we conclude that the drain–to–source resistance R_{DS} is $R_{DS} = V_{DD}/I_{DS} = 10\ V/10\ mA$ or $R_{DS} = 1\ K\Omega$.

Next, let us adjust the voltage source at the gate so that $V_{GG} = -2\ V$, and maintain the voltage at the drain to $V_{DD} = 10\ V$. This reverse–bias condition cause the depletion region to expand and that reduces the effective cross–sectional area of the channel. Let us assume that under those conditions the drain–to–source current through the channel is $I_{DS} = 0.1\ mA$. Therefore, we conclude that the drain–to–source resistance R_{DS} is now $R_{DS} = 10\ V/0.1\ mA$ or $R_{DS} = 100\ K\Omega$.

Now, let us adjust the voltage source at the gate so that $V_{GG} = -4\ V$, and maintain the voltage at the drain to $V_{DD} = 10\ V$. This reverse–bias condition cause the depletion region to expand even more and that reduces further the effective cross–sectional area of the channel. and we assume that under those conditions the drain–to–source current through the channel is $I_{DS} = 0.01\ mA$. Thus, we conclude that the drain–to–source resistance R_{DS} is now $R_{DS} = 10\ V/0.01\ mA$ or $R_{DS} = 1\ M\Omega$, and therefore, we observe that the drain–to–source resistance R_{DS} can be controlled by the voltage applied at the gate. The voltage at the gate which causes the drain–to–source resistance R_{DS} to become infinite, is referred to as the *pinch–off* voltage and it is denoted as V_P. In other words, when $V_{GG} = V_P$, no current flows through the channel.

In our discussion above, we described the operation of an N–channel JFET. A P–channel JFET operates similarly except that the voltage polarities are reversed as shown in Figure 4.4 which also shows the symbols for each.

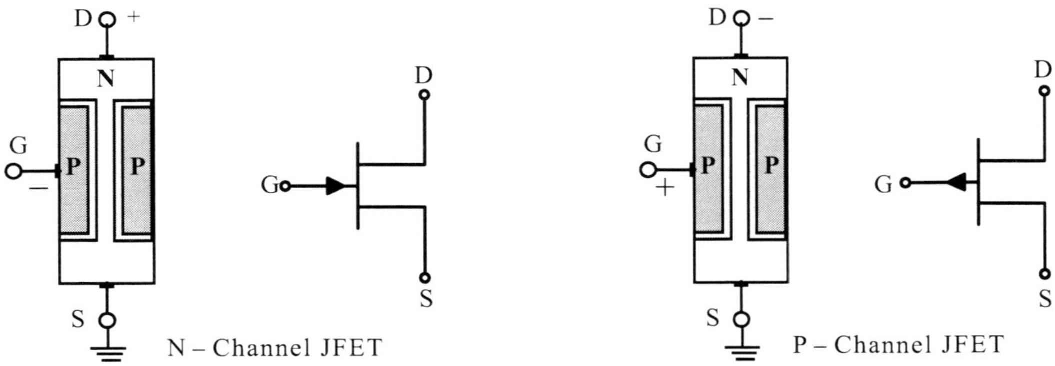

Figure 4.4. N–Channel and P–Channel JFETs

Like in bipolar transistors, one important parameter in FETs is its *transconductance* g_m defined as the ratio of the change in current i_{DS} to the change of voltage v_{GS} which produced it. In other words,

$$g_m = \left.\frac{\partial i_{DS}}{\partial v_{GS}}\right|_{v_{DS} = \text{constant}} \tag{4.1}$$

Example 4.1

Figure 4.5 shows a common–source N–channel JFET amplifier circuit and Table 4.1 shows several values of the current i_{DS} corresponding to the voltage v_{GS}.

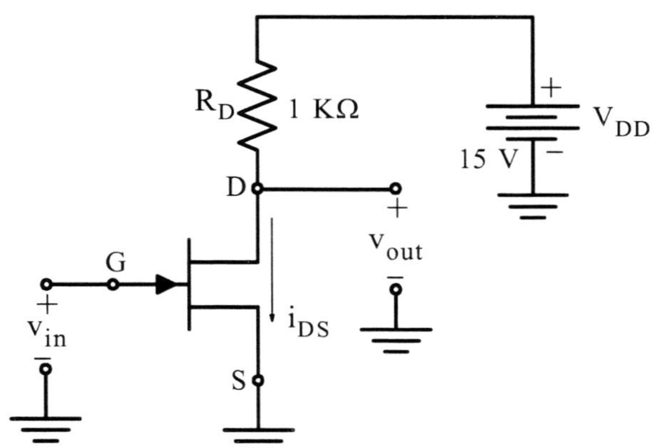

Figure 4.5. Common–source N–channel JFET amplifier for Example 4.1

TABLE 3.1 Current i_{DS} versus voltage v_{GS} for Example 4.1

v_{GS} (V)	0	−1	−2	−3	−4
i_{DS} (mA)	35	20	8	2	0

a. Find the output voltage v_{out} if the input signal is $v_{in} = -2.4$ V.

b. Find the output voltage v_{out} if the input signal is $v_{in} = -1.8$ V.

c. Is this an inverting or a non–inverting amplifier?

d. Find the transconductance using the results of (a) and (b).

e. Plot i_{DS} versus v_{GS} and indicate how the transconductance can be calculated from this plot.

Solution:

a.

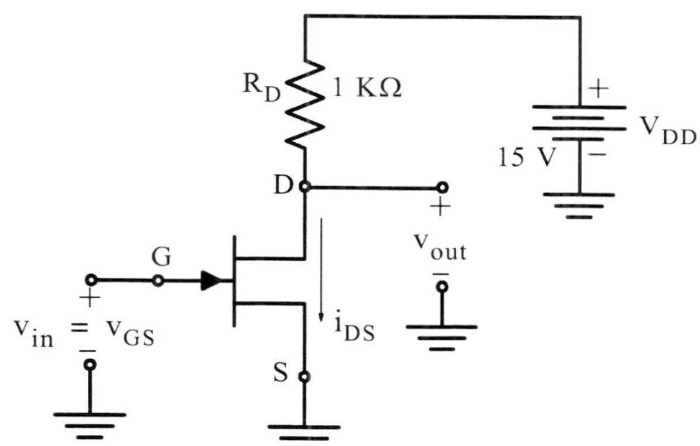

Figure 4.6. Circuit for the solution of Example 4.1

From Figure 4.6,

$$v_{out} = V_{DD} - R_D i_{DS} \tag{4.2}$$

but we do not know all values of i_{DS} for the interval $0 \le i_{DS} \le 35$ mA. Therefore, let us use the following MATLAB script to plot a suitable curve for this interval.

```
vGS=[-4 -3 -2 -1  0]; iDS=[0 2 8 20 35];... % These are the data in Table 4.1
curve=polyfit(vGS,iDS,4);...                % Fits the data to a polynomial of fourth degree
vGSaxis=-4.0:0.1:0.0;...                    % Creates horizontal (vGS) axis
polcurve=polyval(curve,vGSaxis);...         % Computes the polynomial for vGS axis values
plot(vGSaxis,polcurve);...                  % Plots the fourth degree polynomial
xlabel('vGS (volts)'); ylabel('iDS (milliamps)');
title('Plot for Example 4.1'); grid
```

The generated plot is shown in Figure 4.7.

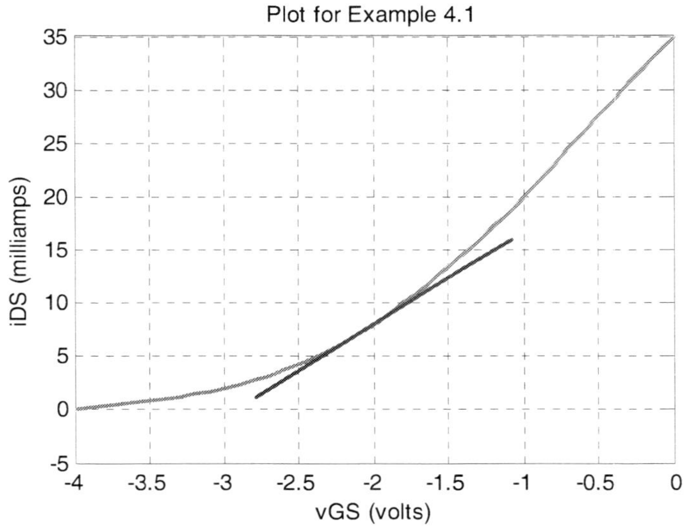

Figure 4.7. MATLAB generated plot for Example 4.1

From the plot of Figure 4.7 we see that for $v_{in} = v_{GS} = -2.4\ V$, $i_{DS} \approx 5\ mA$. Therefore,

$$v_{out}\Big|_{V_{GS} = -2.4\ V} = V_{DD} - R_D i_{DS} = 15 - 10^3 \times 5 \times 10^{-3} = 10\ V \tag{4.3}$$

b.

From the plot of Figure 4.7 we see that for $v_{in} = v_{GS} = -1.8\ V$, $i_{DS} \approx 10\ mA$. Therefore,

$$v_{out}\Big|_{V_{GS} = -1.8\ V} = V_{DD} - R_D i_{DS} = 15 - 10^3 \times 10 \times 10^{-3} = 5\ V \tag{4.4}$$

c.

The results of (a) and (b) indicate that an increase in the input voltage results in a decrease of the output voltage. Therefore, we conclude that the given JFET circuit is an inverting amplifier.

d.

$$g_m = \frac{\Delta i_{DS}}{\Delta v_{GS}} = \frac{(5-10)\times 10^{-3}}{-2.4-(-1.8)} = 0.0083\ \Omega^{-1} \tag{4.5}$$

e. The plot in Figure 4.7 shows the slope of g_m and it is calculated as in (4.5).

4.2 The Metal Oxide Semiconductor Field Effect Transistor (MOSFET)

The most popular type of a FET is the Metal–Oxide–Semiconductor Field Effect Transistor or MOSFET. Another less frequently name for the MOSFET is Insulated–Gate FET or IGFET.

Figure 4.8 shows a cross section and the symbol for an n-channel MOSFET or NMOS FET device. The complementary p-channel MOSFET or PMOS FET is similar but opposite in polarity.

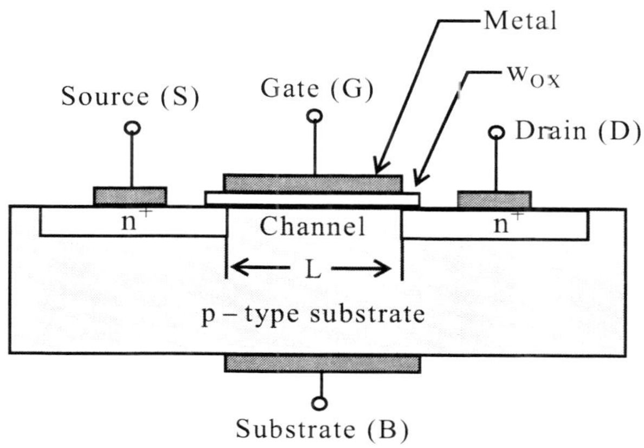

Figure 4.8. Cross section of an n-channel MOSFET

In Figure 4.8, heavily doped N-type regions, indicated by n^+, are diffused into a P-type substrate or base. An n conducting channel may be formed and exist with the gate voltage $V_G = 0$. A negative gate voltage will then drive electrons out of the channel, increasing the resistance from source to drain. This is termed *depletion-mode operation*. The JFET also operates in this manner. Conversely, if no channel exists with $V_G = 0$, one can be formed by applying positive voltage V_G and attracting electrons to a thin surface layer. This is termed *enhancement-mode operation*. The enhancement mode MOSFET has a lightly doped channel and uses forward bias to enhance the current carriers in the channel. A MOSFET can be constructed that will operate in either mode depending upon what type of bias is applied, thus allowing a greater range of input signals.

The symbols for the four basic variations of the MOSFET are shown in Figure 4.9.

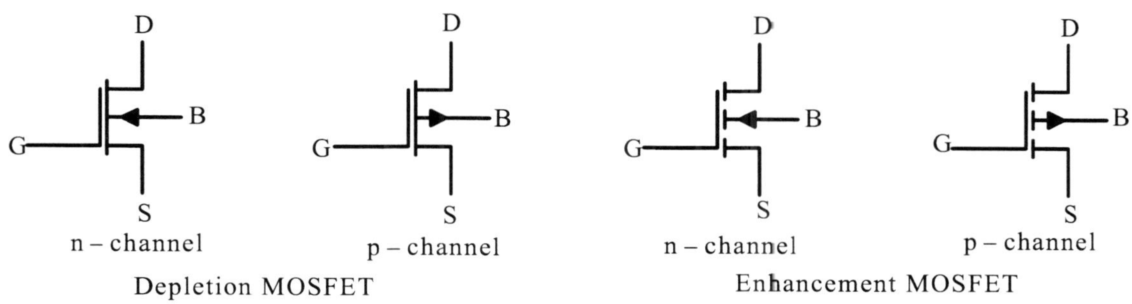

Figure 4.9. MOSFET types and symbols

4.2.1 The N-Channel MOSFET in the Enhancement Mode

Figure 4.10 shows the drain–to–source current i_{DS} versus the drain–to–source voltage v_{DS} for different values of gate–to–source voltage v_{GS}.

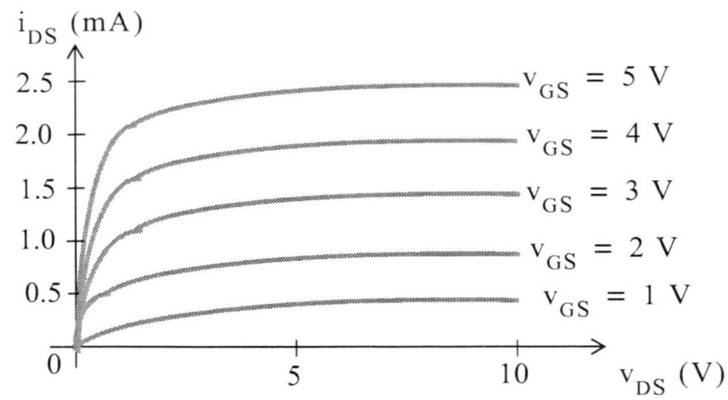

Figure 4.10. Current–voltage characteristics for typical MOSFET

The voltage at which the channel is closed is known as the *pinch–off voltage* V_P, and the minimum voltage required to form a conducting channel between the drain and source is referred to as the *threshold voltage* and it is denoted as V_T. Typical values for V_T are 2 to 4 volts for high voltage devices with thicker gate oxides, and 1 to 2 volts for lower voltage devices with thinner gate oxides. In terms of this, the drain current i_D in a MOSFET can be written approximately as

$$i_D = \frac{\varepsilon\mu W}{Lw_{OX}}\left[(v_{GS} - V_T)v_{DS} - \frac{1}{2}v_{DS}^2\right] \qquad \text{for } v_{DS} \le v_{GS} - V_T \tag{4.6}$$

where L is the channel length as shown in Figure 4.8, W is the width of the structure perpendicular to the paper in Figure 4.8, w_{OX} is the oxide thickness, ε is its dielectric constant, and μ is the mobility of carriers in the channel. The values for the channel length and width depend on the voltage v_{GS} and typical values are $2\ \mu m \le L \le 10\ \mu m$ and $100\ \mu m \le W \le 100\ \mu m$.

It is convenient to represent the quantity $\varepsilon\mu/w_{OX}$ as k_n*, and typical values of k_n are around $20\ \mu A/V^2$. Then (4.6) is expressed as

$$i_D = k_n\frac{W}{L}\left[(v_{GS} - V_T)v_{DS} - \frac{1}{2}v_{DS}^2\right] \qquad \text{for } v_{DS} \le v_{GS} - V_T \tag{4.7}$$

* *The subscript* n *is used here as a reminder that relation (4.7) applies to n–channel MOSFETs.*

where $v_{DS} \le v_{GS} - V_T$ defines the so–called *quadratic* or *triode region**, and

$$i_D = \frac{1}{2} \cdot k_n \frac{W}{L} (v_{GS} - V_T)^2 \qquad \text{for } v_{DS} \ge v_{GS} - V_T{}^{\dagger} \tag{4.8}$$

where $v_{DS} \ge v_{GS} - V_T$ defines the so–called *saturation* or *pentode region*. When $v_{GS} < V_T$, the MOSFET is said to be in the *cutoff region*. The quadratic (triode) and saturation regions are as shown in Figure 4.11.

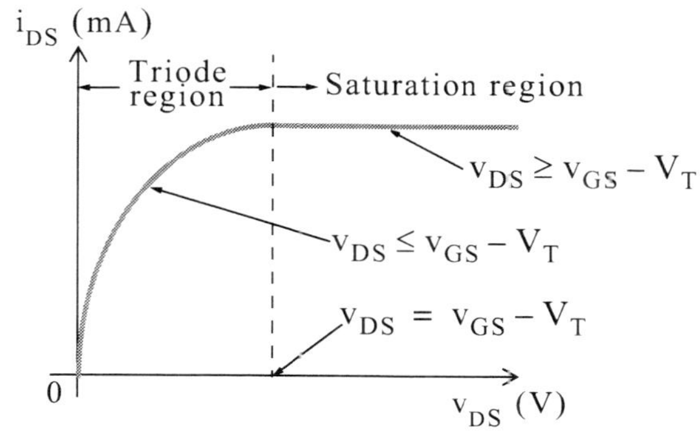

Figure 4.11. Triode and saturation regions for an enhancement type NMOS where $v_{GS} > V_T$

As shown in Figure 4.11, the triode region starts as a linear function where the drain–to–source resistance r_{DS} is linear but it changes to a curve because the resistance changes with changes in the drain–to–source voltage v_{DS}. The saturation region starts where for any further increases in v_{DS} there is no increase in the drain current i_D.

Figures 4.10 and 4.11 reveal that for small values of v_{DS}, say $0.05 \le v_{DS} \le 0.15\ V$, relation (4.7) can be written as

$$i_D \approx k_n \frac{W}{L} [(v_{GS} - V_T) v_{DS}] \qquad \text{for } v_{DS} \le v_{GS} - V_T \tag{4.9}$$

Thus, in that range of v_{DS} the MOSFET behaves as a linear resistor, usually denoted as r_{DS}, and its value can be found from

$$r_{DS} \equiv \frac{v_{DS}}{i_D} = \frac{L}{k_n W (v_{GS} - V_T)} \tag{4.10}$$

---

* *This name is carried over from the old days of the vacuum tube triode whose characteristics are as shown in Figure 4.11. It is also referred to as the quadratic region. Likewise, the saturation region is sometimes referred to as the pentode region.*

† *The drain current* i_D *is not exactly independent of the drain–to–source voltage* v_{DS}. *It increases with increasing* v_{DS} *due to the so–called channel width modulation caused by reduction of the effective channel length, and since* i_D *is inversely proportional to the channel length,* i_D *increases with* v_{DS} *and thus (4.8) is an approximation to the exact drain current* i_D *given by* $i_D = \frac{1}{2} \cdot k_n \frac{W}{L} (v_{GS} - V_T)^2 (1 + \lambda v_{DS})$ *where* λ *is a small quantity, typically* 0.01 .

Example 4.2

Compute and plot the values of r_{DS} for an NMOS device where $k_n = 18\ \mu A/V^2$, $L = 5\ \mu m$, $W = 60\ \mu m$, $V_T = 1\ V$ as v_{GS} varies in the interval $1.5 \le v_{GS} \le 3.5\ V$ in steps of $0.5\ V$. Assume that k_n, L, and W do not change significantly in this interval.

Solution:

The MATLAB script for Example 4.2 is given below.

```
kn=18*10^(-6); L=5*10^(-6); W=60*10^(-6); VT=1;
rD1 = 10.^(-3).*L./(kn.*W.*(1.5-VT)); rD2 = 10.^(-3).*L./(kn.*W.*(2-VT));...% Kilohm values
rD3 = 10.^(-3).*L./(kn.*W.*(2.5-VT)); rD4 = 10.^(-3).*L./(kn.*W.*(3-VT));...
rD5 = 10.^(-3).*L./(kn.*W.*(3.5-VT)); fprintf(' \n');...
disp('vGS (Volts)   rD (KOhms)');...                  % Display vGS and rD values
disp('-------------------------');...
fprintf('%6.1f    %10.2f \n', 1.5,rD1,2.0,rD2,2.5,rD3,3.0,rD4,3.5,rD5)
vDS=(0:0.01:100)*10^(-2); iD1=rD1*vDS; iD2=rD2*vDS; iD3=rD3*vDS; iD4=rD4*vDS;
iD5=rD5*vDS; plot(vDS,iD1, vDS,iD2, vDS,iD3, vDS,iD4, vDS,iD5); xlabel('vDS in V');...
ylabel('iD in mA'); title('Linear region for typical MOSFET'); grid
```

When this program is executed, MATLAB displays the following:

```
vGS (Volts)    rD (KOhms)
---------------------------
   1.5          9.26
   2.0          4.63
   2.5          3.09
   3.0          2.31
   3.5          1.85
```

The plot is shown in Figure 4.12.

Relation (4.8) which is repeated below for convenience, represents an $n-\text{channel}$ MOSFET and we observe that, in saturation, the drain current i_D is independent of the drain voltage v_{DS}

$$i_D = \frac{1}{2}\cdot k_n\frac{W}{L}(v_{GS}-V_T)^2 \qquad \text{for } v_{DS} \ge v_{GS}-V_T \tag{4.11}$$

Therefore, we conclude that in the saturation mode the $n-\text{channel}$ MOSFET behaves as an ideal current source whose value is as in (4.11).

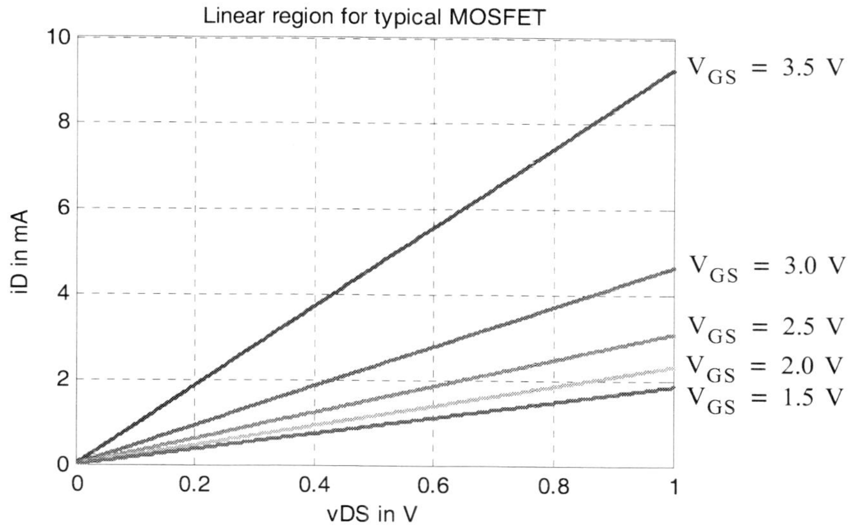

Figure 4.12. Plot for Example 4.2

In analogy with the transconductance in bipolar junction transistors, the MOSFET *transconductance* is defined as

$$g_m = \left.\frac{\partial i_D}{\partial v_{GS}}\right|_{V_{DS} = \text{constant}} \tag{4.12}$$

In other words, the transconductance is a measure of the sensitivity of drain current to changes in gate–to–source bias.

As indicated in the previous chapter, subscripts in upper case represent the sum of the quiescent and small signal parameters, and subscripts in lower case represent just the small signal parameters. Thus, $v_{GS} = V_{GS} + v_{gs}$ and (4.11) can be expressed as

$$i_D = \frac{1}{2}\cdot k_n\frac{W}{L}(V_{GS} + v_{gs} - V_T)^2 = \frac{1}{2}\cdot k_n\frac{W}{L}[(V_{GS} - V_T) + v_{gs}]^2$$

$$= \frac{1}{2}\cdot k_n\frac{W}{L}(V_{GS} - V_T)^2 + k_n\frac{W}{L}(V_{GS} - V_T)v_{gs} + \frac{1}{2}\cdot k_n\frac{W}{L}v_{gs}{}^2$$

For small signals, transconductance is defined in terms of the second term of the above expression and thus

$$i_d = k_n\frac{W}{L}(V_{GS} - V_T)v_{gs}$$

Letting

$$\frac{\partial i_d}{\partial v_{gs}} = \frac{i_{d2} - i_{d1}}{v_{gs2} - v_{gs1}} = \frac{i_d}{v_{gs}}$$

and substituting the last expression above into (4.12) we obtain

$$g_m = \frac{i_d}{v_{gs}} = \frac{k_n(W/L)(V_{GS} - V_T)v_{gs}}{v_{gs}} \tag{4.13}$$

or

$$g_m = \frac{i_d}{v_{gs}} = k_n\frac{W}{L}(V_{GS} - V_T) \tag{4.14}$$

The *output conductance* is defined as

$$g_o = \left.\frac{\partial i_D}{\partial v_{DS}}\right|_{V_{GS} = \text{constant}} \tag{4.15}$$

At saturation the slope of (4.15) is

$$g_o(\text{sat}) = 0 \tag{4.16}$$

From (4.7) for the triode region

$$i_D = k_n\frac{W}{L}\left[(v_{GS} - V_T)v_{DS} - \frac{1}{2}v_{DS}^2\right]$$

Therefore the output conductance is found by differentiation of the last expression above as

$$g_o = \frac{\partial i_D}{\partial v_{DS}} = k_n\frac{W}{L}(v_{GS} - V_T - v_{DS}) \tag{4.17}$$

4.2.2 The N–Channel MOSFET in the Depletion Mode

As we've learned, in the enhancement mode we apply a positive voltage v_{GS} and as this value increases, the channel conductivity increases until we reach saturation. However, a MOSFET can also be fabricated with an implanted channel so that drain current i_D will flow if we apply a voltage v_{DS} even though the voltage v_{GS} is zero. If, however we want to decrease the channel conductivity, we can apply a sufficiently negative v_{GS} to deplete the implanted channel, and this mode of operation is referred to as the *depletion mode*. The negative value of v_{GS} that causes the channel to be entirely depleted is the threshold voltage V_T of the $n-\text{channel}$ MOSFET and obviously has a negative value. At this threshold negative voltage V_T, the drain current i_D is zero although v_{DS} may still be present.

Typical values of drain current i_D versus the gate–to–source voltage v_{GS} characteristics for an $n-\text{channel}$ MOSFET that operates in the enhancement mode only are shown in Figure 4.13(a) and one that operates in both the depletion and enhancement modes are shown in Figure 4.13(b).

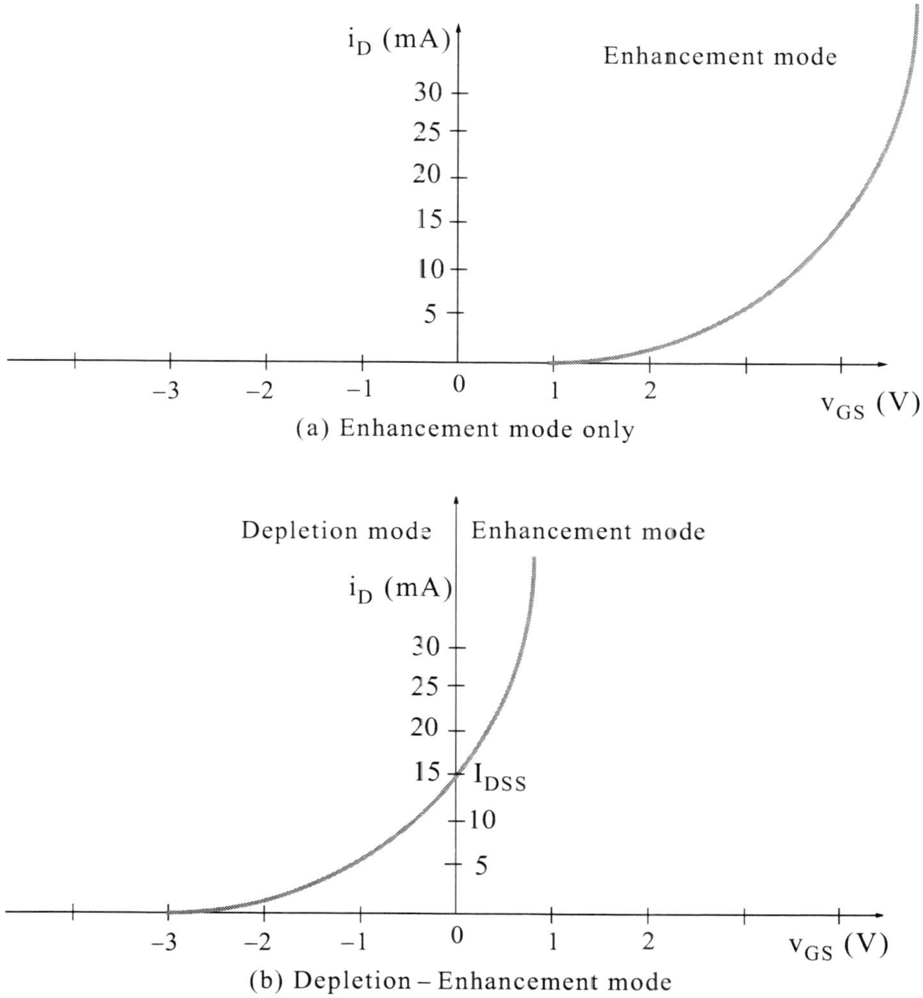

Figure 4.13. Typical n – channel *MOSFET* i_D vs v_{GS} *characteristics*

As noted in Figure 4.13(b), I_{DSS} is the value of the drain current in saturation with $v_{GS} = 0$.

Figure 4.14 shows typical i_D vs v_{DS} characteristics for a depletion–type n – channel MOSFET.

Example 4.3

It is known that for a depletion–type n – channel *MOSFET*, $k_n = 0.3\ mA/V^2$, $W = 100\ \mu m$, $L = 10\ \mu m$, and $V_T = -3\ V$.

a. At what value should the voltage v_{DS} be set so that this device will be operating in the saturation region when $v_{GS} = 2\ V$?

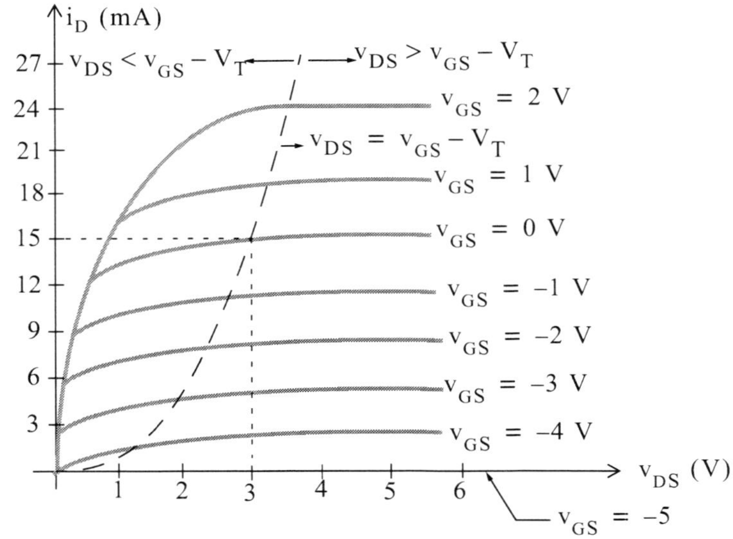

Figure 4.14. Typical i_D vs v_{DS} *characteristics for a depletion–type* $n-channel$ MOSFET

b. What would the drain current i_D be if the voltage v_{DS} is set to its minimum value to keep the device in the saturation region?

Solution:

a. The dotted curve of the i_D vs v_{DS} characteristics of Figure 4.14 separates the saturation region from the triode region. Therefore, the voltage v_{DS} should be such that

$$v_{DS} \ge v_{GS} - V_T \ge 2-(-3) \ge 5 \text{ V}$$

b. Relation (4.11) indicates that at saturation the current i_D is independent of the voltage v_{DS}. Therefore,

$$i_D = \frac{1}{2} \cdot k_n \frac{W}{L}(v_{GS} - V_T)^2 = \frac{1}{2} \cdot 0.3 \cdot \frac{100}{10}[2-(-3)]^2 = 37.5 \text{ mA}$$

4.2.3 The P–Channel MOSFET in the Enhancement Mode

For the $p-channel$ MOSFET the threshold voltage V_T is negative and the drain–to–source voltage v_{DS} must also be negative or the source–to–drain voltage v_{SD} must be positive. Therefore, to create a channel we must have $v_{GS} \le V_T$ and $v_{DS} < 0$, or $v_{SD} > 0$.

In a $p-channel$ MOSFET device operating in triode mode the drain current i_D is found from

$$i_D = k_p \frac{W}{L}\left[(v_{GS} - V_T)v_{DS} - \frac{1}{2}v_{DS}^2\right] \qquad \text{for } (v_{DS} \geq v_{GS} - V_T) \qquad (4.18)$$

and if the device operates in the saturation mode, the drain current i_D is found from

$$i_D = \frac{1}{2} \cdot k_p \frac{W}{L}(v_{GS} - V_T)^2 \qquad \text{for } v_{DS} \leq v_{GS} - V_T^{*} \qquad (4.19)$$

The constant k_p in relations (4.18) and (4.19) is analogous to k_n for the n-channel MOSFET and typical values of k_p are around $10\ \mu A/V^2$.

Example 4.4

It is known that for an enhancement-type p-channel MOSFET, $k_p = 8\ mA/V^2$, $W = 100\ \mu m$, $L = 10\ \mu m$, and $V_T = -1.5\ V$. If $v_G = 0$, and $v_S = 5\ V$, compute the drain current i_D if:

a. $v_D = 4\ V$

b. $v_D = 1.5\ V$

c. $v_D = 0\ V$

d. $v_D = -5\ V$

Solution:

As stated above, to create a channel v_{GS} must be equal or less than V_T, and v_{DS} must be negative and using the conditions specified in (4.18) and (4.19) we will determine whether the device is in the triode or saturation mode. For convenience, we list the following conditions:

$$v_{GS} = v_G - v_S$$

$$v_{GS} < V_T$$

$$v_{DS} = v_D - v_S$$

* *As with the n-channel MOSFET, the drain current i_D is not exactly independent of the drain-to-source voltage v_{DS}. It increases with increasing v_{DS} due to the so-called channel width modulation caused by reduction of the effective channel length, and since i_D is inversely proportional to the channel length, i_D increases with v_{DS} and thus (4.18) is an approximation to the exact drain current i_D given by $i_D = \frac{1}{2} \cdot k_n \frac{W}{L}(v_{GS} - V_T)^2(1 + \lambda v_{DS})$ where v_{GS}, v_{DS}, V_T, and λ are all negative.*

$$v_{DS} > v_{GS} - V_T$$

a.

$$v_{GS} = v_G - v_S = 0 - 5 = -5 \text{ V}$$

$$v_{GS} < V_T \Rightarrow -5 < -1.5$$

$$v_{DS} = v_D - v_S = 4 - 5 = -1 \text{ V}$$

$$v_{GS} - V_T = -5 - (-1.5) = -3.5 \text{ V}$$

$$v_{DS} > v_{GS} - V_T \Rightarrow -1 > -3.5$$

and we conclude that the device operates in the triode mode. Then, with (4.17) we obtain

$$i_D = k_p \frac{W}{L}\left[(v_{GS} - V_T)v_{DS} - \frac{1}{2}v_{DS}^2\right] = 8 \times \frac{10}{1} \times [(-5 - (-1.5)) \times (-1)] - 0.5 \times (-1)^2 = 0.24 \text{ mA}$$

b.

$$v_{DS} = v_D - v_S = 1.5 - 5 = -3.5 \text{ V}$$

$$v_{GS} - V_T = -5 - (-1.5) = -3.5 \text{ V}$$

$$v_{DS} = v_{GS} - V_T = -3.5 \text{ V}$$

and we conclude that the device operates at the point where the triode mode ends and the saturation region begins. Then, with (4.19) we obtain

$$i_D = \frac{1}{2} \cdot k_p \frac{W}{L}(v_{GS} - V_T)^2 = 0.5 \times 8 \times 10[-5 - (-1.5)]^2 = 0.49 \text{ mA}$$

c.

$$v_{DS} = v_D - v_S = 0 - 5 = -5 \text{ V}$$

$$v_{GS} - V_T = -5 - (-1.5) = -3.5 \text{ V}$$

$$v_{DS} < v_{GS} - V_T \Rightarrow -5 < 3.5$$

and we conclude that the device operates well in the saturation region.Then, with (4.19) we obtain

$$i_D = \frac{1}{2} \cdot k_p \frac{W}{L}(v_{GS} - V_T)^2 = 0.5 \times 8 \times 10[-5 - (-1.5)]^2 = 0.49 \text{ mA}$$

d.

$$v_{DS} = v_D - v_S = -5 - 5 = -10 \text{ V}$$

$$v_{GS} - V_T = -5 - (-1.5) = -3.5 \text{ V}$$

$$v_{DS} < v_{GS} - V_T \Rightarrow -10 < 3.5$$

and we conclude that the device operates deeply in the saturation region.Then, with (4.19) we obtain

$$i_D = \frac{1}{2} \cdot k_p \frac{W}{L}(v_{GS} - V_T)^2 = 0.5 \times 8 \times 10[-5-(-1.5)]^2 = 0.49 \text{ mA}$$

4.2.4 The P–Channel MOSFET in the Depletion Mode

Depletion type p – channel MOSFETs operate similarly to depletion type n – channel MOSFETs except that the polarities of all voltages are reversed and the current i_D flows from source to drain. Figure 4. 16 shows typical i_D vs v_{DS} characteristics for a depletion–type p – channel MOSFET.

4.2.5 Voltage Gain

The *voltage gain* in a MOSFET device is defined as the ratio of the small signal quantities v_d to v_{gs}. We will derive it with the aid of the MOSFET in Figure 4.15.

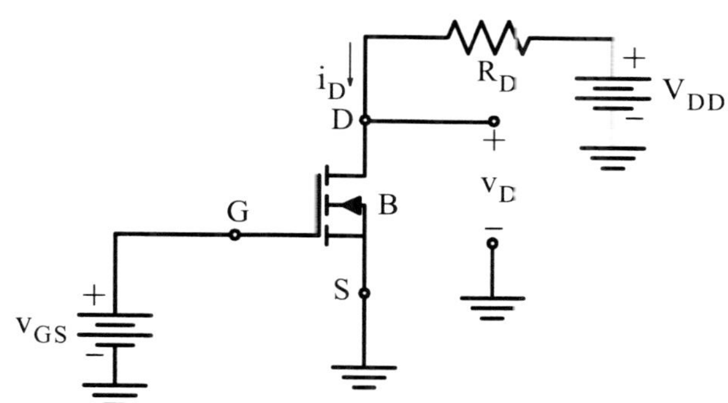

Figure 4.15. Circuit for the derivation of the voltage gain

Figure 4.15 shows that the substrate is connected to the source; this is a common practice with MOSFET devices. From Figure 4.15,

$$v_D = V_{DD} - R_D i_D$$

$$i_D = I_D + i_d$$

$$v_D = V_{DD} - R_D(I_D + i_d)$$

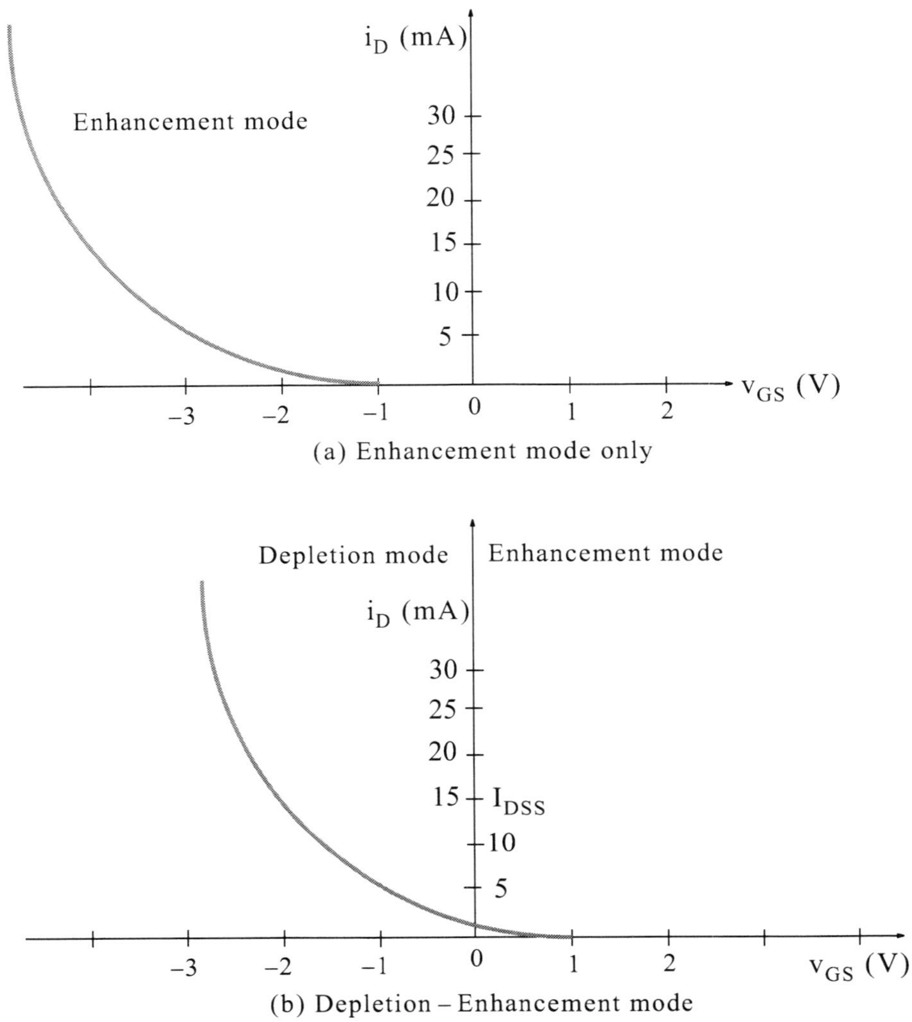

Figure 4.16. Typical i_D vs v_{DS} *characteristics for a depletion–type* $n-channel$ *MOSFET*

$$V_{DD} = V_D + R_D i_D$$

$$v_D = V_{DD} - R_D(I_D + i_d) = V_D + R_D i_D - R_D(I_D + i_d) = V_D - R_D i_d$$

and

$$v_d = -R_D i_d$$

From relation (4.14)

$$g_m = \frac{i_d}{v_{gs}}$$

$$i_d = g_m v_{gs}$$

$$v_d = -R_D i_d = -R_D g_m v_{gs}$$

Then, the voltage gain A_v is

$$A_v = v_d / v_{gs} = -R_D g_m \tag{4.20}$$

and the minus (–) sign indicates 180° phase reversal.

Relation (4.20) reveals that high gains can be achieved with increased drain resistance R_D. However, external resistors are not used in IC MOSFETS because they require large areas and their tolerance may cause some undesirable effects. Instead, constant current sources are used, and the most commonly used current sources are the so–called *current mirrors*. A typical current mirror is shown in Figure 4.17.

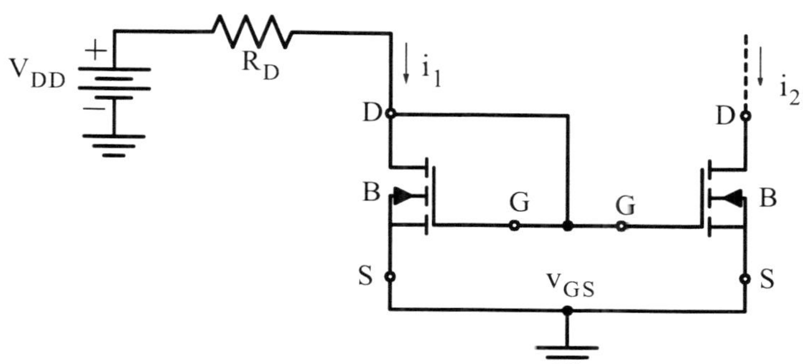

Figure 4.17. Typical circuit of a current mirror

A current mirror is essentially an adjustable current regulator. In the circuit of Figure 4.17, both MOSFETs share the same voltage v_{GS} and the ratio of the currents i_1 and i_2 is directly related to the ratio of the geometry of the transistors, that is,

$$\frac{i_1}{i_2} = \frac{(W/L)_1}{(W/L)_2}$$

and if the two MOSFETs have the same geometry, then $i_1 = i_2$, and hence the name current mirror. A similar current mirror circuit can be implemented with bipolar transistor.

MOSFET devices can be used as amplifiers provided that they operate in the saturation region. However, the CMOS devices, discussed on the next section, are the most common amplifiers in the FET technology.

4.3 Complementary MOS (CMOS)

Complementary MOS or CMOS technology combines one NMOS device and one PMOS device into a single device referred to as CMOS. These devices are used extensively in both analog and digital circuits, and integrated circuits.

In CMOS devices only one of its components is on at any given time, that is, either the NMOS device is on and the PMOS device is off, or vice versa. Thus, CMOS chips require less power than chips using just one type of a MOSFET, and unlike bipolar transistors, a CMOS has almost no static power dissipation. This makes CMOS devices particularly attractive for use in battery–powered devices, such as portable computers. Personal computers also contain a small amount of battery–powered CMOS memory to hold the date, time, and system setup parameters.

In this section we will briefly discuss the three types of CMOS amplifiers, and in subsequent chapters we will see how CMOS devices are used as logic inverters and gates.

4.3.1 The CMOS Common–Source Amplifier

Figure 4.18 shows how a CMOS device is configured as a common–source amplifier where the upper two MOSFETS serve as a current mirror to perform the function of a resistor.

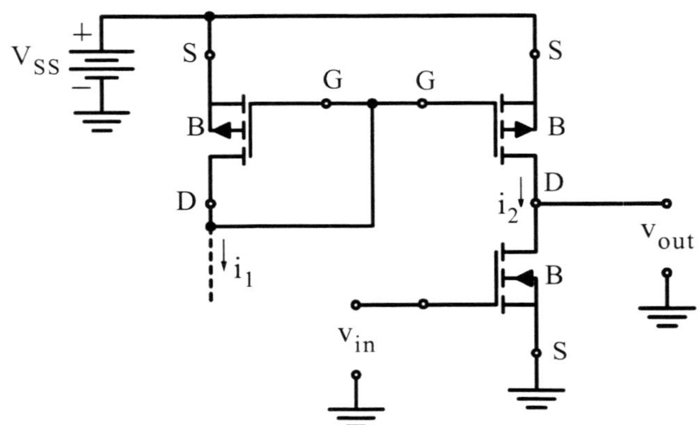

Figure 4.18. Typical CMOS common–source amplifier

A typical common–source CMOS amplifier can provide gains of 20 to 100, and the output is $180°$ out–of–phase with the input. It exhibits both high input and high output resistances.

4.3.2 The CMOS Common–Gate Amplifier

Figure 4.19 shows how a CMOS device is configured as a common–gate amplifier. The DC voltage V_{DC} at the gate provides a bias voltage but the AC signal is zero and this is the reason that the circuit is referred to as CMOS common–gate amplifier. A typical common–source CMOS amplifier can provide gains of 20 to 100, and the output is in–phase with the input. It exhibits a low input resistance, and a high output resistance.

4.3.3 The CMOS Common–Drain (Source Follower) Amplifier

A typical *common–drain* (*source follower*) CMOS amplifier is shown in Figure 4.20. Its voltage gain is less than unity, and the output is in–phase with the input. It exhibits a low output resistance, and thus it can be used as a buffer amplifier. It is referred to as common–drain amplifier because there is no signal at the drain of the upper MOSFET device; the voltage V_{DD} just provides a bias. It is also referred to as source follower because the lower right MOSFET device acts as a load for the upper MOSFET device.

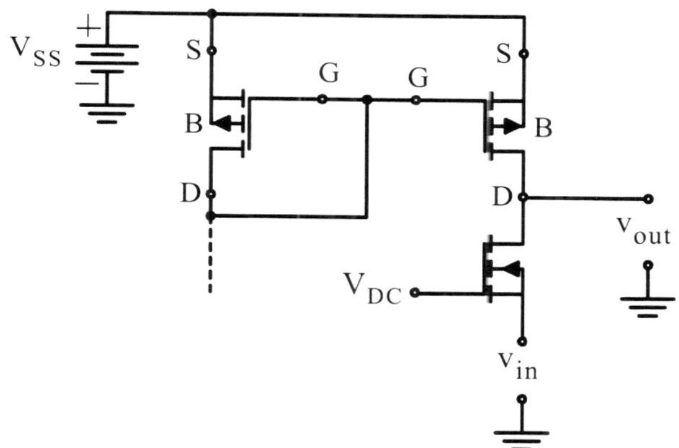

Figure 4.19. Typical CMOS common–gate amplifier

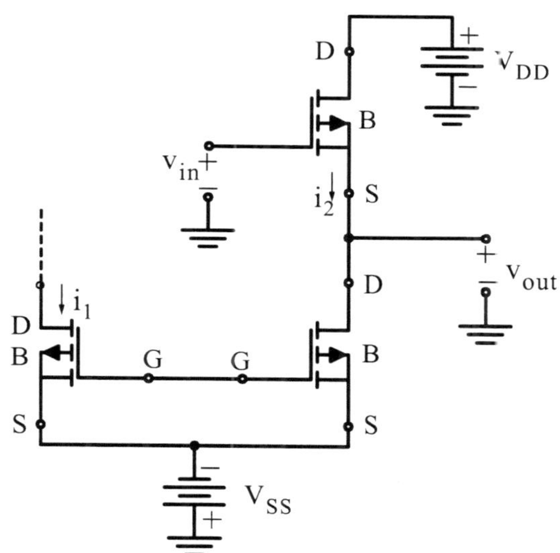

Figure 4.20. Typical common–drain (source follower) amplifier

4.4 The Metal Semiconductor FET (MESFET)

The *Metal–Semiconductor–Field–Effect–Transistor* (MESFET) consists of a conducting channel positioned between a source and drain contact region. The carrier flow from source to drain is controlled by a Schottky metal gate. The control of the channel is obtained by varying the depletion layer width underneath the metal contact which modulates the thickness of the conducting channel and thereby the current. MESFETs use GaAs (gallium arsenide) technology. Only *n – channel* MESFETS are available, because holes have a slower mobility.

The main advantage of the MESFET is the higher mobility – also referred to as surface mobility – of the carriers in the channel as compared to the silicon–type MOSFETs. The higher mobility results in a higher current, transconductance and transit frequency of the device.

However, the presence of the Schottky metal gate limits the forward bias voltage on the gate to the turn–on voltage of the Schottky diode. This turn–on voltage is typically 0.7 V for GaAs Schottky diodes. The threshold voltage therefore must be lower than this turn–on voltage. As a result it is more difficult to fabricate circuits containing a large number of enhancement–mode MESFET.

The higher transit frequency of the MESFET makes it particularly of interest for microwave circuits. While the advantage of the MESFET provides a superior microwave amplifier or circuit, the limitation by the diode turn–on is easily tolerated. Typically depletion–mode devices are used since they provide a larger current and larger transconductance and the circuits contain only a few transistors, so that threshold control is not a limiting factor. The buried channel also yields a better noise performance as trapping and release of carriers into and from surface states and defects is eliminated.

The use of GaAs rather than silicon MESFETs provides two more significant advantages: first of all the room temperature mobility is more than 5 times larger, while the saturation velocity is about twice that in silicon, and second it is possible to fabricate semi–insulating (SI) GaAs substrates which eliminates the problem of absorbing microwave power in the substrate due to free carrier absorption.

4.5 The Unijunction Transistor

The *unijunction transistor* (UJT) is a three–terminal, single–junction device which exhibits negative resistance and switching characteristics totally unlike those of conventional bipolar transistors. As shown in Figure 4.21, the UJT consists of a bar of n–type silicon having ohmic contacts designated Base 1 (B_1) and Base 2 (B_2) on either side of a single PN junction designated as the emitter.

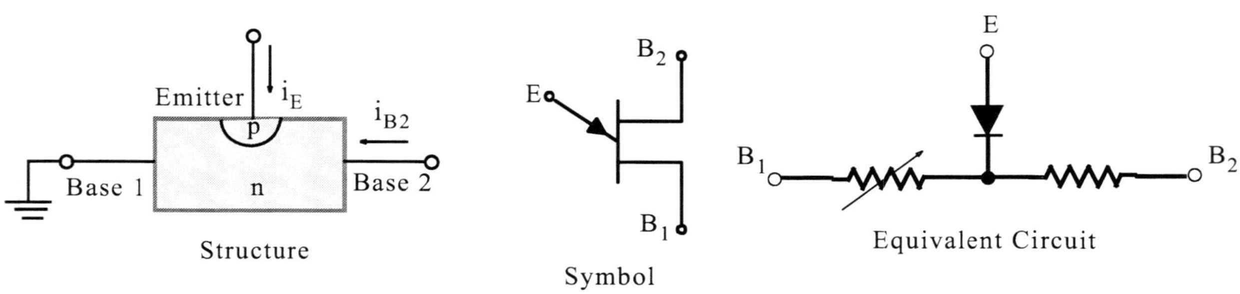

Figure 4.21. Unijunction transistor structure, symbol, and equivalent circuit

In operation, a positive voltage is applied to B_2 and B_1 is placed at ground potential. The $B_2 - E - B_1$ junctions then act like a voltage divider which reverse–biases the emitter junction. An external voltage having a potential higher than this reverse bias will forward–bias the emitter and inject holes into the silicon bar which move toward B_1. The emitter–B_1 resistance then decreases, and this, in turn, causes the emitter voltage to decrease as the emitter current increases, and a negative resistance characteristic is obtained as shown in Figure 4.22.

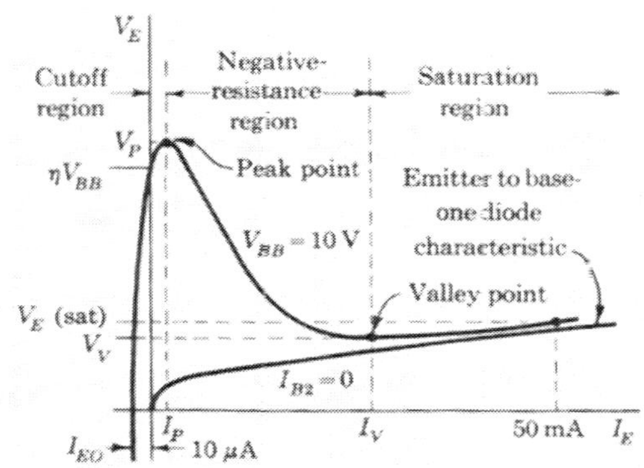

Figure 4.22. UJT emitter characteristics curve with important parameters (Courtesy of General Electric)

On the emitter characteristics curve of Figure 4.22, the points of interest are the *peak point*, and the *valley point*. The region to the left of the peak point is called the *cutoff region*, and in this region the emitter is reverse biased and only a small leakage current flows. The region between the peak point and the valley point is referred to as the negative resistance region. The region to the right of the valley point is the saturation region and as we can see, the resistance in this region is positive.

Device 2N2646 is a popular UJT and can be used for the design of pulse and sawtooth generators, analog–to–digital converters, relay time delay circuits, and frequency dividers.

Other UJT devices, referred to as *programmable UJTs*, can have their parameters set by external components such as resistors and capacitors. Device 2N6027 is known as a programmable UJT.

4.6 The Diac

The *diac* is a two–terminal, transistor–like component which exhibits bistable switching for either polarity of a suitably high applied voltage. As shown in Figure 4.23, the diac closely resembles a PNP transistor without an external base terminal.

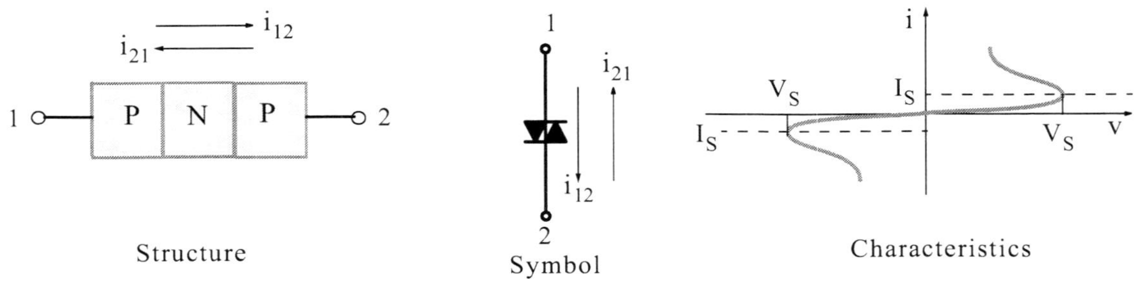

Figure 4.23. Diac structure, symbol, and characteristics

Normally, the diac does not conduct (except for a small leakage current) until the breakover voltage V_S is reached, typically 20 to 40 volts. At that point the diac goes into avalanche conduction also at that point the device exhibits a negative resistance characteristic, and the voltage drop across the diac snaps back, typically about 5 volts, creating a breakover current I_S in the order of 50 to 200 μA sufficient to trigger a triac or SCR.

The negative–resistance characteristic of the diac makes it useful for very simple *relaxation oscillators** and pulse generators, but its major application is in conjunction with a triac, to be discussed in Section 4.8, to produce ac phase–control circuits useful for motor–speed control, light dimming, and other AC power–control functions.

4.7 The Silicon Controlled Rectifier (SCR)

The *silicon controlled rectifier*, usually referred to as an SCR, is one of the family of semiconductors that includes transistors and diodes. Another name for the SCR is *thyristor*†. It is similar to a diode with an additional terminal that is used to turn it on. Once turned on, the SCR will remain on as long as current flows through it. If the current falls to zero, the SCR behaves like an open switch. This device is much larger in size than a transistor or a MOSFET and it is designed to operate at higher voltages and currents, typically $1,000\ V$ or higher, and $100\ A$ or higher

The SCR is a four–layer semiconducting device, with each layer consisting of an alternately N or P–type material, for example N–P–N–P. The main terminals, labelled *anode* and *cathode*, are across the full four layers, and the control terminal, called the *gate*, is attached to one of the middle layers as shown in Figure 4.24.

* *Relaxation oscillators. are circuits that generate non–sinusoidal waveforms such as pulse and sawtooth generators.*

† *An earlier gas filled tube device called a Thyratron provided a similar electronic switching capability, where a small control voltage could switch a large current.*

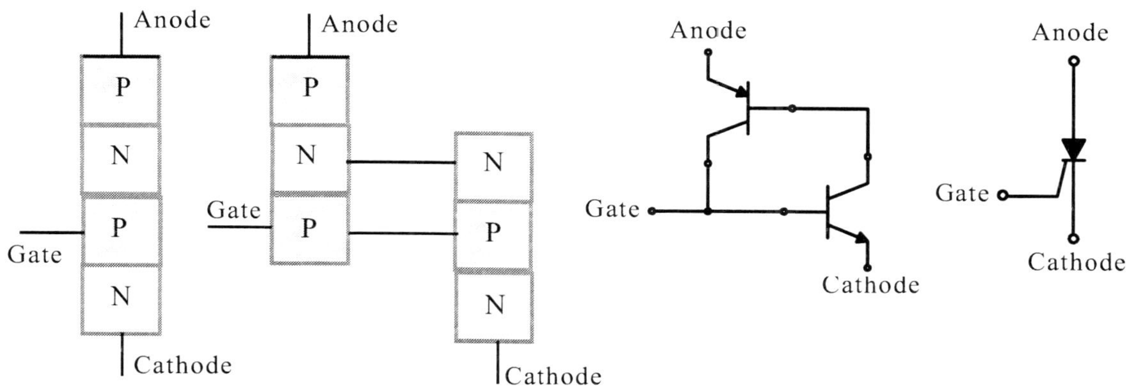

Figure 4.24. Parts of an SCR, the two–transistor equivalent circuit, and its symbol

SCRs are mainly used where high currents and voltages are involved, and are often used to control alternating currents, where the change of sign of the current causes the device to automatically switch off. Like a diode, an SCR conducts only in one direction. A similar 5–layer device, called a *triac*, to be discussed on the next section, conducts current in both directions.

Modern SCRs can switch large amounts of power (up to megawatts). In the realm of very high power applications, they are still the primary choice. However, in low and medium power (from few tens of watts to few tens of kilowatts) they have almost been replaced by other devices with superior switching characteristics like MOSFETs. While an SCR is that is it not a fully controllable switch in the sense that triggering current direction need to be reversed to switch it off, a newer device, the *gate turn–off SCR* (GTO) can be turned on and off with a signal applied to the gate. The turn–on signal is a small positive voltage and the turn–off is a negative small signal. GTOs are used for the output stages of medium–voltage, high horsepower, variable frequency drives. In high–frequency applications, SCRs are poor candidates due to large switching times arising out of bipolar conduction. MOSFETs, on the other hand, has much faster switching capability because of its unipolar conduction (only majority carriers carry the current). Figure 4.25 shows the volt–ampere characteristics of a typical SCR.

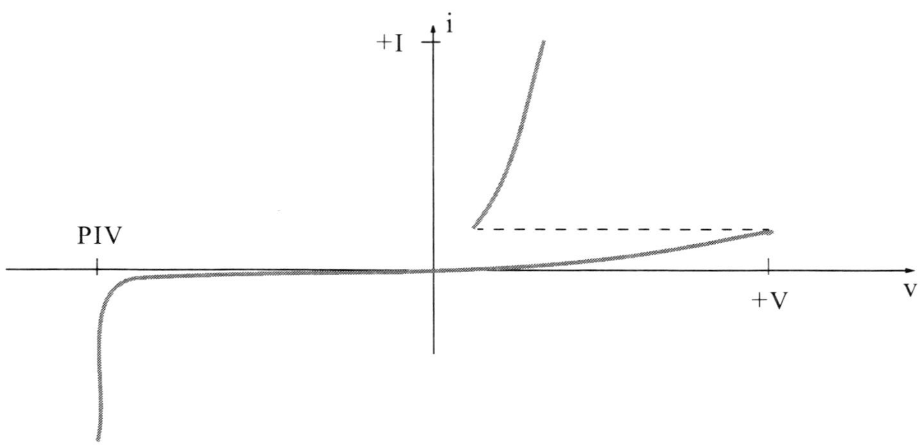

Figure 4.25. The voltage–current characteristics of a typical SCR

As shown in Figure 4.25, when the forward bias voltage from anode to cathode reaches a value indicated as $+V$, and a positive signal is applied to the gate, the device reverts to a low impedance and current flows from the anode to cathode. The dotted line shows the interval of the voltages from anode to cathode in which the SCR will be conducting after the trigger signal has been removed. The current must be limited by the load to the value $+I$, or the SCR will be damaged.

When the forward bias from anode to cathode is reduced to zero or becomes negative, the SCR becomes a non–conductive device and the signal at the gate, even if present, will not change the non–conductive state of the device. If the negative bias exceeds the *peak–inverse voltage* indicated as PIV, the SCR will be damaged. It is recommended that the PIV value should be at least three times the RMS value of the applied voltage.
Figure 4.26 shows sinusoidal waveform applied to the gate of an SCR where during the positive half cycle the SCR is triggered at $45°$ after the sinewave starts from zero and increases in the positive direction.

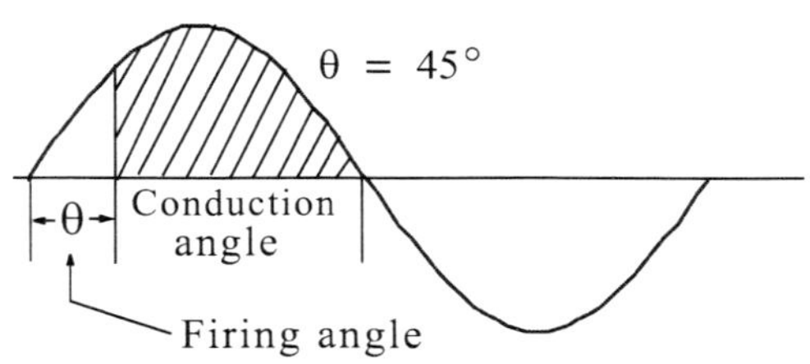

Figure 4.26. SCR gate control

As shown in Figure 4.26, the number of degrees from the beginning of the cycle until the SCR is gated to the on condition is referred to as the *firing angle*, and the number of degrees that the SCR remains conducting is known as the *conduction angle*. For accurate SCR gating, the firing circuit must be synchronized with the AC line voltage being applied anode–to–cathode across the device. Without synchronization, the SCR firing would be random in nature and the system response will be erratic. Also, when the firing angle is greater than zero, the voltage applied to the load is no longer sinusoidal. This presents no problem in the case of motor loads, but for radio and television an interference is created and usually the manufacturer of the SCR equipment will include an *electromagnetic interference* (EMI) filter to rectify the problem.

In closed–loop systems, such as motor control, an *Error Detector Circuit* computes the required firing angle based on the system setpoint and the actual system output. The firing circuit is able to sense the start of the cycle, and, based on an input from the Error Detector, delay the firing pulse until the proper time in the cycle to provide the desired output voltage. An analogy of a firing circuit would be an automobile distributor which advances or retards the spark plug firing based on the action of the vacuum advance mechanism.

In analog control systems the error detector circuit is usually an integrated circuit operational amplifier which takes reference and system feedback inputs and computes the amount of error (difference) between the actual output voltage and the desired setpoint value. Even though the SCR is an analog device, many new control systems now use a microprocessor based, digital, firing circuit to sense the AC line zero–crossing, measure feedback and compare it with the setpoint, and generate the required firing angle to hold the system in a balanced state.

Another consideration is SCR protection. The SCR, like a conventional diode, has a very high one–cycle surge rating. Typically, the device will carry from eight to ten time its continuous current rating for a period of one electrical cycle. It is extremely important that the proper high–speed, current–limiting, rectifier fuses recommended by the manufacturer be employed; one should never substitute with another type fuse. Current limiting fuses are designed to sense a fault in a quarter–cycle and clear the fault in one–half of a cycle, thereby protecting the SCR from damage due to short circuits.

Switching spikes and transients, which may exceed the device PIV rating, are also a serious concern with semiconductor devices. Surge suppressors, such as the GE Metal–Oxide–Varistor (MOV), are extremely effective in absorbing these short–term transients. High voltage capacitors are also often employed as a means of absorbing these destructive spikes and provide a degree of electrical noise suppression as well.

4.7.1 The SCR as an Electronic Switch

The SCR circuit of Figure 4.27 shows a possible arrangement for switching the SCR to the off position. We recall that to switch the SCR to the off position, the anode to cathode voltage must be zero or negative.

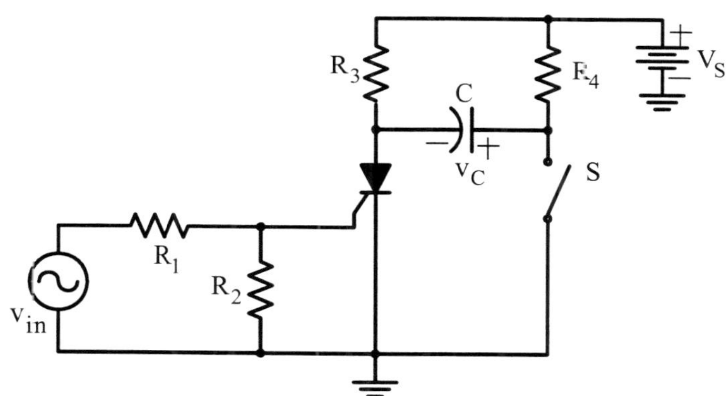

Figure 4.27. A switching circuit using an SCR

The SCR in the circuit of Figure 4.27 is switched to the on (conducting) state by applying a positive triggering pulse at v_{in} and with the SCR on, there is there is a very small voltage drop from anode to cathode. The capacitor then charges to voltage v_C and this voltage is approximately equal to V_S. The SCR will be turned off by closing the switch S, and when this occurs, we observe that the positive side of the capacitor is connected to the ground. Since the capacitor voltage cannot change instantly, the anode of the SCR becomes negative with respect to the cathode. The anode current no longer flows, and the SCR is turned Off. Obviously, for repetitive operation at fast rates the switch S in Figure 4.27 can be replaced by another controlled rectifier as shown in Figure 4.28.

As with the circuit of Figure 4.27, in Figure 4.28 a positive triggering pulse applied at v_{in1} turns the SCR on the left On. Then a subsequent positive pulse applied at v_{in2} turns the SCR on the right On; this action is equivalent to closing the switch in Figure 4.27, and it switches the SCR on the left off. The capacitor now charges with the opposite polarity, and another positive pulse applied at v_{in1} turns the SCR on the left On; with the aid of the capacitor this action turns the SCR on the right Off.

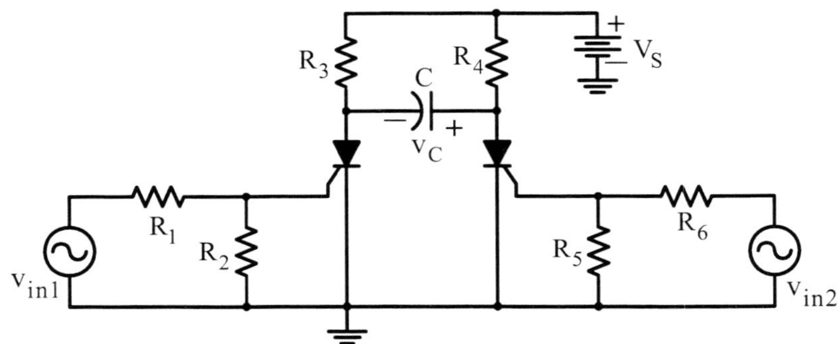

Figure 4.28. SCR circuit for repetitive operation at fast rates

If a train of positive pulses is applied simultaneously at v_{in1} and v_{in2} they have no effect on the rectifier that is conducting, but they turn the nonconducting SCR on. Again, with the aid of the capacitor this action turns the conducting SCR off. Thus a train of positive pulses applied simultaneously to the gates of the SCRs causes them to switch on and off alternately.

4.7.2 The SCR in the Generation of Sawtooth Waveforms

Another important application for SCRs is in circuits for the generation of sawtooth waveforms. Sawtooth generators have many important engineering applications. They are used in oscilloscopes to provide the sweep voltage for horizontal deflection, and they are used in circuits for the measurement of the time interval between the occurrence of two events. Figure 4.29 shows an ideal sawtooth generator.

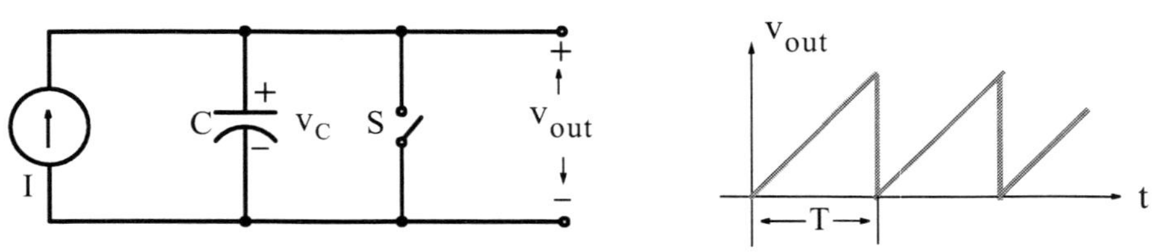

Figure 4.29. An ideal sawtooth generator and its output waveform

When the switch S is open, the current source delivers a constant current to the capacitor, and as a result the voltage across the capacitor increases linearly with time as shown in Figure 4.29.* At a certain instant the switch is closed momentarily, and the capacitor discharges through the switch. The switch is reopened immediately, and the capacitor begins to charge again. If the closing of the switch is periodic, a periodic sawtooth waveform of voltage appears at v_{out}. If the switching is not periodic, the successive peaks in the waveform are not of the same amplitude; however, the amplitude of each peak is directly proportional to the duration of that particular switching interval. A practical sawtooth generator is shown in Figure 4.30.

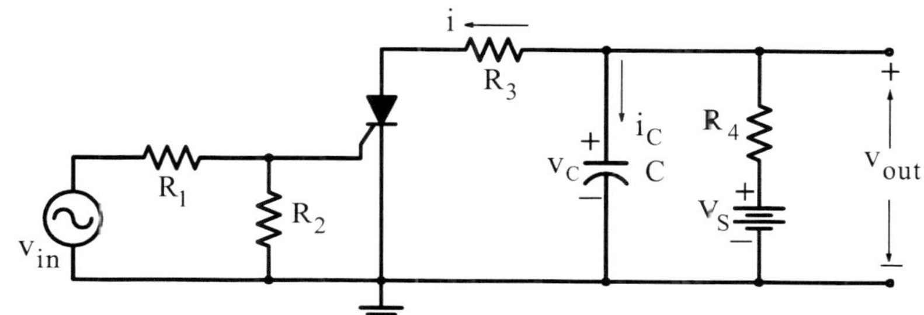

Figure 4.30. A practical sawtooth generator using an SCR as a switch

In the circuit of Figure 4.30 the SCR performs the function of the switch in the ideal generator of Figure 4.29. Resistor R_4 is a large resistor in the order of $1\ M\Omega$ and this with the voltage source V_S approximate the current source in the ideal sawtooth generator; since a true current source is not used, the output voltage shown in Figure 4.31 is not exactly linear.

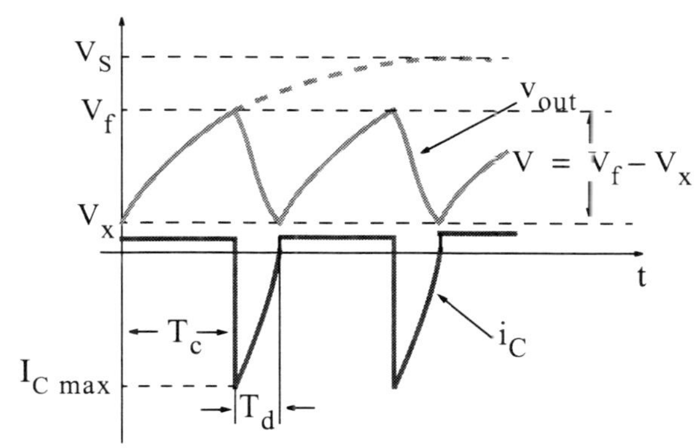

Figure 4.31. Waveforms for the circuit of Figure 4.30

The voltage v_{in} is the control voltage for the rectifier, and R_3 is a resistance that limits the rectifier current to a safe value when the capacitor is discharging.

---

* *We recall that* $v_C = \int I dt = It \text{ (ramp)}$

When the SCR is switched off, the capacitor charges with a current from the DC supply voltage V_S that is almost constant, and the capacitor voltage rises exponentially with time. The waveforms of voltage and current are shown in Figure 4.31. When the capacitor voltage reaches the critical value V_f, known as *firing voltage*, the SCR fires and discharges the capacitor rapidly. When the capacitor is discharged to the critical voltage V_x, referred to as the *extinction voltage*, the SCR switches off, and the cycle repeats itself.

The duration of the charging interval T_c is determined by the firing voltage V_f, the extinction voltage V_x, the capacitor current i_C, and the capacitance C. Since the firing voltage depends on the control voltage v_{in}, the amplitude and frequency, of the sawtooth wave can be adjusted by adjusting v_{in}. For a practical circuit triggering pulses of voltage can be applied at v_{in} to control the firing of the SCR and thereby synchronize the sawtooth with an external signal.

The duration of the discharging interval T_d depends on the amplitude of the sawtooth voltage, the discharging current, and the capacitance C. For rapid discharging, which is highly desired, the discharging current should be as large as possible; however, the current must be limited to a value that is safe for the SCR. Thus, for good performance the SCR should have a high current rating and short switching times.

All applications for the sawtooth generator require a highly linear sawtooth waveform and a short discharge time. These two quantities can be related in a very simple way to the important parameters of the circuit. During the charging interval T_c, the SCR is switched off, and the circuit is a simple series connection of R_4, C, and V_S. Thus, choosing $t = 0$ at the beginning of the charging interval, the equation for the output voltage v_{out} during that interval has the form

$$v_{out} = V_S + Ae^{(-t/R_4C)} \text{*} \tag{4.21}$$

When $t = 0$, v_{out} has its minimum value V_x as shown in Figure 4.31; hence letting $t = 0$ in (4.21) we obtain

$$v_{out} = V_S + A = V_x$$

or

$$A = V_x - V_S$$

and by substitution into (4.21)

$$v_{out} = V_S - (V_S - V_x)e^{(-t/R_4C)} \tag{4.22}$$

* *For an introduction to transient analysis, refer to Circuit Analysis I with MATLAB Applications, ISBN 978–0–9709511–2–0.*

Relation (4.22) defines the output voltage v_{out} during the charging interval T_c. Next, we let

$$(V_S - V_X) = V'_S \tag{4.23}$$

Then

$$V_S = V'_S + V_X \tag{4.24}$$

and by substitution into (4.22)

$$v_{out} = V'_S + V_X - V'_S e^{(-t/R_4C)} \tag{4.25}$$

The exponential factor in (4.25) can now be expanded in a power series to obtain the following more useful expression:

$$v_{out} = V'_S + V_X - V'_S\left[1 - \frac{t}{R_4C} + \frac{1}{2}\left(\frac{t}{R_4C}\right)^2 - \dots\right] = V_X + V'_S \cdot \frac{t}{R_4C} - \frac{1}{2}V'_S\left(\frac{t}{R_4C}\right)^2 + \dots \tag{4.26}$$

Since we want the output voltage v_{out} to be a linear function of time, our sawtooth generator must be designed so that the second term on the right side of (4.26) should be sufficiently small. Then,

$$v_{out} = V_X + V'_S \cdot \frac{t}{R_4C} \tag{4.27}$$

The right–band side of (4.26) is a converging alternating series; hence the error in truncating the series at any point is smaller than the first term in the part out off. Thus, the magnitude of the error at any instant is smaller than

$$\Delta v_{out} = \frac{1}{2}V'_S\left(\frac{t}{R_4C}\right)^2 \tag{4.28}$$

At the end of the charging interval $t = T_c$ the output voltage is

$$v_{out} = V_f = V_X + V'_S \cdot \frac{T_c}{R_4C} \tag{4.29}$$

It now follows from (4.29) and Figure 4.30 that the amplitude of the sawtooth wave is

$$V = V_f - V_X = V'_S \cdot \frac{T_c}{R_4C} \tag{4.30}$$

and the error at the end of the charging interval, which is the amount by which the actual amplitude is less than the amplitude given by (4.30), is smaller than

$$\Delta v_{out} = \frac{1}{2}V'_S\left(\frac{T_c}{R_4C}\right)^2 \tag{4.31}$$

Solving (4.30) for $T_c/(R_4C)$ and substituting the result into (4.31) we obtain

$$\Delta v_{out} = \frac{1}{2}\frac{V^2}{V'_S} \tag{4.32}$$

From (4.30) and (4.32)

$$\frac{\Delta v_{out}}{V} = \frac{V}{2V'_S} = \frac{T_c}{2R_4C} \tag{4.33}$$

Relation (4.33) reveals that the fractional error depends only on the relative sizes of the sawtooth amplitude and the effective DC supply voltage; in order to have good linearity it is necessary to have an effective supply voltage much larger than the required amplitude of the sawtooth. Good linearity is often obtained by the direct expedient of using a large supply voltage or by generating a sawtooth wave of small amplitude. In more sophisticated designs the effect of a very large supply voltage, often thousands of volts, is simulated by the use of additional circuitry in which amplifiers are frequently used. One method of simulating a large supply voltage is illustrated in Example 4.5 at the end of this section. When the capacitor is charged from an ideal current source, as in Figure 4.28, the effective supply voltage is the open–circuit voltage of the source; this voltage is infinite, giving zero error.

In designing a sawtooth generator, V_S and V'_S must be chosen so that the fractional error given by (4.33) is sufficiently small. The time constant R_4C can then be chosen to produce the desired slope to the sawtooth. With the control voltage v_{in} in Figure 4.29 held constant, the firing voltage V_f and the sawtooth amplitude V are also constant, and the slope of the sawtooth determines the length of the charging period. From Equation (4.30),

$$T_c = R_4C\frac{V}{V'_S} \tag{4.34}$$

and when a train of triggering pulses is applied at v_{in}, T_c is determined by the interval between the pulses, and the slope of the sawtooth determines the amplitude V. With a fixed slope, V is a measure of the time interval T_c.

At the end of the charging interval the SCR switches on, and the capacitor discharges exponentially through the SCR at a rate determined by the resistance R_3 in Figure 4.29. During the discharging interval the voltage across the capacitor changes by the amount

$$V = V_f - V_x = \frac{Q_f - Q_x}{C} \tag{4.35}$$

The duration of the discharging interval T_d can be determined from this relation. To simplify the calculation we will approximate the discharging current pulse, shown in Figure 4.30, with a pulse that decreases linearly with time from the peak value $I_{C\ max}$ to zero. With this approximation the average value of the current during the discharging interval is $I_{C\ max}/2$, and the charge removed from the capacitor during this interval is $T_d I_{C\ max}/2$. Thus (4.33) can be written as

$$V \approx \frac{T_d I_{C\ max}}{2C} \tag{4.36}$$

Also, since the capacitor current $I_{C\ max}$ is approximately equal to the current I_{max} through the SCR, we can express (4.36) as

$$V \approx \frac{T_d I_{max}}{2C} \tag{4.37}$$

or

$$T_d = \frac{2CV}{I_{max}} \tag{4.38}$$

Division of (4.38) by (4.34) yields

$$\frac{T_d}{T_c} = \frac{2V'_S}{R_4 I_{max}} \tag{4.39}$$

For good linearity we must have $V \ll V'_S$, and since the voltage drop across R_4 during the charging interval is approximately V'_S, and V'_S/R_4 is very nearly the charging current I_C for the capacitor, (4.37) can be expressed as

$$\frac{T_d}{T_c} = \frac{2I_C}{I_{max}} \tag{4.40}$$

where I_C is the charging current for the capacitor, and I_{max} is the peak discharging current through the SCR. To provide a relatively short discharge time, the peak SCR current must be much greater than the capacitor charging current.

Any load that is connected across the output terminals of the sawtooth generator in Figure 4.30 will affect the operation of the circuit, and any leakage resistance that develops across the capacitor has a similar effect. Figure 4.32(a) shows a sawtooth generator with a load resistor R_L connected across the output terminals.

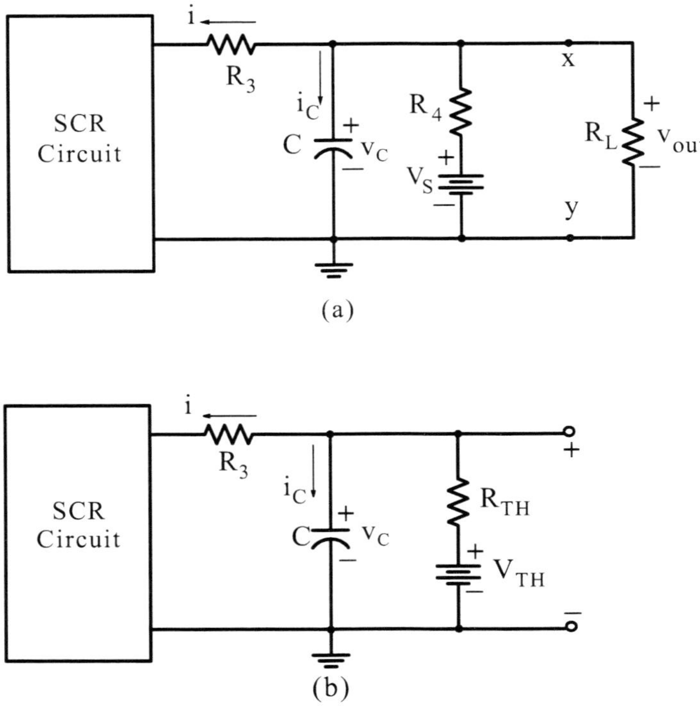

Figure 4.32. A sawtooth generator with resistive load and its Thevenin equivalent

To simplify the circuit of Figure 4.32(a), we apply Thevenin's theorem at points $x - y$ and we obtain the circuit of Figure 4.32(b) where

$$V_{TH} = \frac{R_L}{R_4 + R_L} \cdot V_S \tag{4.41}$$

and

$$R_{TH} = \frac{R_4 R_L}{R_4 + R_L} \tag{4.42}$$

It is then evident that the addition of the load resistor R_L affects both the slope and the linearity of the sawtooth waveform. The presence of R_L affects the effective DC supply voltage, and hence on the linearity, is especially severe. Therefore, it becomes necessary to make R_L greater than $1\ M\Omega$. A load resistor of about $100\ K\Omega$ would reduce the effective supply voltage and will increase the fractional error by a factor of more than 10. It is quite possible in such cases for the effective voltage V_{TH} to be less than the firing voltage V_f, for the SCR; then the rectifier never fires, and no sawtooth wave is generated. Accordingly, the output from the sawtooth generator must be connected to a very high impedance device such as a MOSFET. The output can also be delivered to a bipolar junction transistor, but a very special circuit configuration must be used to provide the required high input resistance.

Example 4.5

A sawtooth generator is shown in Figure 4.33(a). The capacitor is charged through a transistor in order to approximate a current source supply. Figure 4.33(b) shows how an SCR can be used as the discharge switch for the capacitor. We want to design the circuit to produce a 1 KHz sawtooth wave having a fractional error smaller than 1 per cent. The SCR can be assumed ideal with a maximum permissible current rating of 100 mA.

Solution:

With the capacitor in Figure 4.33(b) discharged there is no voltage across the SCR; thus the rectifier is switched off, and its cathode is at a potential of 20 volts. The potential at the gate terminal of the rectifier is fixed by R_2 and R_3, and it is less than 20 volts; thus a reverse bias is applied to the gate. As the capacitor charges through the transistor from the 20–volt supply, the potential at the cathode of the SCR drops, but the SCR remains switched off until the cathode potential drops below the gate potential. At this point the gate.to–cathode voltage becomes positive and the SCR switches on. The capacitor then discharges almost to zero volts, and the SCR switches off, provided the current drawn by the collector of the transistor is less than the holding current for the SCR. The cycle then repeats itself periodically.

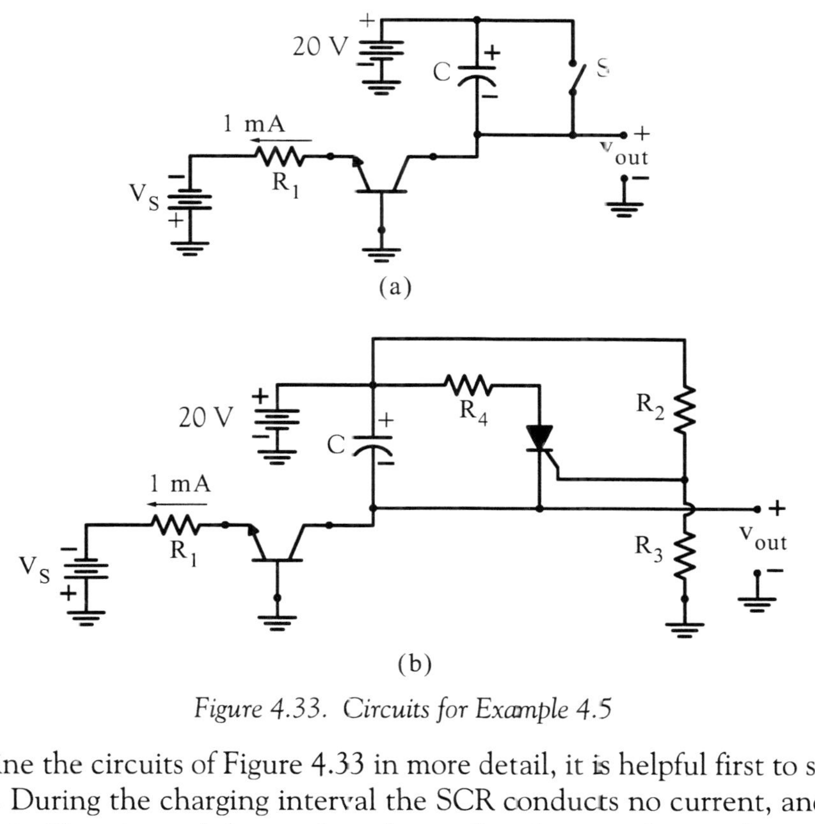

Figure 4.33. Circuits for Example 4.5

In order to examine the circuits of Figure 4.33 in more detail, it is helpful first to simplify and rearrange the circuit. During the charging interval the SCR conducts no current, and it can be omitted from the circuit. If V_S is much larger than the small emitter–to–base voltage drop, the transistor can be represented to a very good approximation by the model shown in Figure 4.34(a) where E, C, and B are the transistor terminals. The resistance r_C in this model is the collector resistance of the transistor.

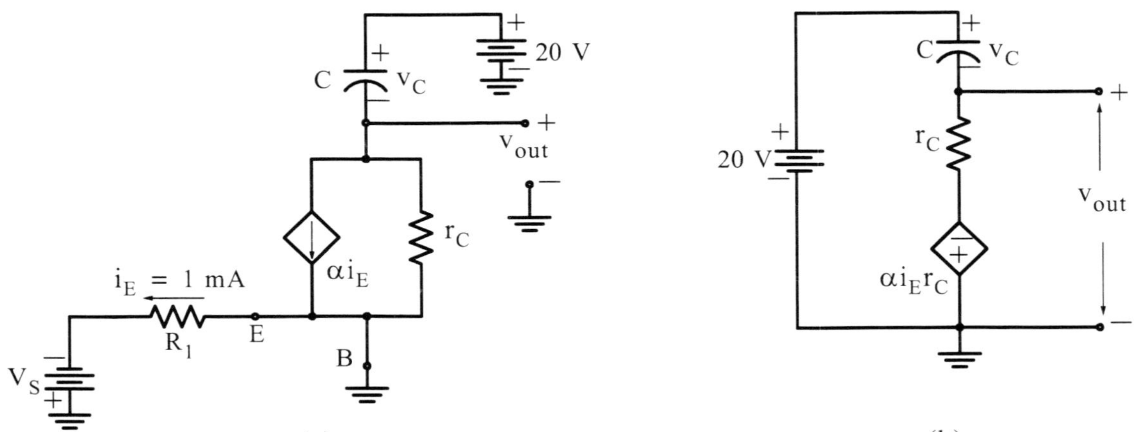

Figure 4.34. Piecewise–linear and equivalent circuits for the sawtooth generator of Figure 4.33

In most circuits the collector resistance is in parallel with a much smaller load resistance and it can be neglected; this is not the case in the circuit of Figure 4.34(a), however, and the collector resistance has an important effect. As the next step in the simplification of the circuit, it is noted that because of the short circuit between base and ground, the emitter circuit affects the collector circuit only through the action of the controlled source αi_E. Thus when i_E is known, the emitter circuit can be ignored. And finally, the current source and its shunt resistance can be converted to an equivalent voltage source with a series resistance. The resulting equivalent circuit is shown in Figure 4.34(b). From relation (4.23), the effective supply voltage is

$$V'_S = V_S - V_X \tag{4.43}$$

and for our example,

$$V_S = 20 + \alpha i_E r_C \tag{4.44}$$

The equivalent circuit of Figure 4.34(b) shows that by charging the capacitor through the transistor an effective supply voltage of

$$V'_S = 20 + \alpha i_E r_C - V_X \tag{4.45}$$

is obtained. The extinction voltage V_X, is a negligible fraction of a volt, and the emitter current i_E is 1 ma. Thus taking typical values of $\alpha = 0.98$ and $r_C = 1\ M\Omega$, relation (4.45) yields

$$V'_S = 20 + 980 = 1000\ V \tag{4.46}$$

For a fractional error of 1 per cent in the amplitude of the sawtooth, relation (4.33) yields

$$\frac{\Delta v_{out}}{V} = 0.01 = \frac{V}{2V'_S} \tag{4.47}$$

and thus

$$V = 20\ V \tag{4.48}$$

Thus, in principle the amplitude of the sawtooth output can be as large as the DC supply voltage; however, to avoid possible non linearities in the transistor characteristics at low voltages and to allow for possible variations in the supply voltage, a value of 15 volts might be chosen for V. Accordingly, the values of resistors R_2 and R_3 should be chosen to make the gate–terminal potential about 5 volts when the SCR is switched off.

To limit the SCR current to the maximum safe value of 100 ma, the value of resistor R_4 should be

$$R_4 = \frac{V}{I_{max}} = \frac{15}{0.1} = 150\ \Omega \tag{4.49}$$

For a sawtooth frequency of $1\ KHz$,

$$T_c + T_d = T_c\left(1 + \frac{T_d}{T_c}\right) = \frac{1}{f} = 0.001\ sec \tag{4.50}$$

From (4.40)

$$\frac{T_d}{T_c} = \frac{2I_C}{I_{max}} \tag{4.51}$$

With $I_C \approx \alpha i_E = 1\ mA$, $I_{max} = 100\ mA$ and using (4.50) and (4.51) we obtain

$$T_c\left(1 + 2\frac{1}{100}\right) = 1.02T_c = 0.001 \tag{4.52}$$

and thus

$$T_c = 0.00098\ sec \tag{4.53}$$

Finally, from (4.34)

$$T_c = R_4 C \frac{V}{V'_S}$$

or

$$C = \frac{T_c}{R_4}\frac{V'_S}{V} = \frac{0.00098}{10^6} \cdot \frac{1000}{15} = 0.0655\ \mu F \tag{4.54}$$

where $R_4 = r_C = 1\ M\Omega$ is as shown in Figure 4.34(b).

4.8 The Triac

The *triac* is a device capable of switching on for either polarity of an applied voltage. As can be seen in Figure 4.35, the triac has a single gate lead and is therefore the AC equivalent of the SCR. A triac is essentially two SCRs connected back to back with a common gate and common terminals where each terminal is the anode of one SCR and the cathode of another.

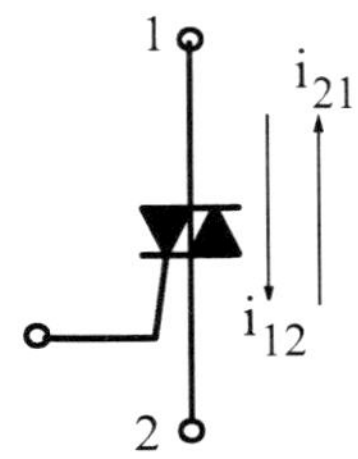

Figure 4.35. Triac symbol

Figure 4.36 shows an SCR circuit and a triac circuit for comparison. The diode D in the SCR circuit ensures a positive trigger voltage. Figure 4.37 shows a comparison of the waveforms at the input, gate, and output of these two devices.

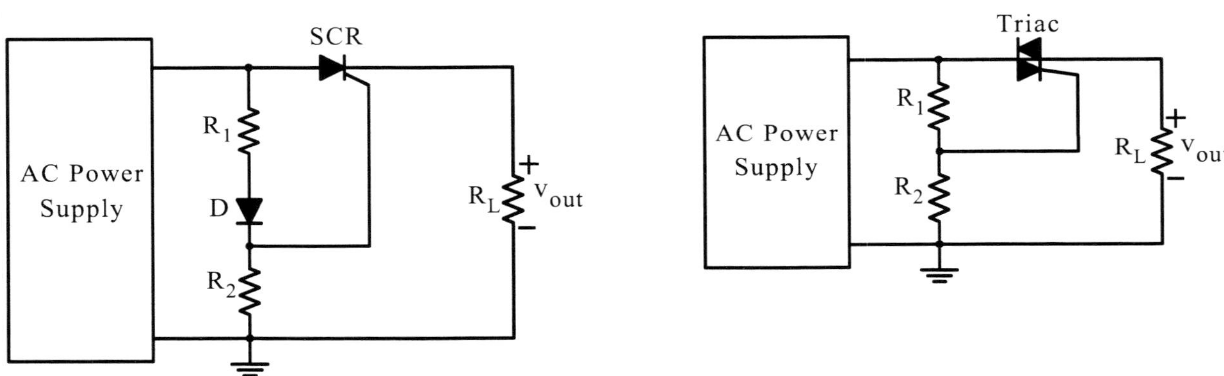

Figure 4.36. Comparison of SCR and Triac circuits

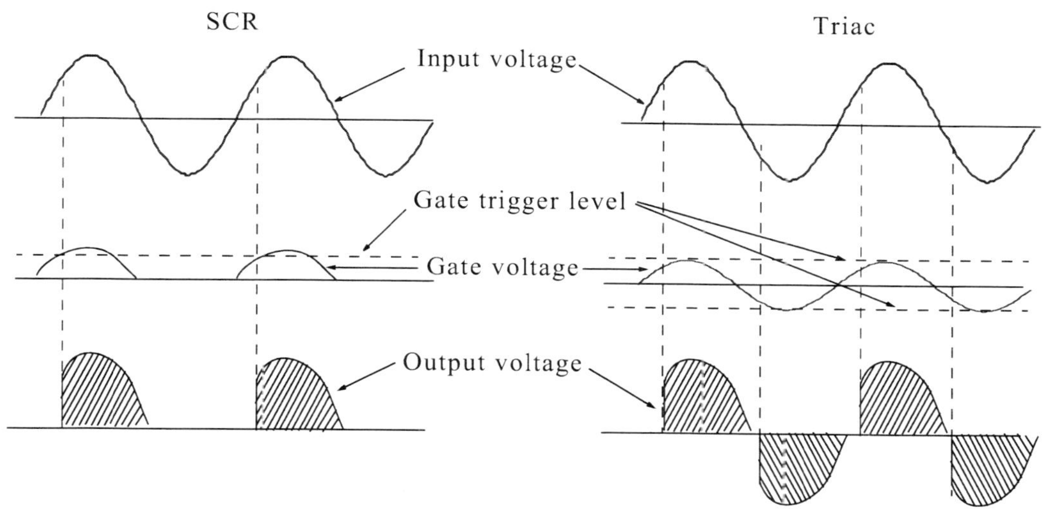

Figure 4.37. Comparison of SCR and Triac waveforms

The triac exhibits voltage–current characteristics similar to those of the SCR. Applications for triacs include AC motor–speed control, AC light dimmers, and general AC power–control applications.

4.9 The Shockley Diode*

The *Shockley diode* or *four–layer diode*, or *PNPN diode*, is a four–layer sandwich of P–N–P–N semiconductor material very similar to the SCR but without a gate as shown in Figure 4.38.

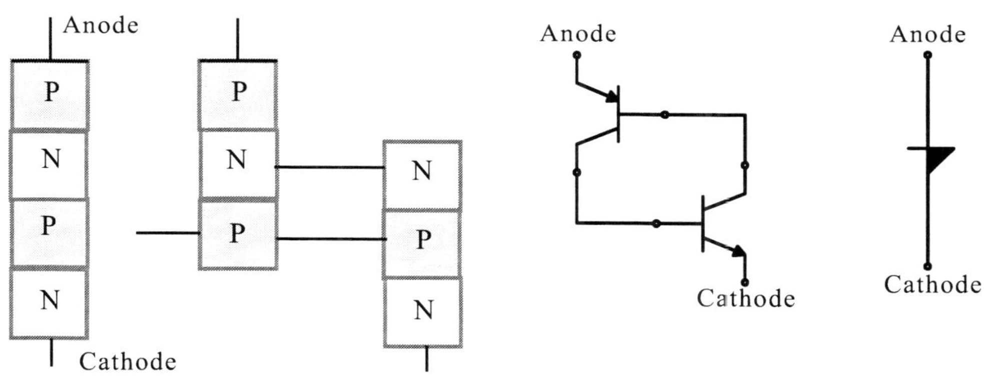

Figure 4.38. Parts of a Shockley diode, the two–transistor equivalent circuit, and its symbol

A Shockley diode can be to turned on by applying sufficient voltage between anode and cathode. This voltage will cause one of the transistors to turn on, which then turns the other transistor on,

* *The Shockley diode should not be confused with the Schottky diode, the two–layer metal–semiconductor device known for its high switching speed. We discussed the Schottky diode in Chapter 2.*

ultimately latching both transistors on where they will tend to remain. The two transistors can be turned off again by reducing the applied voltage to a much lower point where there is too small current to maintain transistor bias, at which point one of the transistors will cutoff, which then halts base current through the other transistor, sealing both transistors in the off state as they were before any voltage was applied at all. In other words, the Shockley diode tends to stay on once it's turned on, and stay off once it's turned off. There is no in–between or active mode in its operation: it is a purely on or off device, as are all thyristors.

There are a few special terms applied to Shockley diodes and all other thyristor devices built upon the Shockley diode foundation. First is the term used to describe its on state: latched. The word latch is reminiscent of a door lock mechanism, which tends to keep the door closed once it has been pushed shut. The term firing refers to the initiation of a latched state. In order to obtain a Shockley diode to latch, the applied voltage must be increased until breakover is attained. Despite the fact that this action is best described in terms of transistor breakdown, the term breakover is used instead because the end result is a pair of transistors in mutual saturation rather than destruction as would be the case with a normal transistor. A latched Shockley diode is reset back into its nonconducting state by reducing current through it until low–current dropout occurs.

It should be noted that Shockley diodes may be fired in a way other than breakover: excessive voltage rise, or `dv/dt`. This is when the applied voltage across the diode increases at a high rate of change. This is able to cause latching (turning on) of the diode due to inherent junction capacitances within the transistors. Capacitors, as we recall, oppose changes in voltage by drawing or supplying current. If the applied voltage across a Shockley diode rises at too fast a rate, those tiny capacitances will draw enough current during that time to activate the transistor pair, turning them both on. Usually, this form of latching is undesirable, and can be minimized by filtering high–frequency (fast voltage rises) from the diode with series inductors and/or parallel resistor–capacitor networks. The voltage rise limit of a Shockley diode is referred to as the *critical rate of voltage rise* and it is provided by the manufacturer.

4.10 Other PNPN Devices

Other PNPN devices include the *light–activated* SCR (LASCR), *silicon unilateral switch* (SUS), *silicon bilateral switch* (SBS), and *light–emitting four–layer diodes.* The LASCR is a conventional SCR installed in a housing having a transparent window or collection lens. Operation of the LASCR is similar to that of a conventional SCR except an optical signal replaces the gate electrical signal.

LASCRs exhibit very–high gain and permit relatively large amounts of current to be controlled by a relatively weak optical signal. The *light–activated silicon controlled switch* (LASCS) is an LASCR with both anode and cathode gate terminals. The SUS is an SCR with an anode gate instead of the usual cathode gate and a self–contained low–voltage avalanche diode between the gate and cathode. The SBS is two SUS devices arranged in inverse–parallel to permit bidirectional switching.

Light–emitting four–layer diodes are gallium–arsenide devices which emit recombination radiation when they are switched on. These devices are unique in that exceptionally simple optical pulse sources can be fabricated since the source is also the switching element. A particularly interesting device of this kind is the PNPN *injection laser*.

The *silicon controlled switch* (SCS) is very similar to the SCR but all four regions are accessible to the external circuit. as shown in Figure 4.39.

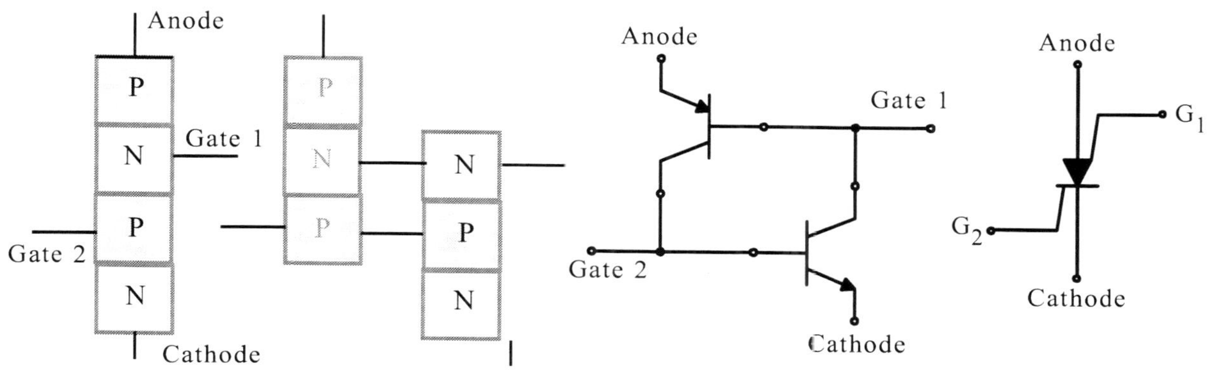

Figure 4.39. Parts of an SCS, the two–transistor equivalent circuit, and its symbol

As shown in Figure 4.39, the basic construction of the SCS is the same as for the SCR, with the addition of a second gate lead. Thus the SCS has an anode, a cathode, an anode gate, and a cathode gate. The SCS has two advantages over the SCR and the Shockley diode. First, because both gate regions are accessible, they can be biased so as to completely cancel the rate effect we described with the four–layer diode. Second, since we can now control both end junctions, we can actively turn the SCS off without the need to reduce the applied voltage or current.

4.11 The Future of Transistors

Presently, a transistor is barely used as an amplifier, as op amps are the most versatile amplifier devices. But transistors are extensively as switching devices also known as latches. Recently, Hewlett–Packard announced a successful simulation of a latch that may eventually replace the transistor in digital circuits. A simplified latch is shown in Figure 4.40, and it is reproduced from the Journal of Applied Physics, 2005.

As shown in Figure 4.40(a), a logic 1 is placed on the line by closing Switch S_1. Likewise, we can place a logic 0 on the line by closing Switch S_0 as shown in Figure 4.40(b). The switches represent bistable two–terminal switches and hopefully their sizes can be made much smaller than the three–terminal transistor.

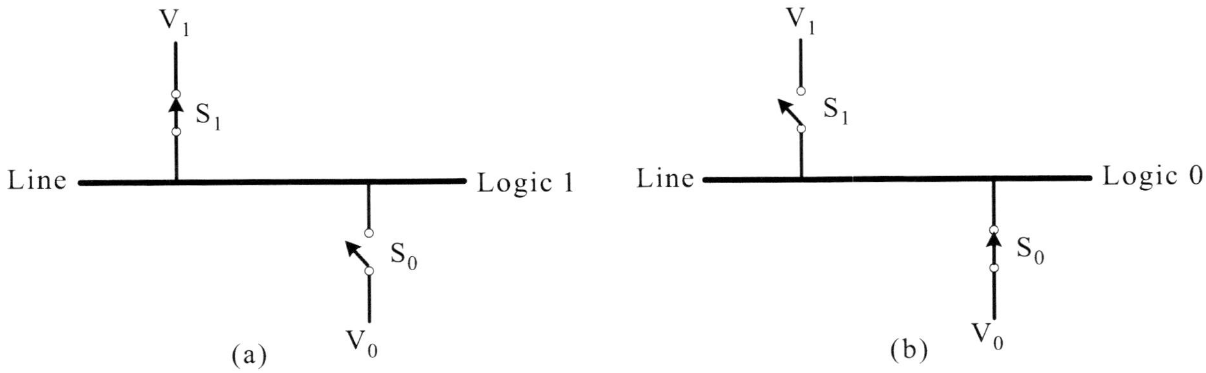

Figure 4.40. A simplified arrangement of a latch as a possible replacement for the transistor

4.12 Summary

- The Field–Effect Transistor (FET) is another semiconductor device. The FET is a voltage–controlled device.The Junction FET (JFET) is the earlier type and the Metal Oxide Semiconductor FET (MOSFET) is now the most popular type. A FET has four terminals, drain, source, gate and substrate. The substrate is normally connected to the source and thus a FET is essentially a three–terminal device.

- One of the most attractive features of the FET is the fact that the input resistance, measured between gate and source, can be made very large, from 1 to 100 megohms.

- N–channel FETs are the most common. P–channel FETs operate similarly except that the voltage polarities are reversed.

- One important parameter in FETs is its transconductance g_m defined as the ratio of the change in current i_{DS} to the change of voltage v_{GS} which produced it. In other words,

$$g_m = \left.\frac{\partial i_{DS}}{\partial v_{GS}}\right|_{v_{DS} = \text{constant}}$$

  In other words, the transconductance is a measure of the sensitivity of drain current to changes in gate–to–source bias.

- A MOSFET is said to operate in the depletion mode when a negative voltage is applied at the gate. The JFET also operates in this manner. A MOSFET is said to operate in the enhancement mode if a positive voltage is applied at the gate. A MOSFET can be constructed that will operate in either mode depending upon what type of bias is applied, thus allowing a greater range of input signals.

- In a MOSFET, the voltage at which the channel is closed is known as the pinch–off voltage V_P, and the minimum voltage required to form a conducting channel between the drain and source is referred to as the threshold voltage and it is denoted as V_T. Typical values for V_T are 2 to 4 volts for high voltage devices with thicker gate oxides, and 1 to 2 volts for lower voltage devices with thinner gate oxides.

- In a MOSFET, the three regions of operation are the cutoff, the quadratic or triode, and the saturation or pentode. In the cutoff mode, $v_{GS} < V_T$ and no drain current flows. In the quadratic region the drain current is determined from the expression

$$i_D = k_n \frac{W}{L}\left[(v_{GS} - V_T)v_{DS} - \frac{1}{2}v_{DS}^2\right] \qquad \text{for } v_{DS} \le v_{GS} - V_T$$

  and in the saturation region from the expression

$$i_D = \frac{1}{2} \cdot k_n \frac{W}{L}(v_{GS} - V_T)^2 \qquad \text{for } v_{DS} \ge v_{GS} - V_T$$

where k_n is a constant depending on electron mobility, and oxide thickness, permittivity, and capacitance for n–channel MOSFETs. For p–channel MOSFETs this parameter is denoted as k_p. The ratio W/L denotes the ratio of the channel width W to channel length L. Typical values of k_n are around $20\ \mu A/V^2$, for k_p, around $10\ \mu A/V^2$, and for W/L about 10.

- For small signals, transconductance is defined as

$$g_m = \frac{i_d}{v_{gs}} = k_n \frac{W}{L}(V_{GS} - V_T)$$

- The output conductance is defined as

$$g_o = \left.\frac{\partial i_D}{\partial v_{DS}}\right|_{V_{GS} = \text{constant}}$$

- The voltage gain A_v in a MOSFET device is defined as

$$A_v = \frac{v_d}{v_{gs}} = -R_D g_m$$

and the minus (–) sign indicates $180°$ phase reversal.

- In high density integrated circuits, current mirrors are normally used in lieu of resistors.
- Complementary MOS or CMOS technology combines one NMOS device and one PMOS device into a single device referred to as CMOS. These devices are used extensively in both analog and digital circuits, and integrated circuits. Also, CMOS devices are the most common amplifiers in the FET technology.
- In CMOS devices only one of its components is on at any given time, that is, either the NMOS device is on and the PMOS device is off, or vice versa. Thus, CMOS chips require less power than chips using just one type of a MOSFET, and unlike bipolar transistors, a CMOS has almost no static power dissipation.
- A CMOS device can be configured as a common–source amplifier, common–gate amplifier, and common–drain or source follower.
- The Metal–Semiconductor–Field–Effect–Transistor (MESFET) consists of a conducting channel positioned between a source and drain contact region. The carrier flow from source to drain is controlled by a Schottky metal gate. The control of the channel is obtained by varying the depletion layer width underneath the metal contact which modulates the thick-

ness of the conducting channel and thereby the current MESFETs use GaAs (gallium arsenide) technology. Only n – channel MESFETS are available because holes have a slower mobility.

- The unijunction transistor (UJT) is a three–terminal, single–junction device which exhibits negative resistance and switching characteristics totally unlike those of conventional bipolar transistors. UJTs can be used for the design of pulse and sawtooth generators, analog–to–digital converters, relay time delay circuits, and frequency dividers. Other UJT devices, referred to as programmable *UJTs*, can have their parameters set by external components such as resistors and capacitors.
- The diac is a two–terminal, transistor–like component which exhibits bistable switching for either polarity of a suitably high applied voltage. The diac closely resembles a PNP transistor without an external base terminal. Its major application is in conjunction with a triac to produce AC phase–control circuits useful for motor–speed control, light dimming, and other ac power–control functions.
- The silicon controlled rectifier (SCR) is a four–layer semiconducting device, with each layer consisting of an alternately N or P–type material, for example N–P–N–P. The main terminals, labelled *anode* and *cathode*, are across the full four layers, and the control terminal, called the *gate*, is attached to one of the middle layers. Another name for the SCR is thyristor. It is similar to a diode with an additional terminal that is used to turn it on. Once turned on, the SCR will remain on as long as current flows through it. If the current falls to zero, the SCR behaves like an open switch.
- The triac is a device capable of switching on for either polarity of an applied voltage; It is the AC equivalent of the SCR.
- The Shockley diode, (not to be confused with the Schottky diode), or four–layer diode, or PNPN diode, is a four–layer sandwich of P–N–P–N semiconductor material very similar to the SCR but without a gate.
- The silicon controlled switch (SCS) is very similar to the SCR but all four regions are accessible to the external circuit.

4.13 Exercises

1. For the JFET amplifier circuit below, prove that the voltage gain A_V depends only on the transconductance g_m and the value of the drain resistor R_D, that is, show that $A_V = -g_m R_D$.

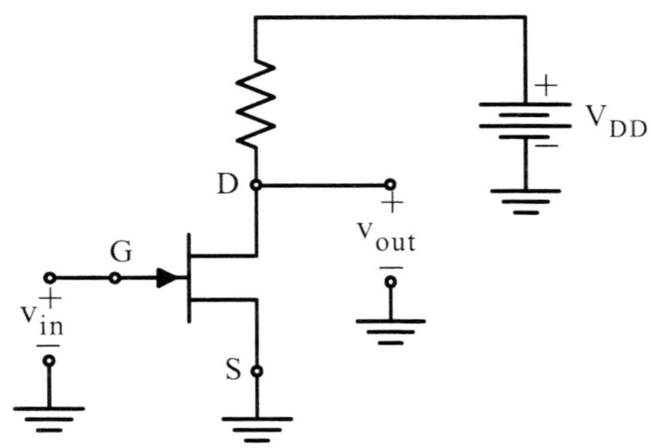

2. Compute the value of r_{DS} for an NMOS device where $k = 18\ \mu A/V^2$, $L = 5\ \mu m$, $W = 60\ \mu m$, $V_T = 2\ V$, and $v_{GS} = 4\ V$.

3. Draw an equivalent circuit that represents the n-channel MOSFET in the saturation mode.

4. An n-channel MOSFET with $k_n = 50 \times 10^{-6}$, $W = 10\ \mu m$, $L = 1\ \mu m$, and $V_T = 1\ V$, is biased with $v_{GS} = 3\ V$ and $v_{DS} = 5\ V$.

   a. Compute the drain current i_D

   b. Compute the transconductance g_m

   c. Compute the output conductance g_o if $V_{DS} = 2\ V$

   d. Compute the output conductance g_o if $V_{DS} = 0\ V$

5. An enhancement-type p-channel MOSFET with $k_p = 10 \times 10^{-6}$, $W = 10\ \mu m$, $L = 1\ \mu m$, and $V_T = -2\ V$, is biased with $v_G = 0\ V$ and $v_S = 5\ V$. What is the highest voltage of v_D that will keep the device in saturation?

6. In the circuit below, the SCR is used to control the power delivered to the $50\ \Omega$ load by the sinusoidal source. As shown, the gate supply V_{GG} is adjustable.

   a. Over what range may the conduction angle of the SCR be continuously varied?

   b. Over what range may the load DC current be continuously varied if the frequency is $60\ Hz$?

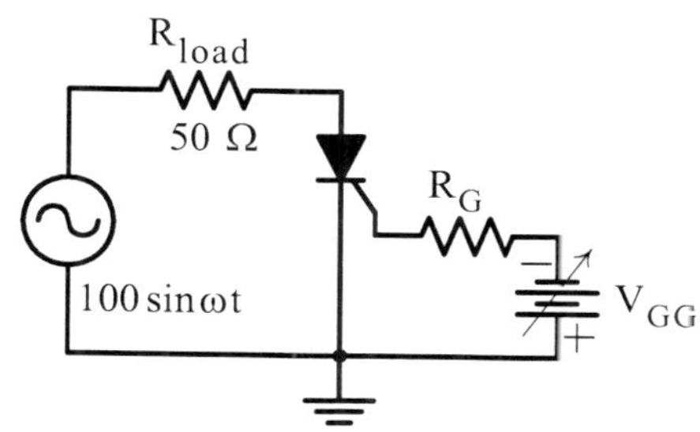

7. The circuit below is in steady state with the switch open and the SCR is in the conducting state. The voltage drop across the SCR is 1 V while it is conducting. The switch is then closed at $t = 0$, and it is assumed that the SCR switches instantly to the non–conducting state and remains non–conducting thereafter.

   a. What are the initial and final values of the voltage across the capacitor with the polarities shown?

   b. If the anode potential of the SCR must remain negative for 10 μs to ensure that the SCR switches to the non–conducting state, what is the minimum value of the capacitor C that should be used?

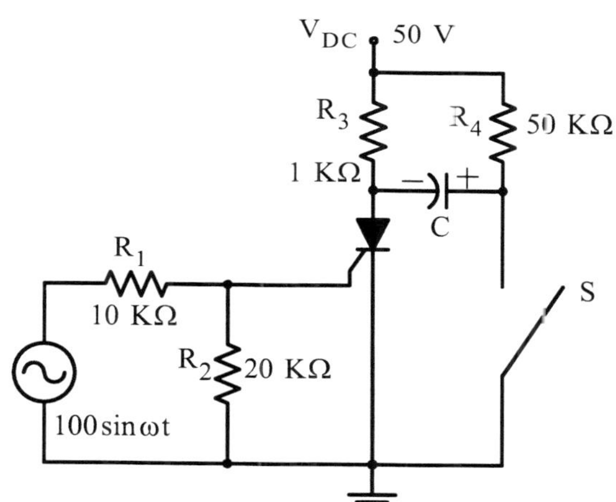

4.14 Solutions to End-of-Chapter Exercises

1.

By definition

$$g_m = di_{DS}/dv_{GS} = di_{DS}/dv_{in} \quad (1)$$

and since

$$v_{out} = V_{DD} - R_D i_{DS}$$

it follows that

$$dv_{out} = -R_D di_{DS}$$

and with (1)

$$dv_{out} = -g_m R_D dv_{in}$$

Thus, the voltage gain is

$$A_v = \frac{dv_{out}}{dv_{in}} = -g_m R_D$$

2. From (4.10),

$$r_{DS} \equiv \frac{v_{DS}}{i_D} = \frac{L}{kW(v_{GS} - V_T)} = \frac{5 \times 10^{-6}}{18 \times 10^{-6} \times 60 \times 10^{-6}(4-2)} = 2.3\ \text{K}\Omega$$

3. In the saturation mode the n – channel MOSFET behaves as an ideal current source whose value is as in (4.11) and it can be represented by the equivalent circuit shown below.

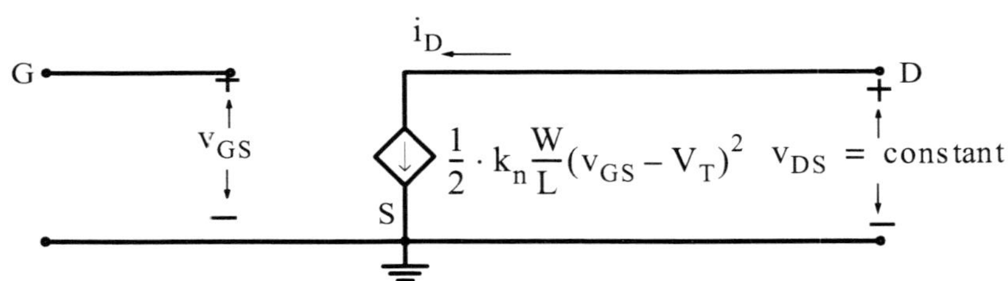

4.

a. Since $v_{DS} > v_{GS} - V_T \Rightarrow 5 > 3 - 1$, the device is in saturation and the drain current i_D is found from (4.11). Thus,

$$i_D = \frac{1}{2} \cdot k_n \frac{W}{L}(v_{GS} - V_T)^2 = \frac{1}{2} \cdot 50 \times 10^{-6} \cdot \frac{10}{1}(3-1)^2 = 1\ \text{mA}$$

b. The transconductance g_m is found from (4.13). Then,

$$g_m = k_n \frac{W}{L}(v_{GS} - V_T) = 50 \times 10^{-6} \cdot \frac{10}{1}(3-1) = 1\ \text{m}\Omega^{-1}$$

c. The output conductance g_m is found from (4.16). Then,

$$g_o = k_n \frac{W}{L}(v_{GS} - V_T - v_{DS}) = 50 \times 10^{-6} \cdot \frac{10}{1}(3 - 1 - 2) = 0$$

This result is not surprising. With reference to Figure 4.14 we observe that for $v_{DS} \geq v_{GS} - V_T$ the slope is zero.

d. If $V_{DS} = 0\ V$, the device is in the triode region since $v_{DS} < v_{GS} - V_T \Rightarrow 0 < 3 - 1$. Then,

$$g_o = k_n \frac{W}{L}(v_{GS} - V_T - v_{DS}) = 50 \times 10^{-6} \cdot \frac{10}{1}(3 - 1 - 0) = 1\ m\Omega^{-1}$$

5.

The device will be in saturation as long as $v_{DS} \leq v_{GS} - V_T$. For this exercise,

$$v_{GS} = v_G - v_S = 0 - (-5) = 5\ V$$

$$v_{DS} = v_D - v_S = v_D - (-5) = v_D + 5$$

$$v_D + 5 = v_{GS} - V_T = 5 - (-2) = 7$$

$$v_D = 2\ V$$

6.

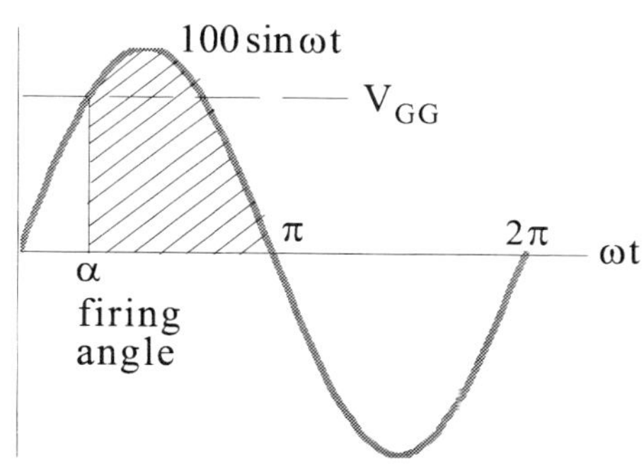

a. The SCR will fire for the interval $0 \leq \omega t \leq \pi/2$ or not at all depending on the value of V_{GG}.

b. The load DC current I_{DC} is the average value for the interval $0 \leq \alpha \leq \pi/2$ and its value is determined by the integral

$$I_{DC} = \frac{1}{2\pi}\int_{\alpha}^{\pi} \frac{100 \sin \omega t}{50} d(\omega t) = \frac{1}{\pi}(-\cos \omega t)\Big|_{\alpha}^{\pi}$$

If $\alpha = 0$, $I_{DC} = 2/\pi = 0.637\ A$

If $\alpha = \pi/2$, $I_{DC} = 1/\pi = 0.318\ A$

If SCR never fires, $I_{DC} = 0$

7.

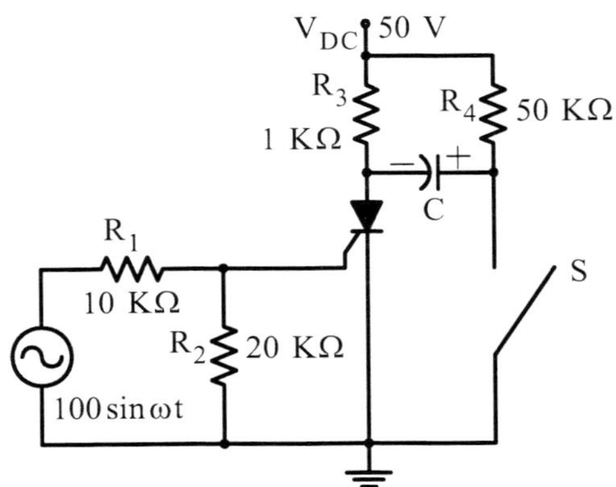

a. Just before the switch closes, the voltage drop across the SCR is 1 V and the voltage drop across the capacitor is 49 V with the polarity as shown. The capacitor voltage cannot change instantaneously so immediately after the switch is closed, the capacitor voltage will be −49 V as shown in the figure below, and the SCR will be Off.

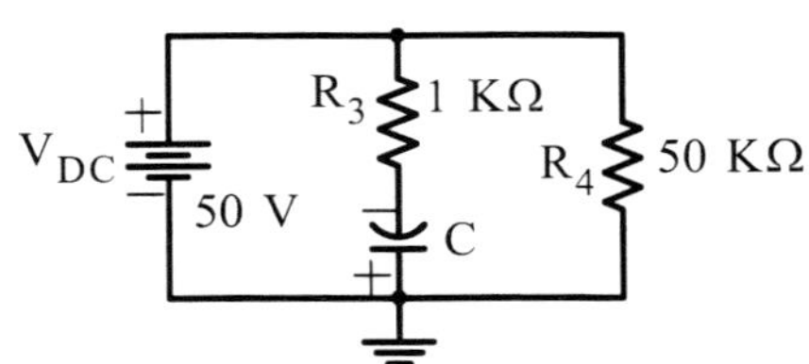

The capacitor voltage will eventually charge to 50 V with respect to the ground and thus it will undergo a change from −49 V (initial value) to +50 V (final value).

b. We want the SCR anode to remain negative for 10 μs, that is, until the voltage across the capacitor reaches the value of 0 V. Using the general formula for the capacitor voltage, we must have

$$v_C(t) = V_\infty - (V_\infty - V_{initial})e^{-t/(RC)}$$

or

$$0 = 50 - [50 - (-49)]e^{-t/(RC)}$$

$$e^{-t/(RC)} = 50/99$$

$$-t/(R_3C) = \ln(50/99)$$

$$C = \frac{-t/R_3}{\ln(50/99)} = \frac{(-10 \times 10^{-6})/10^3}{-0.683} = 14.64 \text{ nF}$$

Chapter 5

Operational Amplifiers

This chapter begins with an introduction to operational amplifiers (op amps), characteristics, and applications. We will discuss the ideal op amp, analysis of circuits in the inverting and non–inverting configurations, and gain and bandwidth on circuit performance. We will also introduce some circuits consisting of op amps and non–linear devices, and analog computers.

5.1 The Operational Amplifier

The *operational amplifier* or simply *op amp*, is the most versatile electronic amplifier. It derives it name from the fact that it is capable of performing many mathematical operations such as addition, multiplication, differentiation, integration, and analog–to–digital conversion or vice versa. It can also be used as a comparator and electronic filter. It is also the basic block in analog computer design. Its symbol and pinouts are shown in Figure 5.1.

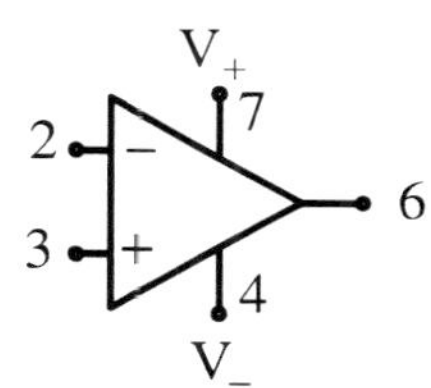

Figure 5.1. Symbol and pinouts for operational amplifier

As shown above the op amp has two inputs but only one output. For this reason it is referred to as *differential input, single ended output* amplifier. Figure 5.2 shows the internal construction of the popular 741 op amp. This figure also shows terminals V_+ and V_-. These are the voltage sources required to power up the op amp. Typically, V_+ is $+15$ volts and V_- is -15 volts. These terminals are not shown in op amp circuits since they just provide power, and do not reveal any other useful information for the op amp's circuit analysis.

5.2 An Overview of the Op Amp

The op amp has the following important characteristics:

1. Very high input impedance (resistance)
2. Very low output impedance (resistance)
3. Capable of producing a very large gain that can be set to any value by connection of external resistors of appropriate values

4. Frequency response from DC to frequencies in the MHz range
5. Very good stability
6. Operation to be performed, i.e., addition, integration etc. is done externally with proper selection of passive devices such as resistors, capacitors, diodes, and so on.

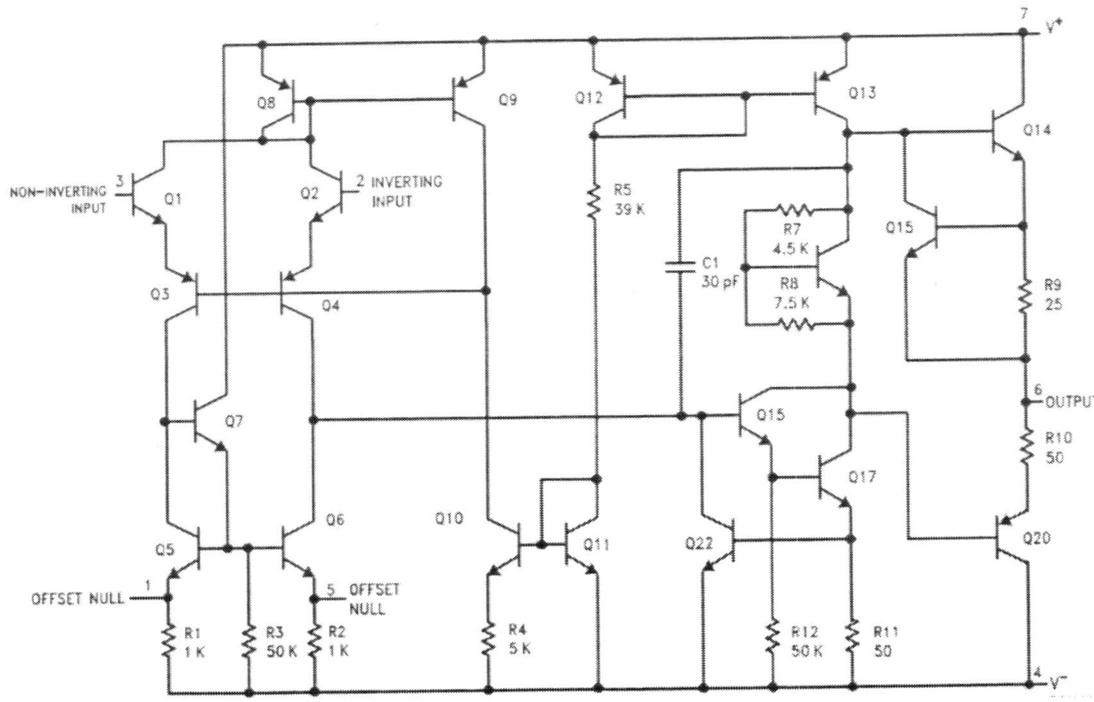

Figure 5.2. The 741 op amp (Courtesy National Semiconductor)

5.3 The Op Amp in the Inverting Mode

An op amp is said to be connected in the *inverting mode* when an input signal is connected to the inverting (–) input through an external resistor R_{in} whose value along with the feedback resistor R_f determine the op amp's gain. The non–inverting (+) input is grounded through an external resistor *R* as shown in Figure 5.3.

For the circuit of Figure 5.3, the voltage gain G_v is

$$G_v = \frac{v_{out}}{v_{in}} = -\frac{R_f}{R_{in}} \tag{5.1}$$

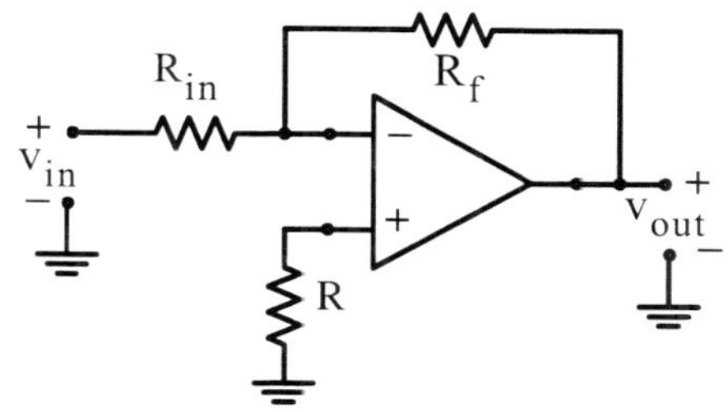

Figure 5.3. Circuit of inverting op amp

Note 1: In the inverting mode, the resistor R connected between the non–inverting (+) input and ground serves only as a current limiting device, and thus it does not influence the op amp's gain. So its presence or absence in an op amp circuit is immaterial.

Note 2: The input voltage v_{in} and the output voltage v_{out} as indicated in the circuit of Figure 5.3, should not be interpreted as open circuits; these designations imply that an input voltage of any waveform may be applied at the input terminals and the corresponding output voltage appears at the output terminals.

As shown in the relation of (5.1), the gain for this op amp configuration is the ratio $-R_f/R_{in}$ where R_f is the feedback resistor which allows portion of the output to be fed back to the input. The minus (–) sign in the gain ratio $-R_f/R_{in}$ implies that the output signal has opposite polarity from that of the input signal; hence the name inverting amplifier. Therefore, when the input signal is positive (+) the output will be negative (–) and vice versa. For example, if the input is +1 volt DC and the op amp gain is 100, the output will be –100 volts DC. For AC (sinusoidal) signals, the output will be 180° out–of–phase with the input. Thus, if the input is 1 volt AC and the op amp gain is 5, the output will be –5 volts AC or 5 volts AC with 180° out–of–phase with the input.

Example 5.1

Compute the voltage gain G_v and then the output voltage v_{out} for the inverting op amp circuit shown in Figure 5.4, given that $v_{in} = 1\ mV$. Plot v_{in} and v_{out} as mV versus time on the same set of axes.

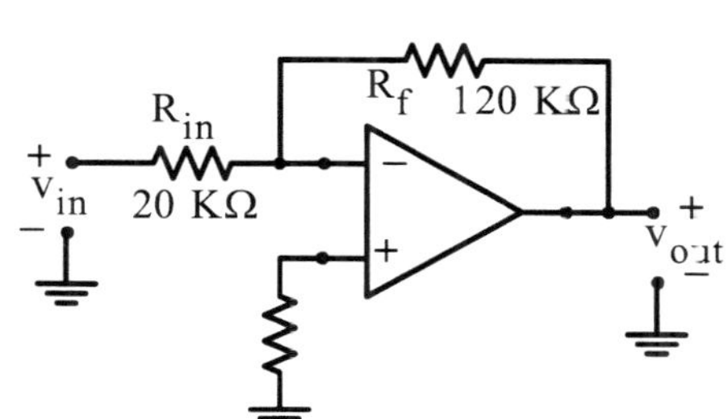

Figure 5.4. Circuit for Example 5.1

Solution:

This is an inverting amplifier and thus the voltage gain G_v is

$$G_v = -\frac{R_f}{R_{in}} = -\frac{120\ K\Omega}{20\ K\Omega} = -6$$

and since

$$G_v = v_{out}/v_{in}$$

the output voltage is

$$v_{out} = G_v v_{in} = -6 \times 1$$

or

$$v_{out} = -6\ mV$$

The voltages v_{in} and v_{out} are plotted as shown in Figure 5.5.

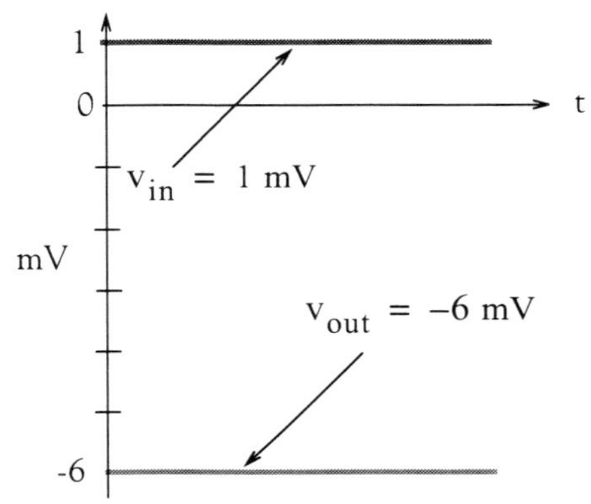

Figure 5.5. Input and output waveforms for the circuit of Example 5.1

Example 5.2

Compute the voltage gain G_v and then the output voltage v_{out} for the inverting op amp circuit shown in Figure 5.6, given that $v_{in} = \sin t\ mV$. Plot v_{in} and v_{out} as mV versus time on the same set of axes.

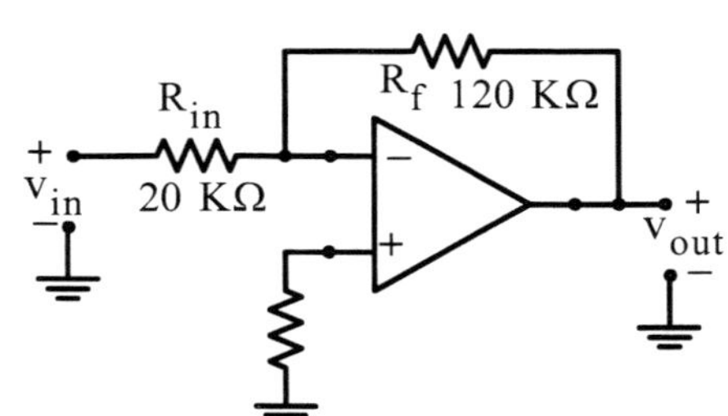

Figure 5.6. Circuit for Example 5.2

Solution:

This is the same circuit as that of the previous example except that the input is a sine wave with unity amplitude and the voltage gain G_v is the same as before, that is,

$$G_v = -\frac{R_f}{R_{in}} = -\frac{120\ K\Omega}{20\ K\Omega} = -6$$

and the output voltage is

$$v_{out} = G_v v_{in} = -6 \times \sin t = -6\sin t\ mV$$

The voltages v_{in} and v_{out} are plotted as shown in Figure 5.7.

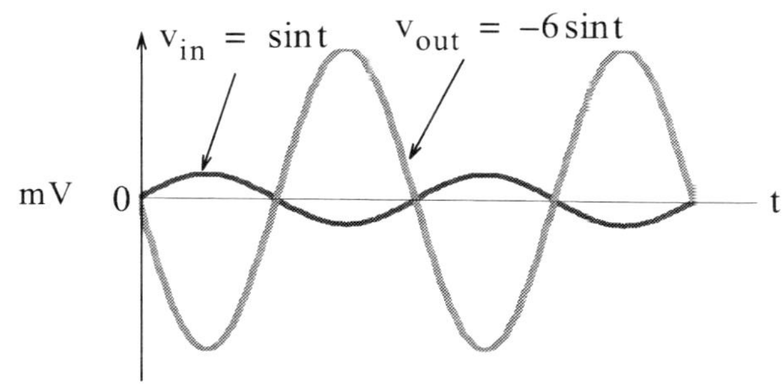

Figure 5.7. Input and output waveforms for the circuit of Example 5.2

5.4 The Op Amp in the Non–Inverting Mode

An op amp is said to be connected in the *non–inverting mode* when an input signal is connected to the non–inverting (+) input through an external resistor R which serves as a current limiter, and the inverting (–) input is grounded through an external resistor R_{in} as shown in Figure 5.8. In our subsequent discussion, the resistor R will represent the internal resistance of the applied voltage v_{in} when this voltage is applied at the non–inverting input.

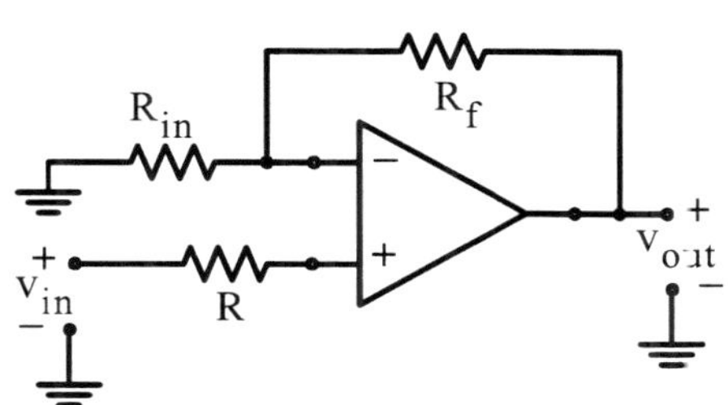

Figure 5.8. Circuit of non–inverting op amp

For the circuit of Figure 5.8, the voltage gain G_v is

$$G_v = \frac{v_{out}}{v_{in}} = 1 + \frac{R_f}{R_{in}} \tag{5.2}$$

As indicated by the relation of (5.2), the gain for the non–inverting op amp configuration is $1 + R_f/R_{in}$ and therefore, in the non–inverting mode the op amp output signal has the same polarity as the input signal; hence, the name non–inverting amplifier. Thus, when the input signal is positive (+) the output will be also positive and if the input is negative, the output will be also negative. For example, if the input is +1 mV DC and the op amp gain is 75, the output will be +75 mV DC. For AC signals the output will be in–phase with the input. For example, if the input is 0.5 V AC and the op amp gain is $G_v = 1 + 19\ K\Omega/1\ K\Omega = 20$, the output will be 10 V AC and in–phase with the input.

Example 5.3

Compute the voltage gain G_v and then the output voltage v_{out} for the non–inverting op amp circuit shown in Figure 5.9, given that $v_{in} = 1\ mV$. Plot v_{in} and v_{out} as mV versus time on the same set of axes.

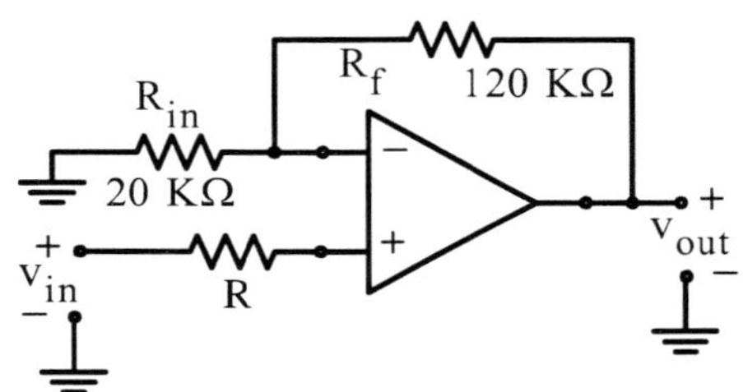

Figure 5.9. Circuit for Example 5.3

Solution:

The voltage gain G_v is

$$G_v = \frac{v_{out}}{v_{in}} = 1 + \frac{R_f}{R_{in}} = 1 + \frac{120\ K\Omega}{20\ K\Omega} = 1 + 6 = 7$$

and thus

$$v_{out} = G_v v_{in} = 7 \times 1\ mV = 7\ mV$$

The voltages v_{in} and v_{out} are plotted as shown in Figure 5.10.

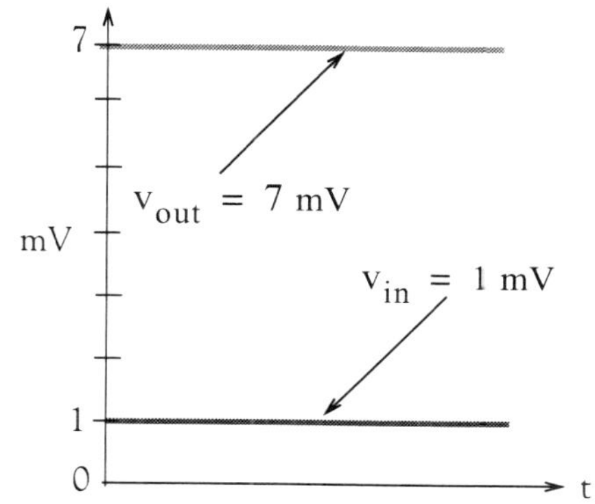

Figure 5.10. Input and output waveforms for the circuit of Example 5.3

Example 5.4

Compute the voltage gain G_v and then the output voltage v_{out} for the non–inverting op amp circuit shown in Figure 5.11, given that $v_{in} = \sin t \text{ mV}$. Plot v_{in} and v_{out} as mV versus time on the same set of axes.

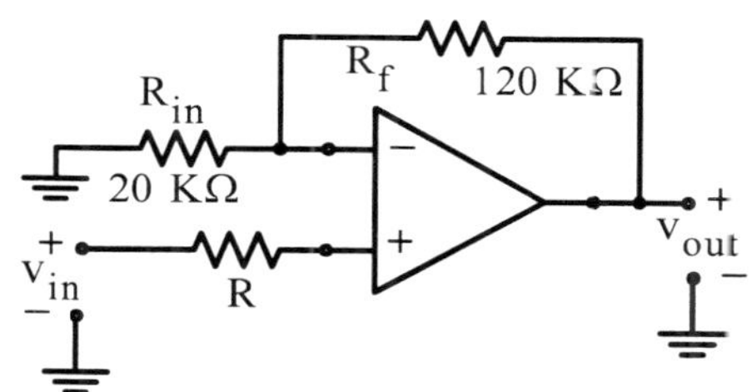

Figure 5.11. Circuit for Example 5.4

Solution:

This is the same circuit as in the previous example except that the input is a sinusoid. Therefore, the voltage gain G_v is the same as before, that is,

$$G_v = \frac{v_{out}}{v_{in}} = 1 + \frac{R_f}{R_{in}} = 1 + \frac{120\ K\Omega}{20\ K\Omega} = 1 + 6 = 7$$

and the output voltage is

$$v_{out} = G_v v_{in} = 7 \times \sin t = 7\sin t \text{ mV}$$

The voltages v_{in} and v_{out} are plotted as shown in Figure 5.12.

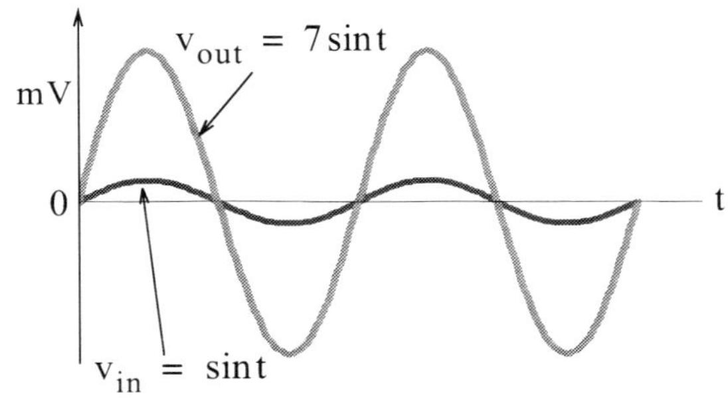

Figure 5.12. Input and output waveforms for the circuit of Example 5.4

Quite often an op amp is connected as shown in Figure 5.13.

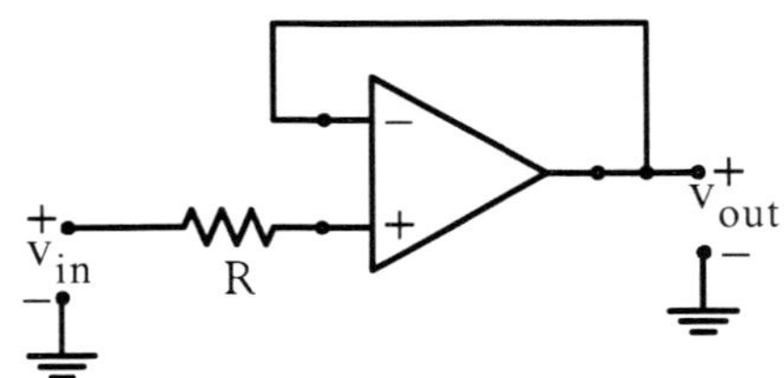

Figure 5.13. Circuit of unity gain op amp

For the circuit of Figure 5.13, the voltage gain G_v is

$$G_v = \frac{v_{out}}{v_{in}} = 1$$

and thus

$$v_{out} = v_{in}$$

For this reason, the op amp circuit of Figure 5.13 it is called *unity gain amplifier*. For example, if the input voltage is 5 mV DC the output will also be 5 mV DC, and if the input voltage is 2 mV AC, the output will also be 2 mV AC. The unity gain op amp is used to provide a very high resistance between a voltage source and the load connected to it. An application is presented in Example 5.8.

5.5 Active Filters

An *active filter* is an electronic circuit consisting of an amplifier and other devices such as resistors and capacitors. In contrast, a *passive filter* is a circuit which consists of passive devices such as resistors, capacitors and inductors. Operational amplifiers are used extensively as active filters.

A *low–pass filter* transmits (passes) all frequencies below a *critical* (*cutoff*) frequency denoted as ω_C, and *attenuates* (blocks) all frequencies above this cutoff frequency. An op amp low–pass filter is shown in Figure 5.14(a) and its amplitude frequency response in Figure 5.14(b).

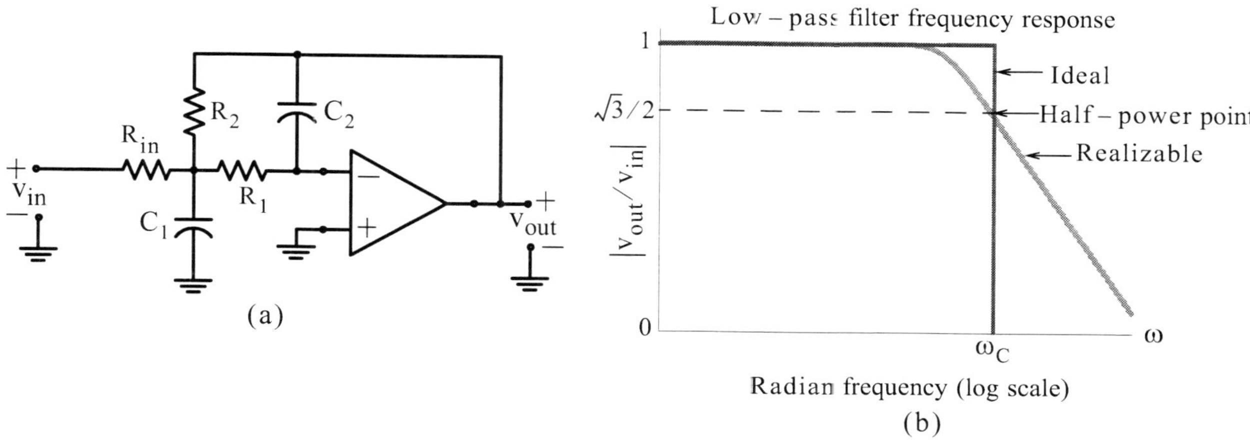

Figure 5.14. An active low–pass filter and its amplitude frequency response

In Figure 5.14(b), the straight vertical and horizontal lines represent the ideal (unrealizable) and the smooth curve represents the practical (realizable) low–pass filter characteristics. The vertical scale represents the magnitude of the ratio of output–to–input voltage v_{out}/v_{in}, that is, the gain G_v. The cutoff frequency ω_c is the frequency at which the maximum value of v_{out}/v_{in} which is unity, falls to $0.707 \times G_v$, and as mentioned before, this is the *half–power* or the −3 dB point.

A *high–pass filter* transmits (passes) all frequencies above a critical (cutoff) frequency ω_c, and attenuates (blocks) all frequencies below the cutoff frequency. An op amp high–pass filter is shown in Figure 5.15(a) and its frequency response in Figure 5.15(b).

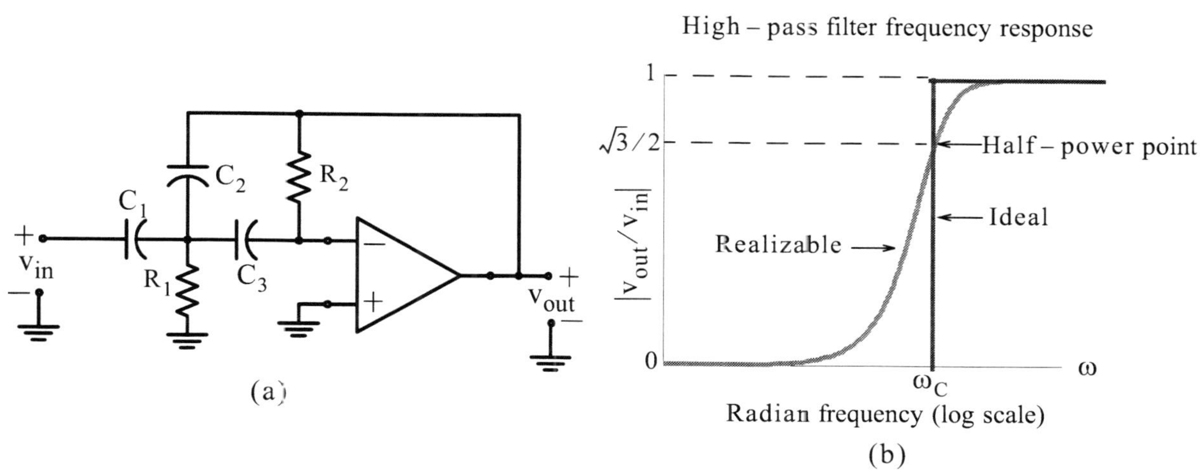

Figure 5.15. An active high–pass filter and its amplitude frequency response

In Figure 5.15(b), the straight vertical and horizontal lines represent the ideal (unrealizable) and the smooth curve represents the practical (realizable) high–pass filter characteristics. The vertical scale represents the magnitude of the ratio of output–to–input voltage v_{out}/v_{in}, that is, the gain G_v. The cutoff frequency ω_c is the frequency at which the maximum value of v_{out}/v_{in} which is unity, falls to $0.707 \times G_v$, i.e., the half–power or the $-3\ \mathrm{dB}$ point.

A *band–pass* filter transmits (passes) the band (range) of frequencies between the critical (cutoff) frequencies denoted as ω_1 and ω_2, where the maximum value of G_v which is unity, falls to $0.707 \times G_v$, while it attenuates (blocks) all frequencies outside this band. An op amp band–pass filter and its frequency response are shown below. An op amp band–pass filter is shown in Figure 5.16(a) and its frequency response in Figure 5.16(b).

A *band–elimination* or *band–stop* or *band–rejection* filter attenuates (rejects) the band (range) of frequencies between the critical (cutoff) frequencies denoted as ω_1 and ω_2, where the maximum value of G_v which is unity, falls to $0.707 \times G_v$, while it transmits (passes) all frequencies outside this band. An op amp band–stop filter is shown in Figure 5.17(a) and its frequency response in Figure 5.17(b).

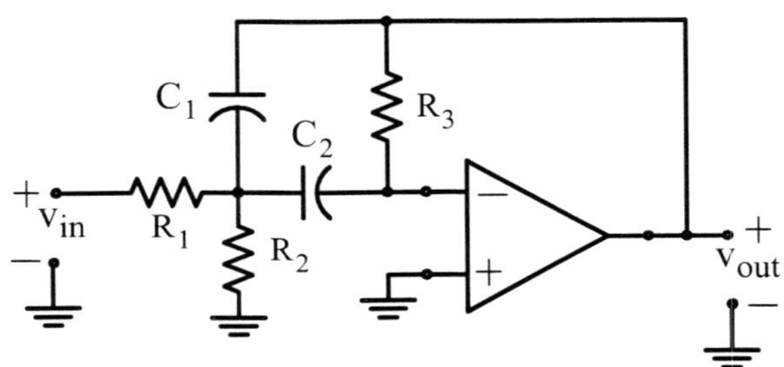

(a)

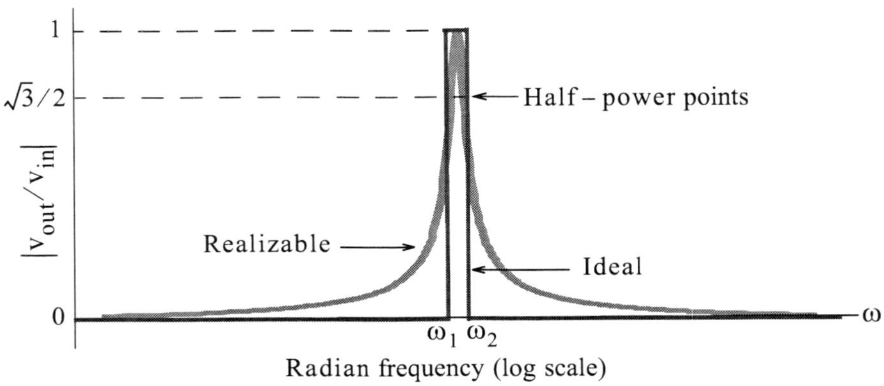

(b)

Figure 5.16. An active band–pass filter and its amplitude frequency response

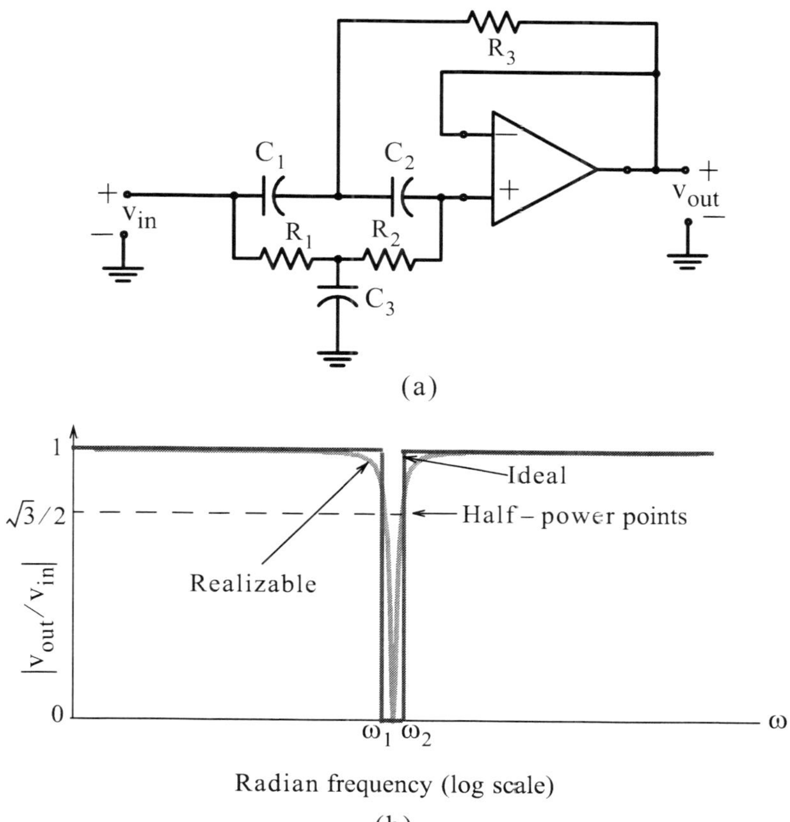

Figure 5.17. An active band−elimination filter and its amplitude frequency response

5.6 Analysis of Op Amp Circuits

The procedure for analyzing an op amp circuit (finding voltages, currents and power) is the same as for the other circuits which we have studied thus far. That is, we can apply Ohm's law, KCL and KVL, superposition, Thevenin's and Norton's theorems. When analyzing an op amp circuit, we must remember that, for all practical purposes, in any op−amp:

a. *The currents into both input terminals are zero*

b. *The voltage difference between the input terminals of an op amp is zero*

c. *For circuits containing op amps, we will assume that the reference (ground) is the common terminal of the two power supplies*. For simplicity, the terminals of the power supplies will not be shown.

In this section we will provide several examples to illustrate the analysis of op amp circuits without being concerned about its internal operation; this will be discussed in a later section.

Example 5.5

The op amp circuit shown in Figure 5.18 is called *inverting op amp*. Prove that the voltage gain G_v is as given in (5.3) below, and draw its equivalent circuit showing the output as a dependent source.

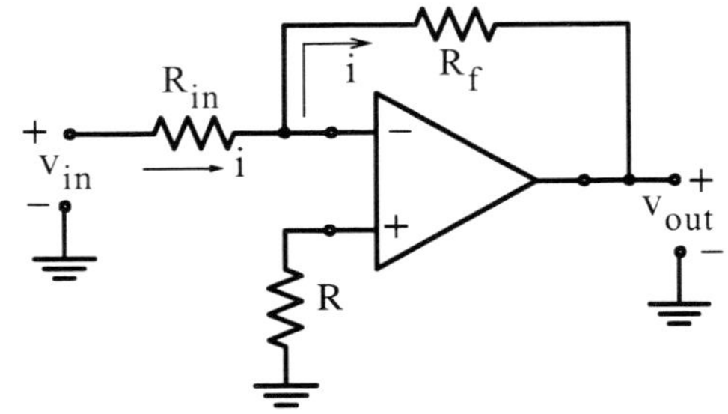

Figure 5.18. Circuit for deriving the gain of an inverting op amp

$$G_v = \frac{v_{out}}{v_{in}} = -\frac{R_f}{R_{in}} \tag{5.3}$$

Proof:

No current flows through the (–) input terminal of the op amp; therefore the current i which flows through resistor R_{in} flows also through resistor R_f. Also, since the (+) input terminal is grounded and there is no voltage drop between the (–) and (+) terminals, the (–) input is said to be at *virtual ground*. From the circuit of Figure 5.18,

$$v_{out} = -R_f i$$

where

$$i = \frac{v_{in}}{R_{in}}$$

and thus

$$v_{out} = -\frac{R_f}{R_{in}} v_{in}$$

or

$$G_v = \frac{v_{out}}{v_{in}} = -\frac{R_f}{R_{in}}$$

The input and output parts of the circuit are shown in Figure 5.19 with the virtual ground being the same as the circuit ground.

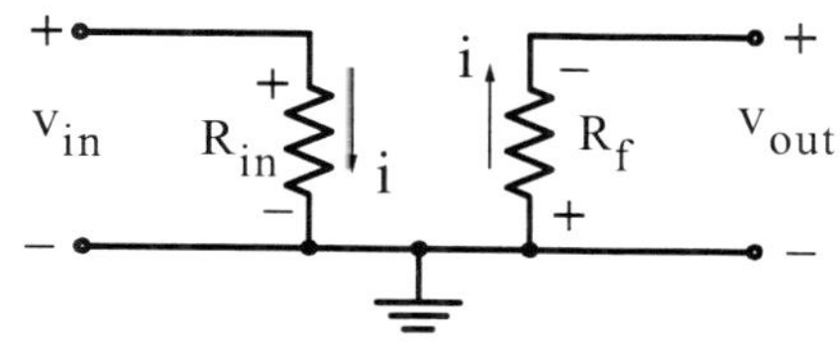

Figure 5.19. Input and output parts for the inverting op amp

These two circuits are normally drawn with the output as a dependent source as shown in Figure 5.20. This is the *equivalent circuit of the inverting op amp* and the dependent source is a Voltage Controlled Voltage Source (VCVS).*

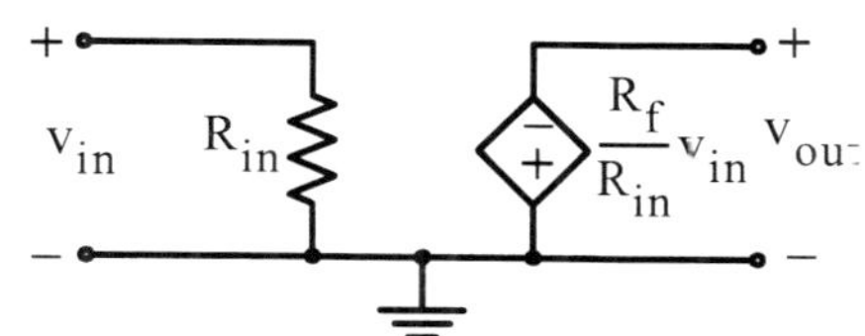

Figure 5.20. Equivalent circuit for the inverting op amp

Example 5.6

The op amp circuit shown in Figure 5.21 is called *non–inverting op amp*. Prove that the voltage gain G_v is as given in (5.4) below, and draw its equivalent circuit showing the output as a dependent source.

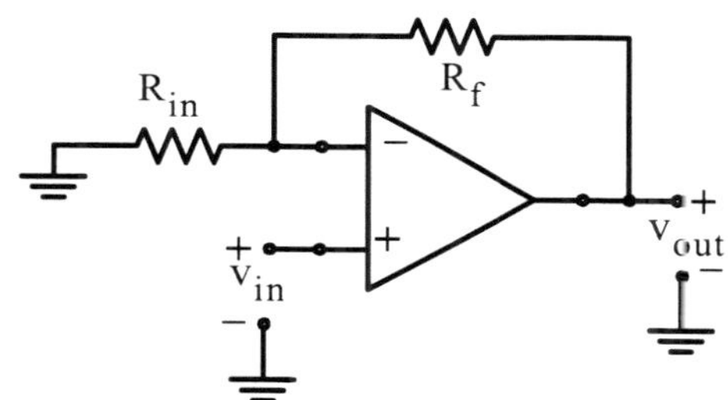

Figure 5.21. Circuit of non–inverting op amp

$$G_v = \frac{v_{out}}{v_{in}} = 1 + \frac{R_f}{R_{in}} \tag{5.4}$$

Proof:

Let the voltages at the (−) and (+) terminals be denoted as v_1 and v_2 respectively as shown in Figure 5.22.

* *For a definition of dependent sources please refer to Circuit Analysis I with MATLAB Applications, ISBN 978–0–9709511–2–0.*

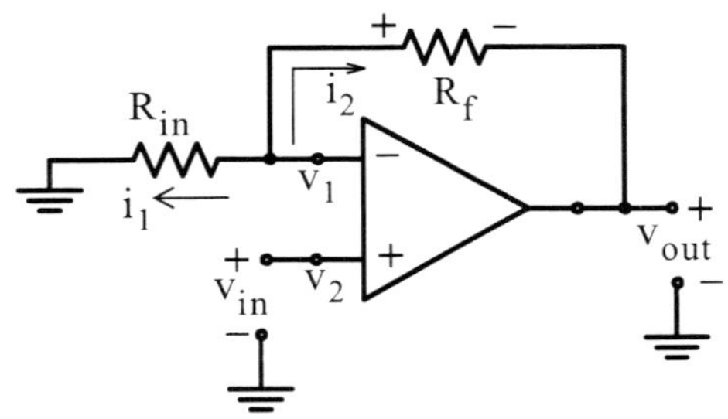

Figure 5.22. Non–inverting op amp circuit for derivation of (5.4)

By application of KCL at v_1

$$i_1 + i_2 = 0$$

or

$$\frac{v_1}{R_{in}} + \frac{v_1 - v_{out}}{R_f} = 0 \tag{5.5}$$

There is no potential difference between the (−) and (+) terminals; therefore, $v_1 - v_2 = 0$ or $v_1 = v_2 = v_{in}$. Relation (5.5) then can be written as

$$\frac{v_{in}}{R_{in}} + \frac{v_{in} - v_{out}}{R_f} = 0$$

$$\left(\frac{1}{R_{in}} + \frac{1}{R_f}\right)v_{in} = \frac{v_{out}}{R_f}$$

Rearranging, we obtain

$$G_v = \frac{v_{out}}{v_{in}} = 1 + \frac{R_f}{R_{in}} \tag{5.6}$$

Figure 5.23 shows the equivalent circuit of Figure 5.22. The dependent source of this equivalent circuit is also a VCVS.

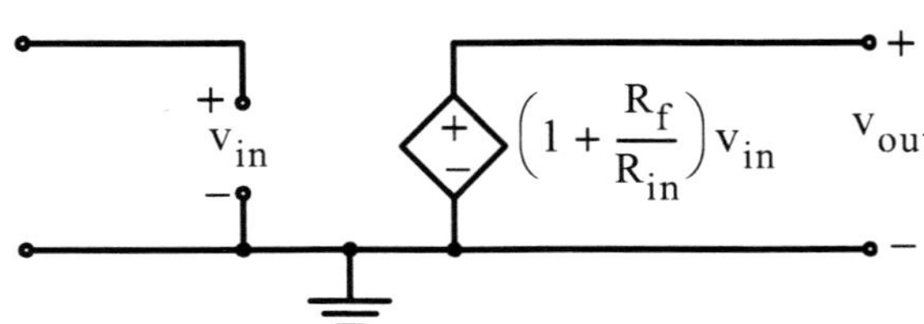

Figure 5.23. Equivalent circuit for the non–inverting op amp

Example 5.7

If, in the non–inverting op amp circuit of Example 5.6, we replace R_{in} with an open circuit ($R_{in} \to \infty$) and R_f with a short circuit ($R_f \to 0$), prove that the voltage gain G_v is

$$G_v = \frac{v_{out}}{v_{in}} = 1 \tag{5.7}$$

and thus

$$v_{out} = v_{in}$$

Proof:

With R_{in} open and R_f shorted, the non–inverting amplifier of the previous example reduces to the circuit of Figure 5.24.

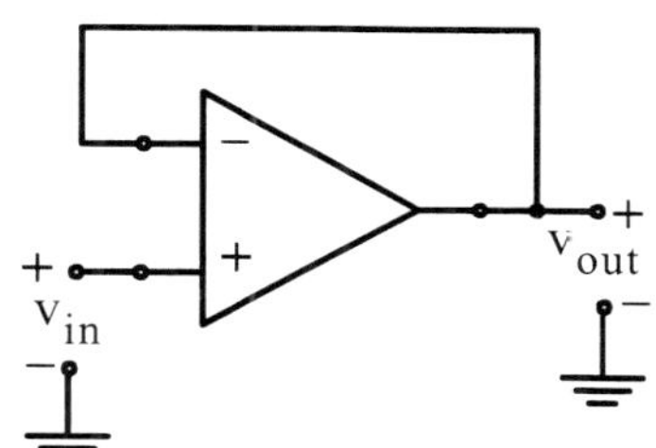

Figure 5.24. Circuit of Figure 5.22 with R_{in} open and R_f shorted

The voltage difference between the (+) and (−) terminals is zero; then $v_{out} = v_{in}$.

We will obtain the same result if we consider the non–inverting op amp gain

$$G_v = \frac{v_{out}}{v_{in}} = 1 + \frac{R_f}{R_{in}}$$

Then, letting $R_f \to 0$, the gain reduces to $G_v = 1$ and for this reason this circuit is called *unity gain amplifier* or *voltage follower*. It is also called *buffer amplifier* because it can be used to "buffer" (isolate) one circuit from another when one "loads" the other as we will see on the next example.

Example 5.8

For the circuit of Figure 5.25:

a. With the load R_{load} disconnected, compute the open circuit voltage v_{ab}

b. With the load connected, compute the voltage v_{load} across the load R_{load}

c. Insert a buffer amplifier between a and b and compute the new voltage v_{load} across the same load R_{load}

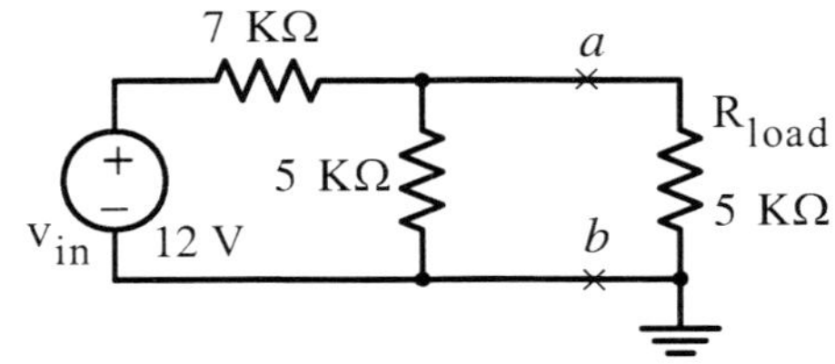

Figure 5.25. Circuit for Example 5.8

Solution:

a. With the load R_{load} disconnected the circuit is as shown in Figure 5.26.

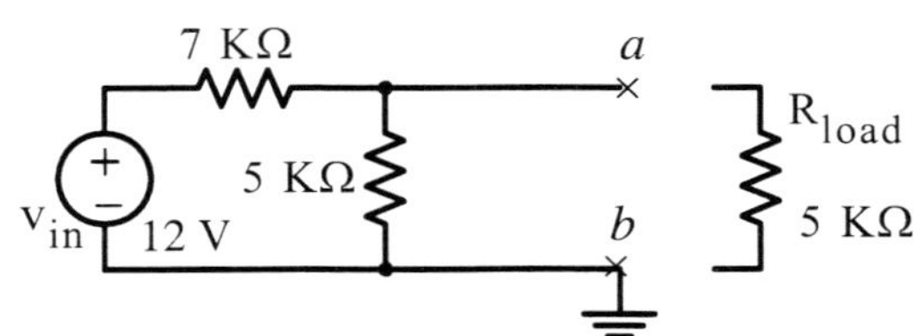

Figure 5.26. Circuit for Example 5.8 with the load disconnected

The voltage across terminals a and b is

$$v_{ab} = \frac{5\ K\Omega}{7\ K\Omega + 5\ K\Omega} \times 12 = 5\ V$$

b. With the load R_{load} reconnected the circuit is as shown in Figure 5.27. Then,

$$v_{load} = \frac{5\ K\Omega \| 5\ K\Omega}{7\ K\Omega + 5\ K\Omega \| 5\ K\Omega} \times 12 = 3.16\ V$$

Here, we observe that the load R_{load} "loads down" the load voltage from 5 V to 3.16 V and this voltage may not be sufficient for proper operation of the load.

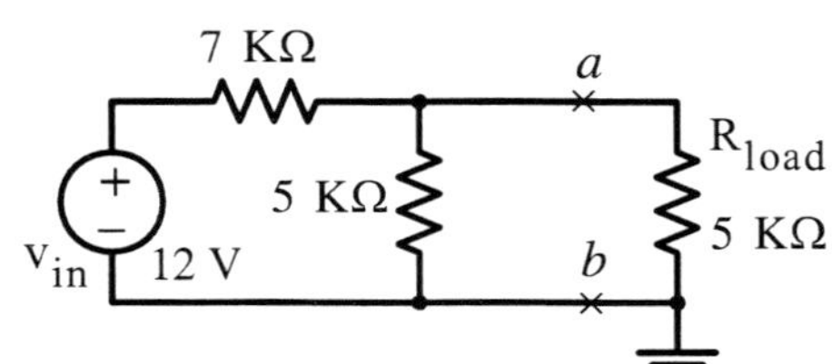

Figure 5.27. Circuit for Example 5.8 with the load reconnected

c. With the insertion of the buffer amplifier between points *a* and *b* and the load, the circuit now is as shown in Figure 5.28.

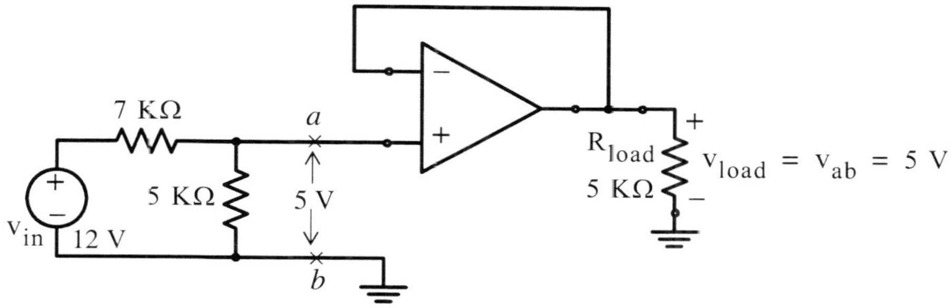

Figure 5.28. Circuit for Example 5.8 with the insertion of a buffer op amp

From the circuit of Figure 5.28, we observe that the voltage across the load is 5 V as desired.

Example 5.9

The op amp circuit shown in Figure 5.29 is called *summing circuit* or *summer* because the output is the summation of the weighted inputs. Prove that for this circuit,

$$v_{out} = -R_f\left(\frac{v_{in1}}{R_{in1}} + \frac{v_{in2}}{R_{in2}}\right) \tag{5.8}$$

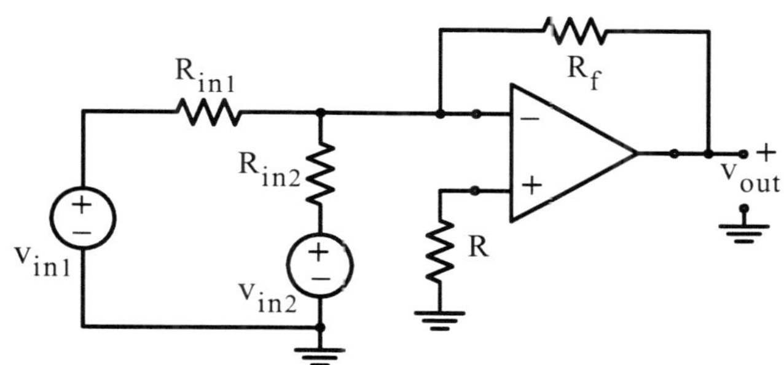

Figure 5.29. Two–input summing op amp circuit

Proof:

We recall that the voltage across the (−) and (+) terminals is zero. We also observe that the (+) input is grounded, and thus the voltage at the (−) terminal is at "virtual ground". Then, by application of KCL at the (−) terminal, we obtain

$$\frac{v_{in1}}{R_{in1}} + \frac{v_{in2}}{R_{in2}} + \frac{v_{out}}{R_f} = 0 \tag{5.9}$$

and solving for v_{out} we obtain (5.8).

Alternately, we can apply the principle of superposition to derive this relation.

Example 5.10

Compute the output voltage v_{out} for the amplifier circuit shown in Figure 5.30.

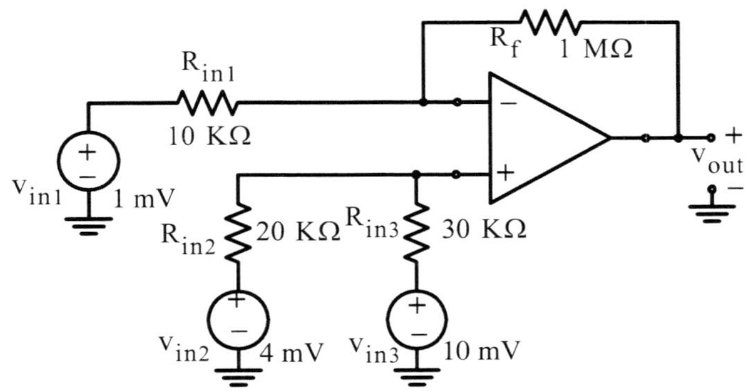

Figure 5.30. Circuit for Example 5.10

Solution:

Let v_{out1} be the output due to v_{in1} acting alone, v_{out2} be the output due to v_{in2} acting alone, and v_{out3} be the output due to v_{in3} acting alone. Then by superposition,

$$v_{out} = v_{out1} + v_{out2} + v_{out3} \tag{5.10}$$

First, with v_{in1} acting alone and v_{in2} and v_{in3} shorted, the circuit becomes as shown in Figure 5.31.

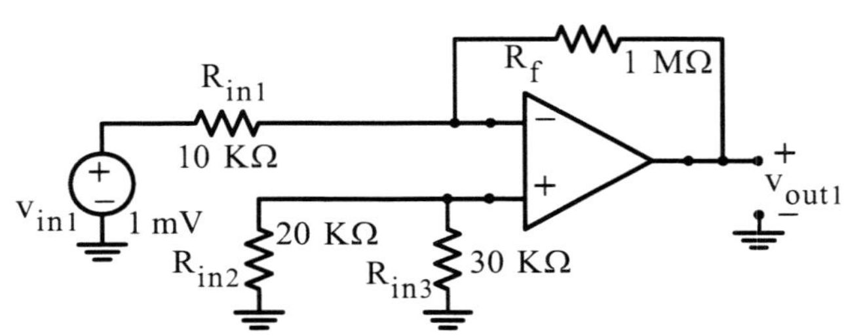

Figure 5.31. Circuit for Example 5.10 with v_{in1} acting alone

We recognize this as an inverting amplifier whose voltage gain G_v is

$$G_v = 1\ M\Omega/10\ K\Omega = 100$$

and thus

$$v_{out1} = (100)(-1\ mV) = -100\ mV \tag{5.11}$$

Next, with v_{in2} acting alone and v_{in1} and v_{in3} shorted, the circuit becomes as shown in Figure 5.32.

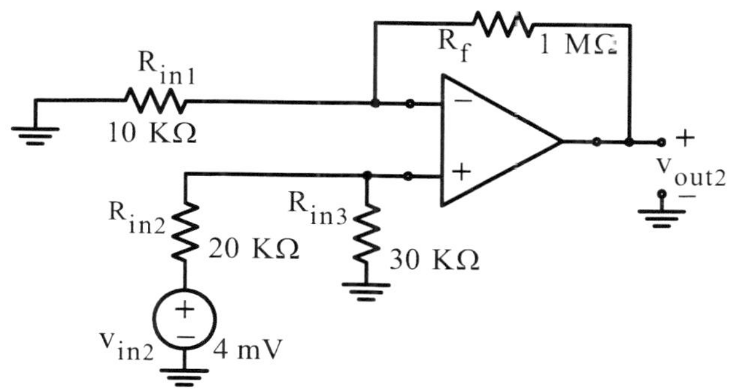

Figure 5.32. Circuit for Example 5.10 with v_{in2} *acting alone*

The circuit of Figure 5.32 as a non–inverting op amp whose voltage gain G_v is

$$G_v = 1 + 1\ M\Omega/10\ K\Omega = 101$$

and the voltage at the plus (+) input is computed from the voltage divider circuit shown in Figure 5.33.

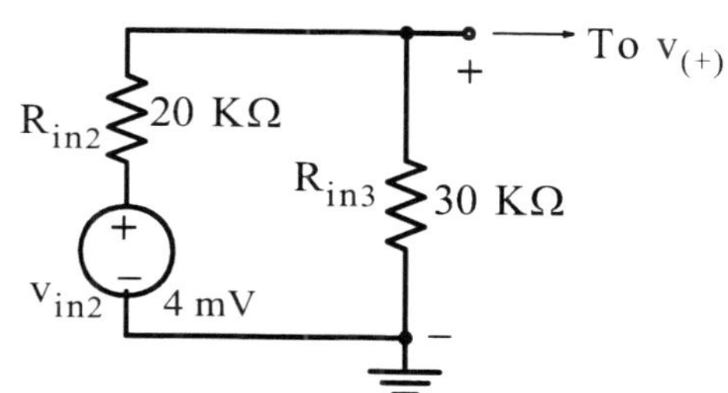

Figure 5.33. Voltage divider circuit for the computation of $v_{(+)}$ *with* v_{in2} *acting alone*

Then,

$$v_{(+)} = \frac{R_{in3}}{R_{in2} + R_{in3}} \times v_{in2} = \frac{30\ K\Omega}{50\ K\Omega} \times 4\ mV = 2.4\ mV$$

and thus

$$v_{out2} = 101 \times 2.4\ mV = 242.4\ mV \tag{5.12}$$

Finally, with v_{in3} acting alone and v_{in1} and v_{in2} shorted, the circuit becomes as shown in Figure 5.34.

The circuit of Figure 5.34 is also a non–inverting op amp whose voltage gain G_v is

$$G_v = 1 + 1\ M\Omega/10\ K\Omega = 101$$

and the voltage at the plus (+) input is computed from the voltage divider circuit shown in Figure 5.35.

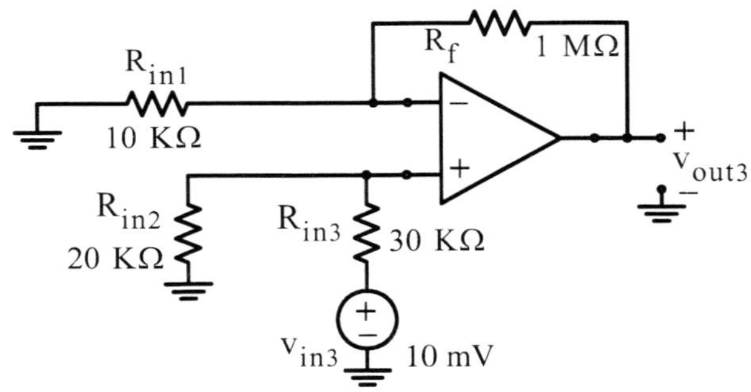

Figure 5.34. Circuit for Example 5.10 with v_{in3} acting alone

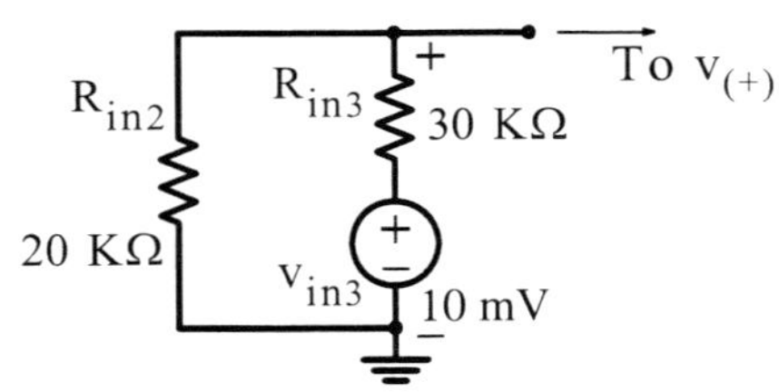

Figure 5.35. Voltage divider circuit for the computation of $v_{(+)}$ with v_{in3} acting alone

From Figure 5.35,

$$v_{(+)} = \frac{R_{in2}}{R_{in2} + R_{in3}} \times v_{in2} = \frac{20\ K\Omega}{50\ K\Omega} \times 10\ mV = 4\ mV$$

and thus

$$v_{out3} = 101 \times 4\ mV = 404\ mV$$

Therefore, from (5.11), (5.12) and (5.13),

$$v_{out} = v_{out1} + v_{out2} + v_{out3} = -100 + 242.4 + 404 = 546.4\ mV$$

Example 5.11

For the circuit shown in Figure 5.36, derive an expression for the voltage gain G_v in terms of the external resistors R_1, R_2, R_3, and R_f.

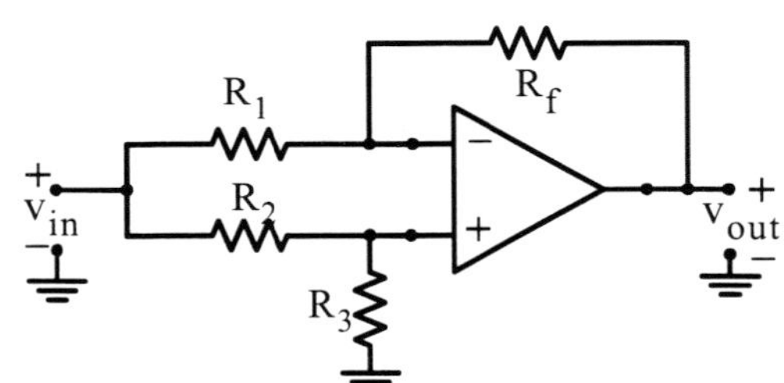

Figure 5.36. Circuit for Example 5.11

Solution:

We apply KCL at nodes v_1 and v_2 as shown in Figure 5.37.

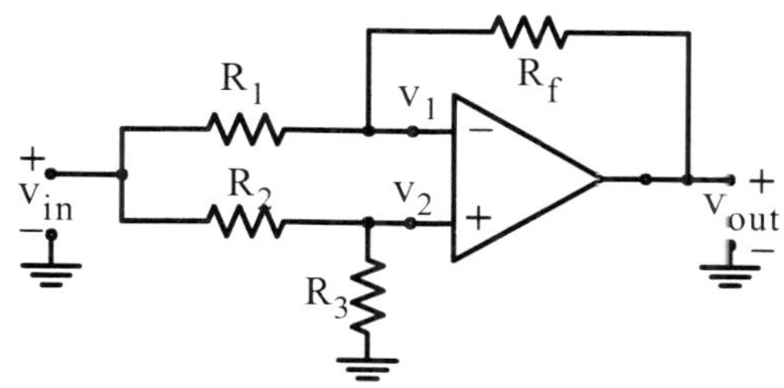

Figure 5.37. Application of KCL for the circuit of Example 5.11

At node v_1:

$$\frac{v_1 - v_{in}}{R_1} + \frac{v_1 - v_{out}}{R_f} = 0$$

$$\left(\frac{1}{R_1} + \frac{1}{R_f}\right)v_1 = \frac{v_{in}}{R_1} + \frac{v_{out}}{R_f}$$

$$\left(\frac{R_1 + R_f}{R_1 R_f}\right)v_1 = \frac{R_f v_{in} + R_1 v_{out}}{R_1 R_f}$$

or

$$v_1 = \frac{R_f v_{in} + R_1 v_{out}}{R_1 + R_f} \qquad (5.13)$$

At node v_2:

$$\frac{v_2 - v_{in}}{R_2} + \frac{v_2}{R_3} = 0$$

or

$$v_2 = \frac{R_3 v_{in}}{R_2 + R_3} \qquad (5.14)$$

and since $v_2 = v_1$, we rewrite (5.14) as

$$v_1 = \frac{R_3 v_{in}}{R_2 + R_3} \qquad (5.15)$$

Equating the right sides of (5.13) and (5.15) we obtain

$$\frac{R_f v_{in} + R_1 v_{out}}{R_1 + R_f} = \frac{R_3 v_{in}}{R_2 + R_3}$$

or

$$R_f v_{in} + R_1 v_{out} = \frac{R_3 v_{in}}{R_2 + R_3}(R_1 + R_f)$$

Dividing both sides of the above relation by $R_1 v_{in}$ and rearranging, we obtain

$$\frac{v_{out}}{v_{in}} = \frac{R_3(R_1 + R_f)}{R_1(R_2 + R_3)} - \frac{R_f}{R_1}$$

and after simplification

$$G_v = \frac{v_{out}}{v_{in}} = \frac{R_1 R_3 - R_2 R_f}{R_1(R_2 + R_3)} \qquad (5.16)$$

5.7 Input and Output Resistances

The input and output resistances are very important parameters in amplifier circuits. The *input resistance* R_{in} of a circuit is defined as the ratio of the applied voltage v_S to the current i_S drawn by the circuit, that is,

$$\boxed{R_{in} = \frac{v_S}{i_S}} \qquad (5.17)$$

Therefore, in an op amp circuit the input resistance provides a measure of the current i_s which the amplifier draws from the voltage source v_s. Of course, we want i_s to be as small as possible; accordingly, we must make the input resistance R_{in} as high as possible.

Example 5.12

Compute the input resistance R_{in} of the inverting op amp amplifier shown in Figure 5.38 in terms of R_1 and R_f.

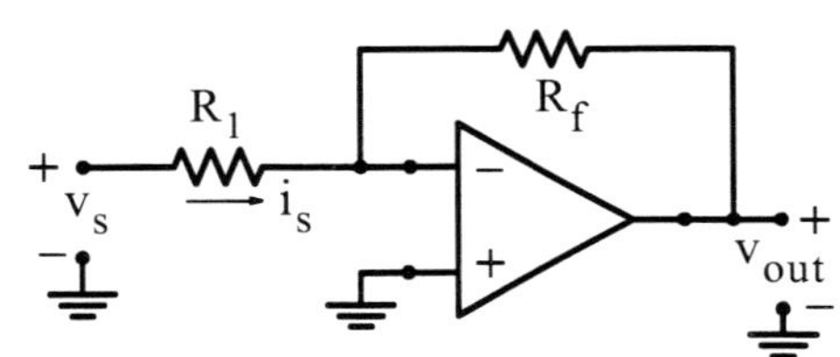

Figure 5.38. Circuit for Example 5.12

Solution:

By definition, $R_{in} = v_s/i_s$ and since no current flows into the minus (–) terminal of the op amp and this terminal is at virtual ground, it follows that

$$i_s = v_s/R_1$$

From the above relations we observe that

$$\boxed{R_{in} = R_1} \tag{5.18}$$

It is therefore, desirable to make R_1 as high as possible. However, if we make R_1 very high such as $10\ M\Omega$, for a large gain, say 100, the value of the feedback resistor R_f should be $1\ G\Omega$. Obviously, this is an impractical value. Fortunately, a large gain can be achieved with the circuit of Exercise 8 at the end of this chapter.

Example 5.13

Compute the input resistance R_{in} of the op amp circuit shown in Figure 5.39.

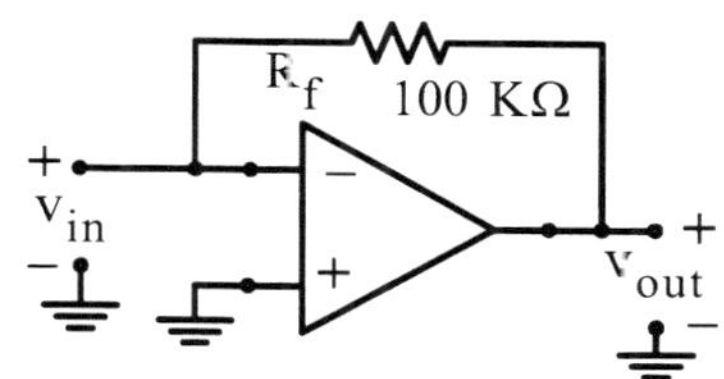

Figure 5.39. Circuit for Example 5.13

Solution:

In the circuit of Figure 5.39, v_{in} is the voltage at the minus (−) terminal; not the source voltage v_S. Therefore, there is no current i_S drawn by the op amp. In this case, we apply a test (hypothetical) current i_X as shown in Figure 5.40, and we treat v_{in} as the source voltage.

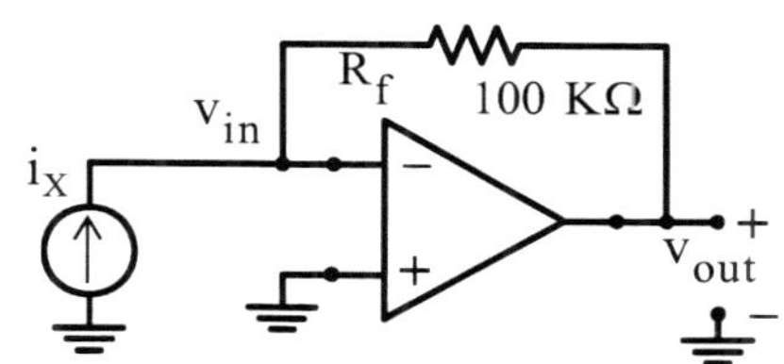

Figure 5.40. Circuit for Example 5.13 with a test current source

We observe that v_{in} is zero (virtual ground). Therefore,

$$R_{in} = \frac{v_{in}}{i_X} = \frac{0}{i_X} = 0$$

By definition, *the output resistance* R_{out} *is the ratio of the open circuit voltage to the short circuit current*, that is,

$$R_{out} = \frac{v_{oc}}{i_{sc}} \tag{5.19}$$

The output resistance R_{out} *is not the same as the load resistance.* The output resistance provides a measure of the change in output voltage when a load which is connected at the output terminals draws current from the circuit. It is desirable to have an op amp with very low output resistance as illustrated by the following example.

Example 5.14

The output voltage of an op amp decreases by 10% when a 5 KΩ load is connected at the output terminals. Compute the output resistance R_{out}.

Solution:

Consider the output portion of the op amp shown in Figure 5.41.

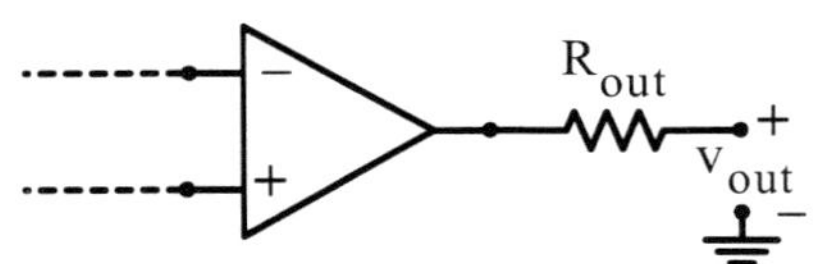

Figure 5.41. Partial circuit for Example 5.14

With no load connected at the output terminals, we observe that

$$v_{out} = v_{oc} = G_v v_{in} \tag{5.20}$$

With a load R_{load} connected at the output terminals, the load voltage v_{load} is

$$v_{load} = \frac{R_{load}}{R_{out} + R_{load}} \times v_{out} \tag{5.21}$$

and from (5.20) and (5.21)

$$v_{load} = \frac{R_{load}}{R_{out} + R_{load}} \times G_v v_{in} \tag{5.22}$$

Therefore,

$$\frac{v_{load}}{v_{OC}} = 0.9 = \frac{5\ \text{K}\Omega}{R_{out} + 5\ \text{K}\Omega}$$

and solving for R_{out} we obtain

$$R_{out} = 555\ \Omega$$

From (5.22) we observe that as $R_{out} \to 0$, $v_{load} = G_v v_{in}$ and with (5.20), $v_{load} = v_{OC}$.

5.8 Op Amp Open Loop Gain

Operational amplifiers can operate either a *closed–loop* or an *open–loop* configuration. The operation, closed–loop or open–loop, is determined by whether or not feedback is used. Without feedback the operational amplifier has an open–loop configuration. This open–loop configuration is practical only when the operational amplifier is used as a *comparator* – a circuit which compares two input signals or compares an input signal to some fixed level of voltage. As an amplifier, the open–loop configuration is not practical because the very high gain amplifies also electrical noise and other unwanted signals, and creates poor stability. Accordingly, operational amplifiers operate in the closed–loop configurations, that is, with *feedback*.

Operational amplifiers are used with negative (degenerative) feedback. Negative feedback has the tendency to oppose (subtract from) the input signal. Although the negative feedback reduces the gain of the operational amplifier, it greatly increases the stability of the circuit. Also, the negative feedback causes the inverting and non–inverting inputs to the operational amplifier will be kept at the same potential. All circuits that we considered in the previous sections of this chapter operate in the closed loop configuration.

The *gain* of any amplifier varies with frequency. The specification sheets for operational amplifiers state the open–loop at DC or 0 Hz. At higher frequencies, the gain is much lower and decreases quite rapidly as frequency increases as shown in Figure 5.42 where both frequency and gain are in logarithmic scales.

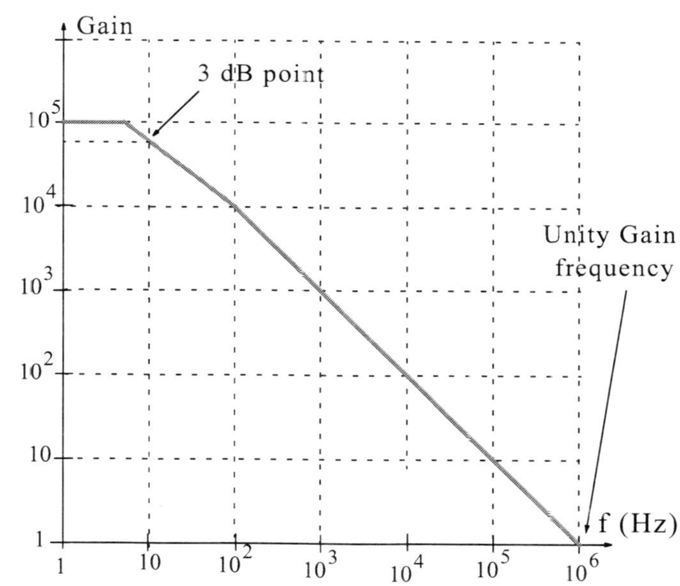

Figure 5.42. Typical op amp open–loop frequency response curve (log scales)

The frequency–response curve of Figure 5.42 shows that the bandwidth is only 10 Hz with this configuration. The *unity gain frequency* is the frequency at which the gain is unity. In Figure 5.46 the unity gain frequency is 1 MHz. We observe that the frequency response curve shows that the gain falls off with frequency at the rate of –20 dB/decade. Figure 5.42 reveals also the gain–bandwidth product is constant at any point of the curve, and this product is equal to 1 MHz, that is, the unity gain frequency. Thus, for any op amp

$$\text{Gain} \times \text{Bandwidth} = \text{Unity Gain Frequency} \tag{5.23}$$

Denoting the open–loop gain as A_{ol}, the 3 dB bandwidth as $BW_{3\ dB}$, and the unity gain frequency as f_{ug}, we can express (5.23) as

$$A_{ol} \cdot BW_{3\ dB} = f_{ug} \tag{5.24}$$

The use of negative feedback increases the bandwidth of an operational amplifier circuit but decreases the gain so that the gain times bandwidth product is always equal to the unity gain frequency of the op amp. The *frequency–response curve* shown in Figure 5.43 is for a circuit in which negative feedback has been used to decrease the circuit gain to 100 (from 100,000 for the operational amplifier without feedback). We observe that the half–power point of this curve is slightly above 10 KHz.

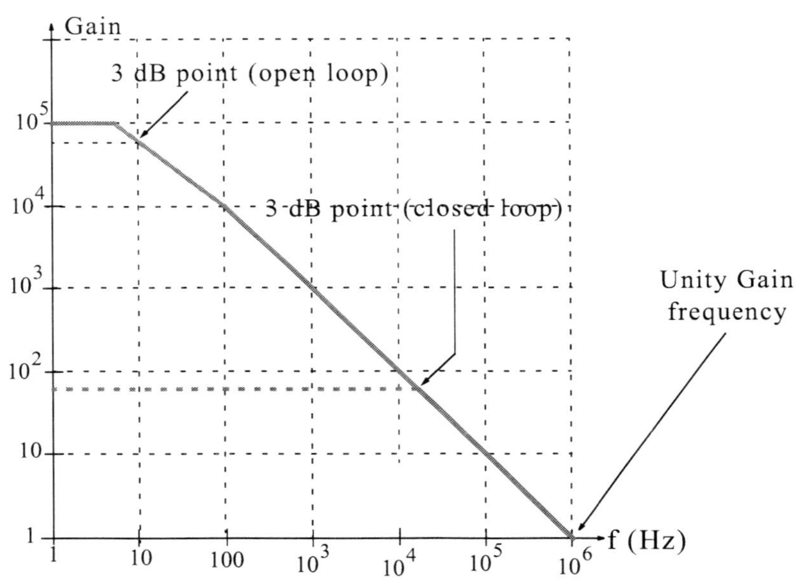

Figure 5.43. Closed–loop frequency response for a gain of 100

5.9 Op Amp Closed Loop Gain

An ideal op amp is shown in Figure 5.44. Of course, an ideal op amp does not exist but the outstanding characteristics of the op amp allow us to treat it as an ideal device. An exact equivalent of the ideal op amp is referred to an a *nullor* consists of two new elements – the *nullator* for the input, i.e., no voltage or current, and the *norator* for the output, i.e., any voltage or current.

The *open–loop gain* A_{ol} of an op amp is very high; for the popular 741 device, $A_{ol} = 200,000$. The external devices, i.e., resistors and capacitors should be chosen for a *closed–loop gain* of about one–tenth to one–twentieth of the open loop gain at a given frequency. This will ensure that the op amp will operate in a stable condition and without distortion.

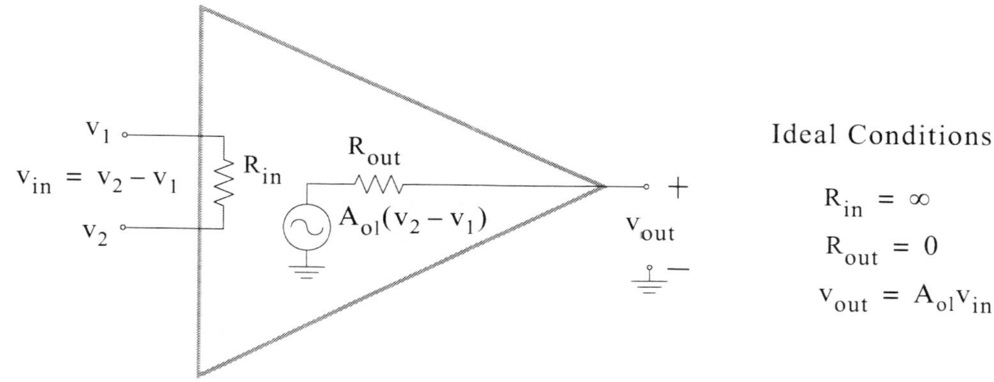

Figure 5.44. The ideal op amp

From Figure 5.44

$$v_{out} = A_{ol}(v_2 - v_1) \tag{5.25}$$

We observe that when $v_2 > v_1$, the output voltage v_{out} is positive and when $v_1 > v_2$, v_{out} is negative. Accordingly, we call the lower terminal v_2 the non–inverting input and the upper terminal v_1 the inverting input. However, if $v_2 = v_1$, $v_{out} = 0$ and we call this condition *common–mode rejection*. In other words, the op amp rejects any signals at its inputs that are exactly the same.

Example 5.15

Figure 5.45 shows a circuit that can be used as a high–pass filter. We want to find the range of the open–loop gain A_{ol} for which the system is stable. Assume that the input impedance is infinite and the output impedance is zero.

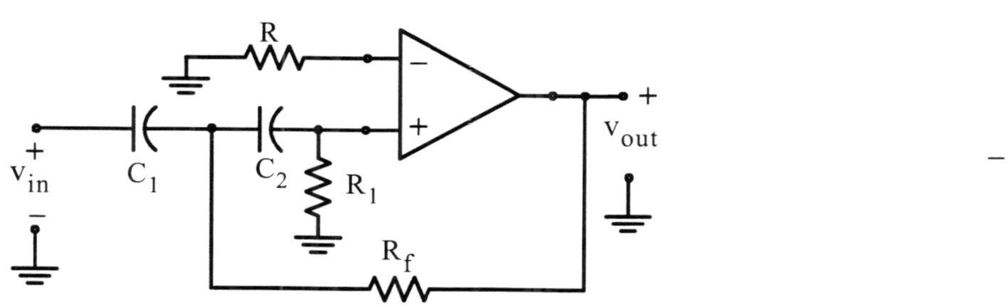

Figure 5.45. Circuit for Example 5.15

Solution:

For *stability*, the coefficients on the denominator of the transfer function must all be positive. To derive the transfer function, we transform the given circuit into its $s-\text{domain}$ equivalent as shown in Figure 5.46.

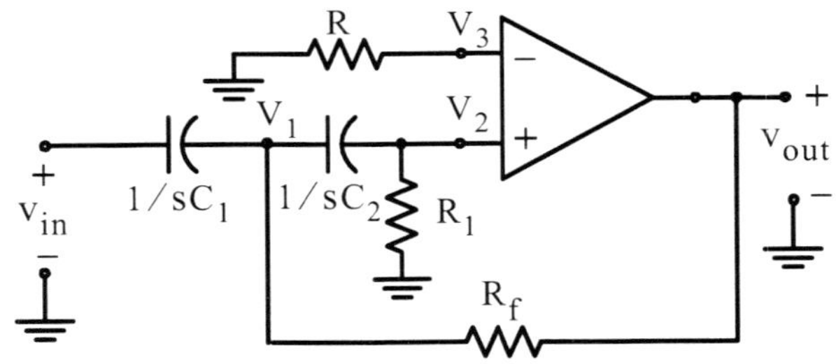

Figure 5.46. The $s-\text{domain}$ *equivalent circuit of Figure 5.49*

Application of KCL at Node V_1 yields

$$sC_1(V_1 - V_{in}) + \frac{V_1 - V_{out}}{R_f} + sC_2(V_1 - V_2) = 0 \tag{5.26}$$

At Node V_2,

$$sC_2(V_2 - V_1) + \frac{V_2}{R_1} \tag{5.27}$$

and since there is no voltage drop across resistor R,

$$V_3 = 0 \tag{5.28}$$

Also,

$$V_{out} = A_{ol}(V_2 - V_3) = A_{ol}(V_2 - 0) = A_{ol}V_2 \tag{5.29}$$

or

$$V_2 = V_{out}/A_{ol} \tag{5.30}$$

Substitution of (5.30) into (5.26) and (5.27) yields

$$sC_1(V_1 - V_{in}) + \frac{V_1 - V_{out}}{R_f} + sC_2(V_1 - V_{out}/A_{ol}) = 0 \tag{5.31}$$

$$sC_2(V_{out}/A_{ol} - V_1) + \frac{V_{out}/A_{ol}}{R_1} \tag{5.32}$$

Solving (5.31) and (5.32) for V_1, equating right sides, and rearranging, we obtain the transfer function

$$G(s) = \frac{V_{out}}{V_{in}} = \frac{A_{ol}[R_1R_fC_1C_2]s^2}{[R_1R_fC_1C_2]s^2 + [R_1C_2(1 - A_{ol}) + R_fC_1 + R_fC_2]s + 1} \tag{5.33}$$

For stability, the coefficient of s in the denominator of (5.33) must be positive, that is,

$$R_1C_2(1 - A_{ol}) + R_fC_1 + R_fC_2 > 0 \tag{5.34}$$

or

$$A_{ol} < 1 + \frac{R_f}{R_1} + \frac{R_fC_1}{R_1C_2} \tag{5.35}$$

Let $R_f = 100\ K\Omega$, $R_1 = 1\ K\Omega$, $C_1 = 1\ \mu F$, and $C_2 = 0.01\ \mu F$. With these values (5.34) becomes

$$A_{ol} < 1 + \frac{10^5}{10^3} + \frac{10^5 \times 10^{-6}}{10^3 \times 10^{-8}}$$

or

$$A_{ol} < 10,101$$

5.10 Transresistance Amplifier

In our previous chapters we introduced voltage gain v_{out}/v_{in}, current gain i_{out}/i_{in}, and transconductance i_{out}/v_{in}. Another term used with amplifiers is the *transresistance gain* $R_m = v_{out}/i_{in}$. The simple op amp circuit shown in Figure 5.47 is known as a *transresistance amplifier*.

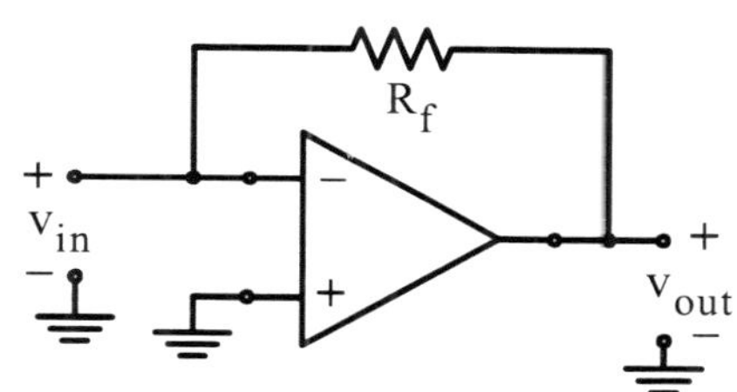

Figure 5.47. Transresistance amplifier

The circuit of Figure 5.47 is the same as that of Figure 5.39, the circuit for Example 5.13, where we found that $R_{in} = 0$. Figure 5.48 shows the circuit model of the transresistance amplifier which was introduced in Exercise 4 in Chapter 1. As in Example 5.13, a test current i_x at the inverting input produces an output voltage v_{out} whose value is $v_{out} = v_{in} - R_fi_x$, and since $v_{in} = 0$, the transresistance R_m of the circuit of Figure 5.47 is

$$R_m = \frac{v_{out}}{i_{in}} = \frac{v_{out}}{i_x} = -R_f \tag{5.36}$$

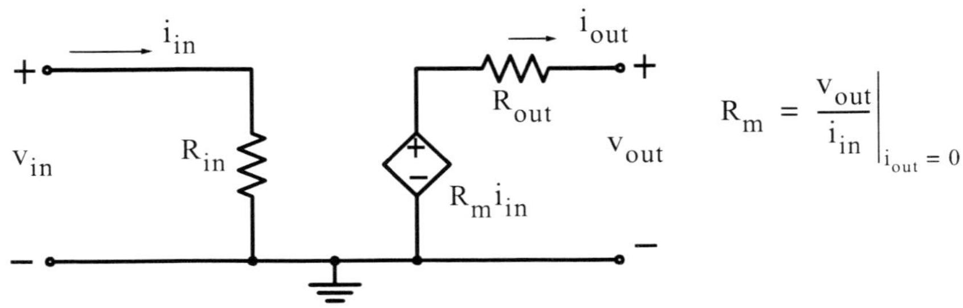

Figure 5.48. Transresistance circuit model

and the minus (–) sign indicates that them output voltage v_{out} and the test current i_x are 180° out–of–phase with each other.

5.11 Closed Loop Transfer Function

In all of the previous sections of this chapter, the external devices in the op amp circuits were resistors. However, several other circuits such as integrators, differentiators, and active filters contain capacitors in addition to resistors. In this case it is convenient to denote devices in series as an impedance in the s – domain as $Z(s)$, and in the $j\omega$ – domain as $Z(j\omega)$. Likewise, it is also convenient to denote devices in parallel as $Y(s)$ or $Y(j\omega)$. Thus, for the inverting input mode, the *closed loop transfer function* $G(s)$ is

$$G(s) = \frac{V_{out}(s)}{V_{in}(s)} = -\frac{Z_f(s)}{Z_1(s)} \tag{5.37}$$

where $Z_f(s)$ and $Z_1(s)$ are as shown in Figure 5.49.

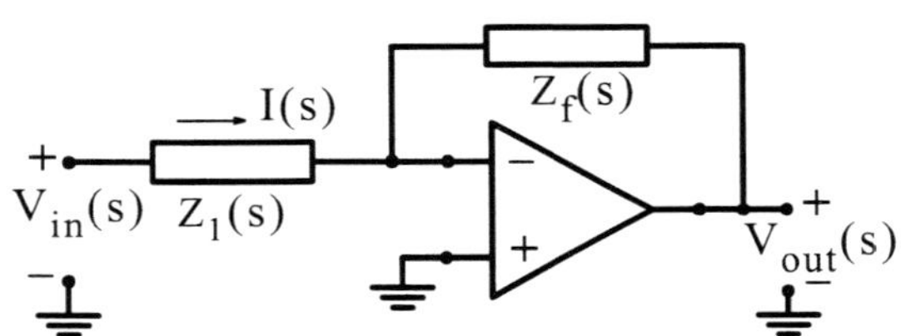

Figure 5.49. The s – domain *inverting amplifier*

Example 5.16

Derive the closed–loop transfer function for the circuit of Figure 5.50.

Solution:

To derive the transfer function, we first convert the given circuit to its s – domain equivalent, and for convenience we denote the series devices as $Z_1(s)$ and the parallel devices as $Y_f(s)$. The circuit then is as shown in Figure 5.51.

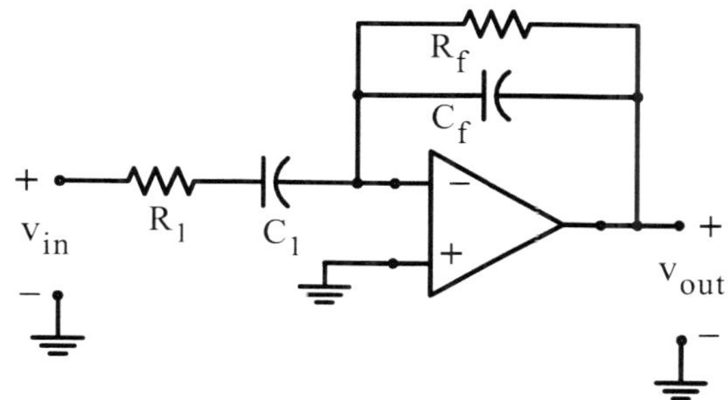

Figure 5.50. Circuit for Example 5.16

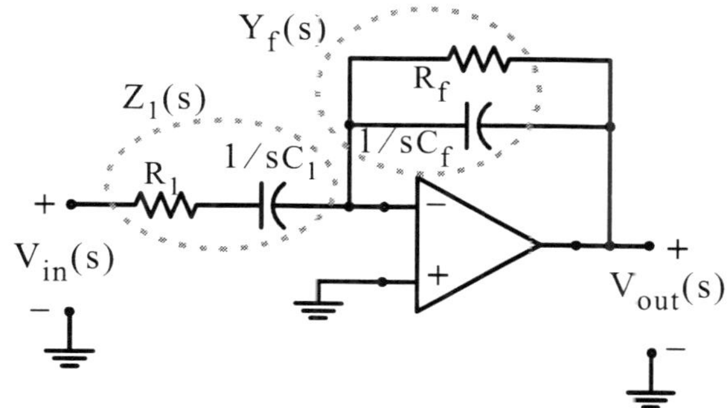

Figure 5.51. The s – domain *equivalent circuit of Figure 5.50*

From Figure 5.51,

$$Z_1(s) = R_1 + 1/sC_1 = (sC_1R_1 + 1)/sC_1 \tag{5.38}$$

$$Y_f(s) = \frac{1}{R_f} + \frac{1}{1/sC_f} = \frac{1}{R_f} + sC_f = (1 + sC_fR_f)/R_f \tag{5.39}$$

$$Z_f(s) = \frac{1}{Y_f(s)} = \frac{R_f}{1 + sC_fR_f} \tag{5.40}$$

From (5.37), (5.38), and (5.40)

$$G(s) = \frac{V_{out}(s)}{V_{in}(s)} = -\frac{Z_f(s)}{Z_1(s)} = -\frac{R_f/(1 + sC_fR_f)}{(sC_1R_1 + 1)/sC_1} = -\frac{sC_1R_f}{(sC_1R_1 + 1)(sC_fR_f + 1)} \tag{5.41}$$

5.12 The Op Amp Integrator

The op amp circuit of Figure 5.52 is known as the *Miller integrator*. For the integrator circuit of Figure 5.52, the voltage across the capacitor is

$$v_C = \frac{1}{C}\int_{-\infty}^{t} i_C dt \tag{5.42}$$

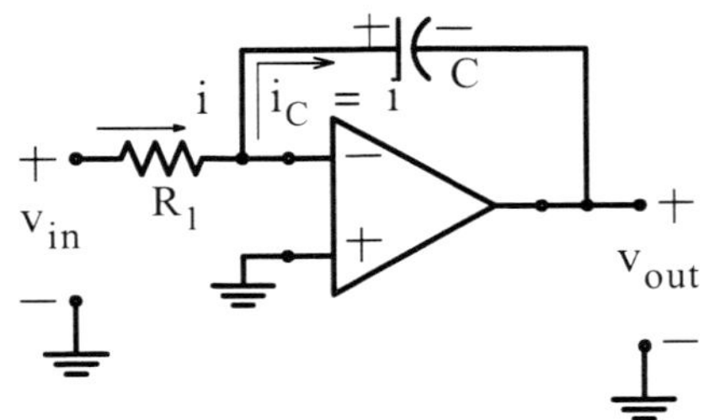

Figure 5.52. The Miller integrator

and assuming the initial condition that at $t = 0$, the voltage across the capacitor is V_0, we can express (5.42) as

$$v_C = \frac{1}{C}\int_0^t i_C dt + V_0 \tag{5.43}$$

Since the inverting input is at virtual ground, the output voltage v_{out} is the negative of the capacitor voltage v_C, that is, $v_{out} = -v_C$, and thus

$$v_{out} = -\frac{1}{C}\int_0^t i_C dt - V_0 \tag{5.44}$$

Also, since

$$i_C = i = \frac{v_{in}}{R_1} \tag{5.45}$$

we rewrite (5.44) as

$$v_{out} = -\frac{1}{R_1 C}\int_0^t v_{in} dt - V_0 \tag{5.46}$$

Example 5.17

The input voltage to the amplifier in Figure 5.53(a) is as shown in Figure 5.53(b). Find and sketch the output voltage assuming that the initial condition is zero, that is, $V_0 = 0$.

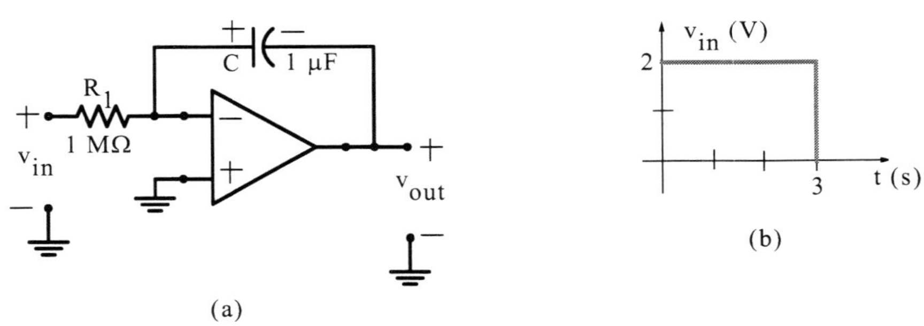

Figure 5.53. Circuit and input waveform for Example 5.17

Solution:

From (5.46)

$$v_{out} = -\frac{1}{R_1C}\int_0^t v_{in}dt - V_0$$

and with $R_1C = 10^6 \times 10^{-6} = 1$ and $V_0 = 0$, the integral above reduces to

$$v_{out} = -\int_0^t v_{in}dt = -\int_0^3 2dt = -2t|_0^3 = -6$$

This result shows that the output voltage v_{out} decreases linearly from zero to -6 V in the time interval $0 \le t \le 3$ s and remains constant at -6 V for $t > 3$ s thereafter as shown in Figure 5.54.

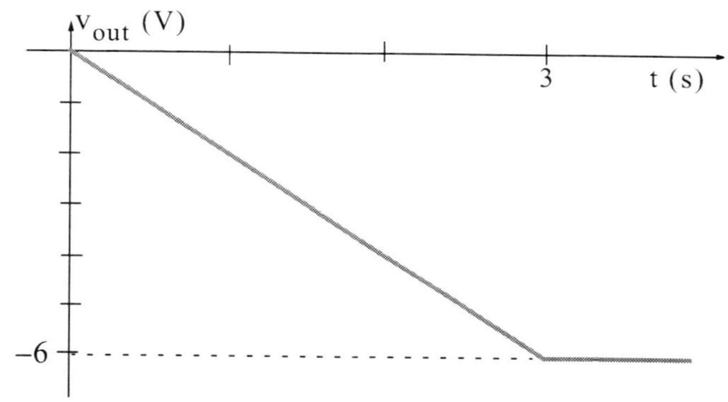

Figure 5.54. Output waveform for the integrator circuit in Figure 5.53

The output voltage waveform in Figure 5.54 indicates that after the capacitor charges to 6 V, it behaves like an open circuit and effectively the negative feedback is an open circuit. Now, let us suppose that the input to the op amp integrator circuit of Figure 5.57(a) is the *unit step function* $u_0(t)$ * as shown in Figure 5.55(a). Then, ideally the output would be a negative ramp towards minus infinity as shown in Figure 5.55(b).

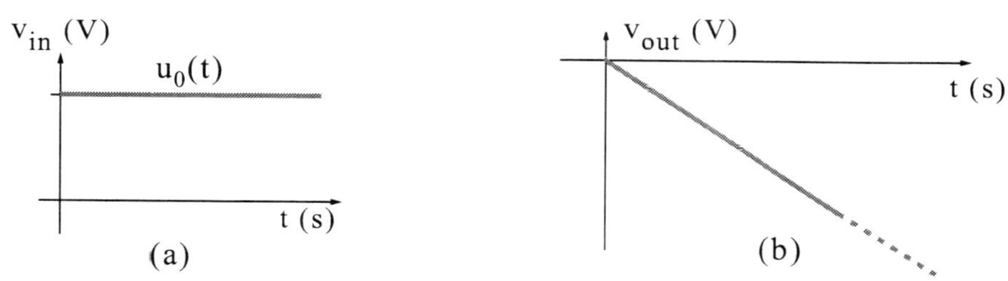

Figure 5.55. The output voltage of the circuit in Figure 5.53 when the input is the unit step function

* *For a detailed discussion on the unit step function refer to Signals and Systems with MATLAB Computing and Simulink Modeling, Fourth Edition, ISBN 978−1−9734404−11−9.*

In reality, the output voltage saturates at the power supply voltage of the op amp, typically ± 15 V depending on the polarity of the input DC signal. This problem can be rectified if we place a feedback resistor R_f in parallel with the capacitor, and in this case the circuit behaves like a low–pass filter as shown in Exercise 13 at the end of this chapter. This feedback resistor should be at least as large as the input resistance R_1.

Example 5.18

The input voltage to the amplifier in Figure 5.56(a) is as shown in Figure 5.56(b). Find and sketch the output voltage for the interval $0 \le t \le 10$ s assuming that the initial condition is zero, that is, $V_0 = 0$.

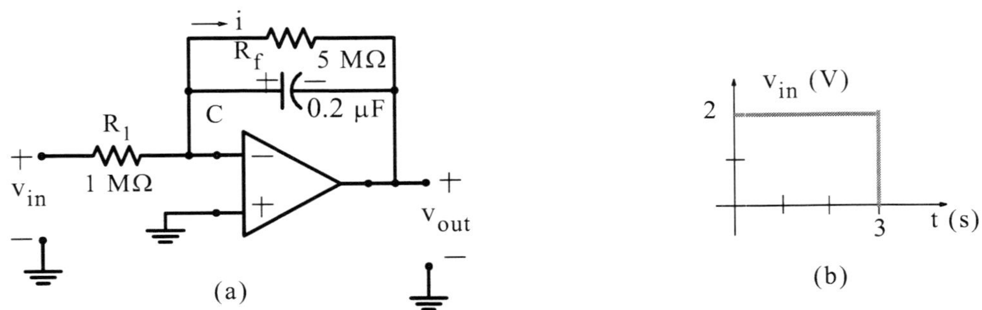

Figure 5.56. Circuit and input waveform for Example 5.18

Solution:

This is the same circuit and input voltage waveform as in Example 5.17 except that a 5 MΩ feedback resistor has been added and the capacitor value was changed to 0.2 μF to simplify the computations. Since it is stated that the initial condition is zero, the capacitor charges in accordance with the relation

$$v_C = V_\infty(1 - e^{-(1/R_fC)t}) \tag{5.47}$$

where

$$V_\infty = -iR_f = -\frac{v_{in}}{R_1}R_f = \frac{-2}{1\ \text{M}\Omega} \times 5\ \text{M}\Omega = -10\text{V}$$

for $0 \le t \le 3$ s. The output voltage for this time interval is

$$v_{out} = -v_C = -10(1 - e^{-(1/R_fC)t})$$

and at $t = 3$ s

$$v_{out}\big|_{t = 3\ s} = -10(1 - e^{-3/1}) = -10(1 - 0.05) = -10(0.95) = -9.5\ \text{V}$$

At $t = 10\ s$

$$v_{out}\Big|_{t=10\ s} = v_{out}\Big|_{t=3\ s} \times e^{-(1/RC)t} = -10e^{-10/1} = -10 \times 4.54 \times 10^{-5} = -45.4\ \mu V$$

The output waveform is shown in Figure 5.57.

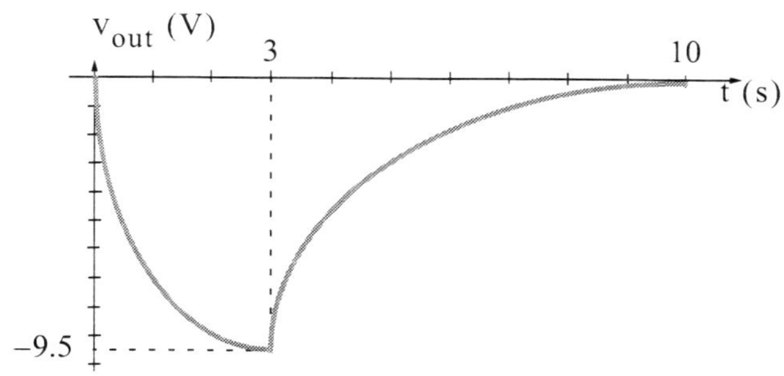

Figure 5.57. Output waveform for the circuit of Example 5.18

As we can see from Figure 5.57, the addition of the feedback resistor makes the circuit of Figure 5.56 somewhat less than an ideal integrator.

5.13 The Op Amp Differentiator

The op amp can also be configured to perform differentiation. The *basic differentiator circuit* is shown in Figure 5.58.

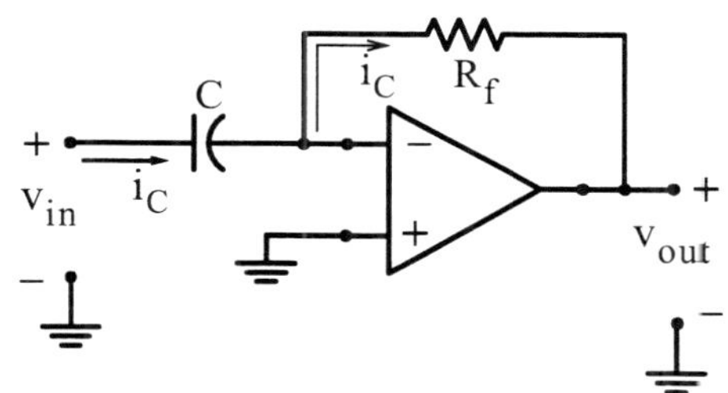

Figure 5.58. Basic differentiator circuit

We observe that the right side of the capacitor is virtually grounded and therefore the current through the capacitor is

$$i_C = C\frac{dv_C}{dt} = C\frac{dv_{in}}{dt}$$

Also,

$$v_{out} = -R_f i_C$$

$$v_{out} = -R_f C\frac{dv_{in}}{dt} \tag{5.48}$$

and we observe that the output voltage v_{out} is the derivative of the input voltage v_{in}.

The circuit of Figure 5.58 is not a practical differentiator because as the frequency increases, the capacitive reactance X_C decreases and the ratio of the feedback resistance R_f to the capacitive reactance increases causing a gain increase without bounds. We could connect a resistor is series with the capacitor but the circuit then becomes a non–ideal differentiator.

Example 5.19

The time constant τ of the differentiator circuit of Figure 5.59 is $\tau = 1\ ms$, and $v_C(0^-) = 0$

a. Find the value of the feedback resistor R_f

b. Derive the transfer function $V_{out}(s)/V_{in}(s)$

c. Find the magnitude and phase at $f = 1\ KHz$

d. If a resistor is added in series with the capacitor to limit the high frequency gain to 100, what should the value of that resistor be?

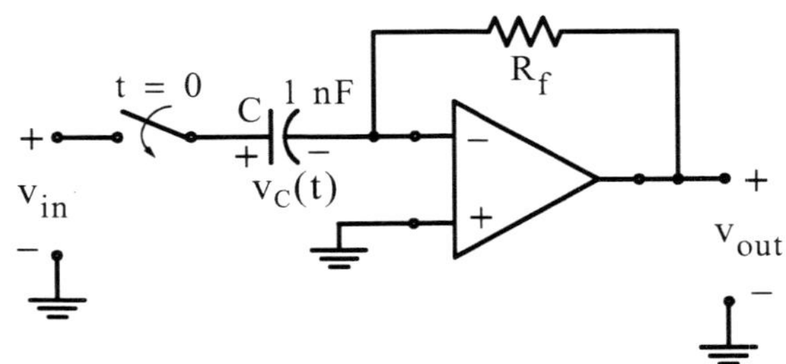

Figure 5.59. Differentiator circuit for Example 5.19

Solution:

a.

$$\tau = R_f C = 10^{-3}\ s$$

and with $C = 10^{-9}$

$$R_f = \frac{\tau}{C} = \frac{10^{-3}}{10^{-9}} = 1\ M\Omega$$

b. Differentiation in the time domain corresponds to multiplication by s in the complex frequency domain, minus the initial value of $f(t)$ at $t = 0^-$.* Thus,

$$\frac{dv_C}{dt} \Leftrightarrow sV_C(s) - v_C(0^-)$$

* For all Laplace transform properties, refer to Circuit Analysis II with MATLAB Applications, ISBN 978–0–9709511–5–1.

and since $v_C(0^-) = 0$, from (5.48)

$$V_{out}(s) = -sR_fCV_{in}(s)$$

or

$$\frac{V_{out}(s)}{V_{in}(s)} = -sR_fC$$

c. With $s = j\omega$, the transfer function can be expressed in magnitude and phase form as

$$\frac{V_{out}(j\omega)}{V_{in}(j\omega)} = -j\omega R_fC$$

$$\left|\frac{V_{out}}{V_{in}}\right| = \omega R_fC$$

$$\theta = -90°$$

$$\left|\frac{V_{out}}{V_{in}}\right|_{f = 1\ KHz} = 2\pi \times 10^3 \times 10^6 \times 10^{-9} = 2\pi \times 10^3 \times 10^{-3} = 2\pi$$

and the phase angle is $-90°$ at all frequencies.

d. As $f \rightarrow \infty$, the capacitor behaves as a short circuit and so with the addition of a resistor R_1 in series with the capacitor, the closed loop voltage gain G_v is

$$G_v = -R_f/R_1$$

and with $R_f = 1\ M\Omega$, for a gain of 100, $R_1 = 10\ K\Omega$.

5.14 Summing and Averaging Op Amp Circuits

The circuit of Figure 5.60 shows the basic *inverting summing* and *averaging op amp* circuit.

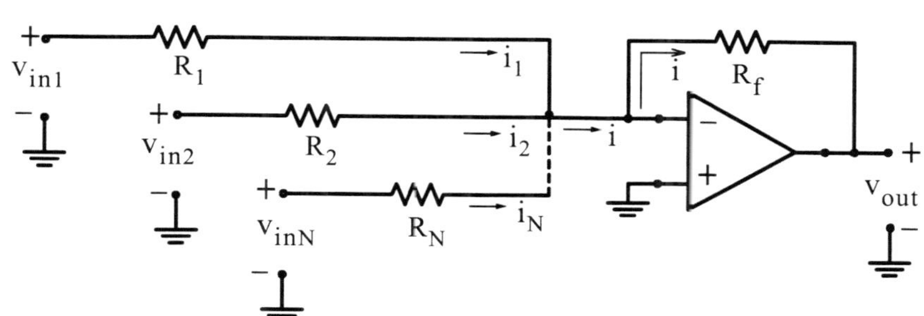

Figure 5.60. Basic inverting summing and averaging op amp circuit

In the circuit of 5.60, the total current is

$$i = i_1 + i_2 + \dots + i_N$$

where

$$i_1 = \frac{v_{in1}}{R_1} \qquad i_2 = \frac{v_{in2}}{R_2} \qquad \dots \qquad i_N = \frac{v_{inN}}{R_N}$$

Also

$$v_{out} = -R_f i = -R_f(i_1 + i_2 + \dots + i_N)$$

Then,

$$v_{out} = -R_f i = -R_f(i_1 + i_2 + \dots + i_N) = -\left(\frac{R_f}{R_1}v_{in1} + \frac{R_f}{R_2}v_{in2} + \dots + \frac{R_f}{R_N}v_{inN}\right) \tag{5.49}$$

If all input resistances are equal, that is, if

$$R_1 = R_2 = \dots = R_N = R$$

the relation of (5.49) reduces to

$$v_{out} = -\frac{R_f}{R}(v_{in1} + v_{in2} + \dots + v_{inN}) \tag{5.50}$$

If $R_f = R$, relation (5.49) reduces further to

$$v_{out} = -(v_{in1} + v_{in2} + \dots + v_{inN}) \tag{5.51}$$

and this indicates that the circuit of Figure 5.60 can be used to find the negative sum of any number of input voltages.

The circuit of Figure 5.60 can also be used to find the average value of all input voltages. The ratio R_f/R is selected such that the sum of the input voltages is divided by the number of input voltages applied at the inverting input of the op amp.

The circuit of Figure 5.61 shows the basic *non–inverting summing* and *non–inverting averaging* op amp circuit. In Figure 5.61 the voltage sources $v_{in1}, v_{in2}, \dots, v_{inN}$ and their series resistances $R_1, R_2, \dots, R_N$ can be replaced by current sources whose values are

$$v_{in1}/R_1, v_{in2}/R_2, \dots, v_{inN}/R_N$$

and their parallel resistances $R_1, R_2, \dots, R_N$.* The circuit of Figure 5.61 can now be represented as in Figure 5.62.

* *For voltage source with series resistance to current source with parallel resistance transformation please refer to Circuit Analysis I with MATLAB Applications, ISBN 978–0–9709511–2–0*

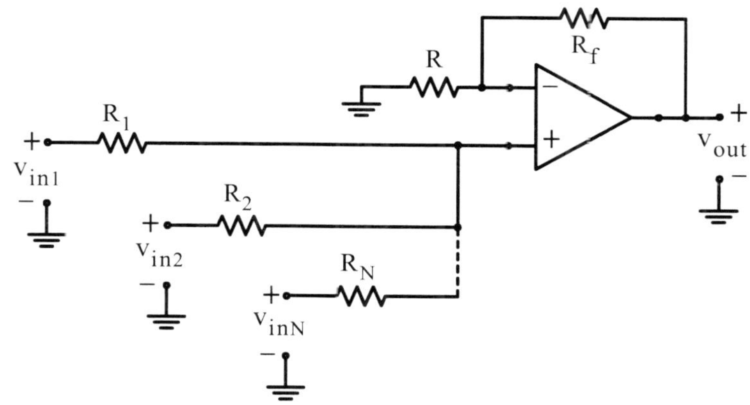

Figure 5.61. Basic non–inverting summing and averaging op amp circuit

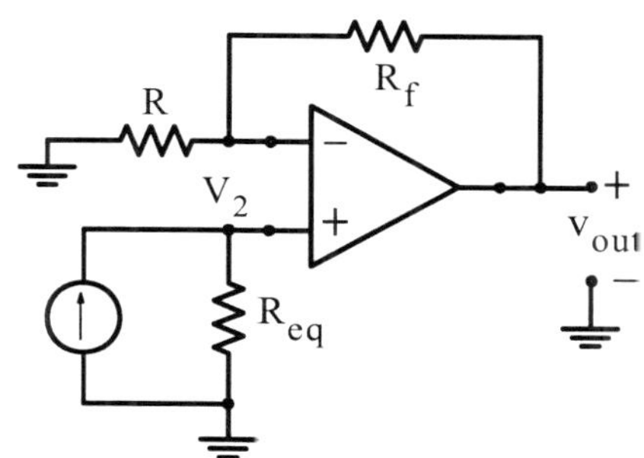

Figure 5.62. Equivalent circuit for the circuit of Figure 5.61

In the circuit of Figure 5.62, the voltage V_2 at the non–inverting input is $V_2 = R_{eq}I_{eq}$ and thus the output voltage is

$$v_{out} = \left(\frac{R_f}{R}+1\right)V_2 = \left(\frac{R_f}{R}+1\right)(R_{eq}I_{eq}) \tag{5.52}$$

5.15 Differential Input Op Amp

The circuit of Figure 5.63 is a *differential input op amp.*

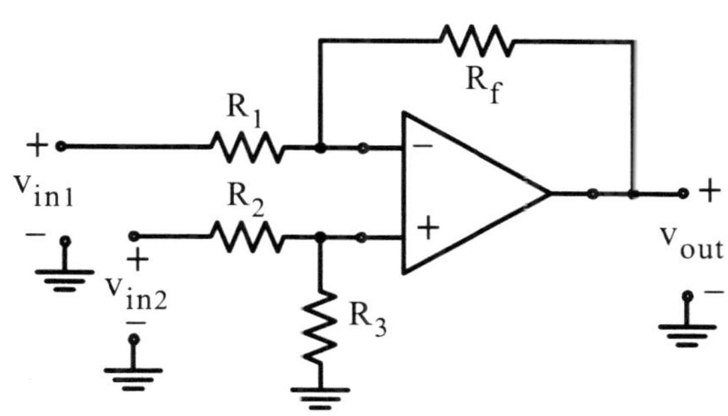

Figure 5.63. Differential input op amp

The differential input configuration allows input signals to be applied simultaneously to both input terminals and produce an output of the difference between the input signals as shown in Figure 5.63. Differential input op amps are used in instrumentation circuits.

We will apply the superposition principle to derive an expression for the output voltage v_{out}. With the input voltage v_{in1} acting alone and v_{in2} grounded, the circuit of Figure 5.63 reduces to that of Figure 5.64.

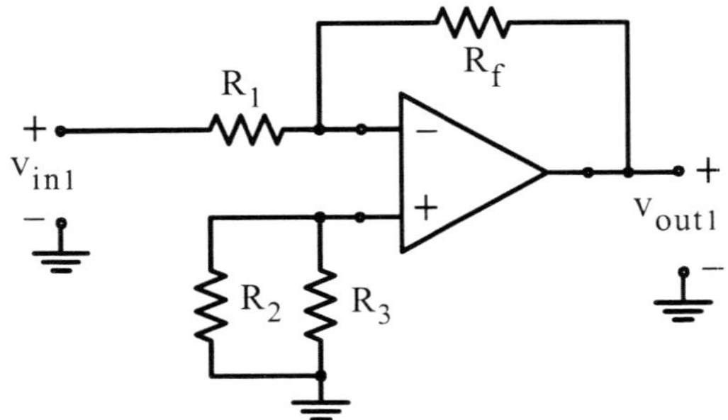

Figure 5.64. The circuit of Figure 5.63 with v_{in1} *acting alone*

The circuit of Figure 5.64 is an inverting amplifier and thus

$$v_{out1} = -\frac{R_f}{R_1}v_{in1} \tag{5.53}$$

Next, with the input voltage v_{in1} grounded and v_{in2} acting alone, the circuit of Figure 5.63 reduces to that of Figure 5.65 and as indicated, we denote the voltage at the non–inverting input as v_2.

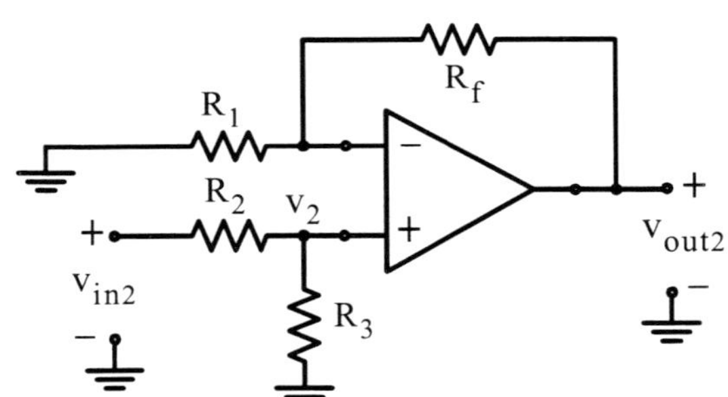

Figure 5.65. The circuit of Figure 5.67 with v_{in2} *acting alone*

Then, by the voltage division expression,

$$v_2 = \frac{R_3}{R_2 + R_3}v_{in2} \tag{5.54}$$

The circuit of Figure 5.65 is a non–inverting amplifier and thus

$$v_{out2} = \left(\frac{R_f}{R_1}+1\right)v_2 \tag{5.55}$$

Then, from (5.54), (5.55), and (5.56),

$$v_{out} = v_{out1} + v_{out2} = -\frac{R_f}{R_1}v_{in1} + \left(\frac{R_f}{R_1}+1\right)\left(\frac{R_3}{R_2+R_3}v_{in2}\right) = -\frac{R_f}{R_1}v_{in1} + \left(\frac{R_fR_3+R_1R_3}{R_1R_2+R_1R_3}\right)v_{in2}$$

or

$$v_{out} = -\frac{R_f}{R_1}v_{in1} + \left(\frac{R_f/R_1+1}{R_2/R_3+1}\right)v_{in2} \tag{5.56}$$

To be useful, a differential input amplifier must have a high *common mode rejection ratio* (CMRR) defined as

$$CMRR = \frac{\text{Differential gain}}{\text{Common mode gain}} = \frac{A_d}{A_{cm}}\,* \tag{5.57}$$

where the *differential gain* A_d is the gain with the input signals applied differentially, and *common mode gain* is the ratio of the output common–mode voltage $V_{cm\ out}$ to the input common–mode voltage $V_{cm\ in}$, that is, $A_{cm} = V_{cm\ out}/V_{cm\ in}$. Ideally, A_{cm} is zero but in reality is finite and much smaller than unity.

It is highly desirable that the differential input op amp of Figure 5.63 produces an output voltage $v_{out} = 0$ when $v_{in2} = v_{in1}$, and as we now know, this is referred to as common–mode rejection. We also want a non–zero output when $v_{in2} \neq v_{in1}$. If the common–mode rejection condition is achieved, that is, when $v_{in2} = v_{in1}$ and $v_{out} = 0$, relation (5.57) above reduces to

$$\frac{R_f}{R_1} = \frac{R_f/R_1+1}{R_2/R_3+1}$$

$$\left(\frac{R_f}{R_1}+1\right)R_1 = \left(\frac{R_2}{R_3}+1\right)R_f$$

$$R_f + R_1 = R_f + \frac{R_2}{R_3}R_f$$

$$R_1 = \frac{R_2}{R_3}R_f$$

and thus for optimum CMMR

$$\frac{R_f}{R_1} = \frac{R_3}{R_2} \tag{5.58}$$

* *The common mode rejection is normally expressed in dB, that is,* $CMR(dB) = 20\log(A_v/A_{cm}) = 20\log CMRR$.

By substitution of (5.58) into (5.56) we obtain

$$v_{out} = \frac{R_f}{R_1}(v_{in2} - v_{in1}) \tag{5.59}$$

Next, we will derive the input resistance for the differential input op amp circuit of Figure 5.63. For convenience in (5.58) we let $R_2 = R_1$. Then $R_3 = R_f$ and with these simplifications the circuit of Figure 5.63 is as shown in Figure 5.66.

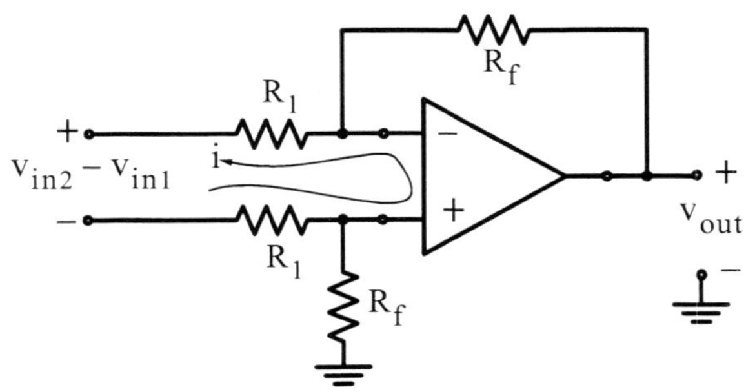

Figure 5.66. Differential input op amp for derivation of the input resistance

Application of KVL around the input circuit starting at the minus (−) input terminal and going counterclockwise, and observing that there is a virtual short between the inverting and non–inverting inputs, we obtain

$$R_1 i + R_1 i - (v_{in2} - v_{in1}) = 0$$

$$v_{in2} - v_{in1} = 2R_1 i \tag{5.60}$$

Also, by definition

$$R_{in} = \frac{v_{in2} - v_{in1}}{i} \tag{5.61}$$

and from (5.60 and (5.61)

$$R_{in} = 2R_1 \tag{5.62}$$

Relation (5.59) reveals that for a large differential gain, we must make the feedback resistor R_f as large as possible and the resistance R_1 as small as possible. But with small R_1 the input impedance will also become small as we can see from (5.62).

5.16 Instrumentation Amplifiers

High input resistance differential input amplifiers are suitable for use in differential measurement applications and the associated circuits are referred to as *instrumentation amplifiers* such as that shown in Figure 5.67.

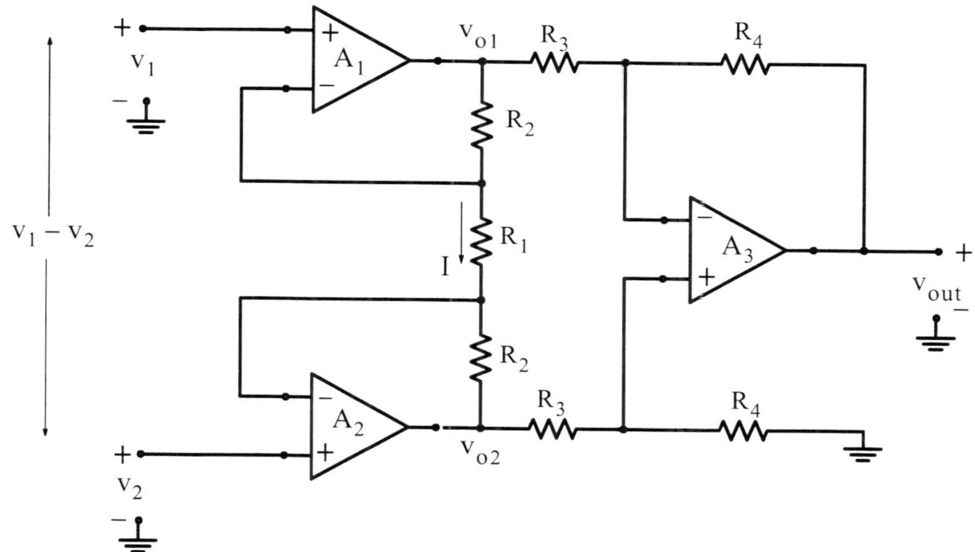

Figure 5.67. High input resistance differential input op amp for use with instrumentation circuits

In the circuit of Figure 5.67 we have made the assumption that there is no current flowing at the inputs of amplifiers A_1 and A_2 and thus the same current flows through the resistive $R_2 - R_1 - R_2$ network. Thus,

$$I = \frac{v_{o1} - v_1}{R_2} = \frac{v_1 - v_2}{R_1} = \frac{v_2 - v_1}{R_2} \tag{5.63}$$

We also have made the assumption that there is no voltage difference between the amplifiers A_1 and A_2 input terminals.

The output voltage v_{o1} of amplifier A_1 is

$$v_{o1} = R_2 I + R_1 I + R_2 I + v_{o2} = (R_1 + 2R_2)I + v_{o2}$$

and with the second term on the right side of (5.63) the above relation can be expressed as

$$v_{o1} = (R_1 + 2R_2)\left(\frac{v_1 - v_2}{R_1}\right) + v_{o2} = \left(1 + \frac{2R_2}{R_1}\right)(v_1 - v_2) + v_{o2}$$

$$v_{o1} - v_{o2} = \left(1 + \frac{2R_2}{R_1}\right)(v_1 - v_2) \tag{5.64}$$

$$v_{o2} - v_{o1} = \left(1 + \frac{2R_2}{R_1}\right)(v_2 - v_1) \tag{5.65}$$

To find the output v_{out} of amplifier A_3 we express (5.59) as

$$v_{out} = \frac{R_4}{R_3}(v_{o2} - v_{o1})$$

and with (5.65)

$$v_{out} = \frac{R_4}{R_3}\left(\frac{2R_2}{R_1} + 1\right)(v_2 - v_1) \tag{5.66}$$

Therefore, the differential gain is

$$A_d = \frac{v_{out}}{(v_2 - v_1)} = \frac{R_4}{R_3}\left(\frac{2R_2}{R_1} + 1\right) \tag{5.67}$$

To make the overall gain of the circuit of Figure 5.67 variable while maintaining CMRR capability, op amp manufacturers recommend that the resistor R_1 be replaced with a fixed value resistor in series with a variable resistor. The fixed resistor will ensure that the maximum gain is limited, while the variable resistor (a potentiometer) can be adjusted for different gains. The interested reader may refer to Exercise 16 at the end of this chapter.

5.17 Offset Nulling

Figure 5.2, Page 5–2, shows that pins 1 and 5 in the 741 op amp are identified as *offset null*. The offset null connections (pins 1 and 5) provide a simple way to balance out the internal variations and zero out the output offset which might be apparent with zero input voltage. It is used simply by connecting a trimmer potentiometer between pins 1 and 5, as shown in Figure 5.68. As shown, the slider on the potentiometer is connected to the negative power supply. To adjust for zero offset, we must set the input voltage to zero and use the offset null potentiometer to set the output voltage precisely to zero.

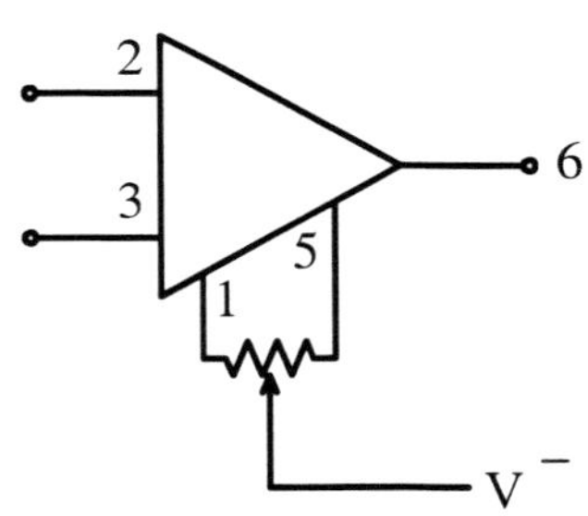

Figure 5.68. Offset nulling terminals of the 741 op amp

According to the 741 op amp specifications, the maximum input offset voltage is ± 6 mv and assuming that the closed loop gain is 100, the output voltage with respect to ground can be $100 \times \pm 6 \times 10^{-3} = \pm 0.6$ V and this value can be either positive or negative even though the input signal is zero volts.

5.18 External Frequency Compensation

General purpose op amps like the 741 op amp, are normally internally frequency compensated so that they will be stable with all values of resistive feedback. Other types of op amps like the 748 op amp, are without *internal frequency compensation* and require external connection of frequency compensating components to the op amp. Typically, the compensating components alter the open loop gain characteristics so that the roll–off is about 20 dB/decade over a wide range of frequencies. Figure 5.69 shows a typical op–amp with *external frequency compensation* where 3 external capacitors can be used as frequency compensating components. With the appropriate selection of capacitors C_1, C_2, and C_3, we can alter the frequency response as shown in Figure 5.70.

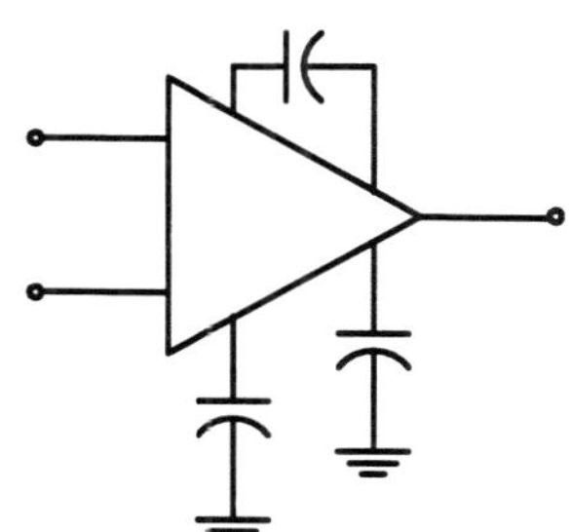

Figure 5.69. Typical op amp with externally connected capacitors for frequency compensation

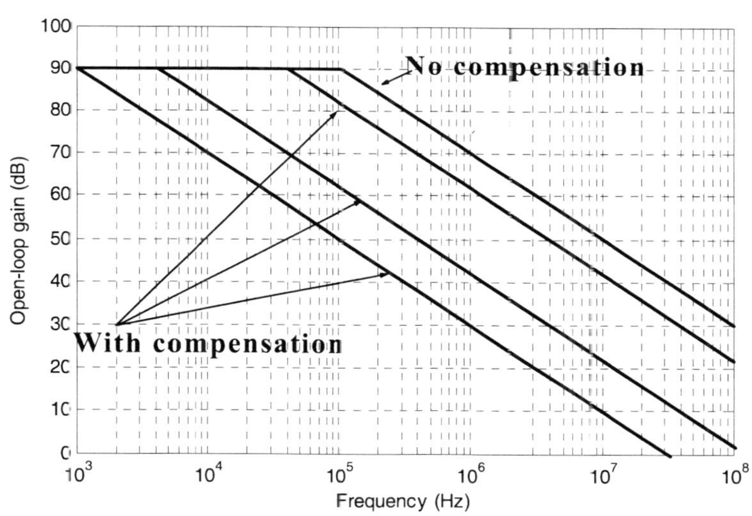

Figure 5.70. Frequency responses with and without external frequency compensation

5.19 Slew Rate

There is a limit to the rate at which the output voltage of an op amp can change. Therefore, manufacturers specify a new parameter referred to as the *slew rate*. By definition, the slew rate (SR) is the maximum rate of change of an output voltage produced in response to a large input step function and it is normally expressed in volts per microsecond, that is,

$$\text{Slew Rate} = SR = \frac{dv_{out}}{dt}_{max} \qquad (5.68)$$

Of course, relation (5.68) is the slope of the output voltage under maximum rate of change conditions. Typical slew rates range from $0.1\ V/\mu s$ to $100\ V/\mu s$, and most internally compensated op amps have slew rates in the order of $1\ V/\mu s$. Figure 5.71 shows a step function of amplitude $10\ V$ applied to the input of a unity gain op amp, and the waveform at the output of this op amp.

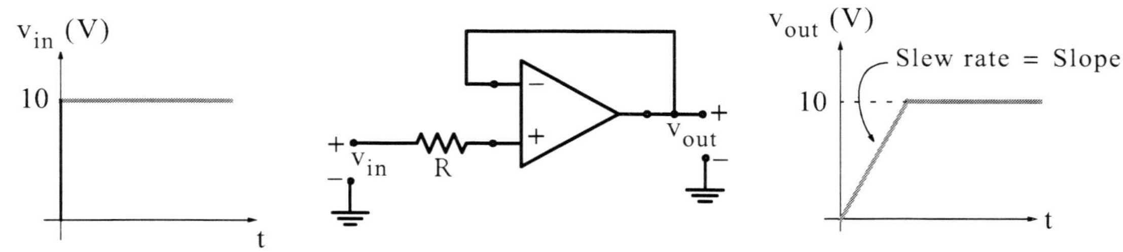

Figure 5.71. The resultant slew rate when a step function is applied to a unity gain op amp

The linearly rising slew rate shown in Figure 5.71 will not be produced if the input voltage is smaller than that specified by the manufacturer. In this case, the slew rate will be a rising exponential such as the rising voltage across a capacitor. In most op amps the slew rate is set by the charging rate of the frequency compensating capacitor and the output voltage is

$$v_{out} = V_f(1 - e^{-\omega_{ug}t}) \qquad (5.69)$$

where V_f is the final value of the output voltage as shown in Figure 5.71, $\omega_{ug} = 2\pi f_{ug}$, and f_{ug} is the unity gain frequency as defined in (5.24), i.e., $A_{ol} \cdot BW_{3\ dB} = f_{ug}$.

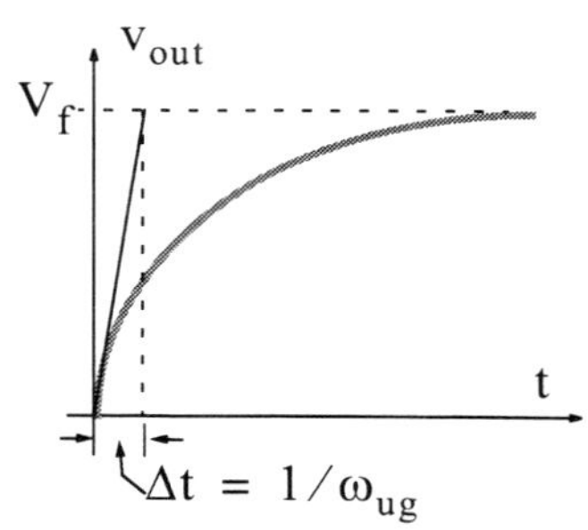

Figure 5.72. Plot for expression $v_{out} = V_f(1 - e^{-\omega_{ug}t})$

5.20 Circuits with Op Amps and Non-Linear Devices

Op amps are often used in circuits with non–linear devices. There are many circuits that can be formed with op amps and non–linear devices such as junction diodes, zener diodes, bipolar transistors, and MOSFETs. In this section we will introduce just a few.

Figure 5.73 shows a *positive and negative voltage limiter* circuit and its transfer characteristics. In the voltage limiting circuit of Figure 5.73 the Zener diodes D_1 and D_2 limit the peak–to–peak value of the output voltage. Thus, when the output voltage is positive, its value is limited to the value $V_{Z_1} + V_F$ where V_F is the voltage drop across the forward–biased Zener diode and it is typically about $0.7\ V$. Likewise, when the output voltage is positive, its value is limited to the value $-(V_{Z_2} + V_F)$.

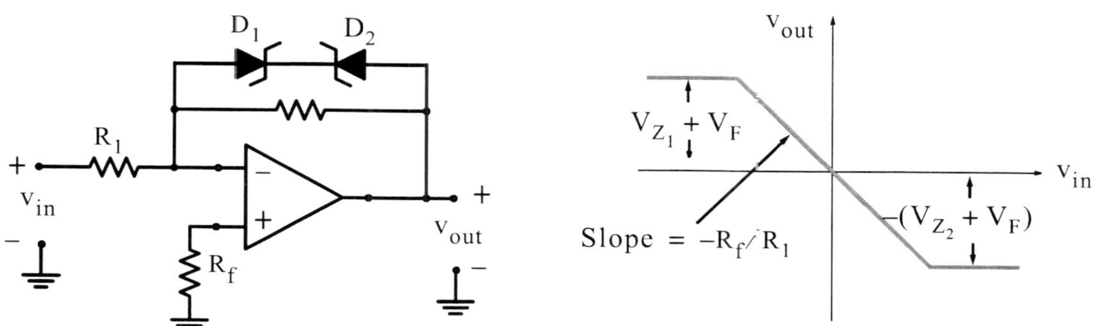

Figure 5.73. A positive and negative voltage limiter circuit and its transfer characteristics

Example 5.20

In the circuit of Figure 5.73, $V_{Z_1} = V_{Z_2} = 6.3\ V$, $V_F = 0.7\ V$, $R_1 = 5\ K\Omega$, and $R_f = 100\ K\Omega$. Describe the output waveforms when

a. $v_{in} = 0.3\sin 10t$

b. $v_{in} = 0.6\cos 100t$

c. $v_{in} = 3\cos(1000t + \pi/6)$

Solution:

Since this is an inverting amplifier, its gain is $R_f/R_1 = 100/5 = 20$* and since we are interested in peak values, the frequencies and phase angles are immaterial for this example. With the given values, the output peaks on positive half–cycles are limited to

$$V_{Z_1} + V_F = 6.3 + 0.7 = 7\ V$$

and the output peaks on negative half–cycles are limited to

* *The gain is always expressed as a positive quantity. The minus sign simply implies inversion.*

$$-(V_{Z_2} + V_F) = -(6.3 + 0.7) = -7\ V$$

a. With $v_{in} = \pm 0.3\ V$ peak,

$$v_{out\ (peak)} = (-R_f/R_1)v_{in} = -20 \times (\pm 0.3) = \mp 6\ V$$

This value is lower than $\pm 7\ V$ and neither Zener diode conducts. Therefore, the output voltage is an unclipped sinusoid.

b. With $v_{in} = \pm 0.6\ V$ peak,

$$v_{out\ (peak)} = (-R_f/R_1)v_{in} = -20 \times (\pm 0.6) = \mp 12\ V$$

and this would be the output peak voltage if the Zener diodes were not present. Since they are, the output peak voltage is clipped to $v_{out\ (peak)} = \mp 7\ V$.

c. With $v_{in} = \pm 3\ V$ peak,

$$v_{out\ (peak)} = (-R_f/R_1)v_{in} = -20 \times (\pm 3) = \mp 60\ V$$

This is indeed a very large voltage for the output of an op amp and even without the Zener diodes, the op amp would saturate. But with the Zener diodes present, $v_{out\ (peak)} = \mp 7\ V$.

Figure 5.74 shows a *positive voltage limiter* and its transfer characteristics where both the Zener diode and the junction diode limit the positive half–cycle of the output voltage. As shown by the transfer characteristics, v_{out} cannot rise above the voltage level $V_{Z_1} + V_F$ because the Zener diode enters the Zener (avalanche) region and the output is clipped. However, the negative half–cycles are not clipped unless the op amp is driven into negative saturation.

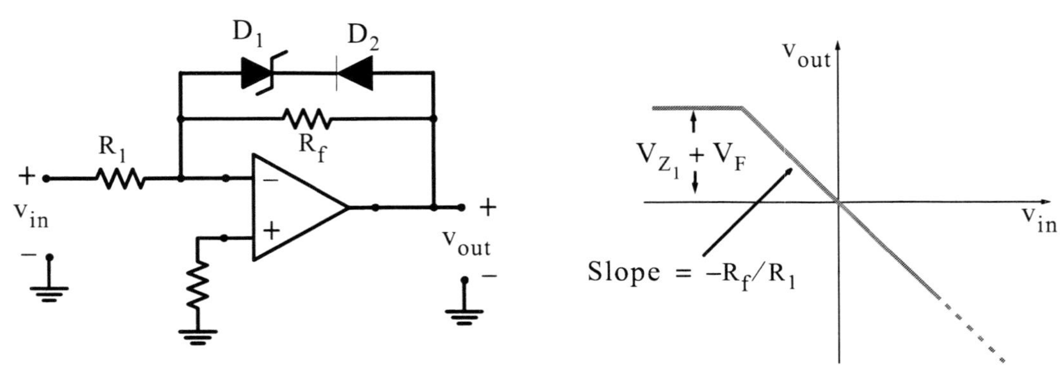

Figure 5.74. A positive voltage limiter circuit and its transfer characteristics

Figure 5.75 shows a *negative voltage limiter* and its transfer characteristics.

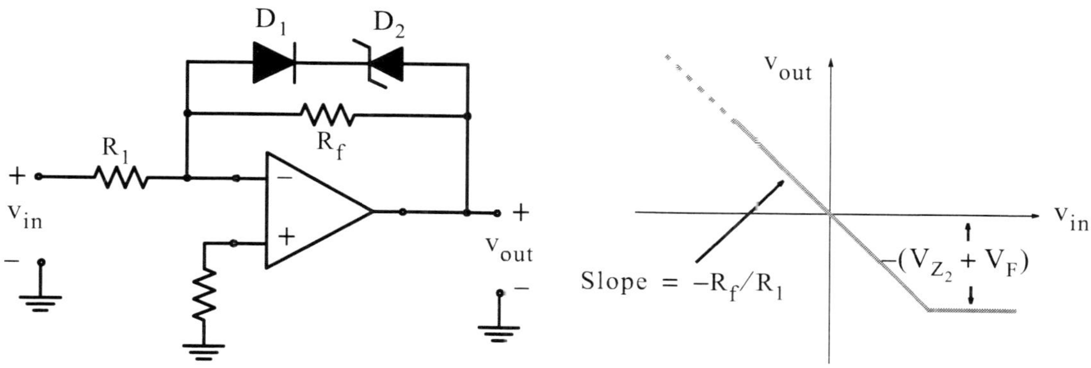

Figure 5.75. A negative voltage limiter circuit and its transfer characteristics

Figure 5.76 shows a limiter where only a single Zener diode is used. This circuit is often referred to as a *half–wave rectifier with limited positive output.*

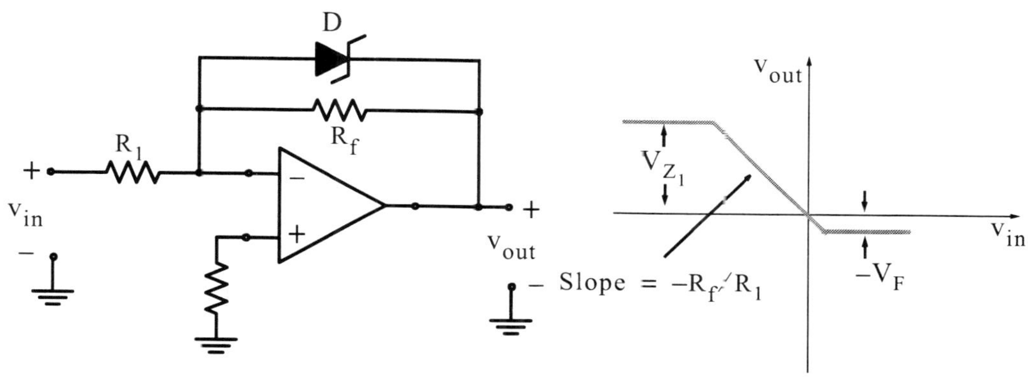

Figure 5.76. A half–wave rectifier with limited positive output and its transfer characteristics

In the circuit of Figure 5.76, the output is limited by the Zener diode during the positive half–cycles of the output voltage, and during the negative half–cycles of the output voltage is limited by the Zener diode forward voltage drop V_F. Figure 5.77 shows another limiter where only a single Zener diode is used.

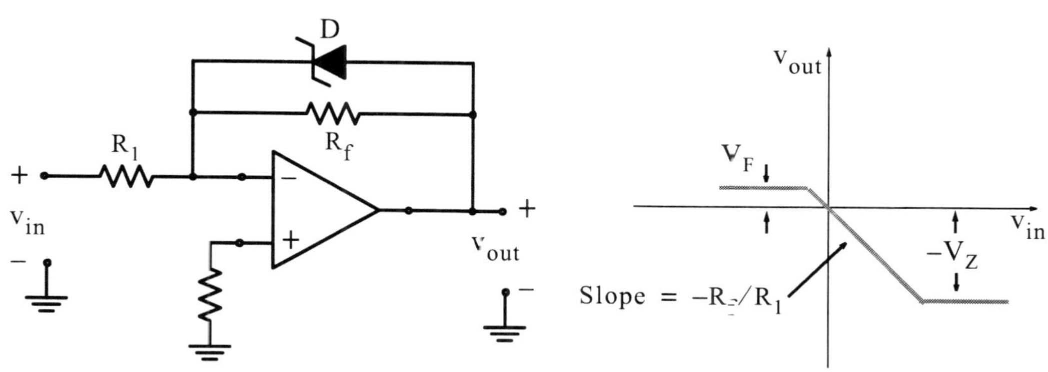

Figure 5.77. A half–wave rectifier with limited negative output and its transfer characteristics

The circuit of Figure 5.77is often referred to as a *half–wave rectifier with limited negative output*. In the circuit of Figure 5.77, the output is limited by the Zener diode during the negative half–cycles of the output voltage, and during the positive half–cycles of the output voltage is limited by the Zener diode forward voltage drop V_F.

5.21 Comparators

A *comparator* is a circuit that senses changes in a varying signal and produces an output when a threshold value is reached. As a comparator, an op amp is used without feedback, that is, the op amp is used in the open loop configuration. Figure 5.78 shows a differential input amplifier without feedback used as a comparator.

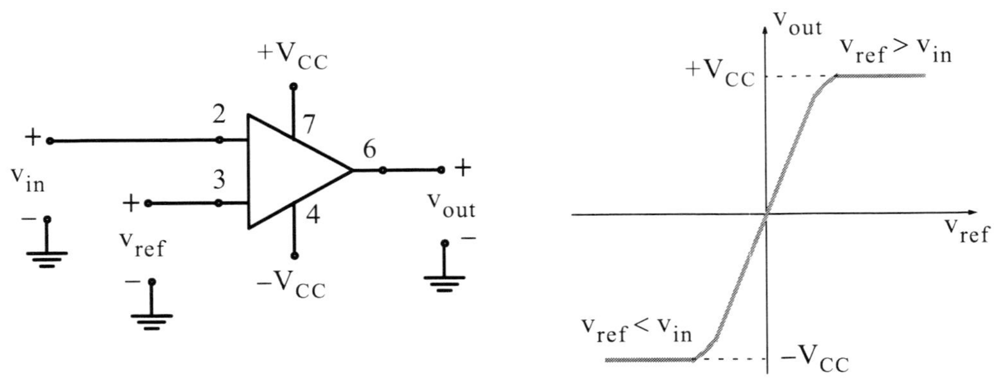

Figure 5.78. A differential input op amp without feedback used as a comparator

As shown in Figure 5.78, $v_{out} = +V_{CC}$ if $v_{ref} > v_{in}$, and $v_{out} = -V_{CC}$ if $v_{ref} < v_{in}$. The switching time from $-V_{CC}$ to $+V_{CC}$ is limited by the slew rate of the op amp. Comparators are used extensively in analog–to–digital conversion as we will see in a subsequent section.

Op amp applications are limitless. It is beyond the scope of this text to describe all. It will suffice to say that other applications include *zero–crossing detectors* also known as *sine–wave to square–wave converters, sample and hold circuits, square–wave generators, triangular–wave generators, sawtooth–wave generators, Twin–T oscillators, Wien bridge oscillators, variable frequency signal generators, Schmitt trigger,* and *multivibrators*. We will discuss the Wien bridge oscillator*, the digital–to–analog converter, and the analog–to–digital converter in the next sections, and the Schmitt trigger and multivibrators in Chapter 7.

5.22 Wien Bridge Oscillator

The circuit shown in Figure 5.79 is known as *Wien bridge oscillator*. This circuit produces a sinusoidal output.

* *We will revisit the Wien bridge oscillator in Chapter 8.*

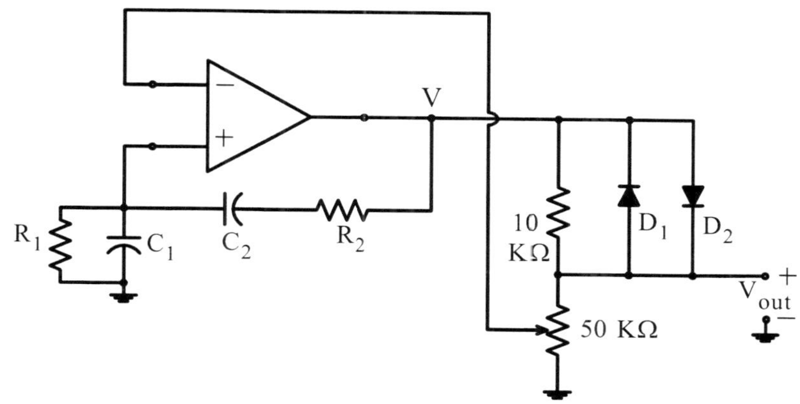

Figure 5.79. The Wien bridge oscillator

Figure 5.79 shows that the Wien bridge oscillator uses two RC networks connected to the non–inverting input of the op amp to form a frequency selective feedback circuit and this causes oscillations to occur. It also amplifies the signal with two negative feedback resistors. The input signal to the non–inverting input is in phase with the output V of the op amp at the particular frequency

$$f_0 = \frac{1}{2\pi RC} \tag{5.70}$$

provided that

$$R = R_1 = R_2$$

and

$$C = C_1 = C_2$$

The Wien bridge oscillator requires precision resistors and capacitors for reliable operation. The feedback signal at the non–inverting input of the op amp leads the output V of the op amp at frequencies below f_0 and lags V at frequencies above f_0. The amount of negative feedback to the inverting input of the op amp and amplitude can be adjusted with the $50\ K\Omega$ potentiometer. The diodes prevent excessive feedback amplitude.

Example 5.21

For the oscillator circuit in Figure 5.79 $R_1 = 100\ K\Omega$. What values of R_2, C_1, and C_2 are required to obtain a frequency of approximately $1\ KHz$?

Solution:

The value of resistor R_2 must be $100\ K\Omega$. The values of capacitors C_1, and C_2 must also be equal. From (5.70),

$$f_0 = 1\ KHz = \frac{1}{2\pi RC} = \frac{1}{2\pi \times 10^5 \times C}$$

Also,

$$C = C_1 = C_2 = \frac{1}{2\pi \times 10^5 \times 10^3} = 15.9\ \mu F$$

and the circuit is as shown in Figure 5.80.

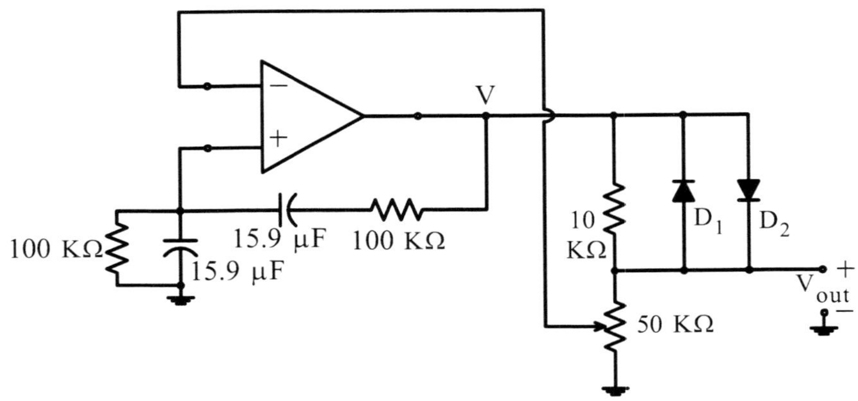

Figure 5.80. Circuit for Example 5.21

5.23 Digital-to-Analog Converters

As we will see in Chapter 6, digital systems* recognize only two levels of voltage referred to as HIGH and LOW signals or as logical 1 and logical 0. This two-level scheme works well with the binary number system. It is customary to indicate the HIGH (logical 1)and LOW (logical 0) by *Single-Pole-Double-Throw* (SPDT) switches that can be set to a positive non-zero voltage like 5 volts for HIGH and zero volts or ground for LOW as shown in Figure 5.81.

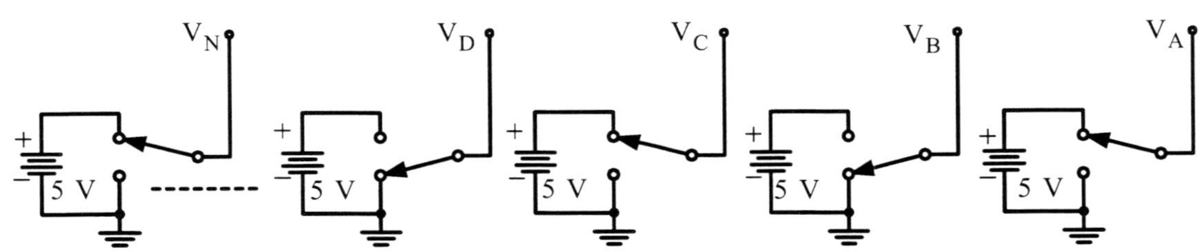

Figure 5.81. Digital circuit represented by SPDT switches

In Figure 5.81 $V_D = 0$, $V_C = 1$, $V_B = 0$, and $V_A = 1$, that is, switches A and C are HIGH (5 volts) and switches B and D are LOW (0 volts). The first 16 binary numbers representing all possible combinations of the four switches with voltage settings V_A (least significant position) through V_D (most significant position), and their decimal equivalents are shown in Table 5.1.

* *Refer also to Digital Circuit Analysis and Design with Simulink Modeling and Introduction to CPLDs and FPGAs, ISBN 978-1-934404-05-8.*

TABLE 5.1 *Voltage levels for the circuit of Figure 5.81 and binary and decimal equivalents*

Voltage Level				Binary Equivalent				Decimal Equivalent
V_D	V_C	V_B	V_A	A	B	C	D	
LOW	LOW	LOW	LOW	0	0	0	0	0
LOW	LOW	LOW	HIGH	0	0	0	1	1
LOW	LOW	HIGH	LOW	0	0	1	0	2
LOW	LOW	HIGH	HIGH	0	0	1	1	3
LOW	HIGH	LOW	LOW	0	1	0	0	4
LOW	HIGH	LOW	HIGH	0	1	0	1	5
LOW	HIGH	HIGH	LOW	0	1	1	0	6
LOW	HIGH	HIGH	HIGH	0	1	1	1	7
HIGH	LOW	LOW	LOW	1	0	0	0	8
HIGH	LOW	LOW	HIGH	1	0	0	1	9
HIGH	LOW	HIGH	LOW	1	0	1	0	10
HIGH	LOW	HIGH	HIGH	1	0	1	1	11
HIGH	HIGH	LOW	LOW	1	1	0	0	12
HIGH	HIGH	LOW	HIGH	1	1	0	1	13
HIGH	HIGH	HIGH	LOW	1	1	1	0	14
HIGH	HIGH	HIGH	HIGH	1	1	1	1	15

A *digital–to–analog (D/A or DAC) converter* is used to convert a binary output from a digital system to an equivalent analog voltage. If there are 16 combinations of the voltages V_D through V_A, the analog device should have 16 possible values. For example, since the binary number 1010 (decimal 10) is twice the value of the binary number 0101 (decimal 5), an analog equivalent voltage of 1010 must be double the analog voltage representing 0101.
Figure 5.82 shows a *DAC with binary–weighted resistors.*

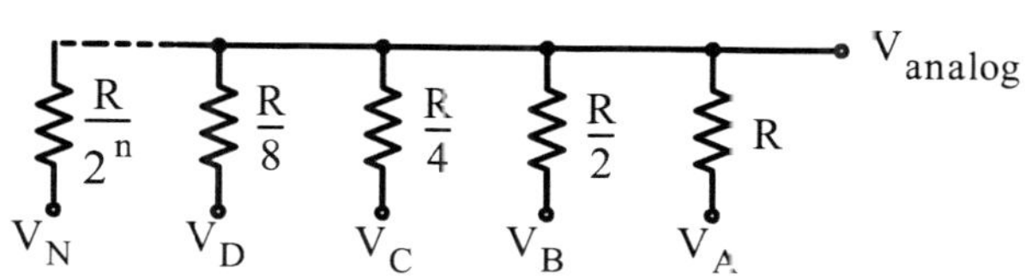

Figure 5.82. Digital–to–analog converter using binary–weighted resistors

We can prove that the equivalent analog voltage V_{analog} shown in Figure 5.82 is obtained from the relation

$$V_{analog} = \frac{V_A + 2V_B + 4V_C + 8V_D + \dots}{1 + 2 + 4 + 8 + \dots} \tag{5.71}$$

The proof is left as an exercise at the end of this chapter.

The DAC with binary–weighted resistors shown in Figure 5.82 has the disadvantage that it requires a large number of precision resistors. The DAC of Figure 5.83, known as *R–2R ladder network*, requires more resistors, but only two sets of precision resistance values, R and 2R.

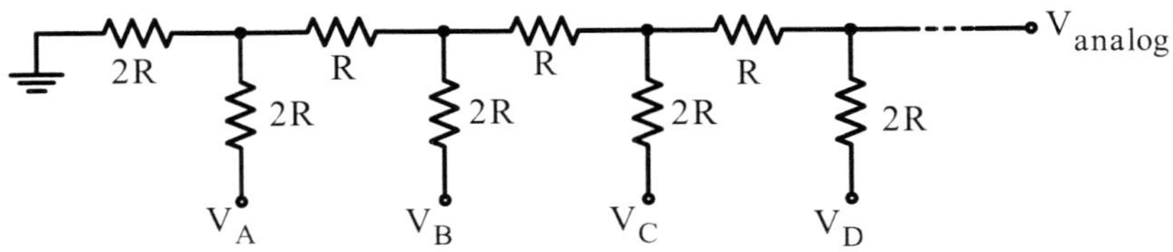

Figure 5.83. Digital–to–analog converter using the R–2R ladder network

We can prove that the equivalent analog voltage V_{analog} shown in Figure 5.83 is obtained from the relation

$$V_{analog} = \frac{V_A + 2V_B + 4V_C + 8V_D + \dots}{2^n} \tag{5.72}$$

where n is the number of digital inputs. The proof is left as an exercise at the end of this chapter.

Figure 5.84 shows a four–bit R–2R ladder network and an op–amp connected to form a DAC. The op amp shown is an inverting amplifier and in this case the reference voltage V_{ref} should be negative so that the amplifier output will be positive. Alternately, a non–inverting op amp could be used with a positive value of V_{ref}.

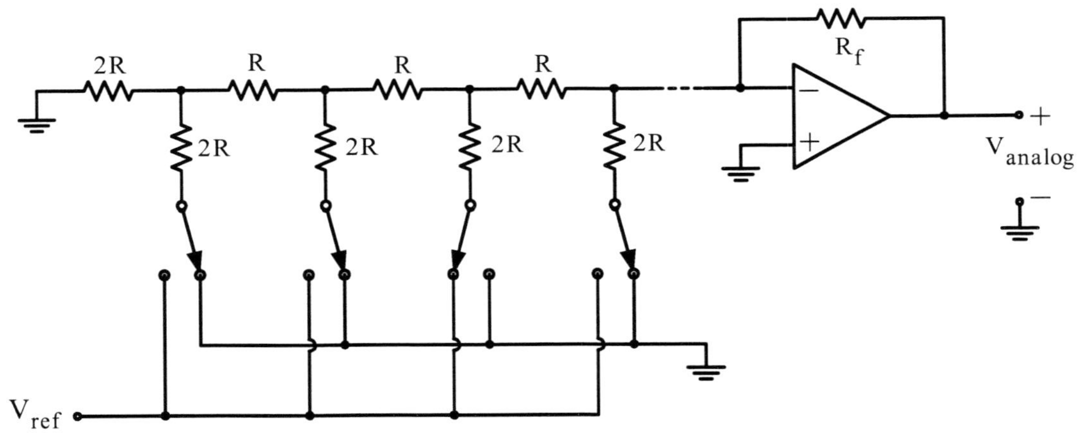

Figure 5.84. Typical R–2R DAC circuit

Example 5.22

Figure 5.85 shows a four–bit DAC where all four switches are set at the ground level. Find the analog voltage value at the output of the unity gain amplifier for each of the sets of the switch positions shown in Table 5.2. Fill–in the right–most column with your answers.

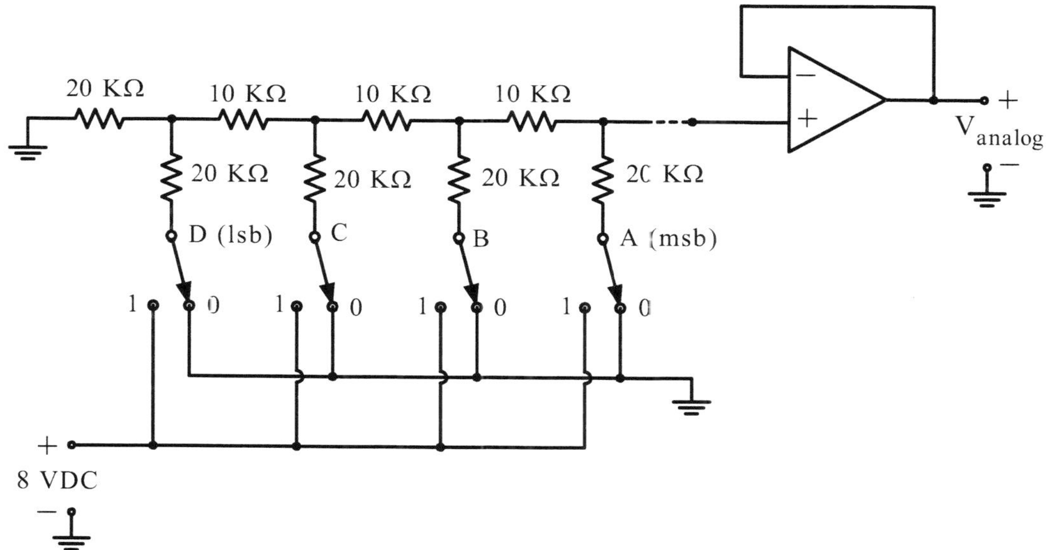

Figure 5.85. DAC circuit for Example 5.22

TABLE 5.2 Switch positions for Example 5.22

	$A\ 2^0$	$B\ 2^1$	$C\ 2^2$	$D\ 2^3$	
(a)	1	1	1	1	
(b)	1	0	0	1	
(c)	1	0	1	0	
(d)	0	1	0	0	

Solution:

This is a 4–bit DAC and thus we have $2^4 = 16$ distinct binary values from 0000 to 1111 corresponding to decimals 0 through 15 respectively. From (5.72):

a.

$$V_{analog} = \frac{V_A + 2V_B + 4V_C + 8V_D}{2^n} = \frac{1 \times 8 + 2 \times 8 + 4 \times 8 + 8 \times 8}{2^4} = 7.5\ V$$

b.

$$V_{analog} = \frac{V_A + 2V_B + 4V_C + 8V_D}{2^n} = \frac{1 \times 8 + 2 \times 0 + 4 \times 0 + 8 \times 8}{2^4} = 4.5\ V$$

c.

$$V_{analog} = \frac{V_A + 2V_B + 4V_C + 8V_D}{2^n} = \frac{1 \times 8 + 2 \times 0 + 4 \times 8 + 8 \times 0}{2^4} = 2.5\ V$$

d.

$$V_{analog} = \frac{V_A + 2V_B + 4V_C + 8V_D}{2^n} = \frac{1 \times 0 + 2 \times 8 + 4 \times 0 + 8 \times 0}{2^4} = 1.0\ V$$

Based on these results, we can now fill–in the right–most column with the values we obtained, and we can plot the output versus inputs of the R–2R network for the voltage levels 0 V and 4 V as shown in Figure 5.86.

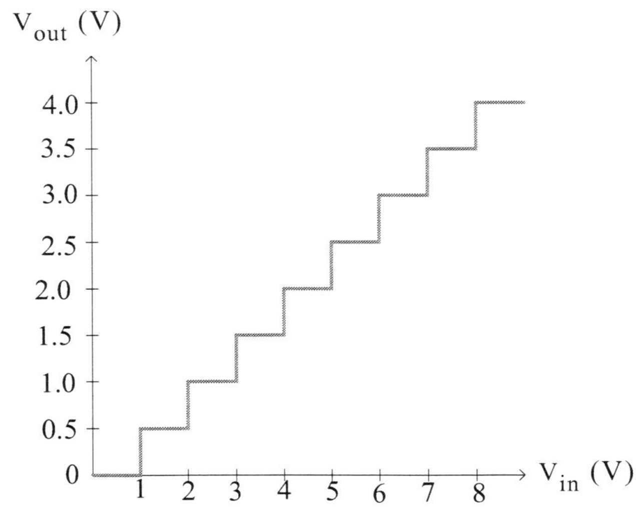

Figure 5.86. Output vs. inputs for an R–2R network with levels 0 to 4 volts

A typical DAC must include an op amp to match the resistive network to a low–resistance load and to provide gain also. Placing an impedance–matching device (the op amp in this case) at the output of the resistive network is called *buffering* the output of the network. Figure 5.87 shows an R–2R type *DAC with buffered output and gain.*

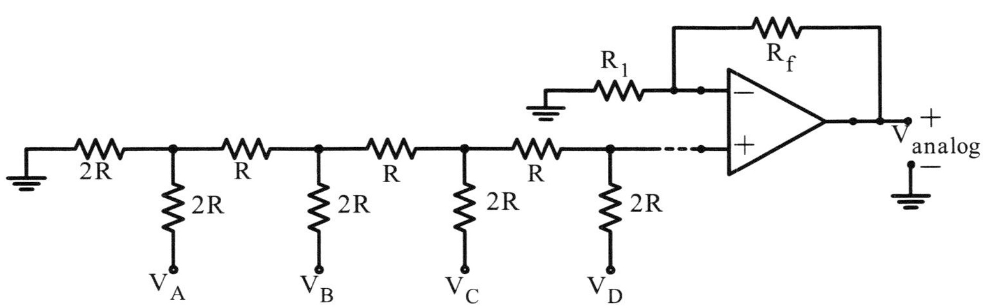

Figure 5.87. Op amp placed between the resistive network and output for buffering and gain

5.24 Analog–to–Digital Converters

Often an analog voltage must be converted to a digital equivalent, such as in a digital voltmeter. In such cases, the principle of the previously discussed digital–to–analog or D/A converter or simply DAC can be reversed to perform analog–to–digital A/D conversion. There are different types of analog–to–digital converters, usually referred to as ADC, such as the flash converter, the successive approximation converter, the dual–slope converter, and the Delta–Sigma algorithmic converter. We will not discuss the latter; the interested reader may refer to *http://www.allaboutcircuits.com/vol_4/chpt_13/9.html*. We begin our discussion with the *flash converter* because of its simplicity.

5.24.1 The Flash Analog-to-Digital Converter

Figure 5.88 shows a typical *flash type ADC* consists of a resistive network, comparators, and an 8–to–3 line encoder.* The flash ADC is so named because of its high conversion speed.

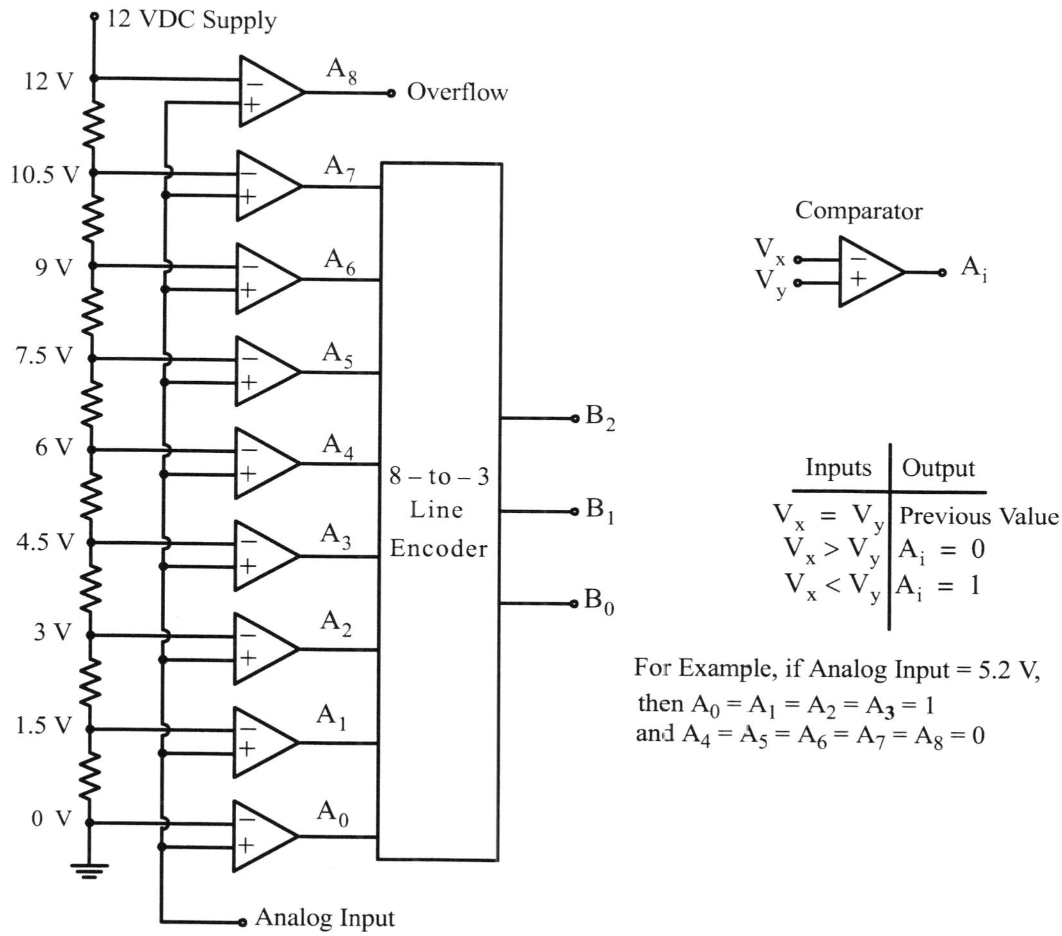

Figure 5.88. A typical circuit for a flash type ADC

The truth table of the 8–to–3 line encoder is shown in Table 5.3. Obviously, a practical ADC would require a circuit with many more comparators and a suitable line encoder.

* *For a detailed discussion of line encoders please refer to Digital Circuit Analysis and Design with Simulink Modeling and Introduction to CPLDs and FPGAs, ISBN 978–1–934404–05–8.*

TABLE 5.3 Truth table for the 8–to–3 line encoder of Figure 7.88

Analog Input	A8	A7	A6	A5	A4	A3	A2	A1	A0	B2	B1	B0
Less than 0 V	0	0	0	0	0	0	0	0	0	x	x	x†
0 to less than 1.5 V	0	0	0	0	0	0	0	0	1	0	0	0
1.5 to less than 3.0 V	0	0	0	0	0	0	0	1	1	0	0	1
3.0 to less than 4.5 V	0	0	0	0	0	0	1	1	1	0	1	0
4.5 to less than 6.0 V	0	0	0	0	0	1	1	1	1	0	1	1
6.0 to less than 7.5 V	0	0	0	0	1	1	1	1	1	1	0	0
7.5 to less than 9.0 V	0	0	0	1	1	1	1	1	1	1	0	1
9.0 to less than 10.5 V	0	0	1	1	1	1	1	1	1	1	1	0
10.5 to 12 V	0	1	1	1	1	1	1	1	1	1	1	1
Greater than 12 V	1	1	1	1	1	1	1	1	1	x	x	x‡

† Underflow
‡ Overflow

5.24.2 The Successive Approximation Analog–to–Digital Converter

Figure 5.89 shows a typical *successive approximation type* ADC.

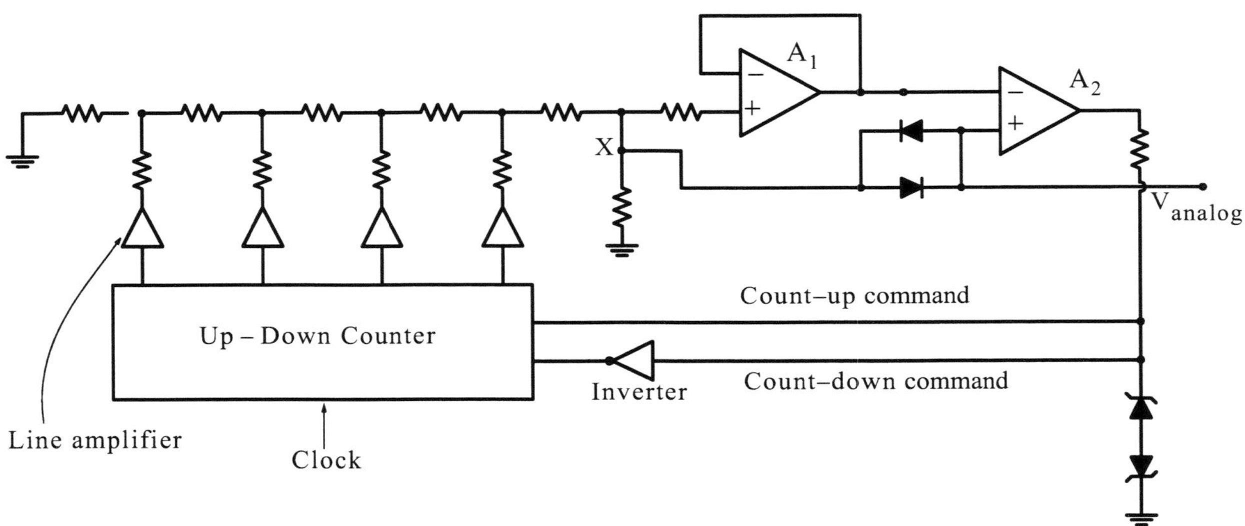

Figure 5.89. A typical circuit for a successive approximation ADC

The up–down counter* has a digital output that increases with each clock pulse when the count–up line is HIGH and the count–down line is LOW. Conversely, it counts down with each clock pulse when its count–up line is LOW and its count–down line is HIGH. Op amp A_2 is a compar-

* *For a detailed discussion on up–down counters, refer to Digital Circuit Analysis and Design with Simulink Modeling and Introduction to CPLDs and FPGAs, ISBN 978–1–934404–05–8.*

ator and its output goes HIGH when its inverting input becomes slightly more negative than its non–inverting input. Conversely, its output goes LOW when its inverting input I becomes slightly more positive than its non–inverting input. When its output is HIGH (positively saturated), the count–up line is also HIGH. Likewise, when the output of A_2 is LOW (negatively saturated), the count–up line is LOW also. Thus, depending on whether the output of A_2 is HIGH or LOW, the up–down counter counts digitally up or down, respectively.

When the up–down counter is counting up, an upward staircase voltage appears at point X. When the counter is counting downward, a downward staircase is present at point X.

Op amp A_1 serves as a voltage follower and it buffers the DAC resistive network.The output voltage of the resistive network is applied to the non–inverting input of A_1 and subsequently to inverting input of A_2 and it is compared with the analog input voltage V_{analog} which is the voltage to be digitized. If V_{analog} exceeds the voltage of the resistive network, the output of A_2 goes HIGH, and the up–down counter counts up, bringing the resistive network's output voltage up, in steps, to the analog input V_{analog}. But if V_{analog} is less than the resistive network's output voltage the output of A_2 goes LOW, and the up–down counter counts down, bringing the voltages at the input of the comparator in line with each other.

Because this system has feedback which keeps the voltage output of the resistive network approximately equal to the analog input voltage V_{analog}, this ADC is also known as the *feedback–type ADC*. In this way the output of the up–down counter is always a digital equivalent of the analog input V_{analog}. In a typical ADC of this type, the output of the counter is fed to a circuit to provide a digital readout.

The line amplifiers are required just in case the maximum value of V_{analog} exceeds the level of output voltages of the up–down counter. The junction diodes prevent excessive differential inputs to the comparator, and the Zener diodes are selected to clip the comparator's output to levels compatible with the up–down counter.

5.24.3 The Dual–Slope Analog–to–Digital Converter

The dual–slope ADC is designed with an integrating dual–slope architecture. The simplest form of an integrating ADC uses a *single–slope architecture* as shown in Figure 5.90(a) where an unknown input DC voltage v_{in} is integrated and the value is compared with a known reference value v_{ref}.It has the advantage of providing a straightforward approach to converting a low bandwidth analog signal into its digital representation.

Op amp A_1 is a Miller integrator and we recall that charging a capacitor with a constant current produces a linear voltage across it. As shown in Figure 5.90(b), the time it takes for the integrator to cause the comparator to change state is proportional to the input voltage v_{in}.

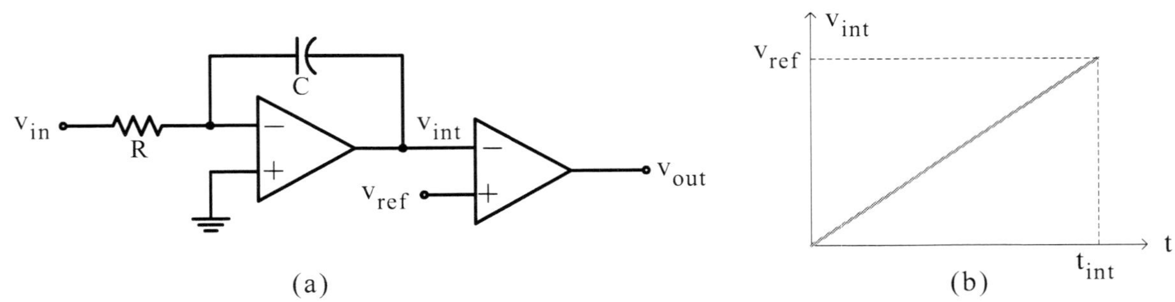

Figure 5.90. An ADC with single–slope architecture

The output voltage v_{out} is fed to a logic circuit which initiates counting in a binary counter. Thus, the circuitry of the this ADC translates a voltage level into a time measurement which can be counted with the binary counter.

The single–slope converter of Figure 5.90(a) is not practical because it is subject to errors caused by variation of the timing components R and C, temperature, and aging. These errors will change the conversion results and make measurement repeatability difficult to attain. To overcome these sensitivities, the dual–slope integrating architecture is used. Figure 5.91 shows a typical dual–slope ADC.

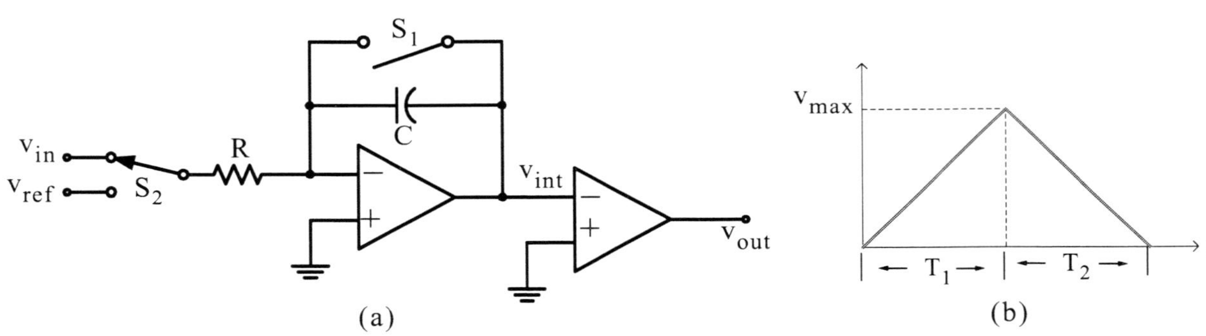

Figure 5.91. An ADC with dual–slope architecture

In the circuit of Figure 5.91(a) Switch S_1 is used to discharge the capacitor before the start of the conversion cycle and thus $v_{int} = 0$. The conversion cycle begins with opening Switch S_1 and placing Switch S_2 to the v_{in} position. A constant current whose value is $I_N = v_{in}/R$ charges the capacitor for a reference time T_1 as shown in Figure 5.91(b). The output voltage v_{out} is fed to a logic circuit which initiates counting in a binary counter. The binary counter counts the pulses from a fixed clock frequency f_C and the counting is proportional to time T_1. At the end of time T_1 the binary counter is reset to zero and Switch S_2 is placed to the v_{ref} position where v_{ref} has a negative value. The current into the integrator reverses direction and its value is

$I_{REF} = v_{ref}/R$. The voltage v_{int} at the output of the integrator decreases linearly during time T_2 as shown in Figure 5.91(b) and when this voltage reaches zero volts, the comparator initiates a signal to stop the binary counter.

Let n_1 be the counter reading at the end of T_1 and n_2 be the counter reading at the end of T_2. Since $v_{max}/T_1 = v_{in}/(RC)$, $v_{max}/T_2 = v_{ref}/(RC)$, and $n_1/T_1 = n_2/T_2$, it follows that

$$n_2 = n_1\left(\frac{v_{in}}{v_{ref}}\right) \tag{5.73}$$

and this expression indicates that the content of the counter at the end of the conversion procedure is directly proportional to the voltage v_{in} to be digitized. The major advantage of the dual–slope architecture ADC over the single–slope architecture ADC is that the entire cycle of the conversion is insensitive to errors since any error introduced by the components R and C during time T_1 will be cancelled out during time T_2.

5.25 Quantization, Quantization Error, Accuracy, and Resolution

In the previous section we've learned how to convert an analog signal to its digital equivalent. We assumed that the analog signal is continuous in both time and amplitude. To express this signal to a digital form we must subdivide both time and amplitude (for instance volts) into a number of equally spaced intervals. The subdivision is referred to as *quantization*. Figure 5.92 shows how an analog voltage is quantized in both time and amplitude. The intervals on the time and voltage axes are equally spaced but need not be the same.

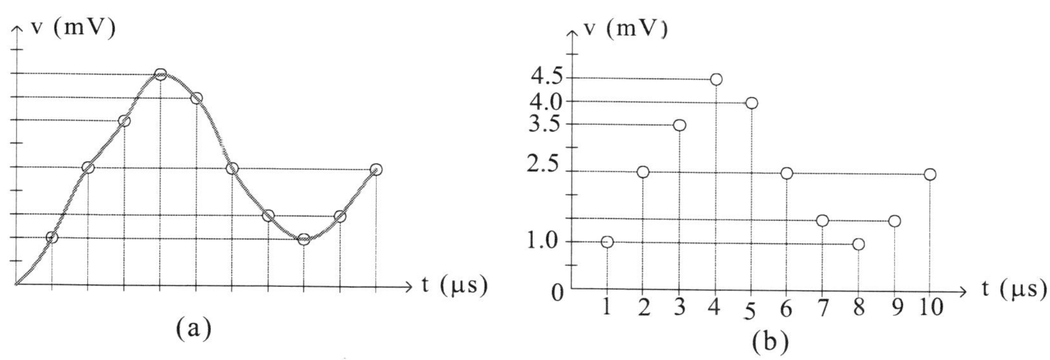

Figure 5.92. An analog voltage quantized in time and amplitude

From Figure 5.92(b) we see that at $t = 1\ \mu s$, $v = 1.0\ mv$, at $t = 2\ \mu s$, $v = 2.5\ mv$, and so on. But with the quantization we chose, we do not know intermediate values, for instance, we do not know what the amplitude of the voltage at $t = 1.6\ \mu s$. Of course, we could quantize the time axis in smaller intervals but no matter how small the intervals are, there will always be a small number that will fall between the time intervals. Subdividing the time axis into a finite number of intervals is known as *sampling*.

We may wonder what sampling rate one must choose to faithfully (exactly) reproduce an analog signal such as that of Figure 5.92(a) from its digital equivalent of Figure 5.92(b). The answer to this question is provided by the well known *sampling theorem* which states that to faithfully reproduce an analog signal from it digital equivalent, the analog signal, before conversion to a digital, must be sampled at least twice the frequency of the highest frequency component of the signal. This minimum frequency is known as the *Nyquist rate.* Thus, to reproduce exactly a 10 KHz sinusoid, theoretically, we must sample it at least at 20 KHz rate and in this case the Nyquist rate is 20 KHz. And how do we know the highest frequency component of a signal? The answer is from Fourier* series.

In practice, the Nyquist rate is not sufficient; it is a theoretical value and because of electronic equipment limitations and component imperfections, this theoretical value cannot be achieved. A general rule is to sample at five to ten times the highest frequency of the analog signal.

The difference between an exact analog value and the closest discrete value after sampling is known as *quantization error*. Obviously, every ADC suffers from a quantization error no matter how small.

The *linearity* of an ADC or DAC is a measure of accuracy and it is an indication of how close the converter's output is to the ideal input–output transfer characteristics. The output of an ADC is accurate to within $\pm 1/2$ of the least significant bit (LSD).

Example 5.23

What is the accuracy of an 8–bit ADC if the converter encodes an analog voltage between 0 and +15 volts.

Solution:

The LSD represents increments of $16/2^8 = 16/256 = 1/16$ so the LSD is $1/16$. With an accuracy of $\pm 1/2$ of the LSD, this ADC produces a digital output that represents the analog input within $\pm 1/32 = 31.3\ \mathrm{mV}$.

Another term used in ADCs and DACs is resolution. Accuracy and resolution have different meaning. *Accuracy* is the degree with which an instrument measures a variable in terms of an accepted standard value or true value; usually measured in terms of inaccuracy but expressed as accuracy; often expressed as a percentage of full–scale range. *Resolution* is the smallest change in the parameter being measured that causes a detectable change in the output of the instrument. Typically, an eight, ten, or at most twelve bits of digital output provides adequate resolution.

* *For a thorough discussion on Fourier series and the Fourier transform, please refer to Signals and Systems with MATLAB Computing and Simulink Modeling, Fourth Edition, ISBN 978–1–934404–11–9.*

For example, a voltmeter is said to have a resolution may be $0.1\ mV$. This means that this value is the step size of the voltmeter's measuring capability. In other words, it is the value with only one decimal place which can be measured and possibly recorded against other voltage records as the voltmeter is not able to detect any smaller step size in voltage changes. By comparison, the accuracy of, say $0.2\ mV$ refers to the degree to which this value may differ from the true value.

Example 5.24

How many clock cycles would we need to obtain a 10–bit resolution with a dual–slope ADC?

Solution:

The integration time T_1 would require $2^{10} = 1024$ clock cycles and also another $2^{10} = 1024$ clock cycles for integration time T_2 for a maximum conversion of $2 \times 2^{10} = 2048$ clock cycles.

5.26 Op Amps in Analog Computers

Present day *analog computers* are build with op amps. In an analog computer the numbers representing the variables are voltages. We will not discuss analog computers in this section. We will simply present two simple examples to illustrate how op amps are used in analog computation.

Example 5.25

Using two op amps and resistors, design an analog computer that will solve the equations

$$\begin{aligned} a_1x + b_1y &= c_1 \\ a_2x + b_2y &= c_2 \end{aligned} \tag{5.74}$$

where a_1, a_2, b_1, b_2, c_1, and c_2 are constant coefficients.

Solution:

We observe that the arithmetic operations involved in (5.74) are addition and multiplication by constant coefficients. The two additions can be performed by summing op amps. Multiplication by a coefficient greater than unity can be performed with an op amp with feedback, and if the coefficient is less than unity, we can use a voltage divider. It is convenient to express the given equations as

$$\begin{aligned} x &= \frac{c_1}{a_1} - \frac{b_1}{a_1}y \\ y &= \frac{c_2}{b_2} - \frac{a_2}{b_2}x \end{aligned} \tag{5.75}$$

and these equations can be solved using two summing amplifiers as shown in Figure 5.93. As shown in Figure 5.93, the output of op amp A_1 is a voltage representing the unknown x and the output of op amp A_2 is a voltage representing the unknown y. The fraction b_1/a_1 of y is obtained from the potentiometer R_{adj4} and this is summed in op amp A_1 with the fraction $-c_1/a_1$ obtained from potentiometer R_{adj1} and voltage source V_{S1}. A similar summation is obtained by amplifier A_2.

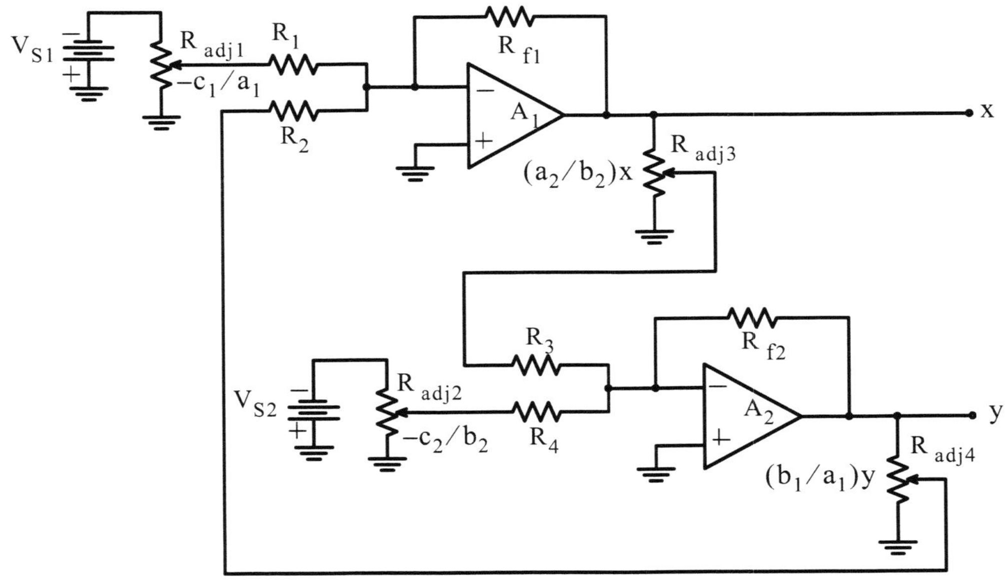

Figure 5.93. Analog computer for the solution of two simultaneous equations with two unknowns

In Figure 5.93 the scaling factors chosen must ensure that the voltages representing the unknowns do not exceed the output capability of the op amps. This procedure can be extended to the solution of simultaneous equations with more than two unknowns.

Before we consider the next example for a circuit to solve a simple differential equation, we need to discuss a practical integrator circuit that provides some means of setting a desired initial value at the beginning of the integration cycle. It is also necessary to provide means to stop the integrator at any time, and for the integrator output to remain constant at the value it has reached at that time. An integrator circuit that provides these means is shown in Figure 5.94.

Initially, the integrator sets the initial condition with the switches as shown in Figure 5.94(a), and denoting the initial condition as V_0. The voltage across the capacitor cannot change instantaneously, and thus

$$v_{out}(t=0) = -\frac{R_2}{R_1}V_0 \tag{5.76}$$

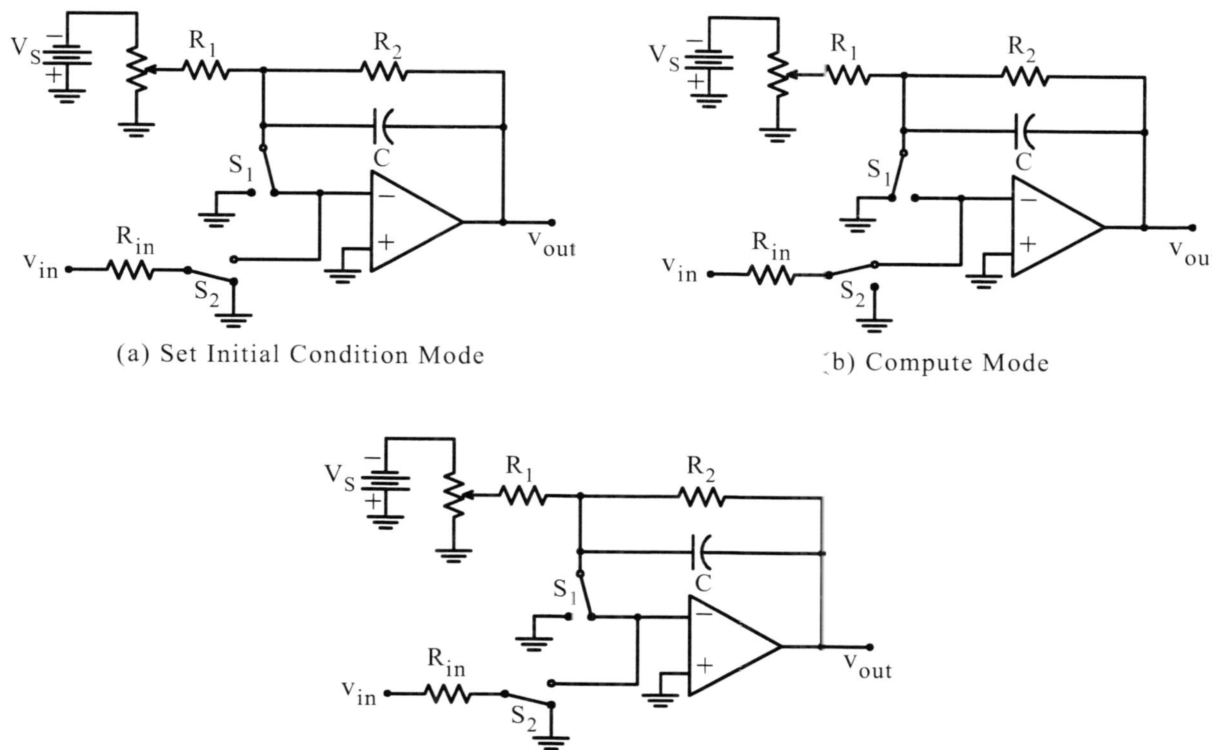

Figure 5.94. Integrator circuit mode control

When switched to the compute mode, the circuit integrates the input voltage and the value of the output voltage is

$$v_{out} = v_{out}(t=0) - \frac{1}{R_{in}C}\int_0^t v_{in}dt \tag{5.77}$$

The integrator is then switched to the hold mode and remains constant at the value reached at the end of the compute mode. We observe that the switches in the hold mode are positioned as in the set initial condition mode.

Example 5.26

Using one op amp, resistors, capacitors, and switches design an analog computer that will solve the simple differential equation

$$\frac{dv_C}{dt} = \frac{1}{C}i_C \tag{5.78}$$

where C is a constant representing the value of a capacitor, dv_C is the voltage across the capacitor and i_C is the current through the capacitor and the initial condition is $v_C(t^{-0}) = V_0$.

Solution:

The given differential equation can be solved with the circuit of Figure 5.95.

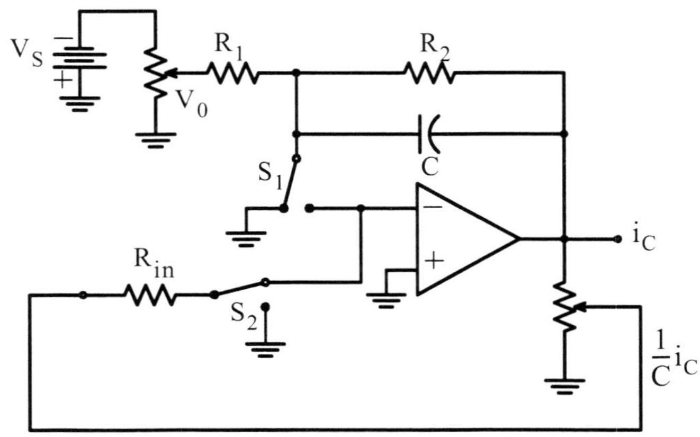

Figure 5.95. Integrator circuit for Example 5.26

Initially, a voltage representing the value of dv_C/dt is applied to the integrator and the output of the integrator will be a voltage representing i_C. Then, the output is multiplied by $1/C$ and fed back to the input.

5.27 Summary

- The operational amplifier (op amp) is the most versatile amplifier and its main features are:
  1. Very high input impedance (resistance)
  2. Very low output impedance (resistance)
  3. Capable of producing a very large gain that can be set to any value by connection of external resistors of appropriate values
  4. Frequency response from DC to frequencies in the MHz range
  5. Very good stability
  6. Operation to be performed, i.e., addition, integration etc. is done externally with proper selection of passive devices such as resistors, capacitors, diodes, and so on.
- The gain of an inverting op amp is the ratio $-R_f/R_{in}$ where R_f is the feedback resistor which allows portion of the output to be fed back to the minus (–) input. The minus (–) sign implies that the output signal has opposite polarity from that of the input signal.
- The gain of an non–inverting op amp is $1 + R_f/R_{in}$ where R_f is the feedback resistor which allows portion of the output to be fed back to the minus (–) input which is grounded through the R_{in} resistor. The output signal has the same polarity from that of the input signal.
- In a unity gain op amp the output is the same as the input. A unity gain op amp is used to provide a very high resistance between a voltage source and the load connected to it.
- Op amps are also used as active filters.
- The input resistance R_{in} is defined as the ratio of the applied voltage v_S to the current i_S drawn by the circuit, that is, $R_{in} = v_S/i_S$.
- The output resistance R_{out} is the ratio of the open circuit voltage to the short circuit current, that is, $R_{out} = v_{OC}/i_{SC}$.
- Operational amplifiers can operate either a closed–loop or an open–loop configuration. The operation, closed–loop or open–loop, is determined by whether or not feedback is used. Most operational amplifiers operate in the closed–loop configurations, that is, with feedback.
- The gain of any amplifier varies with frequency. The specification sheets for operational amplifiers state the open–loop at DC or $0\ Hz$. At higher frequencies, the gain is much lower and decreases as frequency increases at the rate of $-20\ dB/decade$.
- The unity gain frequency, is the frequency at which the gain is unity.
- The gain–bandwidth product is an important op amp parameter and is related to the unity gain

frequency as $\text{Gain} \times \text{Bandwidth} = \text{Unity Gain Frequency}$.

- The use of negative feedback increases the bandwidth of an operational amplifier circuit but decreases the gain so that the gain times bandwidth product is always equal to the unity gain frequency of the op amp.
- Common–mode rejection refers to the condition that if the inputs to the op amp are exactly the same, i.e., if $v_{in2} = v_{in1}$, then $v_{out} = 0$. In other words, the op amp rejects any signals at its inputs that are the same.
- The transresistance gain is defined as the ratio v_{out}/i_{in}.
- The closed loop transfer function G(s) is defined as $G(s) = V_{out}(s)/V_{in}(s)$ and in the inverting input mode $G(s) = -Z_f(s)/Z_1(s)$.
- The output voltage of an op amp integrator is the integral of the input voltage.
- The output voltage of an op amp differentiator is the first derivative of the input voltage.
- A summing amplifier in the inverting mode produces the negative sum of any number of input voltages. Likewise, a summing amplifier in the non–inverting mode produces the positive sum of any number of input voltages.
- A differential input op amp allows input signals to be applied simultaneously to both input terminals and produce an output of the difference between the input signals. A differential input amplifier must have a high common mode rejection ratio (CMRR) defined as

$$\text{CMRR} = \frac{\text{Differential gain}}{\text{Common mode gain}} = \frac{A_d}{A_{cm}}$$

where the differential gain A_v is the gain with the input signals applied differentially, and common mode gain is the ratio of the output common–mode voltage $V_{cm\ out}$ to the input common–mode voltage $V_{cm\ in}$, that is, $A_{cm} = V_{cm\ out}/V_{cm\ in}$. Ideally, A_{cm} is zero but in reality is finite and much smaller than unity.

- High input resistance differential input amplifiers are suitable for use in differential measurement applications and the associated circuits are referred to as instrumentation amplifiers.
- General purpose op amps are normally internally frequency compensated so that they will be stable with all values of resistive feedback. Other types of op amps are without internal frequency compensation and require external connection of frequency compensating components to the op amp. Typically, the compensating components alter the open loop gain characteristics so that the roll–off is about $20\ \text{dB/decade}$ over a wide range of frequencies.
- The slew rate (SR) is defined as the maximum rate of change of an output voltage produced in

response to a large input step function and it is normally expressed in volts per microsecond, that is,

$$SlewRate = SR = \frac{dv_{out}}{dt_{max}}$$

- Op amps can also be used in circuits with non–linear devices. We have introduced the positive and negative voltage limiters, half–wave rectifier with limited positive or negative output.
- A comparator is a circuit that senses changes in a varying signal and produces an output when a threshold value is reached. As a comparator, an op amp is used without feedback, that is, the op amp is used in the open loop configuration.
- A Wien bridge oscillator produces a sinusoidal output. This oscillator uses two RC networks connected to the non–inverting input of the op amp to form a frequency selective feedback circuit and this causes oscillations to occur. It also amplifies the signal with two negative feedback resistors.
- A digital–to–analog converter (D/A converter or DAC) is used to convert a binary output from a digital system to an equivalent analog voltage. If there are 16 combinations of the voltages V_D through V_A, the analog device should have 16 possible values.
- A DAC with binary–weighted resistors has the disadvantage that it requires a large number of precision resistors.
- A DAC formed with an R–2R ladder network, requires more resistors, but only two sets of precision resistance values, R and 2R.
- A DAC must include an op amp to match the resistive network to a low–resistance load and to provide gain also. Placing an impedance–matching device (the op amp in this case) at the output of the resistive network is called buffering the output of the network.
- An analog–to–digital converter (A/D converter or ADC) converts an analog signal to its binary equivalent. Popular types of ADCs are the flash converter, the successive approximation, and the dual–slope.
- Quantization is the process where an analog signal is subdivided into small discrete time and amplitude intervals. Subdividing the time axis into a finite number of intervals is known as sampling. The difference between an exact analog value and the closest discrete value after sampling is known as quantization error.
- The sampling theorem states that to faithfully reproduce an analog signal from it digital equivalent, the analog signal, before conversion to a digital, must be sampled at least twice the frequency of the highest frequency component of the signal. This minimum frequency is known as the Nyquist rate. The Nyquist rate is not sufficient; because of electronic equipment limitations and component imperfections, this theoretical value cannot be achieved. A general rule is to sample at five to ten times the highest frequency of the analog signal.

- The linearity of an ADC or DAC is a measure of accuracy and it is an indication of how close the converter's output is to the ideal input–output transfer characteristics. The output of an ADC is accurate to within $\pm 1/2$ of the least significant bit (LSD).
- Accuracy is the degree with which an instrument measures a variable in terms of an accepted standard value or true value.
- Resolution is the smallest change in the parameter being measured that causes a detectable change in the output of the instrument. Typically, an eight, ten, or at most twelve bits of digital output provides adequate resolution.
- Present day analog computers are build with op amps. In an analog computer the numbers representing the variables are voltages.

5.28 Exercises

1. For the circuit below compute v_{out2}.

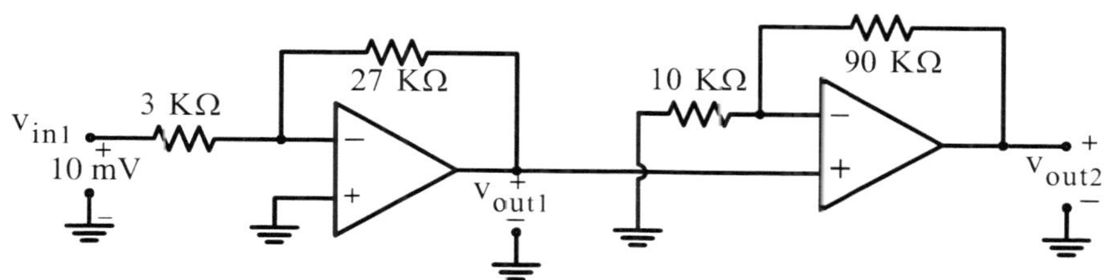

2. For the circuit below compute $i_{5K\Omega}$.

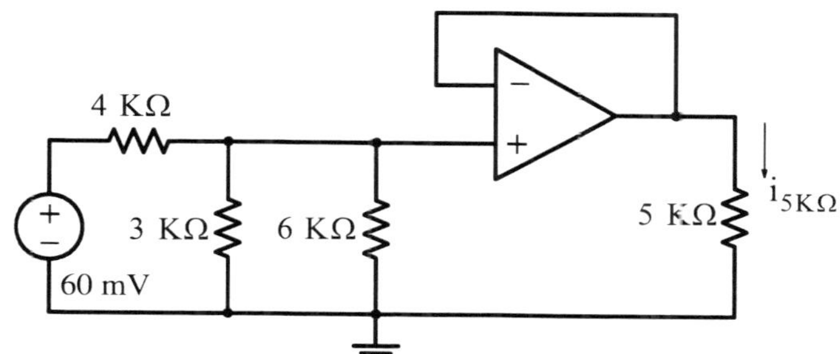

3. For the circuit below, R_{in1}, R_{in2}, and R_{in3} represent the internal resistances of the input voltages v_{in1}, v_{in2}, and v_{in3} respectively. Derive an expression for v_{out} in terms of the input voltage sources, their internal resistances, and the feedback resistance R_f.

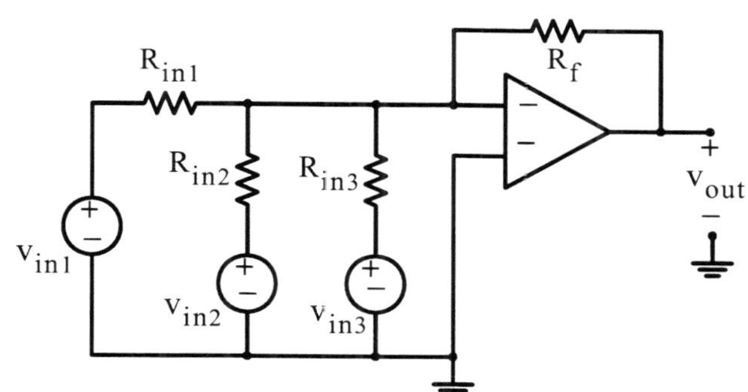

4. For the circuit below compute v_{out}.

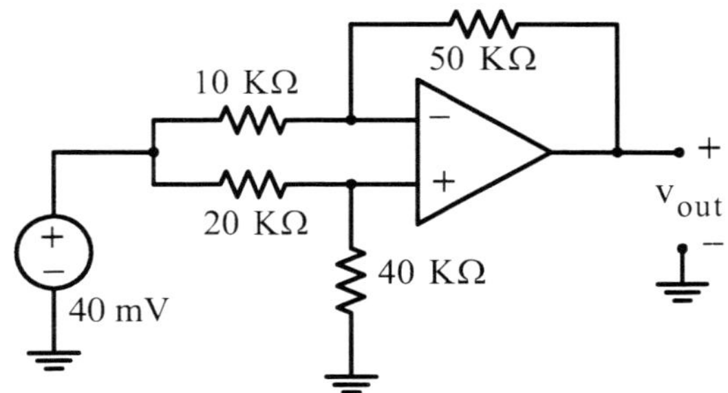

5. The op–amp circuit (a) below can be represented by its equivalent circuit (b). For the circuit (c), compute the value of R_{load} so that it will receive maximum power.

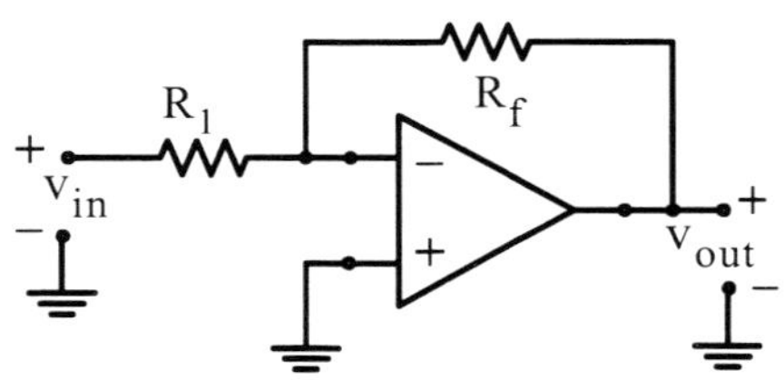

(a)

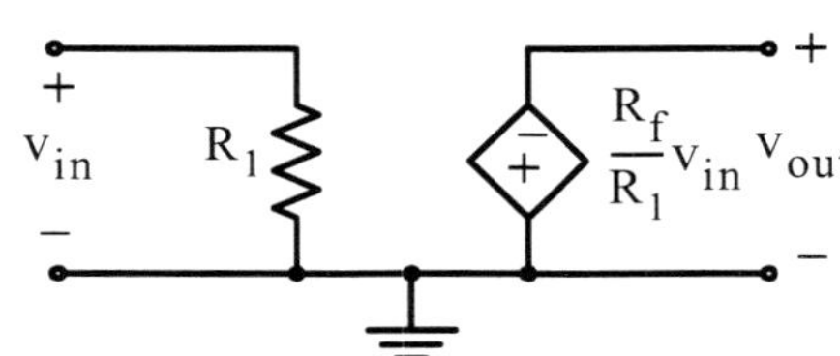

(b)

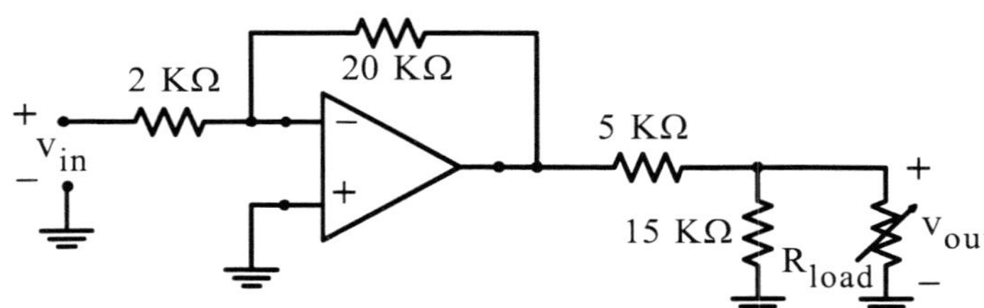

(c)

6. For the circuit below compute $v_{5K\Omega}$ using Thevenin's theorem.

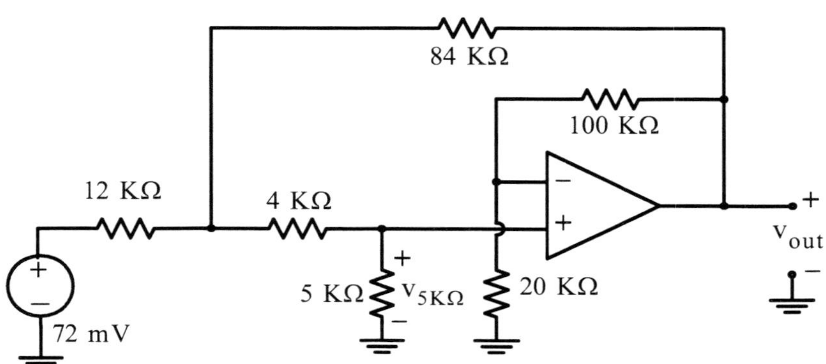

7. For the circuit below compute the gain $G_v = v_{out}/v_{in}$.

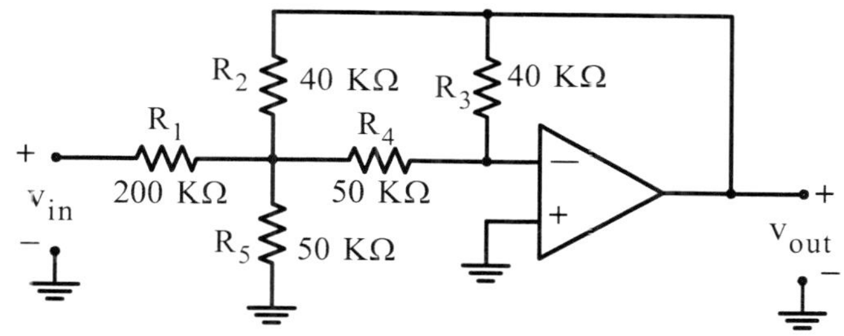

8. For the circuit below show that the gain is given by

$$G_v = \frac{v_{out}}{v_{in}} = -\frac{1}{R_1}\left[R_4 + R_2\left(\frac{R_4}{R_3} + 1\right)\right]$$

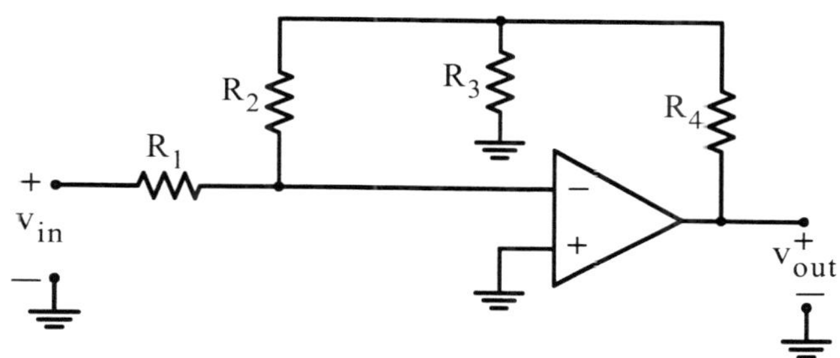

9. The specifications for the popular 741 op amp state that the open–loop gain is 200, 000 and the unity gain frequency is 1 MHz. Plot the frequency response for open–loop gain versus frequency curve and the closed–loop frequency response for a gain of 10.

10. For the op amp shown below the open–loop gain is 100, 000.

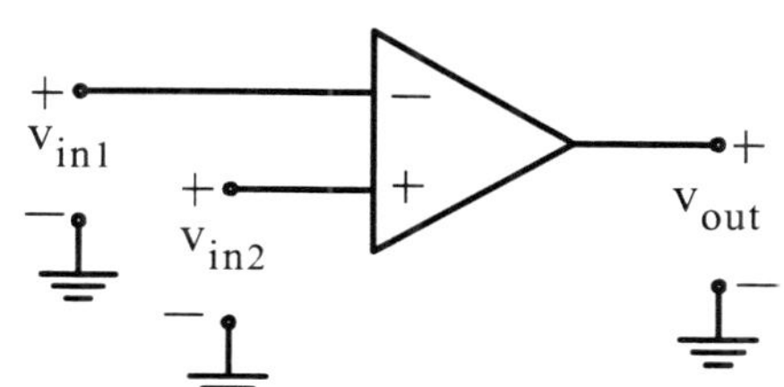

a. Find v_{in1} if $v_{in2} = 3\ mV$ and $v_{out} = 5\ V$

b. Find v_{in2} if $v_{in1} = 2\ mV$ and $v_{out} = -5\ V$

c. Find v_{out} if $v_{in1} = 2\ mV$ and $v_{in2} = -3\ mV$

11. Determine whether the op amp circuit below is stable or unstable. Assume infinite input impedance and zero output impedance.

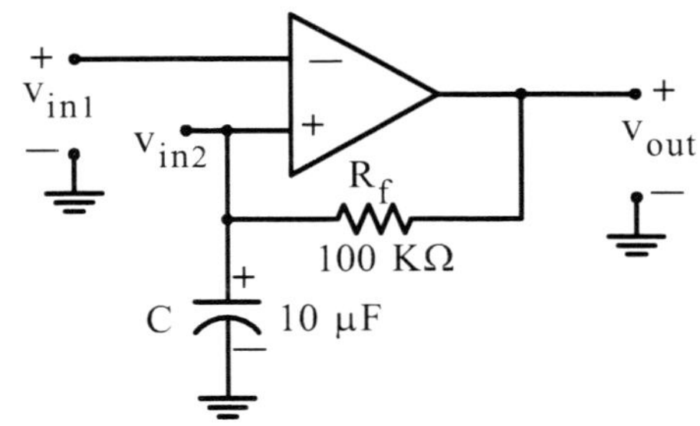

12. For the op amp circuit below, find:

a. the transresistance R_m

b. the output voltage v_{out}

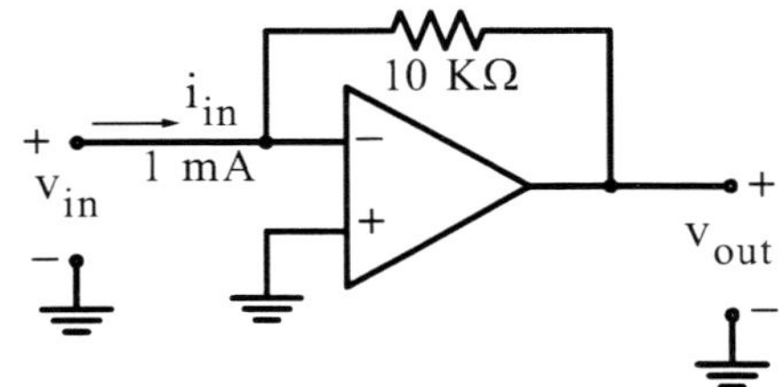

13. For the op amp circuit below:

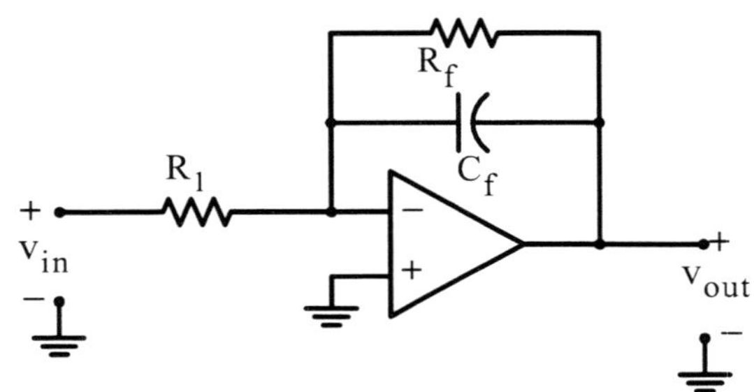

a. Derive the closed–loop transfer function

b. Derive an expression for the DC gain

c. Derive an expression for the 3 dB frequency

d. If $R_1 = 1\ K\Omega$, compute the values of R_f and C_f such that the circuit will have a DC gain of 40 dB and 1 KHz 3 dB frequency

e. Compute the frequency at which the gain is 0 dB

f. The phase angle at the frequency where the gain is 0 dB

14. For the op amp integrator circuit below the time constant is $\tau = 100\ \mu s$.

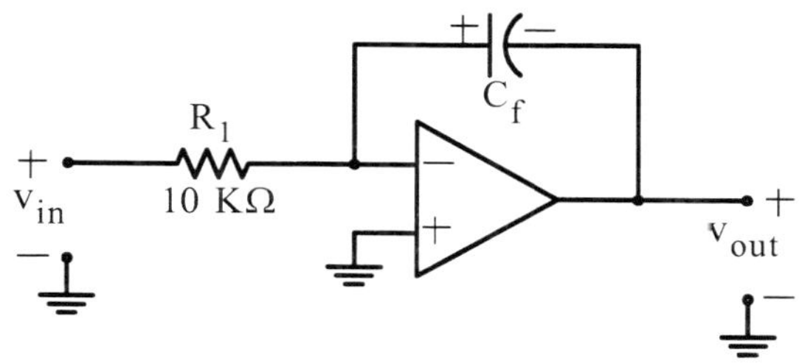

a. Find the value of the capacitor C_f

b. Find the magnitude and phase angle of the gain at $f = 1\ KHz$

c. Find the frequency at which the magnitude of the gain is unity

15. For the circuit below, it is required that the current i_f should not exceed $1\ mA$, and the output voltage should be such that $v_{out} = -(4v_{in1} + 3v_{in2}) \le |12|\ V$. Select appropriate values for resistors R_f, R_1, and R_2 to meet these specifications.

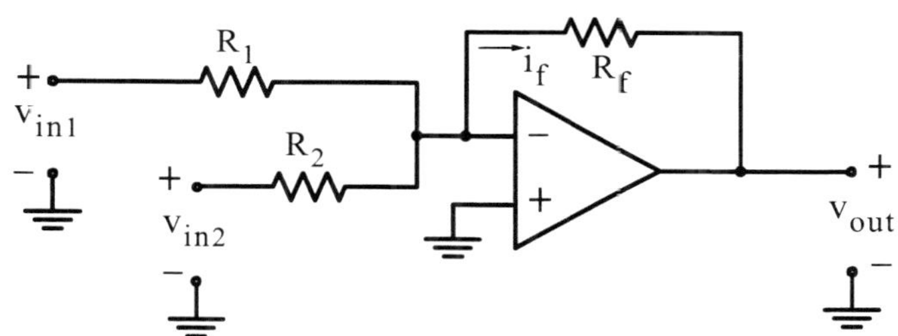

16. As stated earlier in Section 5.16, to make the overall gain of the circuit of Figure 5.67 variable, we replace resistor R_1 with a fixed value resistor R_{fix} in series with a variable resistor R_{var} as shown below.

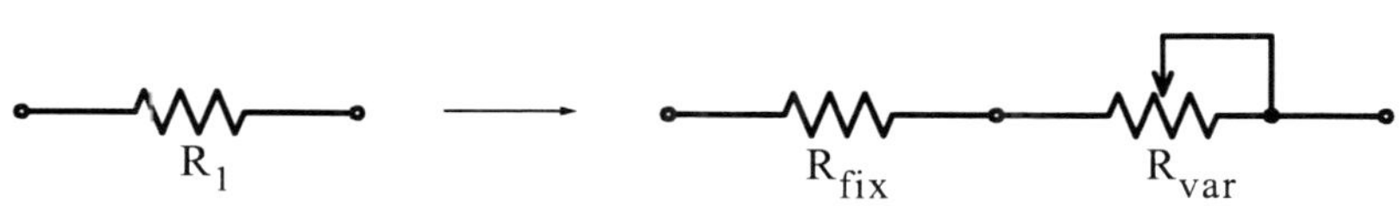

Choose appropriate values for the fixed value resistor R_{fix} and the variable resistor R_{var} so that the overall voltage gain can vary from 10 to 100.

17. Suggest a nulling circuit other than that in Exercise 16 for:

a. an inverting mode op amp

b. a non–inverting mode op amp

18. Find appropriate values for all external resistors so that the differential input op amp circuit below will have an input resistance $R_{in} = 50\ K\Omega$ and voltage gain $G_v = 40$.

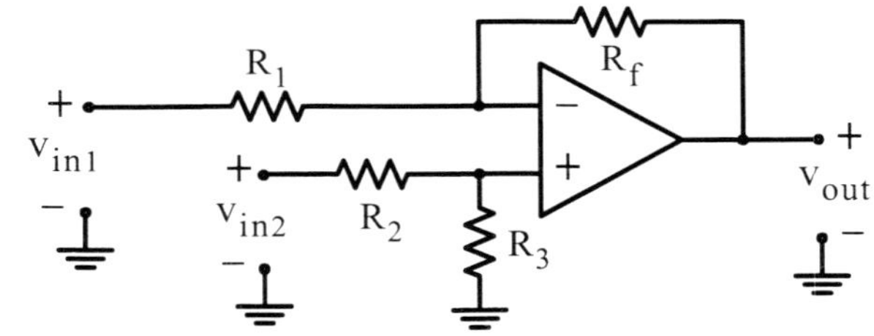

19. For the circuit below, derive the transfer function $G_v(s) = V_{out}(s)/V_{in}(s)$ and the magnitude and phase of $G_v(j\omega) = V_{out}(j\omega)/V_{in}(j\omega)$ at unity gain.

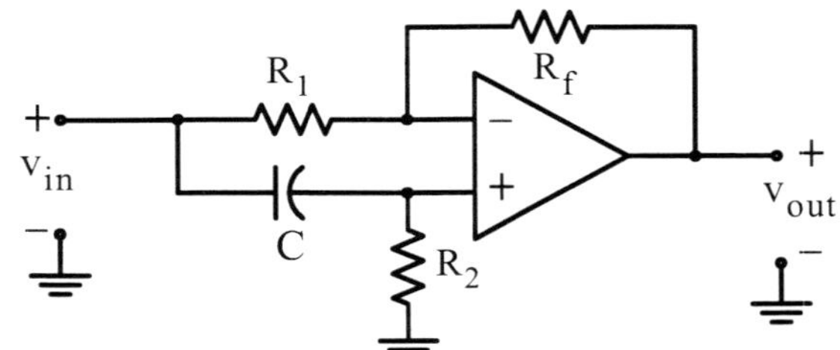

20. The circuit below is referred to as a *differential input integrator*. The resistors and capacitors are of equal value. Derive an expression for $V_{out}(s)$ in terms of the input voltages and the circuit constants.

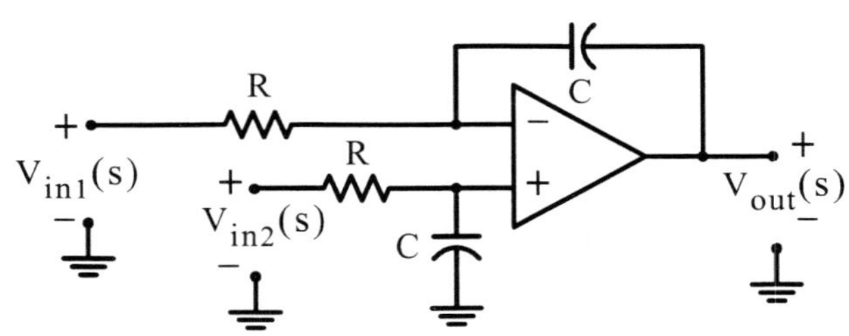

21. The figure below shows a digital–to–analog converter with binary–weighted resistors.

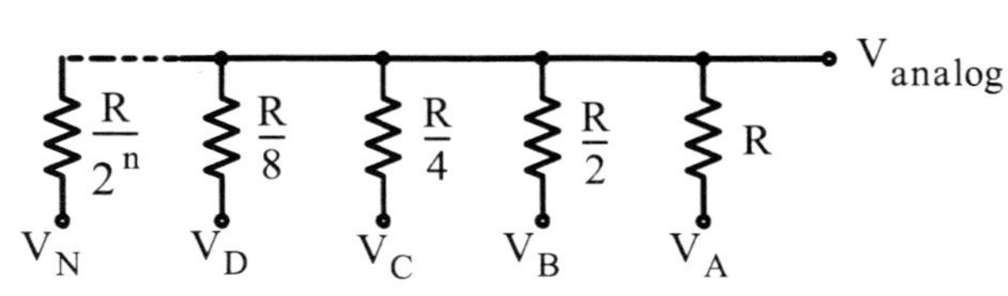

Prove that

$$V_{analog} = \frac{V_A + 2V_B + 4V_C + 8V_D + \ldots}{1 + 2 + 4 + 8 + \ldots}$$

22. The figure below shows a digital–to–analog converter with known as R–2R ladder network.

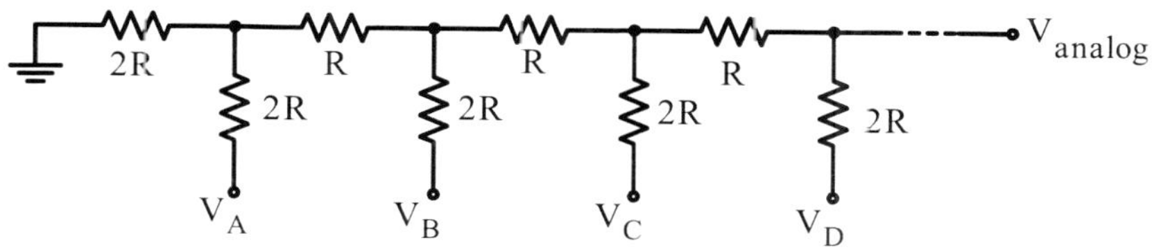

Prove that

$$V_{analog} = \frac{V_A + 2V_B + 4V_C + 8V_D + \ldots}{2^n}$$

where n is the number of digital inputs.

23. Design an analog computer circuit to perform the following integration.

$$v_{out} = -\int_0^t (v_{in1} + 2v_{in2} + 10v_{in3})dt$$

Assume ideal op amp action and that the integrating capacitor has a value of $1\ \mu F$.

24. Design an op amp circuit with input v_{in} and output $v_{out} = -5v_{in} - 3dv_{in}/dt$

25. For the op amp circuit shown below, derive an expression for the output voltage v_{out} in terms of the input voltage v_{in} and the circuit constants.

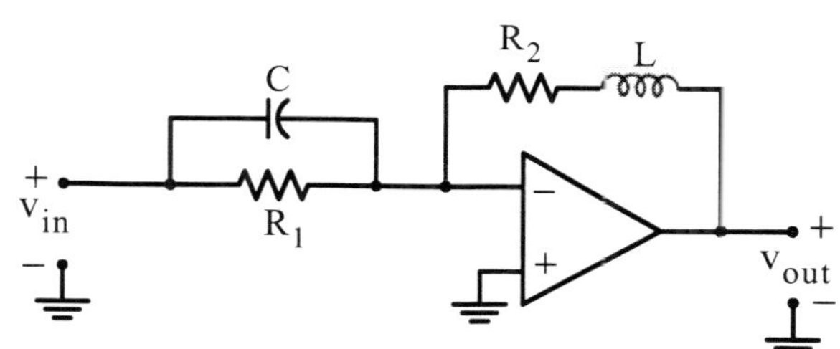

26. For the op amp circuit shown below, derive an expression for the output voltage v_{out} in terms of the input voltage v_{in}.

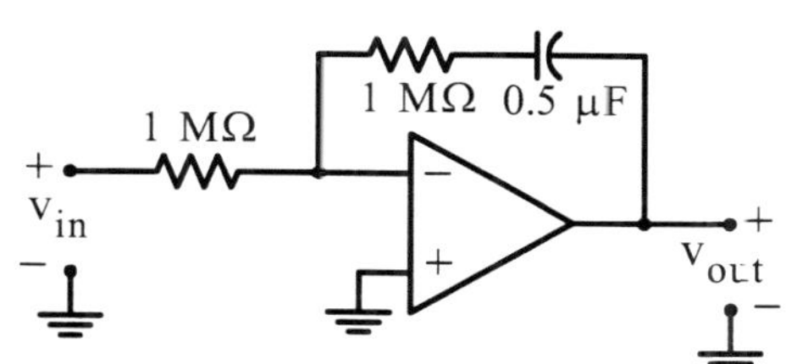

5.29 Solutions to End-of-Chapter Exercises

1.

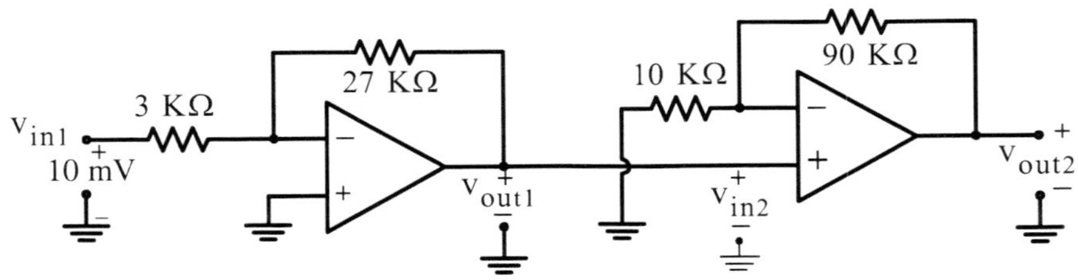

$$v_{out1} = -(27/3) \times 10 = -90 \text{ mV}$$

and thus

$$v_{in2} = v_{out1} = -90 \text{ mV}$$

Then

$$v_{out2} = \left(1 + \frac{90}{10}\right) \times (-90) = -0.9 \text{ V}$$

2. We assign v_1, v_{load}, and R_{load} as shown below.

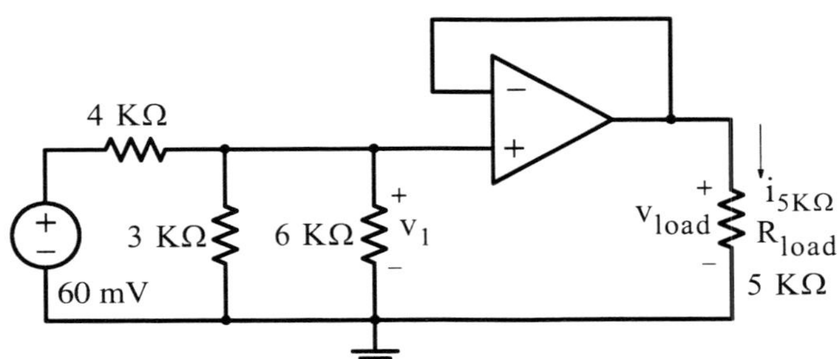

$$3 \text{ K}\Omega \,\|\, 6 \text{ K}\Omega = 2 \text{ K}\Omega$$

and by the voltage division expression

$$v_1 = \frac{2 \text{ K}\Omega}{4 \text{ K}\Omega + 2 \text{ K}\Omega} \times 60 \text{ mV} = 20 \text{ mV}$$

and since this is a unity gain amplifier, we obtain

$$v_{load} = v_1 = 20 \text{ mV}$$

Then

$$i_{5K\Omega} = \frac{v_{load}}{R_{load}} = \frac{20 \text{ mV}}{5 \text{ K}\Omega} = \frac{20 \times 10^{-3}}{5 \times 10^{3}} = 4 \times 10^{-6} \text{ A} = 4 \text{ }\mu\text{A}$$

3.

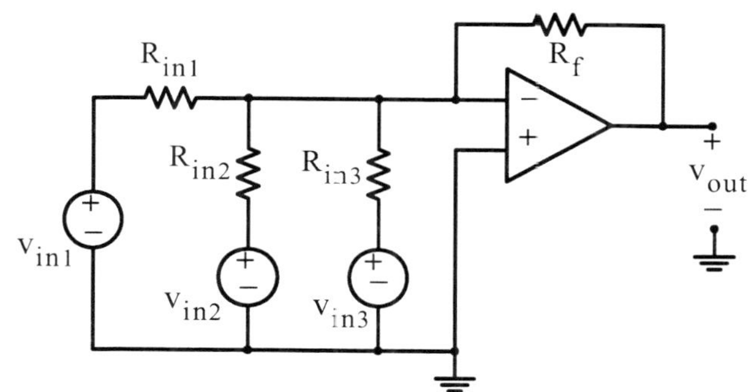

By superposition

$$v_{out} = v_{out1} + v_{out2} + v_{out3}$$

where

$$v_{out1}\Big|_{\substack{v_{in2}=0\\ v_{in3}=0}} = -\frac{R_f}{R_{in1}}v_{in1}$$

We observe that the minus (–) input is a virtual ground and thus there is no current flow in R_{in1} and R_{in2}. Also,

$$v_{out2}\Big|_{\substack{v_{in1}=0\\ v_{in3}=0}} = -\frac{R_f}{R_{in2}}v_{in2}$$

and

$$v_{out3}\Big|_{\substack{v_{in1}=0\\ v_{in2}=0}} = -\frac{R_f}{R_{in3}}(-v_{in3})$$

Then,

$$v_{out} = R_f\left(\frac{v_{in3}}{R_{in3}} - \frac{v_{in2}}{R_{in2}} - \frac{v_{in1}}{R_{in1}}\right)$$

4. We assign voltages v_1 and v_2 as shown below. At the minus (–) terminal

$$\frac{v_1 - 40\ mV}{10\ K\Omega} + \frac{v_1 - v_{out}}{50\ K\Omega} = 0$$

$$\frac{6}{50\times10^3}v_1 - \frac{1}{50\times10^3}v_{out} = 4\times10^{-6}$$

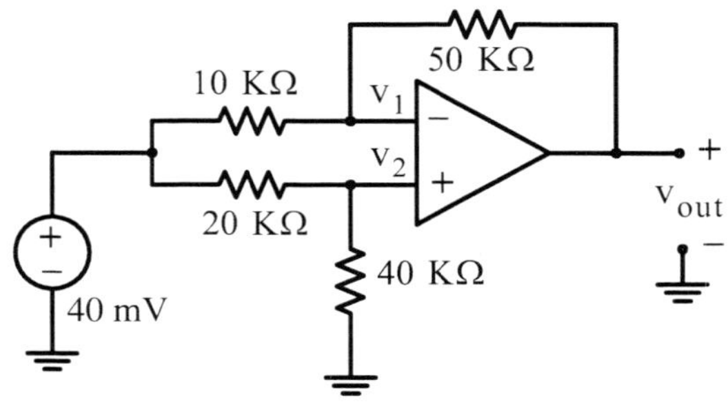

At the plus (+) terminal

$$\frac{v_2 - 40\ mV}{20\ K\Omega} + \frac{v_2}{40\ K\Omega} = 0$$

$$\frac{3}{40 \times 10^3} v_2 = 2 \times 10^{-6}$$

$$v_2 = \frac{80 \times 10^{-3}}{3}$$

Since $v_2 = v_1$ we equate the nodal equations and we obtain

$$\frac{6}{50 \times 10^3}\left(\frac{80 \times 10^{-3}}{3}\right) - \frac{1}{50 \times 10^3} v_{out} = 4 \times 10^{-6}$$

Multiplication by 50×10^3 yields

$$\frac{2 \times 80 \times 10^{-3} \times 50 \times 10^3}{50 \times 10^3} - v_{out} = 4 \times 10^{-6} \times 50 \times 10^3$$

$$v_{out} = -40\ mV$$

Check with (5.16), Page 5-22, using MATLAB:

```
R1=10000; R2=20000; R3=40000; Rf=50000; Vin=40*10^(-3);
Vout=(R1*R3-R2*Rf)*Vin/(R1*(R2+R3))

Vout =
    -0.0400
```

5. We attach the 5 KΩ, 15 KΩ, and R_{load} resistors to the equivalent circuit as shown below.

By Thevenin's theorem,

$$v_{TH} = v_{OC} = v_{ab} = \frac{15\ K\Omega}{5\ K\Omega + 15\ K\Omega}(-10 v_{in})$$

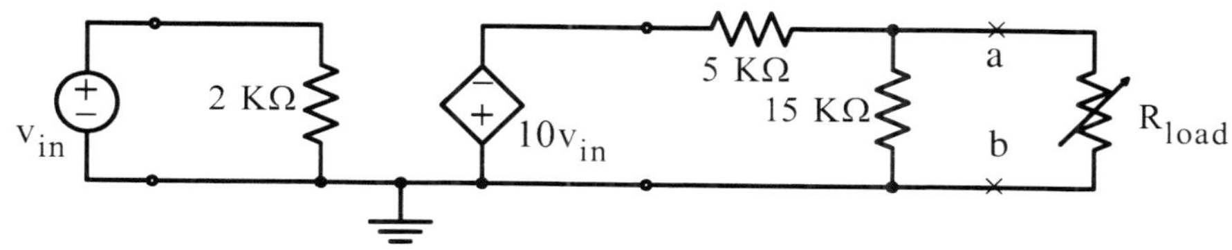

or

$$v_{TH} = -7.5v_{in}$$

Because the circuit contains a dependent source, we must compute the Thevenin resistance using the relation $R_{TH} = v_{TH}/i_{SC}$ where i_{SC} is found from the circuit below.

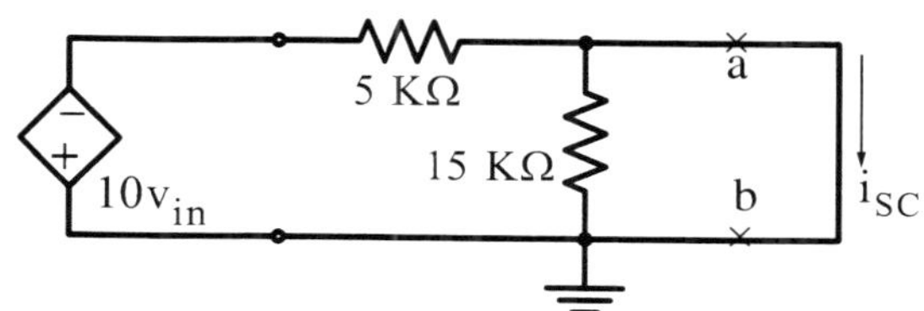

We observe that the short circuit shorts out the 15 KΩ and thus

$$i_{SC} = \frac{-10v_{in}}{5\ K\Omega} = -2\times 10^{-3}v_{in}$$

Then

$$R_{TH} = \frac{-7.5v_{in}}{-2\times 10^{-3}v_{in}} = 3.75\ K\Omega$$

and the Thevenin equivalent circuit is shown below.

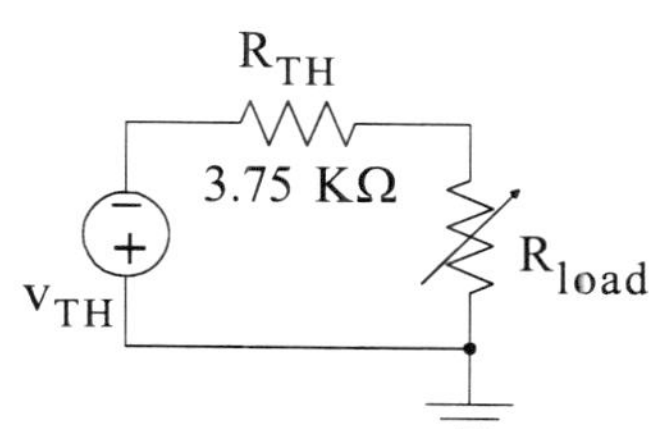

Therefore, for maximum power transfer we must have $R_{load} = R_{TH} = 3.75\ K\Omega$

6. The given circuit is a non–inverting op amp whose equivalent circuit is shown below.

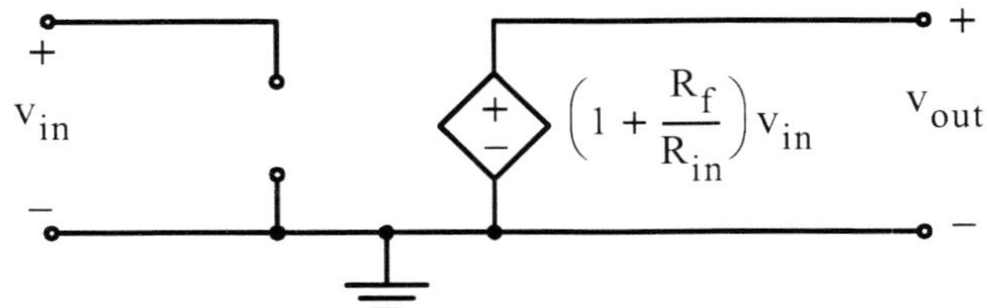

For this circuit $v_{in} = v_{5K\Omega}$ and the output is

$$v_{out} = \left(1+\frac{R_f}{R_{in}}\right)v_{5K\Omega} = \left(1+\frac{100}{20}\right)v_{5K\Omega} = 6v_{5K\Omega}$$

Attaching the external resistors to the equivalent circuit above we obtain the circuit below.

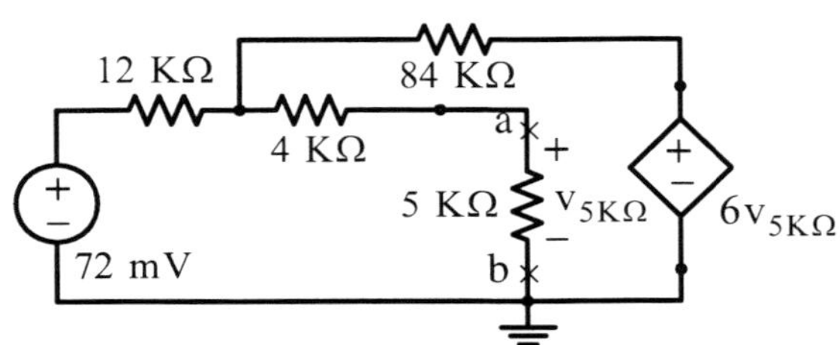

To find the Thevenin equivalent at points *a* and *b* we disconnect the 5 KΩ resistor. When this is done there is no current in the 4 KΩ resistor and the circuit simplifies to the one shown below.

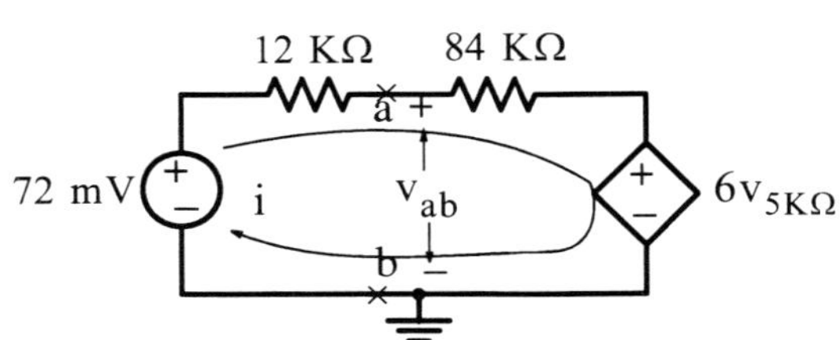

By KVL

$$(12\ K\Omega + 84\ K\Omega)i + 6v_{5K\Omega} = 72\ mV$$

or

$$i = \frac{72\ mV - 6v_{5K\Omega}}{(12\ K\Omega + 84\ K\Omega)}$$

Also

$$v_{TH} = v_{ab} = v_{5K\Omega} = 72\ mV - (12\ K\Omega)i = 72\ mV - 12\ K\Omega\left(\frac{72\ mV - 6v_{5K\Omega}}{96\ K\Omega}\right)$$

$$v_{TH} = 72\ mV - 9\ mV + \frac{3}{4}v_{5K\Omega}$$

or

$$v_{5K\Omega} - \frac{3}{4}v_{5K\Omega} = 63\ \text{mV}$$

and thus

$$v_{TH} = v_{ab} = v_{5K\Omega} = 252\ \text{mV}$$

The Thevenin resistance is found from $R_{TH} = v_{OC}/i_{SC}$ where i_{SC} is computed with the terminals *a* and *b* shorted making $v_{5K\Omega} = 0$ and the circuit is as shown in Figure (a) below. We also perform voltage–source to current–source transformation and we obtain the circuit shown in Figure (b) below.

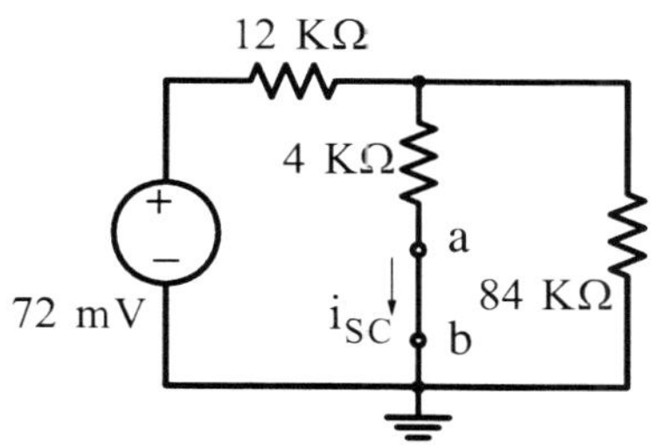

(a)

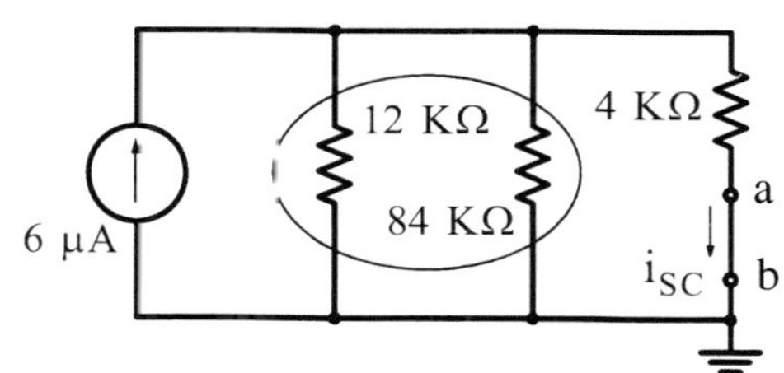

(b)

Now,

$$12\ \text{K}\Omega \,||\, 84\ \text{K}\Omega = 10.5\ \text{K}\Omega$$

and by the current division expression

$$i_{SC} = i_{ab} = \frac{10.5\ \text{K}\Omega}{10.5\ \text{K}\Omega + 4\ \text{K}\Omega} \times 6\ \mu\text{A} = \frac{126}{29}\ \mu\text{A}$$

Therefore,

$$R_{TH} = \frac{v_{OC}}{i_{SC}} = \frac{252}{126/29} = 58\ \text{K}\Omega$$

and the Thevenin equivalent circuit with the 5 KΩ resistor is shown below.

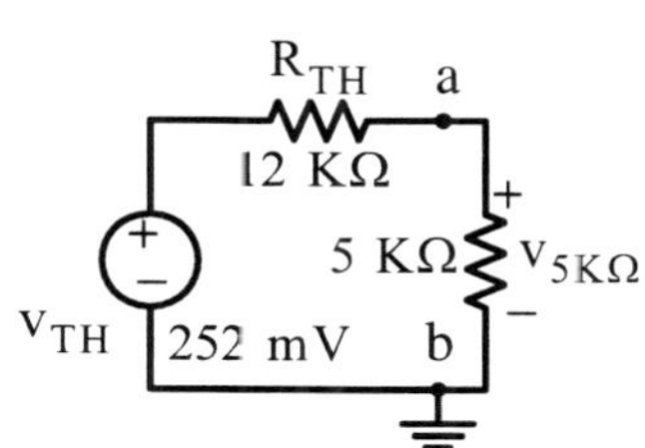

Finally, by the voltage division expression

$$v_{5K\Omega} = \frac{5}{58+5} \times 252 = 20\ \text{mV}$$

7. We assign node voltages v_1 and v_2 as shown below and we write node equations observing that $v_2 = 0$ (virtual ground).

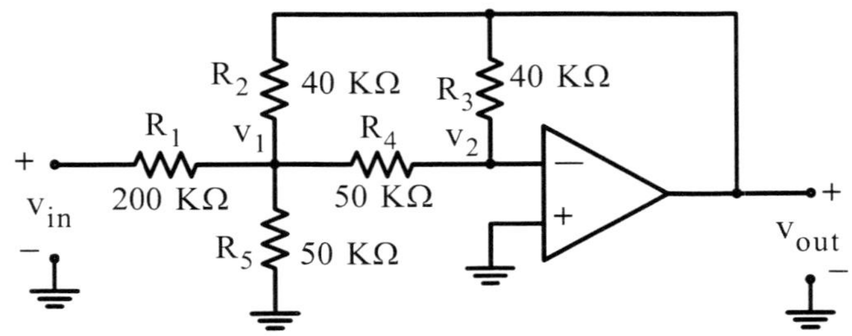

Node 1 at v_1:

$$\frac{v_1 - v_{in}}{200\ K\Omega} + \frac{v_1 - v_{out}}{40\ K\Omega} + \frac{v_1 - 0}{50\ K\Omega} + \frac{v_1}{50\ K\Omega} = 0$$

or

$$\left(\frac{1}{200\ K\Omega} + \frac{1}{40\ K\Omega} + \frac{1}{50\ K\Omega} + \frac{1}{50\ K\Omega}\right)v_1 = \frac{v_{in}}{200\ K\Omega} + \frac{v_{out}}{40\ K\Omega}$$

Multiplication of each term by 200 KΩ and simplification yields

$$v_1 = \frac{1}{14}(v_{in} + 5v_{out}) \quad (1)$$

Node 2 at v_2:

$$\frac{0 - v_1}{50\ K\Omega} + \frac{0 - v_{out}}{40\ K\Omega} = 0$$

or

$$v_1 = -\frac{5}{4}v_{out} \quad (2)$$

Equating the right sides we obtain

$$\frac{1}{14}(v_{in} + 5v_{out}) = -\frac{5}{4}v_{out}$$

or

$$\frac{37}{28}v_{out} = -\frac{1}{14}v_{in}$$

Simplifying and dividing both sides by v_{in} we obtain

$$G_v = \frac{v_{out}}{v_{in}} = -\frac{2}{37}$$

8. We assign node voltages v_1 and v_2 as shown below and we write node equations observing that $v_1 = 0$ (virtual ground).

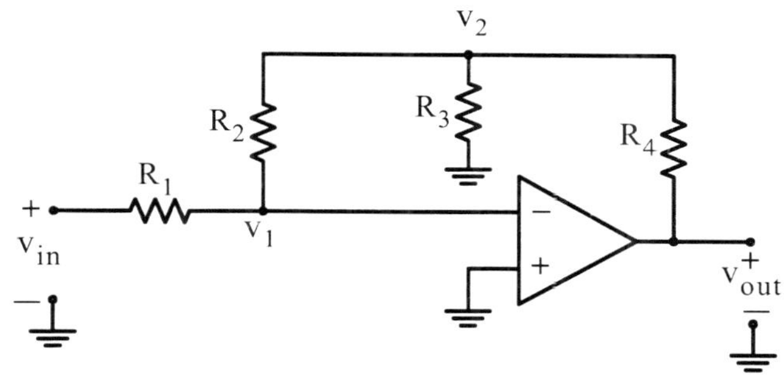

Node 1 at v_1:

$$\frac{0 - v_{in}}{R_1} + \frac{0 - v_2}{R_2} = 0$$

Solving for v_2 we obtain

$$v_2 = -\frac{R_2}{R_1} v_{in} \quad (1)$$

Node 2 at v_2:

$$\frac{v_2 - 0}{R_2} + \frac{v_2}{R_3} + \frac{v_2 - v_{out}}{R_4} = 0$$

$$\left(\frac{1}{R_2} + \frac{1}{R_3} + \frac{1}{R_4}\right) v_2 = \frac{v_{out}}{R_4}$$

$$v_2 = \frac{1}{R_4/R_2 + R_4/R_3 + 1} v_{out} \quad (2)$$

Equating the right sides we obtain

$$\frac{1}{R_5/R_3 + R_5/R_4 + 1} v_{out} = -\frac{R_3}{R_1} v_{in}$$

Simplifying and dividing both sides by v_{in} we obtain

$$G_v = \frac{v_{out}}{v_{in}} = -\frac{1}{R_1}\left[R_4 + R_2\left(\frac{R_4}{R_3} + 1\right)\right]$$

9.

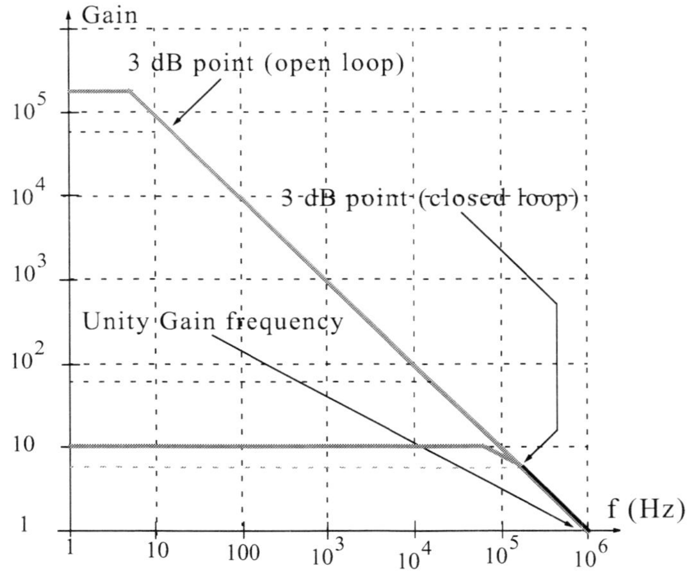

10.

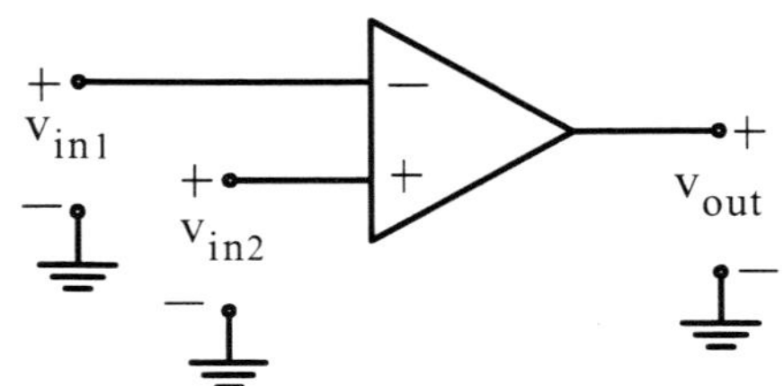

From (5.25), Page 5–27, $v_{out} = A_{ol}(v_{in2} - v_{in1})$, and we are given that $A_{ol} = 100,000$. Then,

a.
$$v_{in1} = v_{in2} - \frac{v_{out}}{A_{ol}} = 3 \times 10^{-3} - \frac{5}{10^5} = 2.95 \text{ mV}$$

b.
$$v_{in2} = v_{in1} + \frac{v_{out}}{A_{ol}} = 2 \times 10^{-3} + \frac{-5}{10^5} = 1.95 \text{ mV}$$

c.
$$v_{out} = A_{ol}(v_{in2} - v_{in1}) = 10^5(-30 - 20) \times 10^{-6} = -1 \text{ V}$$

11.

The s – domain equivalent circuit is shown below.

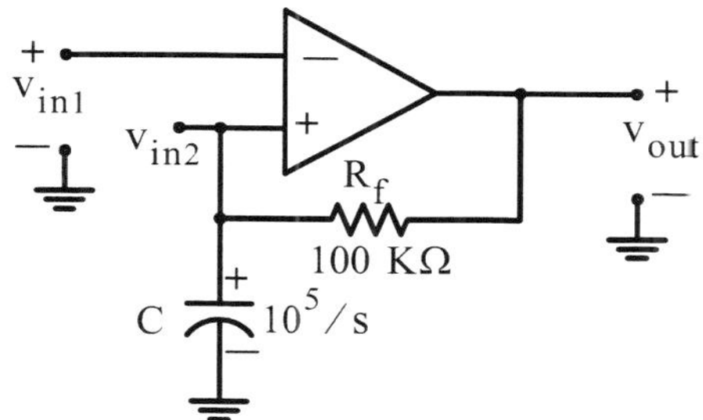

From (5.25), Page 5–27,

$$v_{out} = A_{ol}(v_{in2} - v_{in1}) \quad (1)$$

By KCL

$$\frac{sV_{in2}}{10^5} + \frac{V_{in2} - V_{out}}{10^5} = 0$$

$$V_{out} = (s+1)V_{in2}$$

$$V_{in2} = \frac{V_{out}}{s+1}$$

and by substitution into (1)

$$V_{out} = A_{ol}\left(\frac{V_{out}}{s+1} - V_{in1}\right)$$

$$\left(1 - \frac{A_{ol}}{s+1}\right)V_{out} = -A_{ol}V_{in1}$$

$$G(s) = \frac{V_{out}}{V_{in1}} = \frac{-A_{ol}(s+1)}{s+1-A_{ol}}$$

The pole of $G(s)$ is located at $s = -(1 - A_{ol}) = A_{ol} - 1$ and it is positive for $A_{ol} > 1$. Therefore, the circuit is unstable since the pole is located on the right half–plane.

12.

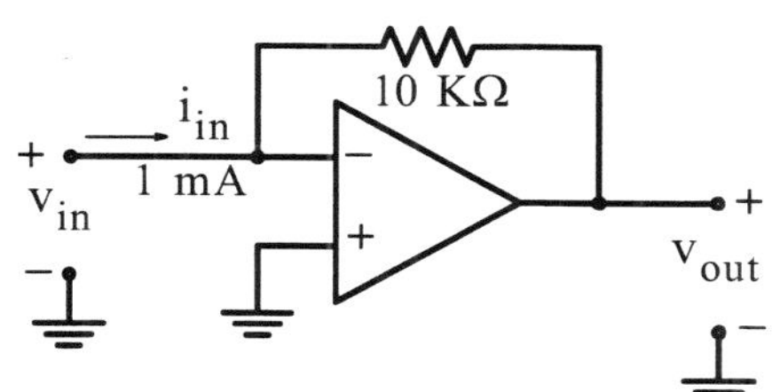

This circuit is the same as that of Figure 5.47, Page 5-29, where we found that $R_m = -R_f$.

Then,

a.
$$R_m = -10\ K\Omega$$

b.
$$R_m = \frac{v_{out}}{i_{in}} = -10\ K\Omega$$

or
$$v_{out} = R_m i_{in} = (-10\ K\Omega)(1\ mA) = -10\ V$$

13.

The s – domain circuit is shown below.

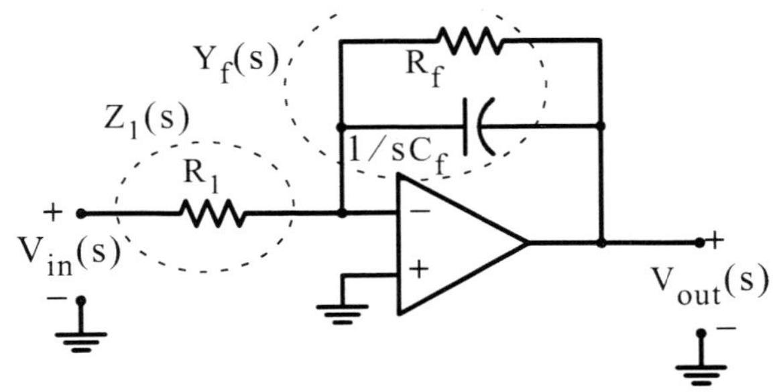

a.
$$Z_1(s) = R_1$$

$$Y_f(s) = \frac{1}{R_f} + \frac{1}{1/sC_f} = \frac{1}{R_f} + sC_f = (1 + sC_fR_f)/R_f$$

$$Z_f(s) = \frac{1}{Y_f(s)} = \frac{R_f}{1 + sC_fR_f}$$

$$G(s) = \frac{V_{out}(s)}{V_{in}(s)} = -\frac{Z_f(s)}{Z_1(s)} = -\frac{R_f/(1 + sC_fR_f)}{R_1} = -\frac{R_f}{R_1(1 + sC_fR_f)} \quad (1)$$

b. The DC gain is defined at the point where the frequency is zero, that is, $s = 0$. Then, (1) above reduces to

$$Gain_{DC} = \frac{V_{out}}{V_{in}} = -\frac{R_f}{R_1}$$

c. We rewrite the transfer function of (1) above as

$$G(s) = -\frac{R_f}{R_1} \cdot \frac{1}{s + 1/C_fR_f}$$

and we recognize this as the transfer function of a low–pass filter whose cutoff 3 dB frequency is $f_{3\ dB} = 1/2\pi C_f R_f$ and falls of at the rate of 20 dB/decade as shown below. Therefore, the 3 dB frequency can be determined by proper selection of the feedback resistor and feedback capacitor.

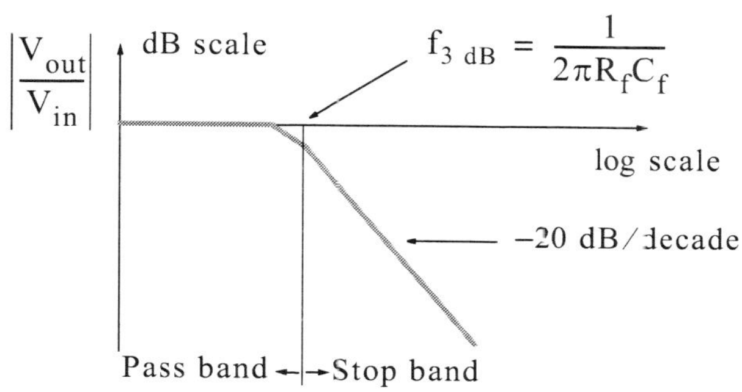

d. 40 dB represents a ratio of 100. Therefore, $R_f/R_1 = 100$ and with $R_1 = 1\ K\Omega$, we find that $R_f = 100\ K\Omega$. Also, since

$$f_{3\ dB} = \frac{1}{2\pi R_f C_f}$$

for $f_{3\ dB} = 1\ KHz$ and $R_f = 100\ K\Omega$, we find that

$$C_f = \frac{1}{2\pi R_f f_{3\ dB}} = \frac{1}{2\pi \times 10^5 \times 10^3} = 1.59\ nF$$

e. The sketch above shows that the gain drops at the rate of 20 dB/decade and thus the drop of the gain from 40 dB to 0 dB represents two decades. And since at 40 dB, $f_{40\ dB} = 1\ KHz$,

$$f_{0\ dB} = 100 \times f_{40\ dB} = 100 \times 1000 = 100\ KHz$$

f. It is shown in Circuit Analysis texts* that for an RC low–pass filter,

$$G(j\omega) = \frac{V_{out}}{V_{in}} = \frac{1}{1 + j\omega RC} = \frac{1}{(\sqrt{1 + \omega^2 R^2 C^2})\angle \mathrm{atan}(\omega RC)} = \frac{1}{\sqrt{1 + \omega^2 R^2 C^2}}\angle -\mathrm{atan}(\omega RC)$$

The MATLAB script below plots the phase angle $\theta = -\mathrm{atan}(\omega RC)$ as a function of frequency ω with $RC = 1$. The script expresses the frequency normalized to the 3 dB frequency where $f_{3\ dB} = 1\ KHz$. Then, the normalized phase angle can be expressed as $\theta_n = -\mathrm{atan}(f/f_{3\ dB})$.

```
f=(0.1:0.01:10); f3dB=1; theta=-atan(f./f3dB).*180./pi; semilogx(f,theta); grid
```

* *Please refer to Chapter 7 in Circuit Analysis I with MATLAB Applications, ISBN 978–0–9709511–2–0*

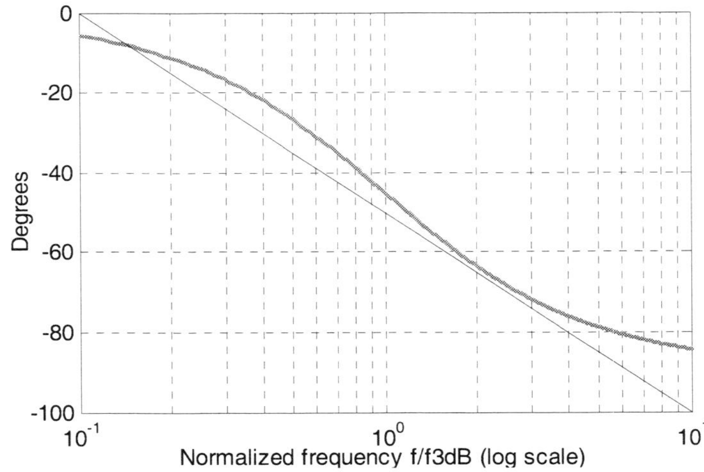

The plot above shows that the $f_{3\ dB} = 1\ KHz$ frequency occurs at $\theta = -45°$ and the asymptotic (straight) line indicates that it drops at the rate of $45°/(\text{decade})$. Therefore, the phase at 100 KHz is $-90°$ and because this is an inverting amplifier, we must add $-180°$ so the total phase shift is $-270°$ or $+90°$.

14.
The s–domain equivalent circuit is shown below.

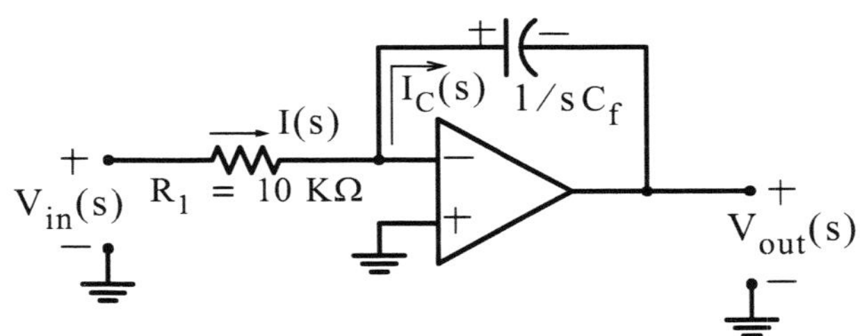

a. The time constant is the product of R_1 and C_f, that is, $\tau = R_1C_f = 10^{-4}\ s$, and thus

$$C_f = \tau/R_1 = 10^{-4}/10^4 = 0.01\ \mu F$$

b.

$$V_{out}(s) = -\frac{1}{sC_f}I_C(s) = -\frac{1}{sC_f}I(s) = -\frac{1}{sC_f}\frac{V_{in}(s)}{R_1}$$

$$\frac{V_{out}(s)}{V_{in}(s)} = -\frac{1}{sR_1C_f}$$

$$\frac{V_{out}(j\omega)}{V_{in}(j\omega)} = -\frac{1}{j\omega R_1 C_f} = -\frac{10^8}{j\omega 10^4} = -\frac{10^4}{j\omega}$$

$$\left|\frac{V_{out}}{V_{in}}\right| = \frac{10^4}{\omega} \quad (1)$$

$$\theta = 180° - 90° = 90°$$

$$\left|\frac{V_{out}}{V_{in}}\right|_{f = 1\ KHz} = \frac{10^4}{2\pi \times 10^3} \approx 1.59$$

$$\theta = 90°$$

c. From (1) above we observe that the gain will be unity when $\omega = 10^4\ r/s$ or

$$f = \frac{10^4}{2\pi} \approx 1.6\ KHz$$

15.

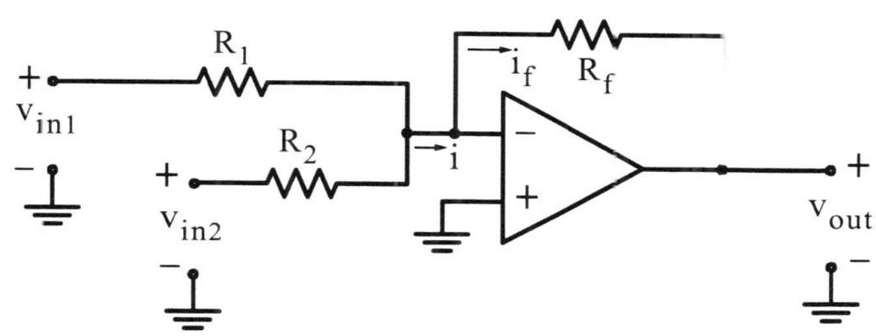

From (5.49), Page 5–38,

$$v_{out} = -\left(\frac{R_f}{R_1}v_{in1} + \frac{R_f}{R_2}v_{in2} + \dots + \frac{R_f}{R_N}v_{inN}\right)$$

and with only two inputs it reduces to

$$v_{out} = -\left(\frac{R_f}{R_1}v_{in1} + \frac{R_f}{R_2}v_{in2}\right) \quad (1)$$

Since the magnitude of v_{out} must 12 V or less and the feedback current i_f must be limited to 1 mA, there must be

$$R_f = \frac{|v_{out}|}{i_f} = \frac{|12|}{10^{-3}} = 12\ K\Omega$$

and with this value (1) above becomes

$$v_{out} = -\left(\frac{12\ K\Omega}{R_1}v_{in1} + \frac{12\ K\Omega}{R_2}v_{in2}\right) \quad (2)$$

Then, the condition $v_{out} = -(4v_{in1} + 3v_{in2})$ can be satisfied if we choose $R_1 = 3\ K\Omega$ and $R_4 = 4\ K\Omega$.

16.

The overall voltage gain can be found from relation (5.67), Page 5–44, that is,

$$A_d = \frac{v_{out}}{(v_{in2} - v_{in1})} = \frac{R_4}{R_3}\left(\frac{2R_2}{R_1} + 1\right)$$

so we need to replace R_1 with $R_{fix} + R_{var}$. For convenience, we can make $R_4 = R_3$ and the above relation reduces to

$$A_d = \frac{2R_2}{R_{fix} + R_{var}} + 1$$

Let us choose a $100\ K\Omega$ potentiometer for R_{var}. For maximum gain, 100 for our case, the potentiometer must be set for zero resistance, in other words, $R_{var} = 0$. Then,

$$A_{d\ max} = \frac{2R_2}{R_{fix}} + 1 = 100 \quad (1)$$

For minimum gain, 10 for our case, the potentiometer must be set for full resistance, in other words, $R_{var} = 100\ K\Omega$. Then,

$$A_{d\ min} = \frac{2R_2}{R_{fix} + 10^5} + 1 = 10 \quad (2)$$

Rearranging (1) and (2) for simultaneous solution, we obtain

$$R_2 - 49.5R_{fix} = 0$$

$$R_2 - 4.5R_{fix} = 4.5 \times 10^5$$

and the solution yields $R_2 = 495\ K\Omega \approx 500\ K\Omega$ and $R_{fix} = 10\ K\Omega$

17.

a.

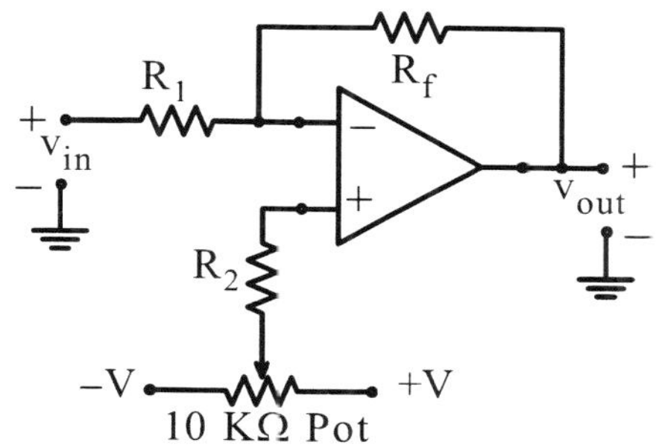

b.

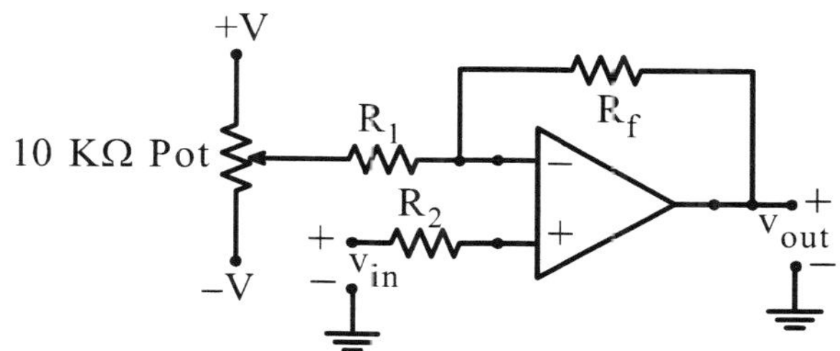

In both circuits of (a) and (b) above, the potentiometer can be adjusted to provide the value and polarity of the DC input voltage required to null the output to zero volts.

18.

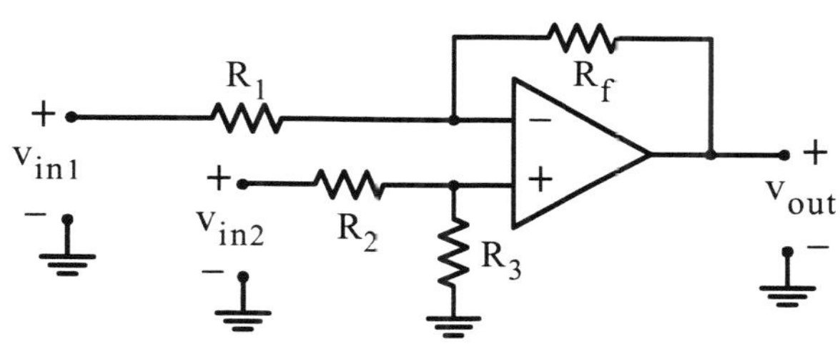

From (5.62), Page 5–42,

$$R_{in} = 2R_1$$

but this relation holds only if $R_2 = R_1$. Since for the circuit above we want $R_{in} = 50\ K\Omega$, we must make $R_1 + R_2 = 50\ K\Omega$, and if $R_2 = R_1$, then $R_1 = R_2 = 25\ K\Omega$.
The circuit above will perform as a differential input op amp if

$$\frac{R_f}{R_1} = \frac{R_3}{R_2}$$

and if we make $R_2 = R_1$, then $R_3 = R_f$, and for $R_1 = R_2 = 25\ K\Omega$ and $G_v = 40$, we must make $R_3 = R_f = 1\ M\Omega$.

19. The s – domain circuit is shown below.

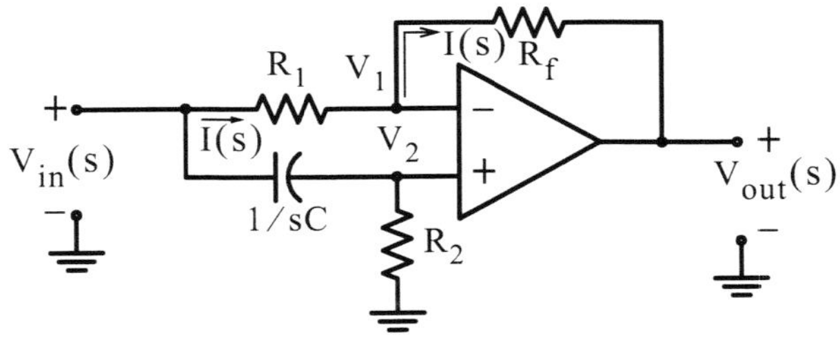

By the voltage division expression

$$V_2 = \frac{R_2}{1/sC + R_2}V_{in}(s)$$

and since

$$V_1 = V_2$$

it follows that

$$V_1 = \frac{R_2}{1/sC + R_2}V_{in}(s)$$

This is a differential input amplifier and the current $I(s)$ is

$$I(s) = \frac{V_{in}(s) - V_1}{R_1}$$

$$I(s) = \frac{1}{R_1}\left(V_{in}(s) - \frac{R_2}{1/sC + R_2}V_{in}(s)\right) = \frac{V_{in}(s)}{R_1}\left(1 - \frac{R_2}{1/sC + R_2}\right)$$

$$I(s) = \frac{V_{in}(s)}{R_1}\left(\frac{1/sC + R_2}{1/sC + R_2} - \frac{R_2}{1/sC + R_2}\right) = \frac{V_{in}(s)}{R_1}\left(\frac{1/sC}{1/sC + R_2}\right)$$

Next,

$$V_{out}(s) = V_1 - R_f I(s)$$

$$V_{out}(s) = \frac{R_2}{1/sC + R_2}V_{in}(s) - R_f\frac{V_{in}(s)}{R_1}\left(\frac{1/sC}{1/sC + R_2}\right)$$

$$G_v(s) = \frac{V_{out}(s)}{V_{in}(s)} = \frac{R_2}{1/sC + R_2} - \frac{R_f}{R_1}\left(\frac{1/sC}{1/sC + R_2}\right) = \frac{R_1R_2 - R_f/sC}{R_1(1/sC + R_2)}$$

$$G_v(s) = \frac{V_{out}(s)}{V_{in}(s)} = \frac{R_2 - R_f/sCR_1}{R_2 + 1/sC} = \frac{sCR_2 - R_f/R_1}{sCR_2 + 1} = \frac{s - R_f/CR_1R_2}{s + 1/CR_2}$$

Now, letting $s = j\omega$, we obtain

$$G_v(j\omega) = \frac{V_{out}(j\omega)}{V_{in}(j\omega)} = \frac{j\omega - R_f/CR_1R_2}{j\omega + 1/CR_2}$$

and for unity gain there must be $R_f = R_1$. Then,

$$G_v(j\omega) = \frac{V_{out}(j\omega)}{V_{in}(j\omega)} = \frac{j\omega - 1/CR_2}{j\omega + 1/CR_2} = \frac{\sqrt{\omega^2 + 1/C^2R_2^2}\ \angle\tan(-\omega CR_2)}{\sqrt{\omega^2 + 1/C^2R_2^2}\ \angle\tan(\omega CR_2)} = 1\angle{-2\tan^{-1}(\omega CR_2)}$$

Therefore,

$$\left|\frac{V_{out}(j\omega)}{V_{in}(j\omega)}\right| = 1$$

and because the magnitude is independent of frequency, this circuit is referred to as *first order all–pass filter*. The phase angle is

$$\theta = 180° - \angle{-2\tan^{-1}(\omega CR_2)}$$

where the angle of $180°$ is due to the fact that the feedback resistor is connected to the inverting input. Because of this phase shift, this circuit is also referred to as a *phase shifter*.

20.

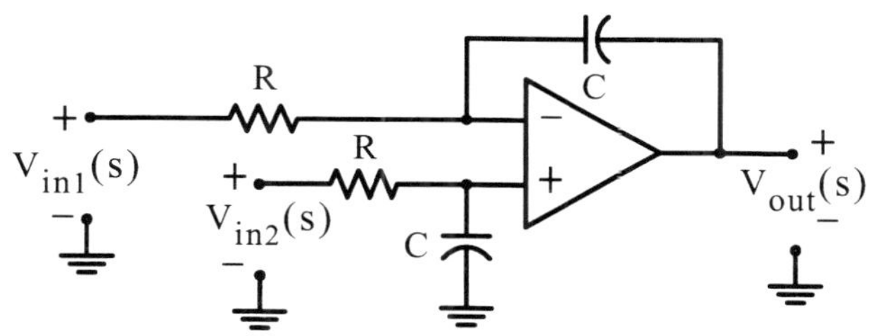

We will apply the superposition principle. With $v_{in1}(s)$ acting alone and $v_{in2}(s) = 0$, the circuit reduces to that shown below.

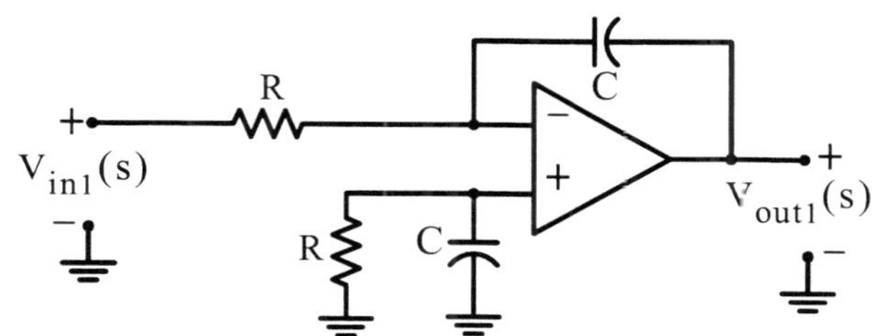

The circuit above is essentially a Miller integrator like that shown in Figure 5.52, Page 5–32, and in Exercise 14 we found that

$$\frac{V_{out}(s)}{V_{in}(s)} = -\frac{1}{sR_1C}$$

Therefore, for the circuit above

$$V_{out1}(s) = -\frac{1}{sRC}V_{in1}(s) \quad (1)$$

Next, with $V_{in2}(s)$ acting alone and $V_{in1}(s) = 0$, the given circuit reduces to that shown below.

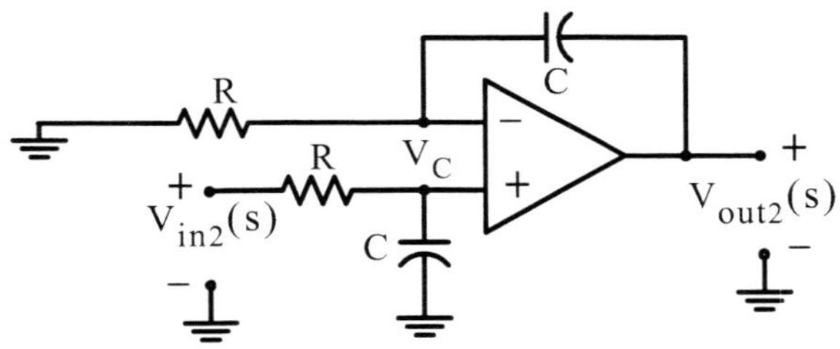

By the voltage division expression

$$V_C = \frac{1/sC}{R+1/sC}V_{in2}(s)$$

and since this is a non–inverting amplifier,

$$V_{out2}(s) = \left(1+\frac{1/sC}{R}\right)V_C = \left(1+\frac{1/sC}{R}\right)\left(\frac{1/sC}{R+1/sC}\right)V_{in2}(s)$$

$$V_{out2}(s) = \left(1+\frac{1}{sCR}\right)\left(\frac{1}{sCR+1}\right)V_{in2}(s) = \left(\frac{sCR+1}{sCR}\right)\left(\frac{1}{sCR+1}\right)V_{in2}(s)$$

$$V_{out2}(s) = \frac{1}{sCR}V_{in2}(s) \quad (2)$$

From (1) and (2)

$$V_{out}(s) = V_{out1}(s) + V_{out2}(s) = \frac{1}{sCR}(V_{in2}(s) - V_{in1}(s))$$

21.

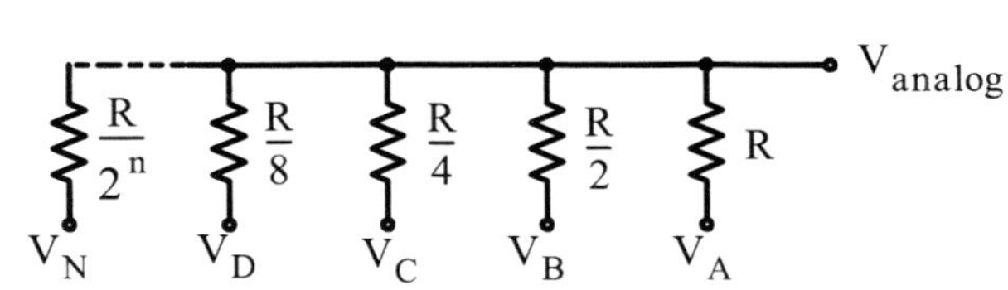

$$I_T = \frac{V_A}{R} + \frac{V_B}{R/2} + \frac{V_C}{R/4} + \frac{V_D}{R/8} + \ldots = \frac{1}{R}(V_A + 2V_B + 4V_C + 8V_D + \ldots)$$

$$\frac{1}{R_{eq}} = \frac{1}{R} + \frac{1}{R/2} + \frac{1}{R/4} + \frac{1}{R/8} + \dots$$

or

$$R_{eq} = \frac{R}{1+2+4+8+\dots}$$

$$V_{analog} = R_{eq}I_T = \frac{R}{1+2+4+8+\dots} \cdot \frac{1}{R}(V_A + 2V_B + 4V_C + 8V_D + \dots)$$

$$V_{analog} = \frac{V_A + 2V_B + 4V_C + 8V_D + \dots}{1+2+4+8+\dots}$$

22.

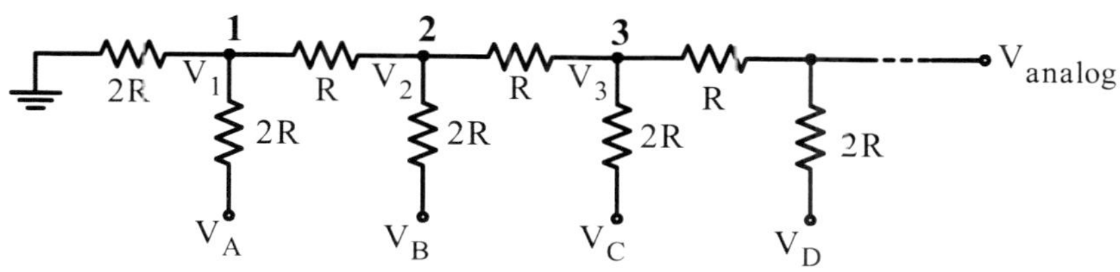

We write nodal equations at Nodes 1, 2, and 3.

$$\frac{V_1}{2R} + \frac{V_1 - V_A}{2R} + \frac{V_1 - V_2}{R} = 0$$

$$\frac{V_2 - V_1}{R} + \frac{V_2 - V_B}{2R} + \frac{V_2 - V_3}{R} = 0$$

$$\frac{V_3 - V_2}{R} + \frac{V_3 - V_C}{2R} + \frac{V_3 - V_D}{R + 2R} = 0$$

Simplifying and collecting like terms we obtain

$$4V_1 - 2V_2 = V_A$$

$$-2V_1 + 5V_2 - 2V_3 = V_B$$

$$-6V_2 + 11V_3 = 3V_C + 2V_D$$

By Cramer's rule*

* *For a review of matrices and determinants please refer to Numerical Analysis Using MATLAB and Excel, ISBN 978–1–934404–03–4.*

$$V_3 = \frac{\begin{bmatrix} 4 & -2 & V_A \\ -2 & 5 & V_B \\ 0 & -6 & (3V_C + 2V_D) \end{bmatrix}}{\begin{bmatrix} 4 & -2 & 0 \\ -2 & 5 & -2 \\ 0 & -6 & 11 \end{bmatrix}} = \frac{1}{128}(12V_A + 24V_B + 48V_C + 32V_D)$$

or

$$V_3 = \frac{1}{32}(3V_A + 6V_B + 12V_C + 8V_D)$$

By the voltage division expression,

$$V_{analog} = \frac{V_3 - V_D}{R + 2R} \cdot 2R + V_D = \frac{2}{3}V_3 + \frac{1}{3}V_D$$

and

$$V_{analog} = \frac{2}{3} \cdot \frac{1}{32}(3V_A + 6V_B + 12V_C + 8V_D) + \frac{1}{3}V_D$$

$$= \frac{1}{16}(V_A + 2V_B + 4V_C + 8V_D)$$

where $16 = 2^4 = 2^n$.

23.

$$v_{out} = -\int_0^t (v_{in1} + 2v_{in2} + 10v_{in3})dt$$

This integration can be performed by a summing integrator such as that shown below where the factor $1/RC$ was chosen to form the coefficients of v_{in1}, v_{in2}, and v_{in3} in the given expression.

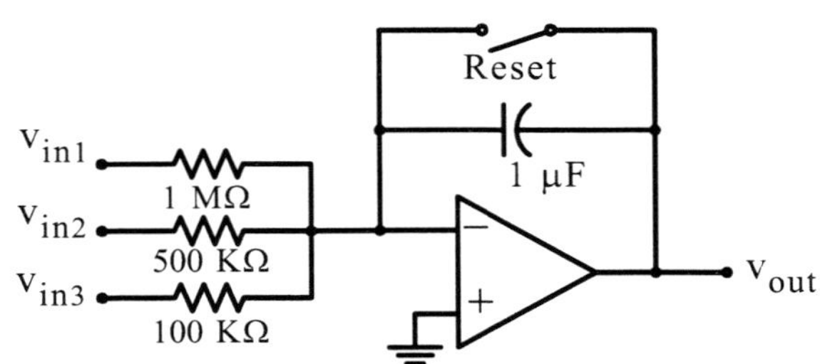

24.

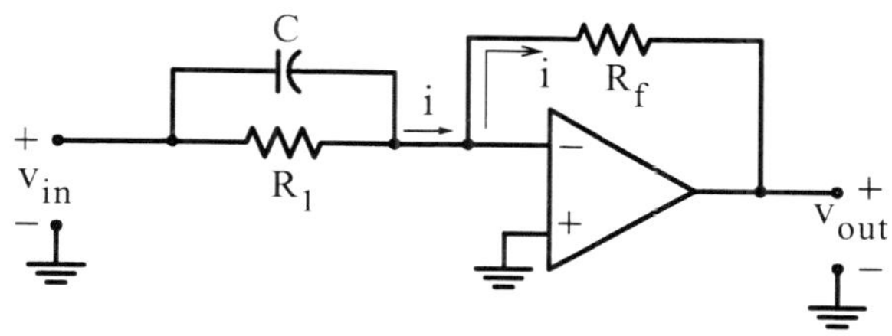

$$i = \frac{v_{in}}{R_1} + C\frac{dv_{in}}{dt} \quad (1)$$

$$v_{out} = -R_f i \quad (2)$$

Substitution of (1) into (2) yields

$$v_{out} = -\frac{R_f}{R_1}v_{in} - R_f C\frac{dv_{in}}{dt}$$

Since we want $v_{out} = -5v_{in} - 3dv_{in}/dt$, we choose $R_f = 5R_1$ and $R_f C = 3$

25.

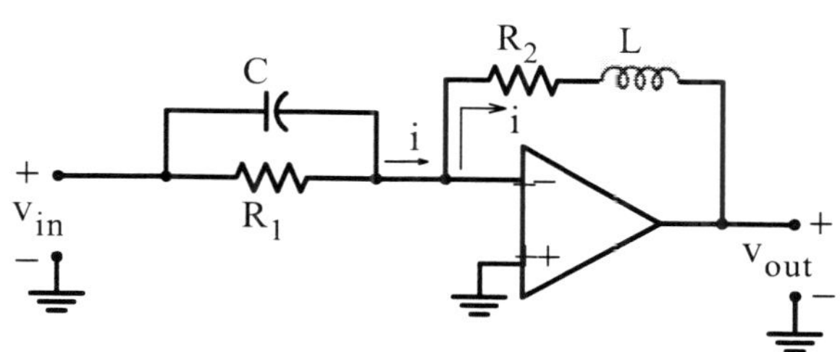

$$i = \frac{v_{in}}{R_1} + C\frac{dv_{in}}{dt} \quad (1)$$

$$v_{out} = -R_2 i - L\frac{di}{dt} \quad (2)$$

Differentiating (1) with respect to *t* we obtain

$$\frac{di}{dt} = \frac{1}{R_1}\frac{dv_{in}}{dt} + C\frac{d^2v_{in}}{dt^2} \quad (3)$$

and from (1), (2), and (3)

$$v_{out} = -\left(\frac{R_2}{R_1}v_{in} + R_2C\frac{dv_{in}}{dt} + \frac{L}{R_1}\frac{dv_{in}}{dt} + LC\frac{d^2v_{in}}{dt^2}\right)$$

26.

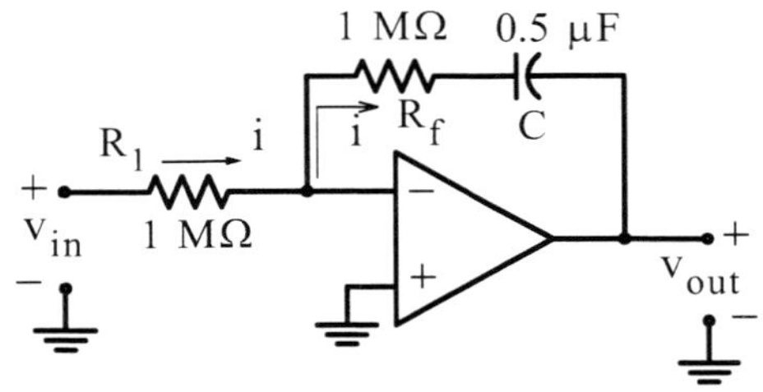

$$i = \frac{v_{in}}{R_1} \quad (1)$$

$$v_{out} = -R_f i - \frac{1}{C}\int i\,dt \quad (2)$$

Substitution of (1) into (2) yields

$$v_{out} = -\frac{R_f}{R_1}v_{in} - \frac{1}{R_1 C}\int v_{in}dt = -v_{in} - 2\int v_{in}dt$$

Alternately,

$$\frac{dv_{out}}{dt} = -\left(\frac{dv_{in}}{dt} + 2v_{in}\right)$$

Chapter 6

Integrated Circuits

This chapter begins with an introduction to electronic logic gates and their function in terms of Boolean expressions and truth tables. Positive and negative logic are defined, and the transistor–transistor logic (TTL), emitter–coupled logic (ECL), CMOS, and BiCMOS logic families are discussed. Earlier logic families are presented in the exercises section.

6.1 The Basic Logic Gates*

Electronic logic gates are used extensively in digital systems and are manufactured as integrated circuits (IC's). The basic logic gates are the inverter or NOT gate, the AND gate, and the OR gate, and these perform the complementation, ANDing, and ORing operations respectively. The symbols for these gates are shown in Figure 6.1.

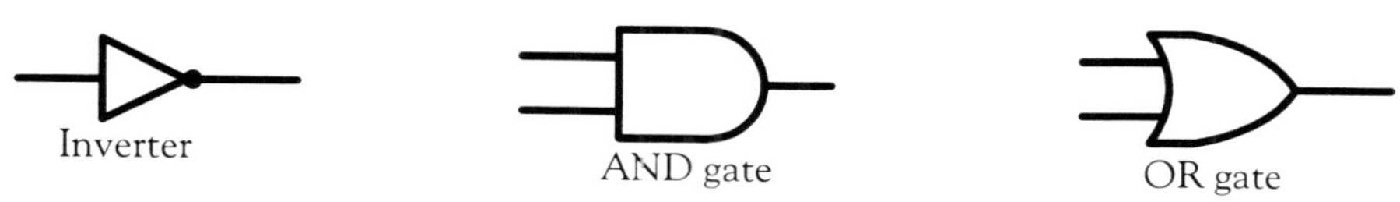

Figure 6.1. The three basic logic gates

Four other logic gates, known as NAND, NOR, Exclusive OR (XOR), and Exclusive NOR (XNOR), are derivatives of the basic AND and OR gates and will be discussed later in this chapter.

6.2 Positive and Negative Logic

Generally, an uncomplemented variable represents a logical 1, also referred to as the *true* condition, and when that variable is complemented, it represents a logical 0, also referred to as the *false* condition. Thus, if $A = 1$ (true), it follows that $\overline{A} = 0$ (false). Of course, digital computers do not understand logical 1, logical 0, true, or false; they only understand voltage signals such as that shown in Figure 6.2.

Figure 6.2. Typical voltage signal for a digital computer

* *For this and the remaining chapters it is assumed that the reader has prior knowledge of the binary, the octal, and hexadecimal number systems, complements of numbers, binary codes, the fundamentals of Boolean algebra, and truth tables. If not, it is strongly recommended that a good book like our Digital Circuit Analysis and Design, ISBN 0–9744239–6–3 is reviewed.*

With reference to the voltage waveform of Figure 6.2, integrated circuit manufacturers assign the letter H (High) to the 5 volt level and the letter L (Low) to the ground level as shown in Figure 6.3.

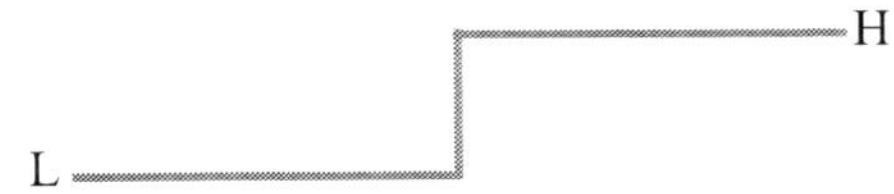

Figure 6.3. High (H) and Low (L) assignments in a voltage waveform

With the H and L assignments as shown in Figure 6.3, the logic circuitry designer has the option of assigning a logical 1 to H and logical 0 to L, or logical 0 to H and logical 1 to L. The former convention is known as *positive logic*, and the latter as *negative logic*. Thus, Figure 6.4(a) represents positive logic and Figure 6.4(b) represents negative logic.

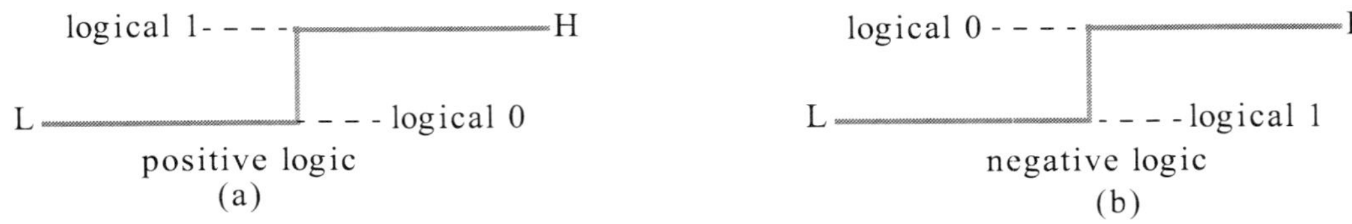

Figure 6.4. Positive and negative logic defined

We will discuss the three common IC logic families in subsequent sections. For our present discussion it will suffice to list the High and Low voltage levels for the Transistor–Transistor Logic (TTL or T^2L), Emitter–Coupled Logic (ECL), and Complementary Metal Oxide Semiconductor (CMOS) logic. Typical values are shown in Table 6.1.

TABLE 6.1 Typical values of High and Low voltage levels for three IC families

IC Family	High voltage (Volts DC)	Low voltage (Volts DC)
TTL	5.0	0.2
ECL	−0.9	−1.75
CMOS	5.0 to 15.0	0

6.3 The Inverter

The symbol for the inverter (NOT gate) is shown in Figure 6.5, and the truth table is shown in Table 6.2.

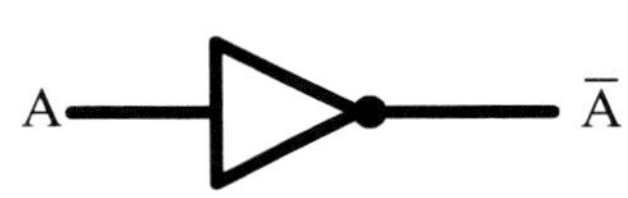

Figure 6.5. Symbol for the inverter

TABLE 6.2 The truth table for the inverter

Input	Output
L	H
H	L

With positive logic, the truth table for the inverter is written as shown in Table 6.3.

TABLE 6.3 The truth table for the inverter when positive logic is assumed

Input	Output
0	1
1	0

With negative logic, the truth table for the inverter is written as shown in Table 6.4.

TABLE 6.4 The truth table for the inverter when negative logic is assumed

Input	Output
1	0
0	1

Figure 6.6 shows the TTL SN7404 Hex Inverter IC where SN identifies the packaging and Hex implies that there are 6 inverters within the IC.

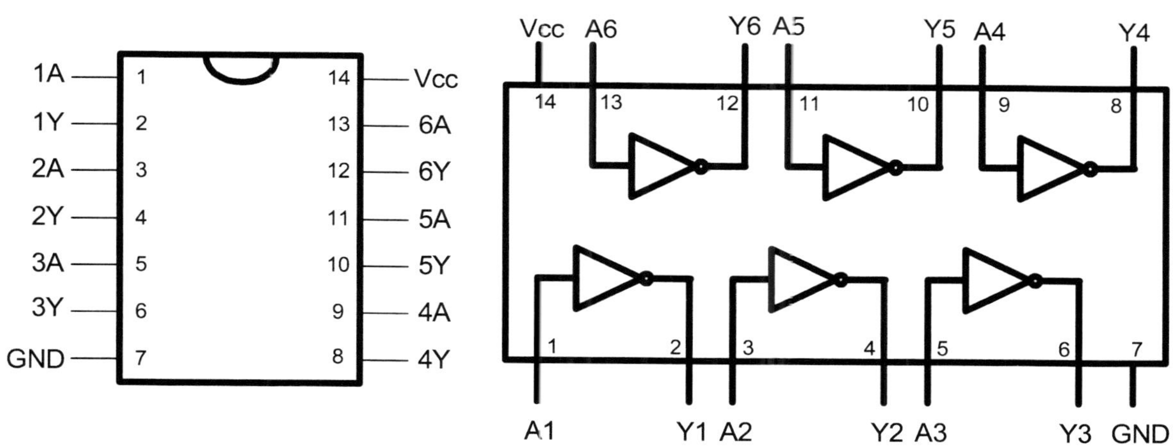

Figure 6.6. The SN7404 Hex Inverter IC

One important parameter is the *input clamp voltage* denoted as V_{IK} and this refers to the maximum negative voltage that may be applied at the input terminals without damaging the IC. A typical value for this parameter is $-1.2\ V$. To insure that this value is not exceeded, diode D_1 is included to clamp the input voltage to less than $-1.2\ V$ with respect to the ground. We recall from Chapter 2 that a practical diode, when forward–biased, may be represented as an ideal diode in series with a $0.7\ V$ source and a small resistance as shown in Figure 6.7.

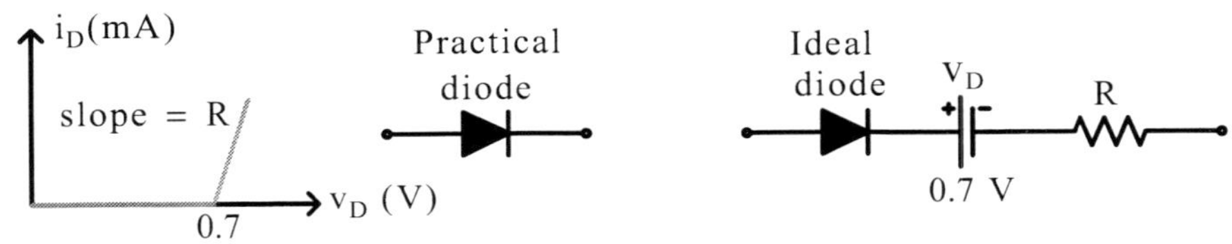

Figure 6.7. Practical and ideal diode representation

The data sheets provided by the manufacturer list several parameters and we will discuss the most important later in this chapter.

Figure 6.8 shows the internal details of the TTL IC 7404 inverter.

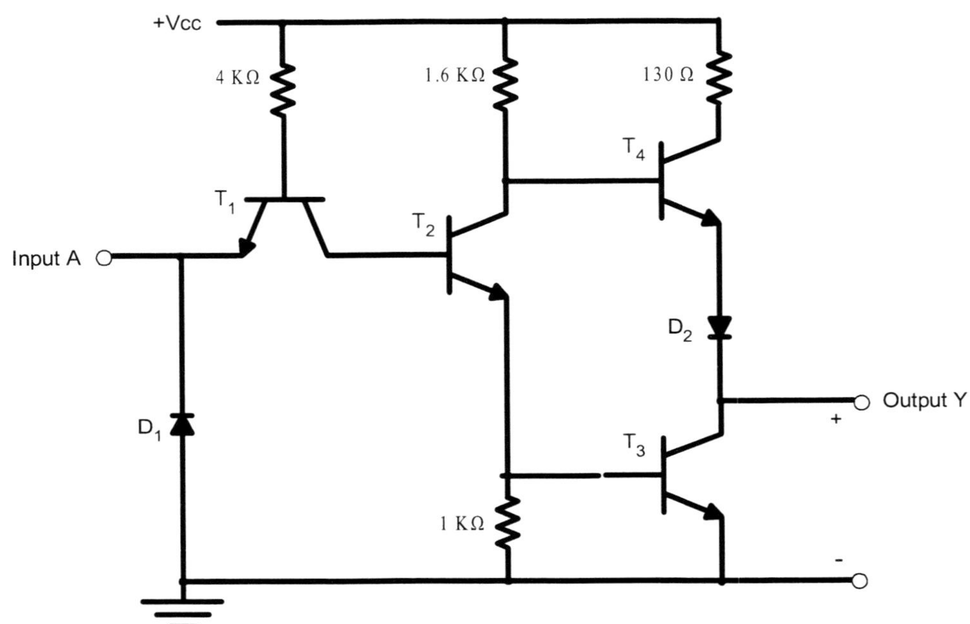

Figure 6.8. Circuit of the SN7404 Inverter

Note: Unless otherwise noted, henceforth Low will mean a nominal 0 V input or output signal and High will mean a nominal 5 V input or output signal.

The inverter circuit of Figure 6.8 functions as follows:

We assume that the input voltage is the output of a previous stage and it is Low at $V_{in} = 0.2\ V$ with respect to the ground as shown in Figure 6.9. With this input, transistor T_1 is ON and the voltage V_1 at the collector of transistor T_1 is $V_1 = V_{CE\ sat} + V_{in} = 0.2 + 0.2 = 0.4\ V$ and under this condition, transistors T_2 and T_3 are both OFF. This is because if transistors T_2 and T_3 were both ON, the voltage V_1 would have to be $V_1 = V_{BE\ T2} + V_{BE\ T3} = 0.7 + 0.7 = 1.4\ V$. Accordingly, we accept the fact that with the input Low, transistors T_2 and T_3 are both OFF. However, the current I that flows through the 1.6 KΩ resistor and then into the base of transistor T_4, is sufficient to turn it ON. Then, $V_{out} = V_{CC} - V_{CE\ T4} - V_{D2} = 5.0 - 0.2 - 0.7 = 4.1\ V$ and thus the output is High and the inversion operation has been performed.

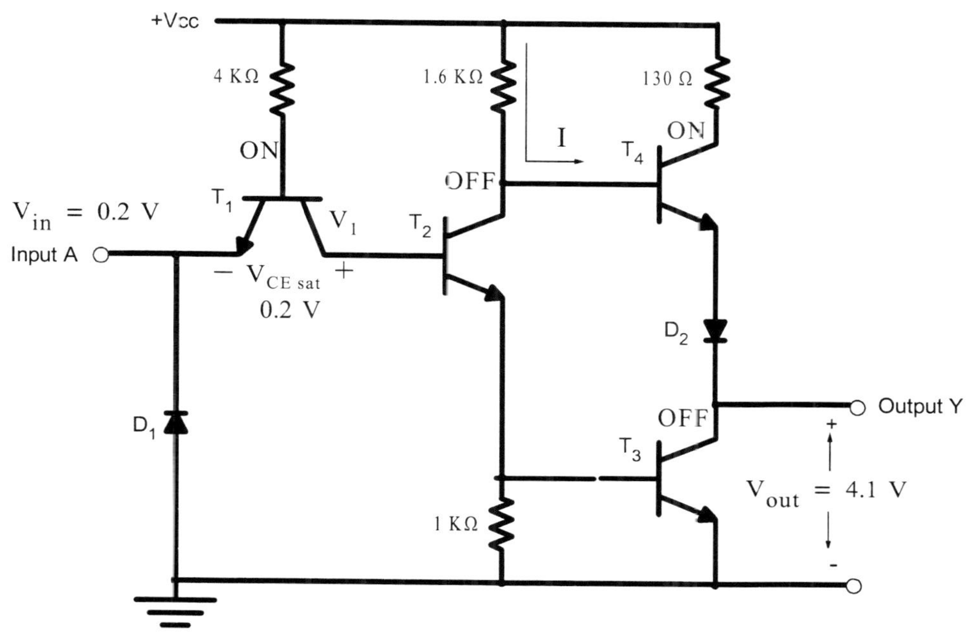

Figure 6.9. Circuit of the SN7404 Inverter for Low–to–High inversion

Next, with the input High the circuit is as shown in Figure 6.10 and the circuit functions as follows:

We assume that the input voltage is the output of a previous stage and it is High at $V_{in} = 5.0\ V$ with respect to the ground. With this input, the base–emitter junction of transistor T_1 is reverse–biased, current flows from V_{CC} through the 4 KΩ resistor and thus the base–collector junction is forward–biased. Under these conditions, transistor T_1 is said to be operating at the *inverse active mode*. The current that flows through the base collector junction of transistor T_1 is sufficient to turn transistor T_2 ON and this causes transistor T_3 to turn ON also. The voltage V_2 at the collector of transistor T_2 is $V_2 = V_{CE\ T2} + V_{BE\ T3} = 0.2 + 0.7 = 0.9\ V$. However, this voltage is not sufficient to turn transistor T_4 ON because for transistor T_4 to be ON there should be $V_2 = V_{BE\ T4} + V_{D2} + V_{out} = 0.7 + 0.7 + 0.2 = 1.6\ V$. Therefore, we conclude that transistor T_4 is OFF, $V_{out} = 0.2\ V$, and thus the output is Low and the inversion operation has been performed.

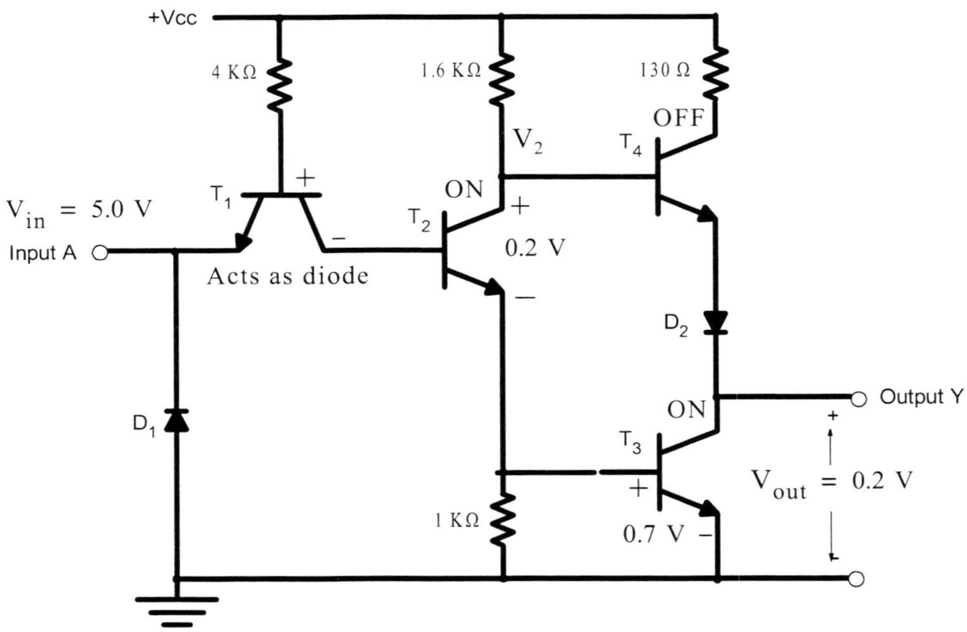

Figure 6.10. Circuit of the SN7404 Inverter for High–to–Low inversion

It should be noted that it is the diode D_2 that prevents transistor T_4 from being ON; if D_2 were not being there, T_4 would be also ON since there would be $V_{BE\ T4} = 0.9 - 0.2 = 0.7\ V$ which is just sufficient to turn T_4 ON.

6.4 The AND Gate

The symbol for a 3–input AND gate is shown in Figure 6.11, and the truth table is shown in Table 6.5.

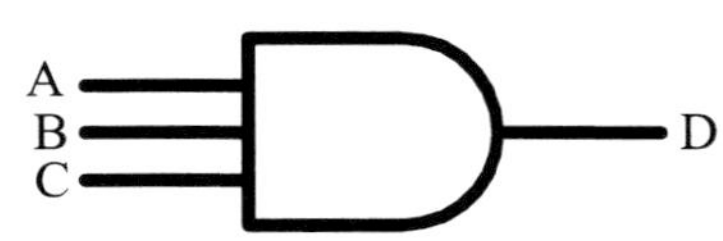

Figure 6.11. 3–input AND gate symbol

With positive logic, the truth table for a 3–input AND gate is written as shown in Table 6.6.

Table 6.6 shows that the output of an AND gate is logical 1 (true) only when all inputs are logical 1 .

Figure 6.12 shows the TTL SN7408 Quad 2–input AND gate where Quad implies that there are 4 AND gates within the IC.

TABLE 6.5 Truth table for 3–input AND gate

Inputs			Output
A	**B**	**C**	**D**
L	L	L	L
L	L	H	L
L	H	L	L
L	H	H	L
H	L	L	L
H	L	H	L
H	H	L	L
H	H	H	H

TABLE 6.6 Truth table of 3–input AND gate when positive logic is used

Inputs			Output
A	**B**	**C**	**D**
0	0	0	0
0	0	1	0
0	1	0	0
0	1	1	0
1	0	0	0
1	0	1	0
1	1	0	0
1	1	1	1

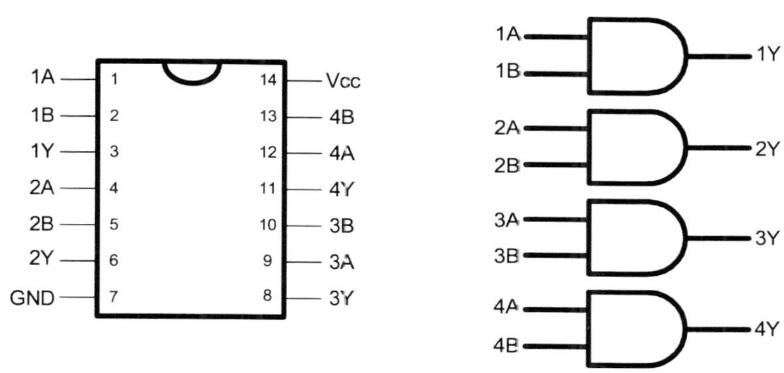

Figure 6.12. The SN7408 Quad 2–input AND gate

Figure 6.13 shows the internal details of a two–input TTL AND gate. We will defer the analysis of this circuit and the reason for deferring the discussion after we introduce the NAND gate.

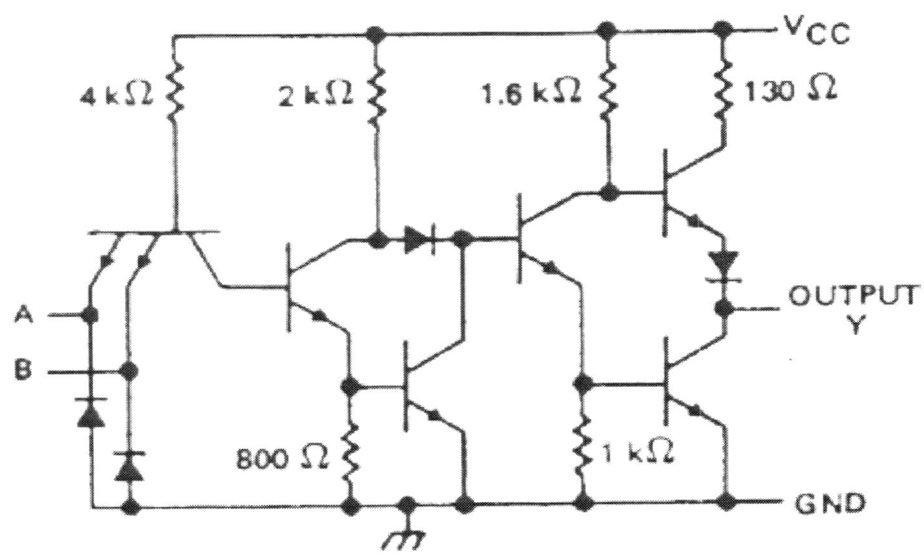

Figure 6.13. Circuit for the SN7408 Quad 2–input AND gate (Courtesy Texas Instruments)

6.5 The OR Gate

The symbol for a 3–input OR gate is shown in Figure 6.14, and the truth table with positive logic is shown in Table 6.7.

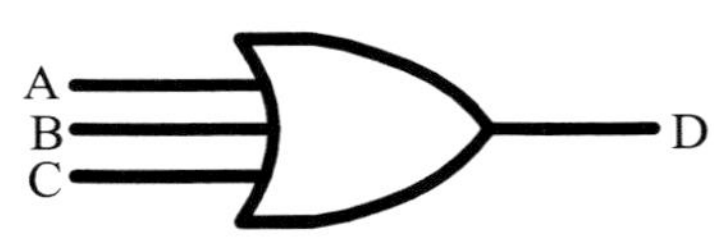

Figure 6.14. Symbol for 3–input OR gate

TABLE 6.7 *Truth table for 3–input OR gate with positive logic*

Inputs			Output
A	**B**	**C**	**D**
0	0	0	0
0	0	1	1
0	1	0	1
0	1	1	1
1	0	0	1
1	0	1	1
1	1	0	1
1	1	1	1

Table 6.7 shows that the output of an OR gate is logical 1 (true) whenever one or more of its inputs are logical 1 .

Figure 6.15 shows the TTL SN7432 Quad 2–input OR gate where Quad implies that there are 4 OR gates within the IC.

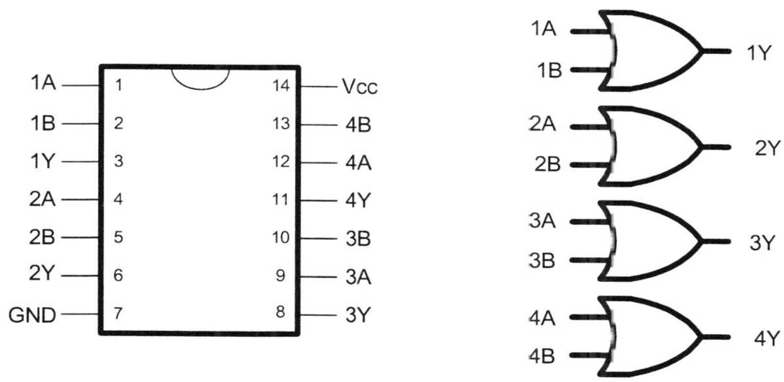

Figure 6.15. The SN7432 Quad 2–input OR gate

Figure 6.16 shows the internal details of the TTL SN7432 Quad 2–input OR gate. We will not describe the functioning of the SN7432 Quad 2–input OR gate at this time. We will defer the circuit operation until we first describe the NOR gate operation in a subsequent section.

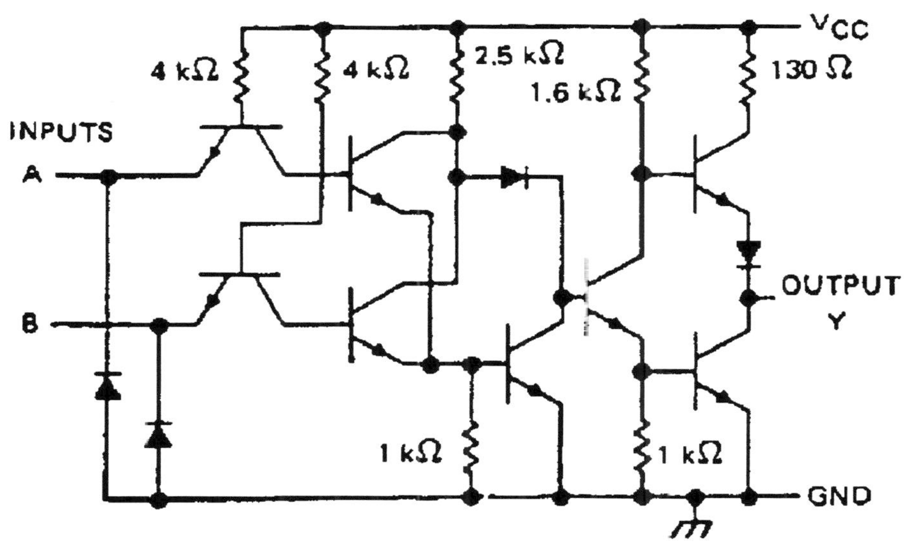

Figure 6.16. Circuit for the SN7432 Quad 2–input OR gate (Courtesy Texas Instruments)

6.6 The NAND Gate

The symbol for a 3–input NAND gate is shown in Figure 6.17, and the truth table with positive logic is shown in Table 6.8.

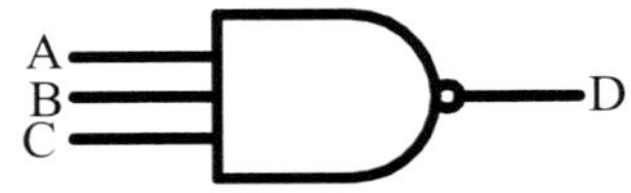

Figure 6.17. Symbol for 3–input NAND gate

TABLE 6.8 Truth table for 3–input NAND gate with positive logic

Inputs			Output
A	**B**	**C**	**D**
0	0	0	1
0	0	1	1
0	1	0	1
0	1	1	1
1	0	0	1
1	0	1	1
1	1	0	1
1	1	1	0

Table 6.8 shows that the output of an NAND gate is logical 0 (false) only when all inputs are logical 1 .

Figure 6.18 shows the TTL SN7400 Quad 2–input NAND gate where Quad implies that there are 4 NAND gates within the IC.

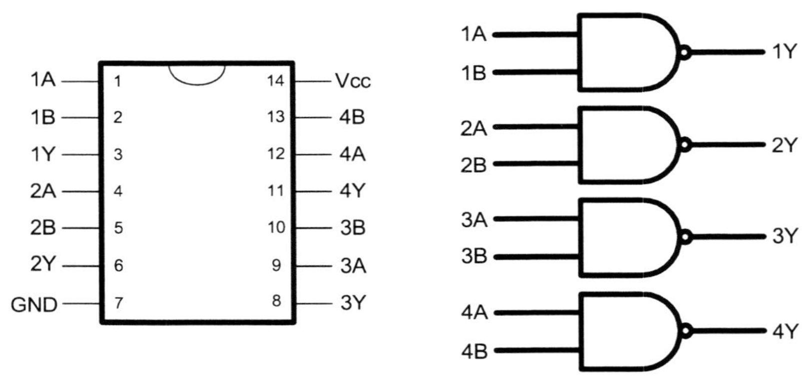

Figure 6.18. The SN7400 Quad 2–input NAND gate

Figure 6.19 shows the internal details of the IC SN7400 NAND gate where transistor T_1 is equivalent to two identical NPN transistors with their bases and collectors tied together; therefore, they are fabricated as a single device with 2 emitters but only one collector and one base as shown in Figure 6.19.

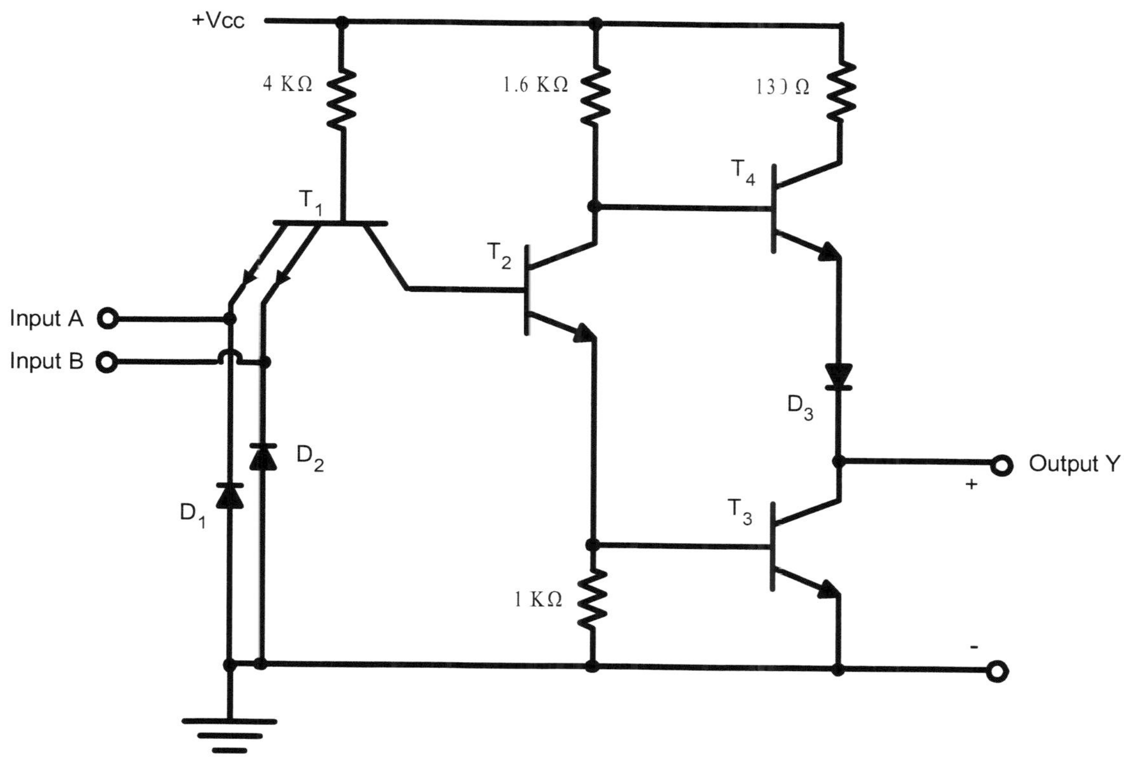

Figure 6.19. Circuit for the TTL SN7400 Quad 2–input NAND gate

We observe that, with the exception of two inputs for the 2–input NAND gate, this circuit is practically the same as that of the inverter circuit of Figure 6.8 where transistor T_4 is stacked on top of transistor T_3 and the operation is complementary, that is, when transistor T_3 is ON transistor T_4 is OFF and vice versa. This arrangement is referred to as *totem–pole* configuration. The 2–input TTL NAND gate circuit of Figure 6.19 functions as follows:

Let us assume that either Input A or B or both are Low at $V_{in} = 0.2\ V$ with respect to the ground as shown in Figure 6.20. Under any of these three conditions, transistor T_1 is ON and the voltage V_1 at the collector of transistor T_1 is $V_1 = V_{CE\ sat} + V_{in} = 0.2 + 0.2 = 0.4\ V$ and thus transistors T_2 and T_3 are both OFF. This is because if transistors T_2 and T_3 were both ON, the voltage V_1 would have to be $V_1 = V_{BE\ T2} + V_{BE\ T3} = 0.7 + 0.7 = 1.4\ V$. Accordingly, we accept the fact that under any of these three conditions, transistors T_2 and T_3 are both OFF. However, The current I that flows through the $1.6\ K\Omega$ resistor and then into the base of transistor T_4 is sufficient to turn it ON. Then, $V_{out} = V_{CC} - V_{CE\ T4} - V_{D2} = 5.0 - 0.2 - 0.7 = 4.1\ V$ and thus the output is High and this satisfies the 2–input NAND gate operation.

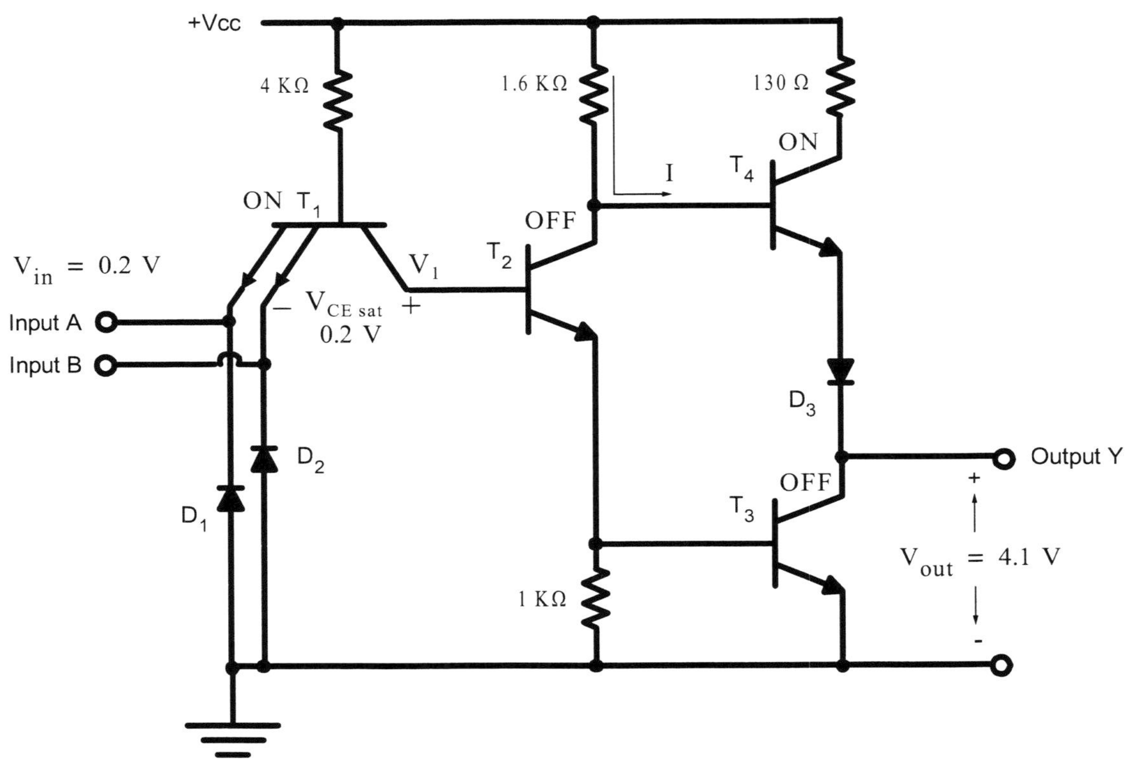

Figure 6.20. Circuit of the SN7400 2–input NAND gate when either Input A of B or both are Low

Next, with both inputs High the circuit is as shown in Figure 6.21 and the circuit functions as follows:

With $V_{in} = 5.0 \text{ V}$ at both inputs, the base–emitter junction of transistor T_1 is reverse–biased, current flows from V_{CC} through the 4 KΩ resistor and thus the base–collector junction is forward–biased. In other words, transistor T_1 is operating at the inverse active mode. The current that flows through the base collector junction of transistor T_1 is sufficient to turn transistor T_2 ON and this causes transistor T_3 to turn ON also. The voltage V_2 at the collector of transistor T_2 is $V_2 = V_{CE\,T2} + V_{BE\,T3} = 0.2 + 0.7 = 0.9 \text{ V}$. However, this voltage is not sufficient to turn transistor T_4 ON because for transistor T_4 to be ON the voltage V_2 should have the value $V_2 = V_{BE\,T4} + V_{D2} + V_{out} = 0.7 + 0.7 + 0.2 = 1.6 \text{ V}$. Therefore, transistor T_4 is OFF, $V_{OUT} = 0.2 \text{ V}$, and this satisfies the 2–input NAND gate operation.

It should be noted that it is the diode D_3 that prevents transistor T_4 from being ON; if D_3 were not being there, T_4 would be also ON since there would be $V_{BE\,T4} = 0.9 - 0.2 = 0.7 \text{ V}$ which is just sufficient to turn T_4 ON.

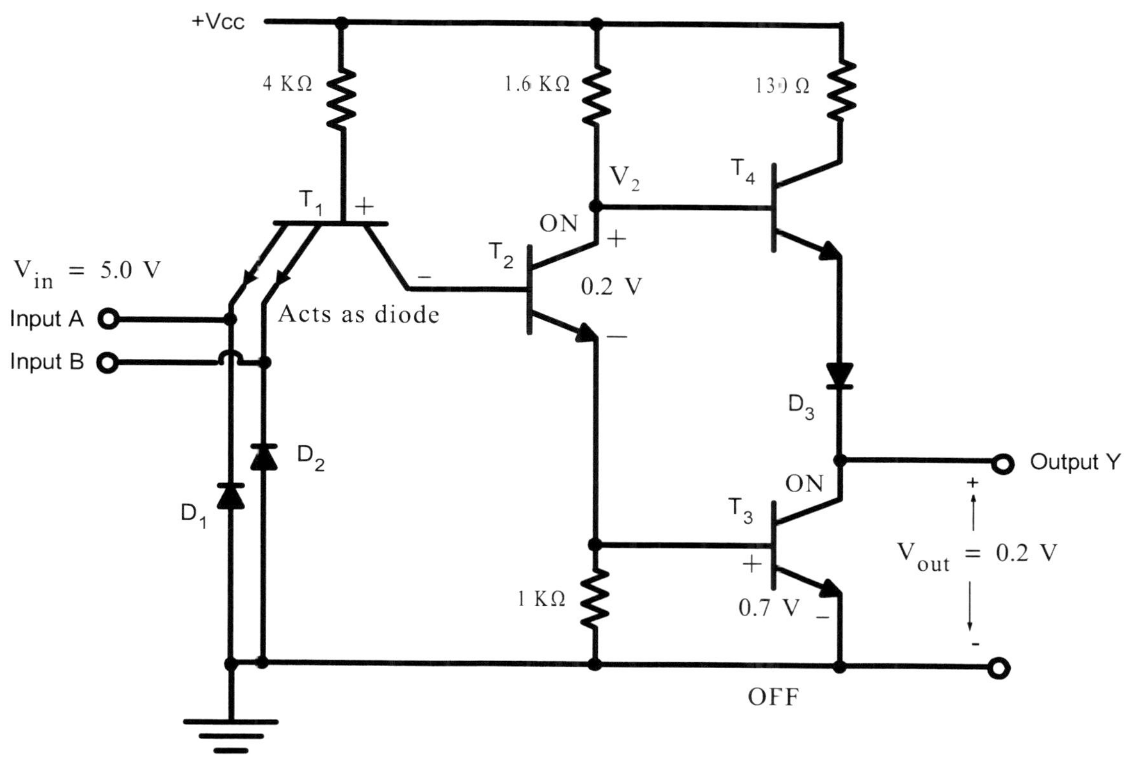

Figure 6.21. Circuit of the SN7400 2–input NAND gate when both inputs are High

Now, referring back to the AND gate circuit of Figure 6.13, we observe that this circuit essentially consists of a NAND gate followed by an inverter as shown in Figure 6.22.

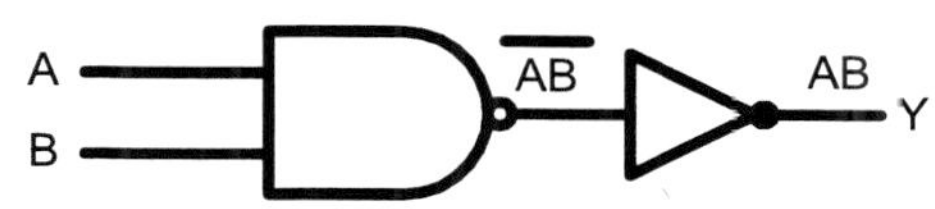

Figure 6.22. The components of a 2–input AND gate

We mentioned earlier that one of the parameters specified by the manufacturer is the *input clamp voltage* denoted as V_{IK} and this refers to the maximum negative voltage that may be applied at the input terminals without damaging the IC. A typical value for this parameter is $-1.2\ V$. To insure that this value is not exceeded, diodes D_1 and D_2 are included in the circuit of the 2–input NAND gate to clamp the input voltage to less than $-1.2\ V$ with respect to the ground.

6.7 The NOR Gate

The symbol for a 3–input NOR gate is shown in Figure 6.23, and the truth table with positive logic is shown in Table 6.9.

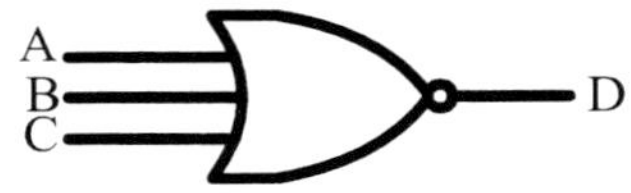

Figure 6.23. Symbol for 3–input NOR gate

TABLE 6.9 Truth table for 3–input NOR gate with positive logic

Inputs			Output
A	**B**	**C**	**D**
0	0	0	1
0	0	1	0
0	1	0	0
0	1	1	0
1	0	0	0
1	0	1	0
1	1	0	0
1	1	1	0

Table 6.9 shows that the output of an NOR gate is logical 1 (true) only when all inputs are logical 0 (false).

Figure 6.24 shows the TTL SN7402 Quad 2–input NOR gate where Quad implies that there are 4 NOR gates within the IC.

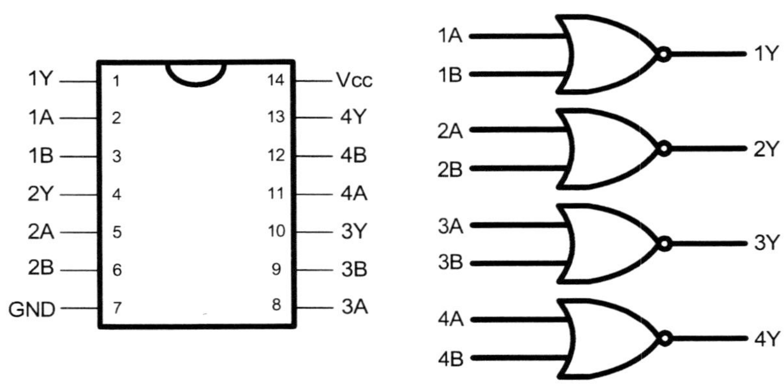

Figure 6.24. The TTL SN7402 Quad 2–input NOR gate

Figure 6.25 shows the internal details of the TTL IC SN7402 NOR gate.

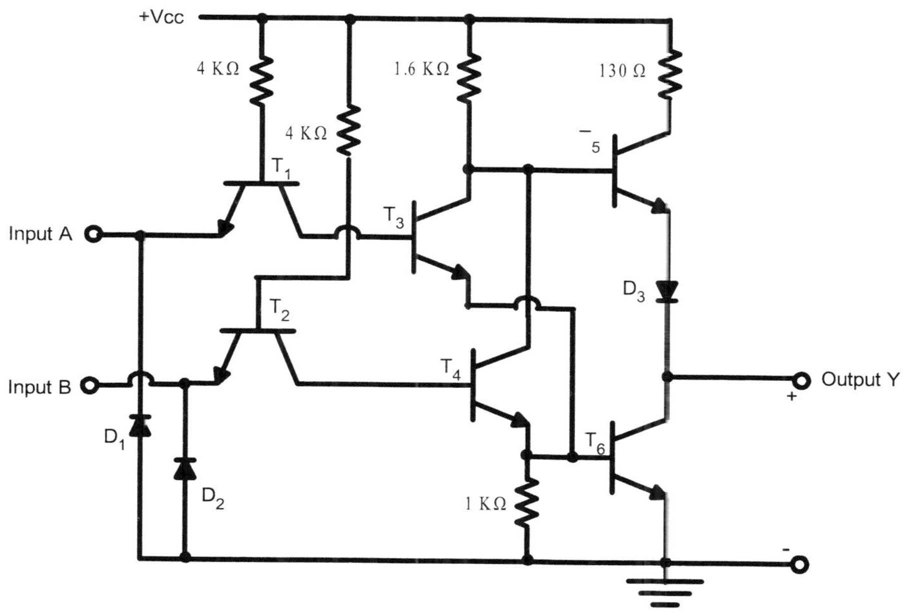

Figure 6.25. Circuit for the TTL SN7402 Quad 2–input NOR gate

The TTL 2–input NOR gate circuit of Figure 6.25 functions as follows:

When both inputs are Low at 0.2 V with respect to the ground, transistors T_1 and T_2 are both ON. The voltages at the base of transistors T_3 and T_4 are the same as the collector voltages of transistors s T_1 and T_2, that is, 0.2 V, and thus transistors T_3 and T_4 are OFF. However, transistor T_5 is ON and therefore, $V_{out} = V_{CC} - V_{CE\ T5} - V_{D5} = 5.0 - 0.2 - 0.7 = 4.1\ V$.

If Input A is Low and Input B is High, transistor T_1 will be ON and transistor T_3 will be OFF. But transistor T_2 will behave as a junction diode turning transistor T_4 ON, and a voltage drop across the $1\ K\Omega$ resistor will be developed and it will be sufficient to turn transistor T_6 ON, and thus $V_{out} = 0.2\ V$. If Input A is High and Input B is Low or both inputs are High, we find that $V_{out} = 0.2\ V$ also.

6.8 The Exclusive OR (XOR) and Exclusive NOR (XNOR) Gates

The exclusive–OR (XOR) logic gate has two inputs and one output. The symbol for the XOR gate is shown in Figure 6.26, and the truth table with positive logic is shown in Table 6.10.

We observe that the output of an XOR gate is logical 1 (true) when only one of the inputs, but not both, is logical 1.

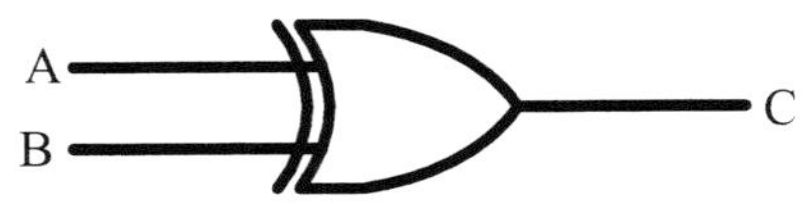

Figure 6.26. Symbol for the XOR gate with positive logic

TABLE 6.10 Truth table for 2–input XOR gate

Inputs		Output
A	**B**	**C**
0	0	0
0	1	1
1	0	1
1	1	0

Figure 6.27 shows the TTL SN7486 Quad XOR gate where Quad implies that there are 4 XOR gates within the IC.

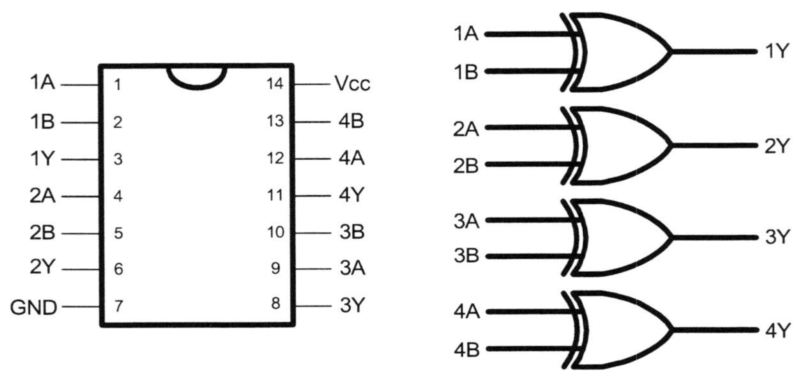

Figure 6.27. The TTL SN7486 Quad XOR gate

Figure 6.28 shows the internal details of the TTL IC SN7486 XOR gate.

The exclusive–NOR (XNOR) logic gate has two inputs and one output. The symbol for the XNOR gate is shown in Figure 6.29. and the truth table with positive logic is shown in Table 6.11.

Table 6.11 shows that the output of a XNOR gate is logical 1 (true) only when the inputs are the same, that is, both logical 0 or both logical 1. For this reason, the XNOR gate is also known as *equivalence gate*.

There is no IC XNOR gate in the TTL family but one can be formed with an XOR gate (SN7486) followed by an inverter (SN7404). We can also implement the XNOR function using the TTL SN7486 IC with negative logic.

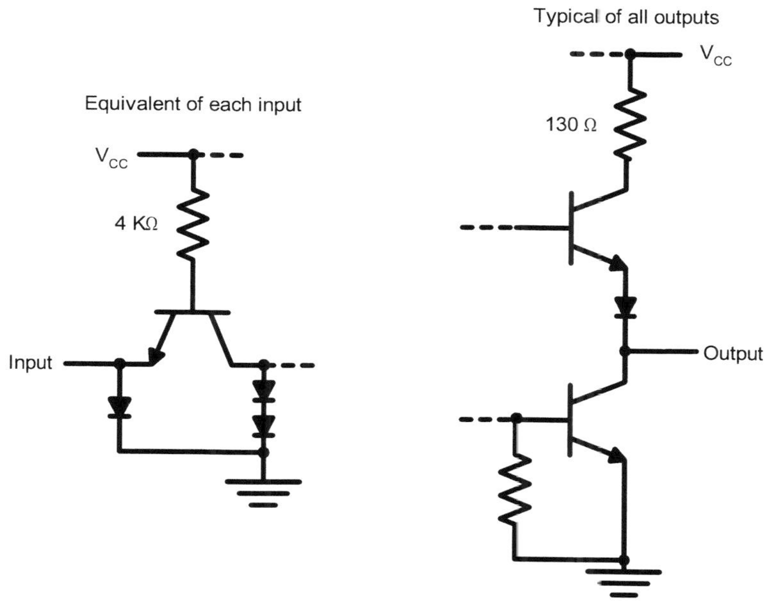

Figure 6.28. Circuit for the SN7486 XOR gate

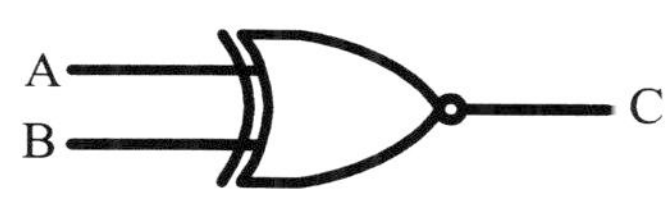

Figure 6.29. Symbol for 2–input XNOR gate

TABLE 6.11 Truth table for 2–input XNOR gate with positive logic

Inputs		Output
A	**B**	**C**
0	0	1
0	1	0
1	0	0
1	1	1

6.9 Fan–In, Fan–Out, TTL Unit Load, Sourcing Current, and Sinking Current

The *fan–in* of a gate is the number of its inputs. Thus, a 3–input NAND gate has a fan–in of 3. *Fan–out* is a term that defines the maximum number of digital inputs that the output of a single logic gate can feed. Generally, TTL gates can feed up to 10 other digital gates or devices. Thus, a typical TTL gate has a fan–out of 10.

In some digital systems, it is necessary for a single TTL logic gate to drive more than 10 other gates or devices. When this is the case, a device called a *buffer* can be used between the TTL gate and the multiple devices it must drive. A buffer of this type has a fan–out of 25 to 30. An inverter (NOT gate) can serve this function in most digital circuits if complementation is also required.

A *unit load* for TTL logic gates is defined as

$$1 \text{ Unit Load } = \begin{cases} 40\ \mu\text{A when output is in High state} \\ 1.6\ \text{mA when output is in Low state} \end{cases} \tag{6.1}$$

Thus, for a fan–out of 10 which is typical for TTL gates, when the output is in High state, there is a current of $10 \times 40\ \mu\text{A} = 0.4\ \text{mA}$, and when the output in Low state there is a current of $0 \times 1.6\ \text{mA} = 16\ \text{mA}$.

We often hear the expressions "passive pull–up" and "active pull–up". Figure 6.30 is an example of *passive pull–up* where the resistor pulls–up the output voltage towards V_{CC}. and the word "passive" is used to indicate that the pull–up device used is passive, i.e., a resistor.

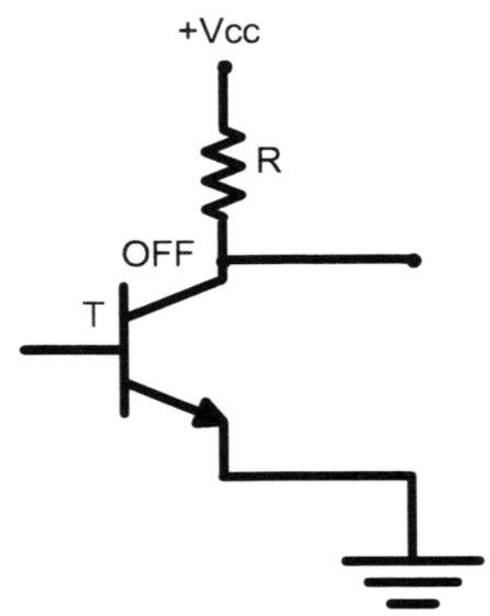

Figure 6.30. A passive pull–up circuit.

The arrangement in Figure 6.31(a) is often referred to as *active pull–up* since when transistor T_1 is ON, it pulls the output up towards V_{CC}. The word "active" is used here to indicate that transistor T_1, like all other transistors, is an active device.

Sourcing and *sinking* currents refer to the current flow in TTL circuits. A driver gate* is said to be sourcing current when it's output is a logic 1 as shown in Figure 6.31(a). A driver gate is sinking current when it's output is a logic 0 as shown in Figure 6.31(b).

The passive pull–up arrangement is often used with *open collector* TTL devices such as the one shown in Figure 6.32. The advantage of the open–collector configuration over the totem–pole configuration is that the outputs of two or more open–collector TTL gates can be tied together to realize the AND function as shown in Figure 6.33.

* *A driver gate is a gate whose output serves as an input to another gate referred to as a unit load.*

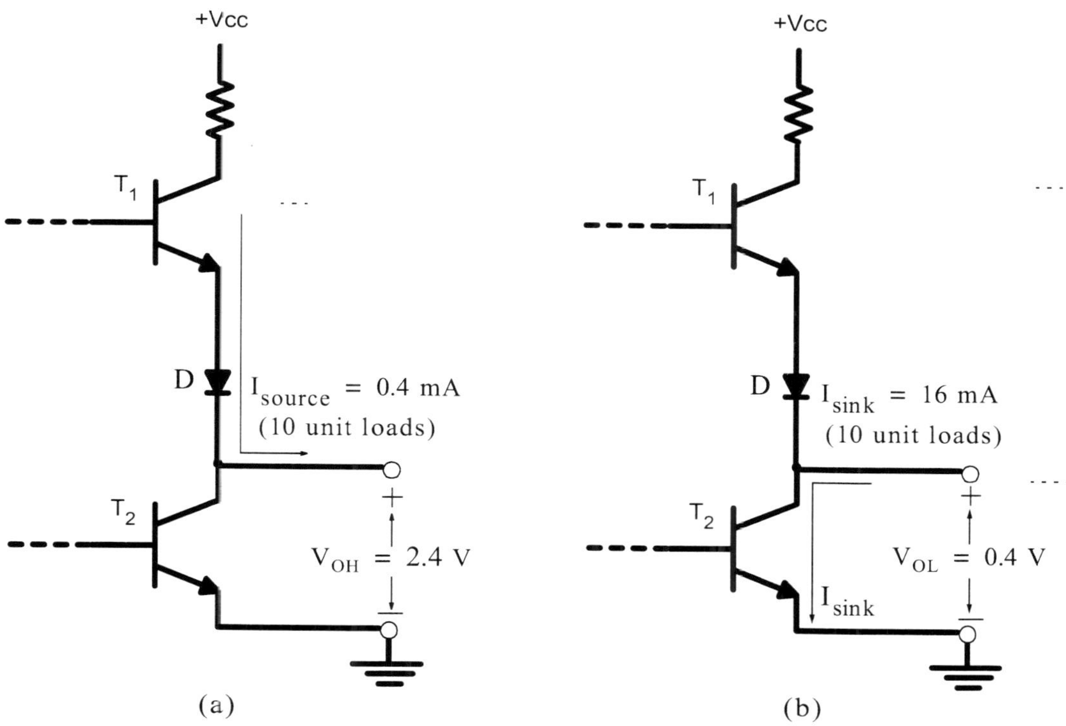

Figure 6.31. Current sourcing and current sinking

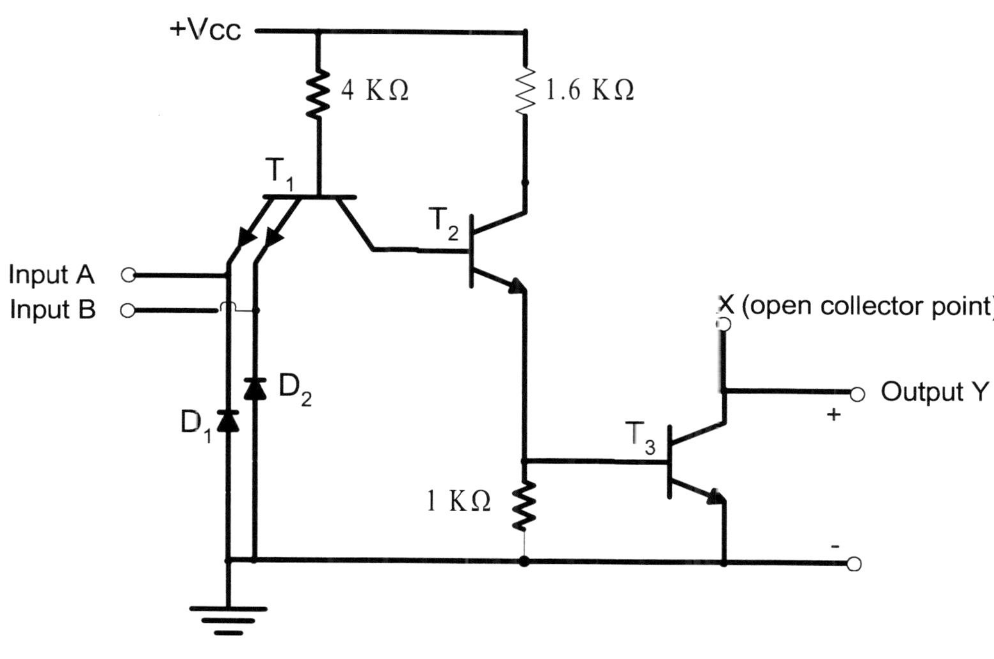

Figure 6.32. Open–collector TTL configuration

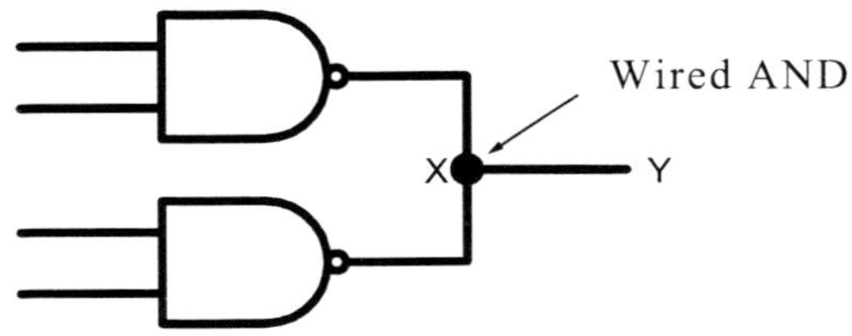

Figure 6.33. Wired AND connection

To see why the outputs of two or more totem pole TTL gates should not be tied together for wired AND operation, let us consider the arrangement shown in Figure 6.34. With the collectors of transistors T_2 and T_4 as shown, the current I can be as high as 55 mA which most likely will damage transistors T_1 and T_4, because per manufacturer's specifications these transistors can sink only 16 mA.

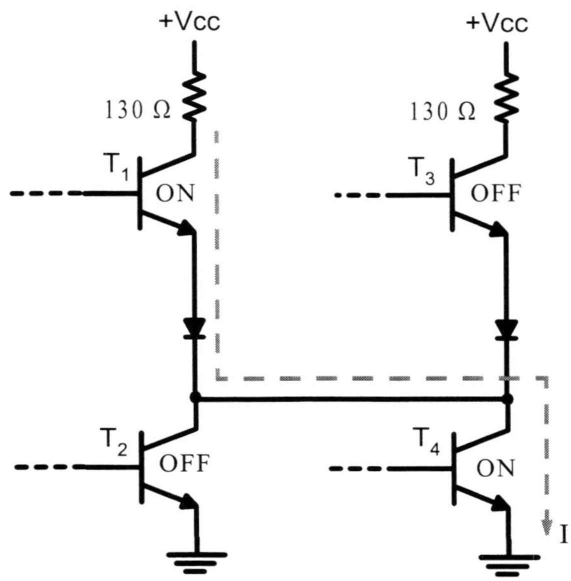

Figure 6.34. Improper wired AND connection

6.10 Data Sheets

Table 6.12 lists some important parameters listed in the data sheets provided by the manufacturer with typical values for a TTL 2−input NAND gate, SN7400. The reader is cautioned that these values are provided just for instructional purposes. For actual values, the manufacturer's latest data sheet should always be used since the values change from time to time.

Item 1 in the table is self−explanatory, and we have already discussed Item 2. Therefore, we will continue with Items 3 through 13.

TABLE 6.12 *Typical parameters for TTL logic circuits provided by IC manufacturers*

	Parameter	Conditions	Min	Typical	Max	Units
1	Supply voltage V_{CC}	$T_A = 25°C$	4.75	5.00	5.25	V
2	Input clamp voltage V_{IK}	$V_{CC} = Min, T_A = 25°C, I_{IN} = -12\ mA$			–1.5	V
3	High Level Input Voltage V_{IH}	$V_{CC} = Min$	2.0			V
4	Low Level Input Voltage V_{IL}	$V_{CC} = Min$			0.8	V
5	High Level Output Voltage V_{OH}	$V_{CC} = Min, V_{IL} = 0.8\ V,$ $I_{OH} = -0.4\ mA$	2.4	3.4		V
6	Low Level Output Voltage V_{OL}	$V_{CC} = Min, V_{IH} = 2.0\ V, I_{OL} = 16\ mA$		0.2	0.4	V
7	High Level Input Current I_{IH}	$V_{CC} = Max, V_{IN} = 2.4\ V$ $V_{CC} = Max, V_{IN} = 5.5\ V$			40 1	μA mA
8	Low Level Input Current I_{IL}	$V_{CC} = Max, V_{IN} = 0.4\ V$			–1.6	mA
9	Output Short Circuit Current I_{OS}	$V_{CC} = Max, V_{IN} = 0\ V, V_{OUT} = 0$	–18		–55	mA
10	Low Level Supply Current I_{CCL}	$V_{CC} = Max, V_{IN} = 4.5\ V$		2.4	4.4	mA
11	High Level Supply Current I_{CCH}	$V_{CC} = Max, V_{IN} = 0$		0.8	1.6	mA
12	Propagation Delay Time – High to Low Level t_{PHL}	$V_{CC} = 5.0\ V, T_A = 25°C, R_L = 400\ \Omega$ $C_L = 15\ pF$		7	15	ns
13	Propagation Delay Time – Low to High Level t_{PLH}	$V_{CC} = 5.0\ V, T_A = 25°C, R_L = 400\ \Omega$ $C_L = 15\ pF$		11	22	ns

3. The high level input voltage V_{IH}, also referred to as logical 1 input voltage is the minimum input voltage level that the IC device will recognize as a valid logical 1 input as shown in Figure 6.35.

4. The low level input voltage V_{IL}, also referred to as logical 0 input voltage is the maximum input voltage level that the IC device will recognize as a valid logical 0 input as shown in Figure 6.35.

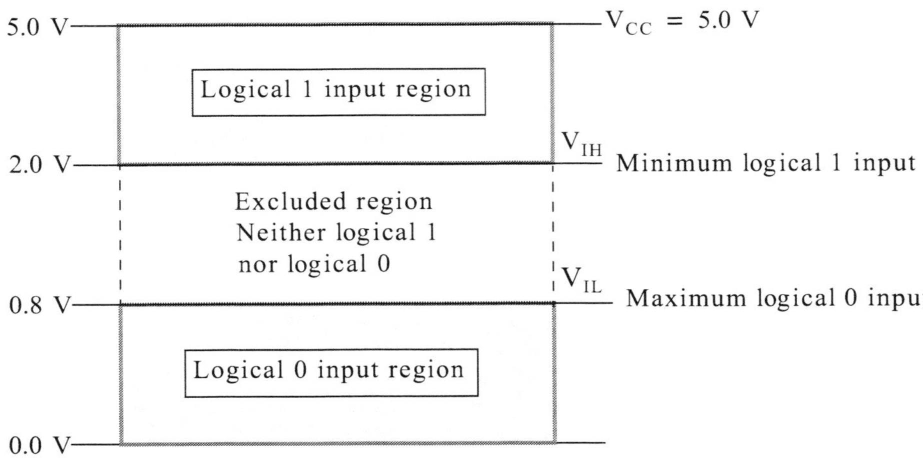

Figure 6.35. Valid logical 1 *and valid logical* 0 *input regions for a typical IC gate*

5. The high level output voltage V_{OH}, also referred to as logical 1 output voltage is the minimum output voltage level that the IC device will recognize as a valid logical 1 output.

6. The low level output voltage V_{OL}, also referred to as logical 0 output voltage is the maximum output voltage level that the IC device will recognize as a valid logical 0 output when a current of 16 mA, representing a fan–out of 10, is sinked by transistor T_2 as indicated in Figure 6.36.

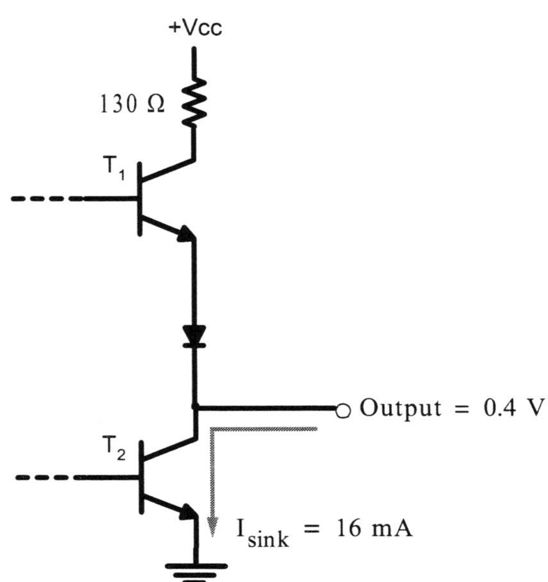

Figure 6.36. Part of a typical NAND gate showing current sinking

7.a. For a 2–input NAND gate, when the input voltage is 2.4 V, the maximum value of the sum of the currents I_{IH} furnished by the input lines should not be greater than $40 \mu A$ for ten unit loads.

7.b. For a 2–input NAND gate, when the input voltage is 5.5 V, the maximum value of the sum of the currents I_{IH} furnished by the input lines should not be greater than 1 mA for ten unit loads.

8. When the input voltage is low at 0.4 V, the maximum value of the current I_{IL} should not exceed -1.6 mA where the minus sign indicates the direction of the current flow as shown in Figure 6.37 where for a logical 0 input voltage the ten gates at the right side "source" (supply) -1.6 mA each, and the single gate on the left "sinks" (accepts) $-1.6 \times 10 = -16 \text{ mA}$.

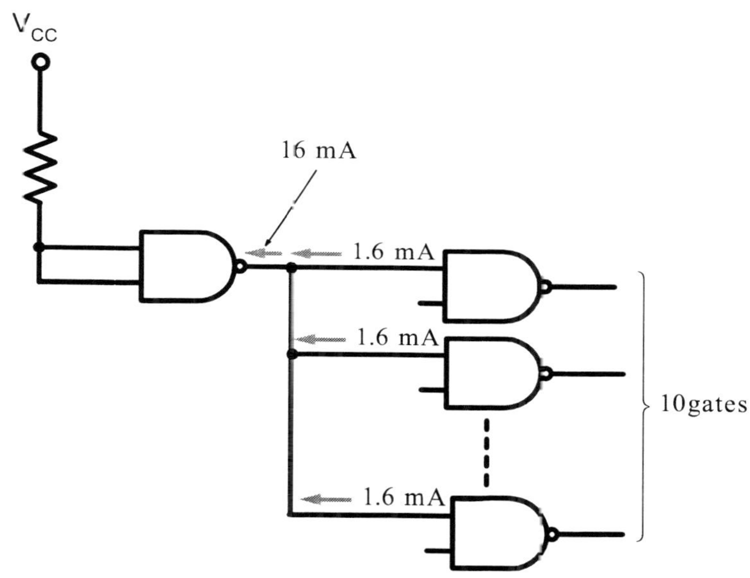

Figure 6.37. An illustration of the input current I_{-L} for a 2–input NAND gate

9. The output short–circuit current I_{OS} is the output current when the output is shorted to ground as shown in Figure 6.38. The data sheet specifies an output short circuit current of 55 mA maximum. When transistor T_1 is ON, transistor T_2 is OFF, and the output voltage is grounded, the resistor will act as a current limiter to limit the current in the range 18 mA to 55 mA. Thus, the resistor provides short circuit protection.

10. The Low Level Supply Current I_{CCL} is the current that the voltage supply V_{CC} must furnish when only one of the four TTL 2–input NAND gates within the IC SN7400 has its output at logical 0. If two outputs are at logical 0, V_{CC} must furnish twice the amount of current specified in the data sheet, that is, $2 \times 4.4 = 8.8 \text{ mA}$.

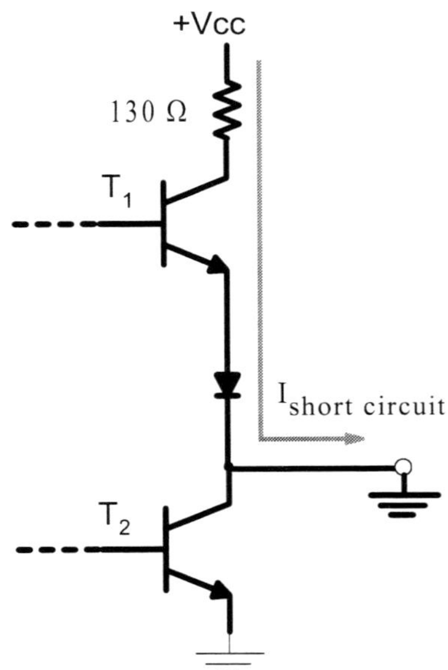

Figure 6.38. The definition of the output short–circuit current

11. The High Level Supply Current I_{CCH} is the current that the voltage supply V_{CC} must furnish when only one of the four 2–input NAND gates within the IC SN7400 has its output at logical 1 . If two outputs are at logical 1 , V_{CC} must furnish twice the amount of current specified in the data sheet, that is, $2 \times 1.6 = 3.2\ mA$.

12. The propagation delay time t_{PHL} is the time required for a logical 1 Input A or logical 1 Input B to appear as logical 0 at Output Y. A typical value is 7 ns .

13. The propagation delay time t_{PLH} is the time required for a logical 0 Input A or logical 0 Input B to appear as logical 1 at Output Y. A typical value is 11 ns .

6.11 Emitter Coupled Logic (ECL)

Let us consider the differential circuit shown in Figure 6.39, assume that transistors T_1 and T_2 are identical, and that $R_{C1} = R_{C2}$. Then,

a. If $V_{B2} = V_{B1}$, then $I_{C2} = I_{C1}$, $R_{C2}I_{C2} = R_{C1}I_{C1}$, $V_{out2} = V_{out1}$ and thus the differential output is zero.

b. If $V_{B2} > V_{B1}$, then $I_{C2} > I_{C1}$, $R_{C2}I_{C2} > R_{C1}I_{C1}$, $V_{out2} < V_{out1}$ and thus the differential output is positive.

c. If $V_{B2} < V_{B1}$, then $I_{C2} < I_{C1}$, $R_{C2}I_{C2} < R_{C1}I_{C1}$, $V_{out2} > V_{out1}$ and thus the differential output is negative.

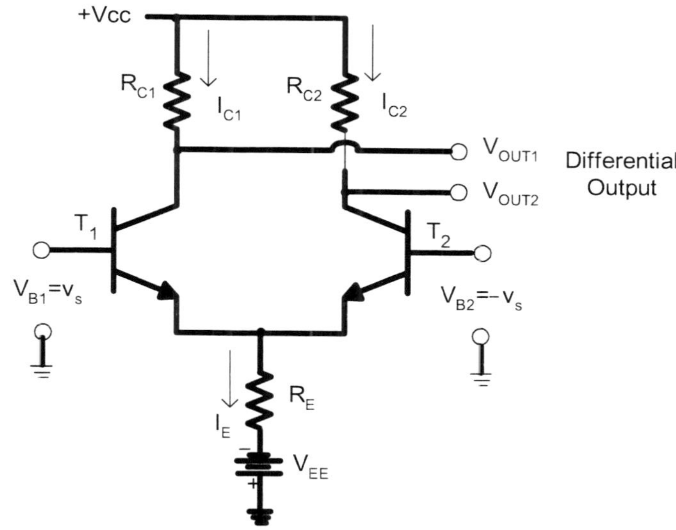

Figure 6.39. Basic differential circuit

Example 6.1

Find the differential voltage gain for the circuit in Figure 6.40.

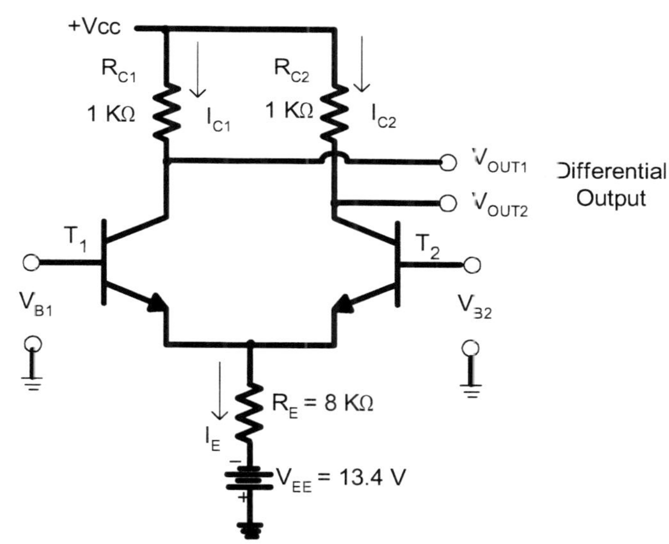

Figure 6.40. Differential circuit for Example 6.1

Solution:

The base–to–emitter resistance is small compared to the external resistor R_E and thus

$$I_E \approx \frac{V_{EE} - V_{BE}}{R_E} = \frac{13.4 - 0.6}{8\ K\Omega} = 1.6\ mA$$

and this current divides equally between transistors T_1 and T_2 assuming that they are identical. Then,

$$I_{C2} = I_{C1} = I_E/2 = 1.6/2 = 0.8 \text{ mA}$$

From Chapter 3, relation (3.78), at room temperature $g_m = 40I_C$ where I_C is in milliamps. Thus,

$$g_m = 40 \times 0.8 \times 10^{-3} = 0.032$$

and from Chapter 3, relation (3.74)

$$g_m = \left.\frac{di_C}{dv_{BE}}\right|_{i_C = I_C}$$

or

$$i_C = g_m v_{BE}$$

and thus

$$v_{out1} = R_{C1} I_{C1} = R_{C1} g_m v_{BE}$$

The small signal voltage gain is

$$A_v = \frac{v_{out1}}{v_{BE}} = R_{C1} g_m = 10^3 \times 0.032 = 32$$

The differential circuit of Figure 6.39 forms the basis for the emitter–coupled logic (ECL) electronic gates. The basic ECL circuit is shown in Figure 6.41.

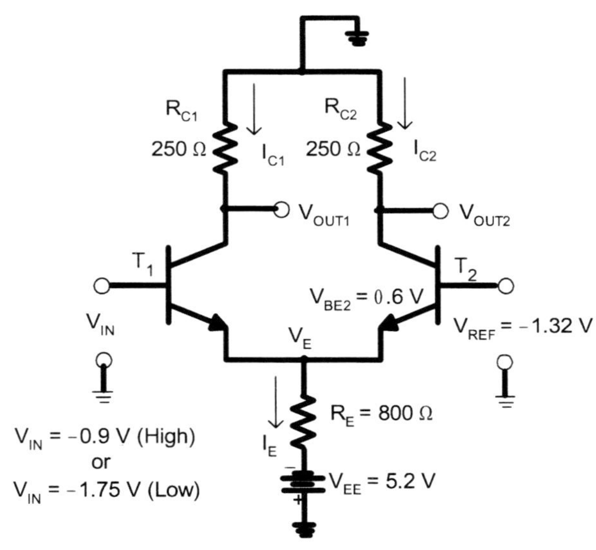

Figure 6.41. Basic ECL circuit

In contrast to TTL where the transistors are either cut off or saturated, depending on the state of the circuit, in ECL the transistors always operate in the active region so they can change state very rapidly. We observe that with $V_{BE2} = 0.6\ V$, the transistor T_2 is barely turned ON or barely turned OFF and this allows a very rapid change of state from ON to OFF or vice versa. The propagation time for this arrangement is less than a nanosecond. However, there is a disadvantage and this is that the transistors are continually drawing current, which means the circuits require high power, and this power is converted to energy in the form of unwanted heat.

In the circuit of Figure 6.41 the resistor values are approximate and for proper operation they must be very precise. The emitter current I_E is almost constant since the base–to–emitter resistance is much smaller than the emitter resistor R_E and this current flows through either transistor T_2 or transistor T_2 depending on the value of V_{in} while V_{REF} is held constant at $-1.32\ V$ with respect to the ground. The condition $V_{in} = -0.9\ V$ is established as logical 1 (High) and $V_{in} = -1.75\ V$ as logical 0 (Low). Also, $V_{BE2} + V_E = V_{REF}$ and thus $V_E = -1.32 - 0.6 = -1.92\ V$.

Let us now assume that $V_{in} = -1.75\ V$, i.e., logical 0. Because $V_E = -1.92\ V$, transistor T_1 does not conduct and V_{out1} is essentially connected to the ground. To find V_{out2} when $V_{in} = -1.75\ V$ (logical 0), we must first find the current I_{C2}. We observe that

$$I_{C2} \approx I_E = \frac{V_E - (-V_{EE})}{R_E} = \frac{-1.92 - (-5.2)}{800} = 4.1\ mA$$

and

$$V_{out2} = -R_{C2} I_{C2} = -250 \times 4.1 \times 10^3 = -1.03\ V$$

Therefore, with $V_{in} = -1.75\ V$ (logical 0), we obtain $V_{out1} = 0\ V$ and $V_{out2} = -1.03\ V$. Next, let $V_{in} = -0.9\ V$, i.e., logical 1. Because $V_E = -1.92\ V$, transistor T_1 conducts hard, and transistor T_2 is OFF, and V_{out2} is essentially connected to the ground. Therefore all current flows through transistor T_1 and the collector current of this transistor is

$$I_{C1} \approx I_E = \frac{V_E - (-V_{EE})}{R_E} = \frac{-1.92 - (-5.2)}{800} = 4.1\ mA$$

and

$$V_{out1} = -R_{C2} I_{C2} = -250 \times 4.1 \times 10^3 = -1.03\ V$$

Therefore, with $V_{in} = -0.9\ V$ (logical 1), we obtain $V_{out1} = -1.03\ V$ and $V_{out2} = 0\ V$. We observe that:

1. V_{out1} and V_{out2} are complements of each other.
2. The output voltage levels are not the same as input voltage levels, that is, the input levels are $-1.75\ V$ and $-0.9\ V$ and the outputs are $0\ V$ and $-1.03\ V$. However, the output levels can be made comparable to the input levels by attaching two emitter followers to the outputs as shown in Figure 6.42.

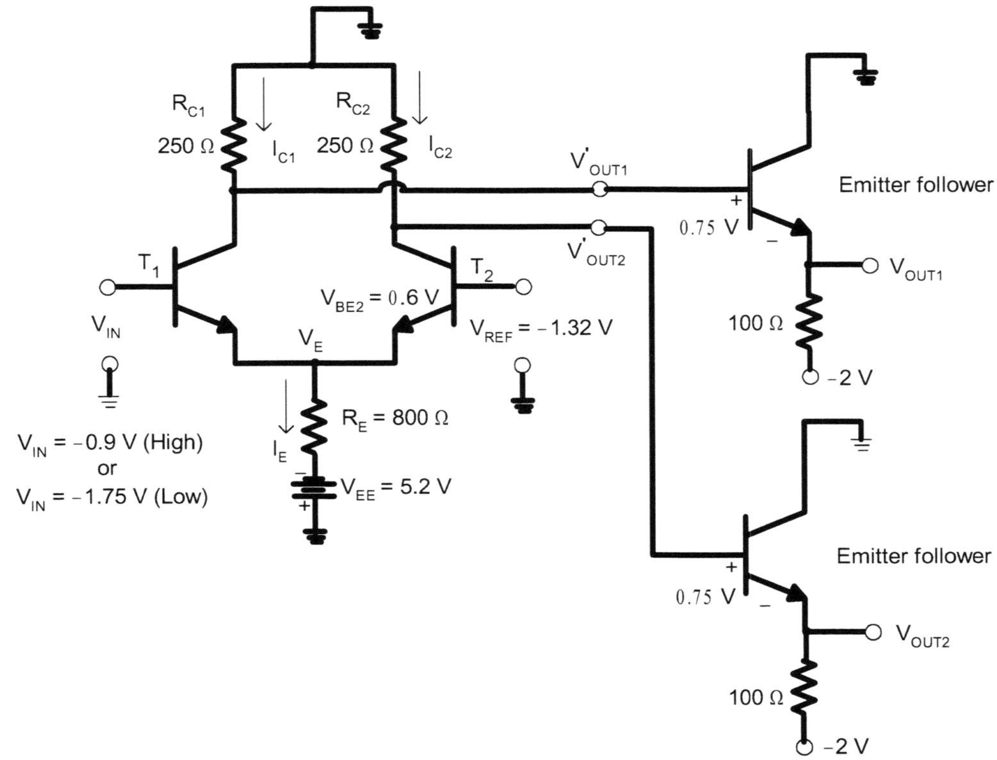

Figure 6.42. The complete ECL circuit

With the emitter–followers connected at the outputs V'_{out1} and V'_{out2} of the basic differential circuit and recalling that in the emitter–follower configuration the output voltage is essentially the same as the input, the outputs at the emitter–followers are

$$V_{out1} = V'_{out1} - V_{BE} = -1.03 - 0.75 = -1.78 \text{ V} \quad (\text{logical } 0)$$

and

$$V_{out2} = V'_{out2} - V_{BE} = 0.00 - 0.45 = -0.75 \text{ V} \quad (\text{logical } 1)$$

Figure 6.43 shows a typical 2–input ECL gate and. as before, the resistor values are approximate. The symbol for a 2–input ECL gate is shown in Figure 6.44. Two types of ECL gates are the ECL 10K and ECL 100K, the latter having a propagation delay time of 0.75 ns and power consumption of 40 mW. It should also be noted that ECL is not a new technology; Motorola introduced the MECL series in the 1980's.

6.12 NMOS Logic Gates

We discussed MOSFETs in Chapter 4 but, for convenience, let us review the basic construction and their operation. We recall that a MOSFET is a 4–terminal device but normally the substrate is connected to the source thus making it a 3–terminal device. Henceforth, the MOSFET will be shown as a 3–terminal block as in Figure 6.45.

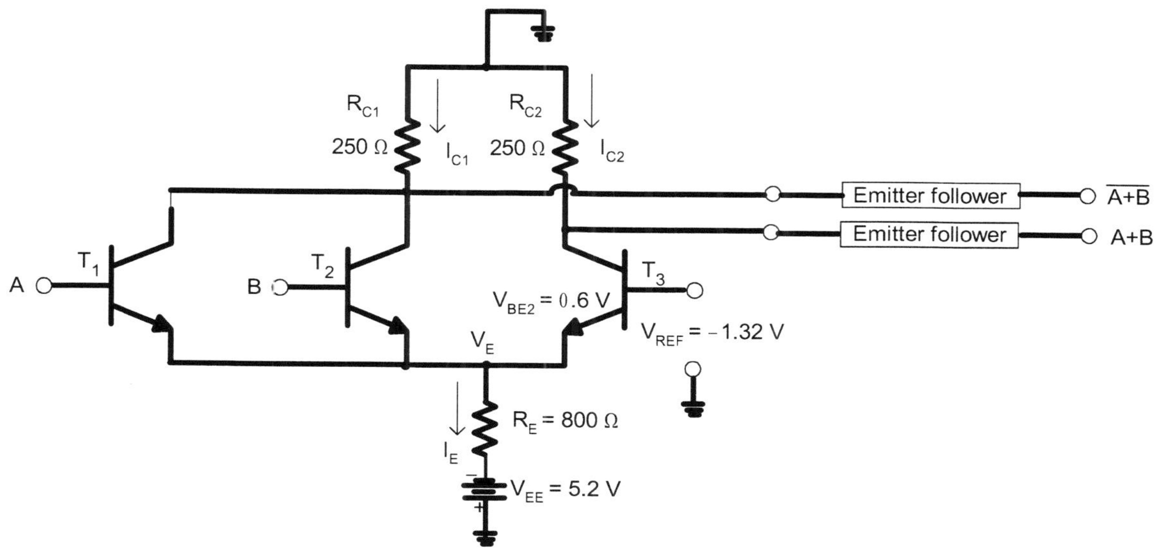

Figure 6.43. Typical 2–input ECL gate

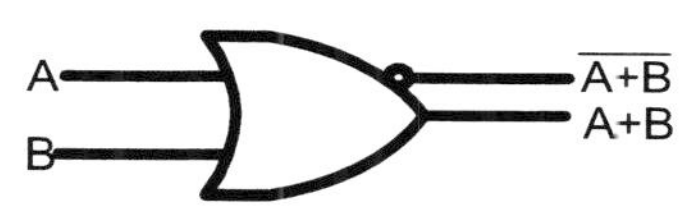

Figure 6.44. Symbol for ECL gate

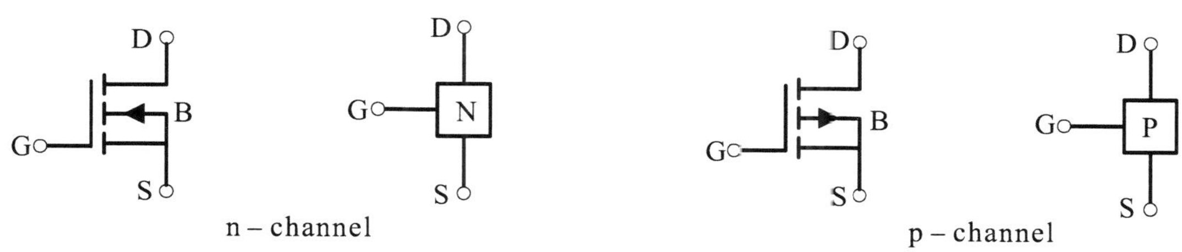

Figure 6.45. Designations for n– and p–channel MOSFETS

An n–channel MOSFET or simply NMOS device, behaves like an open switch, as shown in Figure 6.46, when $V_{GS} < V_T$. where $V_T = 1.5\ V$ and it is referred to as the *threshold voltage*.

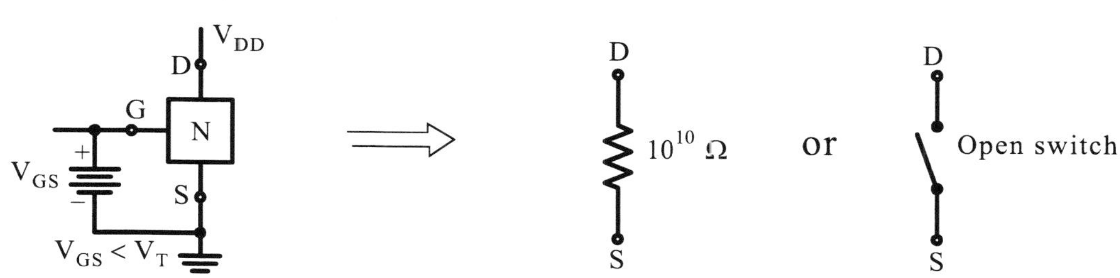

Figure 6.46. Condition under which an NMOS device behaves as an open switch

An NMOS device behaves like a closed switch, as shown in Figure 6.47, when $V_{GS} > V_T$.

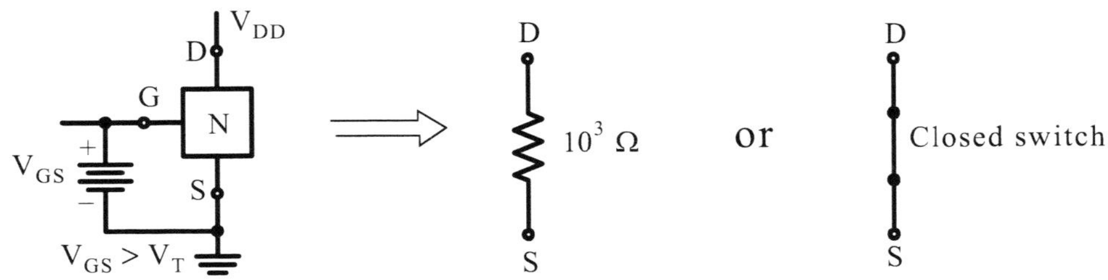

Figure 6.47. Condition under which an NMOS device behaves as a closed switch

A p–channel MOSFET or simply PMOS device, behaves like an open switch, as shown in Figure 6.48, when $V_{GS} > V_T$. where $V_T = 1.5\ V$ is the threshold voltage.

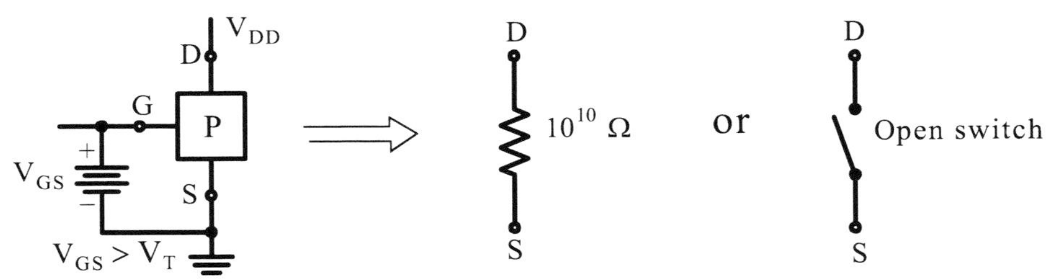

Figure 6.48. Condition under which a PMOS device behaves as an open switch

An PMOS device behaves like a closed switch, as shown in Figure 6.49, when $V_{GS} < V_T$.

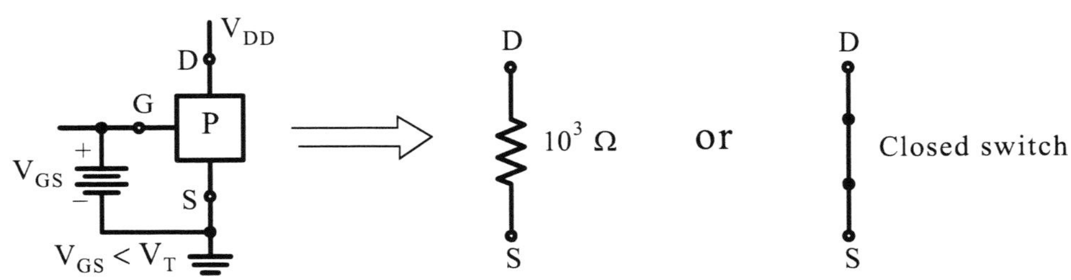

Figure 6.49. Condition under which a PMOS device behaves as a closed switch

For convenience, in our subsequent discussion we will the state of an NMOS or PMOS device ON or OFF (closed or open switch) by a low resistance ($1\ K\Omega$) or a very high resistance ($10^{10}\ \Omega$) respectively. Also, since in integrated circuits resistors are being replaced by NMOS or PMOS devices because these devices require much less space than physical resistors, we will represent these with resistor symbols with a typical value of $100\ K\Omega$. We will denote these as N_L or P_L where the subscript L stands for load.

6.12.1 The NMOS Inverter

A typical NMOS inverter is shown in Figure 6.50 where with $V_{IN} = 0\ V$ (logical 0), the NMOS device N is OFF, $R_X = 10^{10}\ \Omega$, $N_L = R_D = 10^5\ \Omega$, and the output is

$$V_{out} = \frac{R_X}{R_X + R_D} V_{DD} = \frac{10^{10}}{10^{10} + 10^5} \times 5 \approx 5\ V$$

Thus, the inversion operation has been performed. With $V_{in} = 5\ V$ (logical 1), the NMOS device N is ON, $R_X = 10^3\ \Omega$, $N_L = R_D = 10^5\ \Omega$, and the output is

$$V_{out} = \frac{R_X}{R_X + R_D} V_{DD} = \frac{10^3}{10^3 + 10^5} \times 5 \approx 0.05\ V$$

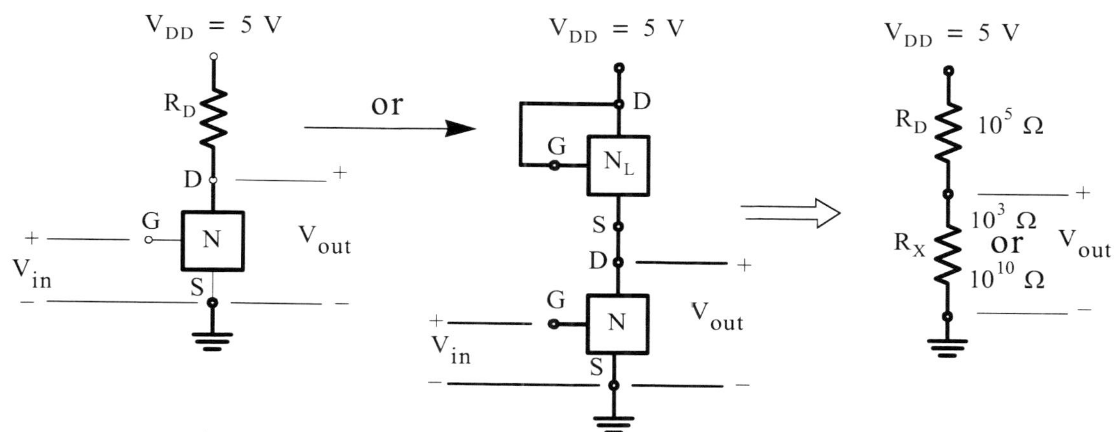

Figure 6.50. Typical NMOS inverter

and again the inversion operation is performed. The truth table for positive logic is shown in Table 6.13.

TABLE 6.13 Truth table for NMOS inverter

V_{in}	V_{out}
0	1
1	0

A PMOS inverter has a similar arrangement where the NMOS devices are replaced by PMOS devices.

6.12.2 The NMOS NAND Gate

A 2–input NMOS NAND gate is shown in Figure 6.51. By application of the voltage division expression as with the NMOS inverter, we find that for positive logic the truth table is as shown in Table 6.14.

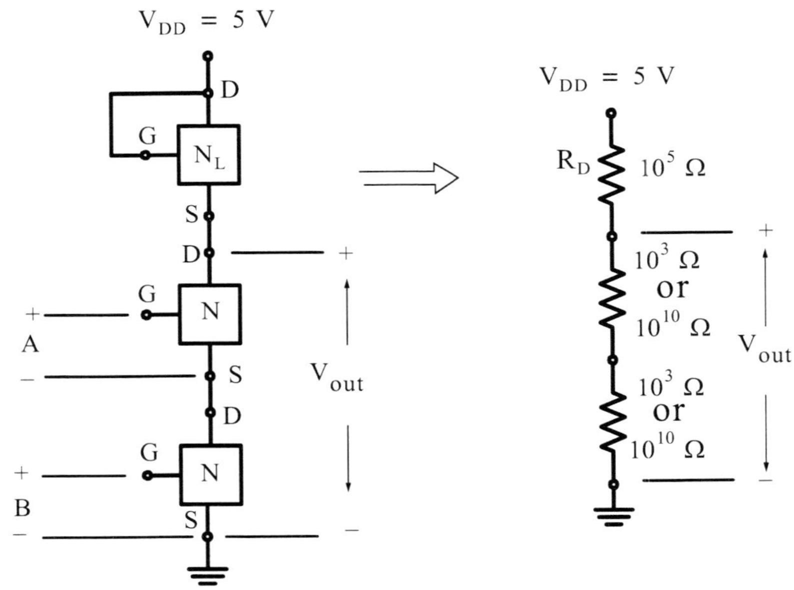

Figure 6.51. 2–input NMOS NAND gate

TABLE 6.14 Truth table for 2–input NMOS NAND gate

A	B	V_{out}
0	0	1
0	1	1
1	0	1
1	1	0

6.12.3 The NMOS NOR Gate

A 2–input NMOS NOR gate is shown in Figure 6.52. By application of the voltage division expression as with the NMOS NAND gate, we find that for positive logic the truth table is as shown in Table 6.15.

6.13 CMOS Logic Gates

CMOS logic uses both NMOS and PMOS devices to form logic functions. CMOS technology is the dominant semiconductor technology for the manufacturing of microprocessors, memories and application specific integrated circuits. The main advantage of CMOS over NMOS and bipolar technology is the much smaller power dissipation. Unlike NMOS or bipolar circuits, a CMOS circuit has almost no static power dissipation. Power is only dissipated in case the circuit actually switches. This allows to integrate many more CMOS gates on an IC than in NMOS or bipolar technology, resulting in much better performance.

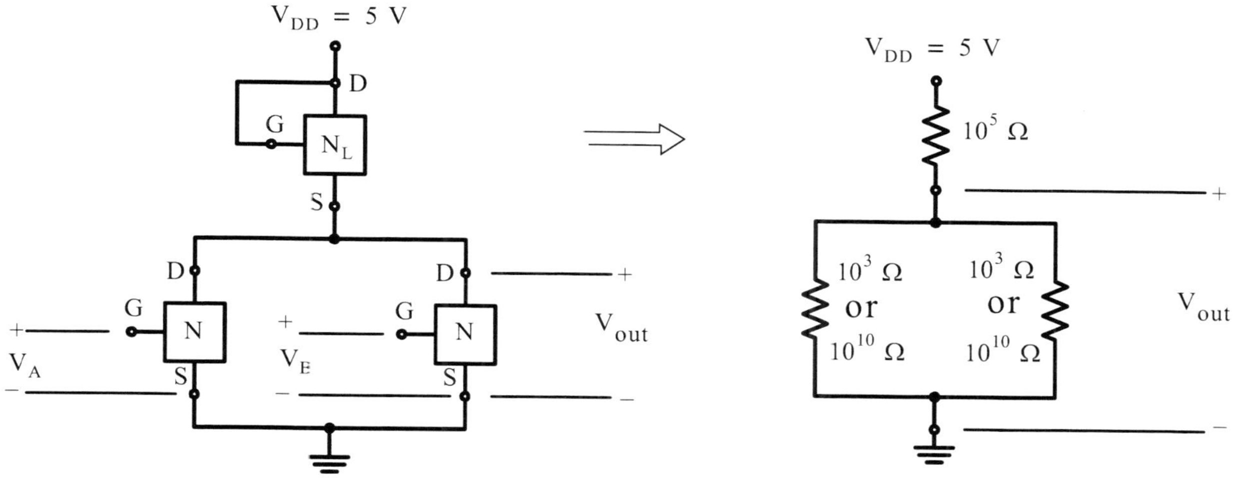

Figure 6.52. 2–input NMOS NOR gate

TABLE 6.15 Truth table for 2–input NMOS NOR gate

A	B	V_{out}
0	0	1
0	1	0
1	0	0
1	1	0

6.13.1 The CMOS Inverter

A typical CMOS inverter is shown in Figure 6.53.

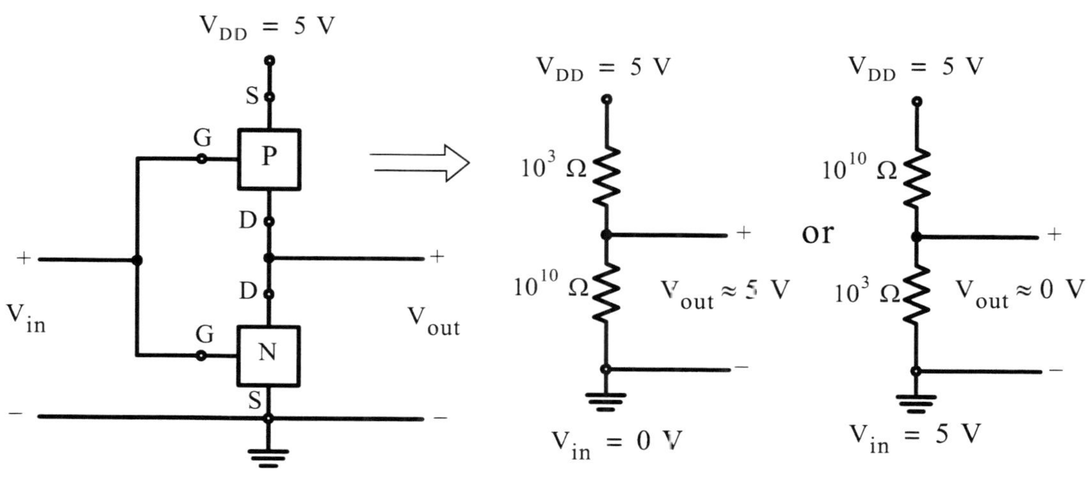

Figure 6.53. Typical CMOS inverter

In Figure 6.53 with $V_{in} = 0\ V$ (logical 0), device P is ON, device N is OFF, and the output is

$$V_{out} = \frac{10^{10}}{10^{10}+10^{3}}V_{DD} = \frac{10^{10}}{10^{10}+10^{3}} \times 5 \approx 5\ V$$

and the inversion operation has been performed. With $V_{in} = 5\ V$ (logical 1), device P is OFF, device N is ON, and the output is

$$V_{out} = \frac{R_X}{R_X+R_D}V_{DD} = \frac{10^{3}}{10^{3}+10^{10}} \times 5 \approx 0\ V$$

and again the inversion operation is performed. The truth table for positive logic is shown in Table 6.16.

TABLE 6.16 Truth table for CMOS inverter

V_{in}	V_{out}
0	1
1	0

6.13.2 The CMOS NAND Gate

A 2–input CMOS NAND gate is shown in Figure 6.54. By application of the voltage division expression as with the CMOS inverter, we find that for positive logic the truth table is as shown in Table 6.17.

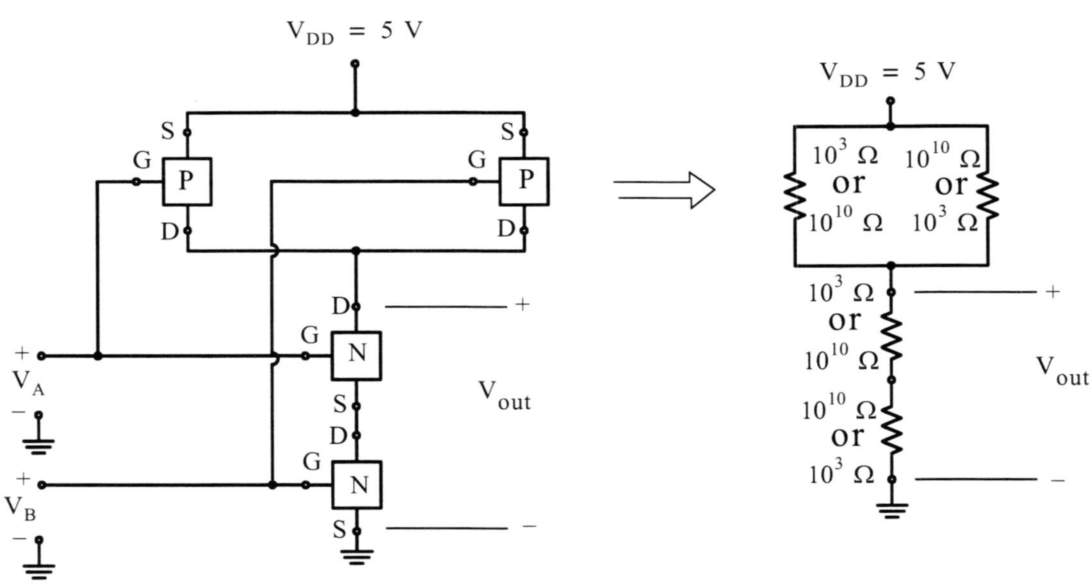

Figure 6.54. 2–input CMOS NAND gate

TABLE 6.17 Truth table for 2–input CMOS NAND gate

V_A	V_B	V_{out}
0	0	1
0	1	1
1	0	1
1	1	0

6.13.3 The CMOS NOR Gate

A 2–input CMOS NOR gate is shown in Figure 6.55. By application of the voltage division expression as with the CMOS inverter, we find that for positive logic the truth table is as shown in Table 6.18.

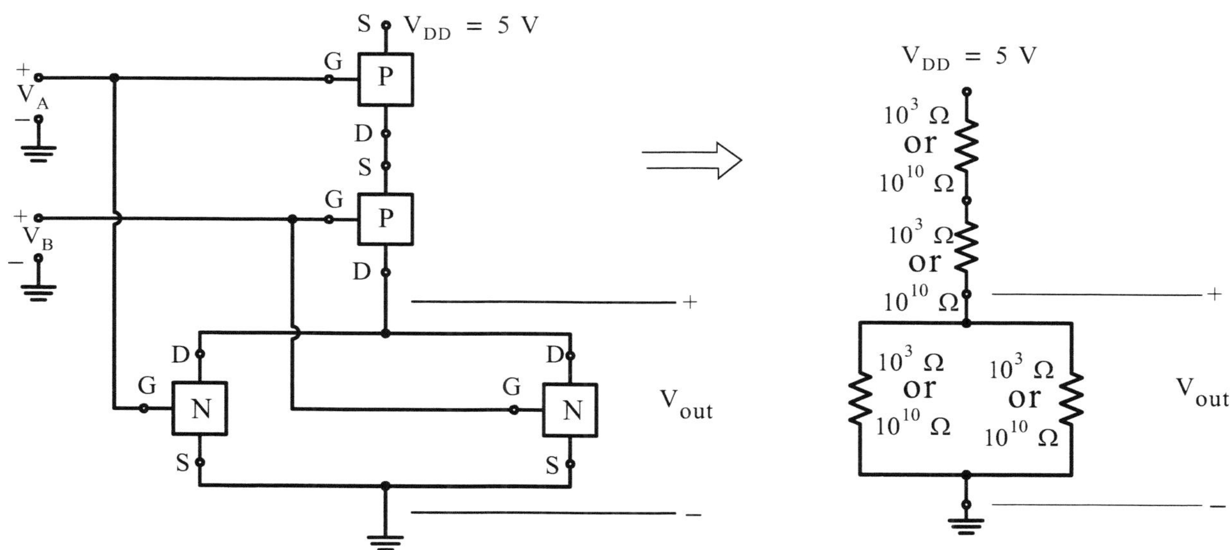

Figure 6.55. 2–input CMOS NOR gate

TABLE 6.18 Truth table for 2–input CMOS NOR gate

V_A	V_B	V_{out}
0	0	1
0	1	0
1	0	0
1	1	0

6.14 Buffers, Tri–State Devices, and Data Buses

In certain applications, the output of a logic circuit needs to be buffered. A buffer or line driver is normally used to change (amplify) the voltage level at the output of a logic circuit as shown in Figure 6.56.

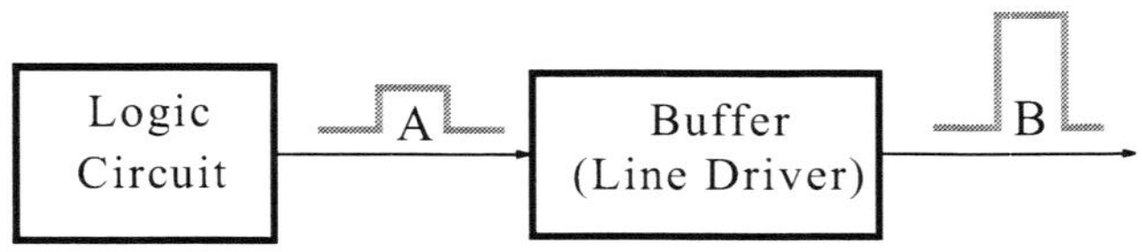

Figure 6.56. Voltage amplification with the use of a buffer

The symbol of a buffer gate is shown in Figure 6.57 and its truth table is shown in Table 6.19.

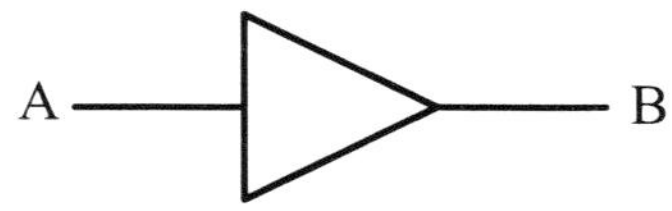

Figure 6.57. Symbol for a buffer gate

TABLE 6.19 Truth table for buffer amplifier with positive logic

A	B
0	0
1	1

The TTL ICs SN7407 and SN7417 are a hex buffers meaning that each contains 6 buffers featuring high–voltage, open–collector outputs for interfacing with high–level circuits (such as CMOS) or for driving high–current loads (such as lamps or relays), and also are characterized for use as buffers for driving TTL inputs. Figure 6.58 shows the circuits for these ICs.

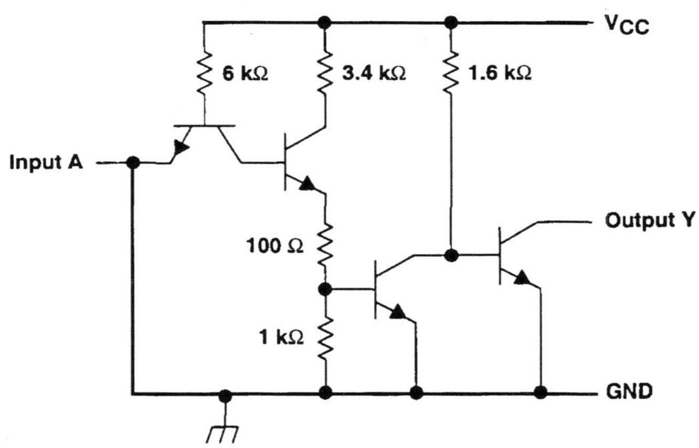

Figure 6.58. Circuit for the IC SN7407/SN7417 buffer (Courtesy of Texas Instruments)

The circuits for the SN7407/SN7417 are completely compatible with most TTL families. The SN7407 has minimum breakdown voltages of 30 V, and the SN7417 has minimum breakdown voltages of 15 V. The maximum sink current is 40 mA for both the SN7407 and SN7417. Inputs are diode clamped to minimize transmission–line effects, which simplifies design. Typical power dissipation is 145 mW, and average propagation delay time is 14 ns.

Other examples of buffered gates are the TTL SN7406/SN7416 Hex Inverter buffers with open–collector high–voltage outputs, the SN7428 quad 2–input positive–nor buffers, and the SN7440 dual–four input positive–NAND buffers.

Some other ICs are *tri–stated buffers* or gates. The output of these devices can assume three states, logic 0, logic 1, and High–Z (high impedance), that is, an open circuit. Figure 6.59 shows the symbols for ICs SN74125 and SN74126 quad bus buffers with 3–state outputs.

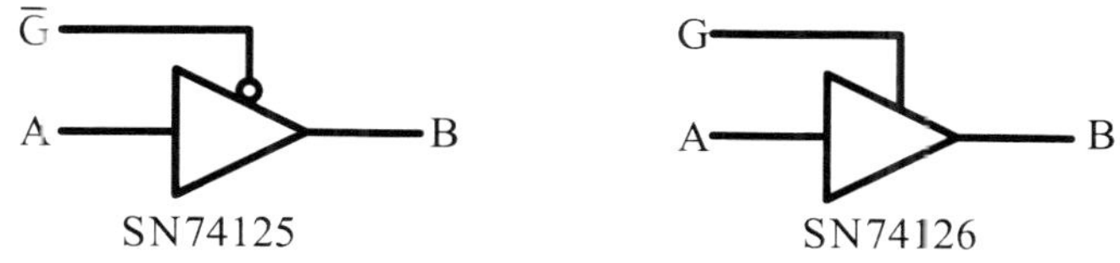

Figure 6.59. Symbols for the TTL SN74125 and SN74126 devices

The output of the SN74125 device is disabled when $\overline{G}$ is High, and the output of the SN74126 is disabled when G is Low. The truth table for the SN74125 device is shown in Table 6.20.

TABLE 6.20 *Truth table for the SN74125 3–state buffer*

$\overline{G}$	A	B
0	0	0
0	1	1
1	0	High Z
1	1	High Z

Figure 6.60 shows the symbol for the TTL SN74366 hex tri–state inverting buffer, and Table 6.21 is the truth table for that device.

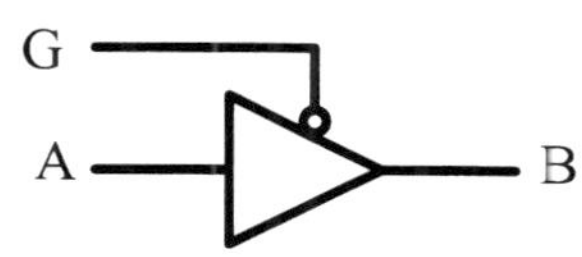

Figure 6.60. Symbol for the SN74366 hex tri–state inverting buffer

TABLE 6.21 *Truth table for the TTL SN74125 3–state buffer*

G	A	B
0	0	1
0	1	0
1	0	High Z
1	1	High Z

Another tri–state device is the CMOS *transmission gate* shown in Figure 6.61 where the control input serves as an open or closed switch. Thus, when Control = 0 the switch is open, both the NMOS and PMOS devices are OFF, and V_{out} = High Z. When Control = 1 the switch is closed, both the NMOS and PMOS devices are ON, and $V_{out} = V_{in}$.

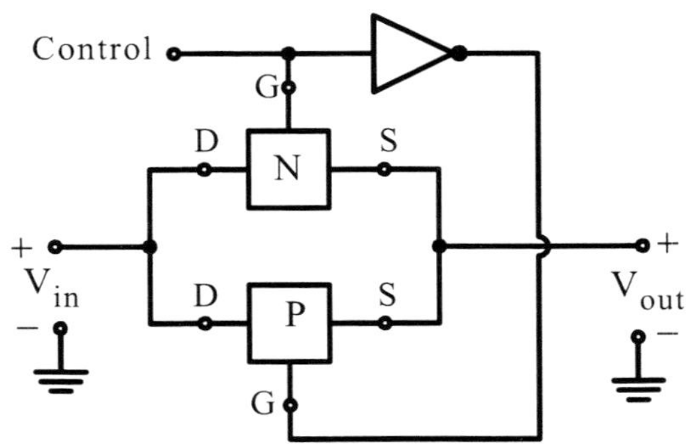

Figure 6.61. The CMOS transmission gate

As shown in Figure 6.61, the CMOS transmission gate circuit consists of one NMOS and one PMOS connected in parallel and controlled by inverted gate voltages. With this arrangement, both NMOS and PMOS can be switched ON of OFF at the same time rather than alternately. If they are ON, the resistance between V_{IN} and V_{OUT} is very low and the signal is transmitted without degradation. If they are OFF, there is no path from the input to the output. Also, the CMOS transmission gate can be used for both analog and digital signals and the only requirement is that the signal does not exceed the power supply voltages. Figure 6.62 shows the symbol for the CMOS transmission gate.

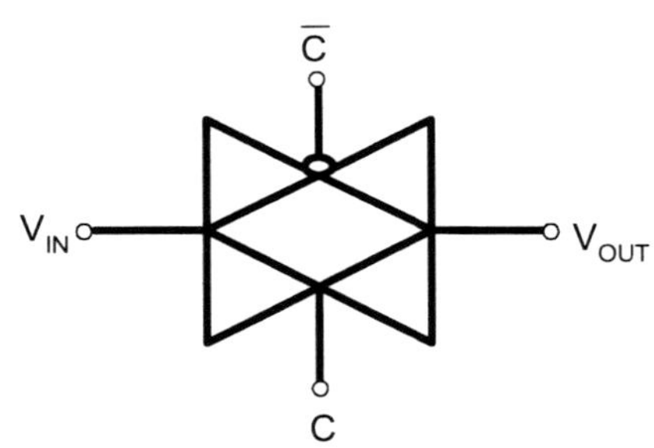

Figure 6.62. Symbol for the CMOS transmission gate

A *data bus* is a group transmission paths such as groups of wires used as a common path to interconnect several devices. A data bus, or simply a bus can be unidirectional, that is, one in which data can flow in one direction only, or *bidirectional* where data can flow in either direction where each direction can be controlled by tri–state devices as shown in Figure 6.63 where the control line $C_{A \to B}$ directs the data flow from left to right and $C_{B \to A}$ from right to left.

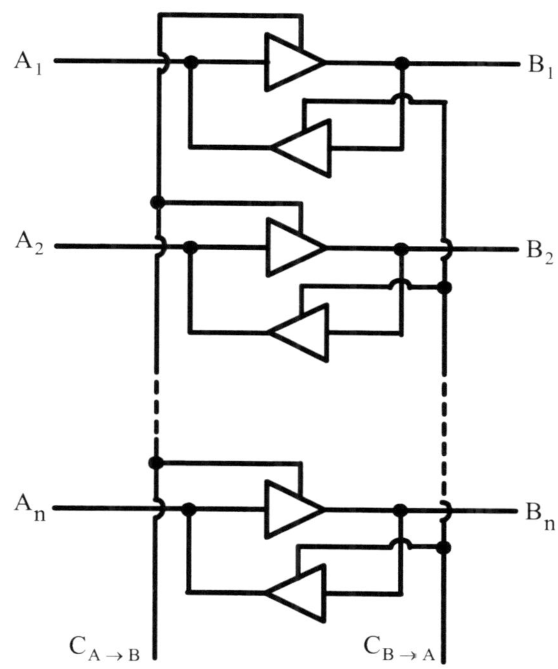

Figure 6.63. Typical bidirectional bus

6.15 Present and Future Technologies

The IC devices that we described in previous sections of this chapter have been around for over forty years. Since speed and low–power consumption are the most desirable characteristics, just about all of these devices have been replaced with newer devices even though the ICs that we described are still in use by experimenters and laboratory work in colleges and universities.

In Chapter 3, we saw how the addition of a Schottky diode can be used to prevent transistor saturation as shown in Figure 6.64.

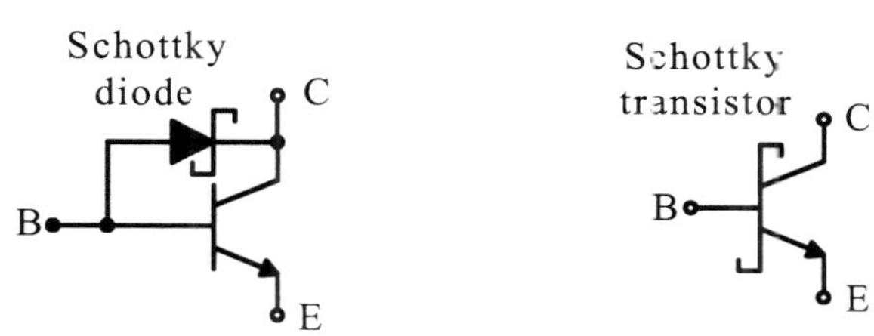

Figure 6.64. Bipolar NPN transistor with Schottky diode and its representation as Schottky transistor

To achieve low–power consumption, the resistances are considerably higher in the upgraded versions of the TTL family and the ICs are identified with the LS designation which stands for Low–power Schottky device. Thus, the Low–power Schottky 2–input NAND gate is identified as SN74LS00 whose internal construction is shown in Figure 6.65.

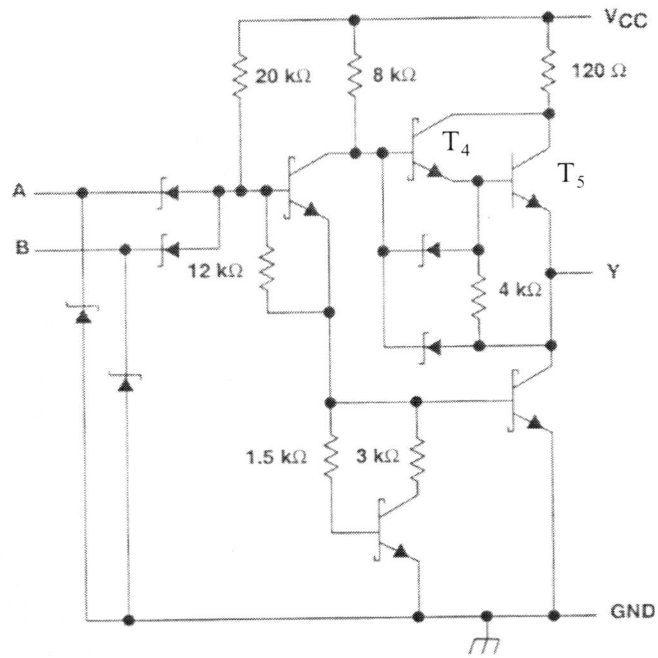

Figure 6.65. Internal construction of the SN74LS00 2–input NAND gate (Courtesy Texas Instruments)

A comparison of the SN7400 2–input NAND gate shown in Figure 6.19 with SN74LS00 2–input NAND gate shown in Figure 6.65, reveals that in the latter all resistances are higher, the conventional diodes have been replaced by Schottky diodes, and all transistors except T_5, have been replaced by Schottky transistors. This is because transistor T_5 never goes into saturation as it is explained below.

Let us consider the terminal voltages of the Schottky transistor shown in Figure 6.66.

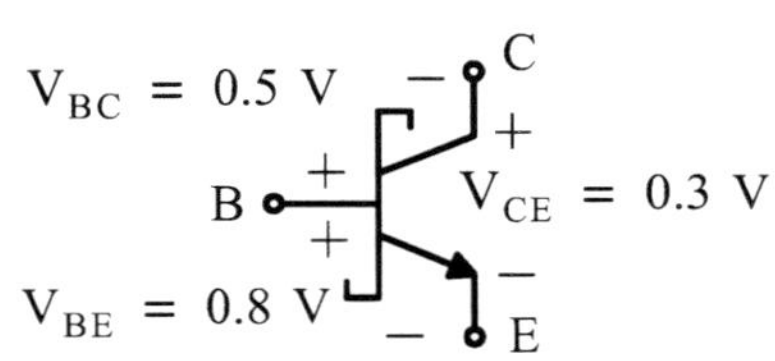

Figure 6.66. Schottky transistor terminal voltages

We observe that these values satisfy KVL since

$$V_{BC} + V_{CE} - V_{BE} = 0 \tag{6.2}$$

From Figure 6.65, by application of (6.2) and noting that $V_{CB} = -V_{BC}$, we find that

$$V_{CET5} = V_{BCT5} + V_{BET5} = V_{CET4} + V_{BET5} = 0.3 + 0.8 = 1.1\ V$$

and therefore we conclude that transistor T_5 never enters the saturation region.

A more recent advancement in TTL technology is the *Advanced Low–power Schottky* denoted as SN74ALSXX. Thus, a 2–input NAND gate in this technology is indicated as SN74ALS00. The propagation delay time is about $4\ ns$ compared to about $11\ ns$ for the standard TTL, and the power consumption is about $1\ mW$ compared to about $10\ mW$ for the standard TTL.

Another recent technology is the *BiCMOS*. It combines bipolar and CMOS to give the best balance between available output current and power consumption. As we know, bipolar transistors offer high speed, high gain, and low output resistance, whereas CMOS technology offers high input resistance, and this combination results in low–power logic gates. Basically, a BiCMOS is a two–stage amplifier that uses a CMOS device in the first stage and a bipolar transistor in the second stage. This device can be used with both analog and digital signals.

We will not discuss the BiCMOS technology in detail. We will only show the BiCMOS inverter and the BiCMOS 2–input NAND gate. The BiCMOS inverter circuit is shown in Figure 6.67. This inverter functions as follows:

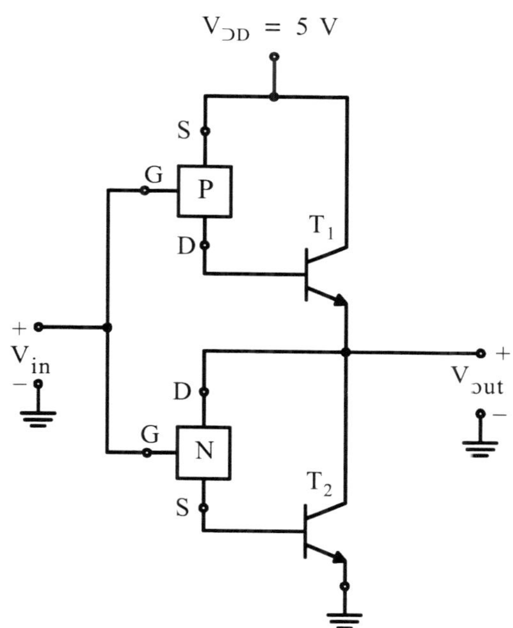

Figure 6.67. Basic BiCMOS inverter circuit.

When $v_{in} = 0\ V$, the NMOS device and transistor T_2 are OFF while the PMOS device and transistor T_1 are ON. Then, $v_{out} = V_{DD} - V_{CET1} \approx 5 - 0.2 = 4.8\ V$. When $v_{in} = 5\ V$, the PMOS device and transistor T_1 are OFF while the NMOS device and transistor T_2 are ON. Then, $v_{out} = V_{CET2} \approx 0.2\ V$.

The BiCMOS 2–input NAND gate has the same arrangement as the 2–input CMOS NAND gate with the addition of two NPN transistors at the outputs of the NMOS and PMOS devices as shown in Figure 6.68.

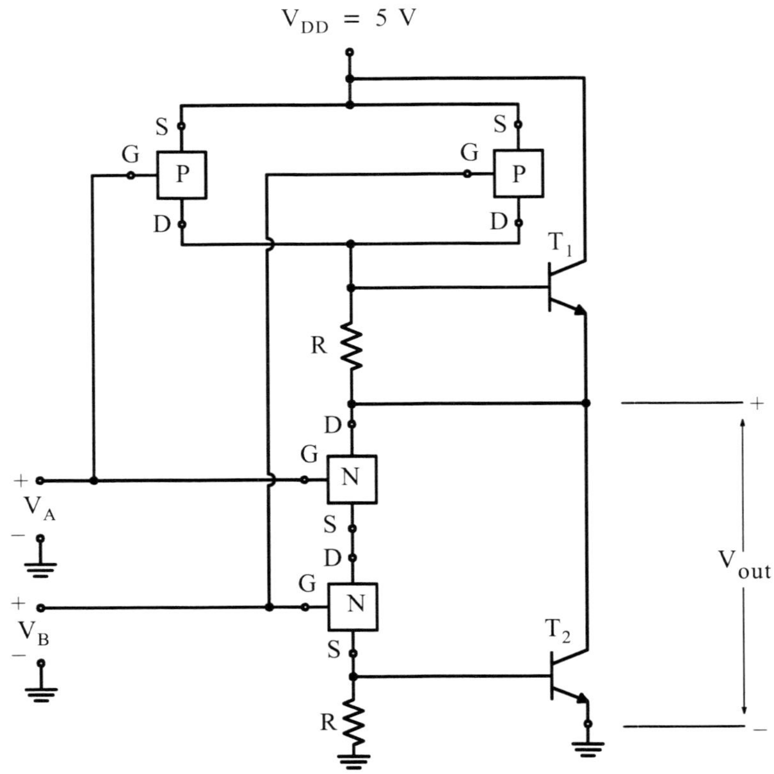

Figure 6.68. The basic BiCMOS NAND gate

As we mentioned in Chapter 3, *gallium arsenide* (GaAs) is the fastest technology with propagation delay times about 20 ps and even though prices for these devices have fallen considerably during the last few years, the cost is still very high in comparison with other IC families.

Undoubtedly, faster and with lower power consumption devices will emerge in the future. The interested reader is urged to check occasionally the Internet for new products developed by Texas Instruments, Motorola, National Semiconductor, Fairchild, and other IC manufacturers for new products.

6.16 Summary

- Electronic logic gates are used extensively in digital systems and are manufactured as integrated circuits (IC's). The basic logic gates are the inverter or NOT gate, the AND gate, and the OR gate, and these perform the complementation, ANDing, and ORing operations respectively.
- Four other logic gates, known as NAND, NOR, Exclusive OR (XOR), and Exclusive NOR (XNOR), are derivatives of the basic AND and OR gates. However, AND and OR gates consist of NAND and NOR gates respectively, followed by an Inverter.
- Generally, an uncomplemented variable represents a logical 1, also referred to as the *true* condition, and when that variable is complemented, it represents a logical 0, also referred to as the *false* condition. Thus, if $A = 1$ (true), it follows that $\overline{A} = 0$ (false).
- Integrated circuit manufacturers assign the letter H (High) to the 5 volt level and the letter L (Low) to the ground level. With the H and L assignments, the logic circuitry designer has the option of assigning a logical 1 to H and logical 0 to L, or logical 0 to H and logical 1 to L. The former convention is known as *positive logic*, and the latter as *negative logic*.
- The Inverter performs the complementation operation. Thus if the input is A the output will be $\overline{A}$ and vice versa. The TTL SN7404 Hex Inverter is a popular transistor–transistor logic (TTL) IC device. All SN74 series devices are TTL devices. The newer SN74ALS04 device is faster and consumes less power. ALS stands for Advanced Low power Schottky transistor.
- The output of an AND gate is logical 1 (true) only when all inputs are logical 1. The earlier TTL SN7408 Quad 2–input AND gate has been superseded with the SN74ALS08 device.
- The output of an OR gate is logical 1 (true) whenever one or more of its inputs are logical 1. The earlier SN7432 Quad 2–input OR gate has been superseded with the SN74ALS32 device.
- The output of an NAND gate is logical 0 (false) only when all inputs are logical 1. The earlier SN7400 Quad 2–input NAND gate has been superseded with the SN74ALS00 device.
- The *input clamp voltage* denoted as V_{IK} is the maximum negative voltage that may be applied at the input terminals of a typical TTL gate without damaging the IC. To insure that this value is not exceeded, diodes are included at the inputs of gates.
- The output of an NOR gate is logical 1 (true) only when all inputs are logical 0 (false). The SN7402 Quad 2–input NOR gate has been superseded with the SN74ALS02 device.
- The exclusive–OR (XOR) logic gate has two inputs and one output. The output of an exclusive–OR gate is logical 1 (true) when only one of the inputs, but not both, is logical 1. The SN7486 XOR gate has been superseded with the SN74ALS86 device.
- The exclusive–NOR (XNOR) logic gate has two inputs and one output. The output of a XNOR gate is logical 1 (true) only when the inputs are the same, that is, both logical 0 or

both logical 1. For this reason, the XNOR gate is also known as *equivalence gate*. There is no IC XNOR gate in the TTL family but one can be formed with an XOR gate (SN7486) followed by an inverter (SN7404). We can also implement the XNOR function using the SN7486 IC with negative logic.

- The *fan–in* of a gate is the number of its inputs. Thus, a 3–input NAND gate has a fan–in of 3.
- *Fan–out* is a term that defines the maximum number of digital inputs that the output of a single logic gate can feed. Generally, TTL gates can feed up to 10 other digital gates or devices. Thus, a typical TTL gate has a fan–out of 10.
- A buffer can be used when it becomes necessary for a single TTL logic gate to drive more than 10 other gates or devices. A typical buffer has a fan–out of 25 to 30. An inverter (NOT gate) can serve this function in most digital circuits if complementation is also required.
- For TTL logic gates, one unit load is defined as $40\ \mu A$ when the output is in High state, and $1.6\ mA$ when the output is in Low state.
- Passive pull–up refers to the condition where a resistor pulls–up the output voltage towards V_{CC}.
- Active pull–up refers to the condition where a transistor rather than a resistor is used to pull–up the output voltage towards V_{CC}.
- *Sourcing* and *sinking* currents refer to the current flow in TTL circuits. A driver gate is said to be sourcing current when it's output is High. A driver gate is sinking current when it's output is Low.
- Totem–pole configuration refers to the arrangement where at the output stage of a TTL gate one transistor is stacked on top of another and their operation is complementary, that is, when one transistor is ON the other transistor is OFF.
- Some TTL logic gates are designed with the collector of transistor at the output stage unconnected and are referred to as open collector TTL devices. The advantage of the open–collector configuration over the totem–pole configuration is that the outputs of two or more open–collector TTL gates can be tied together to realize the AND function. This arrangement is referred to as wired–AND operation. The passive pull–up arrangement is often used with *open collector* TTL devices.
- Totem pole TTL gates should not be tied together for wired AND operation.
- The data sheets provided by IC device manufacturers contain very important parameters that designers must take into consideration.
- Besides the TTL, the emitter–coupled logic (ECL) is another logic family employing bipolar transistors, and with the exception of gallium arsenide technology, is the fastest logic family.

- In contrast to TTL where the transistors are either cut off or saturated, depending on the state of the circuit, in ECL the transistors always operate in the active region so they can change state very rapidly. The basic ECL circuit is essentially a differential amplifier.
- The ECL gate provides two outputs, one which implements the OR function, and the other implements the NOR function.
- Two types of ECL gates are the ECL 10K and ECL 100K, the latter having a propagation delay time of $0.75\ ns$ and power consumption of $40\ mW$. It should also be noted that ECL is not a new technology; Motorola introduced the MECL series in the 1980's.
- The earlier NMOS and PMOS logic gates have now been superseded by CMOS logic gates. CMOS logic uses both NMOS and PMOS devices to form logic functions. CMOS technology is the dominant semiconductor technology for the manufacturing of microprocessors, memories and application specific integrated circuits. The main advantage of CMOS over NMOS and bipolar technology is the much smaller power dissipation.
- A buffer or line driver is normally used to change (amplify) the voltage level at the output of a logic circuit.
- Some other ICs are *tri–stated buffers* or gates. The output of these devices can assume three states, logic 0, logic 1, and High–Z (high impedance).
- The CMOS *transmission gate* is another tri–state device where the control input serves as an open or closed switch.
- A data bus is a group transmission paths such as groups of wires used as a common path to interconnect several devices. A data bus, or simply a bus can be unidirectional, that is, one in which data can flow in one direction only, or bidirectional where data can flow in either direction where each direction can be controlled by tri–state devices.
- A more recent advancement in TTL technology is the *Advanced Low–power Schottky* denoted as ALS. Thus, a 2–input NAND gate in this technology is indicated as SN74ALS00. The propagation delay time is about $4\ ns$ compared to about $11\ ns$ for the standard TTL, and the power consumption is about $1\ mW$ compared to about $10\ mW$ for the standard TTL.
- The BiCMOS technology combines bipolar and CMOS to give the best balance between available output current and power consumption. Bipolar transistors offer high speed, high gain, and low output resistance, whereas CMOS technology offers high input resistance, and this combination results in low–power logic gates. Basically, a BiCMOS is a two–stage amplifier that uses a CMOS device in the first stage and a bipolar transistor in the second stage. This device can be used with both analog and digital signals.
- *Gallium arsenide* (GaAs) still remains the fastest technology with propagation delay times about $20\ ps$ and even though prices for these devices have fallen considerably during the last few years, the cost is still very high in comparison with other IC families.

6.17 Exercises

1. The circuit below is known as Resistor Logic (RL) gate.

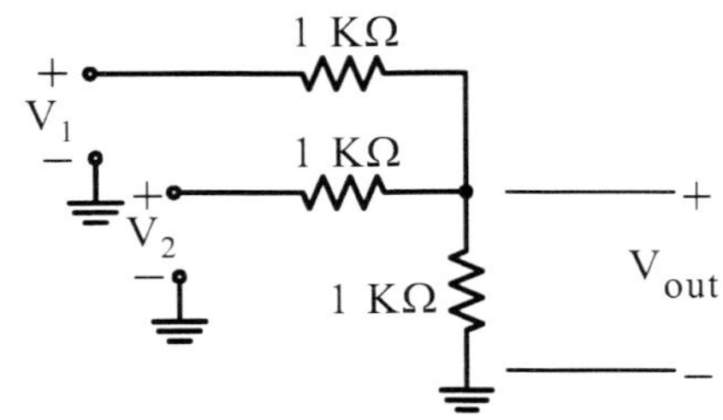

a. Write the truth table for the input combinations of 0 V and 5 V with respect to the ground.

b. State the conditions under which this circuit can be classified as an AND gate.

c. State the conditions under which this circuit can be classified as an OR gate.

d. State some of the deficiencies of this circuit when used as a logic gate.

2. The circuit below is known as Diode Logic (DL) gate.

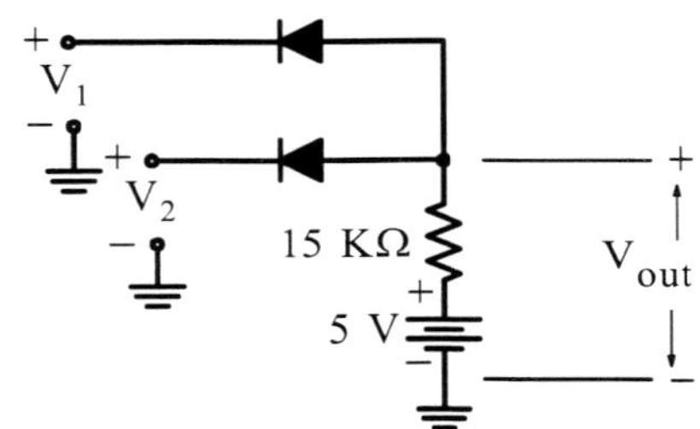

Write the truth table for the input combinations of 0 V and 3 V with respect to the ground. Which type of logic gate does this circuit represent?

3. The circuit below is another form of a Diode Logic (DL) gate.

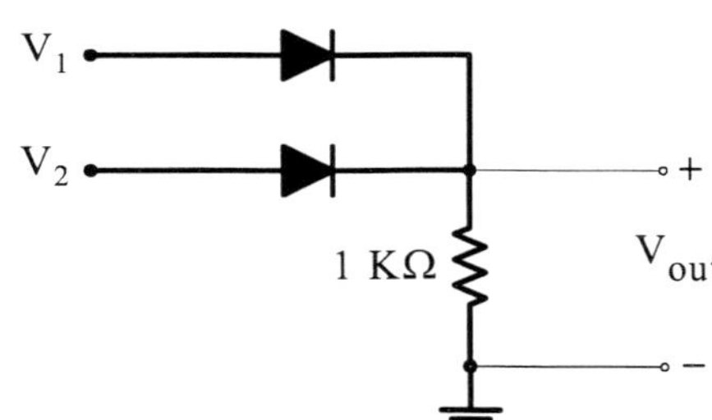

Write the truth table for the input combinations of 0 V and 5 V with respect to the ground. Which type of logic gate does this circuit represent?

4. The circuit below is a 3–input Resistor–Transistor Logic (RTL) gate.

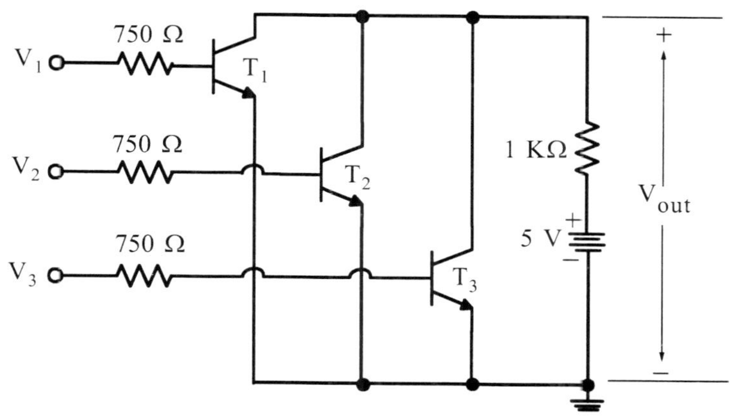

Write the truth table for the input combinations of 0 V and 5 V with respect to the ground. Which type of logic gate does this circuit represent?

5. For the circuit below find V_{out1} and V_{out2} when each of V_1 through V_4 assumes the values of 0 V and 5 V with respect to the ground.

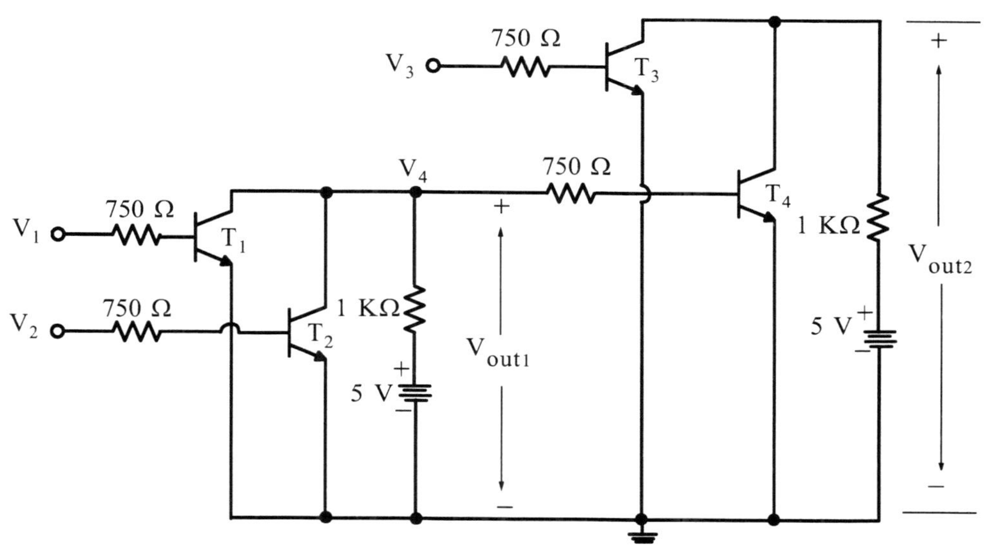

6. For the circuit of Exercise 5, derive an expression for V_{out1} High level when the driving gate has a fan–out of 2. Hint: Start with an equivalent circuit.

7. The circuit below is a 3–input Diode–Transistor Logic (DTL) gate.

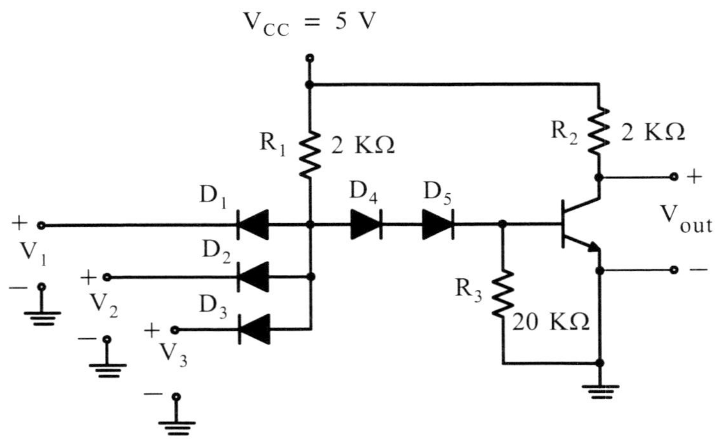

Write the truth table for the input combinations of 0 V and 5 V with respect to the ground. Which type of logic gate does this circuit represent?

8. It is known that the inverter circuit shown below has the value $h_{FE} = 30$ minimum. Calculate the levels of the output voltage v_{out} when the levels of the input voltage v_{in} are 0 and 12 volts. You may first assume that $v_{out}(sat) = 0$ V to obtain approximate values and then assume that $v_{out}(sat) = 0.2$ V for exact values.

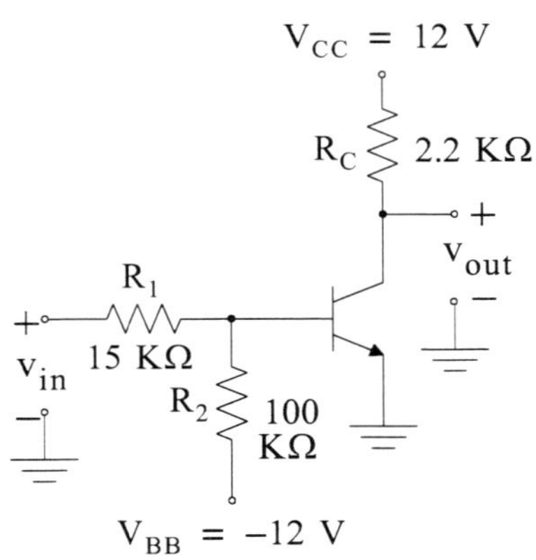

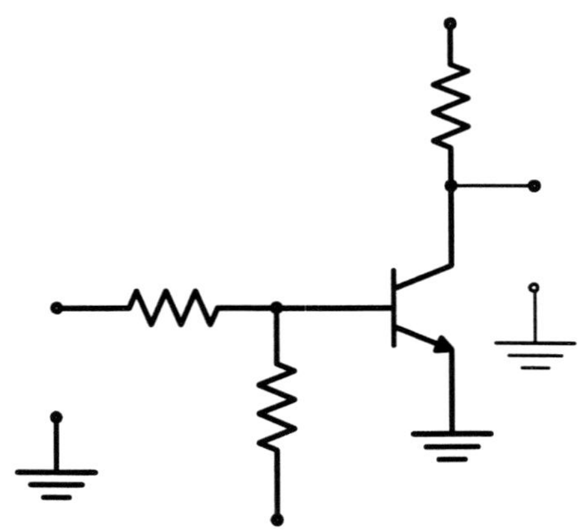

6.18 Solutions to End–of–Chapter Exercises

1.

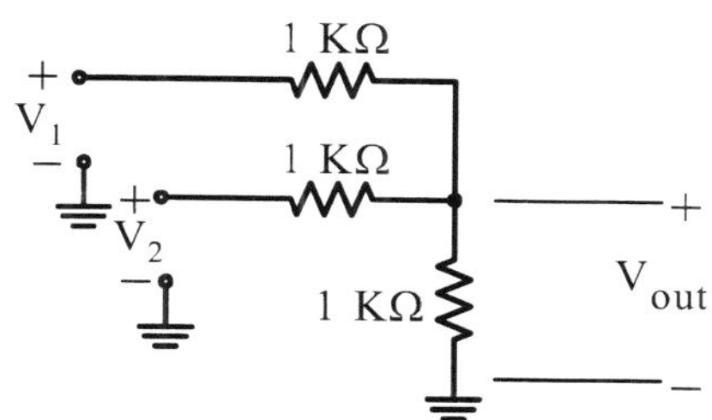

a. With $V_1 = V_2 = 0\ V$, $V_{out} = 0\ V$. With $V_1 = 0\ V$, and $V_2 = 5\ V$, R_1 is in parallel with R_3 and by the voltage division expression

$$V_{out}\Big|_{V_1 = 0\ V} = \frac{R_1 \| R_3}{R_2 + R_1 \| R_3} \cdot V_2 = \frac{0.5\ K\Omega}{1\ K\Omega + 0.5\ K\Omega} \cdot 5\ V = 1.67\ V$$

With $V_1 = 5\ V$, and $V_2 = 0\ V$, R_2 is in parallel with R_3 and by the voltage division expression

$$V_{out}\Big|_{V_2 = 0\ V} = \frac{R_2 \| R_3}{R_1 + R_2 \| R_3} \cdot V_1 = \frac{0.5\ K\Omega}{1\ K\Omega + 0.5\ K\Omega}\ 5\ V = 1.67\ V$$

With $V_1 = V_2 = 5\ V$, we apply the principle of superposition and we find that

$$V_{out}\Big|_{V_1 = V_2 = 5\ V} = V_{out}\Big|_{V_1 = 0\ V} + V_{out}\Big|_{V_2 = 0\ V} = 1.67 + 1.67 = 3.34\ V$$

With these values we construct the truth table below.

V_1	V_2	V_{out}
0 V	0 V	0 V
0 V	5 V	1.67 V
5 V	0 V	1.67 V
5 V	5 V	3.34 V

b. This circuit can be classified as an AND gate if $V_{out} \ge 3.34\ V$

c. This circuit can be classified as an OR gate if $V_{out} \ge 1.67\ V$

d. This circuit is not practical for use as a logic gate because if more than two inputs exist, more different output levels result. Moreover, there is no compatibility between the input and output levels. With the above arrangement for instance, with $V_1 = V_2 = 5\ V$ the output is $V_{out} = 3.34\ V$ and if this output becomes the input of another gate shown below as Gate 3, the outputs drop to $2.72\ V$ as shown in parentheses.

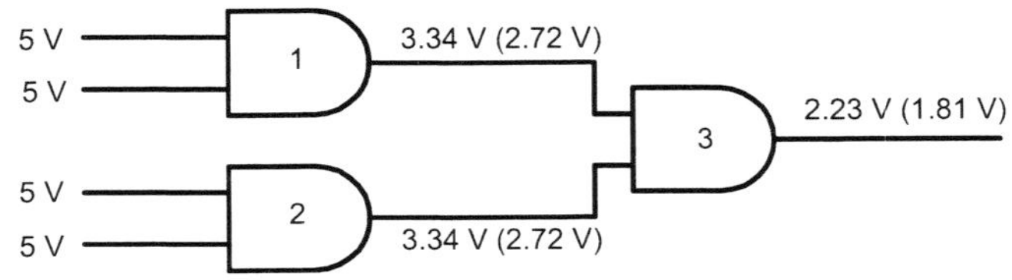

2.

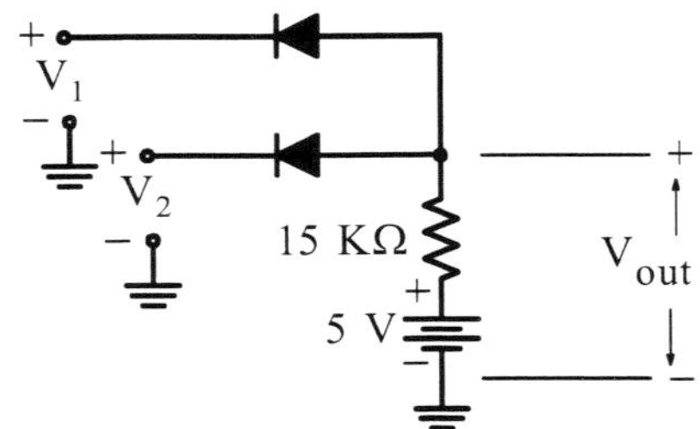

Both diodes are forward–biased for any combination of the inputs since the value of V_{CC} is greater than 3 V and can be represented as in the circuit below.

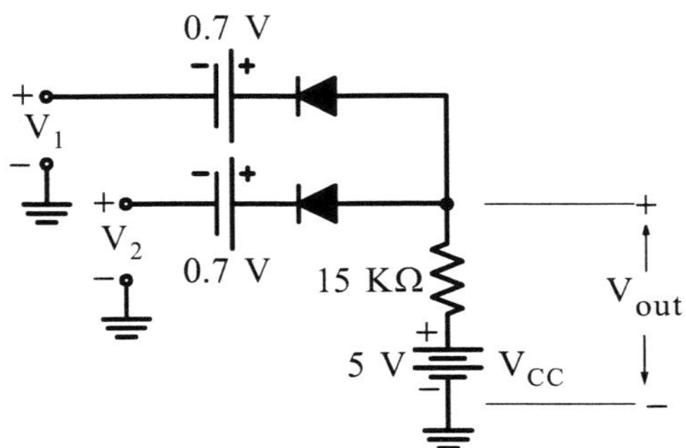

By inspection, when $V_1 = V_2 = 0\ V$, or $V_1 = 0\ V$ and $V_2 = 3\ V$, or $V_1 = 3\ V$ and $V_2 = 0\ V$, the output is $V_{out} = 0.7\ V$. When $V_1 = V_2 = 3\ V$, the output is $V_{out} = 3.7\ V$ and thus the truth table is as shown below.

V_1	V_2	V_{out}
0 V	0 V	0.7 V
0 V	3 V	0.7 V
3 V	0 V	0.7 V
3 V	3 V	3.7 V

The truth table indicates that this circuit could be used as an AND gate.

3.

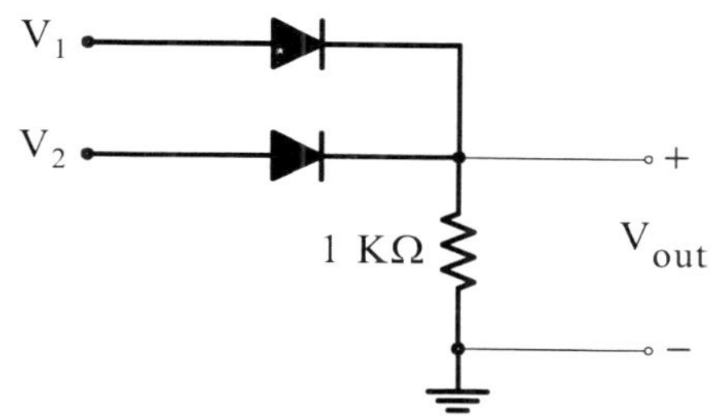

By inspection, when $V_1 = V_2 = 0\ V$, the output is $V_{out} = 0\ V$. When $V_1 = 0\ V$ and $V_2 = 5\ V$, or $V_1 = 5\ V$ and $V_2 = 0\ V$, or $V_1 = V_2 = 5\ V$, one or both diodes conduct and the output is $V_{out} = 5.0 - 0.7 = 4.3\ V$ and thus the truth table is as shown below.

V_1	V_2	V_{out}
0 V	0 V	0 V
0 V	5 V	4.3 V
5 V	0 V	4.3 V
5 V	5 V	4.3 V

The truth table indicates that this circuit could be used as an OR gate.

4.

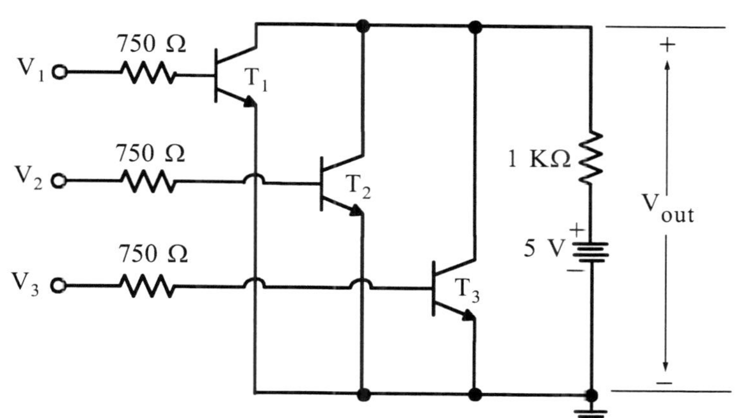

By inspection, when $V_1 = V_2 = V_3 = 0\ V$, none of the three transistors conduct and thus $V_{out} = 5\ V$. When one or more of the input voltages is $5\ V$ one or more of the transistors will saturate and the output will be $V_{out} = 0.2\ V \approx 0\ V$. The truth table is as shown below and thus this circuit behaves as a 3–input NOR gate.

Inputs			Output
V_1	V_2	V_3	V_{out}
0 V	0 V	0 V	5 V
0 V	0 V	5 V	0 V
0 V	5 V	0 V	0 V
0 V	5 V	5 V	0 V
5 V	0 V	0 V	0 V
5 V	0 V	5 V	0 V
5 V	5 V	0 V	0 V
5 V	5 V	5 V	0 V

5.

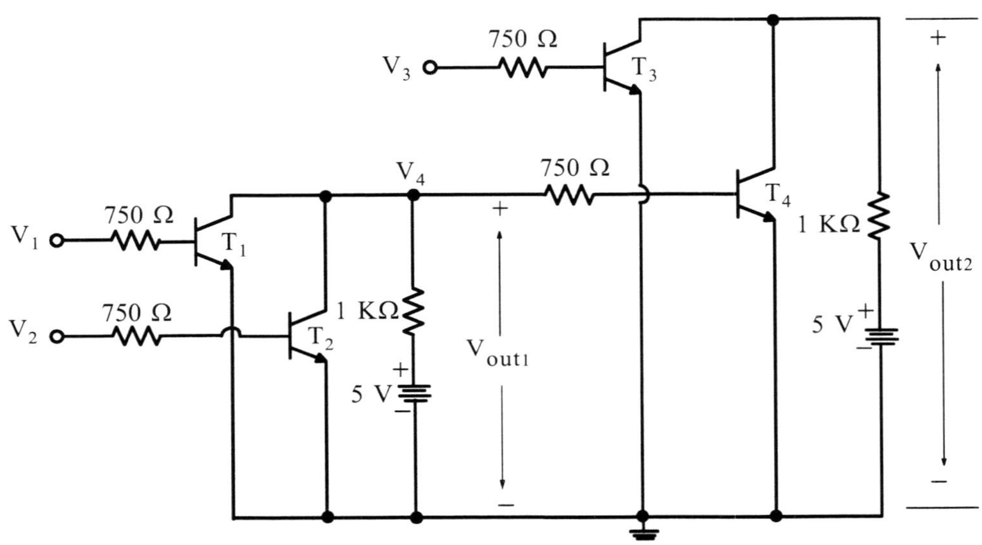

This circuit is a 2–input NOR gate with fan–out of 1. We recall that fan–out is the number of gate inputs that can be driven by a single gate output. Thus, in the circuit above the output of the first 2–input NOR gate drives (is connected to) one of the inputs of the second 2–input NOR gate. By inspection, when $V_1 = 0\ V$ and $V_2 = 5\ V$, or $V_1 = 5\ V$ and $V_2 = 0\ V$, or $V_1 = V_2 = 5\ V$, one of both transistors T_1 and T_2 conduct and thus $V_{out1} = 0.2\ V \approx 0\ V$. When $V_3 = 0\ V$ and $V_{out1} = 0\ V$, $V_{out2} = 5\ V$.

Next, let us examine the case where $V_1 = V_2 = 5\ V$. Under this condition $V_{out1} < 5\ V$ because current flows through the $750\ \Omega$ at the base of transistor T_4 which now conducts, and and $V_{BET4} = 0.7\ V$. We can now find that current from the equivalent circuit below.

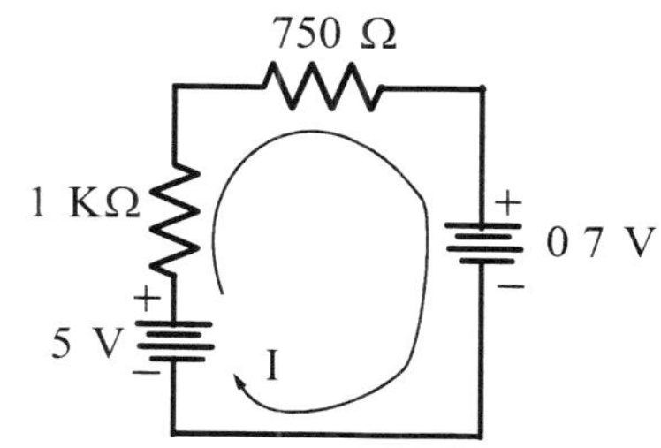

$$I = \frac{5 - 0.7}{1000 + 750} = \frac{4.3}{1750} = 2.46\ mA$$

The actual value of V_{out1} is 5 V minus the voltage drop across the 1 KΩ resistor. Thus,

$$V_{out1} = 5 - 10^3 \times 2.46 \times 10^{-3} = 2.54\ V$$

We see then that by connecting one input of a gate to the output of another gate causes the high level to decrease from its open circuit value of 5 V to the value of 2.54 V.

6.

For a fan–out of 2, the equivalent circuit is as shown in Figure (a) below and its Thevenin equivalent is as shown in Figure (b).

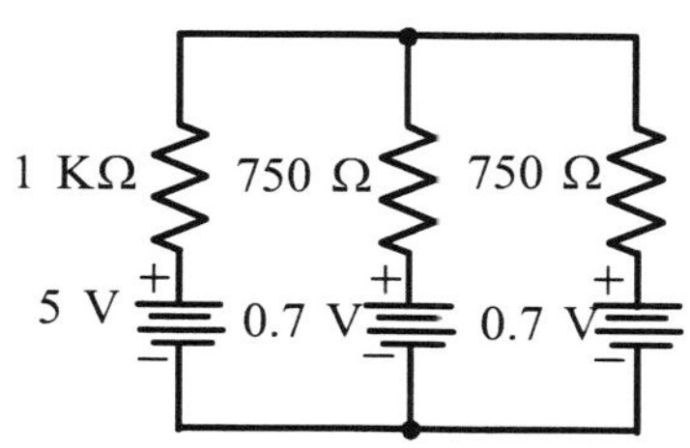

(a)

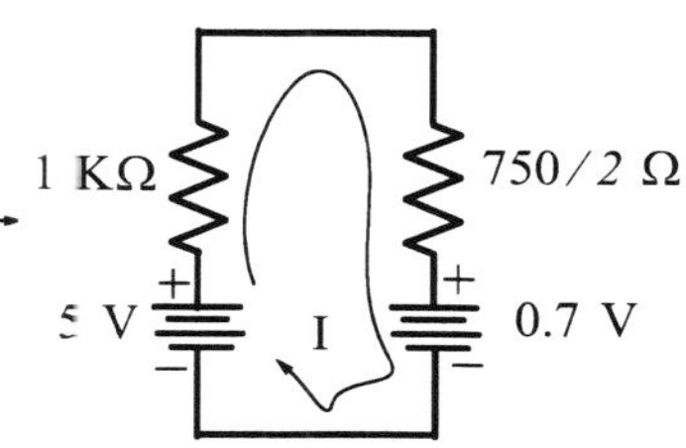

(b)

From Figure (b)

$$I = \frac{5 - 0.7}{1000 + 750/2} = \frac{4.3}{1000 + 750/2}$$

and

$$V_{out1} = 5 - \frac{1000}{1000 + 750/2} 4.3 = 1.87\ V$$

The equivalent circuit of Figure (a) can be extended to the situation where there are N gates driven by one output of a gate, that is a fan–out of N, by replacing the resistance 750/2 by 750/N. Then,

$$V_{out1} = 5 - \frac{1000}{1000 + 750/N} 4.3$$

Note: RTL is no longer used as a logic family. It is one of the earlier families employed years ago with a maximum fan–out of 5 and fan–in of 4.

7.

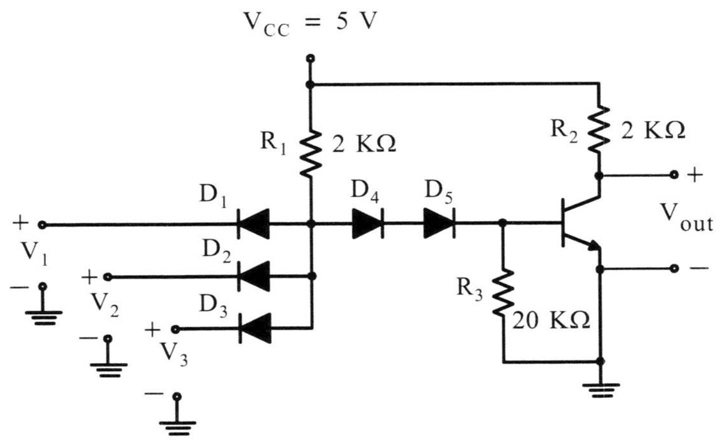

With one or more of the inputs V_1, V_2, or V_3 at $0\ V$, the given circuit becomes as shown below where the $0.2\ V$ is assumed to be the output of a previous logic gate with logical 0.

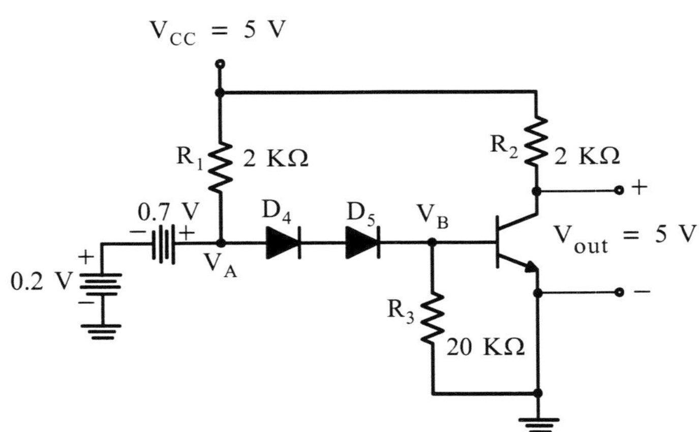

With $V_A = 0.2\ V + 0.7\ V = 0.9\ V$, the transistor is OFF. For the transistor to be ON, the voltage V_A with respect to the ground would have to be

$$V_A = V_{D4} + V_{D5} + V_{BE} = 0.7 + 0.7 + 0.7 = 2.1\ V$$

above ground. Therefore, $V_{out} = 5\ V$.

Next, we will consider the case where all three inputs are high, that is, $V_1 = V_2 = V_3 = 5\ V$. In this case, the three input diodes D_1, D_2, and D_3 are reverse–biased and thus do not conduct. However, diodes D_4 and D_5 are forward–biased and the equivalent circuit is as shown below.

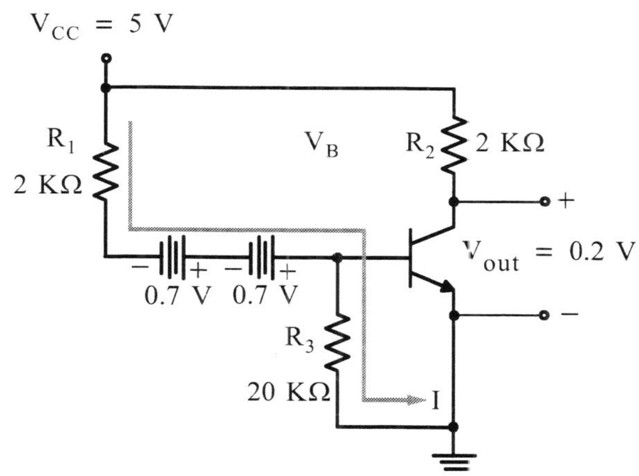

To determine the output voltage V_{out} we need to find out whether the transistor is ON or OFF. Let us assume that it is OFF. Then, by KVL

$$R_1 I - 0.7\ V - 0.7\ V + R_3 I = V_{CC} = 5\ V$$

$$V_B = V_{BE} = R_3 I = 20\ K\Omega \times \frac{5 - (0.7 + 0.7)}{2\ K\Omega + 20\ K\Omega} = 3.27\ V$$

This voltage is more than sufficient to turn the transistor on and thus $V_{out} = 0.2\ V$.

The two diodes D_4 and D_5 in series are both necessary for the following reason:

If only one diode were used, when at least one of the input voltages is Low, the voltages V_A and V_B would be as shown below.

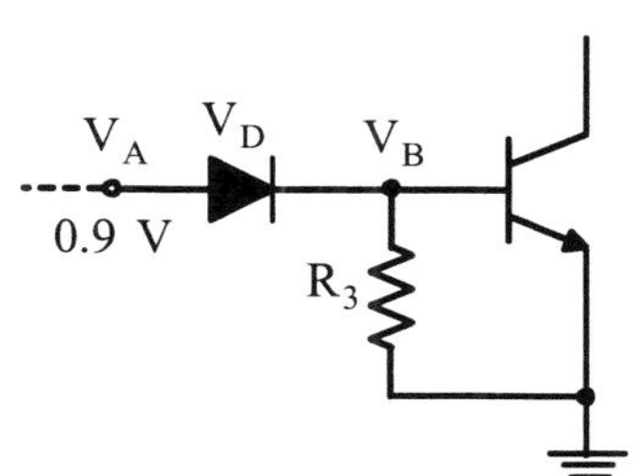

From the circuit above, $V_{BE} = V_B = V_B - V_B = 0.9 - 0.7 = 0.2\ V$ and thus the transistor will be OFF. But this would not be a good design since any small noise voltage such as $0.5\ V$ when added to $0.2\ V$ would result to $0.2 + 0.5 = 0.7\ V$ and this would turn the transistor ON.

The truth table for this 3–input DTL gate is shown below and we see that it behaves like a 3–input NAND gate.

Inputs			Output
V_1	V_2	V_3	V_{out}
0 V	0 V	0 V	5 V
0 V	0 V	5 V	5 V
0 V	5 V	0 V	5 V
0 V	5 V	5 V	5 V
5 V	0 V	0 V	5 V
5 V	0 V	5 V	5 V
5 V	5 V	0 V	5 V
5 V	5 V	5 V	0 V

Note: Like the RTL, DTL is no longer used as a logic family. It is one of the earlier families employed years ago with power consumption of about $10\ mW$ and propagation delay time of approximately $30\ ns$.

8.

With $v_{in} = 0\ V$ the circuit is as shown below.

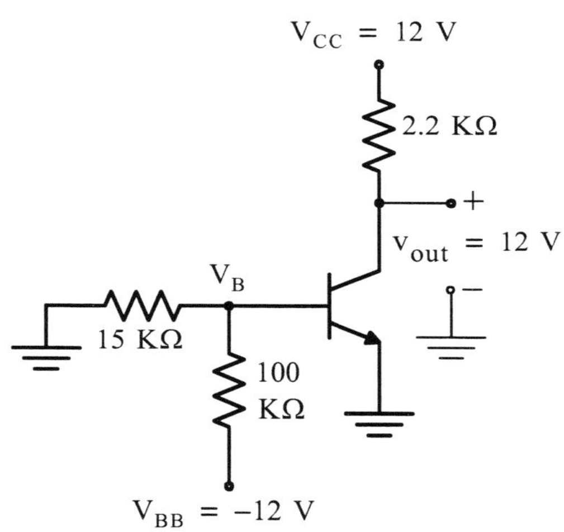

By the voltage division expression

$$V_B = V_{BE} = \frac{15\ K\Omega}{100\ K\Omega + 15\ K\Omega} \cdot (-12\ V) = -1.57\ V$$

and this voltage will certainly keep the transistor at cutoff. Therefore, for $v_{in} = 0\ V$, $v_{out} = 12\ V$ and the inversion operation has been performed.

With $v_{in} = 12\ V$ the circuit is as shown below. First let us assume that $v_{out}(sat) = 0\ V$ to obtain approximate values and verify the assumption that the transistor is indeed in saturation.

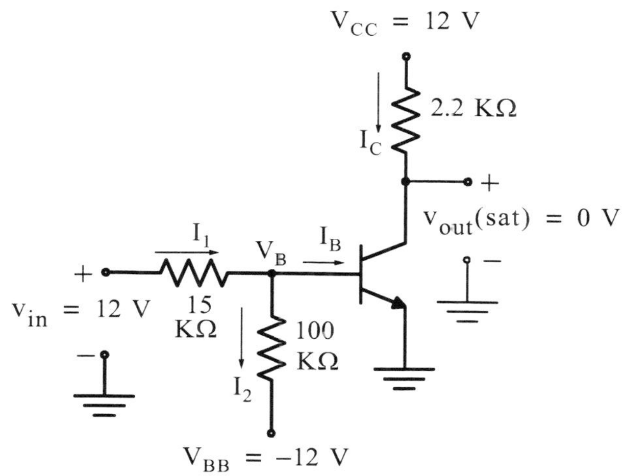

Let the minimum base current for saturation be denoted as $I_B(min)$. Then, $I_B(min) = I_C/h_{FE}$ where

$$I_C = \frac{12\ V}{2.2\ K\Omega} = 5.45\ mA$$

and thus

$$I_B(min) = \frac{I_C}{h_{FE}} = \frac{5.45\ mA}{30} = 0.182\ mA$$

The actual base current can be found from the circuit above. We find that

$$I_1 = \frac{12\ V}{15\ K\Omega} = 0.80\ mA$$

$$I_2 = \frac{12\ V}{100\ K\Omega} = 0.12\ mA$$

and

$$I_B = I_1 - I_2 = 0.80 - 0.12 = 0.68\ mA$$

Since this value is considerably larger than $I_B(min)$, we accept the fact that the transistor is in saturation.

Next, we assume that $v_{out}(sat) = 0.2\ V$ to obtain more exact values. The circuit then is as shown below, and since the transistor is in saturation, there must be

$$v_{BE}(sat) = 0.7\ V$$

and with these values,

$$I_C = \frac{12\ V - 0.2}{2.2\ K\Omega} = 5.1\ mA$$

$$I_B(min) = \frac{I_C}{h_{FE}} = \frac{5.1\ mA}{30} = 0.17\ mA$$

$$I_1 = \frac{12\ V - 0.7\ V}{15\ K\Omega} = 0.75\ mA$$

$$I_2 = \frac{0.7\ V - (-12\ V)}{100\ K\Omega} = 0.13\ mA$$

and

$$I_B = I_1 - I_2 = 0.75 - 0.13 = 0.62\ mA$$

We observe that the more exact computations do not differ significantly from those obtained by assuming that the transistor in saturation is essentially a short circuit. Therefore, when $v_{in} = 12\ V$, $v_{out}(sat) = 0.2\ V$ and again the inversion operation is performed.

Chapter 7

Pulse Circuits and Waveform Generators

This chapter is an introduction to several pulsed circuits and waveform generators also known as relaxation oscillators. These circuits are part of a family of circuits that include differentiators, integrators, clipping and clamping circuits, pulse–timing and delay circuits, logic circuits, and switching circuits. In this chapter we discuss the three types of multivibrators, the 555 Timer, and the Schmitt trigger. Sinusoidal oscillators are discussed in Chapter 10.

7.1 Astable (Free–Running) Multivibrators

An *astable* or *free–running multivibrator*, shown in Figure 7.1, is essentially a digital clock which has two momentarily stable states which alternate continuously thus producing square pulses as shown in Figure 7.2. We have assumed that all devices in the circuit of Figure 7.1 are identical.

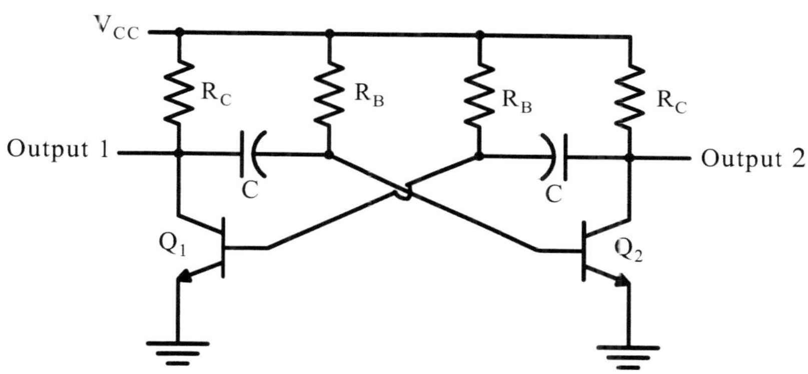

Figure 7.1. Astable multivibrator

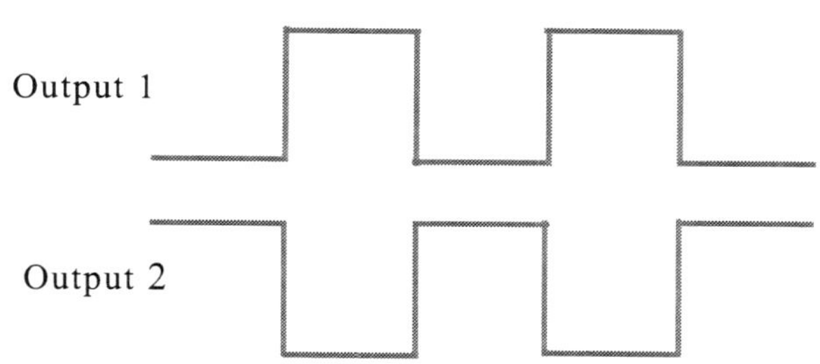

Figure 7.2. Output waveforms of the astable multivibrator

Let us now consider Output 1 and assume that the time ON T_1 and time OFF T_2 are equal and the period is T as shown in Figure 7.3.

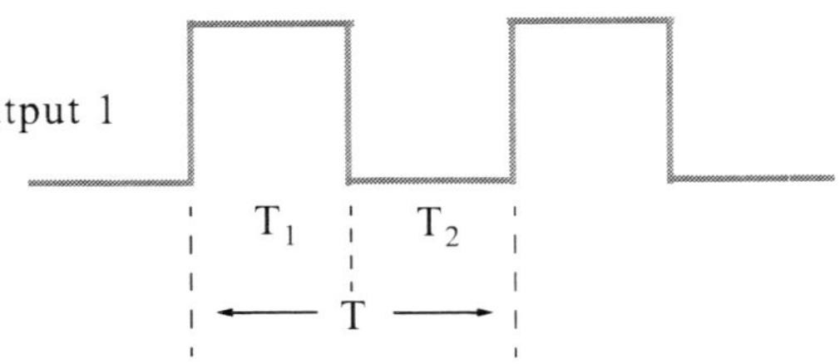

Figure 7.3. Waveform showing period T and times ON and OFF for the multivibrator circuit of Figure 7.1

If in Figure 7.3 $T_1 = T_2$, the pulse repetition frequency f is given by

$$f = \frac{1}{T} = \frac{1}{T_1 + T_2} = \frac{1}{\ln 2 \times R_B \times C} = \frac{1}{0.69 \times R_B \times C}$$

where R_B and C are as shown in Figure 7.1.

7.2 The 555 Timer

The 555 *Timer* circuit is a widely used IC for generating waveforms. A simplified diagram is shown in Figure 7.4.

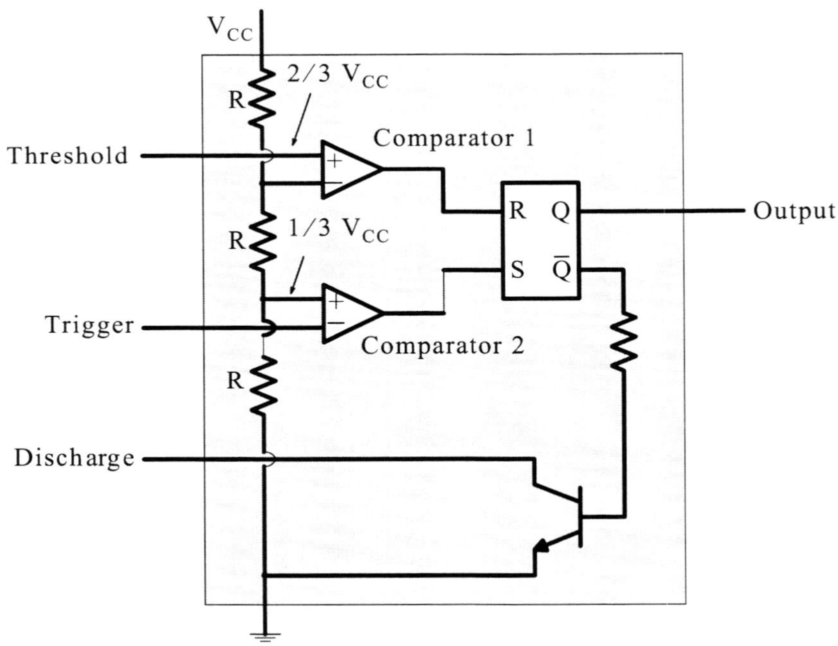

Figure 7.4. Simplified circuit for the 555 Timer

We observe that this IC includes a resistive voltage divider consisting of three identical resistors and this divider sets the voltage at the plus (+) input of the lower comparator at $1/3\ V_{CC}$ and at plus (+) input of the upper comparator at $2/3\ V_{CC}$. The outputs of the comparators determine the state of the SR flip–flop whose output is either Q or $\overline{Q}$. Thus, if Q is High (Set state), $\overline{Q}$ is Low, and if Q is Low, $\overline{Q}$ will be High (Reset state) or vice versa.

The SR flip–flop is Set when a High level is applied to the S input, and it is Reset when a High level is applied to the R input. Accordingly, the flip–flop is Set or Reset depending on the outputs of the two comparators, and these outputs are determined by the inputs Threshold at the plus (+) input of Comparator 1, and Trigger at the minus (–) input of Comparator 2. The output of the 555 Timer is the Q output of the SR flip–flop and when it is Low, $\overline{Q}$ will be High, and if the input Discharge is High, the transistor will be saturated and it will provide a path to the ground.

7.3 Astable Multivibrator with the 555 Timer

A useful application of the 555 Timer is as an astable multivibrator. From our previous discussion, we recall that the astable multivibrator is a pulse generator producing waveforms such as that shown in Figure 7.5.

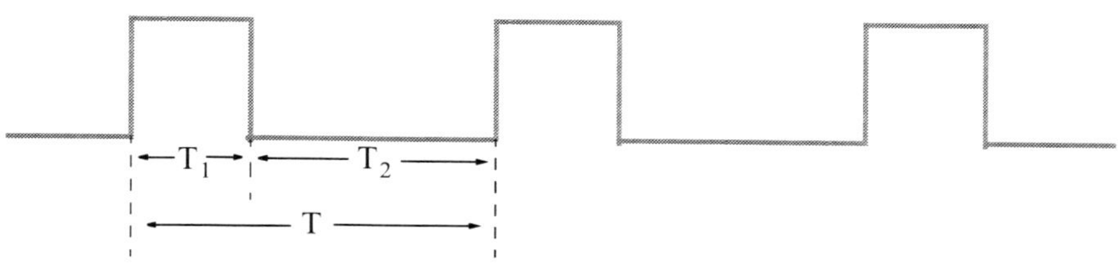

Figure 7.5. Typical waveform for an astable multivibrator

For the waveform of Figure 7.5,

$$\text{Period} = T = T_1 + T_2$$

$$\text{Pulse Repetition Frequency} = f = \frac{1}{T}$$

$$\text{Duty Cycle} = \frac{T_1}{T}$$

Figure 7.6 shows an astable multivibrator employing a 555 Timer. We will see that with this circuit the duty cycle will always be greater than 0.5 as shown in Figure 7.7.

For the circuit of Figure 7.6 let us first assume that the capacitor is uncharged and the SR flip–flop is Set. In this case the output is High and the transistor is does not conduct. The capacitor then will charge towards V_{CC} through the resistors R_A and R_B. When the capacitor reaches the value $v_C = 1/3\ V_{CC}$ the output of Comparator 2 goes Low and the SR flip–flop remains Set.

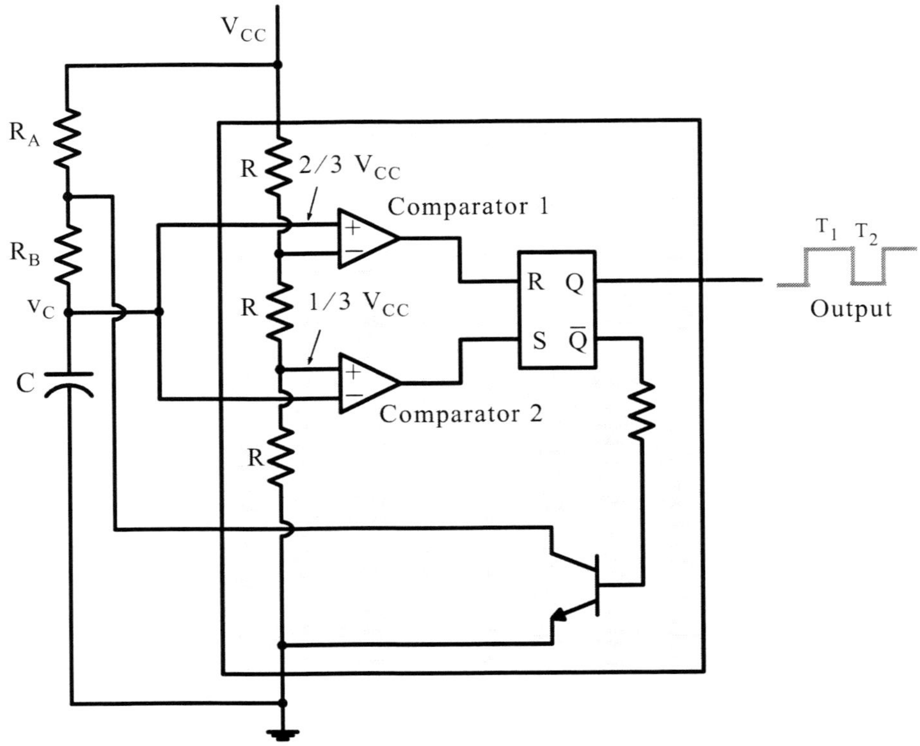

Figure 7.6. Implementation of an astable multivibrator with a 555 Timer

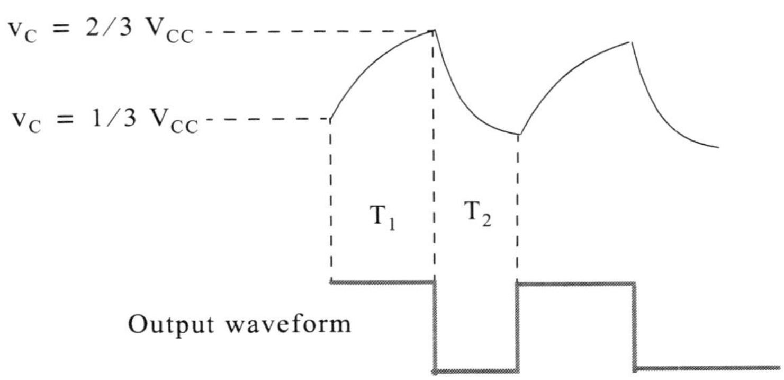

Figure 7.7. Input and output waveforms for the astable multivibrator of Figure 7.6

When the capacitor reaches the value $v_C = 2/3\ V_{CC}$, the output of Comparator 1 goes High and resets the SR flip–flop and thus the output Q goes Low, $\overline{Q}$ goes High, the transistor becomes saturated, its collector voltage becomes almost zero, and since it appears at the common node of resistors R_A and R_B, the capacitor begins to discharge through resistor R_B and the collector of the transistor. The capacitor voltage v_C decreases exponentially with time constant $t_d = R_BC$

and when it reaches the value $v_C = 1/3\ V_{CC}$, the output of Comparator 2 goes High and Sets the SR flip–flop. The output Q goes High, $\overline{Q}$ goes Low, the transistor is turned OFF, the capacitor begins to charge through the series combination of resistors R_A and R_B, and its voltage rises exponentially with time constant $t_r = (R_A + R_B)C$ and when it reaches the value $v_C = 2/3\ V_{CC}$ it resets the flip–flop and the cycle is repeated.

We are interested in the pulse repetition frequency $f = 1/T$ and the duty cycle T_1/T where $T = T_1 + T_2$ and the desired values are dependent on appropriate values of resistors R_A and R_B and the capacitor C. As we know, the capacitor voltage as a function of time is given by

$$v_C(t) = V_\infty - (V_\infty - V_{in})e^{-t/R_{eq}C} \tag{7.1}$$

where V_∞ is the final value and V_{in} is the initial value of the voltage across the capacitor.

For the charging interval T_1 the circuit is as shown in figure 7.8.

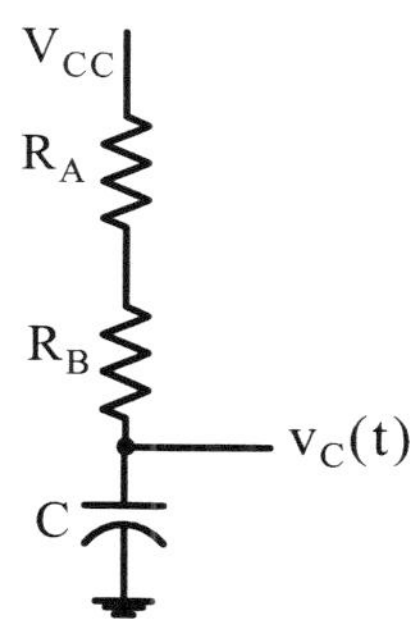

Figure 7.8. Circuit for the charging interval

For this interval, $V_\infty = V_{CC}$, $V_{in} = 1/3\ V_{CC}$, and $R_{eq} = R_A + R_B$. Then, relation (7.1) becomes

$$v_C(t) = V_{CC} - (V_{CC} - 1/3\ V_{CC})e^{-t/(R_A + R_B)C} = V_{CC} - (2/3\ V_{CC})e^{-t/(R_A + R_B)C} \tag{7.2}$$

At $t = T_1$, $v_C(t) = 2/3\ V_{CC}$ and relation (7.2)becomes

$$2/3\ V_{CC} = V_{CC} - (2/3\ V_{CC})e^{-T_1/(R_A + R_B)C}$$

$$(2/3\ )e^{-T_1/(R_A + R_B)C} = 1/3$$

$$e^{-T_1/(R_A + R_B)C} = 1/2$$

$$-T_1/(R_A + R_B)C = \ln(1/2) = \ln 1 - \ln 2 = 0 - \ln 2$$

or

$$T_1 = \ln 2 \cdot (R_A + R_B)C = 0.69(R_A + R_B)C \tag{7.3}$$

For the discharging interval T_2 the circuit is as shown in figure 7.9.

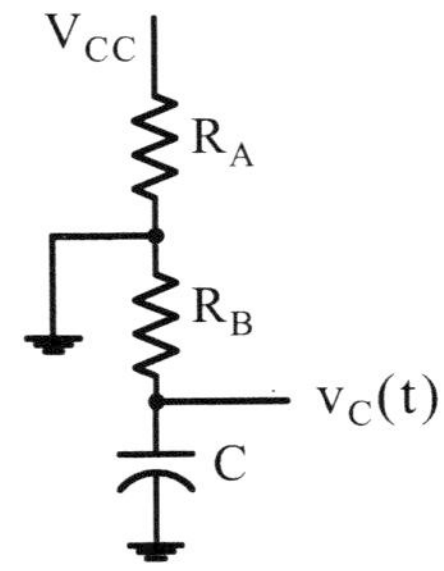

Figure 7.9. Circuit for the discharging interval

For this interval, $V_\infty = 0\ V$, $V_{in} = 2/3\ V_{CC}$, and $R_{eq} = R_B$. Then, relation (7.1) becomes

$$v_C(t) = 0 - (0 - 2/3\ V_{CC})e^{-t/R_BC} = (2/3\ V_{CC})e^{-t/R_BC} \tag{7.4}$$

At $t = T_2$, $v_C(t) = 1/3\ V_{CC}$ and relation (7.4)becomes

$$1/3\ V_{CC} = (2/3\ V_{CC})e^{-T_2/R_BC}$$

$$(2/3\ )e^{-T_2/R_BC} = 1/3$$

$$e^{-T_2/R_BC} = 1/2$$

$$-T_2/R_BC = \ln(1/2) = \ln 1 - \ln 2 = 0 - \ln 2$$

or

$$T_2 = \ln 2 \cdot (R_A + R_B)C = 0.69R_BC \tag{7.5}$$

The period T is the summation of (7.3) and (7.5). Thus,

$$T = T_1 + T_2 = 0.69(R_A + R_B)C + 0.69R_BC = 0.69(R_A + 2R_B)C \tag{7.6}$$

The right side of (7.6) cannot be negative; accordingly, the circuit of Figure 7.9 cannot achieve the condition $T_2 > T_1$. Moreover, from (7.3) and (7.5) we observe that T_1 and T_2 cannot be equal. Therefore, with the circuit of Figure 7.9 T_1 will always be greater than T_2 and the duty cycle will always be greater than 0.5 of 50 percent. But we can achieve a duty cycle close to 50 percent if we choose resistor R_A to be much smaller than resistor R_B.

Example 7.1

For the astable multivibrator of Figure 7.10 the capacitor has the value of $10\ nF$. Determine appropriate values for the resistors R_A and R_B so that the circuit will produce a pulse repetition frequency of $200\ KHz$ with a duty cycle of 60%.

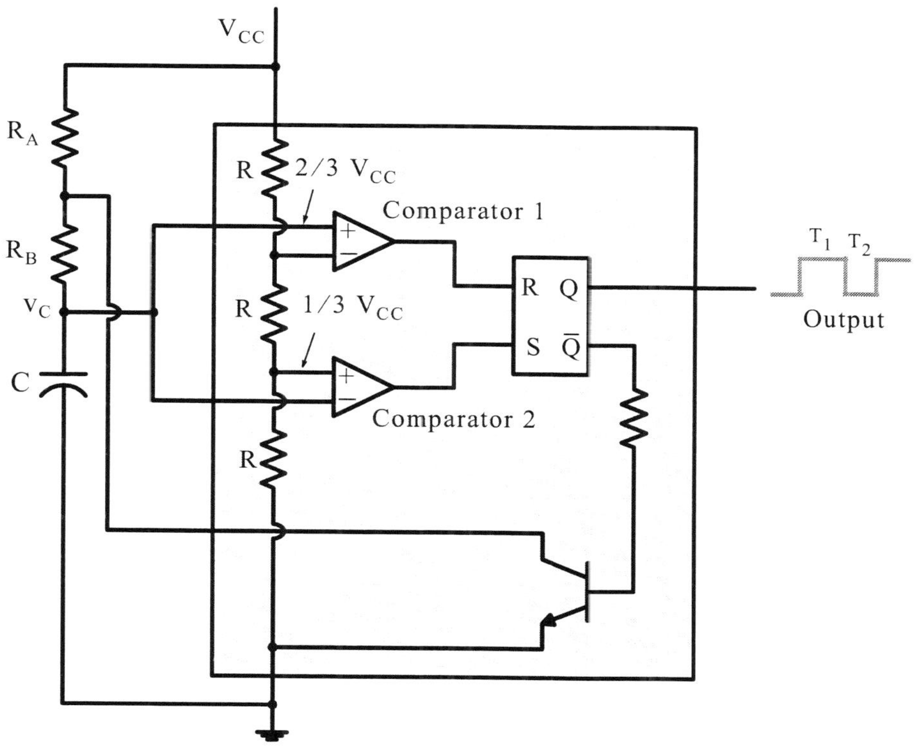

Figure 7.10. Astable multivibrator for Example 7.1

Solution:

The period $T = T_1 + T_2$ is a function of the capacitor and resistor values as shown in relation (7.6), that is,

$$T = T_1 + T_2 = 0.69(R_A + 2R_B)C$$

The pulse repetition frequency is $f = 1/T$ and thus

$$f = 10^5 = \frac{1}{0.69(R_A + 2R_B)C} = \frac{1.443 \times 10^8}{R_A + 2R_B}$$

$$R_A + 2R_B = \frac{1.443 \times 10^8}{10^5} = 14.43\ K\Omega \tag{7.7}$$

The duty cycle is T_1/T and from relations (7.3) and (7.6) we obtain

$$\text{Duty cycle} = 0.6 = \frac{T_1}{T} = \frac{0.69(R_A + R_B)C}{0.69(R_A + 2R_B)C} = \frac{R_A - R_B}{R_A + 2R_B} = \frac{R_A + R_B}{14.43\ K\Omega}$$

$$R_A + R_B = 0.6 \times 14.43 = 8.66\ K\Omega \tag{7.8}$$

Subtraction of (7.8) from (7.7) yields

$$R_B = 14.43 - 8.66 = 5.77\ K\Omega \tag{7.9}$$

and from (7.8)

$$R_A = 8.66 - 5.77 = 2.89\ K\Omega$$

Figure 7.11 shows an alternate form of an astable multivibrator using the 555 Timer.

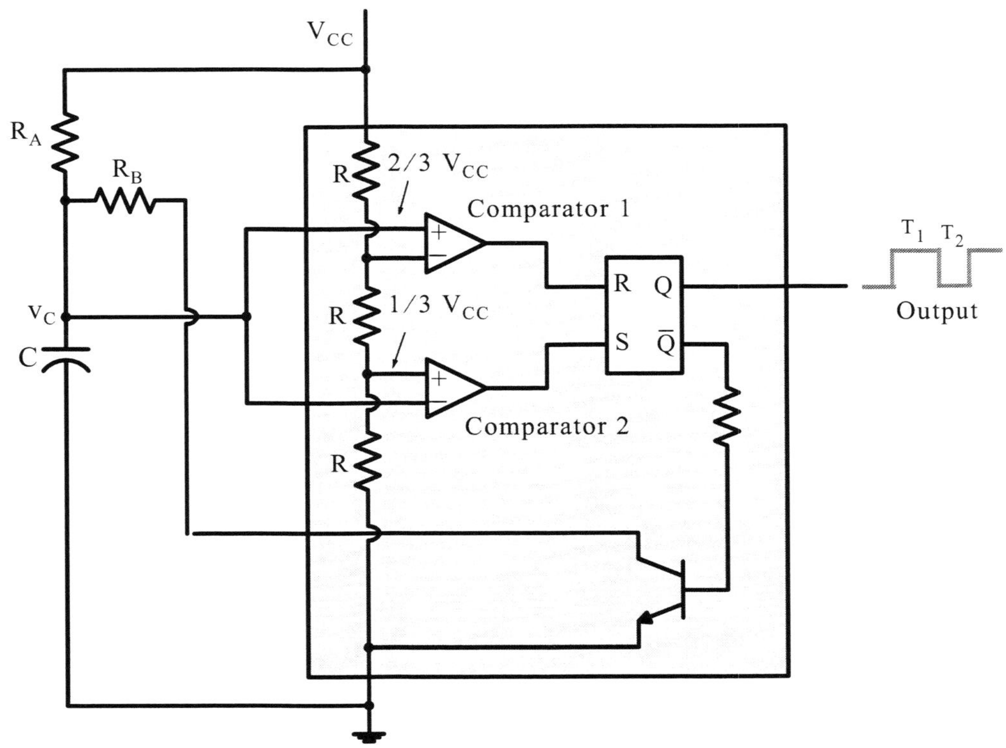

Figure 7.11. Alternate form of an astable multivibrator with a 555 Timer

With the circuit of Figure 7.11 we can obtain any desired value for the duty cycle as we will see from the relations below.

For the charging interval T_1 the circuit is as shown in figure 7.12.

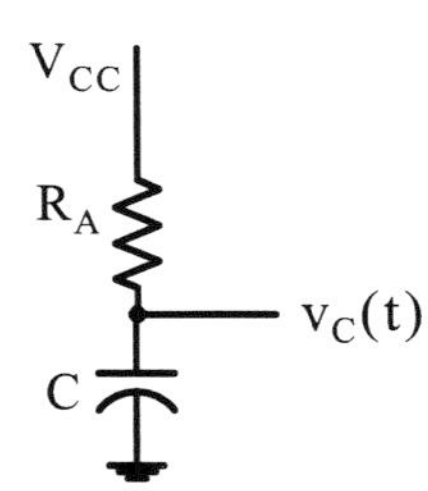

Figure 7.12. Circuit for the charging interval

For this interval, $V_\infty = V_{CC}$, $V_{in} = 1/3\ V_{CC}$, and $R_{eq} = R_A$. Then, relation (7.1) becomes

$$v_C(t) = V_{CC} - (V_{CC} - 1/3\ V_{CC})e^{-t/R_AC} = V_{CC} - (2/3\ V_{CC})e^{-t/R_AC} \tag{7.10}$$

At $t = T_1$, $v_C(t) = 2/3\ V_{CC}$ and relation (7.10) becomes

$$2/3\ V_{CC} = V_{CC} - (2/3\ V_{CC})e^{-t/R_AC}$$

$$(2/3\ )e^{-t/R_AC} = 1/3$$

$$e^{-t/R_AC} = 1/2$$

$$-T_1/R_AC = \ln(1/2) = \ln 1 - \ln 2 = 0 - \ln 2$$

$$T_1 = \ln 2 \cdot R_AC = 0.69R_AC \tag{7.11}$$

For the discharging interval T_2 the circuit is as shown in figure 7.13.

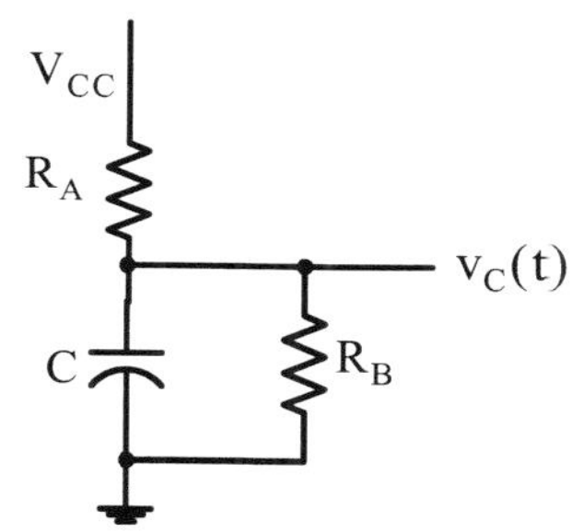

Figure 7.13. Circuit for the discharging interval

For this interval, $V_\infty = (\ R_B/(R_A+R_B)\ )V_{CC}$, $V_{in} = 2/3\ V_{CC}$, and $R_{eq} = R_A \| R_B$. Then, relation (7.1) becomes

$$v_C(t) = \left(\frac{R_B}{R_A+R_B}\right)V_{CC} - \left[\left(\frac{R_B}{R_A+R_B}\right)V_{CC} - 2/3\ V_{CC}\right]e^{-t/(R_A \| R_B)C} \tag{7.12}$$

At $t = T_2$, $v_C(t) = 1/3\ V_{CC}$ and relation (7.12)becomes

$$1/3\ V_{CC} = \left(\frac{R_B}{R_A+R_B}\right)V_{CC} - \left[\left(\frac{R_B}{R_A+R_B}\right)V_{CC} - 2/3\ V_{CC}\right]e^{-T_2/(R_A \| R_B)C}$$

$$e^{-T_2/(R_A \| R_B)C} = \frac{[1/3 - \ R_B/(R_A+R_B)\ ]V_{CC}}{[2/3 - \ R_B/(R_A+R_B)\ ]V_{CC}} = \left(\frac{R_A - 2R_B}{2R_A - R_B}\right) \tag{7.13}$$

$$e^{-T_2/R_BC} = 1/2$$

$$-T_2/(R_A \| R_B)C = \ln\left(\frac{R_A - 2R_B}{2R_A - R_B}\right) = \ln(R_A - 2R_B) - \ln(2R_A - R_B)$$

$$T_2/(R_A \| R_B)C = \ln(2R_A - R_B) - \ln(R_A - 2R_B) = \ln\left(\frac{2R_A - R_B}{R_A - 2R_B}\right)$$

or

$$T_2 = \frac{R_A \cdot R_B}{R_A + R_B} C \cdot \ln\left(\frac{2R_A - R_B}{R_A - 2R_B}\right) \tag{7.14}$$

However, there is one restriction as we can see from relation (7.13), that is, the term $R_B/(R_A + R_B)$ must be less than $1/3$. Therefore, the period T is the summation of (7.11) and (7.14) from which we obtain the relation

$$T = T_1 + T_2 = 0.69R_AC + \frac{R_A \cdot R_B}{R_A + R_B} C \cdot \ln\left(\frac{2R_A - R_B}{R_A - 2R_B}\right)$$

$$T = \left[0.69R_A + \frac{R_A \cdot R_B}{R_A + R_B}\right] \cdot C \cdot \ln\left(\frac{2R_A - R_B}{R_A - 2R_B}\right) \tag{7.15}$$

subject to the condition that $R_B/(R_A + R_B) < 1/3$, and this condition can be simplified as shown below.

$$\frac{R_B}{R_A + R_B} < 1/3$$

$$3R_B < R_A + R_B$$

$$2R_B < R_A$$

$$R_B < R_A/2 \tag{7.16}$$

Example 7.2

Using the astable multivibrator of Figure 7.11 find an appropriate ratio for resistors R_A and R_B so that the output will be a square waveform, i.e., a waveform with duty cycle at $50\ \%$.

Solution:

To achieve a duty cycle at $50\ \%$, we must make $T_1 = T_2$ and this condition will be satisfied if we equate relations (7.11) and (7.14) subject to the constraint of (7.16). Then,

$$\ln 2R_AC = \frac{R_A \cdot R_B}{R_A + R_B} C \cdot \ln\left(\frac{2R_A - R_B}{R_A - 2R_B}\right)$$

$$\frac{R_B}{R_A + R_B} \cdot \ln\left(\frac{2R_A - R_B}{R_A - 2R_B}\right) = \ln 2$$

$$\ln 2 \cdot \left(\frac{R_A + R_B}{R_B}\right) = \ln\left(\frac{2R_A - R_B}{R_A - 2R_B}\right)$$

$$\ln 2 \cdot \left(\frac{R_A}{R_B} + 1\right) = \ln\left(\frac{2R_A/R_B - 1}{R_A/R_B - 2}\right)$$

For convenience, we let $x = R_A/R_B$ and the relation above is written as

$$\ln 2 \cdot (x + 1) = \ln\left(\frac{2x - 1}{x - 2}\right) \tag{7.17}$$

This is a non–linear equation and we could try to solve it with the MATLAB symbolic expression **solve**('eqn') as

```
syms x;  solve('(x+1)*log(2)=log((2*x–1)/(x–2))')
```

but this returns the value of -1 which obviously is meaningless for this example. Therefore, let us plot relation (7.17) with the following MATLAB script to find the x–axis zero crossing.

```
x=2.1:0.01:2.5; y=(x+1)*log(2)–log((2.*x–1)./(x–2)); plot(x,y);grid
```

Figure 7.14 indicates that x–axis zero crossing occurs at approximately $x = R_A/R_B = 2.363$ and thus to achieve a duty cycle at $50\ \%$, we should choose that the ratio R_A/R_B should be close to this value.

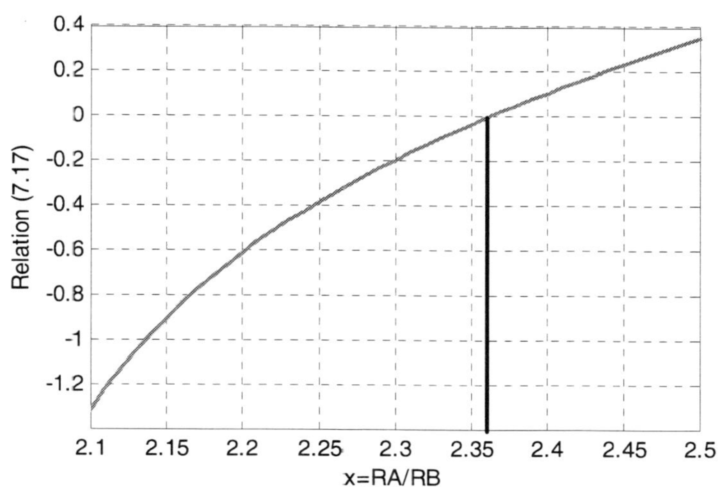

Figure 7.14. Plot for finding the ratio R_A/R_B *to achieve* $50\ \%$ *duty cycle for the circuit of Figure 7.14*

Let us choose $R_A = 5.1\ K\Omega$; then $R_B = 5.1/2.363 = 2.16\ K\Omega$. The value of the capacitor can be found from either expression (7.11) or (7.14) but (7.11) is simpler. Thus, if we require that $T_1 = 5\ \mu s$, then,

$$T_1 = 5 \times 10^{-6} = 0.69 R_A C = 0.69 \times 5.1 \times 10^3 C$$

from which

$$C = \frac{5 \times 10^{-6}}{0.69 \times 5.1 \times 10^3} = 1.42\ nF$$

An unstable multivibrator can also be constructed with an op amp as shown in Figure 7.15.

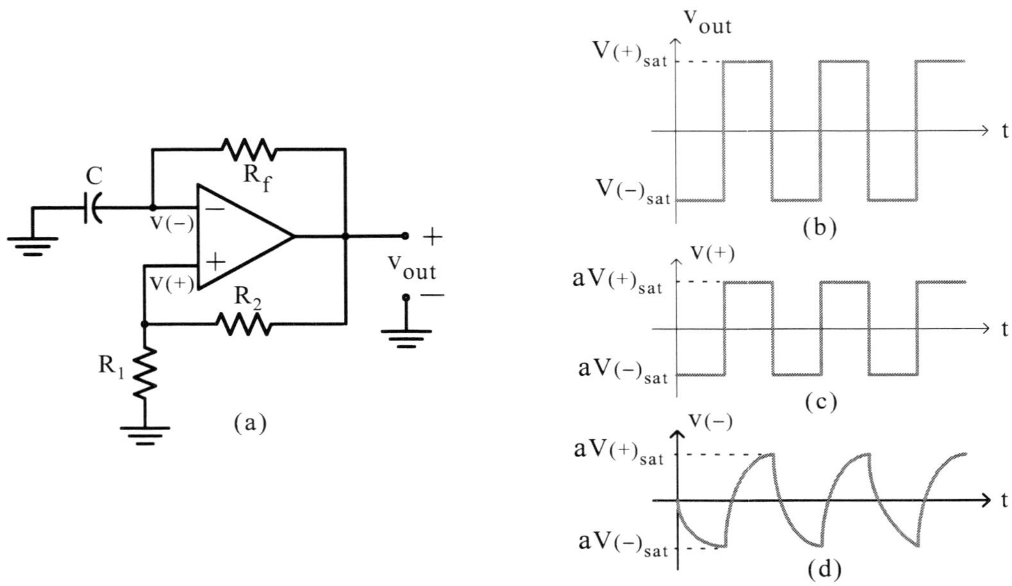

Figure 7.15. An unstable multivibrator using an op amp

From the circuit of Figure 7.15(a), application of KCL at the non–inverting input of the op amp yields

$$\frac{v(+) - V_{out}}{R_2} + \frac{v(+)}{R_1} = 0$$

$$\frac{(R_1 + R_2)v(+)}{R_1 R_2} = \frac{V_{out}}{R_2}$$

$$\frac{v(+)}{V_{out}} = \frac{R_1}{(R_1 + R_2)} = a \tag{7.18}$$

where a is the *attenuation factor* since it is less than unity. Accordingly, the positive and negative saturation values are denoted as $aV(+)_{sat}$ and $aV(-)_{sat}$ respectively, and are shown in Figures 7.15(c) and 7.15(d). Assuming that the initially the op amp is in negative saturation, the voltage at the non–inverting input is $aV(-)_{sat}$. The capacitor C and the feedback resistor R_f form a simple RC circuit and when the voltage $v(-)$ at the inverting input is at a higher than the voltage $v(+)$ at the non–inverting input, the capacitor charges through R_f towards the negative saturation value $aV(-)_{sat}$.

When the potential difference between the inverting and non–inverting inputs of the op amp approaches the zero value the op amp comes out of saturation and subsequently the positive feedback from the output to the inverting input of the op amp drives the op amp to positive saturation. The voltage across the capacitor cannot change instantaneously and rises exponentially as shown in Figure 7.15(d) and when it reaches the value $aV(+)_{sat}$ the output switches back to the negative saturation state.

For the astable multivibrator of the op amp circuit of Figure 7.15(a) we can derive the period of the pulse repetition frequency as follows:

From relation (7.1)

$$v_C(t) = V_\infty - (V_\infty - V_{in})e^{-t/R_fC}$$

Taking the natural log of both sides and solving for the time t we obtain

$$t = R_fC\ln\frac{(V_\infty - V_{in})}{(V_\infty - v_C(t))} \tag{7.19}$$

Substitution of the voltages shown in Figures 7.15(a) and 7.15(b) yields the appropriate expressions for the timing periods as indicated below.

$$t_1 = R_fC\ln\frac{V(+)_{sat} - aV(-)_{sat}}{V(+)_{sat} - aV(+)_{sat}} = R_fC\ln\frac{V(+)_{sat} - aV(-)_{sat}}{V(+)_{sat}(1 - a)} \tag{7.20}$$

$$t_2 = R_fC\ln\frac{V(-)_{sat} - aV(+)_{sat}}{V(+)_{sat} - aV(+)_{sat}} = R_fC\ln\frac{V(-)_{sat} - aV(+)_{sat}}{V(-)_{sat}(1 - a)} \tag{7.21}$$

Relations (7.20) and (7.21) indicate that if the positive and negative values of the saturation voltage have the same amplitude, then $t_1 = t_2$ and the period of the pulse repetition becomes

$$T = t_1 + t_2 = 2R_fC\ln\frac{1 + a}{1 - a} \tag{7.22}$$

and with (7.18)

$$T = t_1 + t_2 = 2R_fC\ln\left(1 + \frac{2R_1}{2R_2}\right) \tag{7.23}$$

Example 7.3

For the astable multivibrator circuit of Figure 7.16 compute the timing periods given that $C = 0.1\ \mu F$, $R_f = 50\ K\Omega$, $R_1 = 10\ K\Omega$, $R_2 = 50\ K\Omega$, $V(+)_{sat} = 10\ V$, and $V(-)_{sat} = -5\ V$.

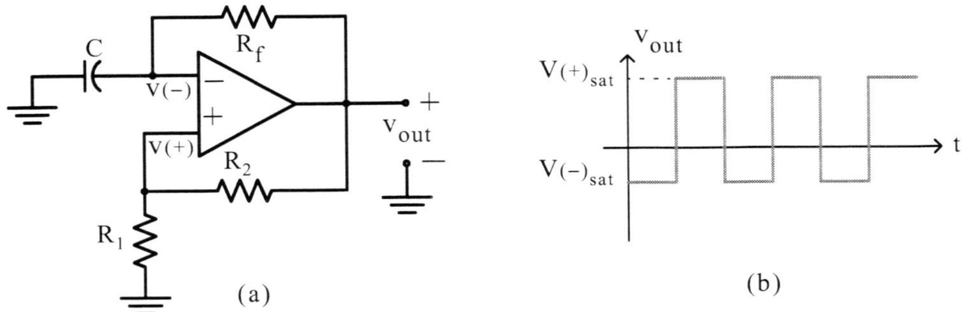

Figure 7.16. Unstable multivibrator circuit for Example 7.3

Solution:

From (7.18) the attenuation factor is

$$a = \frac{R_1}{(R_1 + R_2)} = \frac{10}{10 + 500} = \frac{1}{6}$$

and with (7.20) and (7.21)

$$t_1 = R_f C \ln\frac{V(+)_{sat} - aV(-)_{sat}}{V(+)_{sat}(1-a)} = 50 \times 10^4 \times 0.1 \times 10^{-6} \ln\left(\frac{10 - (1/6)(-5)}{10 \times (1 - 1/6)}\right)$$

$$= 5 \times 10^{-2} \ln\left(\frac{65/6}{50/6}\right) = 1.31\ ms$$

$$t_2 = R_f C \ln\frac{V(-)_{sat} - aV(+)_{sat}}{V(-)_{sat}(1-a)} = 50 \times 10^4 \times 0.1 \times 10^{-6} \ln\left(\frac{-5 - (1/6)(10)}{-5 \times (1 - 1/6)}\right)$$

$$= 5 \times 10^{-2} \ln\left(\frac{-40/6}{-25/6}\right) = 2.35\ ms$$

7.4 Monostable (One-Shot) Multivibrators

A *monostable* or *one–shot multivibrator* is an electronic circuit with two output states in which only one state is stable. This circuit stays in its normal state (usually a logical 0) until a signal is applied to its input. When triggered by an input signal, the output reverses states, that is, changes from 0 to 1 for a short period of time, depending upon the circuit parameters which are discussed below, and then returns to its normal state.

Figure 7.17 shows a block diagram of a typical monostable and its input and output waveforms.

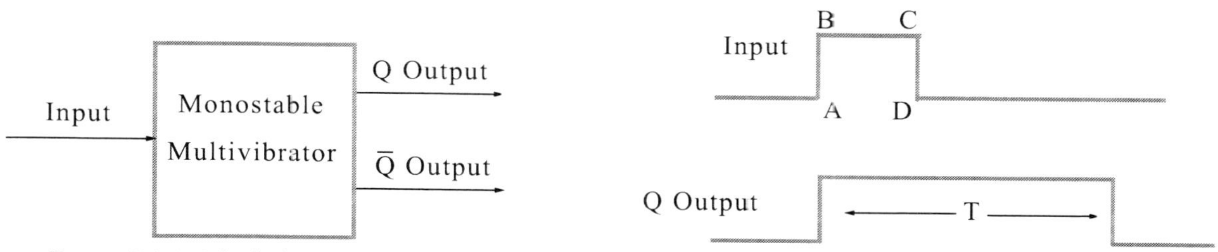

Figure 7.17. Block diagram and typical input and output waveforms for a monostable multivibrator

We observe that a positive going pulse at the input signal (transition from A to B) which is often referred to as the *leading edge* of the pulse (the transition from C to D is referred to as the *trailing edge* of the pulse) makes the 0 output to change state momentarily from state 0 to state 1 for a time period designated as T which is determined by an external resistor R_X and an external capacitor C_X as shown in Figure 7.18. Device SN74121 is a typical monostable multivibrator circuit and it is shown in Figure 7.18.

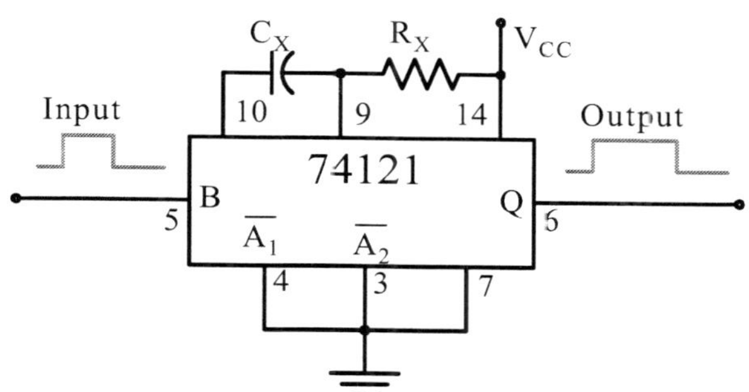

Figure 7.18. The 74121 monostable multivibrator

As shown in Figure 7.18, the 74121 has three trigger inputs: A_1, A_2, and B. Depending on the desired circuit operation, any or all of these three pins may be connected to the input trigger signal. In Figure 7.5 A_1 and A_2 are both active low and the B input is active high. Under those conditions, the monostable multivibrator will produce the output pulse shown.

As stated above, time period that the output will stay High is determined by an external resistor R_X and an external capacitor C_X. Typically, the capacitor value is greater than $1\ \mu F$ and the resistor value is greater than $10\ K\Omega$ and the pulse width period T at the output is determined from the relation

$$T = 0.33 \times R_X \times C_X$$

For instance, if $R_X = 20\ K\Omega$ and $C_X = 5\ \mu F$, the pulse width will be $T = 100\ ms$. With the 74121 device, the output pulse width can be varied from $40\ ns$ to $28\ s$. Also the 74121 is retriggable meaning that the input may be a train of periodic pulses.

The monostable multivibrator is used primarily in pulse shaping applications. It can be used to widen a narrow pulse as shown in Figure 7.18, and to block unwanted input pulses as shown in Figure 7.19.

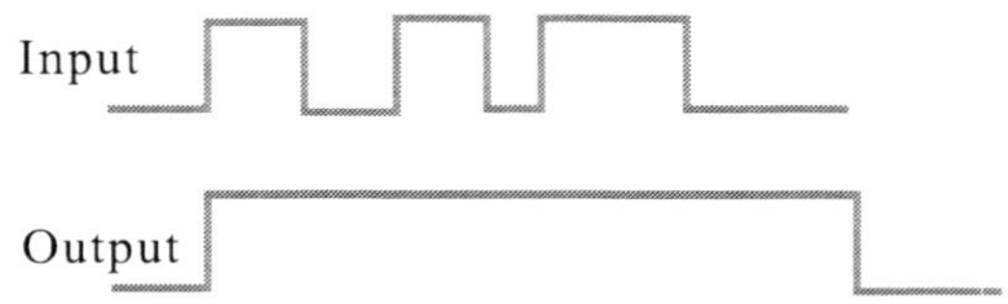

Figure 7.19. The use of a monostable multivibrator to block unwanted input pulses

A monostable multivibrator can also be constructed with an op amp as shown in Figure 7.20.

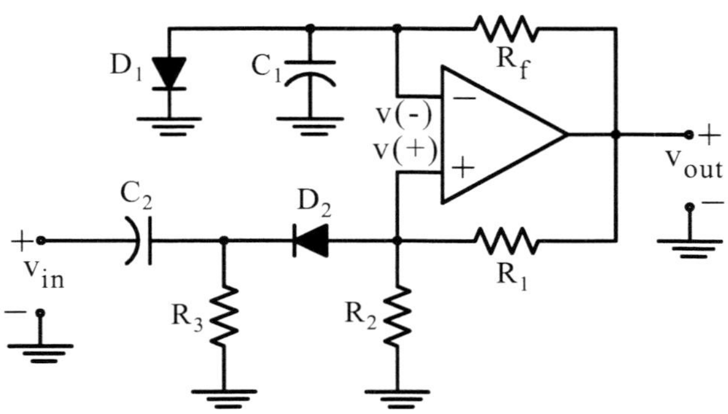

Figure 7.20. A monostable multivibrator using an op amp

In the circuit of Figure 7.20, while the input pulse is zero, i.e., $v_{in} = 0$, the op amp remains in positive saturation, that is, $v_{out} = V(+)_{sat}$, and this is the stable state of the circuit, hence the name monostable. At this state, diode D_1 is forward biased and the voltage across the capacitor C_1 is the same as that of diode D_1, about 0.7 V and thus the voltage across the inverting input of the op amp is $v(-) = 0.7 \text{ V}$.

By the voltage division expression, the voltage at the non–inverting input of the op amp is

$$v(+) = \frac{R_2}{R_1 + R_2} V(+)_{sat} \tag{7.24}$$

Usually, resistors R_1 and R_2 are chosen such that $R_2/R_1 = 10$ and thus $v(+) \gg v(-)$. Accordingly, the output of the op amp is held at positive saturation.

Next, let us assume that a negative going pulse whose amplitude exceeds the voltage $v(+)$. This voltage at the non–inverting of the op amp will go momentarily negative and will drive the output of the op amp into negative saturation $V(-)_{sat}$. So, with the output at negative saturation, the voltage $v(+)$ at the non–inverting input will also become negative. But this condition is only temporary because diode D_1 is now reverse–biased and allows capacitor C_1 to charge negatively, and when $v_{(-)}$ becomes slightly more negative $v_{(+)}$, and the output of the op amp is once again driven into positive saturation.

We see then that the output of the op amp remains at negative saturation $V_{(-)sat}$ only long enough for capacitor C_1 to charge to approximately the value of $v_{(+)}$, and the time t is determined by the time constant of capacitor C_1 and feedback resistor R_f, and also by the ratio R_1/R_2. The expression for the timing period can be derived from relations (7.18) or (7.19) which for the monostable multivibrator is appropriately expressed as

$$t = R_f C_1 \ln \frac{(V_{(-)sat} - 0)}{(V_{(-)sat} - aV_{(-)sat})} = R_f C_1 \ln \frac{1}{1-a} \tag{7.25}$$

where a is the attenuation factor defined as before as

$$a = \frac{R_1}{(R_1 + R_2)}$$

Alternately, we may express (7.25) as

$$t = R_f C_1 \ln\left(1 + \frac{R_1}{R_2}\right) \tag{7.26}$$

and if we choose $R_2/R_1 = 10$, relation (7.26) simplifies to the approximate expression

$$t \approx 0.1 R_f C_1 \tag{7.27}$$

As with astable multivibrators, we can form a monostable multivibrator using a 555 timer circuit as shown in Figure 7.21. As shown in Figure 7.21, in the stable state the SR flip–flop will be in the Reset state keeping the transistor in saturation and thus the capacitor voltage will be close to zero volts. The voltage at the input terminal is sufficiently high to keep the output of Comparator 2 Low. The output Q will also be Low.

Now, if a negative going pulse is applied at the input, the output of Comparator 2 will go High, the flip–flop will be Set and its output Q will go High and output $\bar{Q}$ will go Low and the transistor will be at cutoff. At this time, the capacitor will begin charging through resistor R_1 and will rise exponentially towards the supply voltage V_{CC}. Once the capacitor voltage exceeds the threshold voltage of Comparator 1, the output of this comparator will go High, the flip–flop will be Reset, the transistor will go in saturation, and the capacitor will discharge. Thus, the multivibrator will be returned to it stable state at zero volts level.

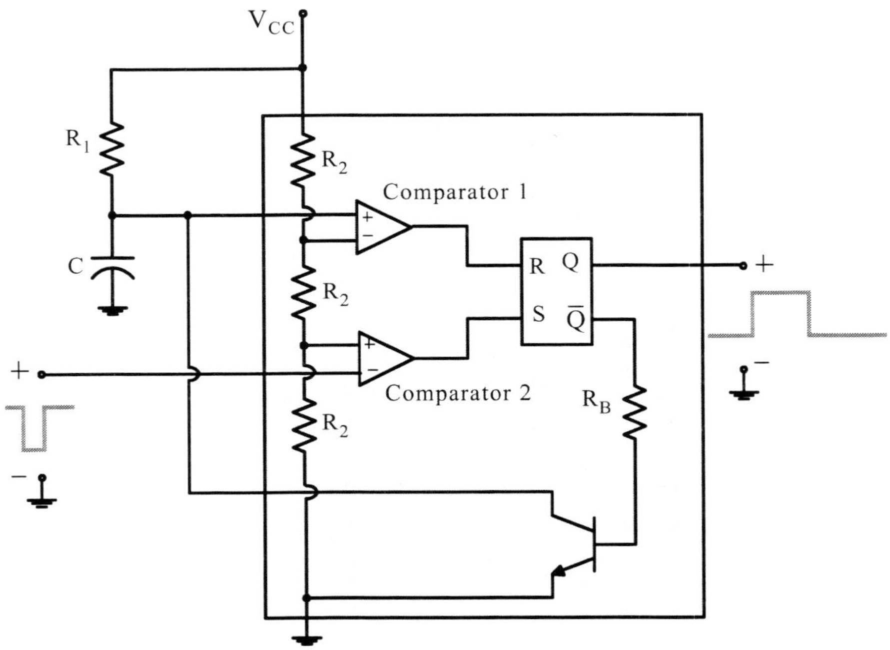

Figure 7.21. A monostable multivibrator using a 555 timer circuit.

We can find the pulse width T for the time that the monostable multivibrator changes momentarily from its stable state by observing that the capacitor voltage, when charging, reaches the threshold voltage at the minus (–) input of Comparator 1. Because the three resistors have the same value, this threshold voltage is $2/3\ V_{CC}$. From (7.1),

$$v_C(t) = V_\infty - (V_\infty - V_{in})e^{-t/R_1C}$$

and assuming that the negative going pulse is applied at the input at $t = 0$ when $V_{in} = 0$, at $t = T$ the above relation becomes

$$(2/3)V_{CC} = V_{CC} - V_{CC}e^{-T/R_1C}$$

$$e^{-T/R_1C} = 1/3$$

$$-T/R_1C = \ln(1/3) = \ln 1 - \ln 3$$

$$T \approx 1.1R_1C \tag{7.28}$$

7.5 Bistable Multivibrators (Flip–Flops)

Bistable multivibrators, or more commonly referred to as *flip–flops*,* are electronic circuits with two stable outputs one of which is the complement of the other. The outputs will change only when directed by an input command. In this section we will introduce the fixed–bias flip–flop, the self–biased flip–flop, and the Schmitt trigger circuit.

7.5.1 The Fixed–Bias Flip–Flop

The *fixed–bias circuit* shown in Figure 7.22 is one of the earlier types. For this circuit it is assumed that the value of the collector resistance R_C is chosen such that the collector current $I_C \approx V_{CC}/R_C$ does not exceed the maximum permissible current. It is also assumed that the values of resistors R_1 and R_2, and the bias voltage V_{BB} are selected so that in one state the base current is adequate to drive one transistor into saturation and in the second state the base–emitter junction must be below cutoff. The cutoff and saturation characteristics are specified by the manufacturer.

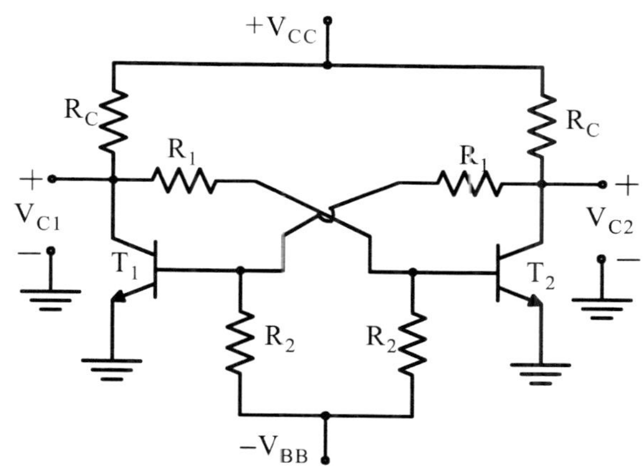

Figure 7.22. Fixed–bias flip–flop circuit

We will illustrate the operation of the circuit of Figure 7.22† with the following example. The triggering signal which is used to induce a transition from one state to another will be discussed later in this section.

Example 7.4

Compute the stable–state currents and voltages for the flip–flop circuit of Figure 7.23. Assume that the transistors have a minimum h_{FE} value of 50.

---

* *The logic diagrams and characteristic tables for the Set–Reset, Data, J–K, and Toggle flip–flops are described in Digital Circuit Analysis and Design with Simulink Modeling and Introduction to CPLDs and FPGAs, ISBN 978–1–934404–05–8.*

† *This circuit is also known as Ecless–Jordan configuration*

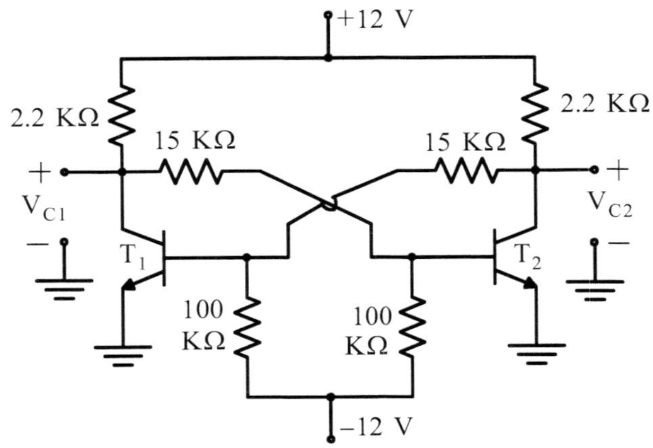

Figure 7.23. Fixed–bias flip–flop circuit for Example 7.4.

Solution:

This circuit consists of two cross–coupled inverter circuits such as that of Exercise 8 in Chapter 6. The analysis is facilitated by breaking the given circuit into two parts, the first part indicating the connections between the base of transistor T_1 and the collector of transistor T_2 as shown in Figure 7.24(a), and the second part indicating the connection between the collector of T_1 and the base of T_2 as shown in Figure 7.24(b).

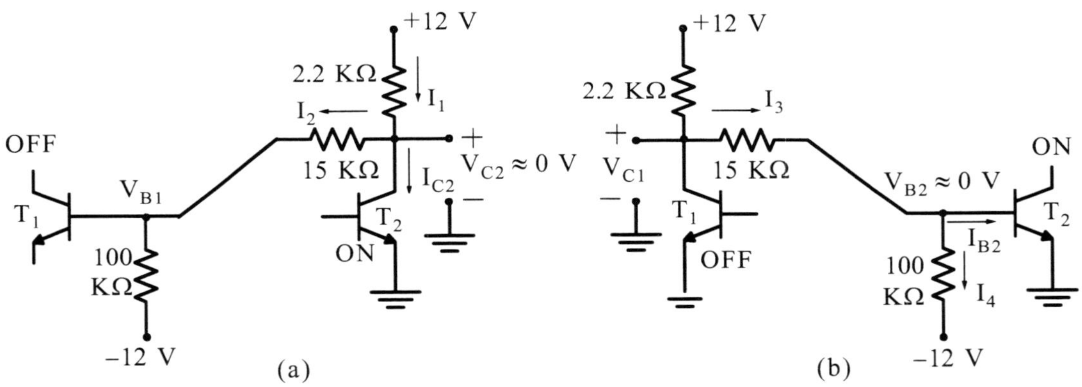

Figure 7.24. Circuits for the computation of the stable states for Example 7.4

Let us first assume that transistor T_1 is OFF and transistor T_2 is ON. Since the saturation voltages are small (about 0.2 V), we will initially neglect them and we let $V_{C2}(sat) \approx 0$ V and $V_{B2}(sat) \approx 0$ V as shown in Figures 7.20(a) and 7.20(b) respectively.

By the voltage division expression

$$V_{B1} = \frac{15\ K\Omega}{100\ K\Omega + 15\ K\Omega} \cdot (-12\ V) = -1.57\ V$$

and this voltage will certainly keep transistor T_1 at cutoff. To verify that with transistor T_1

beyond cutoff transistor T_2 is in saturation, we calculate I_{C2} by first finding I_1 and I_2 as follows:

$$I_1 = \frac{12\ V}{2.2\ K\Omega} = 5.45\ mA$$

Also,

$$I_2 = \frac{0-(-12)}{15\ K\Omega + 100\ K\Omega} = 0.10\ mA$$

Then,

$$I_{C2} = I_1 - I_2 = 5.45 - 0.10 = 5.35\ mA$$

We observe that the value of I_2 is much less than the value of I_1 and in a practical flip–flop design this may not be the case.

Usually, the minimum base current I_{B2} is read from the collector characteristics curve for a particular transistor and since no curves were given, we will use the given minimum value of h_{FE} .for the minimum value of I_{B2} for saturation. Thus,

$$I_B(min) = \frac{I_{C2}}{h_{FE}} = \frac{5.35\ mA}{50} = 0.11\ mA$$

Let us now compute the value of I_{B2} from Figure 7.24(b). We find that

$$I_3 = \frac{12}{2.2\ K\Omega + 15\ K\Omega} = 0.70\ mA$$

and

$$I_4 = \frac{12}{100\ K\Omega} = 0.12\ mA$$

Then,

$$I_{B2} = I_3 - I_4 = 0.70 - 0.12 = 0.58\ mA$$

Since $(I_{B2} = 0.58\ mA) > (I_B(min) = 0.11\ mA)$, that is, the value of I_{B2} exceeds the minimum base current value of $I_B(min)$ required for saturation, we have shown that transistor T_2 is deeply into saturation.

Finally, the collector voltage V_{C1} if found from Figure 7.20(b) as

$$V_{C1} = 12 - 2.2 \times 10^3 \times I_3 = 12 - 2.2 \times 10^3 \times 0.70 \times 10^{-3} = 10.5\ V$$

and this value is close to the collector supply voltage of 12 volts.

With the assumption that the saturation are zero volts, the voltages and currents for a stable state for the flip–flop of this example are summarized in Table 7.1.

The second stable state is the one in which transistor T_1 is ON and transistor T_2 is OFF. The analysis is the same as above where the voltages and currents are interchanged between transistors T_1 and T_2, and $V_{C2} = 10.5\ V$.

TABLE 7.1 Voltages and currents for a stable state for the flip–flop of Example 7.4

Parameter	Value
I_{C1}	0 mA
V_{C1}	10.5 V
I_{C2}	5.35 mA
V_{C2}	0 V
I_{B1}	0 mA
V_{B1}	−1.56 V
I_{B2}	0.58 mA
V_{B2}	0 V

7.5.2 The Self–Bias Flip–Flop

This is also one of the earlier types of flip–flops. In a *self–bias flip–flop*, a common emitter resistor R_E, as shown in Figure 7.25, is used to provide self–bias and thus the need for a negative power supply is eliminated.

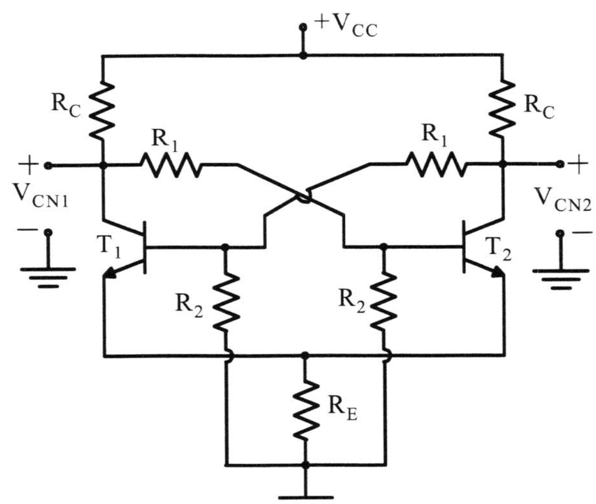

Figure 7.25. Self–bias flip–flop circuit

The procedure for calculating the stable states is similar to that for a fixed–bias flip–flop and it is illustrated in Example 7.5.

Example 7.5

Compute the stable–state currents and voltages for the flip–flop circuit in Figure 7.26 which uses P–N–P identical transistors. Find the minimum value of h_{FE} which will keep the ON transistor in saturation.

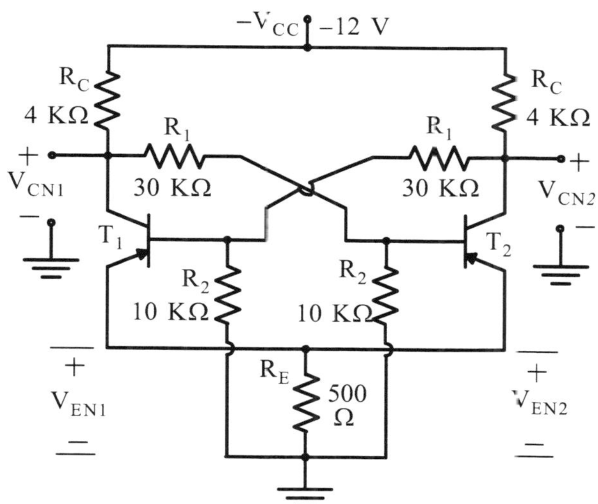

Figure 7.26. Self–bias flip–flop circuit for Example 7.5

Solution:

This circuit consists of two cross–coupled inverter circuits such as that of Exercise 8 in Chapter 6. The analysis is facilitated by breaking the given circuit into two parts, the first part indicating the connections between the base of transistor T_1 and the collector of transistor T_2 as shown in Figure 7.27, and the second part indicating the connection between the collector of T_1 and the base of T_2 as shown in Figure 7.28.

Let us first assume that transistor T_1 is OFF and transistor T_2 is ON. We must first compute the voltage V_{EN1} shown in Figure 7.27.

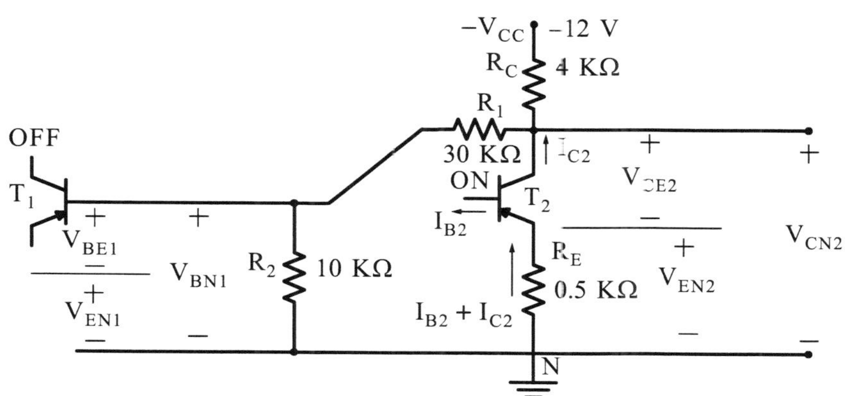

Figure 7.27. The circuit of Example 7.5 showing the connections between the base of T_1 and collector of T_2

From the circuit of Figure 7.27 we observe that $V_{EN2} = V_{EN1}$ and from Figure 7.28

$$V_{EN2} = -(I_{B2} + I_{C2})R_E \tag{7.29}$$

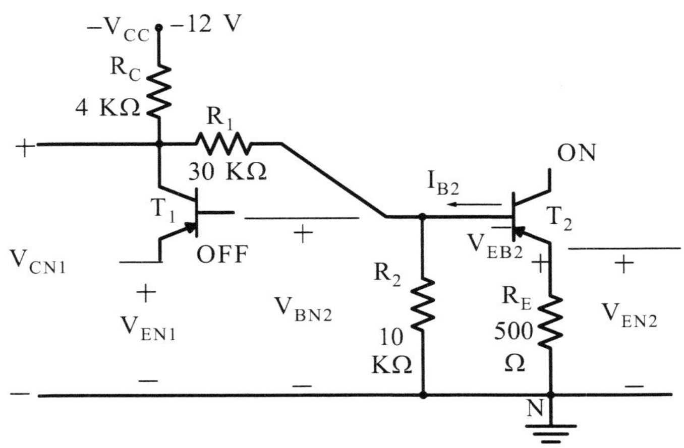

Figure 7.28. The circuit of Example 7.5 showing the connections between the base of T_2 *and collector of* T_1

where the minus (–) sign indicates that the neutral point (ground) is at a higher potential than the emitter. To find the saturation currents I_{B2} and I_{C2} we will apply Thevenin's theorem by first replacing the collector circuit of transistor T_2 by its equivalent, then by replacing the base circuit of T_2 by its equivalent, combining these two, and inserting the transistor T_2 between them.

With transistor T_2 and emitter resistor R_E disconnected, the Thevenin equivalent is as shown in Figure 7.29.

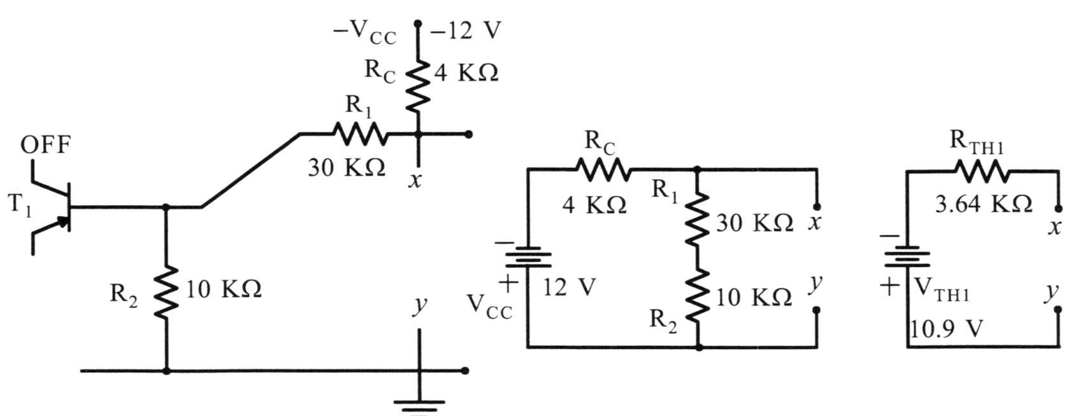

Figure 7.29. Collector circuit for the computation of Thevenin equivalent, Example 7.5

From Figure 7.29

$$V_{TH1} = V_{xy} = \frac{(R_1 + R_2)}{(R_1 + R_2 + R_C)}(-V_{CC}) = \frac{(30 + 10)}{(30 + 10 + 4)}(-12) = -10.9 \text{ V} \tag{7.30}$$

$$R_{TH1} = (R_1 + R_2) \| R_C = \frac{(R_1 + R_2)R_C}{R_1 + R_2 + R_C} = \frac{40 \times 4}{44} = 3.64 \text{ K}\Omega \tag{7.31}$$

Next, we replace the base circuit of T_2 by its equivalent as shown in Figure 7.30.

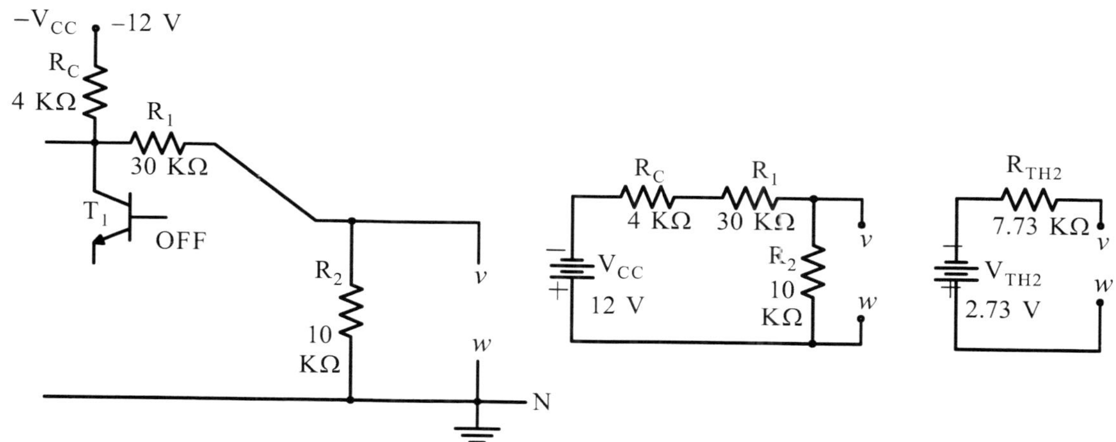

Figure 7.30. Base circuit for the computation of Thevenin equivalent, Example 7.5

From Figure 7.30

$$V_{TH2} = V_{vw} = \frac{R_2}{(R_1 + R_2 + R_C)}(-V_{CC}) = \frac{10}{(30 + 10 + 4)}(-12) = -2.73 \text{ V} \tag{7.32}$$

$$R_{TH2} = (R_C + R_1) \| R_2 = \frac{(R_C + R_1)R_2}{R_C + R_1 + R_2} = \frac{34 \times 10}{44} = 7.73 \text{ K}\Omega \tag{7.33}$$

The Thevenin equivalent voltages and resistances of V_{TH1}, R_{TH1}, V_{TH2}, and R_{TH2}, combined with transistor T_2 and resistor R_E yield the circuit of Figure 7.31 where have assumed that $V_{EC}(sat) = 0.2 \text{ V}$ and $V_{EB}(sat) = 0.7 \text{ V}$.

We find the values of I_{E2} and I_{C2} by application of KVL around the loops indicated in Figure 7.31.

$$0.5(I_{B2} + I_{C2}) + 0.7 + 7.73I_{B2} = 2.73$$

$$0.5(I_{B2} + I_{C2}) + 0.2 + 3.64I_{C2} = 10.9$$

Rearranging and simplifying, we obtain

$$8.23I_{B2} + 0.5I_{C2} = 2.03$$

$$0.5I_{B2} + 4.14I_{C2} = 10.7$$

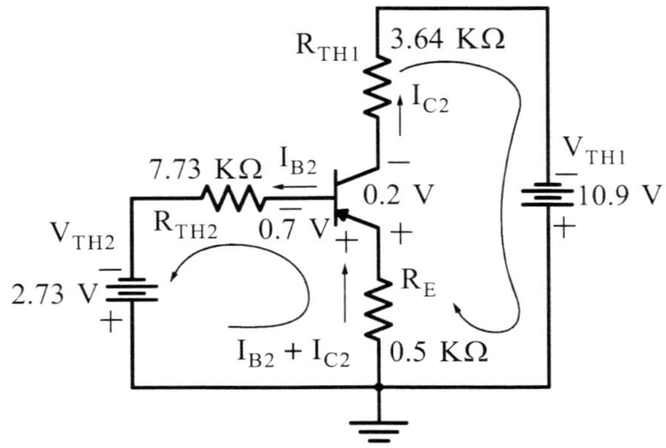

Figure 7.31. The composite Thevenin equivalent circuit of Example 7.5 when transistor T_2 *is in saturation*

Using MATLAB with the matrix left division operator, we obtain write and execute the following script which produces the values of the currents I_{B2} and I_{C2}.

```
A=[8.23 0.5; 0.5 4.14]; B=[2.03 10.7]'; x=A\B;...
fprintf(' \n'); fprintf('IB2=%2.3f mA \t',x(1)); fprintf('IC2=%2.3f mA \t',x(2)); fprintf(' \n')
```

```
IB2=0.090 mA IC2=2.574 mA
```

Thus,

$$I_{C2} = 2.57 \tag{7.34}$$

and

$$I_{B2} = 0.09 \tag{7.35}$$

With these values, we now find the value of h_{FE}.

$$h_{FE} = \frac{I_{C2}}{I_{B2}} = \frac{2.57}{0.09} = 28.5 \tag{7.36}$$

The voltages can now be found from the circuits of Figures 7.27 and 7.28. From Figure 7.26 we observe that $V_{EN2} = V_{EN1}$ and from relation (7.18)

$$V_{EN1} = V_{EN2} = -(I_{B2} + I_{C2})R_E = -(0.09 + 2.57) \times 0.5 = -1.33 \text{ V} \tag{7.37}$$

From Figure 7.27,

$$V_{CN2} = V_{CE2} + V_{EN2} = -0.2 - 1.33 = -1.53 \text{ V} \tag{7.38}$$

From Figure 7.28,

$$V_{BN2} = V_{BE2} + V_{EN2} = -0.7 - 1.33 = -2.03 \text{ V} \tag{7.39}$$

From Figure 7.27,

$$V_{BN1} = \frac{R_2}{R_1 + R_2} V_{CN2} = \frac{10}{30 + 10}(-1.53 \text{ V}) = -0.38 \text{ V} \tag{7.40}$$

and

$$V_{BE1} = V_{BN1} - V_{EN1} = -0.38 - (1.33) = +0.95 \text{ V} \tag{7.41}$$

With this value, the PNP transistor is OFF.

Finally, to find the value of V_{CN1} we apply superposition to the circuit of Figure 7.28 which, for simplification is redrawn into two simple resistive networks as shown in Figure 7.32.

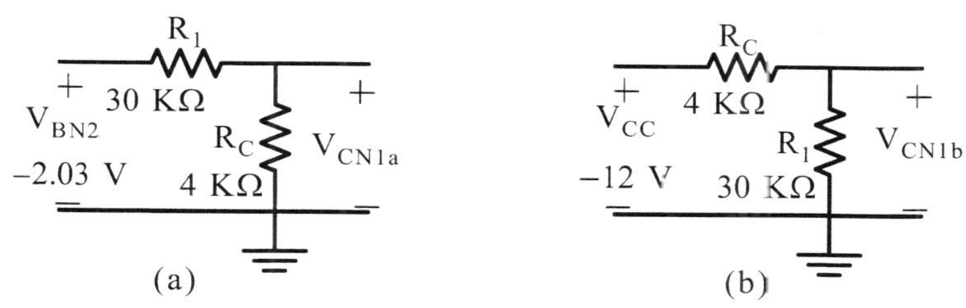

Figure 7.32. Circuits for the computation of V_{CN1}*, Example 7.4*

From the circuit of Figure 7.32(a) we find that

$$V_{CN1a} = \frac{R_C}{R_1 + R_C} V_{BN2} = \frac{4}{30+4}(-2.03) = -0.24$$

and from the circuit of Figure 7.32(b) we find that

$$V_{CN1b} = \frac{R_1}{R_C + R_1} V_{CC} = \frac{30}{4+30}(-12) = -10.59 \text{ V}$$

Therefore,

$$V_{CN1} = V_{CN1a} + V_{CN1b} = -10.83 \text{ V} \tag{7.42}$$

The stable state values of the circuit of Example 7.5 are summarized in Table 7.2

TABLE 7.2 Voltages and currents for a stable state for the flip–flop of Example 7.4

Parameter	Value
I_{C1}	0 mA
V_{CN1}	−10.83 V
I_{C2}	2.57 mA
V_{CN2}	−1.53 V
I_{B1}	0 mA
V_{BN1}	−1.56 V
I_{B2}	0.09 mA
V_{BN2}	−2.03 V
$V_{EN1} = V_{EN2}$	−1.33 V

We observe that the output changes from $V_{CN2} = -1.53$ V to $V_{CN1} = -10.83$ V or vice versa.

7.5.3 Triggering Signals for Flip–Flops

A flip–flop will remain in one of its stable states indefinitely until a triggering signal is applied to make a transition. In some applications it is desired to have a change of state occur immediately after the application of an abrupt triggering signal. The interval time during which conduction transfers from one transistor to the other is known as the *transition time*. It is always desirable to reduce the transition time and this can be accomplished by inserting small capacitors, referred to as *speed–up capacitors*, in parallel with the coupling resistors R_1 as shown in Figure7.33.

Let us assume that in the flip–flop circuit of Figure 7.33 transistor T_1 if OFF, transistor T_2 is ON, and to initiate a transition, a negative pulse is applied at the base of T_2 denoted as point X_2. The collector voltage of T_2, denoted as point Y_2 will rise rapidly and it is desirable that this rapid rise will be transmitted with minimum delay to the base of T_1 denoted as point X_1. Transistor T_1 has an input capacitance C_i and in the absence of the parallel capacitor C_1, the circuit configuration consisting of resistors R_1 and R_2, and input capacitor C_i constitutes an *uncompensated attenuator.**

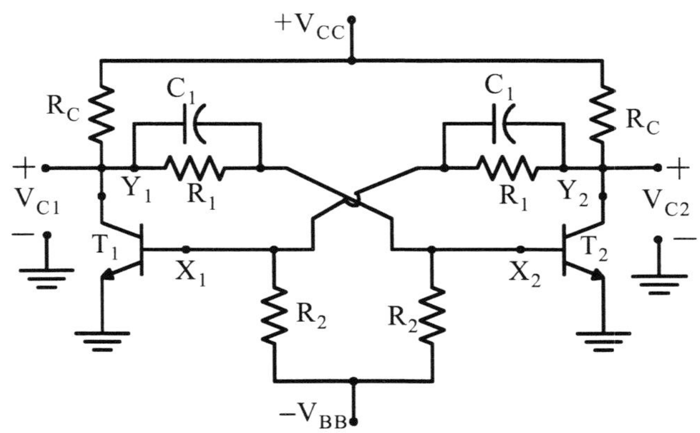

Figure 7.33. Fixed–bias flip–flop circuit including speed–up capacitors

The smallest allowable interval between triggers is referred to as the *resolving time* of the flip–flop, and its reciprocal is the maximum frequency at which the flip–flop will respond. In practice, the maximum frequency f_{max} of operation is given by the relation

$$f_{max} = \frac{1}{2\tau} = \frac{R_1 + R_2}{2C_1R_1R_2} \tag{7.43}$$

The triggering signal which is usually employed to initiate a transition from one state to the other is either a pulse of short time duration or a step voltage. This pulse or step may be used in a man-

* *For a discussion on compensated attenuators, please refer to Appendix B.*

ner that will produce either symmetrical or unsymmetrical triggering. In *unsymmetrical triggering* signal is effective in initiating a transition in only one direction (polarity). A second triggering signal from a separate source must be initiated in a different manner to achieve the reverse direction. The circuits of Figure 7.34 show a method of triggering unsymmetrically an NPN and a PNP bipolar transistor.

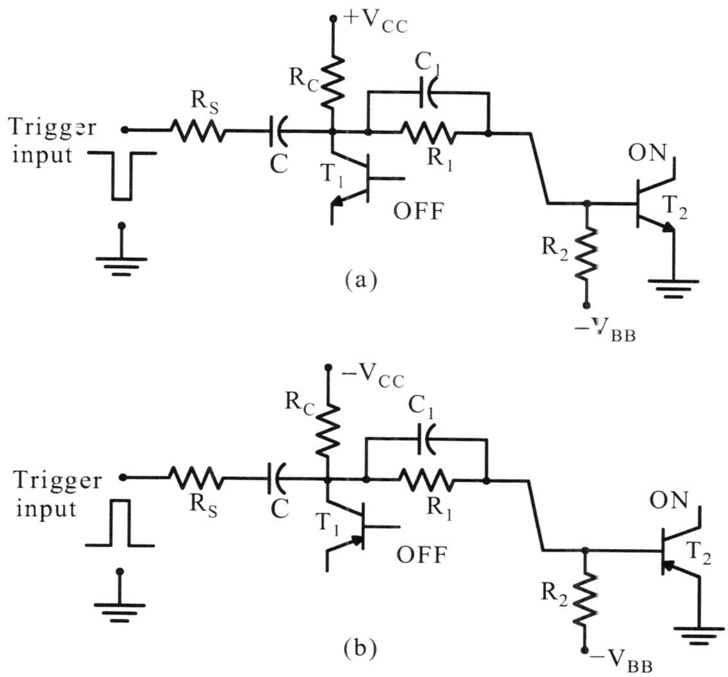

Figure 7.34. A method of triggering unsymmetrically an NPN or a PNP transistor

In *symmetrical triggering*, each successive triggering signal initiates a transition, regardless of the state in which the flip–flop happens to be. The circuits of Figure 7.35 show how a flip–flop may be triggered in a symmetrical manner.

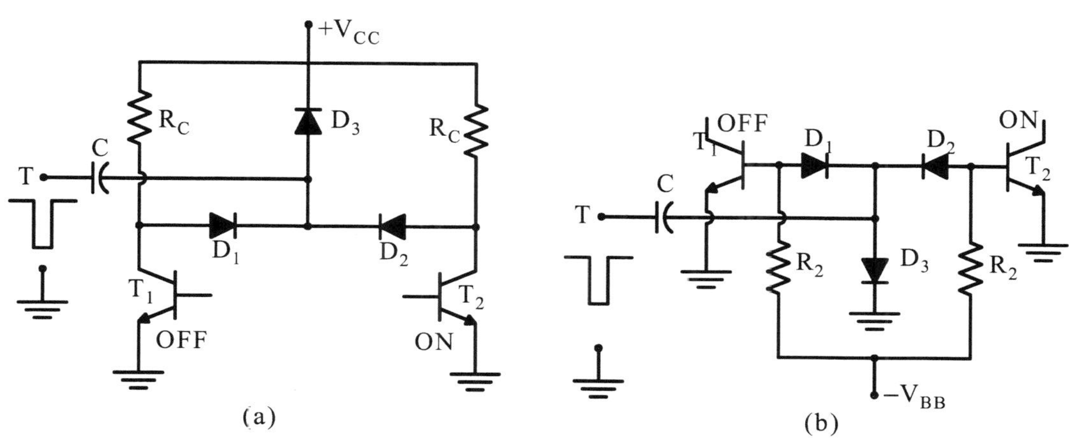

Figure 7.35. Methods of symmetrical triggering through diodes at the collectors or the bases of the transistors

7.5.4 Present Technology Bistable Multivibrators

The bistable multivibrator (flip–flop) circuits we've discussed thus far are the original circuits and were in use in the 1960s. We have included them in this text because they are the circuits from which present technology bistable multivibrators such as those with CMOS technology and op amps have evolved. We will briefly discuss a bistable multivibrator with an op amp in this subsection.

The circuit of Figure 7.36 shows how an op amp can be configured to behave as a bistable multivibrator.

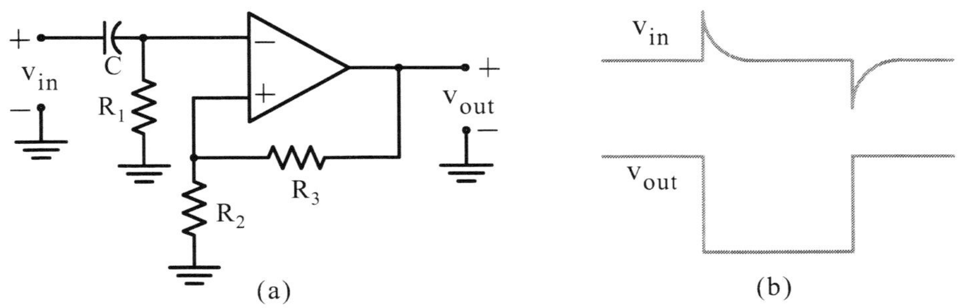

Figure 7.36. Op amp configured as a bistable multivibrator

The stable states for the bistable multivibrator of Figure 7.36(a) are the conditions where the output is at positive or negative saturation. It assumes either positive or negative saturation by the positive feedback formed by resistors R_2 and R_3. A positive or negative going pulse as shown in Figure 7.36(b) causes the circuit to switch states.

7.6 The Schmitt Trigger

Another bistable multivibrator circuit is the *Schmitt trigger* shown in Figure 7.37(a).

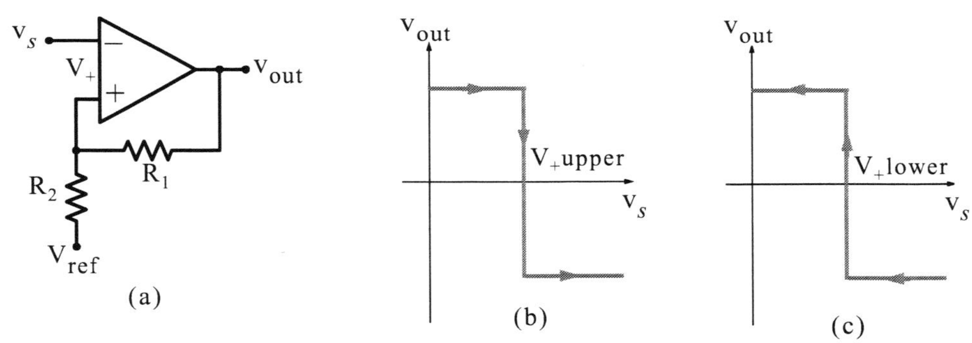

Figure 7.37. Schmitt trigger circuit and waveforms for increasing and decreasing input signals

The *Schmitt trigger* circuit and transfer characteristics are similar to the comparator. It provides an voltage when its input signal v_s reaches some predetermined value set at the non–inverting input of the op amp. The output of the op amp changes from the positive saturation voltage $V_{out}(max)$ to its negative saturation voltage $-V_{out}(max)$ and vice versa. As shown in Figure 7.37(b), the output is positively saturated as long as the input signal v_S is less than the upper threshold V_{+upper}. If the input signal v_S rises slightly above this threshold voltage, the output drop abruptly to $-V_{out}(max)$ and stays there until v_S drops below a lower threshold voltage V_{+lower}. The threshold voltages V_{+upper} and V_{+lower} are determined by the resistors R_1 and R_2, and the reference voltage V_{ref}. These threshold voltages can be found by application of KCL at the non–inverting input of the op amp. Thus,

$$\frac{V_+ - V_{ref}}{R_2} + \frac{V_+ - V_{out}}{R_1} = 0$$

$$\left(\frac{1}{R_1} + \frac{1}{R_2}\right)V_+ = \frac{V_{out}}{R_1} + \frac{V_{ref}}{R_2}$$

$$\left(\frac{R_1 + R_2}{R_1 \cdot R_2}\right) \cdot V_+ = \frac{R_1 V_{ref} + R_2 V_{out}}{R_1 \cdot R_2}$$

$$V_+ = \frac{R_1 V_{ref} + R_2 V_{out}}{R_1 + R_2} \tag{7.44}$$

In (7.44), when V_{out} is the maximum positive output voltage, $V_+ = V_{+upper}$, and when V_{out} is the maximum negative output voltage, $V_+ = V_{+lower}$.

Normally, the peak–to–peak output voltage of the Schmitt trigger of Figure 7.37 is often limited by the use of back–to–back zeners across the output terminal and ground. The zener voltages are chosen so that the output swing from positive to negative or vice versa, is compatible with commercially available IC digital devices.

Example 7.6

For the Schmitt trigger circuit of Figure 7.38(a), the input signal v_S is as shown in Figure 7.38(b).

Find and sketch V_{+upper} and V_{+lower}.

Solution:

It is given that the Schmitt trigger saturates positively at $+12$ V. Then, with (7.44)

$$V_{+upper} = \frac{R_1 V_{ref} + R_2 V_{out}}{R_1 + R_2} = \frac{10^4 \times 2 + 250 \times 12}{10250} = 2.24 \text{ V}$$

That is, when the input signal v_S exceeds this value, the output abruptly swings to negative saturation, i.e., $-12\ V$ as shown in Figure 7.39.

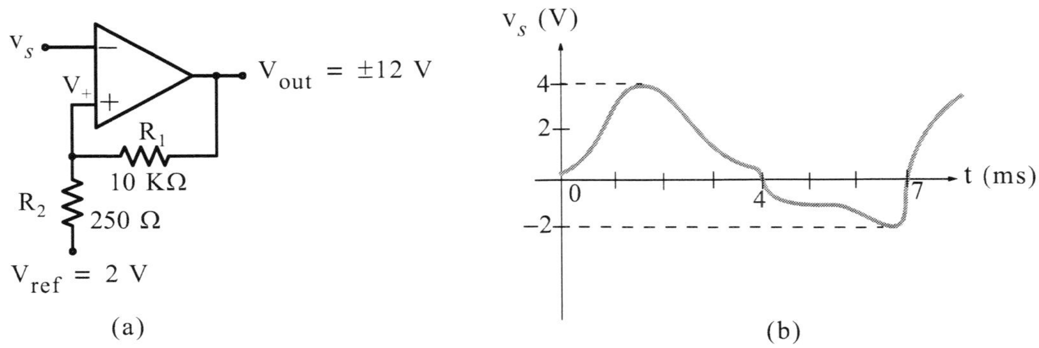

Figure 7.38. Schmitt trigger circuit and input signal waveform for Example 7.6

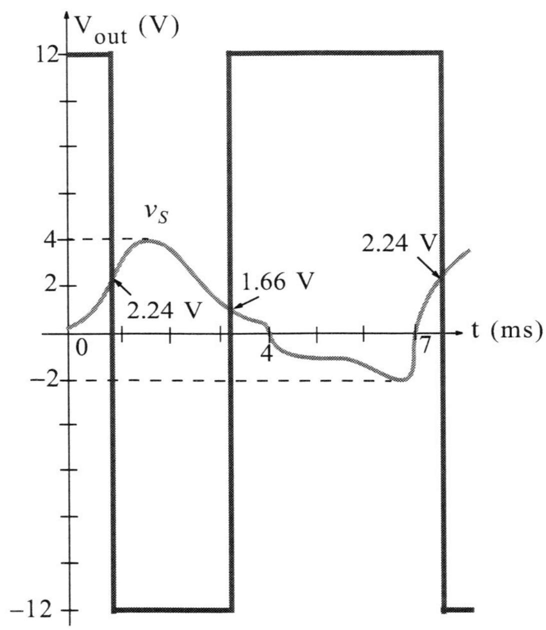

Figure 7.39. Waveforms for Example 7.3

When the output saturates negatively at $-12\ V$ the lower threshold voltage is

$$V_{+}lower = \frac{R_1 V_{ref} + R_2 V_{out}}{R_1 + R_2} = \frac{10^4 \times 2 + 250 \times (-12)}{10250} = 1.66\ V$$

That is, when the input signal v_S drops below this value, the output abruptly swings back to positive saturation, i.e., $+12\ V$, and when the input signal v_S rises again to $2.24\ V$, the output drops again to $-12\ V$ as shown in Figure 7.39.

7.7 Summary

- An astable multivibrator is essentially a digital clock which has two momentarily stable states which alternate continuously producing square pulses.
- The 555 Timer circuit is a widely used IC for generating waveforms.
- A monostable multivibrator is an electronic circuit with two output states in which only one state is stable. This circuit stays in its normal state until a signal is applied to its input. When triggered by an input signal, the output reverses states for a short period of time, depending upon the circuit parameters. A monostable multivibrator is used primarily in pulse shaping applications and to widen narrow pulses.
- Both astable and monostable multivibrators can constructed with a 555 timer circuit.
- IC device SN74121 is a popular monostable multivibrator where the time period that the output will stay High is determined by an external resistor R_X and an external capacitor C_X.
- A bistable multivibrator, commonly referred to as flip–flop, is an electronic circuits with two stable outputs one of which is the complement of the other. The outputs will change only when directed by an input command.
- All three types of the multivibrators, i.e., astable, monostable, and bistable, can be constructed with op amps employed as comparators.
- The Schmitt trigger is another bistable multivibrator circuit.

7.8 Exercises

1. For the astable multivibrator below $V_{CC} = 5\ V$ and $C = 1\ nF$. Find the values for the resistors R_A and R_B so that the circuit will produce the waveform shown.

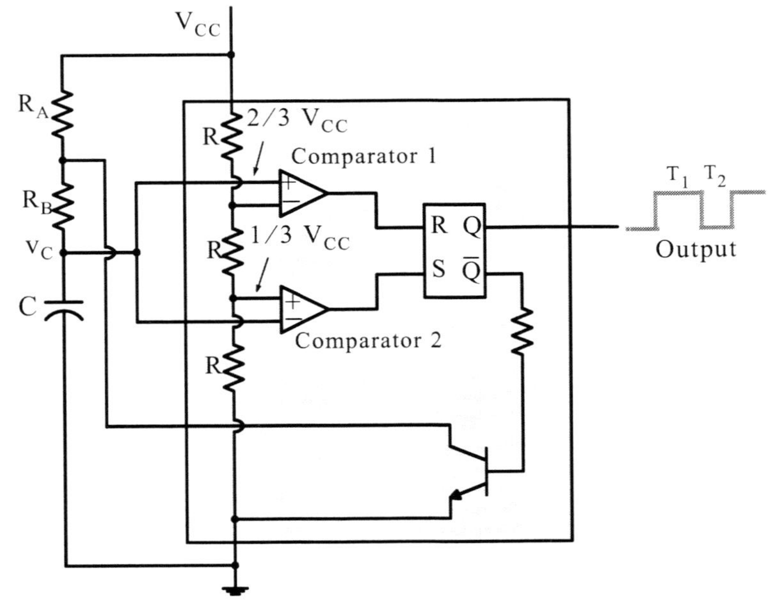

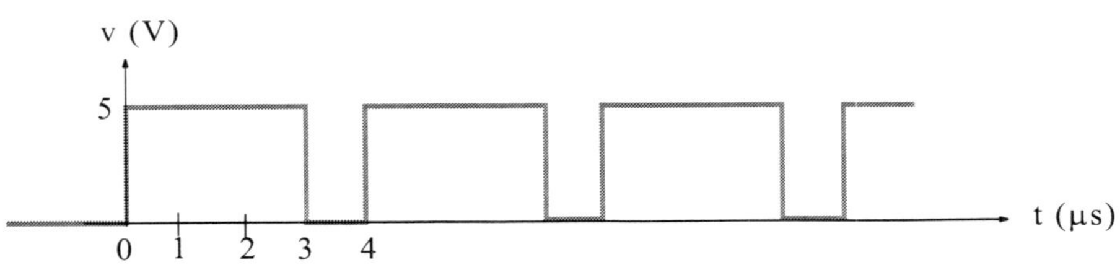

2. For the astable multivibrator below $V_{CC} = 5\ V$ and $C = 1\ nF$. Find the values for the resistors R_A and R_B so that the circuit will produce the waveform shown.

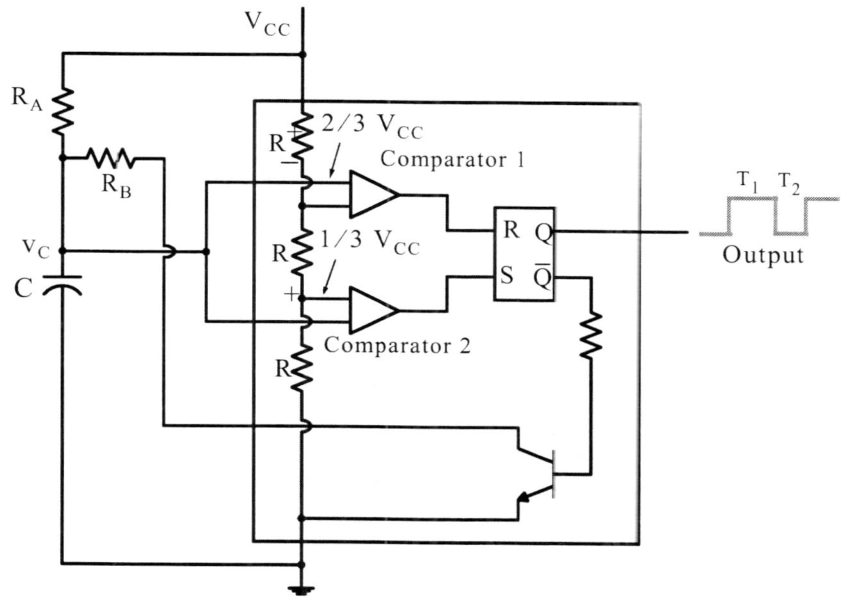

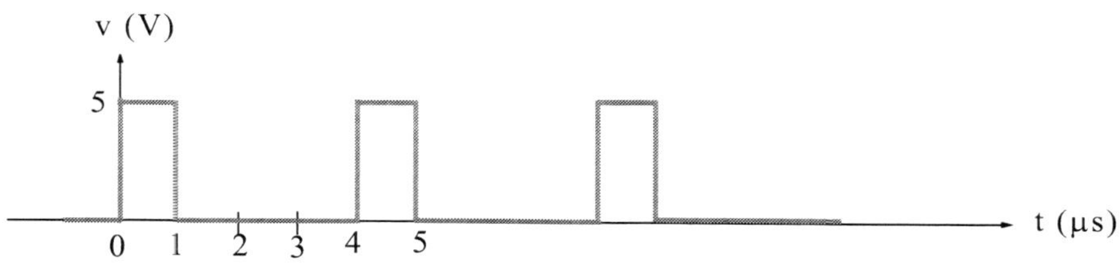

3. Repeat Exercise 2 to design a circuit that will produce the following waveform.

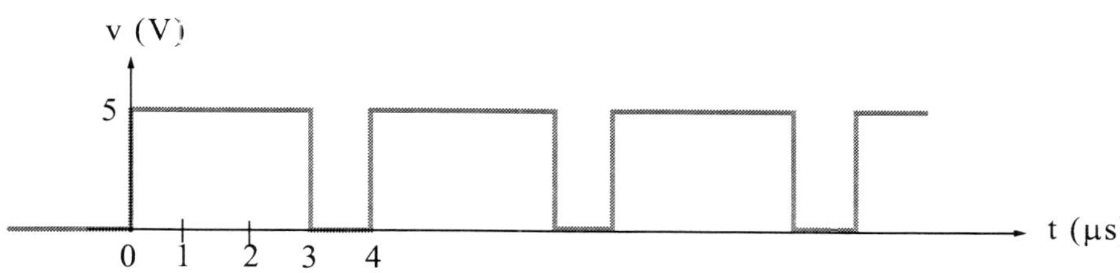

4. Design a monostable multivibrator using a 555 timer, a capacitor with value $C = 1\ nF$ and appropriate resistor values to produce an output pulse of $20\ \mu s$ duration.

5. The fixed–bias flip–flop circuit below is the same as that of Example 7.4.

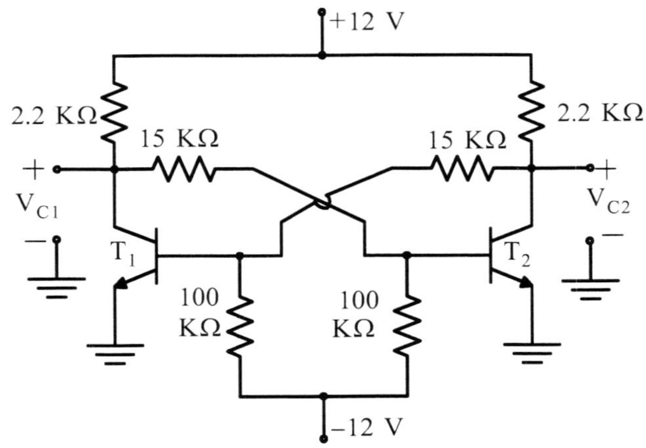

The transistors are identical and the curves supplied by the manufacturer specify that $V_{CE2}(sat) = 0.15\ V$ and $V_{BE2}(sat) = 0.7\ V$. Assume that the transistors have a minimum h_{FE} value of 50. Using these voltages, recalculate the stable–state currents and voltages.

6. For the Schmitt trigger circuit of Figure (a) below, the input signal v_S is as shown in Figure (b).

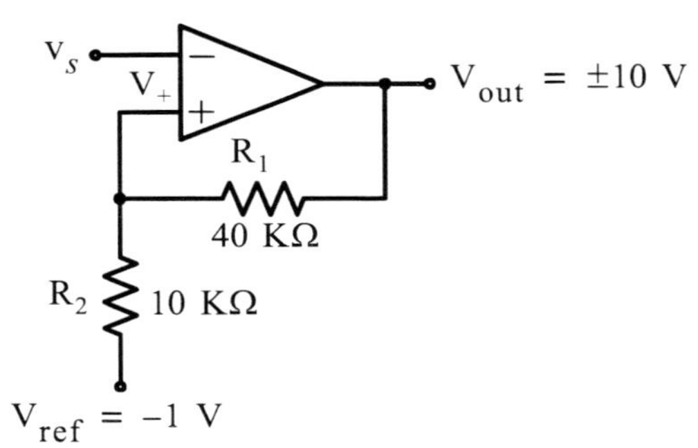

(a)

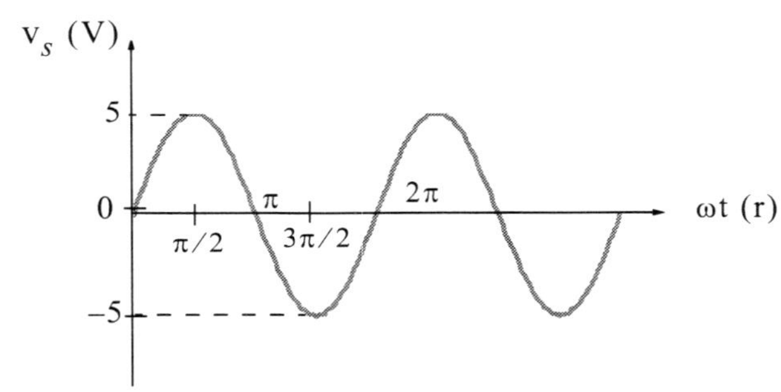

(b)

Find and sketch $V_{+}upper$ and $V_{+}lower$.

7.9 Solutions to End–of–Chapter Exercises

1.

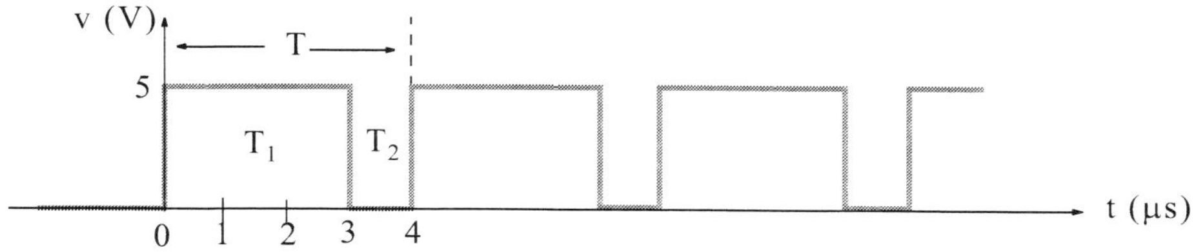

The period is $T = T_1 + T_2 = 3 + 1 = 4\ \mu s$ and the duty cycle is

$$T_1/T = 3/4 = 0.75 = 75\ \%$$

From relation (7.6)

$$T = 4 \times 10^{-6} = 0.69(R_A + 2R_B)C = 0.69 \times 10^{-9}(R_A + 2R_B)$$

$$R_A + 2R_B = \frac{4 \times 10^{-6}}{0.69 \times 10^{-9}} = 5.77\ K\Omega \quad (1)$$

The duty cycle is T_1/T and from relations (7.3) and (7.6) we obtain

$$\text{Duty cycle} = 0.75 = \frac{T_1}{T} = \frac{0.69(R_A + R_B)C}{0.69(R_A + 2R_B)C} = \frac{R_A + R_B}{R_A + 2R_B} = \frac{R_A + R_B}{5.77\ K\Omega}$$

$$R_A + R_B = 0.75 \times 5.77 = 4.33\ K\Omega \quad (2)$$

Subtraction of (2) from (1) yields

$$R_B = 5.77 - 4.33 = 3.44\ K\Omega$$

and from (2)

$$R_A = 4.33 - 3.44 = 890\ \Omega$$

2.

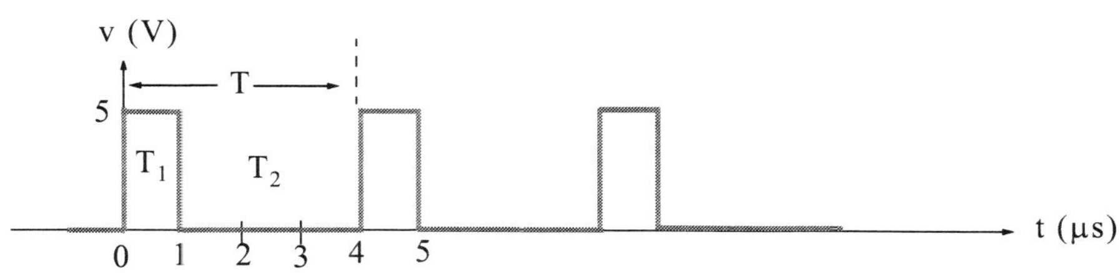

The period is $T = T_1 + T_2 = 1 + 3 = 4\ \mu s$. The duty cycle is $T_1/T = 1/4 = 0.25 = 25\ \%$. To achieve a duty cycle at $25\ \%$, we must make $T_2 = 3T_1$ and this condition will be satisfied if we multiply relation (7.11) by 3 and equate it to relation (7.14) subject to the constraint of (7.16). Then,

$$3\ln 2R_AC = \frac{R_A \cdot R_B}{R_A + R_B}C \cdot \ln\left(\frac{2R_A - R_B}{R_A - 2R_B}\right)$$

$$\frac{R_B}{R_A + R_B} \cdot \ln\left(\frac{2R_A - R_B}{R_A - 2R_B}\right) = 3\ln 2$$

$$3\ln 2 \cdot \left(\frac{R_A + R_B}{R_B}\right) = \ln\left(\frac{2R_A - R_B}{R_A - 2R_B}\right)$$

$$3\ln 2 \cdot \left(\frac{R_A}{R_B} + 1\right) = \ln\left(\frac{2R_A/R_B - 1}{R_A/R_B - 2}\right)$$

For convenience, we let $x = R_A/R_B$ and the relation above is written as

$$3\ln 2 \cdot (x + 1) = \ln\left(\frac{2x - 1}{x - 2}\right)$$

Therefore, let us plot the above relation with the following MATLAB script to find the x–axis zero crossing.

```
x=2.005:0.0001:2.006; y=3.*log(2).*(x+1)–log((2.*x–1)./(x–2)); plot(x,y);grid
```

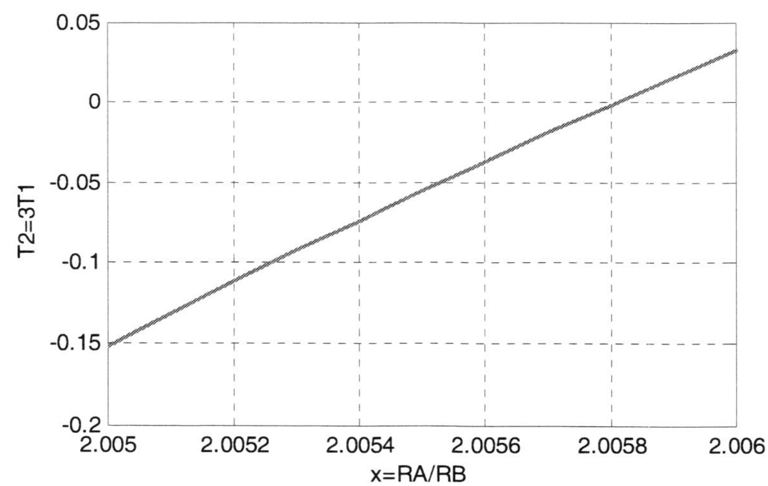

The plot above indicates that the x–axis zero crossing occurs at approximately

$$x = R_A/R_B = 2.0058$$

and thus to achieve a duty cycle at 25 %, we should choose that the ratio R_A/R_B should be close to this value but not exactly equal or less than 2. The value of the capacitor can be found from either expression (7.11) or (7.14) but (7.11) is simpler. We require that $T_1 = 1\ \mu s$ and assuming $R_A = 5.1\ K\Omega$, we obtain

$$T_1 = 10^{-6} = 0.69 R_A C = 0.69 \times 5.1 \times 10^3 C$$

$$C = \frac{10^{-6}}{0.69 \times 5.1 \times 10^3} = 284\ pF$$

3.

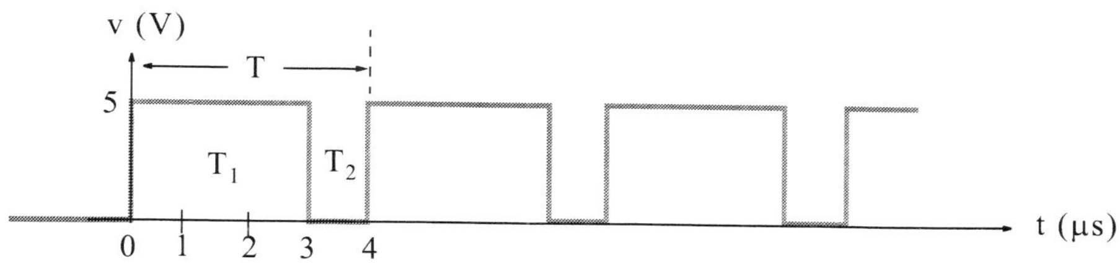

The period is $T = T_1 + T_2 = 3 + 1 = 4\ \mu s$ and the duty cycle is

$$T_1/T = 3/4 = 0.75 = 75\ \%$$

Here, we must make $T_1 = 3T_2$ and this condition will be satisfied if we multiply relation (7.14) by 3 and equate it to relation (7.11) subject to the constraint of (7.16). Then,

$$\ln 2 \cdot R_A C = 3 \cdot \frac{R_A \cdot R_B}{R_A + R_B} C \cdot \ln\left(\frac{2R_A - R_B}{R_A - 2R_B}\right)$$

$$\frac{R_B}{R_A + R_B} \cdot \ln\left(\frac{2R_A - R_B}{R_A - 2R_B}\right) = \frac{\ln 2}{3}$$

$$\frac{\ln 2}{3} \cdot \left(\frac{R_A + R_B}{R_B}\right) = \ln\left(\frac{2R_A - R_B}{R_A - 2R_B}\right)$$

$$\frac{\ln 2}{3} \cdot \left(\frac{R_A}{R_B} + 1\right) = \ln\left(\frac{2R_A/R_B - 1}{R_A/R_B - 2}\right)$$

For convenience, we let $x = R_A/R_B$ and the relation above is written as

$$\ln 2 \cdot (x + 1) = 3\ln\left(\frac{2x - 1}{x - 2}\right)$$

Therefore, let us plot above relation with the following MATLAB script to find the x–axis zero crossing.

```
x=2.1:0.01:6; y=log(2).*(x+1)–3.*log((2.*x–1)./(x–2)); plot(x,y);grid
```

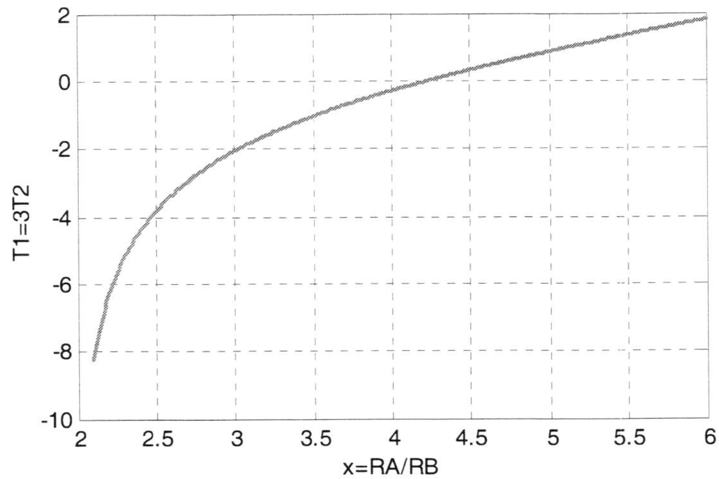

The plot above indicates that x–axis zero crossing occurs at approximately

$$x = R_A/R_B = 4.25$$

and thus to achieve a duty cycle at 75 %, we should choose that the ratio R_A/R_B should be close to this value. The value of the capacitor can be found from either expression (7.11) or (7.14) but (7.11) is simpler. We require that $T_1 = 3\ \mu s$ and assuming $R_A = 5.1\ K\Omega$, we obtain

$$T_1 = 3 \times 10^{-6} = 0.69R_AC = 0.69 \times 5.1 \times 10^3C$$

$$C = \frac{3 \times 10^{-6}}{0.69 \times 5.1 \times 10^3} = 853\ pF$$

Example 7.2 and Exercises 2 and 3 reveal that as the duty cycle decreases the ratio R_A/R_B approaches the value of 2, and as the duty cycle increases, this ratio approaches the value of 11.

4. We will use the monostable multivibrator of Figure 7.21 which is repeated below for convenience. From relation (7.28),

$$T \approx 1.1R_1C$$

With a capacitor of value $C = 1\ nF$ to produce an output pulse of $T = 20\ \mu s$ duration, resistor R_1 must have the value of

$$R_1 = \frac{T}{1.1C} = \frac{20 \times 10^{-6}}{1.1 \times 10^{-9}} = 18.2\ K\Omega$$

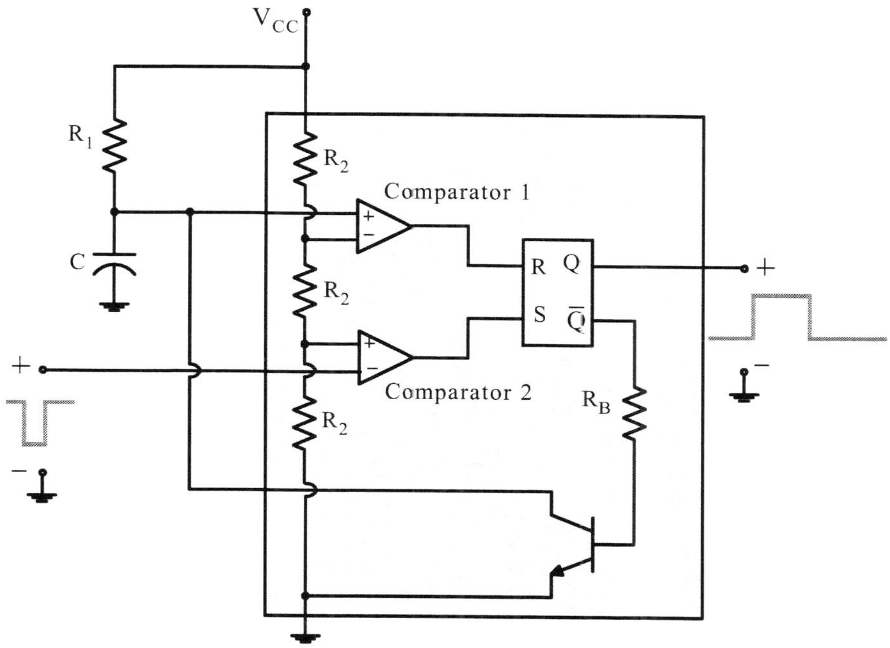

5.

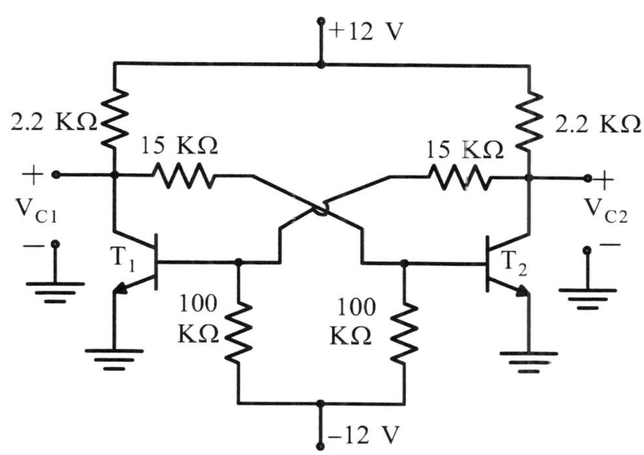

Figure 7.40. Fixed–bias flip–flop circuit for Example 7.3.

As in Example 7.4, we break the given circuit into two parts, the first part indicating the connections between the base of transistor T_1 and the collector of transistor T_2 as shown in Figure (a), and the second part indicating the connection between the collector of T_1 and the base of T_2 as shown in Figure (b).

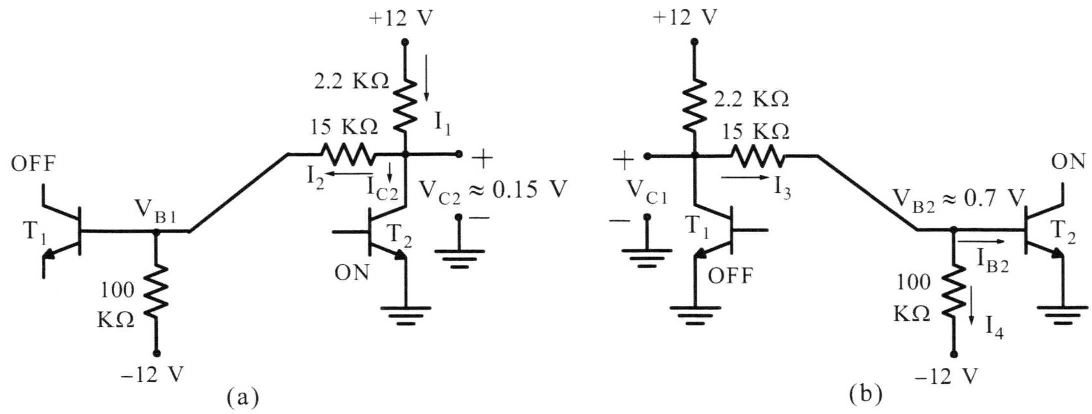

Let us first assume that transistor T_1 is OFF and transistor T_2 is ON. By the voltage division expression and the superposition principle,

$$V_{B1} = \frac{15\ K\Omega}{100\ K\Omega + 15\ K\Omega} \cdot (-12\ V) + \frac{100\ K\Omega}{100\ K\Omega + 15\ K\Omega} \cdot 0.15 = -1.43\ V$$

and this voltage will certainly keep transistor T_1 at cutoff. To verify that with transistor T_1 beyond cutoff transistor T_2 is in saturation, we calculate I_{C2} by first finding I_1 and I_2 as follows:

$$I_1 = \frac{12\ V - 0.15\ V}{2.2\ K\Omega} = 5.39\ mA$$

$$I_2 = \frac{0.15 - (-12)}{15\ K\Omega + 100\ K\Omega} = 0.11\ mA$$

Then,

$$I_{C2} = I_1 - I_2 = 5.39 - 0.11 = 5.28\ mA$$

and

$$I_B(min) = \frac{I_{C2}}{h_{FE}} = \frac{5.28\ mA}{50} = 0.11\ mA$$

Let us now compute the value of I_{B2} from Figure (b). We find that

$$I_3 = \frac{12 - 0.7}{2.2\ K\Omega + 15\ K\Omega} = 0.66\ mA$$

and

$$I_4 = \frac{0.7 - (-12)}{100\ K\Omega} = 0.13\ mA$$

Then,

$$I_{B2} = I_3 - I_4 = 0.66 - 0.13 = 0.53 \text{ mA}$$

Since $(I_{B2} = 0.53 \text{ mA}) > (I_B(min) = 0.11 \text{ mA})$, that is, the value of I_{B2} exceeds the minimum base current value of $I_B(min)$ required for saturation, we have shown that transistor T_2 is deeply into saturation.

Finally, the collector voltage V_{C1} if found from Figure (b) as

$$V_{C1} = 12 - 2.2 \times 10^3 \times I_3 = 12 - 2.2 \times 10^3 \times 0.66 \times 10^{-3} = 10.5 \text{ V}$$

and this value is close to the collector supply voltage of 12 volts.

Therefore, with the values given the more accurate voltages and currents for a stable state for the flip–flop of this example are summarized in the table below.

Parameter	Value
I_{C1}	0 mA
V_{C1}	10.5 V
I_{C2}	5.28 mA
V_{C2}	0.15 V
I_{B1}	0 mA
V_{B1}	−1.43 V
I_{B2}	0.53 mA
V_{B2}	0.7 V

A comparison of the values of Table 7.1 and those of the table above reveals that the error in assuming that a transistor in saturation behaves as an ideal short circuit is very small and this error can be neglected.

The second stable state is the one in which transistor T_1 is ON and transistor T_2 is OFF. The analysis is the same as above where the voltages and currents are interchanged between transistors T_1 and T_2, and $V_{C2} = 10.5 \text{ V}$.

6.

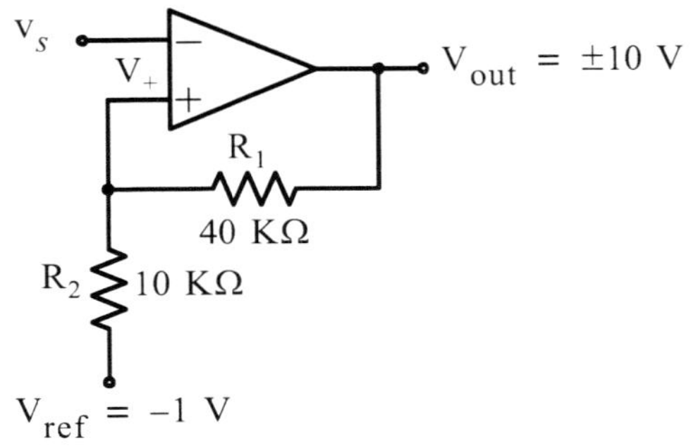

(a)

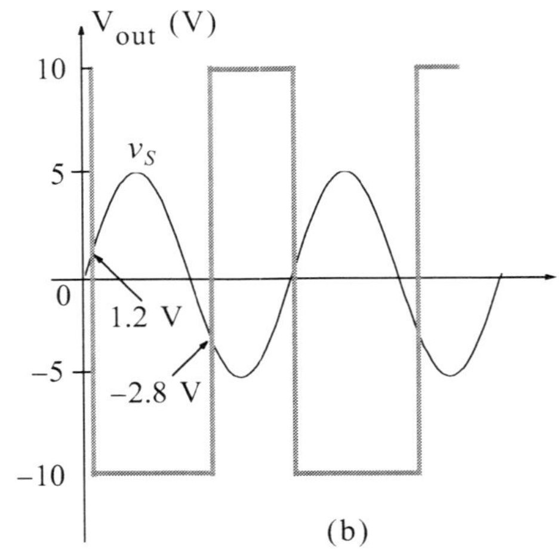

(b)

$$V_{+}upper = \frac{R_1 V_{ref} + R_2 V_{out}}{R_1 + R_2} = \frac{4 \times 10^4 \times (-1) + 10^4 \times 10}{5 \times 10^4} = 1.2 \text{ V}$$

That is, when the input signal v_s exceeds this value, the output abruptly swings to negative saturation, i.e., -10 V as shown.

$$V_{+}lower = \frac{R_1 V_{ref} + R_2 V_{out}}{R_1 + R_2} = \frac{4 \times 10^4 \times (-1) + 10^4 \times (-10)}{5 \times 10^4} = -2.8 \text{ V}$$

That is, when the input signal v_S drops below this value, the output abruptly swings back to positive saturation, i.e., $+10$ V, and this cycle repeats.

Chapter 8

Frequency Characteristics of Single-Stage and Cascaded Amplifiers

This chapter presents certain basic concepts and procedures that are applicable to frequency dependent single–stage and cascaded amplifiers. The intent is to provide the basic principles of the frequency dependence that is an essential prerequisite for the effective design and utilization of most electronic circuits.

8.1 Properties of Signal Waveforms

In the study of electronic systems we encounter an extremely wide variety of signal waveforms. The simplest are sinusoidal with frequencies of 60 Hz or 400 Hz but in most cases we are concerned with a more complicated class of periodic but non–sinusoidal waveforms. For these waveforms we attempt to represent the signals as a superposition of sinusoidal components of appropriate amplitudes, frequencies, and phases. Fortunately, very powerful techniques for this purpose are available such as the Fourier series,* Fourier transform, and the Laplace transformation.

Periodic signals can be represented as the superposition of sinusoidal components by a Fourier series of the form

$$v(t) = V + V_1\cos(\omega t + \theta_1) + V_2\cos(2\omega t + \theta_2) + V_3\cos(3\omega t + \theta_3) + \ldots \tag{8.1}$$

where V is the DC component of the composite signal, ω is the *fundamental frequency* or *first harmonic*, 2ω, is the *second harmonic*, 3ω is the *third harmonic*, and so on.

Unfortunately, not all signals consist of harmonically related components. In the case of speech or music, for instance, the sinusoidal composition of the sound is continuously changing. In this case, the composite signal can be approximated by the superposition of a number of sinusoidal components expressed as

$$v(t) = V + V_1\cos(\omega_1 t + \theta_1) + V_2\cos(\omega_2 t + \theta_2) + V_3\cos(\omega_3 t + \theta_3) + \ldots \tag{8.2}$$

where V is the DC component of the composite signal, and ω_1, ω_2, ω_3, and so on, are not harmonically related.

Figure 8.1, referred to as the *frequency spectrum* of the signal reveals information about the frequencies and amplitude, but it contains no information about the phase angles of the components.

* *For a thorough discussion on Fourier series, the Fourier transform, and the Laplace transformation please refer to Signals and Systems with MATLAB Computing and Simulink Modeling, ISBN 978–1–934404–11–9.*

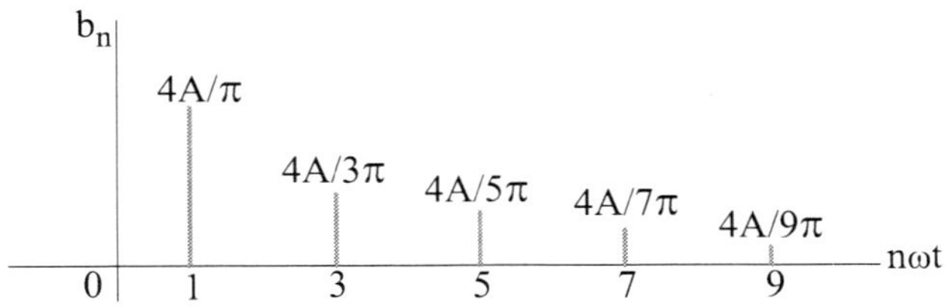

Figure 8.1. Typical frequency spectrum

The frequency spectrum of Figure 8.1 represents the square waveform shown in Figure 8.2.*

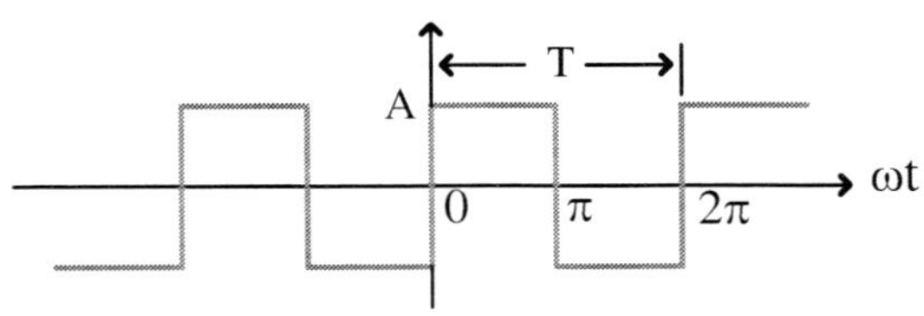

Figure 8.2. Waveform that produces the frequency spectrum in Figure 8.1

Example 8.1

The frequency components of the frequency spectrum of Figure 8.1 are as shown in relation (8.3) below.

$$v(t) = \frac{4A}{\pi}\left(\sin\omega t + \frac{1}{3}\sin 3\omega t + \frac{1}{5}\sin 5\omega t + \dots\right) = \frac{4A}{\pi}\sum_{n\,=\,odd}\frac{1}{n}\sin n\omega t \tag{8.3}$$

What percentage of the average power is absorbed by

a. the fundamental frequency

b. the fundamental and the third harmonic

c. the fundamental, the third, and the fifth harmonic

Solution:

The instantaneous power is $p = v^2/R$ and for simplicity let us assume that $R = 1$. The average power P over a period T is

$$P = \frac{1}{T}\int_0^T v^2 dt = \frac{1}{T}v^2 t\Big|_0^T = v^2 \tag{8.4}$$

We recall that the RMS value of a sinusoid is the amplitude of the sinusoid divided by $\sqrt{2}$. Thus,

* *Please refer to Chapter 7, Signals and Systems with MATLAB Computing and Simulink Modeling, ISBN 978–1–934404–11–9.*

the average power in the harmonics of relation (8.3) is

$$P_1 + P_3 + P_5 + \dots = V_1^2 + V_3^2 + V_5^2 + \dots = \left(\frac{4A}{\sqrt{2}\pi}\cdot 1\right)^2 - \left(\frac{4A}{\sqrt{2}\pi}\cdot\frac{1}{3}\right)^2 + \left(\frac{4A}{\sqrt{2}\pi}\cdot\frac{1}{5}\right)^2 + \dots$$

$$= \frac{8A^2}{\pi^2}\left(1 + \frac{1}{9} + \frac{1}{25} + \dots\right)$$

a. Percentage of power absorbed by the fundamental frequency component:

$$\left(\frac{8}{\pi^2}\cdot 1\right) \times 100 = 81\%$$

b. Percentage of power absorbed by the fundamental frequency and the third harmonic components:

$$\left[\frac{8}{\pi^2}\cdot\left(1 + \frac{1}{9}\right)\right] \times 100 = 90\%$$

c. Percentage of power absorbed by the fundamental frequency, the third, and the fifth harmonic components:

$$\left[\frac{8}{\pi^2}\cdot\left(1 + \frac{1}{9} + \frac{1}{25}\right)\right] \times 100 = 93\%$$

This example indicates that most of the power is absorbed by the fundamental frequency and the first two non–zero harmonics. Of course, the power absorbed would be considerably higher if the signal would include a DC component.

Figure 8.3 shows a typical curve of amplification versus frequency.

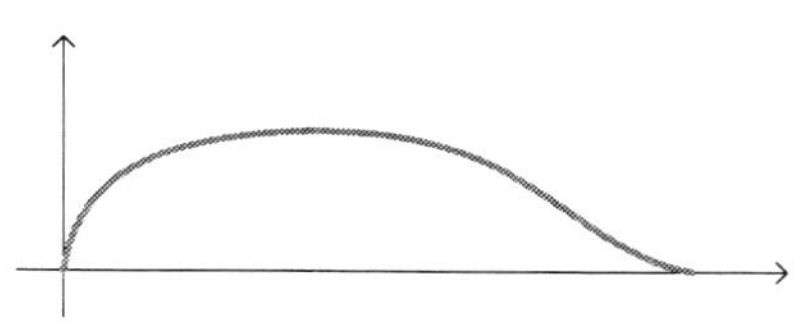

Figure 8.3. Typical frequency characteristics of an amplifier

As shown in Figure 8.3, the amplification decreases at high frequencies as a result of the parasitic capacitances in the circuit, and decreases at low frequencies because of the presence of coupling capacitors. Also, for a good designed amplifier, the amplification near the lower and near the higher band of frequencies should not be less than 70 percent of the amplification in the middle of the band. It turns out that if an amplifier retards the phase of each sinusoidal component of the signal in direct proportion to the frequency of that component, then the waveform of the output signal is an exact copy of the input waveform, but it is delayed in time. This can be illustrated by expressing relation (8.2) with the introduction of a constant time delay T. Then,

$$v(t-T) = V + V_1\cos(\omega_1(t-T)+\theta_1) + V_2\cos(\omega_2(t-T)+\theta_2) + V_3\cos(\omega_3(t-T)+\theta_3) + \dots$$

Letting $\phi_k = \omega_k T$ we obtain

$$v(t-T) = V + V_1\cos(\omega_1 t - \phi_1 + \theta_1) + V_2\cos(\omega_2 t - \phi_2 + \theta_2) + V_3\cos(\omega_3 t - \phi_3 + \theta_3) + \ldots \quad (8.5)$$

Relation (8.5) reveals that the introduction of a constant delay adds a phase lag ϕ_k to each component, directly proportional to the frequency of that component. Reversing the above process we can see that if an amplifier introduces a phase lag that increases linearly with frequency, then $\phi = K\omega$, and the amplifier produces a time delay $T = K$, and if the amplification is the same for all frequencies, the signal is not distorted; it is merely delayed in time. In conclusion, the analysis and design of linear amplifiers is concerned with the characteristics of amplification and phase shift versus frequency.

For the remaining of this chapter and Chapter 9, we will develop procedures for constructing these characteristics. Subsequently, an easy method to begin with the design of amplifiers at low, medium, and high frequencies is to first define the transfer function $G(s)$ gain and the frequencies as the poles of the transfer function and draw the Bode plot using MATLAB.*

The results of the following example will be useful in our analysis and development of equivalent circuits.

Example 8.2

Prove that the circuits of Figure 8.4(a) and 8.4(b) are equivalent.

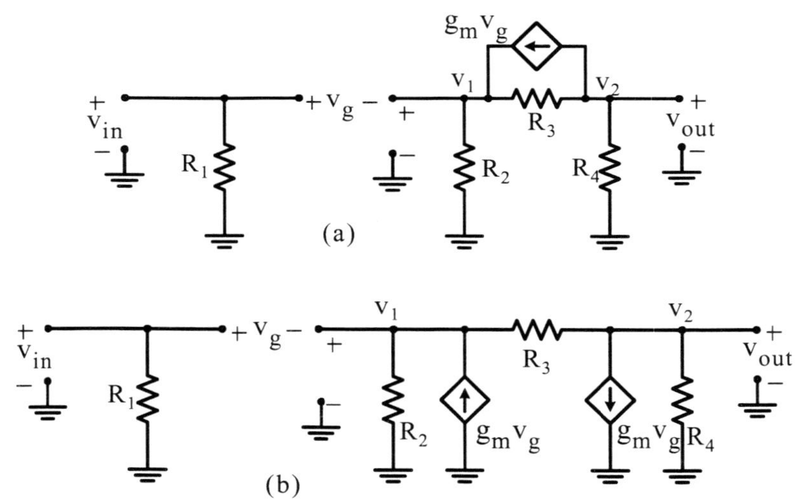

Figure 8.4. Equivalent circuits for Example 8.2

* *Frequency response and Bode plots are discussed in detail in Circuit Analysis II with MATLAB Applications, ISBN 978–0–9709511–5–1.*

Proof:

Application of KCL at v_1 and v_2 of Figure 8.4(a) yields

$$\begin{aligned} \frac{v_1}{R_2} + \frac{v_1 - v_2}{R_3} - g_m v_g &= 0 \\ \frac{v_2 - v_1}{R_3} + \frac{v_2}{R_4} + g_m v_g &= 0 \end{aligned} \tag{8.6}$$

Application of KCL at v_1 and v_2 of Figure 8.4(b) yields

$$\begin{aligned} \frac{v_1}{R_2} - g_m v_g + \frac{v_1 - v_2}{R_3} &= 0 \\ \frac{v_2 - v_1}{R_3} + g_m v_g + \frac{v_2}{R_4} &= 0 \end{aligned} \tag{8.7}$$

We observe that equations (8.6) and (8.7) are the same and thus the circuits of Figures 8.4(a) and 8.4(b) are equivalent.

8.2 The Transistor Amplifier at Low Frequencies

Figure 8.5 shows a transistor amplifier and its equivalent circuits that can be used as a low frequency amplifier.

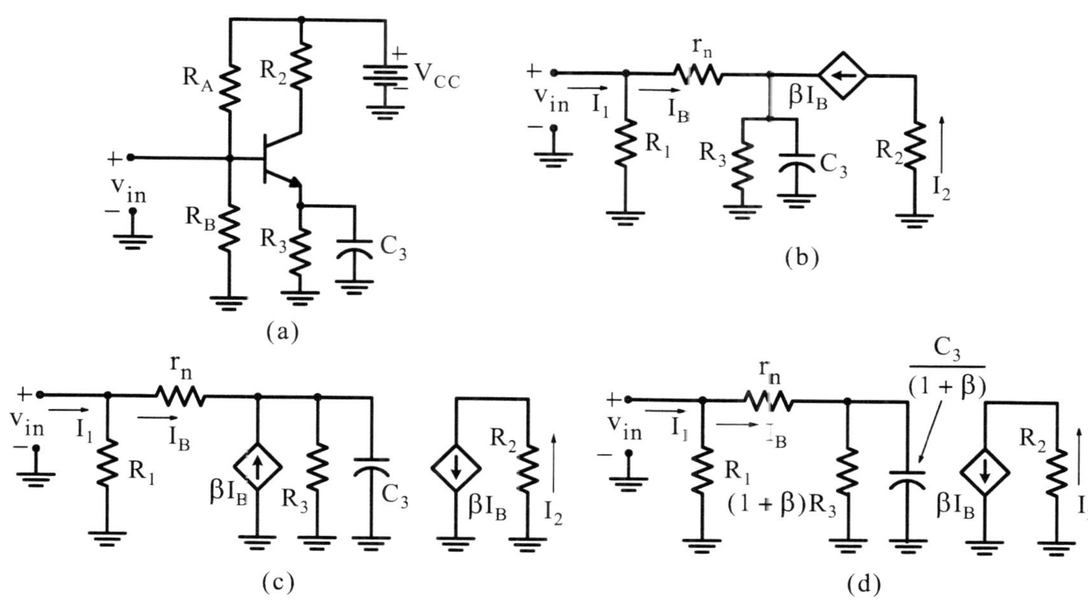

Figure 8.5. Transistor amplifier and equivalent circuits at low frequencies

The behavior of the amplifier of Figure 8.5(a) at low frequencies is affected by the action of capacitor C_3. An incremental model for the amplifier that is valid at low and medium frequencies is shown in Figure 8.5(b). The parameters shown are approximations and this circuit should not be used for precision calculations; they are used primarily for estimating how large C_3 must be for satisfactory performance. For such purposes, it is good engineering practice to use coarse approximations whenever they contribute a substantial simplification of the problem without destroying the usefulness of the results.

As we've learned in Example 8.2, the current source βI_B in the circuit of Figure 8.5(b) can be replaced by an equivalent pair of sources to obtain the equivalent circuit shown in Figure 8.5(c). The reduction theorem* is then applied to obtain the simplified equivalent circuit of Figure 8.5(d). In a typical circuit R_1 may be 5 or 10 kilohms, and $(1+\beta)R_3$ may be 50 or 100 kilohms; hence at very low frequencies, where C_3 acts as an open circuit, most of the signal current flows through R_1 rather than into the base of the transistor. The current amplification is reduced accordingly. It follows that for large, uniform amplification, C_3 should be chosen to act as a short circuit at all frequencies in the band occupied by the sinusoidal components of the signal.

By inspection of the circuit in Figure 8.5(d), the current gain is

$$A_C = I_2/I_1 = \beta I_b/I_1 \tag{8.8}$$

and

$$I_b = \frac{R_1}{R_1 + r_n + Z_{eq}} \cdot I_1 \tag{8.9}$$

where Z_{eq} is the parallel combination of resistance $(1+\beta)R_3$ and capacitance $C_3/(1+\beta)$. Thus,

$$Z_{eq} = \frac{(1+\beta)R_3 \times 1/j\omega(C_3/(1+\beta))}{(1+\beta)R_3 + 1/j\omega(C_3/(1+\beta))} = \frac{(1+\beta)R_3}{1+j\omega R_3 C_3}$$

$$I_b = \frac{R_1}{R_1 + r_n + \dfrac{(1+\beta)R_3}{1+j\omega R_3 C_3}} \cdot I_1 \tag{8.10}$$

We now recall that the amplitude of a low–pass filter is given by†

$$|G(j\omega)| = \left|\frac{V_{out}}{V_{in}}\right| = \frac{1}{\sqrt{1+\omega^2 R^2 C^2}}$$

and the half–power (3 dB) frequency is defined at $|G(j\omega)| = 1/\sqrt{2}$ which occurs when $\omega = 1/RC$. Accordingly, in (8.10) let us define the frequency

* *The reduction theorem is discussed in Appendix E.*

† *Please refer to Chapter 11, Signals and Systems with MATLAB Computing and Simulink Modeling, ISBN 978–1–934404–11–9.*

$$\omega_3 = 1/R_3C_3 \tag{8.11}$$

Factoring $R_1 + r_n$ out of the denominator in (8.10) and multiplying both numerator and denominator by $1 + j\omega/\omega_3$, we obtain

$$I_b = \frac{R_1 I_1}{R_1 + r_n} \cdot \frac{1 + j\omega/\omega_3}{1 + \frac{(1+\beta)R_3}{R_1 + r_n} + \frac{j\omega}{\omega_3}} \tag{8.12}$$

and with (8.8) we find that the current gain is given by

$$A_C = \frac{I_2}{I_1} = \frac{\beta I_b}{I_1} = \frac{\beta R_1}{R_1 + r_n} \cdot \frac{1 + j\omega/\omega_3}{1 + \frac{(1+\beta)R_3}{R_1 + r_n} + \frac{\omega}{\omega_3}} \tag{8.13}$$

In the medium–frequency range the term ω/ω_3 is very large, and the second factor on the right side in (8.13) approaches unity; hence the first factor is the current gain at medium frequencies. Relation (8.13) can be expressed in a more convenient form by defining two new terms as

$$A_m = \frac{\beta R_1}{R_1 + r_n} \tag{8.14}$$

and

$$k_3 = \frac{(1+\beta)R_3}{R_1 + r_n} \tag{8.15}$$

Substitution of (8.14) and (8.15) in (8.13) and factoring $1 + k_3$ out the denominator yields

$$A_C = \frac{A_m}{1 + k_3} \cdot \frac{1 + j\omega/\omega_3}{1 + j\omega/(1 + k_3)\omega_3} \tag{8.16}$$

Relation (8.16) can be used as a guide in the design of transistor amplifiers for low–frequency responses once the quiescent operating point has been established. The bypass capacitor is then chosen to make the break in the amplitude characteristic at $\omega = (1 + k_3)\omega_3$ occur at a suitable point below the band of frequencies occupied by the signal. Changing the value of C_3 does not change the shape of the ramp in the amplitude characteristic; it only shifts the ramp parallel to the frequency axis.

Example 8.3

For the transistor amplifier in Figure 8.5, it is known that $R_1 = 3\ K\Omega$, $r_n = 12\ K\Omega$, $R_3 = 1\ K\Omega$, $\beta = 80$, and $C_3 = 2.5\ \mu F$. Construct the logarithmic amplitude characteristic for this amplifier in the low– and medium–frequency range.

Solution:

The amplification at medium frequencies is given by (8.40). Thus

$$A_m = \frac{\beta R_1}{R_1 + r_n} = \frac{80 \times 3}{3 + 12} = 18$$

and in decibels

$$A_m(dB) = 20\log 18 = 20 \times 1.25 = 25 \text{ dB}$$

The break frequency ω_3 in (8.16) is

$$\omega_3 = \frac{1}{R_3 C_3} = \frac{1}{1000 \times 2.5 \times 10^{-6}} = 400 \text{ rad/sec}$$

or

$$f_3 = \frac{\omega_3}{2\pi} = \frac{400}{2\pi} \approx 64 \text{ Hz}$$

The factor k_3 is found from (8.15), that is,

$$k_3 = \frac{(1+\beta)R_3}{R_1 + r_n} = \frac{(1+80) \times 1}{3 + 12} = 5.4$$

and the break frequency for the factor $(1 + k_3)\omega_3$ in the denominator of relation (8.16) is

$$(1 + k_3)\omega_3 = (1 + 5.4) \times 400 = 2560 \text{ rad/sec}$$

or

$$(1 + k_3)f_3 = (1 + 5.4) \times 64 = 410 \text{ Hz}$$

The amount by which the amplification at low frequencies is less than its value at medium frequencies is

$$20\log(1 + k_3) = 20 \times 0.8 = 16 \text{ dB}$$

Therefore, from DC to the break frequency at 64 Hz, the gain is $25 - 16 = 9 \text{ dB}$. The amplitude characteristic is shown in Figure 8.6 where the frequency is shown in logarithmic scale and the gain is in dB scale.

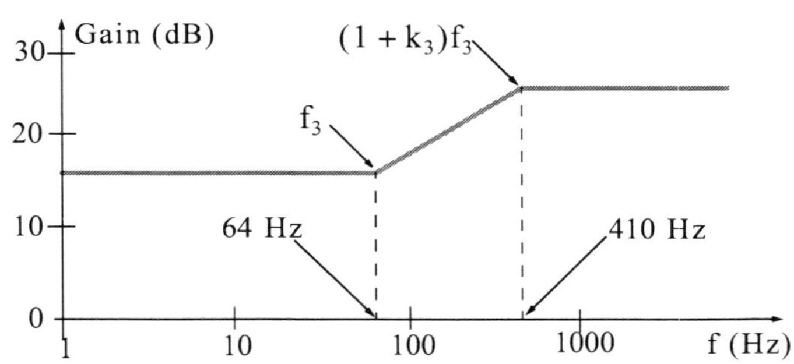

Figure 8.6. Amplitude characteristic for the amplifier of Example 8.3

A smooth curve shown in Figure 8.7 can be obtained with the following MATLAB script:

```
f=1:10:10^(6); w=2.*pi.*f; beta=80; R1=3000; R3=1000; C3=2.5.*10.^(-6); rn=12000;...
w3=1./(R3.*C3); Am=(beta.*R1)./(rn+R1); k3=((1+beta).*R3)./(rn+R1);...
Ac=(Am.*(1+j.*w./w3))./((1+k3)*(1+j.*w./((1+k3).*w3))); y=20*log10(abs(Ac)); semilogx(f,y); grid
```

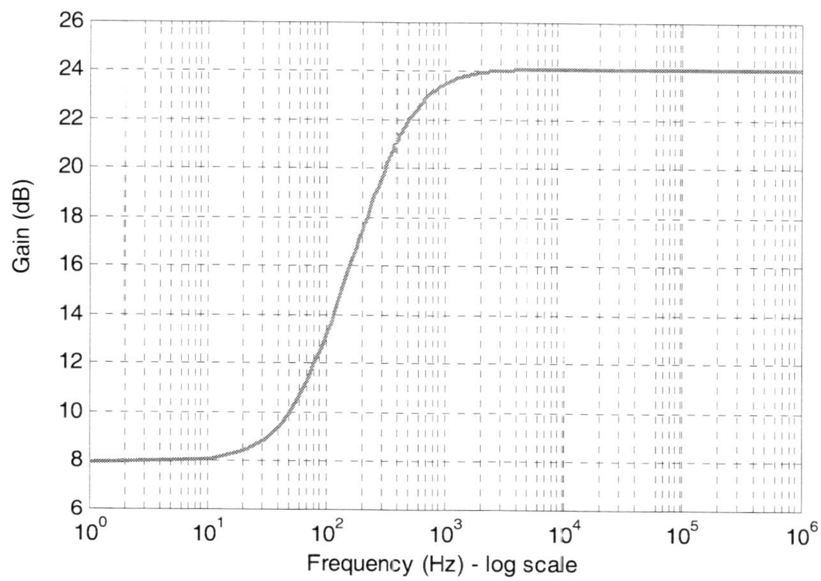

Figure 8.7. Gain versus frequency plot for Example 8.3 with MATLAB

8.3 The Transistor Amplifier at High Frequencies

In Chapter 3, Section 3.13, we introduced the hybrid−π model for a transistor at high frequencies and this model is repeated in Figure 8.8 for convenience.

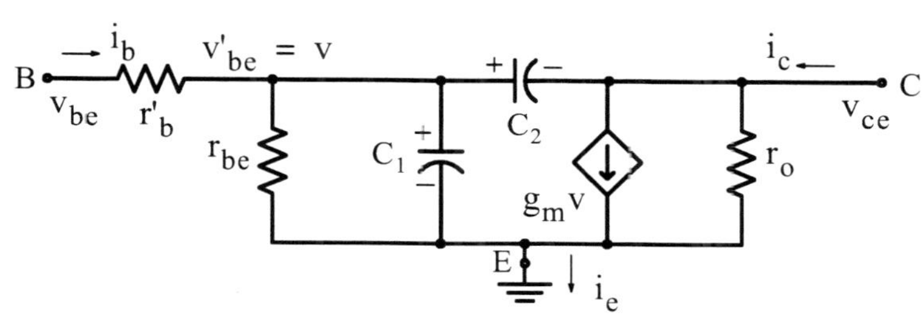

Figure 8.8. The hybrid−π model for the transistor at high frequencies

An incremental model for the transistor amplifier in Figure 8.9(a) that is valid at medium and high frequencies is shown in Fig. 8.9(b).

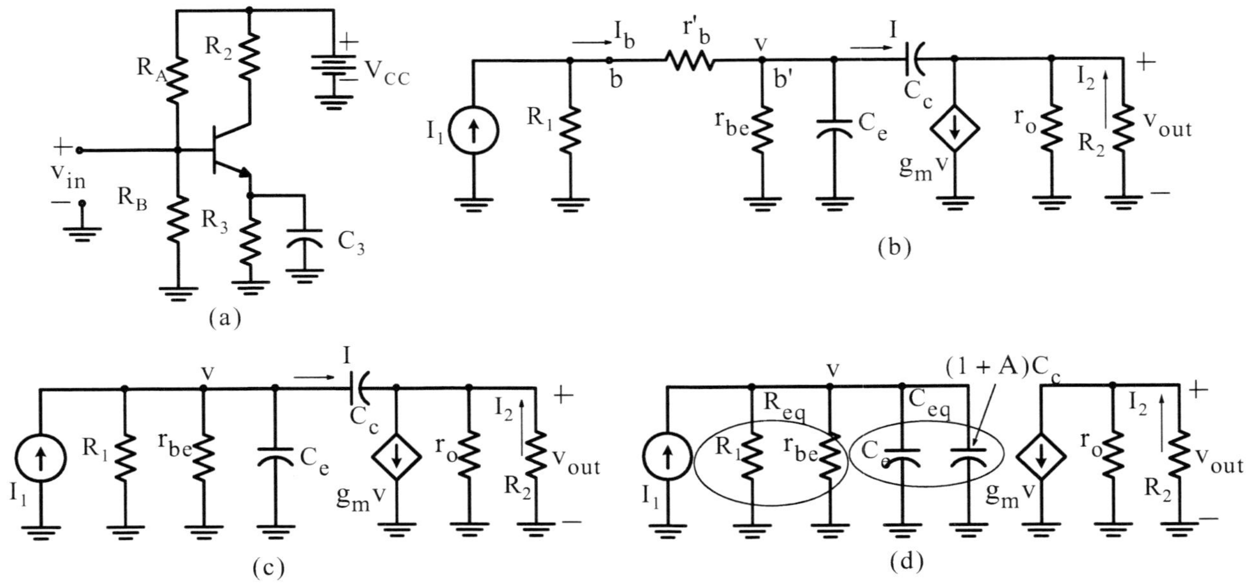

Figure 8.9. Transistor amplifier and equivalent circuits at high frequencies

The transistor circuit of Figure 8.9(a) is represented by the hybrid–π model of Figure 8.9(b), where the resistance R_1 represents resistances R_A, R_B, and the resistance associated with the source I_1. The behavior of this amplifier is affected at high frequencies by the parasitic capacitances C_e and C_c. At high frequencies, these capacitances tend to short–circuit node b' to ground and this results in a significant reduction of the voltage v and the current $g_m v$. In the circuit of Figure 8.9(b), resistance r'_b is typically in the order of 50 to 100 ohms while resistance R_1 is in the order of several kilohms and thus we can eliminate r'_b by replacing the portion of the circuit on the left of r_{be} with a Norton equivalent circuit and Figure 8.9(c) shows the circuit after this modification. The current I flowing through capacitor C_c is given by

$$I = j\omega C_c(v - v_{out}) \tag{8.17}$$

and for the useful frequency range this current is much smaller than the current $g_m v$. Therefore, with the assumption that $I \ll g_m v$, we find that

$$v_{out} = \frac{r_o R_2}{r_o + R_2}(-g_m v) \tag{8.18}$$

and letting

$$A = \frac{-g_m r_o R_2}{(r_o + R_2)} \tag{8.19}$$

we obtain

$$v_{out} = -Av \tag{8.20}$$

Typically, the gain A in (8.20) is in the range of 10 to 1000. Also, $r_o \gg R_2$, and thus $A \approx g_m R_2$. Therefore, we can express (8.17) as

$$I = j\omega C_c(v + Av) = j\omega C_c(1 + A)v \approx j\omega C_c Av \tag{8.21}$$

This relation enable us to replace the current I by a capacitor of the value $(1 + A)C_c$ as shown in Figure 8.9(d), and as we know from Appendix C, this amplifying property of the transistor which increases the apparent value of the collector capacitance is known as the *Miller effect*.

From the circuit of Figure 8.9(d),

$$I_2 = \frac{r_o}{r_o + R_2} g_m v \tag{8.22}$$

$$C_{eq} = C_e + (1 + A)C_c \tag{8.23}$$

$$R_{eq} = \frac{r_{be} R_1}{r_{be} + R_1} \tag{8.24}$$

$$v = \left(\frac{1/j\omega C_{eq}}{R_{eq} + 1/j\omega C_{eq}} \cdot I_1 \right) R_{eq} = \frac{R_{eq}}{1 + j\omega R_{eq} C_{eq}} \cdot I_1 \tag{8.25}$$

The current gain A_c is now found by division of (8.22) by (8.25). Thus,

$$A_c = \frac{I_2}{I_1} = \frac{r_o}{r_o + R_2} \cdot \frac{g_m R_{eq}}{1 + j\omega R_{eq} C_{eq}} \tag{8.26}$$

Next, we let

$$A_m = \frac{r_o}{r_o + R_2} g_m R_{eq} \tag{8.27}$$

and we define the half–power (3 dB) frequency as

$$\omega_1 = \frac{1}{R_{eq} C_{eq}} \tag{8.28}$$

Substitution of (8.27) and (8.28) into (8.26) yields

$$A_c = A_m \frac{1}{1 + j\omega/\omega_1} \tag{8.29}$$

At medium frequencies ω is small, that is, $\omega \ll \omega_1$, and the second factor on the right side in

(8.26) is close to unity; hence the quantity A_m in relation (8.27) is the current gain at medium frequencies. The useful frequency range for the amplifier is considered to extend up to the half–power frequency given by (8.26). Therefore, the magnitude of the frequency characteristics for the transistor amplifier at medium and high frequencies have the form shown in Figure 8.10.

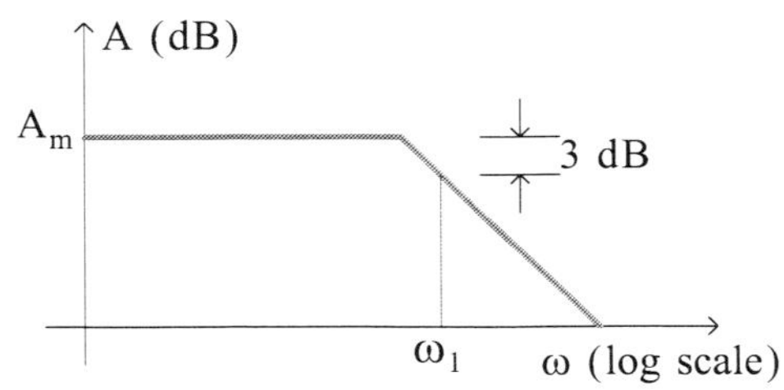

Figure 8.10. Typical logarithmic amplitude characteristics for the transistor amplifier at high frequencies

Substitution of (8.23) and (8.24) into (8.28) yields

$$\omega_1 = \frac{r_{be} + R_1}{[C_e + (1 + A)C_c]r_{be}R_1} \tag{8.30}$$

and since A is much larger than unity, (8.30) can be expressed as

$$\omega_1 = \frac{1}{r_{be}R_1} \cdot \frac{1 + r_{be}/R_1}{1 + AC_c/C_e} \tag{8.31}$$

In Chapter 3 we defined the β cutoff frequency (relation 3.88) as $\omega_\beta = 1/r_{be}C_e$; hence we can express (8.31) as

$$\omega_1 = \omega_\beta \cdot \frac{1 + r_{be}/R_1}{1 + AC_c/C_e} \tag{8.32}$$

or

$$f_1 = f_\beta \cdot \frac{1 + r_{be}/R_1}{1 + AC_c/C_e} \tag{8.33}$$

Thus, relation (8.33) provides us the half–power point for the current gain of a transistor at high frequencies. Obviously, the half–power point frequency f_1 can be greater or smaller than the β cutoff frequency f_β depending on the values of resistor R_1 and the gain $A = g_m R_2$.

Example 8.4

The model for a transistor amplifier at high frequencies is shown in Figure 8.11.

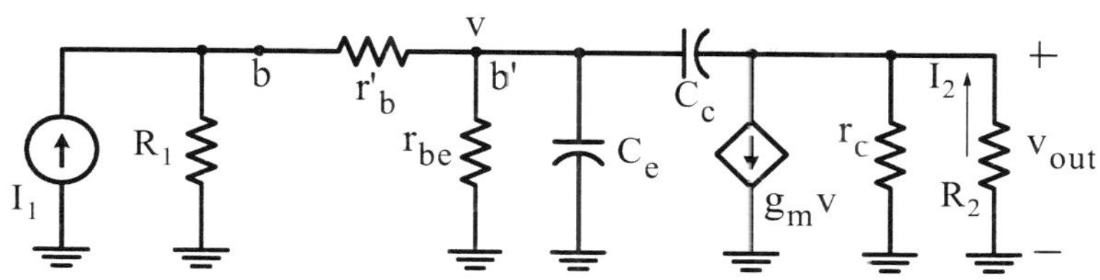

Figure 8.11. Model of the transistor amplifier for Example 8.6

For this circuit it is known that $R_1 = 5\ K\Omega$, $R_2 = 1\ K\Omega$, the quiescent collector current is $I_C = 5\ mA$, $\beta = 80$, $r'_b = 50\ \Omega$, $C_c = 3\ pF$, $r_o = 50\ K\Omega$, and the current gain–bandwidth frequency is $f_T = 400\ MHz$. Determine:

a. the Miller effect $(1 + A)C_c$ at high frequencies

b. the medium frequency current amplification A_m

c. the half–power frequency f_1

Solution:

Since $R_1 \gg r'_b$ and $r_o \gg R_2$, the given model circuit can be simplified to that shown in Figure 8.12.

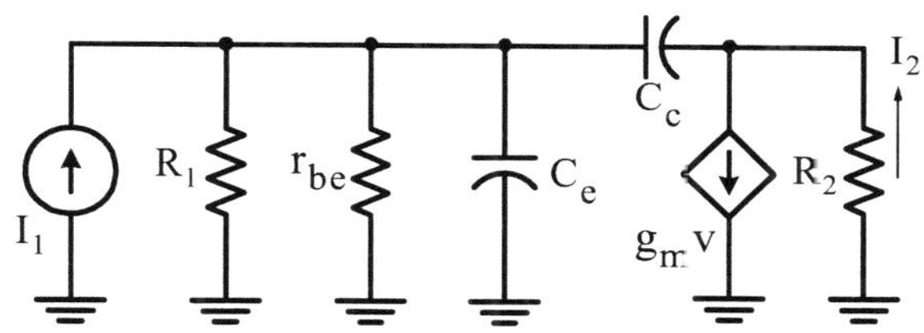

Figure 8.12. Simplified model of the transistor amplifier for Example 8.6

From Chapter 3, relation (3.78),

$$g_m = 40I_C = 40 \times 5 \times 10^{-3} = 0.2\ \Omega^{-1}$$

From relation (3.83),

$$r_{be} = \frac{\beta}{g_m} = \frac{80}{0.2} = 400\ \Omega$$

and from (3.92),

$$f_\beta = \frac{f_T}{\beta} = \frac{400}{80} = 5\ MHz$$

This indicates that large current amplification is provided at frequencies up to $5\ MHz$.

From (3.93),

$$C_e = \frac{g_m}{\omega_T} = \frac{0.2}{2\pi \times 4 \times 10^8} = 79.6\ pF$$

a. The Miller effect is $(1 + A)C_c$ where

$$A = g_m R_2 = 0.2 \times 1000 = 200$$

Thus,

$$(1 + A)C_c = 201 \times 3 \times 10^{-12} = 603 \times 10^{-12} = 603 \text{ pF}$$

and this indicates that the dominant effect at high frequencies is the Miller effect.

b. From Figure 8.12,

$$R_{eq} = \frac{r_{be} R_1}{r_{be} + R_1} = \frac{400 \times 5 \times 10^3}{5.4 \times 10^3} = 370 \ \Omega$$

and the medium frequency current amplification A_m is given by (8.27) as

$$A_m = \frac{r_o}{r_o + R_2} g_m R_{eq}$$

and since $r_o \gg R_2$,

$$A_m \approx g_m R_{eq} = 0.2 \times 370 = 74$$

This value is close to the value of β of the transistor and thus it is near the maximum value that can be obtained by this transistor.

c. The half–power frequency f_1 is found from (8.33). Then,

$$f_1 = f_\beta \cdot \frac{1 + r_{be}/R_1}{1 + AC_c/C_e} = 5 \times 10^6 \cdot \frac{1 + 400/5000}{1 + 200 \times 3 \times 10^{-12}/(80 \times 10^{-12})} = 0.635 \text{ MHz}$$

8.4 Combined Low– and High–Frequency Characteristics

The frequency dependence of various amplifier circuits is examined by treating the low–frequency characteristics separately from the high–frequency characteristics. However, the combined low– and high–frequency characteristics can be displayed on a single set of coordinates as shown in Figure 8.13

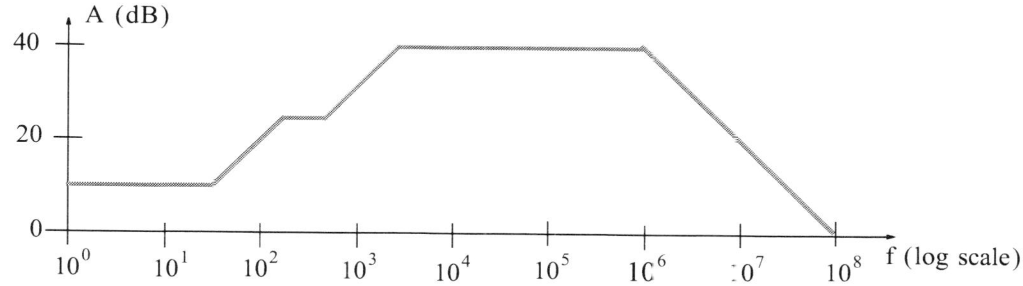

Figure 8.13. Combined low– and high–frequency amplitude versus frequency characteristics

8.5 Frequency Characteristics of Cascaded Amplifiers

The magnitude of the signal amplification obtainable from single–stage voltage and current amplifiers is limited to a factor of a less than 100 depending on the type of transistor employed. Since there are many applications requiring greater signal amplifications than this, it is common practice to connect a number of stages in cascade so that each stage amplifies the signal in succession; the total amplification is then the product of the amplifications of the individual stages. However, cascading amplifiers in this manner introduces some additional considerations in circuit design, and it requires the use of additional circuit components that affect the behavior of the circuit.

In most cases, it is possible to analyze, or design, each stage in a multistage amplifier as a separate part of the overall circuit and to determine the overall properties of the amplifier by combining the results of these separate analyses. Thus, the results obtained in the previous sections of this chapter are to a large extent directly applicable in the analysis of cascaded amplifiers; they are usually modified in some respects, however, by the networks used to couple the individual stages in cascade.

The properties of cascaded amplifiers are usually such that the signal transmission tends to zero at both low and high frequencies; hence these amplifiers are referred to as *bandpass networks*. In some cases cascaded amplifiers are required to amplify signals occupying a wide band of frequencies; the principal design problem is then to obtain uniform amplification over a sufficiently wide band of frequencies. It is also possible to design a particular circuit to amplify signals in a certain frequency range and to exclude of all other signals at other undesirable frequencies; the design problem in these cases is to obtain uniform amplification in the desired band with sufficiently strong discrimination against signals in adjacent bands.

Figure 8.14 shows an RC–coupled transistor amplifier circuit and its high–frequency equivalents.

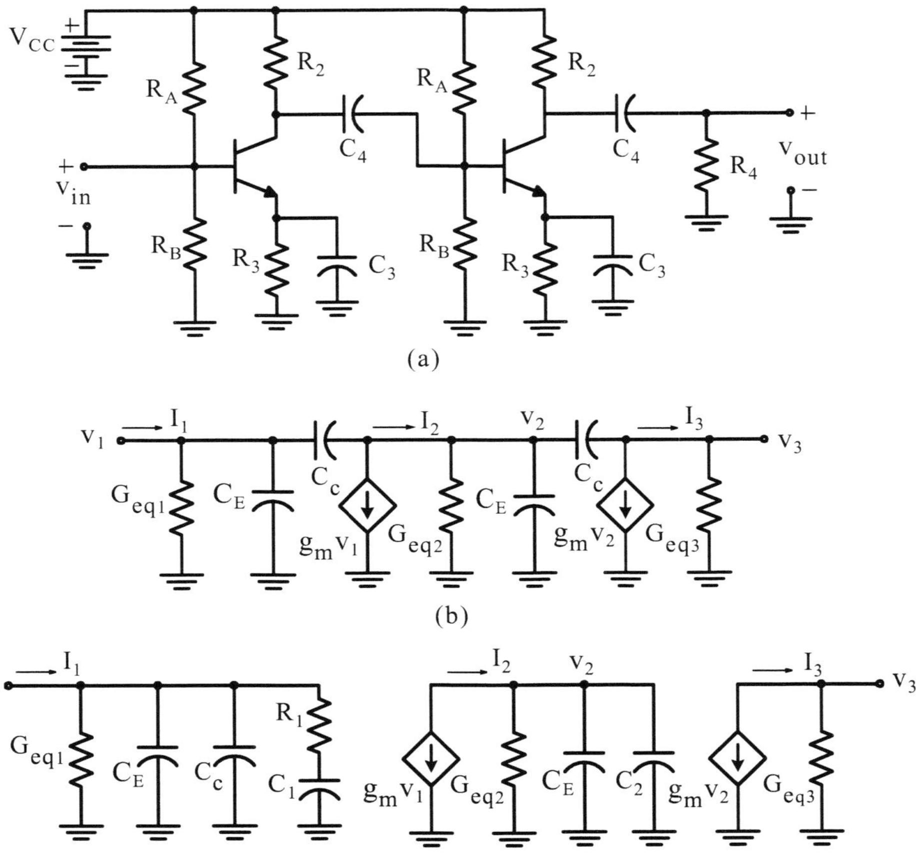

Figure 8.14. Circuit of a cascaded RC–coupled transistor amplifier and its high frequency models

In Figure 8.14(a), capacitor C_4 is referred to as the *interstage coupling capacitor* and its function is to block the DC component through the RC coupling network. Transistors can also be direct coupled under certain conditions.

A high–frequency model for the cascaded transistor stages is shown in Figure 8.14(b) where we have assumed that the base spreading resistance r'_{be} is negligible. The conductances G_{eq1}, G_{eq2}, and G_{eq3}, are the equivalent conductances connected between the three nodes v_1, v_2, v_3, and ground. Thus,

$$G_{eq1} = \frac{1}{R_A} + \frac{1}{R_B} + \frac{1}{r_{be}} \tag{8.34}$$

$$G_{eq2} = \frac{1}{R_A} + \frac{1}{R_B} + \frac{1}{r_{be}} + \frac{1}{R_2} \tag{8.35}$$

and

$$G_{eq3} = \frac{1}{R_2} + \frac{1}{R_4} \tag{8.36}$$

The model for the second stage of the amplifier has the same form as the model for a single–stage amplifier shown in Figure 8.9(c); hence the effect of the collector capacitance in this stage can be accounted for by a Miller capacitance as shown in Figure 8.9(d). When this is done, the second stage has the simplified representation shown in Figure 8.14(c). The Miller capacitance has the value

$$C_2 = (1 + A_2)C_C \tag{8.37}$$

where

$$A_2 = |v_3/v_2| \tag{8.38}$$

Similarly,

$$A_1 = |v_2/v_1| \tag{8.39}$$

The model for the first stage of the amplifier differs from that of the second stage in that its load has a capacitive component as well as a resistive component; hence the Miller effect is slightly different in the first stage. The first stage is as shown in Figure 8.14(c) where C_1 is the shunt capacitance that is present in the load of the first stage transistor, and the resistor R_1 is negligible in comparison to the reactance of capacitor C_1. Typically, R_1 is one kilohm or less.

The value of R_1 depends on the mid–band value of the gain A_1 as defined in (8.39), the collector capacitance C_c, and the half–power frequency for the output circuit of the first stage defined as

$$\omega_2 = \frac{G_{eq2}}{C_E - C_2 + C_C} \tag{8.40}$$

In terms of relations (8.39) and (8.40), the values of R_1 and C_1 can be found from

$$R_1 = \frac{1}{\omega_2 A_1 C_C} = \frac{C_E + C_2 + C_C}{g_m C_C} \tag{8.41}$$

and

$$C_1 = A_1 C_C \tag{8.42}$$

We can also express R_1 in terms of C_1 using (8.42), that is,

$$R_1 = \frac{1}{\omega_2 C_1} \tag{8.43}$$

If the resistance R_1 in the circuit of Figure 8.24(c) were negligible in comparison with the reactance of C_1, then the simplified representation for the first stage would reduce to the same form as that for the second stage, and the process could be repeated for each stage in a cascade of any number of stages. However, relation (8.43) shows that at $\omega = \omega_2$, $R_1 = 1/\omega_2 C_1$ where ω_2 is the half–power frequency for the output circuit of the first stage. Thus, it is convenient to neglect R_1 at frequencies up to ω_2. If R_1 is not neglected, and if the amplifier of Figure 8.24(c) is preceded by yet another transistor stage, the Miller effect in this stage has a still more complicated form. Of course, the Miller effect computations become more and more complicated as more stages are cascaded. To avoid these complications, it is practical to neglect R_l and consider all computations as approximations. Then with these approximations, the representation for each stage in the cascade takes a simple form, and stage–by–stage analysis is easy.

From Figure 8.14(c),

$$I_3 = -g_m v_2 \tag{8.44}$$

$$I_2 = [G_{eq2} + j\omega(C_E + C_2)]v_2 \tag{8.45}$$

and division of (8.44) by (8.45) yields

$$A_{c2} = \frac{I_3}{I_2} = \frac{-g_m}{G_{eq2} + j\omega(C_E + C_2)} = -\frac{g_m}{G_{eq2}} \cdot \frac{1}{1 + j\omega(C_E + C_2)/G_{eq2}} \tag{8.46}$$

Also, from Figure 8.14(c),

$$I_2 = -g_m v_1 \tag{8.47}$$

and neglecting R_l,

$$I_1 = [G_{eq1} + j\omega(C_E + C_c + C_1)]v_1 \tag{8.48}$$

division of (8.47) by (8.48) yields

$$A_{c1} = \frac{I_2}{I_1} = \frac{-g_m}{G_{eq1} + j\omega(C_E + C_c + C_1)} = -\frac{g_m}{G_{eq1}} \cdot \frac{1}{1 + j\omega(C_E + C_c + C_1)/G_{eq1}} \tag{8.49}$$

Example 8.5

For the cascaded amplifier circuit of Figure 8.15, the transistors are identical with $R_A = 15\ K\Omega$, $R_B = 3.3\ K\Omega$, $R_2 = 2.6\ K\Omega$, $R_3 = 0.68\ K\Omega$, and $R_4 = 0.5\ K\Omega$. The quiescent collector current in each stage is $I_C = 4\ mA$, and at this operating point $\beta = 100$, $r'_b = 50\ \Omega$, $C_C = 4\ pF$, $f_T = 100\ MHz$, and the output resistance r_o is very large and it can be neglected. Compute the overall current gain A_c at medium and high frequencies. Sketch and dimension the asymptotes for the high–frequency amplitude characteristics.

Solution:

From Chapter 3, relation (3.78),

$$g_m = 40I_C = 40 \times 4 \times 10^{-3} = 0.16\ \Omega^{-1}$$

From relation (3.83),

$$r_{be} = \frac{\beta}{g_m} = \frac{100}{0.16} = 625\ \Omega$$

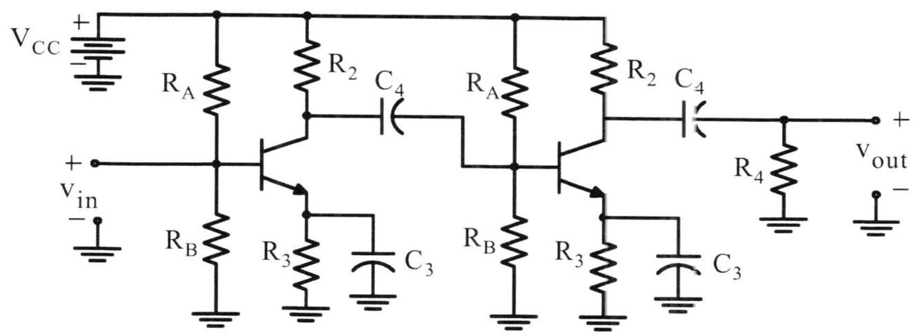

Figure 8.15. Cascaded amplifier for Example 8.5

and from (3.93),

$$C_E = \frac{g_m}{\omega_T} = \frac{0.16}{2\pi \times 10^8} = 255 \text{ pF}$$

For the second stage of the amplifier, from (8.36),

$$G_{eq3} = \frac{1}{R_2} + \frac{1}{R_4} = \frac{1}{2.6 \text{ K}\Omega} + \frac{1}{0.5 \text{ K}\Omega} = 2.4 \text{ m}\Omega^{-1}$$

From (8.27)

$$A_m = \frac{r_o}{r_o + R_2} g_m R_{eq}$$

and since $r_o \gg R_2$, the current gain of the second stage amplifier at medium frequencies is

$$A_{m2} = g_m R_{eq} = \frac{g_m}{G_{eq3}} = \frac{0.16}{2.4 \times 10^{-3}} = 67$$

From (8.37) the Miller effect capacitance is

$$C_2 = (1 + A_{m2})C_C = (1 + 67)4 = 272 \text{ pF}$$

From (8.35),

$$G_{eq2} = \frac{1}{R_A} + \frac{1}{R_B} + \frac{1}{r_{be}} + \frac{1}{R_2} = \frac{1}{15 \text{ K}\Omega} + \frac{1}{3.3 \text{ K}\Omega} + \frac{1}{50 \text{ }\Omega} + \frac{1}{2.6 \text{ K}\Omega} = 2.4 \text{ m}\Omega^{-1}$$

From (8.46),

$$A_{c2} = \frac{I_3}{I_2} = -\frac{g_m}{G_{eq2}} \cdot \frac{1}{1 + j\omega(C_E + C_2)/G_{eq2}}$$

$$= -\frac{0.16}{2.4 \times 10^{-3}} \cdot \frac{1}{1 + j\omega(255 + 272) \times 10^{-12}/2.4 \times 10^{-3}} = -67\frac{1}{1 + j\omega 2.2 \times 10^{-7}}$$

and expressing ω in megaradians per second, we find that

$$A_{c2} = -67\frac{1}{1 + j\omega/4.5}$$

With the frequency expressed in MHz, we obtain

$$A_{c2} = -67\frac{1}{1+jf/0.72}$$

For the first stage of the amplifier we found above that

$$G_{eq2} = 2.4 \ \ m\Omega^{-1}$$

From (8.29)

$$A_m = \frac{r_o}{r_o+R_2}g_m R_{eq}$$

and since $r_o \gg R_2$, the current gain of the first stage amplifier at medium frequencies is

$$A_{m1} = g_m R_{eq} = \frac{g_m}{G_{eq2}} = \frac{0.16}{2.4\times 10^{-3}} = 67$$

From (8.37) the Miller effect capacitance is

$$C_1 = (1+A_{m1})C_C = (1+67)4 = 272 \ pF$$

From (8.34),

$$G_{eq1} = \frac{1}{R_A}+\frac{1}{R_B}+\frac{1}{r_{be}} = \frac{1}{15 \ K\Omega}+\frac{1}{3.3 \ K\Omega}+\frac{1}{625 \ \Omega} = 2 \ \ m\Omega^{-1}$$

From (8.49),

$$A_{c1} = \frac{I_2}{I_1} = -\frac{g_m}{G_{eq1}}\cdot\frac{1}{1+j\omega(C_E+C_c+C_1)/G_{eq1}}$$

$$= -\frac{0.16}{2\times 10^{-3}}\cdot\frac{1}{1+j\omega(255+4+272)\times 10^{-12}/2\times 10^{-3}} = -80\frac{1}{1+j\omega 2.66\times 10^{-7}}$$

and expressing ω in megaradians per second, we find that

$$A_{c1} = -80\frac{1}{1+j\omega/3.8}$$

With the frequency expressed in MHz, we obtain

$$A_{c1} = -80\frac{1}{1+jf/0.60}$$

The overall current gain A_c is obtained by multiplying A_{c1} by A_{c2}. Thus,

$$A_c = A_{c1}A_{c2} = \left(-80\frac{1}{1+jf/0.60}\right)\left(-67\frac{1}{1+jf/0.72}\right) = 5360\cdot\frac{1}{1+jf/0.60}\cdot\frac{1}{1+jf/0.72}$$

Figure 8.16 shows a sketch showing the asymptotes for the high–frequency amplitude characteristics where the frequency is shown in a log scale and the gain in dB where $A_c(dB) = 20\log 5360 \approx 75$, and the slope of each high–frequency asymptote is –20 dB per decade.

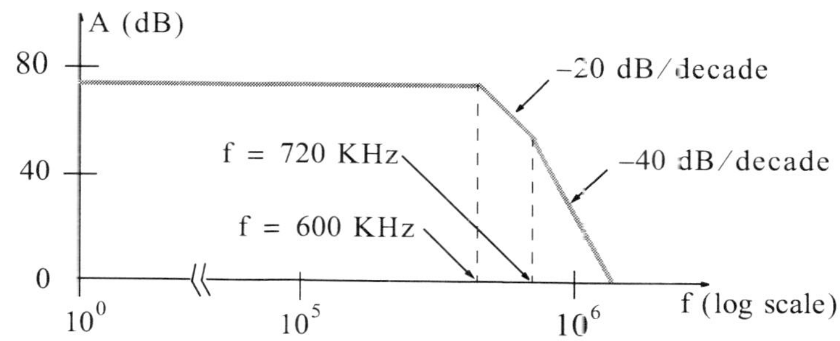

Figure 8.16. Asymptotes for the high–frequency characteristics of the amplifier for Example 8.5

Figure 8.17 shows a cascaded transistor amplifier circuit and its model that is valid at low and medium frequencies. With r_o assumed to be very large, we used the same procedure as in the simplification of Figure 8.5(a) to derive the simple equivalent circuit of Figure 8.5(d). The characteristics of the amplifier are affected at low frequencies by the coupling and bypass capacitors.

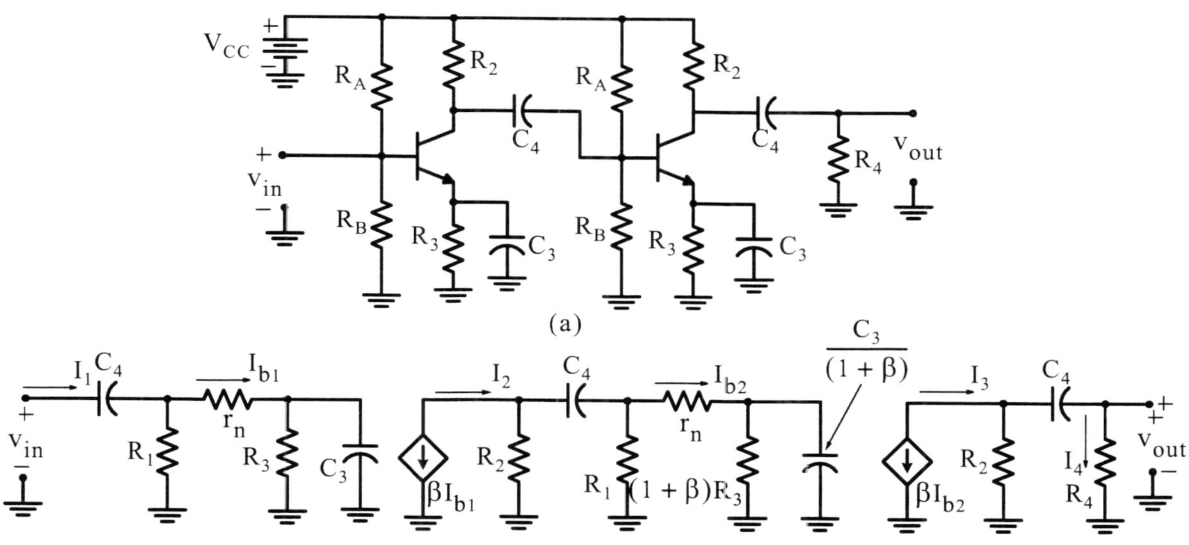

Figure 8.17. RC–coupled amplifier circuit and its model at low frequencies

The effect of the coupling capacitor at the output of the amplifier is determined from right side part of Figure 8.17(b) with the use of the current division expression as follows:

$$I_4 = \frac{R_2}{R_2 + R_4 + 1/j\omega C_4} \cdot I_3$$

or

$$\frac{I_4}{I_3} = \frac{R_2}{R_2 + R_4 + 1/j\omega C_4} = \frac{R_2}{R_2 + R_4} \cdot \frac{j\omega C_4(R_2 + R_4)}{1 + j\omega C_4(R_2 + R_4)} \tag{8.50}$$

Defining the frequency ω_4 as

$$\omega_4 = \frac{1}{C_4(R_2 + R_4)} \tag{8.51}$$

we express (8.50) as

$$\frac{I_4}{I_3} = \frac{R_2}{R_2 + R_4} \cdot \frac{j\omega/\omega_4}{1 + j\omega/\omega_4} \tag{8.52}$$

To simplify the procedure of deriving an expression for the gain I_3/I_2 we make use of the fact that the value required for the bypass capacitor C_3 is much larger than the value required for the coupling capacitor C_4. Thus, we choose the capacitors so that the low–frequency half–power point is determined by C_3, and with this choice C_4 acts as a short circuit while C_3 is making the transition from a short circuit to an open circuit, and with this transition, the interstage network has the same form as the amplifier shown in Figure 8.5(d), and the current gain for each stage has the same form as (8.16). Thus, with C_4 acting as a short circuit we find that

$$A_C = -\frac{A_m}{1 + k_3} \cdot \frac{1 + j\omega/\omega_3}{1 + j\omega/(1 + k_3)\omega_3}\ ^* \tag{8.53}$$

where

$$\omega_3 = 1/R_3C_3 \tag{8.54}$$

and

$$A_m = \frac{\beta R'_1}{R_1 + r_n} \tag{8.55}$$

$$k_3 = \frac{(1 + \beta)R_3}{R'_1 + r_n} \tag{8.56}$$

where, with C_4 acting as a short circuit, R'_1 is the parallel combination of R_1 and R_2 in Figure 8.17(b). Thus,

$$R'_1 = \frac{R_1 R_2}{R_1 + R_2} \tag{8.57}$$

At very low frequencies C_3 acts as an open circuit and the interstage coupling capacitor C_4 contributes the factor

$$\frac{j\omega/\omega_5}{1 + j\omega/\omega_5} \tag{8.58}$$

* *The minus (–) sign stems from the fact that the current I_3 in Figure 8.17(b) has a direction opposite to the current I_2 in Figure 8.5(d).*

where

$$\omega_5 = \frac{1}{R_{eq}C_4} \tag{8.59}$$

and R_{eq} is the equivalent resistance when C_4 acts as a short circuit and C_3 as an open circuit. Thus,

$$R_{eq} = R_1 \| R_2 + r_n + (1+\beta)R_3 \tag{8.60}$$

Usually, $(1+\beta)R_3$ is much larger than the other resistors in that group, and thus

$$\omega_5 \approx \frac{1}{(R_1 + R_2)C_4} \tag{8.61}$$

and with the factor contributed by the interstage coupling capacitor C_4, the complete expression for the current gain is

$$A_C = -\frac{A_m}{1+k_3} \cdot \frac{1+j\omega/\omega_3}{1+j\omega/(1+k_3)\omega_3} \cdot \frac{j\omega/\omega_5}{1+j\omega/\omega_5} \tag{8.62}$$

Normally, the design process begins with the selection of the proper resistor values to establish a suitable quiescent point. Then, C_3 is chosen to make the break at $(1+k_3)\omega_3$ occur at a desired point below the band of the high frequencies of the signal. Finally C_4 is chosen to make the break at ω_5 occur at some frequency below ω_3 where C_3 acts as an open circuit.

Example 8.6

A cascaded transistor amplifier circuit and its model that is valid at low and medium frequencies are shown in Figure 8.18.

The transistors are identical with $R_A = 56\ K\Omega$, $R_B = 12\ K\Omega$, $R_2 = 6.8\ K\Omega$, $R_3 = 2.2\ K\Omega$, and $R_4 = 1\ K\Omega$. The quiescent collector current in each stage is $I_C = 1\ mA$, and the transistor parameters are $\beta = 200$, $r'_b = 200\ \Omega$, $C_C = 20\ pF$, $f_T = 8\ MHz$, and the output resistance r_o is very large in comparison with R_2 and thus it can be neglected. The coupling and bypass capacitors are to be chosen to yield suitable performance at low frequencies.

a. Find the values of the resistances in the simplified low–frequency model in Figure 8.28(b).

b. The break frequency associated with the coupling capacitor C_4 at the output of the amplifier is to occur at $16\ Hz$. Compute the value of C_4 required in the output circuit.

c. The highest break frequency associated with the interstage coupling network is to be $(1+k_3)f_3 = 160\ Hz$. Find ω_3, and, treating C_4 in the interstage network as a short circuit, find the required value for the bypass capacitor C_3.

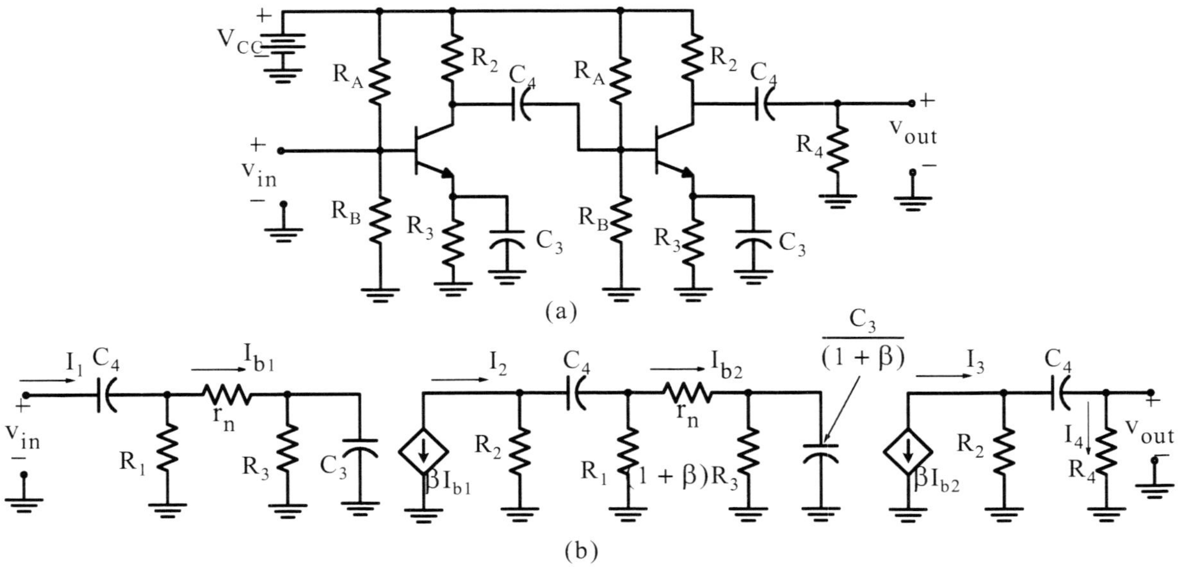

Figure 8.18. Cascaded amplifier circuit and its low–frequency equivalent for Example 8.6.

d. The break frequency associated with the coupling capacitor C_4 in the interstage network, denoted as ω_5 in relation (8.87), is to be at least as small as ω_3. Treating C_3 as an open circuit, find the minimum value for C_4 in the interstage network.

Solution:

a. The values of $R_2 = 6.8\ K\Omega$, $R_3 = 2.2\ K\Omega$, and $R_4 = 1\ K\Omega$ are given. Also,

$$R_1 = \frac{R_A R_B}{R_A + R_A} = \frac{56 \times 12}{56 + 12} = 9.9\ K\Omega$$

$$(1+\beta)R_3 = (1+200)2.2 = 442\ K\Omega$$

From Chapter 3, relation (3.78),

$$g_m = 40I_C = 40 \times 1 \times 10^{-3} = 40\ m\Omega^{-1}$$

From relation (3.83),

$$r_{be} = \frac{\beta}{g_m} = \frac{200}{40 \times 10^{-3}} = 5\ K\Omega$$

and from (3.79),

$$r_n = r'_b + r_{be} = 200\ \Omega + 5\ K\Omega = 5.2\ K\Omega$$

b. From relation (8.51),

$$C_4 = \frac{1}{(R_2 + R_4)\omega_4} = \frac{1}{(6.8+1) \times 10^3 \times 2\pi \times 16} = 1.28\mu F$$

c. From relation (8.57),

$$R'_1 = \frac{R_1 R_2}{R_1 + R_2} = \frac{9.9 \times 6.8}{9.9 + 6.8} = 4.0\ K\Omega$$

and from (8.56),

$$k_3 = \frac{(1+\beta)R_3}{R'_1 + r_n} = \frac{(1+200) \times 2.2}{4.0 + 5.2} = 48.1$$

We are given that $(1 + k_3)f_3 = 160\ Hz$. Then,

$$f_3 = \frac{160}{48.1} = 3.33\ Hz$$

or

$$\omega_3 = 2\pi f_3 = 2\pi \times 3.33 = 21\ rad/s$$

and from (8.54),

$$C_3 = \frac{1}{\omega_3 R_3} = \frac{1}{21 \times 2.2 \times 10^3} = 21.6\ \mu F$$

d. With $\omega_5 = \omega_3 = 21\ rad/s$ and with (8.61),

$$C_4 = \frac{1}{(R_1 + R_2)\omega_5} = \frac{1}{(9.9 + 6.8) \times 10^3 \times 21} = 2.85\ \mu F$$

The computations for all examples and the end–of–chapter exercises, can be greatly simplified by writing a MATLAB script such as that shown below and plot the overall current gain.

```
w=1:10:10^5; RA=56000; RB=12000; R2=6800; R3=2200; R4=1000;...
beta=200; rbprime=200; Ic=10^(-3); f4=16; w4=2*pi*f4;...
R1=RA*RB/(RA+RB); R1prime=R1*R2/(R1+R2); gm=40*Ic; rbe=beta/gm;...
rn=rbprime+rbe; k3=(1+beta)*R3/(R1prime+rn); f3=160/(1+k3);...
w3=2*pi*f3; C3=1/(w3*R3); w5=w3; C4b=1/((R2+R4)*w4);...
C4d=1/((R1+R2)*w5); Am=beta*R1prime/(rn+R1prime);...
Acnum=(Am.*(1+j.*w./w3).*(j.*w./w5));...
Acden=(1+k3).*(1+j.*w./((1+k3).*w3)).*(1+j.*w./w5);...
Ac=Acnum./Acden; AcdB=20.*log10(abs(Ac));...
fprintf(' \n'); fprintf('RA = %2.2f K \t',RA/1000); fprintf(' \n');...
fprintf('RB = %2.2f K \t',RB/1000); fprintf(' \n');...
fprintf('R2 = %2.2f K \t',R2/1000); fprintf(' \n');...
fprintf('R3 = %2.2f K \t',R3/1000); fprintf(' \n');
fprintf('R4 = %2.2f K \t',R4/1000); fprintf(' \n');...
fprintf('R1 = %2.2f K \t',R1/1000); fprintf(' \n');...
fprintf('R1prime = %2.2f K \t',R1prime/1000); fprintf(' \n');...
fprintf('rbprime = %2.2f K \t',rbprime/1000); fprintf(' \n');...
fprintf('rbe = %2.2f K \t',rbe/1000); fprintf(' \n');...
fprintf('rn = %2.2f K \t',rn/1000); fprintf(' \n');...
fprintf('Ic = %2.2f mA \t',Ic*1000); fprintf(' \n');...
```

```
fprintf('beta = %2.0f \t',beta); fprintf(' \n');...
fprintf('f4 = %2.2f Hz \t',f4); fprintf(' \n');...
fprintf('w4 = %2.2f rps \t',w4); fprintf(' \n');...
fprintf('C4b = %2.2f microF \t',C4b*10^6); fprintf(' \n');...
fprintf('gm = %2.2f mmho \t',gm*1000); fprintf(' \n');...
fprintf('k3 = %2.2f \t',k3); fprintf(' \n');...
fprintf('f3 = %2.2f Hz \t',f3); fprintf(' \n');...
fprintf('w3 = %2.2f rps \t',w3); fprintf(' \n');...
fprintf('C3 = %2.2f microF \t',C3*10^6); fprintf(' \n');...
fprintf('w5 = %2.2f rps \t',w5); fprintf(' \n');...
fprintf('C4d = %2.2f microF \t',C4d*10^6); fprintf(' \n');...
fprintf('Am = %2.2f \t',Am); fprintf(' \n');...
f=w./(2.*pi); semilogx(f, AcdB); grid;...
xlabel('Frequency in Hz (log scale)'); ylabel('Current gain (dB scale)');...
title('Low–frequency characteristics for multistage amplifier, Example 8.8')
```

```
RA = 56.00 K
RB = 12.00 K
R2 = 6.80 K
R3 = 2.20 K
R4 = 1.00 K
R1 = 9.88 K
R1prime = 4.03 K
rbprime = 0.20 K
rbe = 5.00 K
rn = 5.20 K
Ic = 1.00 mA
beta = 200
f4 = 16.00 Hz
w4 = 100.53 rps
C4b = 1.28 microF
gm = 40.00 mmho
k3 = 47.92
f3 = 3.27 Hz
w3 = 20.55 rps
C3 = 22.12 microF
w5 = 20.55 rps
C4d = 2.92 microF
Am = 87.30
```

The plot for the current gain (in dB scale) versus frequency (in log scale) is shown in Figure 8.19.

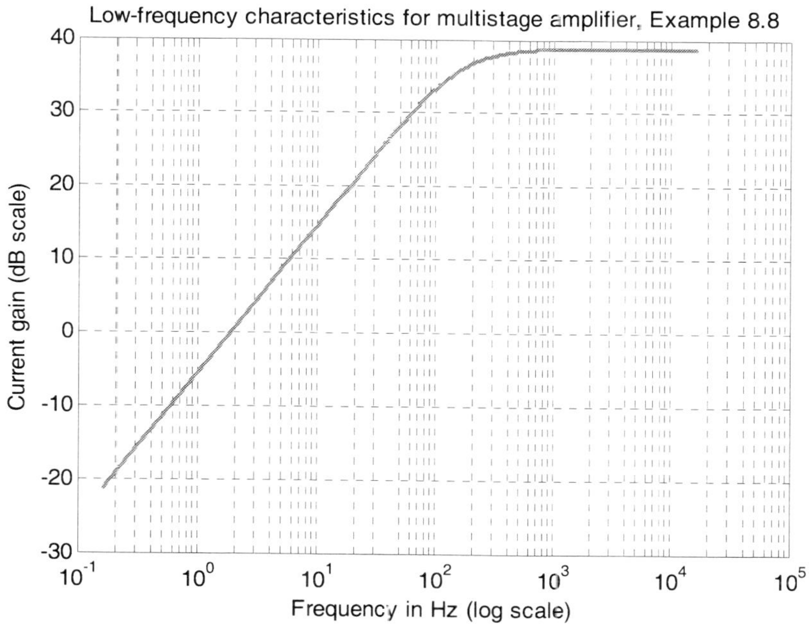

Figure 8.19. Amplitude of current gain versus frequency for Example 8.8

8.6 Overall Characteristics of Multistage Amplifiers

The asymptotes for the amplitude and phase characteristics for a typical RC–coupled stage amplifier are shown in Figure 8.20 where the symbols f_L and f_H define the low– and high–frequency ranges respectively. At low frequencies the voltage gain A_{VL} is obtained from the relation

$$A_{VL} = -A_{mL}\frac{j\omega/\omega_L}{1 + j\omega/\omega_L} \tag{8.63}$$

where A_{mL} depends on the circuit parameters, and ω_L is the low–frequency half–power point. For instance, for the transistor amplifier circuit of Figure 8.5 at low frequencies, $\omega_L = \omega_3$ is as defined in relation (8.11), and A_{mL} is as defined in relation (8.14).

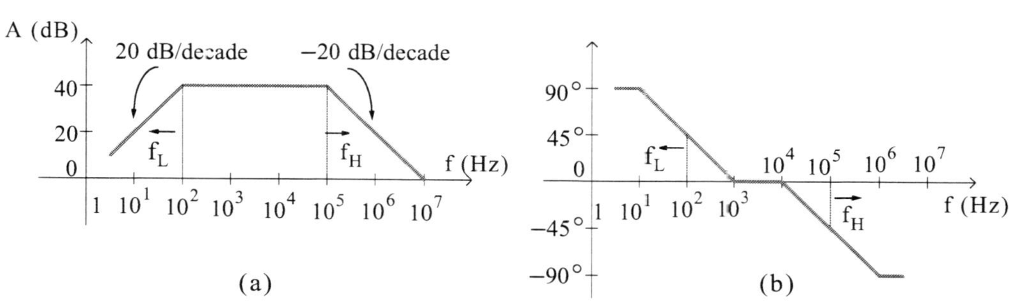

Figure 8.20. Frequency characteristics for a typical single stage amplifier

At high frequencies the voltage gain A_{vH} is obtained from the relation

$$A_{vH} = -A_{mH}\frac{1}{1+j\omega/\omega_H} \tag{8.64}$$

where A_{mH} depends on the circuit parameters, and ω_H is the high–frequency half–power point. For instance, for the transistor amplifier circuit of Figure 8.9 at high frequencies, $\omega_H = \omega_1$ is as defined in relation (8.28), and A_{mH} is as defined in relation (8.27).

We can obtain the combined low– and high–frequency characteristics for the voltage gain A_v by multiplying relation (8.89) by (8.90); thus,

$$A_v = A_{mL}A_{mH}\frac{j\omega/\omega_L}{(1+j\omega/\omega_L)(1+j\omega/\omega_H)} \tag{8.65}$$

If two identical stages such as that shown in Figure 8.18, are connected in cascade, the overall voltage gain A_v is obtained from the relation

$$A_v = A_m^2\frac{(j\omega/\omega_L)^2}{(1+j\omega/\omega_L)^2(1+j\omega/\omega_H)^2} \tag{8.66}$$

The overall amplitude characteristics for two identical stages in cascade is shown in Figure 8.21.

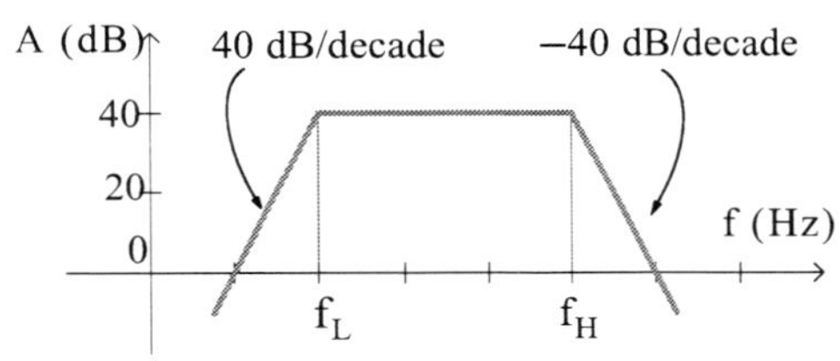

Figure 8.21. Typical overall amplitude characteristics for two identical stages in cascade

By comparing the amplitude characteristics of Figures 8.20 and 8.21, we observe that the bandwidth between the half–power points decreases as more stages are connected in cascade. It can be shown* that the half–power bandwidth of the cascade of n identical stages is given by

$$B_n = B_1\sqrt{2^{1/n}-1} \tag{8.67}$$

where $B_1 = \omega_H$ and ω_H is the half–power frequency for on stage amplifier at medium and high frequencies. The quantity $\sqrt{2^{1/n}-1}$ is referred to as the *bandwidth reduction factor* for n identical stages in cascade. For two stages the bandwidth reduction factor is 0.64 , for three stages is 0.51 , and for four stages it is 0.43 .

* *The proof is left as an exercise for the reader at the end of this chapter.*

If two non–identical stages are connected in cascade, the overall voltage gain is given by

$$A_v = A_{m1}A_{m2}\frac{(j\omega/\omega_{L1})(j\omega/\omega_{L2})}{(1+j\omega/\omega_{L1})(1+(j\omega/\omega_{L2}))(1+j\omega/\omega_{H1})(1+j\omega/\omega_{H2})} \quad (8.68)$$

and the overall amplitude and phase characteristics are obtained by addition of the constituent characteristics. The asymptotes for a typical amplitude characteristic are shown in Figure 8.22 where the notations 20 and 40 stand for 20 dB/decade and 40 dB/decade respectively.

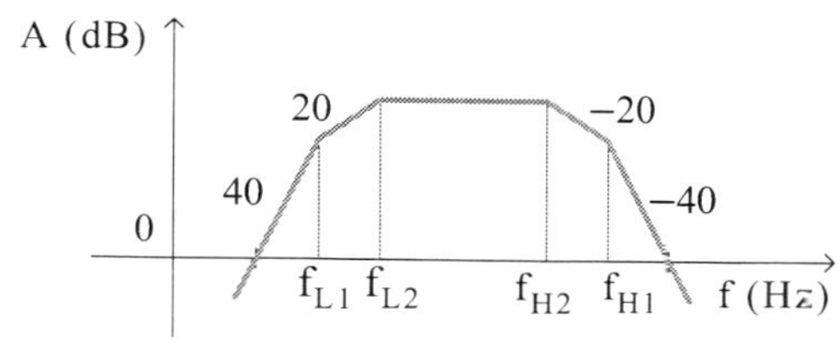

Figure 8.22. Typical overall amplitude characteristics for two non-identical stages in cascade

Relation (8.65) is more conveniently expressed in a pole–zero pattern by letting

$$K = -A_{mL}A_{mH}\omega_H \quad (8.69)$$

Then,

$$A_v(j\omega) = K\frac{j\omega}{(j\omega+j\omega_L)(j\omega+j\omega_H)} \quad (8.70)$$

The roots of the denominator of an expression for $A_v(j\omega)$ may be purely imaginary such as $j\omega_L = -100$ or complex conjugates such as $j\omega_H = -3+j\sqrt{2}$. However, we are interested on the values that lie on the $j\omega$ axis of the complex frequency plane. Therefore, it is convenient to substitute the symbol s for $j\omega$ and after the roots of a quadratic equation in s are found, we can let $s = j\omega$. Replacing $j\omega$ with s and defining $s_L = -\omega_L$, and $s_H = -\omega_H$, relation (8.70) is written as

$$A_v(s) = K\frac{s}{(s-s_L)(s-s_H)} \quad (8.71)$$

and we can obtain the steady–state voltage gain for any frequency ω by letting $s = j\omega$.

For all practical electric and electronic circuits, voltage and current gains can be expressed as the ratio of two polynomials in the variable s, and when the highest power of the denominator is greater than the highest degree of the numerator, the expression is referred to as a *rational function*. The values of s that make a rational function zero are called *zeros* of the function; hence $s = 0$ is a zero of the function of (8.71). The values of s that make rational functions infinite are called *poles* of the function; hence s_L and s_H are poles of the function of (8.71).

As indicated by the form of (8.71), if all the poles and zeros of a rational function are known, the function is completely specified except for a constant multiplier. For example, if the zeros of a certain voltage gain are -1 and -3, and if the poles are -2 and -4, then the voltage gain is

$$A_V(s) = K\frac{(s+1)(s+3)}{(s+2)(s+4)} = K\frac{s^2+4s+3}{s^2+6s+8} \tag{8.72}$$

and for steady–state sinusoidal conditions,

$$A_V(j\omega) = K\frac{(j\omega+1)(j\omega+3)}{(j\omega+2)(j\omega+4)} = \frac{3K}{8}\cdot\frac{(1+j\omega/1)(1+j\omega/3)}{(1+j\omega/2)(1+j\omega/4)} \tag{8.73}$$

Thus if the poles and zeros of a rational function are known, the logarithmic amplitude and phase characteristics can be constructed, except for the effect of the constant multiplier. If the constant multiplier is not unity, it adds a constant number of decibels to the amplitude characteristic; if it is not positive, it adds an angle of $180°$ to the phase shift. It is also interesting to note that the poles and zeros of $A_V(j\omega)$ correspond to the break frequencies expressed in radians per second. A zero corresponds to a break of $20\ dB/decade$ upward in the asymptotic characteristic, and a pole corresponds to a break of $20\ dB/decade$ downward.

The relation between the poles and zeros of the voltage or current gain and the frequency characteristics is displayed in a useful way by the *pole–zero diagram*. For instance, Figure 8.23 shows the pole–zero diagram for the voltage gain of a single–stage amplifier obtained from relation (8.71).

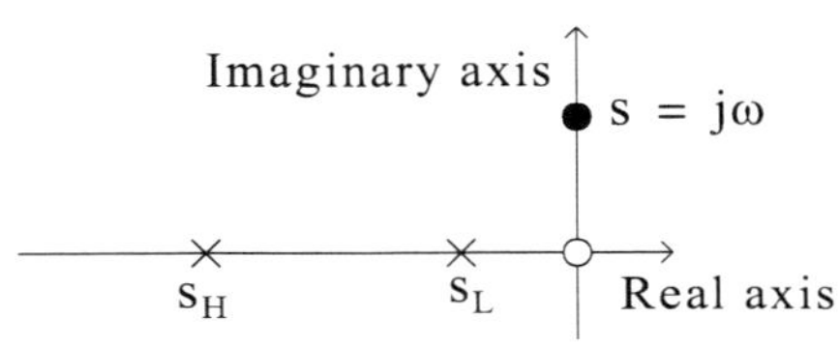

Figure 8.23. Pole–zero pattern for relation (8.71)

As shown in Figure 8.23, the pole–zero diagram is a set of rectangular coordinates known as the complex plane. The zero of $A_V(s)$ which occurs when $s = 0$, is denoted by a circle at the origin of the complex plane. The poles of $A_V(s)$ which occur when $s = s_L = -\omega_L$ and $s = s_H = -\omega_H$ are represented by crosses on the negative real axis. For steady–state sinusoidal operation, the variable $s = j\omega$ corresponds to some point on the imaginary axis shown as a filled circle.

The design of networks to provide a desired frequency characteristic begins with the choice of a pole–zero pattern that yields an approximation to the desired performance. The design is then completed by finding a network configuration and the parameter values that yield the desired pol–zero pattern.

Example 8.7

The pole–zero pattern for the voltage gain of a certain amplifier is shown in Figure 8.24. The amplification at high frequencies is $40\ dB$. Construct the asymptotes for the logarithmic amplitude characteristic.

Solution:

With the given values of the poles and zeros, the voltage gain is

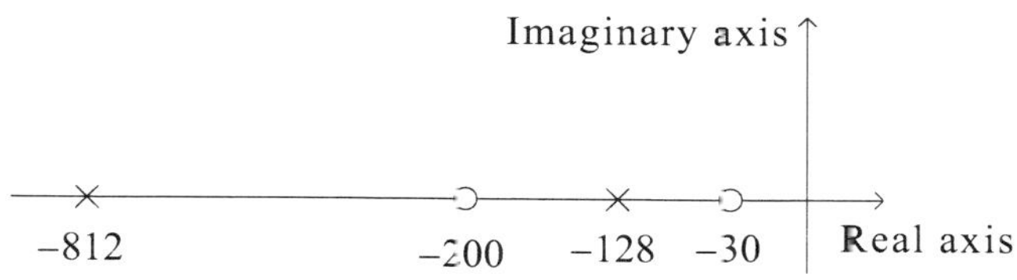

Figure 8.24. Pole–zero pattern for Example 8.9

$$A_v(j\omega) = K\frac{(j\omega+30)(j\omega+200)}{(j\omega+128)(j\omega+812)}$$

For very large values of the frequency ω, the above relation reduces to

$$A_v(j\omega)|_{\omega\to\infty} = K$$

It is given that at high frequencies the voltage gain is $40\ dB$. Recalling that

$$A_{dB} = 20\log 100 = 40$$

it follows that $K = 100$ and

$$A_v(j\omega) = 100\frac{(j\omega+30)(j\omega+200)}{(j\omega+128)(j\omega+812)} = 100\frac{(1+j\omega/30)(1+j\omega/200)\times 30\times 200}{(1+j\omega/128)(1+j\omega/812)\times 128\times 812}$$

$$A_v(j\omega) = 5.77\frac{(1+j\omega/30)(1+j\omega/200)}{(1+j\omega/128)(1-j\omega/812)}$$

The asymptotes for the logarithmic amplitude characteristic with

$$A_v(dB) = 20\log(5.77) = 15.2\ dB$$

is shown in Figure 8.25 where the radian frequencies $30,\ 128,\ 200,$ and 812 are shown in Hz as $4.8,\ 20,\ 32,$ and 129 respectively.

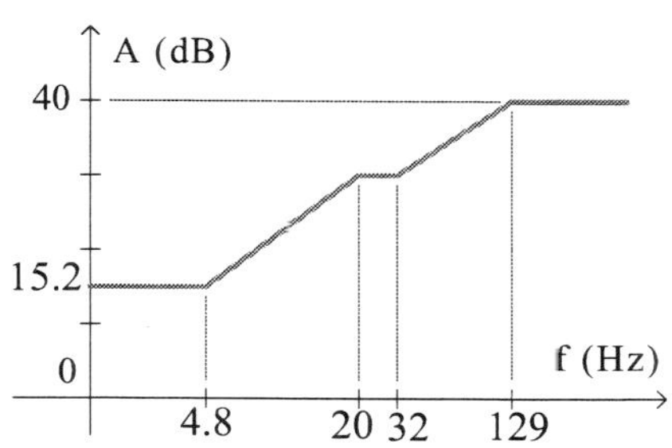

Figure 8.25. Logarithmic amplitude characteristic for Example 8.7

8.7 Amplification and Power Gain in Three or More Cascaded Amplifiers

Transistor amplifiers are conveniently characterized as current amplifiers. A typical transistor radio many employ a cascade of five amplifier stages. The first four are designed to yield the greatest possible current amplification, and the fifth stage is a power amplifier designed to produce the greatest possible power output. Let us consider the equivalent circuit of a voltage amplifier shown in Figure 8.26.

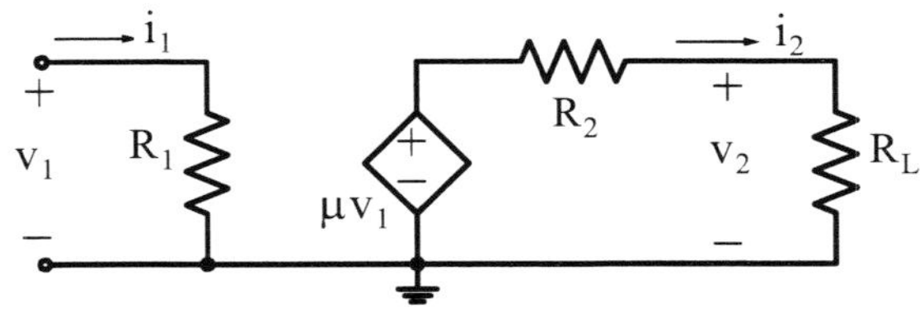

Figure 8.26. Equivalent circuit of a typical voltage amplifier

The instantaneous input power to this amplifier is

$$p_{in} = \frac{v_1^2}{R_1} \tag{8.74}$$

and the instantaneous output power is

$$p_{out} = \frac{v_2^2}{R_L} \tag{8.75}$$

The power gain in dB is defined as

$$G_p = 10\log\frac{p_{out}}{p_{in}} \tag{8.76}$$

Thus, the power gain of the circuit of Figure 8.26 is

$$G_p = 10\log\frac{v_2^2/R_L}{v_1^2/R_1} = 10\log\frac{v_2^2 R_1}{v_1^2 R_L} = 20\log\frac{v_2}{v_1} + 10\log\frac{R_1}{R_L} \tag{8.77}$$

In communications circuits it is desirable to make $R_L = R_1$, and in this case the second term on the right side of (8.77) is zero. Denoting the voltage gain as A_v we obtain

$$A_v = 20\log\frac{v_2}{v_1} \tag{8.78}$$

Example 8.8

For the three–stage amplifier shown in Figure 8.27, find:

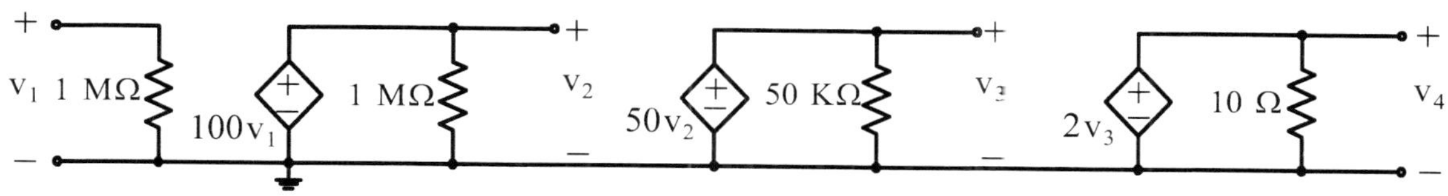

Figure 8.27. Three−stage amplifier for Example 8.8

a. the voltage amplification and power gain of each stage in dB

b. the overall voltage amplification and overall power gain of each stage in dB

Solution:

a. First stage:

$$A_{v1} = 20\log\frac{v_2}{v_1} = 20\log\frac{100v_1}{v_1} = 40 \text{ dB}$$

$$G_{p1} = A_{v1} + 10\log\frac{1\ \text{M}\Omega}{1\ \text{M}\Omega} = A_{v1} + 0 = 40 \text{ dB}$$

Second stage:

$$A_{v2} = 20\log\frac{v_3}{v_2} = 20\log\frac{50v_2}{v_2} = 34 \text{ dB}$$

$$G_{p2} = A_{v2} + 10\log\frac{1\ \text{M}\Omega}{50\ \text{K}\Omega} = A_{v2} + 10\log 20 = 34 + 13 = 47 \text{ dB}$$

Third stage:

$$A_{v3} = 20\log\frac{v_4}{v_3} = 20\log\frac{2v_3}{v_3} = 6 \text{ dB}$$

$$G_{p3} = A_{v3} + 10\log\frac{50\ \text{K}\Omega}{10\ \Omega} = A_{v3} + 10\log 5000 = 6 + 37 = 43 \text{ dB}$$

b. Overall voltage gain:

$$v_4 = 2v_3 = 2 \times 50v_2 = 2 \times 50 \times 100v_1 = 10000v_1$$

$$A_v = 20\log\frac{v_4}{v_1} = 20\log\frac{10000v_1}{v_1} = 80 \text{ dB}$$

Overall power gain:

$$G_p = A_v + 10\log\frac{1\ \text{M}\Omega}{10\ \Omega} = A_v + 10\log 100000 = 80 + 50 = 130 \text{ dB}$$

8.8 Summary

- Periodic signals can be represented as the superposition of sinusoidal components by a Fourier series of the form

$$v(t) = V + V_1\cos(\omega t + \theta_1) + V_2\cos(2\omega t + \theta_2) + V_3\cos(3\omega t + \theta_3) + \ldots$$

  where V is the DC component of the composite signal, ω is the *fundamental frequency* or *first harmonic*, 2ω, is the *second harmonic*, 3ω is the *third harmonic*, and so on.

- Signals that are not can be approximated by the superposition of a number of sinusoidal components expressed as

$$v(t) = V + V_1\cos(\omega_1 t + \theta_1) + V_2\cos(\omega_2 t + \theta_2) + V_3\cos(\omega_3 t + \theta_3) + \ldots$$

  where V is the DC component of the composite signal, and ω_1, ω_2, ω_3, and so on, are not harmonically related.

- The frequency spectrum of a signal reveals information about the frequencies and amplitude, but it contains no information about the phase angles of the components.
- Amplification decreases at high frequencies as a result of the parasitic capacitances in the circuit, and decreases at low frequencies because of the presence of coupling capacitors.
- For a good designed amplifier, the amplification near the lower and near the higher band of frequencies should not be less than 70 percent of the amplification in the middle of the band.
- If an amplifier retards the phase of each sinusoidal component of the signal in direct proportion to the frequency of that component, then the waveform of the output signal is an exact copy of the input waveform, but it is delayed in time.
- The behavior of transistor amplifier at low frequencies is affected by the action of the bypass capacitor that is in parallel with the emitter resistor.
- Transistor amplifier circuits designed for low frequencies are discussed in Section 8.2.
- Transistor amplifier circuits designed for high frequencies are discussed in Section 8.3.
- The amplifying property of the transistor which increases the apparent value of the collector capacitance is known as the Miller effect.
- The frequency dependence of various amplifier circuits is examined by treating the low–frequency characteristics separately from the high–frequency characteristics. However, the combined low– and high–frequency characteristics can be displayed on a single set of coordinates.
- To achieve higher gains, we use multistage amplifier circuits. The overall characteristics of multistage amplifiers can be easily approximated and displayed with the use of asymptotes.

- The bandwidth between the half–power points decreases as more stages are connected in cascade. The half–power bandwidth of the cascade of n identical stages is given by

$$B_n = B_1\sqrt{2^{1/n}-1}$$

where $B_1 = \omega_H$ and ω_H is the half–power frequency for on stage amplifier at medium and high frequencies. The quantity $\sqrt{2^{1/n}-1}$ is referred to as the *bandwidth reduction factor* for n identical stages in cascade. For two stages the bandwidth reduction factor is 0.64, for three stages is 0.51, and for four stages it is 0.43.

- If two non–identical stages are connected in cascade, the overall voltage gain is given by

$$A_v = A_{m1}A_{m2}\frac{(j\omega/\omega_{L1})(j\omega/\omega_{L2})}{(1+j\omega/\omega_{L1})(1+(j\omega/\omega_{L2}))(1+j\omega/\omega_{H1})(1+j\omega/\omega_{H2})}$$

- For all practical electric and electronic circuits, voltage and current gains can be expressed as the ratio of two polynomials in the variable s, and when the highest power of the denominator is greater than the highest degree of the numerator, the expression is referred to as a *rational function*. The values of s that make a rational function zero are called *zeros* of the function; the values of s that make rational functions infinite are called *poles* of the function. If all the poles and zeros of a rational function are known, the function is completely specified except for a constant multiplier. For example, if the zeros of a certain voltage gain are -1 and -3, and if the poles are -2 and -4, then the voltage gain is expressed as

$$A_v(s) = K\frac{(s+1)(s+3)}{(s+2)(s+4)}$$

and for steady–state sinusoidal conditions,

$$A_v(j\omega) = K\frac{(j\omega+1)(j\omega+3)}{(j\omega+2)(j\omega+4)}$$

- A zero corresponds to a break of 20 dB/decade upward in the asymptotic characteristic, and a pole corresponds to a break of 20 dB/decade downward.

- The relation between the poles and zeros of the voltage or current gain and the frequency characteristics are displayed in a useful way by the pole–zero diagram.

- Some amplifiers require a cascade of three or more applications to achieve higher voltage amplification and power gain. The last stage is used to produce the greatest possible power output, and the previous stages are used for voltage amplification.

8.9 Exercises

1. For the transistor amplifier below, $R_A = 33\ K\Omega$ and $R_B = 15\ K\Omega$, $R_2 = 2\ K\Omega$, $R_3 = 1\ K\Omega$, the input resistance $r_n = 500\ \Omega$, and $\beta = 80$. Find the value of capacitor C_3 so that the highest break frequency for the low–frequency model will be $1\ KHz$.

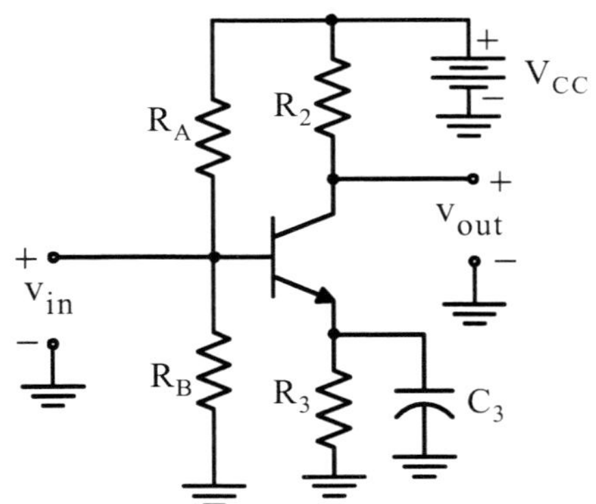

2. A transistor is used as an amplifier for video signals. The high–frequency model has the form shown below.

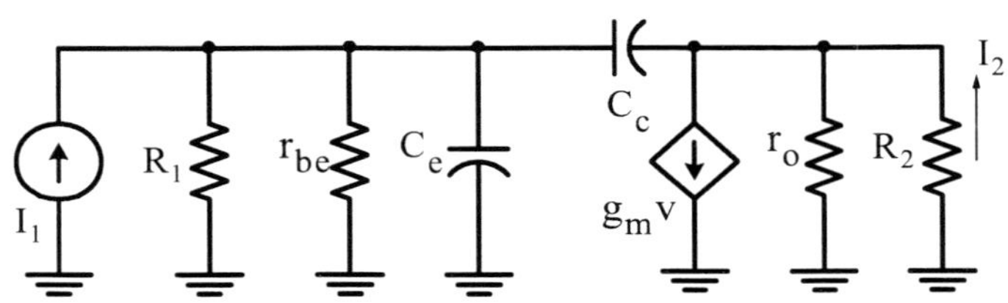

The quiescent collector current is $I_C =$ mA and the collector load resistor is $R_2 = 1\ K\Omega$. The transistor parameters are $\beta = 50$, $r'_b = 50\ \Omega$, $C_c = 3\ pF$, $r_o = 30\ K\Omega$, and the current gain–bandwidth frequency is $f_T = 200\ MHz$. The amplifier circuit is to be designed so that the high–frequency half–power point for the current gain is $3\ MHz$.

a. Determine the values of g_m, r_{be}, C_e, and the Miller effect

b. Determine the value of R_1

c. With the value of R_1 found in part (b) what is the mid–band current amplification?

3. For the cascaded amplifier circuit below, the transistors are identical with $R_A = 56\ K\Omega$, $R_B = 12\ K\Omega$, $R_2 = 6.8\ K\Omega$, $R_3 = 2.2\ K\Omega$, and $R_4 = 1\ K\Omega$. The quiescent collector current in each stage is $I_C = 1\ mA$, and the transistor parameters are $\beta = 200$, $r'_b = 200\ \Omega$,

$C_C = 20\ pF$, $f_T = 8\ MHz$, and the output resistance r_o is very large in comparison with R_2 and thus it can be neglected.

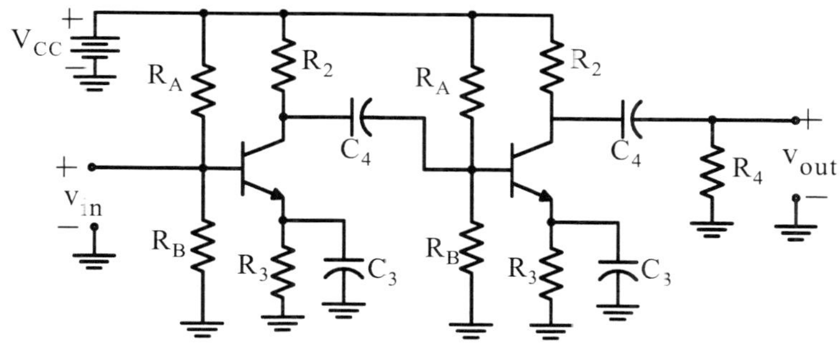

a. Find the parameter values of the simplified high–frequency model circuit shown below.

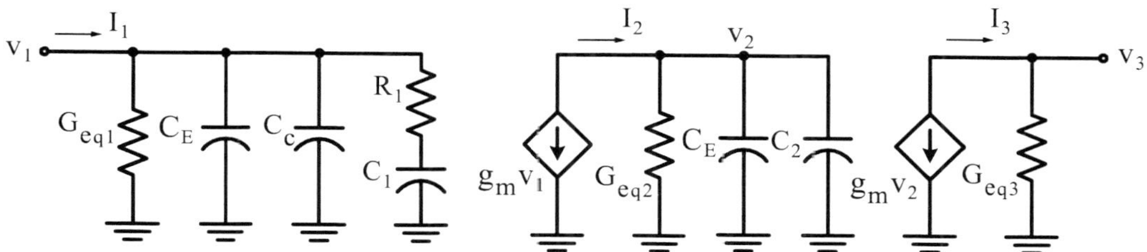

b. Neglecting the effect of the resistance R_1 in the model circuit above, find the overall current gain $A_c = I_3/I_1$.

c. Sketch and dimension the asymptotes for the high–frequency amplitude characteristics.

4. Prove that the half–power bandwidth of the cascade of n identical stages is given by

$$B_n = B_1\sqrt{2^{1/n} - 1}$$

where $B_1 = \omega_H$ and ω_H is the half–power frequency for one stage amplifier at medium and high frequencies. Hint: Begin with the assumption that the amplitude of the voltage gain of a one–stage amplifier is given by

$$|A_v| = A = A_m/(\sqrt{1 + (\omega/\omega_H)^2})$$

5. The pole–zero patterns for the voltage gain of two transistor amplifiers are shown below.

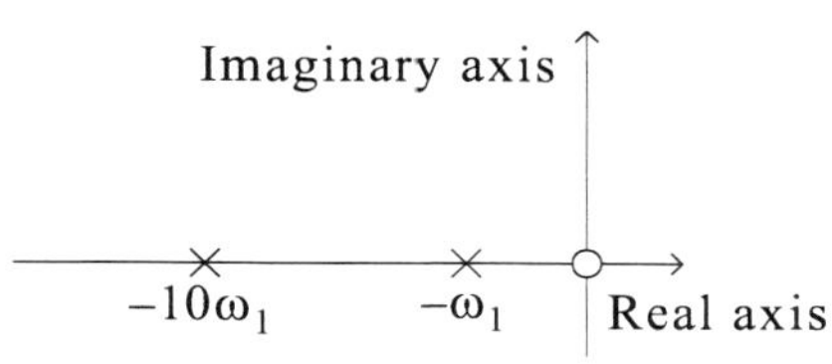

(a)

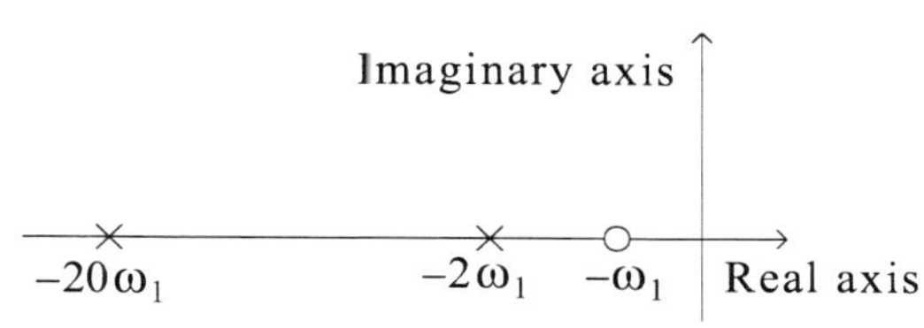

(b)

a. Will both of these amplifiers transmit DC signals?

b. Use MATLAB to obtain the plot for the amplitude characteristic for the pole–zero pattern of Figure (b) assuming that the function is multiplied by the constant $K = 40,000$.

6. A certain stereo amplifier delivers a sinusoidal voltage of 10 mV RMS on open circuit and has an internal resistance of $500\ \Omega$. This amplifier is to deliver a signal to a loudspeaker whose resistance is $10\ \Omega$.

a. If the amplifier is connected directly to the loudspeaker, how much power does it deliver to the loudspeaker?

b. A cascade of amplifier stages similar to that shown in Figure (a) below is to be used to amplify the signal from the stereo amplifier and deliver 10 w to the loudspeaker. Two types of amplifiers shown in Figures (b) and (c) below are available for the cascade connection. Which of these amplifiers can deliver 10 w to the loudspeaker with the smallest input voltage?

c. Which of the amplifiers in Figures (b) and (c) yields the greatest no–load voltage amplification?

d. Draw the circuit diagram for a cascade of stages like those in Figures (b) and (c) that will meet the specifications of part (b) above. Use the smallest possible number of stages and show how a $1\ M\Omega$ potentiometer can be used in the first stage as a volume control.

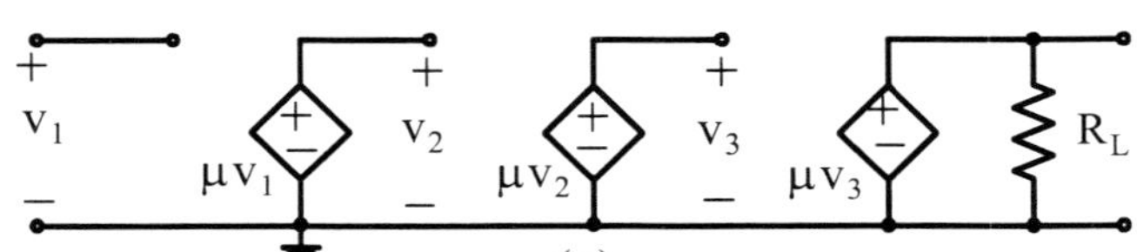

(a)

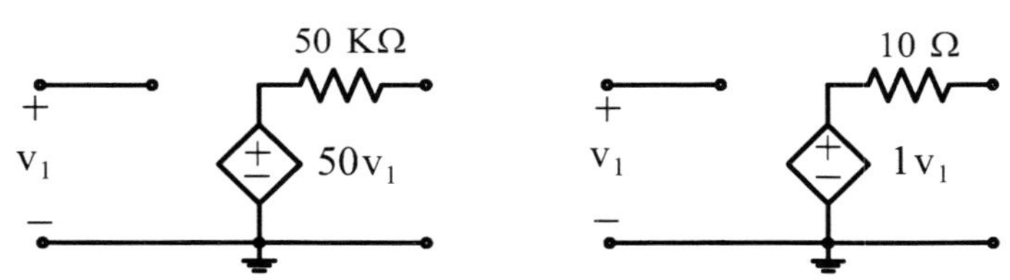

(b) (c)

8.10 Solutions to End–of–Chapter Exercises

1.

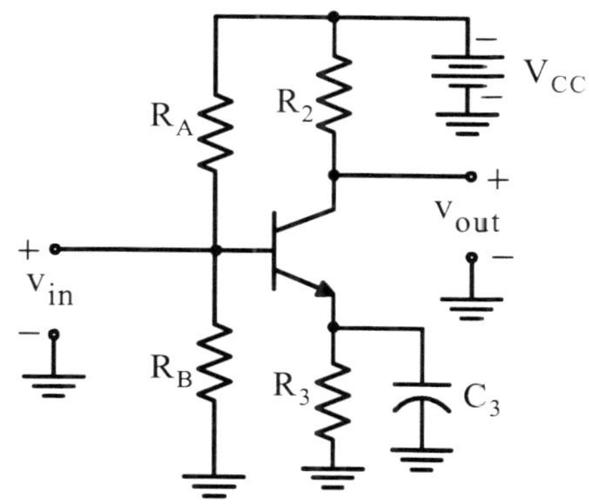

The low–frequency model is shown below where $R_1 \parallel (R_A + R_B)$

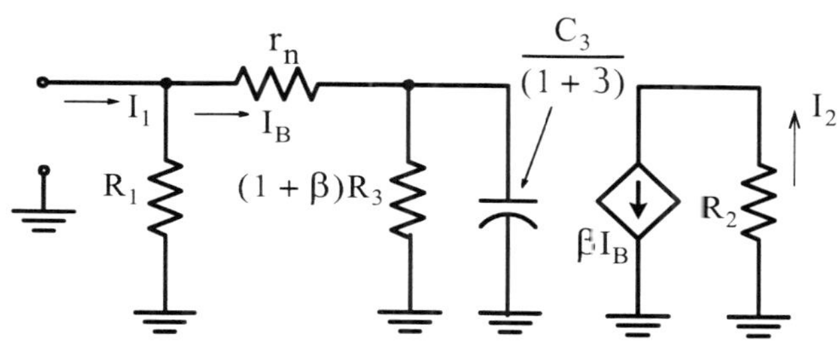

where

$$R_1 = \frac{R_A R_B}{R_A + R_B} = \frac{33 \times 15}{33 + 15} = 10.3\ K\Omega$$

For this exercise we must find k_3 and f_3 such that

$$(1 + k_3)f_3 = 1\ KHz$$

From (8.15),

$$k_3 = \frac{(1+\beta)R_3}{R_1 + r_n} = \frac{(1+80) \times 1}{10.3 + 0.5} = 7.5$$

and thus,

$$f_3 = \frac{1000}{(1+7.5)} \approx 118\ Hz$$

From (8.11)

$$\omega_3 = \frac{1}{R_3 C_3}$$

or

$$f_3 = \frac{1}{2\pi R_3 C_3}$$

Then,

$$C_3 = \frac{1}{2\pi R_3 f_3} = \frac{1}{2\pi \times 10^3 \times 118} = 1.35\ \mu F$$

The amplification at medium frequencies is given by (8.14). Thus

$$A_m = \frac{\beta R_1}{R_1 + r_n} = \frac{80 \times 10.3}{10.3 + 0.5} = 76.3$$

and in decibels

$$A_m(dB) = 20\log 76.3 = 20 \times 1.88 = 37.6 \text{ dB}$$

The amount by which the amplification at low frequencies is less than its value at medium frequencies is

$$20\log(1 + k_3) = 20 \times 0.93 = 18.6 \text{ dB}$$

Therefore, from DC to the break frequency at 118 Hz, the gain is $37.6 - 18.6 = 19$ dB. The amplitude characteristics are shown below where the frequency is shown in logarithmic scale and the gain is in dB scale.

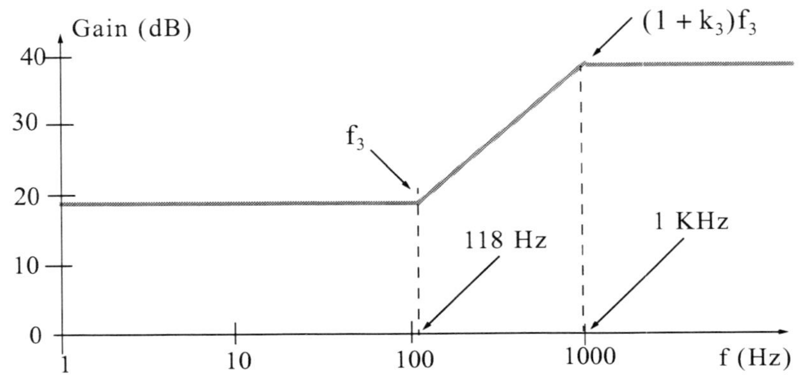

2.

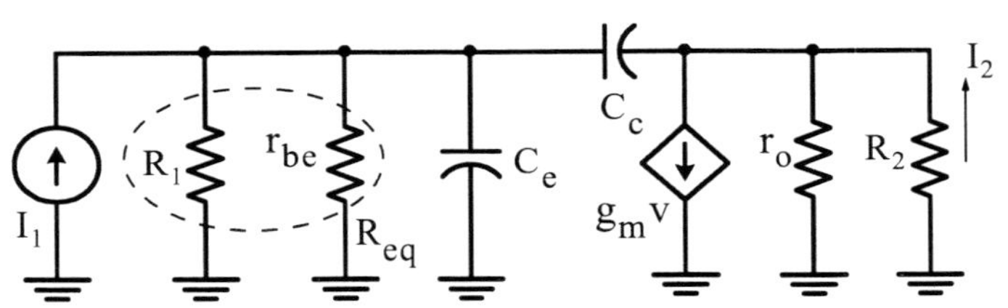

Figure 8.28. Simplified model of the transistor amplifier for Example 8.6

a. From Chapter 3, relation (3.78),

$$g_m = 40I_C = 40 \times 1 \times 10^{-3} = 0.04 \ \Omega^{-1}$$

From relation (3.83),

$$r_{be} = \frac{\beta}{g_m} = \frac{50}{0.04} = 1.25 \text{ K}\Omega$$

and from (3.92),

$$f_\beta = \frac{f_T}{\beta} = \frac{200}{50} = 4 \text{ MHz}$$

This indicates that large current amplification is provided at frequencies up to 4 MHz.

From (3.93),

$$C_e = \frac{g_m}{\omega_T} = \frac{0.04}{2\pi \times 2 \times 10^8} = 31.8 \text{ pF}$$

The Miller effect is $(1 + A)C_c$ where

$$A \approx g_m R_2 = 0.04 \times 1000 = 40$$

Thus,

$$(1 + A)C_c = 41 \times 3 \times 10^{-12} = 123 \times 10^{-12} = 123 \text{ pF}$$

and this indicates that the dominant effect at high frequencies is the Miller effect.

b. From (8.19),

$$|A| = \left|\frac{g_m r_o R_2}{(r_o + R_2)}\right| = \frac{0.04 \times 30 \times 10^6}{31 \times 10^3} = 38.7$$

and by substitution into (8.33) with $f_1 = 3$ MHz

$$3 \times 10^6 = 4 \times 10^6 \cdot \frac{1 + 1.25 \times 10^3/R_1}{1 + 38.7 \times 3 \times 10^{-12}/31.8 \times 10^{-12}}$$

$$4\left(1 + 1.25 \times \frac{10^3}{R_1}\right) = 3(1 + 3.65) = 14$$

$$\frac{5 \times 10^3}{R_1} = 10$$

$$R_1 = 500\ \Omega$$

c. The medium frequency current amplification A_m is given by (8.27) as

$$A_m = \frac{r_o}{r_o + R_2} g_m R_{eq} = \frac{r_o}{r_o + R_2} g_m \frac{R_1 r_{be}}{R_1 + r_{be}} = \frac{30}{31} \times 0.04 \times \frac{500 \times 1250}{1750} = 13.8$$

3.

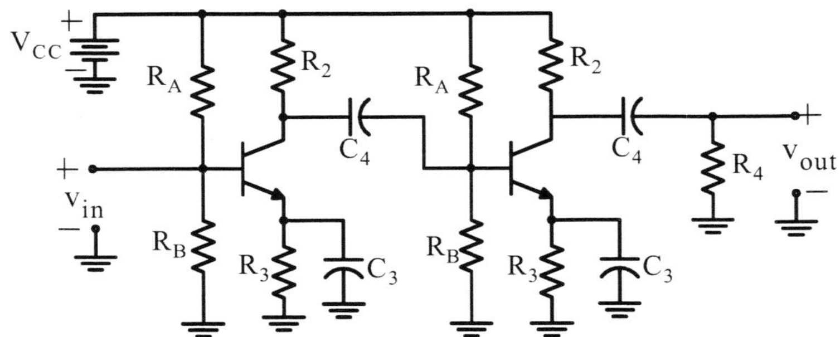

a. The parameter values of the simplified high–frequency model circuit shown below are as follows:

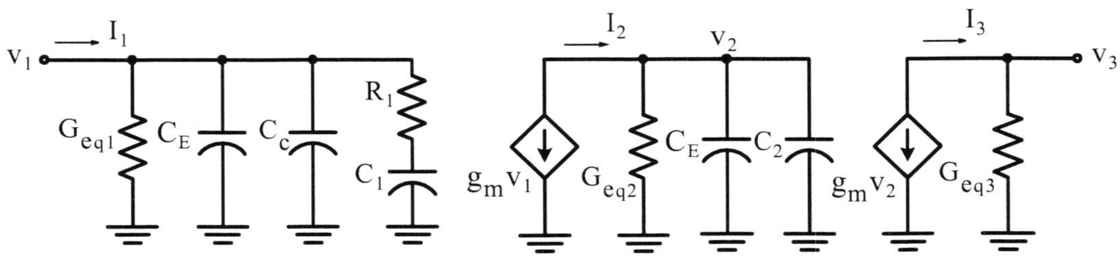

From Chapter 3, relation (3.78),

$$g_m = 40I_C = 40 \times 1 \times 10^{-3} = 0.04\ \Omega^{-1}$$

From relation (3.83),

$$r_{be} = \frac{\beta}{g_m} = \frac{200}{0.04} = 5\ \text{K}\Omega$$

and from (3.93),

$$C_E = \frac{g_m}{\omega_T} = \frac{0.04}{2\pi \times 8 \times 10^6} = 796\ \text{pF}$$

For the second stage of the amplifier, from (8.36),

$$G_{eq3} = \frac{1}{R_2} + \frac{1}{R_4} = \frac{1}{6.8\ \text{K}\Omega} + \frac{1}{1\ \text{K}\Omega} = 1.15\ \ \text{m}\Omega^{-1}$$

From (8.27)

$$A_m = \frac{r_o}{r_o + R_2} g_m R_{eq}$$

and since $r_o \gg R_2$, the current gain of the second stage amplifier at medium frequencies is

$$A_{m2} = g_m R_{eq} = \frac{g_m}{G_{eq3}} = \frac{0.04}{1.15 \times 10^{-3}} = 34.8$$

From (8.37) the Miller effect capacitance is

$$C_2 = (1 + A_{m2})C_C = (1 + 34.8)20 = 716 \text{ pF}$$

From (8.35),

$$G_{eq2} = \frac{1}{R_A} + \frac{1}{R_B} + \frac{1}{r_{be}} + \frac{1}{R_2} = \frac{1}{56 \text{ K}\Omega} + \frac{1}{12 \text{ K}\Omega} + \frac{1}{5 \text{ K}\Omega} + \frac{1}{6.8 \text{ K}\Omega} = 0.45 \text{ m}\Omega^{-1}$$

From (8.46),

$$A_{c2} = \frac{I_3}{I_2} = -\frac{g_m}{G_{eq2}} \cdot \frac{1}{1 + j\omega(C_E + C_2)/G_{eq2}}$$

$$= -\frac{0.04}{0.45 \times 10^{-3}} \cdot \frac{1}{1 + j\omega(796 + 716) \times 10^{-12}/0.45 \times 10^{-3}} = -89\frac{1}{1 + j\omega 3.36 \times 10^{-6}}$$

and expressing ω in megaradians per second, we find that

$$A_{c2} = -89\frac{1}{1 + j\omega/0.3}$$

With the frequency expressed in KHz, we obtain

$$A_{c2} = -57\frac{1}{1 + jf/47.4}$$

For the first stage of the amplifier we found above that

$$G_{eq2} = 0.45 \text{ m}\Omega^{-1}$$

From (8.27)

$$A_m = \frac{r_o}{r_o + R_2} g_m R_{eq}$$

and since $r_o \gg R_2$, the current gain of the first stage amplifier at medium frequencies is

$$A_{m1} = g_m R_{eq} = \frac{g_m}{G_{eq2}} = \frac{0.04}{0.45 \times 10^{-3}} = 89$$

From (8.37) the Miller effect capacitance is

$$C_1 = (1 + A_{m1})C_C = (1 + 89)20 = 1800 \text{ pF}$$

b. Neglecting R_1, from (8.34),

$$G_{eq1} = \frac{1}{R_A} + \frac{1}{R_B} + \frac{1}{r_{be}} = \frac{1}{56 \text{ K}\Omega} + \frac{1}{12 \text{ K}\Omega} + \frac{1}{5 \text{ K}\Omega} = 0.30 \text{ m}\Omega^{-1}$$

From (8.49),

$$A_{c1} = \frac{I_2}{I_1} = -\frac{g_m}{G_{eq1}} \cdot \frac{1}{1 + j\omega(C_E + C_c + C_1)/G_{eq1}}$$

$$= -\frac{0.04}{0.30 \times 10^{-3}} \cdot \frac{1}{1 + j\omega(796 + 20 + 1800) \times 10^{-12}/0.30 \times 10^{-3}} = -133.3\frac{1}{1 + j\omega 8.72 \times 10^{-6}}$$

and expressing ω in megaradians per second, we find that

$$A_{c1} = -133.3\frac{1}{1 + j\omega/0.115}$$

With the frequency expressed in KHz, we obtain

$$A_{c1} = -133.3\frac{1}{1 + jf/18.3}$$

c. The overall current gain A_c is obtained by multiplying A_{c1} by A_{c2}. Thus,

$$A_c = A_{c1}A_{c2} = \left(-133.3\frac{1}{1 + jf/47.4}\right)\left(-89\frac{1}{1 + jf/18.3}\right) = 11864 \cdot \frac{1}{1 + jf/47.4} \cdot \frac{1}{1 + jf/18.3}$$

The sketch below shows the asymptotes for the high–frequency amplitude characteristics with the gain in dB where $A_c(dB) = 20\log 11864 \approx 82$, the frequency in log scale, and the slope of each high–frequency asymptote is –20 dB per decade.

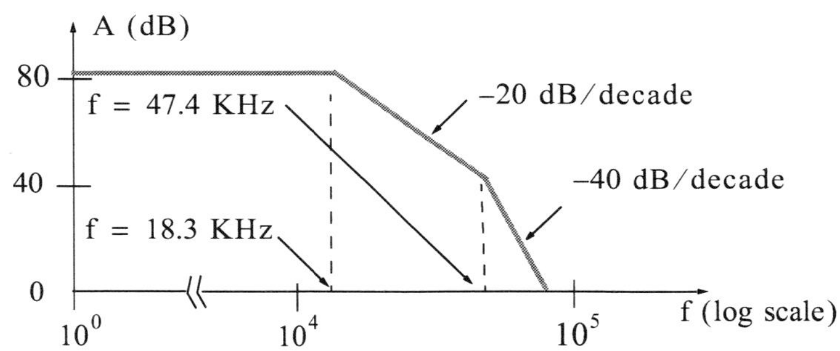

4.

Since the amplitude of the voltage gain of a one–stage amplifier is given by

$$|A_v| = A = \frac{A_m}{\sqrt{1 + (\omega/\omega_H)^2}}$$

for *n* identical stages of this type connected in cascade, the overall amplification will be

$$A_n = \frac{A_m^n}{(\sqrt{1 + (\omega/\omega_H)^2})^n}$$

The half–power frequency for the cascade of n identical stages is the value of ω that will make the denominator of the above expression equal to $\sqrt{2}$. Hence, since $\omega_H = B_1$,

$$\left[1+\left(\frac{\omega}{B_1}\right)^2\right]^n = 2$$

$$\left[1+\left(\frac{\omega}{B_1}\right)^2\right]^n = 2^{1/n}$$

from which

$$\omega = B_1\sqrt{2^{1/n}-1}$$

This value of ω is the half–power bandwidth of the cascade of n identical stages. Therefore,

$$B_n = B_1\sqrt{2^{1/n}-1}$$

5.

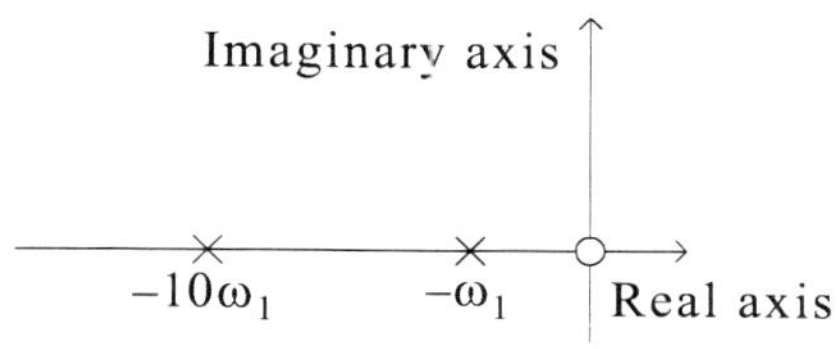

(a)

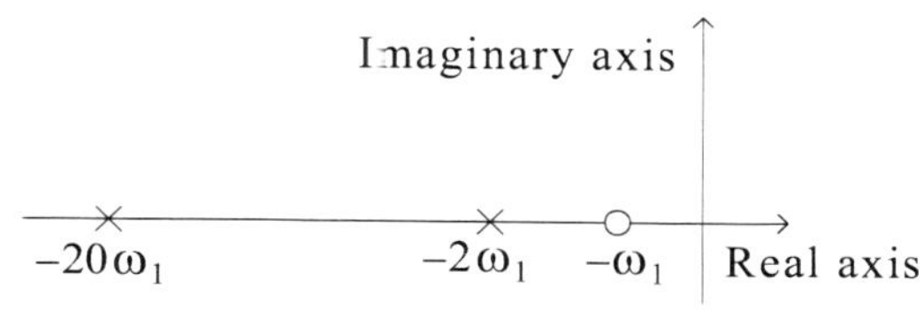

(b)

a. For the amplifier whose pole–zero pattern is shown in (a),

$$A_V = \frac{j\omega}{(j\omega+\omega_1)(j\omega+10\omega_1)}$$

and when $\omega = 0$ (DC condition), $A_V = 0$. Therefore, this amplifier will not transmit DC signals.

For the amplifier whose pole–zero pattern is shown in (b),

$$A_V = \frac{(j\omega+\omega_1)}{(j\omega+2\omega_1)(j\omega+20\omega_1)}$$

and when $\omega = 0$ (DC condition),

$$A_V = \frac{\omega_1}{(2\omega_1)(20\omega_1)} = \frac{1}{40\omega_1}$$

Therefore, this amplifier will transmit DC signals.

b. The MATLAB script and resulting plot are shown below.

```
w=1:10:100000; N=j.*w+100; D=(j.*w).^2+2200.*(j.*w)+400000;...
Av=4.*10.^4.*N./D; AvdB=20.*log10(abs(Av)); semilogx(w AvdB); grid;...
xlabel('Frequency in rad/s (log scale)'); ylabel('Gain (dB scale)');...
title ('Amplitude characteristic for Exercise 5')
```

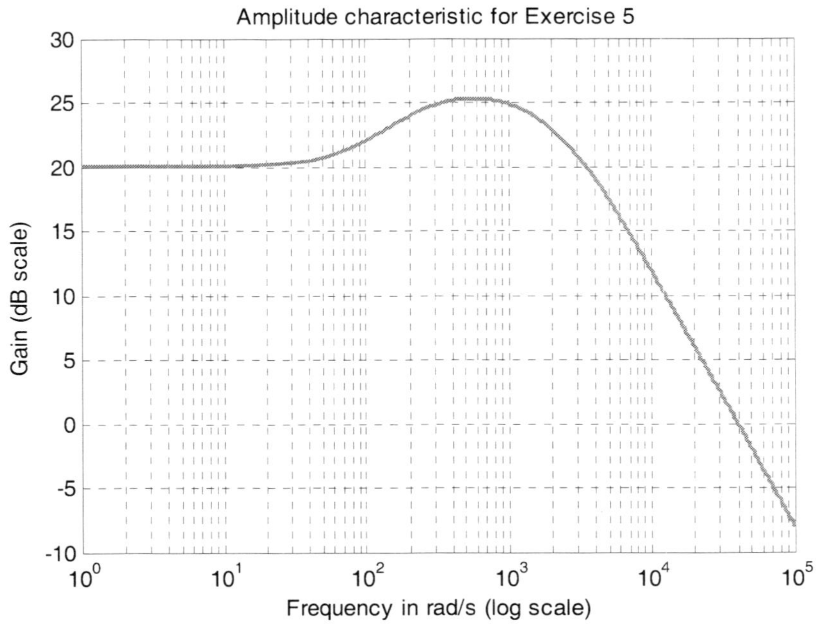

6.

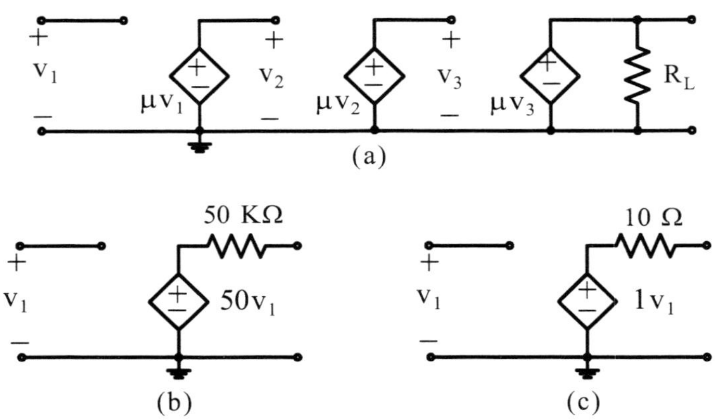

a. If the amplifier is connected directly to the loudspeaker, the total series resistance will be

$$R_{eq} = 500 + 10 = 510\ \Omega$$

and the current through the loudspeaker will be

$$i = \frac{10\ \text{mV}}{510\ \Omega} = 19.6\ \mu\text{A}$$

Thus, the power delivered to the loudspeaker will be

$$p = i^2R = (19.6\times10^{-6})^2\times10 = 3.84\times10^{-9}\text{ w}$$

b. Let us assume 1 v input to each of the amplifiers in Figures (b) and (c).

1. Current in amplifier of Figure (b) will be

$$I_b = \frac{50\times1}{10000+10} = 5\text{ mA}$$

and power will be

$$P_b = I_b^2R_b = (5\times10^{-3})^2\times10\times10^3 = 0.25\text{ mw}$$

2. Current in amplifier of Figure (c) will be

$$I_c = \frac{1\times1}{10+10} = 50\text{ mA}$$

and power will be

$$P_c = I_c^2R_c = (50\times10^{-3})^2\times10 = 25\text{ mw}$$

Therefore, the amplifier of Figure (c) will deliver 10 w to the loudspeaker with the smallest input voltage.

c.

1. Voltage amplification in amplifier of Figure (b) will be

$$A_b = 20\log\frac{50v_1}{v_1} = 34\text{ dB}$$

2. Voltage amplification in amplifier of Figure (c) will be

$$A_c = 20\log\frac{1v_1}{v_1} = 0\text{ dB}$$

Therefore, the amplifier of Figure (b) will produce the greatest voltage amplification.

d.

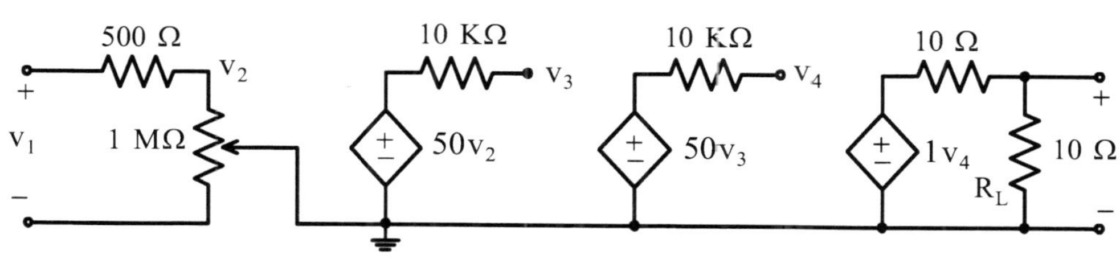

For last stage we must have $P_L = I_L^2R_L = 10\text{ w}$ and since $R_L = 10\ \Omega$, $I_L = 1\text{ A}$, and $V_L = 10\text{ V}$. Therefore, $v_4 = 10+10 = 20\text{ V}$, and

$$v_2 = \frac{1\ \text{M}\Omega}{1\ \text{M}\Omega + 500\ \Omega} \cdot 10\ \text{mV} \approx 10\ \text{mV}$$

The first and second stages were so chosen as to provide the greatest voltage amplification.

Chapter 9

Tuned Amplifiers

This chapter begins with an introduction to tuned amplifiers. We will examine the properties of various tuned amplifiers that find applications to telecommunications systems employing narrowband modulated signals. New tools for the analysis and design of tuned amplifiers are developed.

9.1 Introduction to Tuned Circuits

The amplifiers discussed in previous chapters are sufficient for most applications in which it is not required that the signals be transmitted over long distances. Thus they are adequate for audio amplifiers used in public address and home entertainment systems, servomechanisms, automatic pilots, electronic instruments, and a host of similar applications. However, when the signals must be transmitted over long distances, as between two cities or between a space vehicle and a ground station, effective use of the transmission medium requires the use of narrowband systems operating at high frequencies. Various systems of this kind operate throughout the frequency spectrum from 10 KHz to about 10 GHz although at the higher end of this range ordinary transistors are not used. For these higher frequencies it is necessary to use tuned amplifiers with RLC coupling to overcome the effects of parasitic capacitances and to provide a filtering operation in the form of frequency–selective amplification. In this section we will introduce the properties of various tuned amplifiers that find application in telecommunication systems.

A *tuned amplifier* is essentially a bandpass filter.* A passive bandpass filter is constructed with passive devices, i.e., resistors, inductors, and capacitors, and thus it provides no gain. For small signals, active filters with op amps are very popular. Tuned amplifiers can also be designed with bipolar junction transistors and MOSFETs, and our subsequent discussion will be based on these devices.

When signals must be transmitted over long distances, either by wire or wireless, efficient utilization of the transmission medium requires the use of high–frequency, narrowband signals. To generate these signals a high–frequency carrier wave, usually a sinusoid, is caused to change instant by instant in accordance with the information signal to be transmitted. The process by which the carrier wave is made to change in accordance with the information signal is called *modulation*. A sinusoid has three characteristics that can be modulated; they are the amplitude, the frequency, and

* *For a thorough discussion on passive and active filters, please refer to Signals and Systems with* MATLAB *Applications, ISBN 978–1–934404–11–9.*

the phase, and they give rise to *amplitude–modulated* (AM), *frequency–modulated* (FM), and *phase–modulated* (PM) signals. To understand the requirements that must be met by tuned amplifiers it is necessary to examine the nature of these waves and to understand the way in which they are used in telecommunication systems. The properties of AM waves are developed in the paragraphs that follow; although they are not discussed specifically, FM* and PM waves have similar properties.

If the information, or modulating, wave is a voltage $v_m(t)$, then an AM carrier can be expressed as

$$v_s(t) = [V_c + v_m(t)]\cos\omega_C t \tag{9.1}$$

where ω_C is the carrier frequency, and V_C is the amplitude of the unmodulated carrier. The waveform of such an AM wave is shown in Figure 9.1.

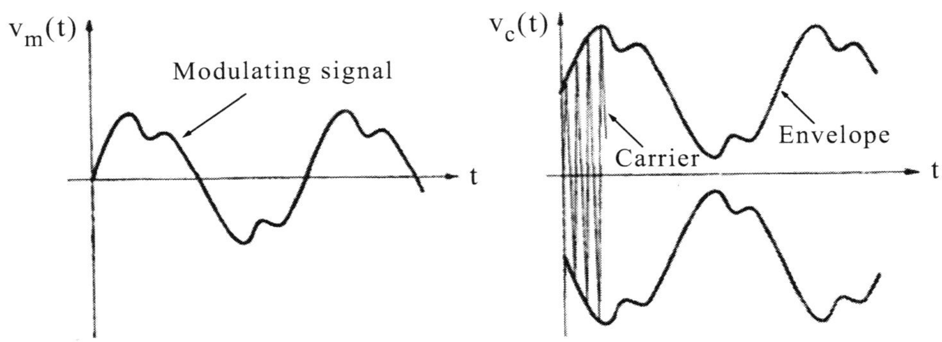

Figure 9.1. Signal and carrier wave that is amplitude–modulated by the signal.

In an AM system the magnitude of $v_m(t)$ is always less than V_C; hence the envelope of the modulated wave never drops to zero. From relation (9.1) and Figure 9.1 we observe that the waveform of the envelope of the modulated signal is the same as the waveform of the modulating signal $v_m(t)$. Thus if the information in the modulated carrier is to be preserved, the waveform of the envelope must be preserved. The information can be recovered by recovering the waveform of the envelope; the peak rectifier circuit that we discussed in Chapter 2 is used for this purpose. The peak rectifier is also known as *diode detector*, and in this application it is frequently called an *envelope detector*.

To examine the properties of the AM wave further, it is convenient to assume that the modulating signal is a sinusoid expressed as

$$v_m(t) = V_m\cos(\omega_m t + \theta_m) \tag{9.2}$$

* *FM and PM systems are best described by Bessel functions. For an introduction to these functions, please refer to Numerical Analysis Using MATLAB and Excel, ISBN 978–1–934404–03–4.*

Substitution of (9.2) into (9.1) yields

$$v_s(t) = [V_c + V_m \cos(\omega_m t + \theta_m)]\cos\omega_c t \tag{9.3}$$

and letting

$$m = V_m/V_c \tag{9.4}$$

we express (9.3) as

$$v_s(t) = V_c[1 + m\cos(\omega_m t + \theta_m)]\cos\omega_c t \tag{9.5}$$

and the quantity m is known as *the modulation index in an AM system*, and it is always less than unity.

In (9.2) and (9.5) we have assumed that the modulating signal $v_m(t)$ is sinusoidal and as such, it predictable for all times; hence, it conveys no useful information. In practice, $v_m(t)$ is non–sinusoidal but if it is periodic in nature as most physical phenomena are, it can be expanded in a Fourier series,* and the analysis of an AM system can be performed by the superposition of the effects of each sinusoidal component known as *harmonics*. Relation (9.5) can be simplified by using the trigonometric identity for the product of two cosines. Thus,

$$v_s(t) = V_c\cos\omega_c t + \frac{m}{2}V_c\cos[(\omega_c + \omega_m)t + \theta_m]t + \frac{m}{2}V_c\cos[(\omega_c - \omega_m)t - \theta_m] \tag{9.6}$$

Relation (9.6) reveals that a sinusoidally modulated AM wave is the superposition of three sinusoidal components, a carrier wave at the carrier frequency, and two side waves at frequencies on either side of the carrier frequency. The frequency spectrum of a typical sinusoidally modulated AM wave is shown in Figure 9.2.

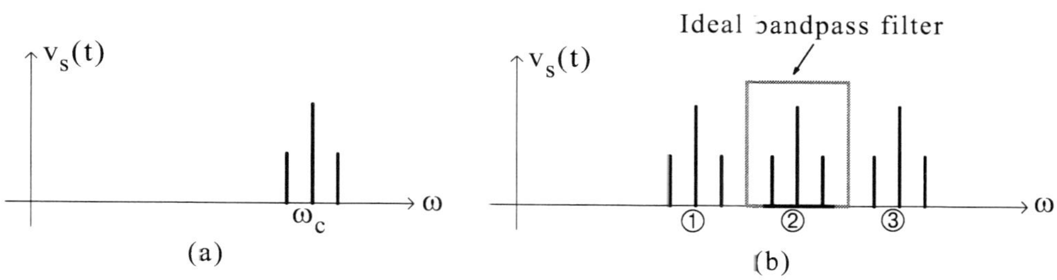

Figure 9.2. Frequency spectra of AM waves

Figure 9.2(a) shows the relative amplitudes and frequencies of the carrier and side waves, except that usually the side waves are much closer to the carrier wave than it is possible to show in the diagram. In the more general case there are many pairs of side waves, and these are referred to as the upper and lower sidebands of the AM wave. In most AM waves the carrier frequency is much larger than the highest frequency in the modulating signal. Thus, a typical AM radio station broadcasting speech and music with frequencies up to 5 KHz has a carrier frequency of 1 MHz

* *For a detailed discussion and applications in Fourier series, please refer to Signals and Systems with MATLAB Computing and Simulink Modeling, ISBN 978–1–934404–11–9.*

and a total bandwidth of 10 KHz centered at 1 MHz. The advantage of AM signals is that they are *narrowband signals*; that is, their total bandwidths are small compared to their center frequencies. A narrowband allows the transmission of many simultaneous telephone conversations on a single pair of conductors in a coaxial cable and for the simultaneous use of space by all radio transmitters.

Figure 9.2(b) shows the frequency spectra of three AM signals indicated as ①, ②, and ③. Each signal uses a different carrier frequency, and normally the difference between these frequencies is great enough so that the spectra of the signals are completely separated and occupy different frequency bands. These signals can be separated by frequency–selective amplification; the idealized gain characteristic for such an amplifier is indicated in Figure 9.2(b). Amplifiers of this kind are called *bandpass amplifiers*. Tuning a radio receiver is the operation by which the passband of the amplifier is shifted from one place to another in the frequency spectrum. The study of tuned amplifiers in the sections that follow is concerned with the response of the amplifiers to modulated waves and with design techniques for obtaining optimum performance. Optimum performance is concerned with the rejection of adjacent channels as well as with faithful amplification of the desired channel.

Relation (9.6) shows that a symmetry exists between the upper and lower sidebands of an AM wave. The sidebands have equal amplitudes, and they have phase angles with respect to the carrier that are equal in magnitude but opposite in sign. If these components are to add up to give the original AM wave, this symmetry must be preserved. The significance and importance of this symmetry is illustrated further by the phasor diagrams in Figure 9.3, in which each sinusoidal component of the wave is represented by a rotating vector. If the carrier vector is taken as the reference and is assumed to be stationary, then the upper–sideband vector rotates counterclockwise at ω_m rad/sec, and the lower–sideband vector rotates clockwise at the same speed.

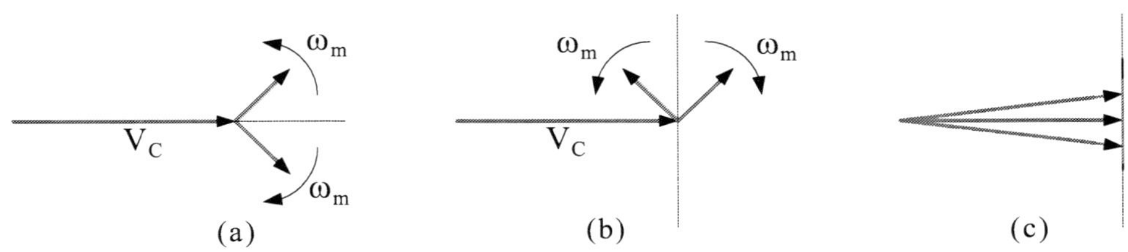

Figure 9.3. Phasor representation of an AM wave

In Figure 9.3(a) the sum of the two sideband vectors (represented by the dotted line) is at every instant colinear with the carrier vector and it gives a sinusoidal variation in the amplitude of the total voltage. However, that this result depends on the symmetry between the sidebands. To illustrate what may happen when the symmetry is destroyed, Figure 9.3(b) shows the case in which the sidebands have been shifted 90° out–of–phase in phase with respect to the carrier. In this case, the sum of the sideband vectors (represented by the dotted line) is at every instant normal to the carrier vector, and the amplitude of the total voltage becomes almost constant as indicated in Figure 9.3(c) where the dotted vectors represent the vector sum of the carrier and the sidebands. When the condition of Figure 9.3(c) occurs, the signal recovered by an envelope detector at the receiving end is very small, and it is badly distorted. Just about all radio signals, when transmitted over long distances, are subject to this kind of fading and distortion as a result of the properties of the transmission medium.

System performance can be improved in this respect by transmitting only one of the two sidebands; with single–sideband systems the need for coherence between two sidebands is eliminated. Single sideband systems are discussed in communications systems text, but our intent here is to illustrate that poorly designed amplifiers can degrade the signal. Ideally, then, amplifiers for AM signals should provide uniform amplification over the band of frequencies occupied by the signal, and they should provide no phase shift in this band. The amplification decreases at high frequencies as a result of the parasitic capacitances in the tuned amplifier. It also decreases at low frequencies because of the presence of coupling capacitors. In a properly designed tuned circuit, the amplification at the ends of the band of frequencies occupied by the signal should not be less than 70 percent of the amplification in the center of the band.

The no phase shift requirement stated above can be somewhat controlled by the fact that a linear phase–shift characteristic produces a pure delay in the envelope waveform without distorting it. This fact can be illustrated by considering a modulated signal consisting of the sum of a number of sinusoids of frequencies $\omega_1, \omega_2, \omega_3, \ldots$, not necessarily harmonically related, such that

$$v_m(t) = V_1\cos(\omega_1 t - \theta_1) + V_2\cos(\omega_2 t - \theta_2) + V_3\cos(\omega_3 t - \theta_3) + \ldots \tag{9.7}$$

Substitution of (9.7) into (9.1) yields

$$\begin{aligned} v_S(t) = {} & V_c\cos\omega_c t + \frac{1}{2}V_1\cos[(\omega_c + \omega_1)t - \theta_1] + \frac{1}{2}V_1\cos[(\omega_c - \omega_1)t - \theta_1] \\ & + \frac{1}{2}V_2\cos[(\omega_c + \omega_2)t - \theta_2] + \frac{1}{2}V_2\cos[(\omega_c - \omega_2)t - \theta_2] + \ldots \end{aligned} \tag{9.8}$$

The frequency spectrum for this signal is similar to the spectra shown in Figure 9.2 except that in this case each sideband contains a sinusoidal component corresponding to each sinusoidal component of the modulating signal; each pair of sideband components exhibits the symmetry discussed above.

If the modulating signal is allowed a constant time delay T, then (9.7) becomes

$$v_m(t) = V_1\cos[\omega_1(t - T) - \theta_1] + V_2\cos[\omega_2(t - T) - \theta_2] + V_3\cos[\omega_3(t - T) - \theta_3] + \ldots \tag{9.9}$$

or

$$v_m(t) = V_1\cos(\omega_1 t - \omega_1 T - \theta_1) + V_2\cos\cos(\omega_2 t - \omega_2 T - \theta_2) + V_3\cos(\omega_3 t - \omega_3 T - \theta_3) + \ldots \tag{9.10}$$

Since T is a constant, it follows from (9.10) that imposing a constant time delay on $v_m(t)$ is equivalent to imposing a lagging phase shift on each sinusoidal component, proportional to the frequency of that component. Thus if we let $\omega_1 T = \phi_1$, $\omega_2 T = \phi_2$ and so on, the corresponding AM wave of (9.8) with the delayed envelope can be expressed as

$$\begin{aligned} v_s(t) &= V_C \cos\omega_C t + \frac{1}{2}V_1 \cos[(\omega_C + \omega_1)t - \phi_1 - \theta_1] + \frac{1}{2}V_1 \cos[(\omega_C - \omega_1)t + \phi_1 + \theta_1] \\ &+ \frac{1}{2}V_2 \cos[(\omega_C + \omega_2)t - \phi_2 - \theta_2] + \frac{1}{2}V_2 \cos[(\omega_C - \omega_2)t + \phi_2 + \theta_2] + \dots \end{aligned} \tag{9.11}$$

where $\phi_1, \phi_2, \dots$ are directly proportional to the separation of the side frequency from the carrier frequency. Thus, if an AM signal is transmitted through an amplifier having a linear phase characteristic with a negative slope, it can be shown by going through this process in reverse that the result is a pure delay in the envelope. It can also be shown that a constant phase shift added to the linear phase characteristic produces a delay in the carrier. Alternatively, a constant phase shift rotates all vectors in the diagrams of Figure 9.3 by the same amount and leaves the relations among them unchanged.

The results obtained in the preceding discussion are useful guides in judging the merits of various tuned amplifiers and in designing these amplifiers for optimum performance. A good amplifier for use with modulated signals should have uniform gain in the frequency band occupied by the signal, and it should have a linear phase characteristic in that frequency band.

Example 9.1

A carrier

$$v_C(t) = 10\cos 2\pi \times 10^6 t$$

is amplitude–modulated with a signal

$$v_m(t) = 3\cos 1000\pi t + 5\cos 2000\pi t$$

Sketch a diagram of the frequency spectrum of the amplitude–modulated wave similar to that of Figure 9.2. Show the amplitude and frequency for each component of the AM wave.

Solution:

From relation (9.6)

$$\begin{aligned} v_S(t) &= 10\cos 2\pi \times 10^6 t + \frac{3}{2}\cos[(2\pi \times 10^6 + \pi \times 10^3)t + \theta_m]t + \frac{3}{2}\cos[(2\pi \times 10^6 - \pi \times 10^3)t] \\ &+ \frac{5}{2}\cos[(2\pi \times 10^6 + \pi \times 2 \times 10^3)t + \theta_m]t + \frac{5}{2}\cos[(2\pi \times 10^6 - \pi \times 2 \times 10^3)t] \end{aligned}$$

and the frequency spectrum is as shown in Figure 9.4.

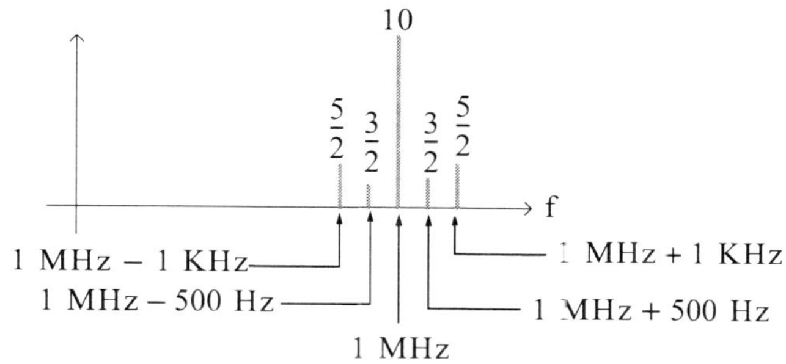

Figure 9.4. Frequency spectrum for Example 9.1

Example 9.2

A radio receiver* with a parallel GLC circuit whose inductance is $L = 0.5\ mH$ is tuned to a radio station transmitting at $810\ KHz$ resonant frequency. The resonant frequency is given by $\omega_0 = 1/\sqrt{LC}$ and the quality factor at parallel resonance is given by $Q_{0P} = \omega_0 C/G$

a. What is the value of the capacitor of this circuit at this resonant frequency?

b. What is the value of conductance G if $Q_{0P} = 75$?

c. If a nearby radio station transmits at $740\ KHz$ and both signals picked up by the antenna have the same current amplitude $I\ (\mu A)$, what is the ratio of the voltage at $810\ KHz$ to the voltage at $740\ KHz$?

Solution:

a.

$$\omega_0^2 = 1/(LC)$$

or

$$f_0^2 = 1/4\pi^2 LC$$

Then,

$$C = \frac{1}{4\pi^2 0.5 \times 10^{-3} \times (810 \times 10^3)^2} = 77.2\ pF$$

b.

$$Q_{0P} = \omega_0 C/G$$

or

$$G = \frac{2\pi f_0 C}{Q_{0P}} = \frac{2\pi \times 8.1 \times 10^5 \times 77.2 \times 10^{-12}}{75} = 5.4\ \mu\Omega^{-1}$$

c.

$$|V_{810\ KHz}| = \frac{I}{|Y_{810\ KHz}|} = \frac{I}{Y_0} = \frac{I}{G} = \frac{I}{5.24 \times 10^{-6}} \qquad (9.12)$$

* *For a block diagram of a typical AM radio receiver and description, please refer to Circuit Analysis II with MATLAB Applications, ISBN 978–0–9709511–5–1, Chapter 2, Figure 2.16.*

Also,

$$|V_{740\ KHz}| = \frac{I}{|Y_{740\ KHz}|} \tag{9.13}$$

where

$$|Y_{740\ KHz}| = \sqrt{G^2 + (\omega C - 1/(\omega L))^2}$$

or

$$|Y_{740\ KHz}| = \sqrt{(5.24 \times 10^{-6})^2 + \left(2\pi \times 740 \times 10^3 \times 77.2 \times 10^{-12} - \frac{1}{2\pi \times 740 \times 10^3 \times 0.5 \times 10^{-3}}\right)^2}$$

or

$$|Y_{740\ KHz}| = 71.2\ \mu\Omega^{-1}$$

and

$$|V_{740\ KHz}| = \frac{I}{71.2 \times 10^{-6}} \tag{9.14}$$

Then from (9.12) and (9.14),

$$\frac{|\mathbf{V}_{810\ KHz}|}{|\mathbf{V}_{740\ KHz}|} = \frac{I/5.24 \times 10^{-6}}{I/71.2 \times 10^{-6}} = \frac{71.2 \times 10^{-6}}{5.24 \times 10^{-6}} = 13.6 \tag{9.15}$$

that is, the voltage developed across the parallel circuit when it is tuned at $f = 810\ KHz$ is 13.6 times larger than the voltage developed at $f = 740\ KHz$.

9.2 Single-tuned Transistor Amplifier

The small signal model for a single–tuned transistor amplifier is shown in Figure 9.5.

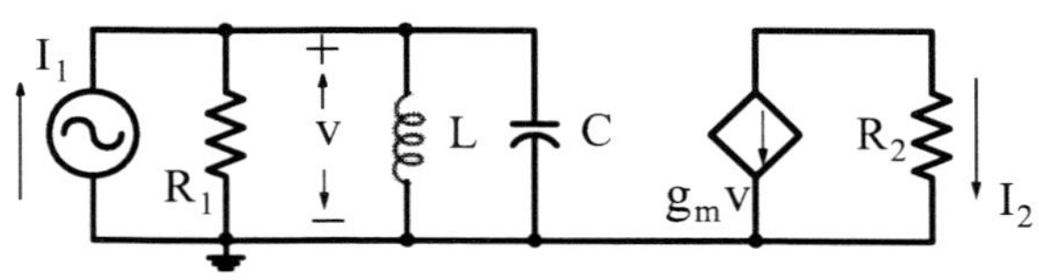

Figure 9.5. Small signal tuned transistor amplifier circuit

In the circuit of Figure 9.5, we assume that the parasitic capacitances are negligible, and with this assumption this circuit is good at frequencies up to a few hundred KHz. The resistance R_1 is in the order of 1 kilohm or less and represents the internal resistance of the current source, the input resistance of the transistor, the biasing transistors, and the losses in the inductor. The admittance $Y(s)$ on the left side of the circuit is

$$Y(s) = \frac{1}{R_1} + \frac{1}{sL} + sC \tag{9.16}$$

and thus

$$I_1(s) = Y(s)V(s) \tag{9.17}$$

Also,

$$I_2(s) = -g_m V(s) \tag{9.18}$$

From (9.12), (9.13), and (9.14) the current gain A_c is

$$A_c(s) = \frac{I_2(s)}{I_1(s)} = \frac{-g_m}{1/R_1 + 1/(sL) + sC}$$

or

$$A_c(s) = -\frac{g_m}{C} \cdot \frac{s}{s^2 + (1/RC)s + 1/LC} \tag{9.19}$$

For convenience, we define two new symbols as

$$\omega_0^2 = \frac{1}{LC} \tag{9.20}$$

and

$$2\alpha = \frac{1}{RC} \tag{9.21}$$

Substitution of (9.16) and (9.17) into (9.15) yields

$$A_c(s) = -\frac{g_m}{C} \cdot \frac{s}{s^2 + 2\alpha s + \omega_0^2} \tag{9.22}$$

The quadratic denominator in (9.18) can be expressed as the product of two linear factors, i.e.,

$$A_c(s) = -\frac{g_m}{C} \cdot \frac{s}{(s - s_1)(s - s_2)} \tag{9.23}$$

and we find that

$$s_1 = -\alpha + \sqrt{\alpha^2 - \omega_0^2} \tag{9.24}$$

and

$$s_2 = -\alpha - \sqrt{\alpha^2 - \omega_0^2} \tag{9.25}$$

Depending on the values of the circuit parameters, the poles of (9.18) may be either real or complex. If they are real, relation (9.19) has the same form as relation (8.71) in Chapter 8; hence the pole–zero pattern and the frequency characteristics of the tuned amplifier with real poles are the same as those of the RC–coupled amplifier which cannot provide a truly narrow–band characteristic. Therefore, in our subsequent discussion we will be concerned with circuit parameters which will yield complex poles.

We are also interested in the current gain at the resonant frequency denoted as A_{c0}. From the parallel network of Figure 9.5, we observe that at parallel resonance inductive susceptance* and

* *The inductive susceptance is defined as* $B_L = 1/j\omega L$ *and the capacitive susceptance as* $B_C = j\omega C$

capacitive susceptance cancel each other and the admittance of the parallel R_1LC circuit reduces to R_1 and thus

$$A_{c0} = I_2/I_1 = -g_m R_1 \tag{9.26}$$

As in the case of the RC amplifier in Chapter 8, the half–power frequencies provide a measure of the bandwidth of the circuit, and the *half–power frequency* is the frequency at which the amplification is $A = A_c(max)/\sqrt{2}$. Also, there are two half–power frequencies, one above and the other below the resonance frequency and thus the bandwidth between the two half–power frequencies is given by

$$BW = 2\alpha = \frac{1}{RC} \tag{9.27}$$

Example 9.3

The equivalent circuit of single–tuned amplifier is shown in Figure 9.6 and has a transconductance $g_m = 40\ m\Omega^{-1}$ and $R_1 = 500\ \Omega$.

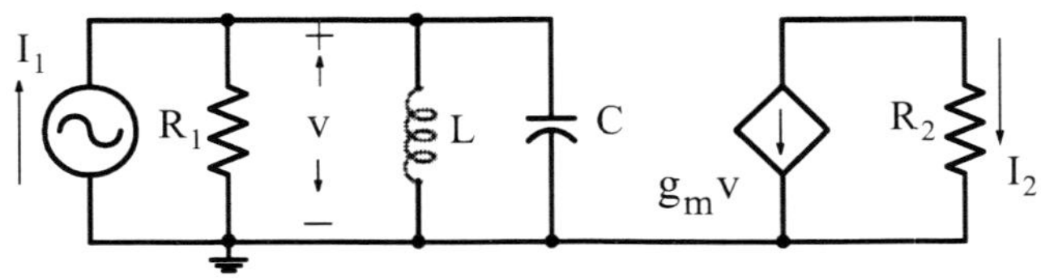

Figure 9.6. Transistor equivalent for the tuned amplifier of Example 9.2

a. Find the values of L, C, and the amplification at resonance so that the half–power bandwidth will be $BW = 10\ KHz$ and the passband will be centered at $1\ MHz$.

b. The quality factor Q_0* at parallel resonance.

Solution:

a. For the specified bandwidth it is necessary that

$$BW = \frac{1}{R_1 C} = 2\pi \times 10^4$$

and with $R_1 = 500\ \Omega$,

$$C = \frac{1}{2\pi \times 10^4 \times 500} = 32\ nF$$

* *For a detailed discussion on quality factor Q, please refer to Circuit Analysis II with MATLAB Applications, ISBN 978–0–9709511–5–1*

For the specified center frequency

$$\omega_0^2 = \frac{1}{LC} = (2\pi \times 10^6)^2 = 39.5 \times 10^{12}$$

and with $C = 32 \times 10^{-9}$,

$$L = \frac{1}{39.5 \times 10^{12} \times 32 \times 10^{-9}} = 0.8\ \mu H$$

The current gain at the resonant frequency is

$$|A_{c0}| = g_m R_1 = 40 \times 10^{-3} \times 500 = 20$$

b. The quality factor Q_0 at parallel resonance is

$$Q_0 = \omega_0 C R_1 = 2\pi \times 10^6 \times 32 \times 10^{-9} \times 500 = 100$$

In Example 9.2, the very low value of $R_1 = 500\ \Omega$ makes a large value of C necessary to obtain the narrow bandwidth $BW = 10\ \text{KHz}$, and a relatively high quality factor Q_0. Moreover, the very low value of L causes the inductive reactance at $1\ \text{MHz}$ to have a very low value of

$$\omega_0 L = 2\pi \times 10^6 \times 0.8 \times 10^{-6} = 5\ \Omega$$

Fortunately, we can transform values required by the specifications into practical values of the circuit parameters by the use of a transformer. Figure 9.7(a) shows a practical tuned transistor amplifier in which a tapped inductor is used as a step–down autotransformer.

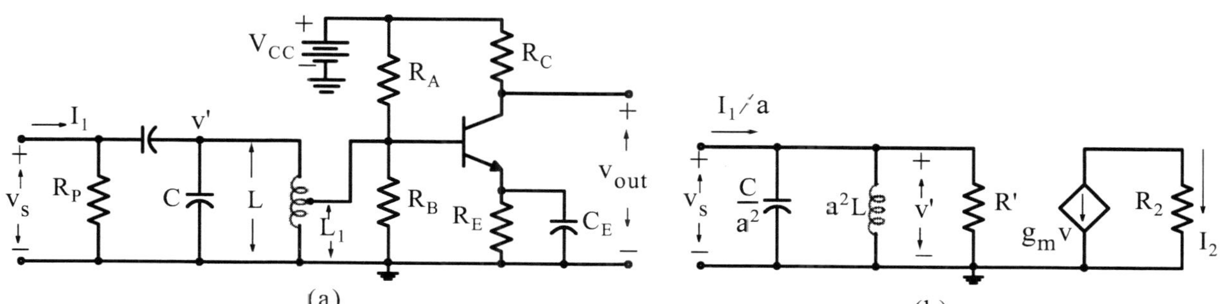

Figure 9.7. Impedance transformation in a tuned transistor amplifier and its model

Resistors R_A and R_B provide the necessary biasing but they also include coils in series with these resistors (not shown) and these coils are known as *radio frequency chokes* (RFCs). An RFC blocks radio frequencies but passes audio frequencies when both frequencies are applied to the same circuit. The inductors used in the autotransformer have ferrite cores, the coefficient of coupling* among the turns is close to unity, and the transformation ratio is

$$\frac{V_1}{V} = \sqrt{\frac{L_1}{L}} = \frac{N_1}{N} = a < 1 \tag{9.28}$$

where L is the total inductance, N is the number of turns in the coil, L_1 is the inductance portion of the coil below the tap, and N_1 is the number of turns in the coil below the tap.

When the portion of the circuit on the left of the transformer coil in Figure 9.7(a) is transferred to the right side, the small signal model assumes the form shown in Figure 9.7(b), and the capacitance C and the total inductance L are transformed by the factor a^2.

In Figure 9.7(a), the primary side of the step–down autotransformer is the total inductance and this represents the primary side of the autotransformer, and the secondary side is the portion of the coil below the tap. As we know, a practical inductor contains some resistance in series with it. Let the resistance in series with the coil in the primary be denoted as R_P and the resistance in series with the coil in the secondary be denoted as R_S. The resistance R_P includes the internal resistance of the source supplying the input signal, and when it is referred to the secondary, it assumes the value a^2R_P.

The resistance R' shown in Figure 9.7(b) represents the parallel combination of a^2R_P with R_S, that is,

$$R' = \frac{a^2R_PR_S}{a^2R_P + R_S} \tag{9.29}$$

The input signal current is also transformed to I_1/a as shown in Figure 9.7(b) and since a is less than unity, we observe that the current amplification is increased by the transformer action. From Figure 9.7(b)

$$Y'(s) = \frac{1}{R'} + \frac{1}{a^2Ls} + \frac{C}{a^2}s \tag{9.30}$$

$$\frac{I_1(s)}{a} = Y'(s)V'(s) \tag{9.31}$$

Also,

$$I_2(s) = -g_mV'(s) \tag{9.32}$$

$$A_c(s) = \frac{I_2(s)}{I_1(s)/a} = \frac{-g_m}{1/R' + 1/a^2Ls + Cs/a^2}$$

* *For a detailed discussion on transformers, please refer to Circuit Analysis II with MATLAB Applications, ISBN 978–0–9709511–5–1*

$$A_c(s) = -\frac{ag_m}{C} \cdot \frac{s}{s^2 + (a^2/R'C)s + 1/LC} \tag{9.33}$$

Therefore, the transformer action leaves the resonant frequency ω_0 unchanged, but the bandwidth becomes

$$BW = \frac{a^2}{R'C} = \frac{a^2 R_P + R_S}{R_P R_S C} \tag{9.34}$$

The LC combination in Figure 9.7(a) forms a *lossless coupling network* and if R_P and R_S are fixed, the coupling network can be adjusted for maximum amplification, and this condition occurs when maximum power is delivered to R_S. Therefore, maximum amplification is obtained when the value of a is chosen so that

$$a^2 R_P = R_S \tag{9.35}$$

At resonance, the inductive susceptance and capacitive susceptance cancel out and thus the admittance is just R'. Therefore, relation (9.33) reduces to

$$A_{co} = -\frac{g_m R'}{a} \tag{9.36}$$

The quality factor Q_0 at parallel resonance is

$$Q_0 = \omega_0 C R' \tag{9.37}$$

Example 9.4

The equivalent circuit of single–tuned amplifier shown in Figure 9.8 has the transformation ratio $a = 1/20$, transconductance $g_m = 40\ m\Omega^{-1}$, and $R' = 500\ \Omega$.

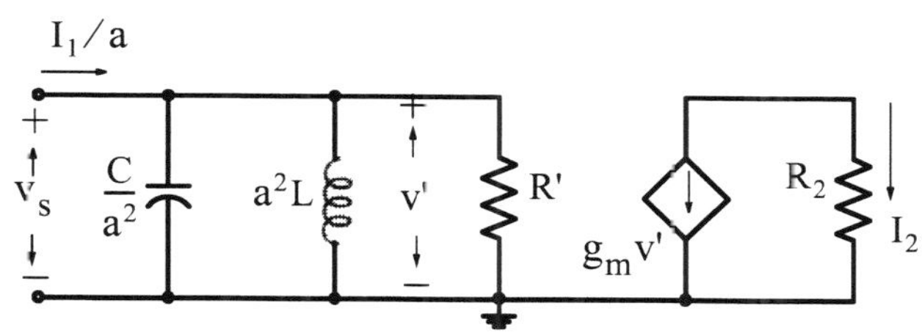

Figure 9.8. Circuit for Example 9.4

a. Find the values of L, C, and the amplification at resonance so that the half–power bandwidth will be $BW = 10\ KHz$ and the passband will be centered at $1\ MHz$.

b. The quality factor Q_0 at parallel resonance.

Solution:

a. For the specified bandwidth it is necessary that

$$BW = \frac{a^2}{R'C} = 2\pi \times 10^4$$

and with $R' = 500\ \Omega$,

$$C = \frac{1/400}{2\pi \times 10^4 \times 500} = 80\ pF$$

For the specified center frequency

$$\omega_0^2 = \frac{1}{LC} = (2\pi \times 10^6)^2 = 39.5 \times 10^{12}$$

and with $C = 80 \times 10^{-12}$,

$$L = \frac{1}{39.5 \times 10^{12} \times 80 \times 10^{-12}} = 0.32\ mH$$

The current gain at the resonant frequency is

$$|A_{c0}| = \frac{g_m R'}{a} = \frac{40 \times 10^{-3} \times 500}{1/20} = 400$$

b. The quality factor Q_0 at parallel resonance is

$$Q_0 = \omega_0 CR' = 2\pi \times 10^6 \times 80 \times 10^{-12} \times 500 = 100$$

9.3 Cascaded Tuned Amplifiers

To achieve higher selectivity it is necessary that tuned amplifiers are connected in cascade by means of a coupling capacitor or transformer coupling. When coupling capacitors are used, they are chosen to act as short circuits in the passband of the amplifier. When two identical stages like the one shown in Figure 9.5 are cascaded, the overall current gain is given by the square of relation (9.23); hence the zero at the origin and the poles s_1 and s_2 each appear twice in the overall current gain. The corresponding pole–zero diagram has the form shown in Figure 9.9 where the numbers in parentheses indicate double poles and zeros. In this form, the two stages are said to be *synchronously tuned*. We have assumed that the Miller effect is negligible.

9.3.1 Synchronously Tuned Amplifiers

The amplitude characteristic for two *synchronously tuned amplifiers* is the characteristic for one stage but with the amplification squared at each point. However, the half–power bandwidth is less than the bandwidth of one stage. The phase characteristic is the characteristic for one stage with the angle multiplied by 2.

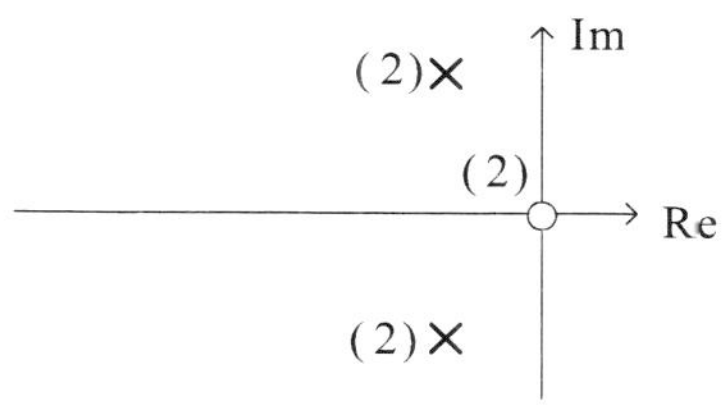

Figure 9.9. Pole–zero pattern for two synchronously tuned amplifiers in cascade

In the study of tuned amplifiers, two important relations are *the narrowband approximations*. They are very useful in the analysis and design of practically all narrowband amplifiers and filters. For the typical narrowband amplifier they are very good approximations. To understand these approximations, we refer back to relations (9.24) and (9.25), that is,

$$s_1 = -\alpha + \sqrt{\alpha^2 - \omega_0^2} \tag{9.38}$$

$$s_2 = -\alpha - \sqrt{\alpha^2 - \omega_0^2} \tag{9.39}$$

Since we are concerned with complex numbers, we express these as

$$s_1 = -\alpha + j\sqrt{\alpha^2 - \omega_0^2} \tag{9.40}$$

$$s_2 = -\alpha - j\sqrt{\alpha^2 - \omega_0^2} \tag{9.41}$$

and letting

$$\beta^2 = \alpha^2 - \omega_0^2 \tag{9.42}$$

we obtain

$$s_1 = -\alpha + j\beta \tag{9.43}$$

$$s_2 = -\alpha - j\beta \tag{9.44}$$

From relation (9.42) we observe that ω_0 is the hypotenuse of a right triangle, and in the complex frequency plane it represents the distance of the poles s_1 and s_2 from the origin as shown in Figure 9.10(a).

Figure 9.10. Graphical representation of the relations $s_1 = -\alpha + j\beta$, $s_2 = -\alpha - j\beta$, *and* $\beta^2 = \alpha^2 - \omega_0^2$

A typical AM signal has a center frequency that is 100 times the bandwidth of the signal to be transmitted. An amplifier suitable for use with this signal will have a resonant frequency of $\omega_0 = 200\alpha$ and as we can see from Figure 9.10(b), the poles s_1 and s_2 are extremely close to the imaginary axis. Now, let us consider the sinusoidal operation where $s = j\omega$ and as $j\omega$ moves along the imaginary axis near the pole s_1, the vector $s - s_1$ gets smaller and experiences large percentage variations. However, since the vector $s - s_2$ is relatively further away, the percentage variations are much smaller and the variations in s and $s - s_2$ tend to cancel. Therefore, for sinusoidal operation in the passband we obtain the narrowband approximations

$$s = j\omega \approx j\omega_0 \tag{9.45}$$

and

$$s - s_2 \approx 2j\omega_0 \tag{9.46}$$

When these approximations are substituted into relation 9.23, we obtain the simplified expression

$$A(s) = -\frac{g_m}{C} \cdot \frac{j\omega_0}{(s - s_1)2j\omega_0} = -\frac{g_m}{2C} \cdot \frac{1}{(s - s_1)} \tag{9.47}$$

Relation (9.47) shows that in narrowband operation the ratio $s/(s - s_2)$ can be expressed as the constant $1/2$ and that the narrowband approximations reduce the quadratic factor in the denominator to a linear factor. This reduction is very important when more complex circuits are to be analyzed. Also, for sinusoidal operation $s = j\omega$ and relation (9.47) can be expressed as

$$A(j\omega) = -\frac{g_m}{2C} \cdot \frac{1}{j\omega - s_1} \tag{9.48}$$

and since $\omega_0 = 200\alpha$, we see from Figure 9.10(a) that $\beta \approx \omega_0$, and thus

$$s_1 = -\alpha + j\beta \approx -\alpha + j\omega_0 \tag{9.49}$$

and by substitution into (9.48)

$$A(j\omega) = -\frac{g_m}{2C} \cdot \frac{1}{j\omega + \alpha - j\omega_0} = -\frac{g_m}{2C} \cdot \frac{1}{\alpha + j(\omega - \omega_0)} \tag{9.50}$$

Letting $\omega - \omega_0 = \Delta\omega$ we can express (9.50) as

$$A(j\omega) = -\frac{g_m}{2C} \cdot \frac{1}{\alpha + j\Delta\omega} = -\frac{g_m}{2\alpha C} \cdot \frac{1}{1 + j\Delta\omega/\alpha} \tag{9.51}$$

Also, since $\beta \approx \omega_0$, at resonance the amplification is given by

$$A_0 = |A_c(j\omega_0)| = \frac{g_m}{2C} \cdot \frac{1}{j\beta + \alpha - j\beta} = \frac{g_m}{2\alpha C} \tag{9.52}$$

We also recall that the bandwidth is given by

$$BW = 2\alpha \tag{9.53}$$

Substitution of (9.52) and (9.53) into (9.51) yields

$$A(j\omega) = -A_0 \cdot \frac{1}{1 + j2\Delta\omega/BW} \tag{9.54}$$

This relation can also be expressed in terms of the quality factor at resonance, i.e., $BW = \omega_0/Q_0$. Then,

$$A(j\omega) = -A_0 \cdot \frac{1}{1 + j2Q_0\Delta\omega/\omega_0} \tag{9.55}$$

In terms of the bandwidth (relation9.54), the magnitude of the amplification for one stage is

$$A_1 = |A| = A_{01} \cdot \frac{1}{\sqrt{1 + (2\Delta\omega/BW_1)^2}} \tag{9.56}$$

where the subscript unity denotes a single stage. For n stages connected in cascade, the magnitude of the amplification is

$$A_1^n = |A| = A_{01}^n \cdot \frac{1}{\sqrt[n]{1 + (2\Delta\omega/BW_1)^2}} \tag{9.57}$$

The half–power frequency for n identical stages connected in cascade occurs when the square of the denominator in (9.57) equals 2. Thus, at the half–power frequency

$$\left[1 + \left(\frac{2\Delta\omega}{BW_1}\right)^2\right] = 2 \tag{9.58}$$

from which

$$\Delta\omega = \pm\frac{BW_1}{2}\sqrt{2^{1/n} - 1} \tag{9.59}$$

There are two half–power frequencies, one above and one below the resonant frequency. The bandwidth between these two half–power frequencies is

$$BW_n = 2\Delta\omega = BW_1\sqrt{2^{1/n} - 1} \tag{9.60}$$

The square root of this expression is the bandwidth reduction factor for n identical stages connected in cascade; for two stages it is 0.64, for three stages it is 0.51, and for four stages it is 0.43.

Example 9.5

Two synchronously tuned stages connected in cascade are to designed so that the resonant frequency will be $f_0 = 25\ \text{MHz}$, and the capacitors will have a value of $C = 80\ \text{pF}$ for each stage. Determine the values of L and R and the half–power bandwidth for each stage so that the overall half–power bandwidth will be $100\ \text{KHz}$.

Solution:

The half–power bandwidth of each stage can be determined from relation (9.60), the values of R from the relation $BW_1 = 1/RC$ and the values of L from $\omega_0^2 = 1/LC$. For convenience, we will let the following MATLAB script do the calculations for us.

```
BWn=2*pi*10^5; C=25*10^(–12); L=1/((2*pi*25*10^6)^2*C); n=2;...
BW1= BWn/sqrt(2^(1/n)–1); BWf=BW1/(2*pi); R=1/(BW1*C); fprintf(' \n');...
fprintf('C = %5.2e F \t',C);...
fprintf('L = %5.2e H \t',L);...
fprintf('BW1 = %5.2e rad/sec \t', BW1); fprintf(' \n');...
fprintf('BWf = %5.2e Hz \t', BWf);...
fprintf('R = %5.2e Ohms \t', R); fprintf(' \n')
```

Upon execution of this script, MATLAB displays the following results:

```
C = 2.50e-011 F         L = 1.62e-006 H     BW1 = 9.76e+005 rad/sec
BWf = 1.55e+005 Hz       R = 4.10e+004 Ohms
```

The half–power bandwidth of each stage is $155\ \text{KHz}$ and, as expected, is larger than the overall half–power bandwidth of $100\ \text{KHz}$.

Synchronous tuning for two stages connected in cascade does not provide the best approximation to the ideal bandpass. As stated above, the bandwidth reduction for synchronous tuning of two stages connected in cascade is about 0.64 and this is confirmed by the ratio $100\ \text{KHz}/155\ \text{KHz}$. Figure 9.11 shows the frequency response of a synchronous tuning for two stages in cascade in comparison to a single stage.

9.3.2 Stagger–Tuned Amplifiers

We can obtain a much better approximation by staggering the tuning of the stages to produce a pole–zero configuration like the one shown in Figure 9.12.

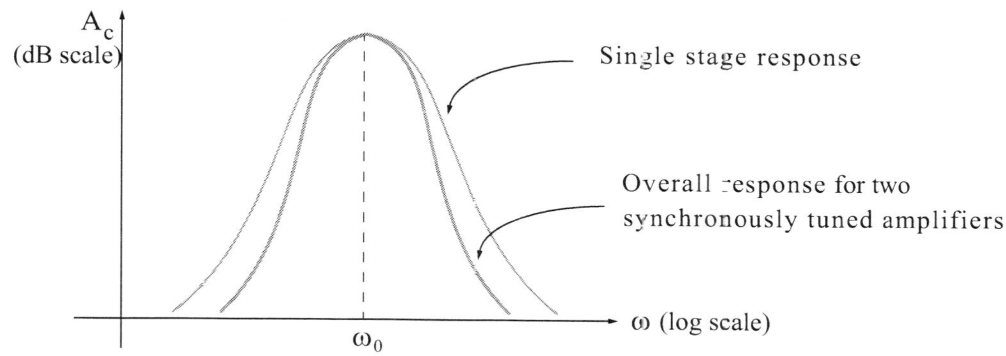

Figure 9.11. Frequency responses for single stage and two synchronously tuned amplifiers.

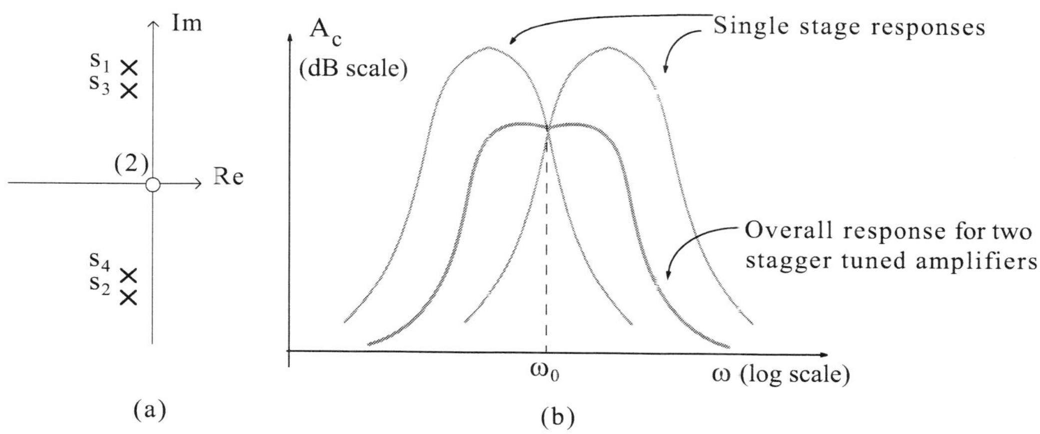

Figure 9.12. Pole–zero pattern and amplitude characteristic for two stagger–tuned amplifiers

Poles s_1 and s_2 in Figure 9.12(a) are the poles one of the stages, and poles s_3 and s_4 are the poles of the other stage; all four poles lie on a line parallel to the imaginary axis. With this arrangement, the amplitude characteristic of the stagger–tuned amplifier will be shown in Figure 9.12(b). We observe that the passband of such stagger–tuned amplifier is wider and flatter than the passband of the synchronously tuned amplifier.

The top flatness of the characteristic depends on the spacing between the poles in Figure 9.12(a), and it is under the control of the designer. Consider, for example the relations (9.20) and (9.21) which are repeated below for convenience, and the pole–zero pattern of Figure 9.10(a).

$$\omega_0^2 = \frac{1}{LC} \tag{9.61}$$

$$2\alpha = \frac{1}{RC} \tag{9.62}$$

From relation (9.62) we observe that if R and C remain constant, the spacing α from the imaginary axis remains constant also. From relation (9.61) we observe that a change in L will result in a change in ω_0 but since α remains constant, the poles must move on a path parallel to the imaginary axis. Hence, the staggering of the poles along the vertical line can be adjusted by adjusting the inductances in the two stages for slightly different values.

We found that the current gain of one stage under sinusoidal conditions and after the narrowband approximations have been applied, is given by relation (9.48), that is,

$$A(j\omega) = -\frac{g_m}{2C} \cdot \frac{1}{j\omega - s_1} \tag{9.63}$$

Therefore, the current gain of a two stagger–tuned amplifier is

$$A(j\omega) = \frac{g_m^2}{4C_1C_2} \cdot \frac{1}{(j\omega - s_1)(j\omega - s_3)} \tag{9.64}$$

where $s_1 = -\alpha + j\beta_1$, $s_3 = -\alpha + j\beta_3$, β is as shown in Figure 9.10(a), C_1 and C_2 are the capacitances of the first and second stages respectively, and the narrowband approximations have reduced a fourth–degree polynomial in the denominator to a second–degree polynomial.

The narrowband approximations and relation (9.64) reveal that the frequency characteristics of the amplifier in the passband are determined by the two poles s_1 and s_3, and the poles s_2 and s_4 together with the zeros at the origin contribute the constant $1/4$ in the scale factor of that relation. Figure 9.13(a) shows an enlarged view of the pole–zero pattern in the vicinity of the poles s_1 and s_3.

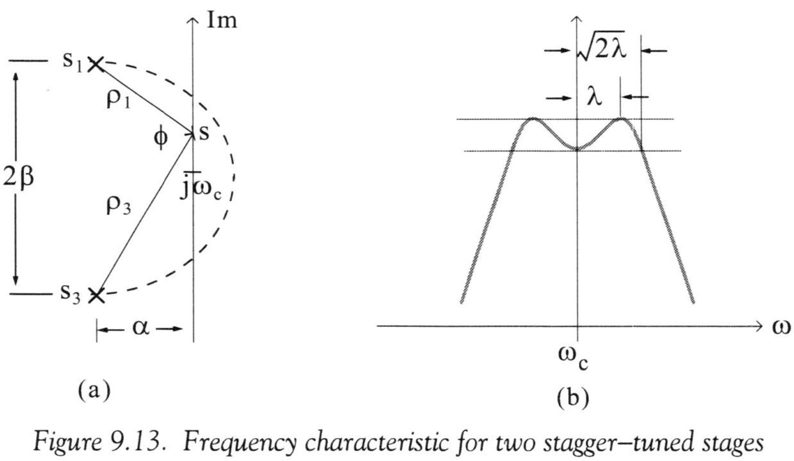

Figure 9.13. Frequency characteristic for two stagger–tuned stages

The frequency ω_c, which is the center frequency of the passband, corresponds to a point on the imaginary axis equidistant from s s_1 and s_3, and the vectors ρ_1 and ρ_3 correspond to the linear factors $j\omega - s_1$ and $j\omega - s_3$. The amplification, given by (9.64), can be expressed as

$$A = \frac{g_m^2}{4C_1C_2} \cdot \frac{1}{\rho_1\rho_3} \tag{9.65}$$

As the variable $s = j\omega$ moves along the imaginary axis, the area of the triangle $s_1\, s\, s_3$ remains constant and its value is

$$a = \alpha\beta \tag{9.66}$$

The area can also be expressed in trigonometric form as

$$a = \frac{1}{2}\rho_1\rho_3\sin\phi \tag{9.67}$$

and from (9.66) and (9.67)

$$\frac{1}{\rho_1\rho_3} = \frac{\sin\phi}{2\alpha\beta} \tag{9.68}$$

Substitution of (9.68) into (9.65) yields

$$A = \frac{g_m^2}{4C_1C_2} \cdot \frac{\sin\phi}{2\alpha\beta} \tag{9.69}$$

The poles for the two stages have the same real parts, $\alpha = 1/2R_1C_1 = 1/2R_2C_2$ and thus relation (9.69) can be expressed as

$$A = g_m^2 R_1 R_2 \frac{\alpha}{2\beta}\sin\phi \tag{9.70}$$

where $g_m^2 R_1 R_2$ is the resonant amplification of two stages with synchronous tuning. Therefore, in relation (9.70), $\sin\phi$ is the only factor that varies with changes in ω.

If α is less than β, a circle with center at the middle of the line joining poles s_1 and s_3 that is parallel to the imaginary axis and with radius β, this circle will intersect the imaginary axis at two points, as shown by the dotted semicircle in Figure 9.13(a), and the amplitude characteristic has two peaks as shown in Figure 9.13(b). The amplification at the peaks occurs when $\sin\phi = 1$ and thus

$$A_p = g_m^2 R_1 R_2 \frac{\alpha}{2\beta} \tag{9.71}$$

In general,

$$A = A_p \sin\phi \tag{9.72}$$

If α is greater than β, the semicircle does not intersect the imaginary axis and the amplitude characteristic has only one peak; it is located at $\omega = \omega_c$ and it is less than A_p.
The frequencies at which the double peaks occur are denoted as

$$\omega_p = \omega_c \pm \lambda \tag{9.73}$$

and since the radius of the semicircle is β, it follows that

$$\lambda^2 = \beta^2 - \alpha^2 \tag{9.74}$$

The *triple–point frequencies* ω_t are the frequencies outside the double peaks at which the amplification has the same value as at the center frequency are obtained by

$$\omega_t = \omega_c \pm \sqrt{2}\lambda \tag{9.75}$$

Let ϕ_c be the value of the angle ϕ when $\omega = \omega_c$. Then, the amplification at the center frequency, denoted as A_c is

$$A_c = A_p \sin\phi_c \tag{9.76}$$

The *peak–to–valley ratio* for the double–peaked amplitude characteristic is

$$\frac{A_p}{A_c} = \frac{1}{\sin\phi_c} \tag{9.77}$$

The angle ϕ_c is related to the pole positions by

$$\tan\frac{\phi_c}{2} = \frac{\beta}{\alpha} = r \tag{9.78}$$

When the amplifier is adjusted so that $\alpha = \beta$, the semicircle is tangent to the imaginary axis, and the amplitude characteristic is maximally flat. In this case, the amplifier is also said to be *flat–staggered,* and this condition marks the transition from the *overstaggered* (double peaks) to the *understaggered* adjustment. We are mostly interested on the flat–staggered and the overstaggered cases.

In all of the amplifiers studied up to this point, the half–power bandwidth has been related in a simple way to the circuit parameters. In the case of the overstaggered pair of tuned stages, however, a simpler and in some respects more useful measure of the bandwidth is the so–called *triple–point bandwidth* shown by the dotted lines in Figure 9.13(b). It is defined as

$$W = 2\sqrt{2}\lambda^{*} \tag{9.79}$$

From relations (9.74) and (9.79)

$$W^2 = 8\lambda^2 = 8[\beta^2 - \alpha^2] \tag{9.80}$$

It is convenient to express relation (9.80) in terms of the design parameter $r = \beta/\alpha$ which is related to the peak–to–valley ratio in relation (9.77). Rearranging relation (9.80) we obtain

$$\left(\frac{W}{2\alpha}\right)^2 = 2\left[\left(\frac{\beta}{\alpha}\right)^2 - 1\right] = 2(r^2 - 1) \tag{9.81}$$

* *We reserve the notation* BW *for the* 3 – dB *bandwidth and we will use* W *for the triple–point bandwidth.*

When the amplifier is flat–staggered or overstaggered, the peak amplification is A_p. In these cases the half–power frequencies can be taken as the frequencies at which $A = A_p/\sqrt{2}$, and it follows from relation (9.76) that this condition occurs when $\phi = 45°$. The corresponding half–power bandwidth in terms of the design parameter $r = \beta/\alpha$ can be determined from the geometrical constructions in Figure 9.14.

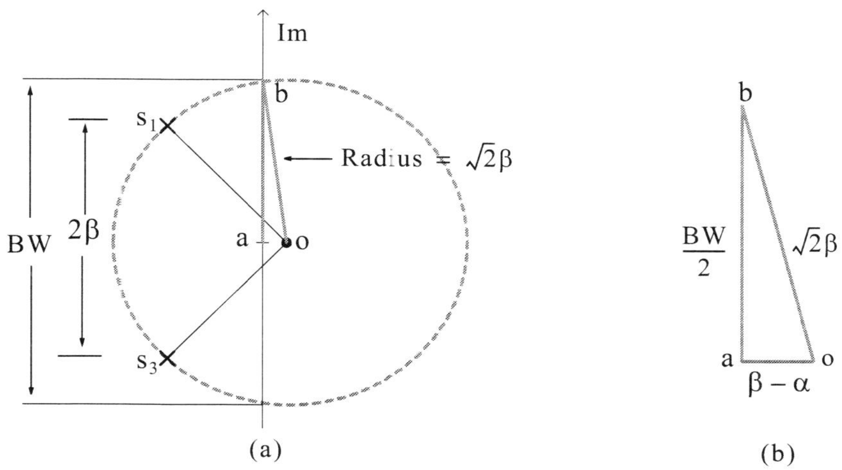

Figure 9.14. Determination of the half–power frequencies of flat–staggered and overstaggered amplifiers

In Figure 9.14(a) the line segment joining the poles s_1 and s_3 subtends the angle $2\phi = 90°$, the radius of the circle is $\sqrt{2}\beta$, and the center of the circle lies a distance $\beta - \alpha$ to the right of the imaginary axis. Figure 9.14(b) is an enlarged version of the small right triangle oab and by the Pythagorean theorem

$$\left(\frac{BW}{2}\right)^2 = 2\beta^2 - (\beta - \alpha)^2 \tag{9.82}$$

or

$$\left(\frac{BW}{2\alpha}\right)^2 = \left(\frac{\beta}{\alpha}\right)^2 + \frac{2\beta}{\alpha} - 1 \tag{9.83}$$

Letting $r = \beta/\alpha$, we can express (9.83) as

$$\left(\frac{BW}{2\alpha}\right)^2 = r^2 + 2r - 1 \tag{9.84}$$

Using relation (9.84), flat–staggered and overstaggered amplifiers can be designed for a specified half–power bandwidth. Also, a comparison of (9.84) with (9.81) reveals that the half–power bandwidth and the triple–point bandwidth are nearly equal except for cases of very slight overstaggering.

The relations derived above provide a simple, direct design procedure for overstaggered stages.

The design requirements specify the center frequency ω_c, the triple–point bandwidth W, and the peak–to–valley ratio A_p/A_p. These specifications are sufficient to define the poles of the amplitude gain A, and hence the circuit parameters. Specifically, the ratio A_p/A_p defines $r = \beta/\alpha$ through relations (9.77) and (9.78). Then, relation (9.81) gives the required value of α, and the required value of β is obtained from the relation $\beta = \alpha r$. Thus, the two of the three parameters R, L, and C are determined, and the third can be chosen as desired to satisfy the other requirement as it was illustrated in the previous examples.

Example 9.6

A transistorized two–stage stagger–tuned amplifier has a pole–zero pattern like the one shown in Figure 9.13. The pertinent dimensions are $\omega_p = 10^6$ rad/s, $\alpha = 1500$ rad/s, and $\beta = 2500$ rad/s. Both transistors have a transconductance of $g_m = 40\ \text{m}\Omega^{-1}$, and the total shunt resistance in each stage is $R_1 = R_2 = 0.75\ \text{K}\Omega$. Find the peak amplification A_p, the peak–to–valley ratio A_p/A_c, the triple–point bandwidth W, the half–power bandwidth BW, and the triple–point frequencies.

Solution:

The peak amplification A_p is found from (9.71), i.e.,

$$A_p = g_m^2 R_1 R_2 \frac{\alpha}{2\beta} = (40\times 10^{-3})^2 \times (0.75\times 10^3)^2 \times \frac{1.5}{5} = 270$$

From relation (9.78)

$$\tan\frac{\phi_c}{2} = \frac{\beta}{\alpha} = r = \frac{2.5}{1.5} = \frac{5}{3}$$

$$\phi_c/2 = \tan^{-1}(5/3) = 59°$$

$$\phi_c = 118°$$

The peak–to–valley ratio A_p/A_c is given by relation (9.77)

$$\frac{A_p}{A_c} = \frac{1}{\sin\phi_c} = \frac{1}{\sin 118°} = 1.13$$

Therefore, the amplifiers are overstaggered.

The triple–point bandwidth W, is given by relation (9.79), i.e., $W = 2\sqrt{2}\lambda$ where from relation (9.74)

$$\lambda^2 = \beta^2 - \alpha^2 = (2.5^2 - 1.5^2) \times 10^6 = 4 \times 10^6 \text{ (rad/s)}^2$$

$$\lambda = 2 \times 10^3 \text{ rad/s}$$

$$W = 2\sqrt{2}\lambda = 4\sqrt{2} \times 10^3 = 5.66 \times 10^3 \text{ rad/s}$$

The half–power bandwidth BW is given by relation (9.83), i.e.,

$$\left(\frac{BW}{2\alpha}\right)^2 = \left(\frac{\beta}{\alpha}\right)^2 + \frac{2\beta}{\alpha} - 1$$

$$BW = 2\alpha\sqrt{\left(\frac{\beta}{\alpha}\right)^2 + \frac{2\beta}{\alpha} - 1} = 2 \times 1.5 \times 10^3\sqrt{\left(\frac{2.5}{1.5}\right)^2 + \frac{5}{1.5} - 1} = 6.78 \times 10^3 \text{ rad/s}$$

The triple–point frequencies are given by relation (9.75). Thus,

$$\omega_t = \omega_c \pm \sqrt{2}\lambda = 10^6 \pm \sqrt{2} \times 2 \times 10^3$$

$$\omega_{t1} = 1.003 \times 10^6 \text{ rad/s} \qquad \omega_{t2} = 0.997 \times 10^6 \text{ rad/s}$$

The procedure for finding all pertinent data, except the values of L and C, was well illustrated in Example 9.6. The following example illustrates the procedure for finding of L and C to meet all design specifications.

Example 9.7

The pole locations for a two–stage stagger–tuned amplifier and the values of R_1 and R_2 are as given in Example 9.6. Determine the values of L and C required for each stage.

Solution:

From relation (9.27)

$$BW = 2\alpha = \frac{1}{RC}$$

and thus

$$2\alpha = \frac{1}{R_1C_1} = \frac{1}{R_2C_2}$$

With $\omega_c = 10^6$ rad/s, $\alpha = 1500$ rad/s and $R_1 = R_2 = 0.75\ K\Omega$, we find that

$$C_1 = C_2 = \frac{1}{2\alpha R_1} = \frac{1}{2\alpha R_2} = \frac{1}{3 \times 10^3 \times 0.75 \times 10^3} = 0.44\ \mu F$$

and

$$L_1 = \frac{1}{\omega_{t1}^2 C_1} = \frac{1}{0.997^2 \times 10^{12} \times 0.44 \times 10^{-6}} = 2.28\ \mu H$$

$$L_2 = \frac{1}{\omega_{t2}^2 C_2} = \frac{1}{1.003^2 \times 10^{12} \times 0.44 \times 10^{-6}} = 2.27\ \mu H$$

Figure 9.15 shows the normalized amplitude characteristics for two tuned stages under four different adjustments.

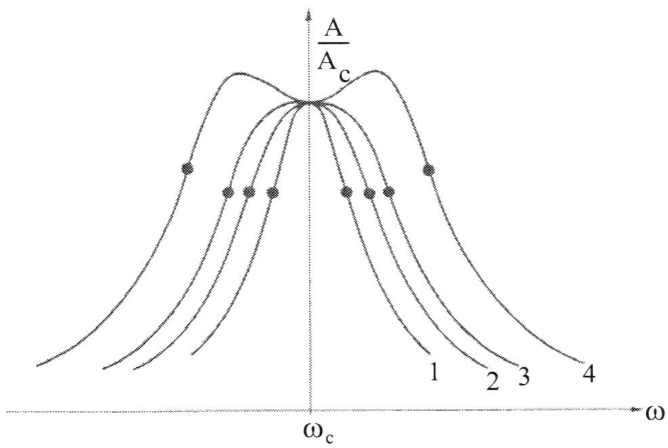

Figure 9.15. Amplitude characteristics for two tuned stages with different degrees of staggering

For the curves in Figure 9.15, α is held constant while β is increased in steps. Curve 1 is for synchronous tuning, Curve 2 is for understaggering with $\phi_c = 70°$, Curve 3 is for flat staggering, and Curve 4 is for overstaggering with $A_p/A_c = 1.1$. The half–power bandwidths are indicated by the solid dots on these curves.

For the flat–staggered case $\beta = \alpha$, $r = 1$, and $\sin\phi_c = 1$. Thus, from relations (9.71) and (9.72) we obtain

$$A_c = A_p = \frac{1}{2} g_m^2 R_1 R_2 \tag{9.85}$$

Letting B_{fs} denote the bandwidth of two stages with flat staggering and BW the $3-dB$ bandwidth of one of the stages, from (9.84) we obtain

$$B_{fs} = 2\sqrt{2}BW \tag{9.86}$$

The gain–bandwidth product for the flat–staggered amplifier is given by

$$A_c B_{fs} = (1/\sqrt{2})(g_m^2 R_1 R_2)BW \tag{9.87}$$

For two synchronously tuned stages we denote the bandwidth of two stages as B_{syn}. Then, for the gain–bandwidth product, using relation (9.60) we obtain

$$A_c B_{syn} = A_1^2 BW_1 \sqrt{2^{1/2} - 1} = 0.64(g_m^2 \mathcal{R}_1 R_2)(BW) \qquad (9.88)$$

The foregoing discussion of stagger–tuned amplifiers considers only the case in which the poles are staggered along a line parallel to the imaginary axis. If the poles do not lie on the same vertical line, skewed frequency characteristics result, and, in accordance with the discussion in Section 9.1, this fact may cause distortion of the envelopes of AM signals.

9.3.3 Three or More Tuned Amplifiers Connected in Cascade

When three or more tuned stages are to be connected in cascade, the relations encountered are similar in many respects to those encountered in the two–stage case. In particular, for narrowband amplifiers the frequency characteristics are determined entirely by the poles lying near the segment of the imaginary axis corresponding to the passband. Thus, in the usual case, the zeros at the origin and the poles in the lower half plane contribute only a constant to the gain in the narrow passband. By a more extensive treatment* it can be shown that for a maximally flat amplitude characteristic, the poles of the signal gain must be uniformly spaced on a semicircle having its center on the imaginary axis. The amplitude characteristics for one, three, and five stages are shown in Figure 9.16.

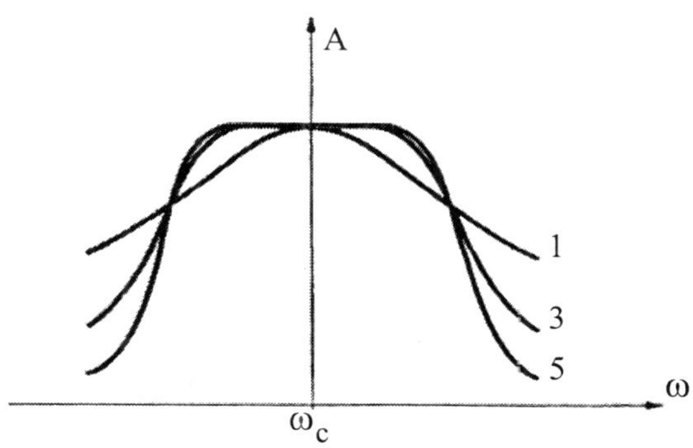

Figure 9.16. Maximally flat multistage tuned amplifiers

In Figure 9.16, The center frequency ω_c of the passband corresponds to the center of the circle on which the poles lie, and no matter how many stages are cascaded, the half–power frequencies are the frequencies at which the circle intersects the imaginary axis. The more stages cascaded in this way, the flatter the amplitude characteristic in the passband and the steeper the characteristic at the edges of the band. The distribution of poles producing these characteristics is known as the *Butterworth configuration.*

* *For a detailed discussion on this topic, please refer to Signals and Systems with MATLAB Computing and Simulink Modeling, ISBN 978–1–934404–11–9.*

9.4 Summary

- A tuned amplifier is essentially a bandpass filter. A passive bandpass filter is constructed with passive devices, i.e., resistors, inductors, and capacitors, and thus it provides no gain. For small signals, active filters with op amps are very popular. Tuned amplifiers can also be designed with bipolar junction transistors and MOSFETs.
- The analysis and design of tuned amplifiers is greatly facilitated with the small signal models.
- For a single stage tuned amplifier the current gain at the resonant frequency is given by

$$|A_{c0}| = I_2/I_1 = g_m R_1$$

and the bandwidth BW between the two half–power frequencies is given by

$$BW = 2\alpha = 1/RC$$

- In the design of tuned amplifiers, we can transform values required by the specifications into practical values of the circuit parameters by the use of a transformer.
- A radio frequency choke (RFC) blocks radio frequencies but passes audio frequencies when both frequencies are applied to the same circuit.
- Tuned amplifiers are connected in cascade to achieve higher selectivity.
- When two identical stages are cascaded, the two stages are said to be synchronously tuned.
- The amplitude characteristic for two synchronously tuned amplifiers is the characteristic for one stage but with the amplification squared at each point.
- In the study of tuned amplifiers, two important relations are the narrowband approximations. They are very useful in the analysis and design of practically all narrowband amplifiers and filters.
- For synchronously tuned amplifiers the bandwidth between the half–power points is given by

$$BW_n = 2\Delta\omega = BW_1\sqrt{2^{1/n}-1}$$

- The passband of such stagger–tuned amplifier is wider and flatter than the passband of the synchronously tuned amplifier and thus provides a better approximation to the ideal bandpass.
- The amplification at resonance for a two stagger–tuned amplifier is given by

$$A = g_m^2 R_1 R_2 \frac{\alpha}{2\beta} \sin\phi$$

where $g_m^2 R_1 R_2$ is the resonant amplification of two stages with synchronous tuning.

- For stagger–tuned amplifiers, the frequencies at which the double peaks occur given by

$$\omega_p = \omega_c \pm \lambda$$

- For stagger–tuned amplifiers, the frequencies outside the double peaks at which the amplification has the same value as at the center frequency are obtained by

$$\omega_t = \omega_c \pm \sqrt{2}\lambda$$

- For stagger–tuned amplifiers adjusted so that $\alpha = \beta$, the amplifier is also said to be flat–staggered, and this condition marks the transition from the overstaggered (double peaks) to the understaggered adjustment. We are mostly interested on the flat–staggered and the overstaggered cases.
- For an overstaggered pair of tuned stages the triple–point bandwidth is defined as

$$W = 2\sqrt{2}\lambda$$

and half–power bandwidth BW is given by

$$\left(\frac{BW}{2\alpha}\right)^2 = \left(\frac{\beta}{\alpha}\right)^2 + \frac{2\beta}{\alpha} - 1$$

- For uniform amplification over a wide band of frequencies we can cascade 3 or more amplifiers where the distribution of poles produce characteristics is known as the Butterworth configuration.

9.5 Exercises

1. The small–signal model for a single–tuned amplifier is shown below, and for this model, it is known that $g_m = 40\ m\Omega^{-1}$ and $R_1 = 1\ K\Omega$.

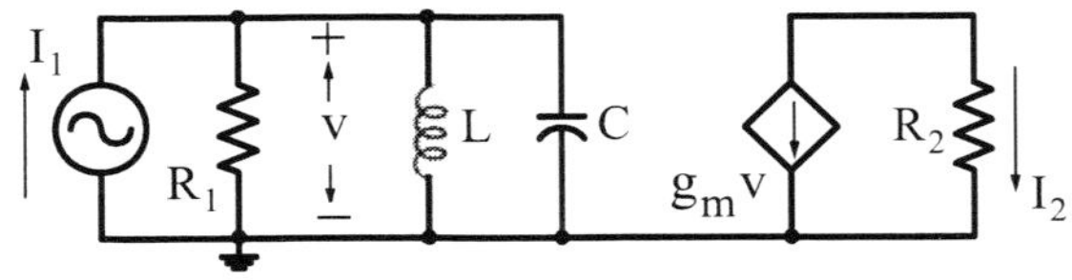

a. Determine the values of L and C so that the amplifier will have a resonant frequency of $100\ KHz$ and a half–power bandwidth of $5\ KHz$.

b. Determine the current gain $A_{c0} = I_2/I_1$, i.e., the current gain at the resonant frequency.

2. For the incremental model of a typical MOSFET tuned circuit below, $g_m = 5\ m\Omega^{-1}$

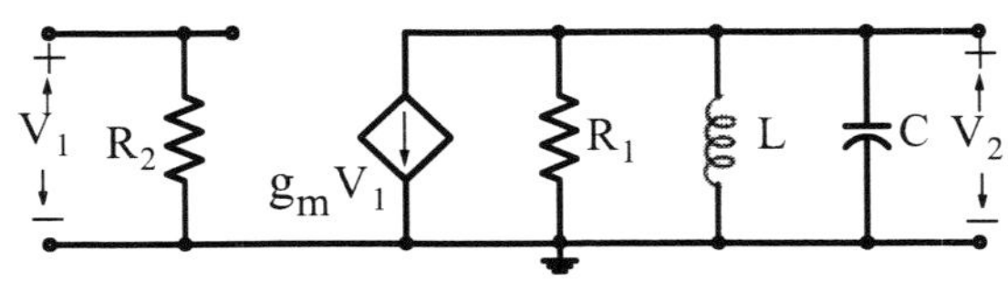

By a procedure similar to that in Section 9.2, we can show that the resonant frequency, the half–power frequency, and the voltage gain at resonance are given by the relations

$$\omega_0^2 = \frac{1}{LC} \qquad BW = 2\alpha = \frac{1}{R_1C} \qquad |A_{v0}| = g_mR_1$$

a. Find suitable values for circuit parameters so that the half–power bandwidth will be the $10\ KHz$, and the passband will be centered at $1\ MHz$.

b. Find the voltage amplification at resonance.

c. Find the quality factor at resonance.

3. Two synchronously tuned stages are connected in cascade and the transconductance of both transistors used is $g_m = 40\ m\Omega^{-1}$ where $R_1 = 1\ K\Omega$ and $R_2 = 1.5\ K\Omega$ are to designed so that the resonant frequency will be $f_c = 25\ MHz$. Find the resonant amplification.

4. A two–stage flat–staggered tuned amplifier is to designed with two transistors each with transconductance $g_m = 40\ m\Omega^{-1}$ where $R_1 = 1\ K\Omega$ and $R_2 = 1.5\ K\Omega$ are to designed so that the resonant frequency will be $f_c = 25\ MHz$. Find the resonant amplification.

9.6 Solutions to End–of–Chapter Exercises

1.

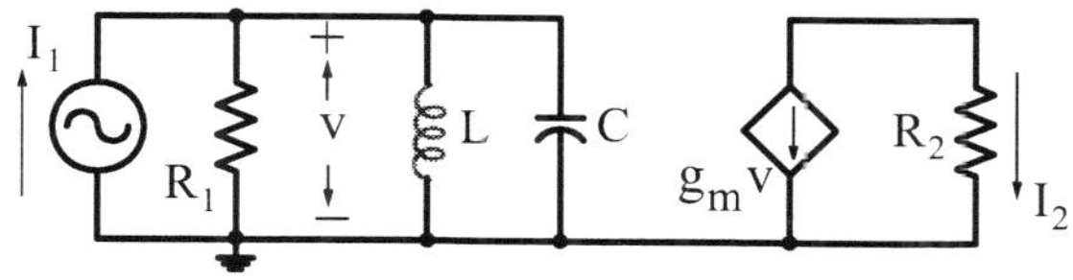

a. The bandwidth in radians per second is

$$BW = 2\alpha = \frac{1}{RC} = 2\pi \times 5000 = 3.14 \times 10^4$$

and with $R_1 = 1\ K\Omega$,

$$C = \frac{1}{BW \cdot R_1} = \frac{1}{3.14 \times 10^4 \times 10^3} = 31.8\ nF$$

The value of L is found from $\omega_0^2 = 1/LC$, or $f_0^2 = 1/2\pi LC$. Thus,

$$L = \frac{1}{(2\pi \times 10^5)^2 \times 31.8 \times 10^{-9}} = 80\ \mu H$$

b. The current gain at resonance is

$$A_{c0} = g_m R_1 = 40\ m\Omega^{-1} \times 1\ K\Omega = 40$$

2.

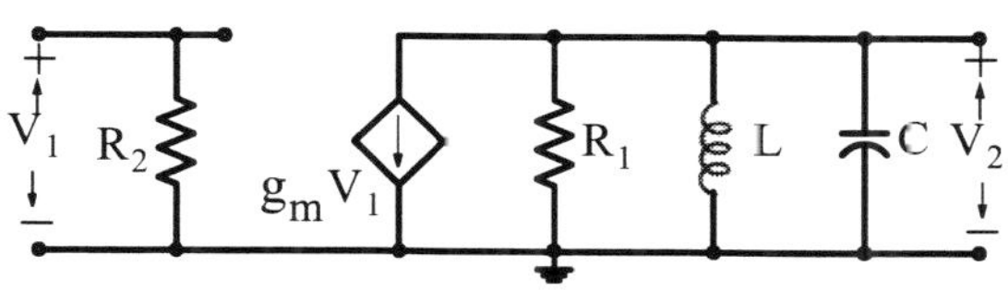

a. For the required bandwidth it is necessary that

$$BW = 2\alpha = \frac{1}{R_1 C} = 2\pi \times 10^4$$

and for the specified center frequency we must have

$$\omega_0^2 = 1/LC = (2\pi \times 10^5)^2$$

For high gain we can choose a large value for R_1. Or, we can choose a large value for C to overcome the parasitic capacitances. Let us assume that the parasitic capacitances are of more concern to us and choose a 200 pF capacitor. Then, the bandwidth requirement yields

$$R_1 = \frac{1}{BWC} = \frac{10^{12}}{2\pi \times 10^4 \times 200} = 80\ K\Omega$$

The value of L to meet the center–frequency requirement is

$$L = \frac{1}{\omega_0^2 C} = \frac{10^{12}}{(2\pi \times 10^6)^2 \times 200} = 0.125\ mH$$

b. The voltage gain at the center–frequency is

$$|A_{v0}| = g_m R_1 = 5 \times 10^{-3} \times 80 \times 10^3 = 400$$

c. The quality factor at resonance is

$$Q_0 = \omega_0 C R_1 = 2\pi \times 10^6 \times 200 \times 10^{-12} \times 80 \times 10^3 = 100$$

3. By relation (9.26) the resonant amplification of the first stage is $|A_1| = g_m R_1$, and that of the second stage is $|A_2| = g_m R_2$. Therefore, the overall gain is

$$A = g_m^2 R_1 R_2 = (40 \times 10^{-3})^2 \times 1 \times 1.5 \times 10^6 = 60$$

4. By relation (9.85)

$$A_c = A_p = \frac{1}{2} g_m^2 R_1 R_2 = \frac{1}{2}(40 \times 10^{-3})^2 \times 1 \times 1.5 \times 10^6 = 30$$

As expected this gain is half of that of Exercise 3 where the circuit was designed with synchronously tuned stages.

Chapter 10

Sinusoidal Oscillators

This chapter begins with an introduction to sinusoidal oscillators. Subsequently, we will discuss the RC phase–shift and LC oscillator families, and we will describe the Armstrong, Hartley, Colpitts, crystal, and the crystal–controlled Pierce oscillators.

10.1 Introduction to Oscillators

An oscillator may be defined as a class of wave generators that produce sinusoidal, square, triangular, and sawtooth waveforms. Non–sinusoidal oscillators are known as *relaxation oscillators*. In Chapter 5 we introduced the Wien bridge oscillator, and in Chapter 7 the family of multivibrators. In this chapter we will be concerned with oscillators that produce sinusoidal waveforms. An oscillator is essentially an amplifier circuit that provides its own input signal. Oscillators are classified in accordance with the waveshapes they produce and the circuitry required to produce the desired oscillations. Henceforth, in this chapter, unless otherwise specified, the term oscillator will mean a sinusoidal oscillator.

10.2 Sinusoidal Oscillators

As the name implies, a sinusoidal oscillator produces a sine–wave output signal. An ideal oscillator should produce an output signal with constant amplitude with no variation in frequency. But a practical oscillator cannot meet these criteria, the degree to which the ideal is approached depends on the class of amplifier operation, amplifier characteristics, frequency stability, and amplitude stability. Sinusoidal oscillators generate signals ranging from low audio frequencies to ultrahigh radio and microwave frequencies. Many low–frequency oscillators use resistors and capacitors to form their frequency–determining networks and are referred to as *RC oscillators*. These are used in the audio–frequency range.

Other types of employ inductors and capacitors for its frequency–determining network. These are known as the LC oscillators and use *tank circuits**, are commonly used for the higher radio frequencies. They are not suitable for use as extremely low–frequency oscillators because the inductors and capacitors would be large in size, heavy, and costly to manufacture.

A third type of sinusoidal oscillator is the *crystal–controlled oscillator*. The crystal–controlled oscillator provides excellent frequency stability and is used from the middle of the audio range through the radio frequency range.

* *The term tank circuit stems from the fact that a network consisting of inductors and capacitors store energy same way as liquids and gases are stored in tanks.*

An oscillator must provide amplification where the amplification of signal power occurs from the input to the output of the oscillator and a portion of the output is fed back to the input to sustain a constant input. We will briefly discuss *positive* and *negative feedback* with reference to the block diagram of a feedback amplifier shown in Figure 10.1.

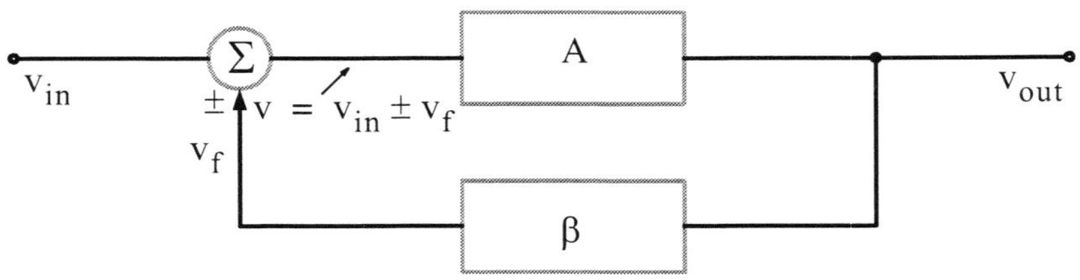

Figure 10.1. Block diagram of a typical feedback amplifier

From Figure 10.1,

$$v_{out} = Av \tag{10.1}$$

where A denotes the *open–loop gain* of the amplifier. Also,

$$v_f = \beta v_{out} \tag{10.2}$$

where β is known as the feedback factor. If v_f subtracts from v_{in}, then $v = v_{in} - v_f$ and the feedback is known as *degenerative* or *negative*, and the amplifier is provided with *negative feedback*. If v_f is added to v_{in}, then $v = v_{in} + v_f$ and the feedback is known as *regenerative* or *positive*, and the amplifier is provided with *positive feedback* From (10.1), (10.2), and $v = v_{in} - v_f$, we obtain the general expression for a feedback loop as

$$A_f = \frac{v_{out}}{v_{in}} = \frac{A}{1 \pm \beta A} \tag{10.3}$$

where the quantity βA is known as the *loop gain*, and $1 \pm \beta A$ is known as the *amount of feedback*.

In a practical amplifier, the open loop gain A is much greater than unity and thus the gain with feedback A_f is for all practical purposes equal to $\pm 1/\beta$ and this means that A_f is effectively dependent on the feedback network, and since this network can be constructed with accurate devices such as precision resistors and capacitors, the gain A_f can be designed precisely.

With negative feedback, relation (10.3) is expressed as

$$A_f = \frac{v_{out}}{v_{in}} = \frac{A}{1 + \beta A} \tag{10.4}$$

and with positive feedback as

$$A_f = \frac{v_{out}}{v_{in}} = \frac{A}{1 - \beta A} \tag{10.5}$$

Practically, all amplifiers employ negative feedback, but as we will see shortly, sinusoidal oscillators use positive feedback.

Example 10.1

A certain amplifier has an open–loop gain A of 70 at an angle of zero degrees, distortion of 10%, and normal input voltage of $1\ V$. Determine the effects of negative feedback if the feedback factor β is 0.1.

Solution:

For negative feedback $\beta A = 0.1 \times 70 = 7$ and in accordance with relation (10.4) the gain with negative feedback is

$$A_f = \frac{A}{1+\beta A} = \frac{70}{1+7} = 8.75$$

The amount of distortion D without feedback is 10% of the normal input voltage v_{in} of $1\ V$, and with feedback is

$$D_f = \frac{0.1 \times 1}{1+\beta A} = \frac{0.10}{1+7} = 0.0125 = 1.25\%$$

New input voltage v'_{in} with feedback is

$$v'_{in} = v_{in}(1+\beta A) = 1 \times (1+7) = 8\ V$$

The output voltage v_{out} without feedback is

$$v_{out} = Av_{in} = 70 \times 1 = 70\ V$$

and the output voltage v'_{out} with feedback is

$$v'_{out} = A_f v'_{in} = 8.75 \times 8 = 70\ V$$

Actual input voltage v to amplifier is

$$v = v'_{in} - \beta v'_{out} = 8 - 0.1 \times 70 = 1\ V$$

and the reduction in gain expressed in dB is

$$dB = 20\log\frac{A}{A_f} = 20\log\frac{70}{8.75} = 18.06\ dB$$

Thus, the introduction of negative feedback has reduced the amplifier gain by $18\ dB$ or we can say that $18\ dB$ of feedback has been applied to the input of the amplifier.

Example 10.1 above and Exercise 1 at the end of this chapter illustrate the effects of degenerative (negative) feedback. In our subsequent discussion we will be concerned with regenerative (positive) feedback. Let us again consider relation (10.5) which is repeated below for convenience.

$$A_f = \frac{v_{out}}{v_{in}} = \frac{A}{1-\beta A} \tag{10.6}$$

In an oscillator circuit the positive feedback must be large enough to compensate for circuit losses so that oscillations will be sustained. Moreover, a practical oscillator must oscillate at a predetermined frequency and thus the oscillator must include a frequency–determining device which essentially is a band–pass filter allowing only the desired frequency to pass. Let us now assume that at that predetermined frequency the loop gain βA in relation (10.6) above is very close to unity and in this case the closed loop A_f is very large. Accordingly, it is possible to obtain a finite output with a very small input and this is the principle behind the operation of an oscillator.

To find the predetermined frequency ω_0 to produce sustained oscillations, we express relation (10.6) in the s–domain, that is,

$$A_f(s) = \frac{V_{out}(s)}{V_{in}(s)} = \frac{A(s)}{1-\beta(s)A(s)} \tag{10.7}$$

and letting $s = j\omega$ we obtain

$$A_f(j\omega) = \frac{V_{out}(j\omega)}{V_{in}(j\omega)} = \frac{A(j\omega)}{1-\beta(j\omega)A(j\omega)} \tag{10.8}$$

To sustain oscillations at $\omega = \omega_0$ the denominator of (10.8) must meet the *Barkhausen criterion* which states that to provide sinusoidal oscillations the magnitude and phase angle of the term $\beta(j\omega)A(j\omega)$ must be such that

$$|\beta(j\omega_0)A(j\omega_0)| \geq 1 \tag{10.9}$$

and

$$\angle\beta(j\omega_0) + \angle A(j\omega_0) = 2\pi n \qquad n = 1, 2, 3, \ldots \tag{10.10}$$

10.3 RC Oscillator

An amplifier and an *RC network* can form an oscillator as shown in Figure 10.2. This oscillator is also known as *phase–shift oscillator*. In the oscillator circuit of Figure 10.2 the transistor, being in the common–emitter configuration, provides $180°$ phase shift and the three RC sections can be selected that each section will provide $60°$ shift. Therefore, the total phase shift in that oscillator will be $3 \times 60° + 180° = 360°$ and the condition of relation (10.10) will be satisfied. The condition of relation (10.9) can also be satisfied by making the product βA equal to unity. For instance, if $A = 10$ we can make $\beta = 1/10$ and the condition of relation (10.9) will also be satisfied.

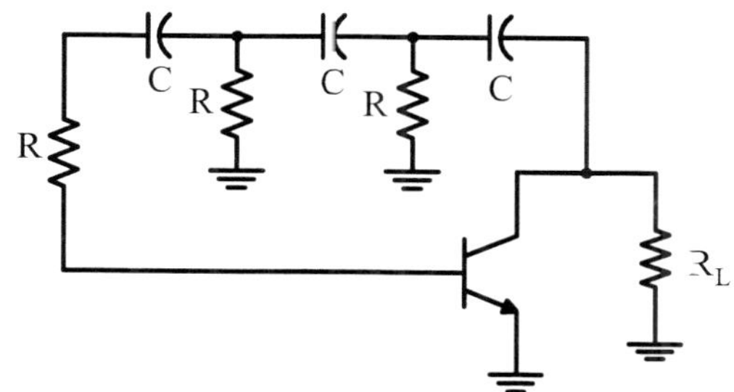

Figure 10.2. Simple form of an RC oscillator

The output of the oscillator contains only a single sinusoidal frequency. When the oscillator is powered on, the loop gain βA is greater than unity and the amplitude of the oscillations will increase. Eventually, a level is reached where the gain of the amplifier decreases, and the value of the loop gain decreases to unity and constant amplitude oscillations are sustained. The frequency of oscillations is determined by the values of resistance and capacitance in the three sections. Variable resistors and capacitors are usually employed to provide tuning in the feedback network for variations in phase shift. For an RC phase–shift oscillator, the amplifier is biased for Class A operation to minimize distortion of the signal.

For the RC oscillator of Figure 10.2, the frequency of oscillation is given by

$$\omega_0 = \frac{1}{C\sqrt{4RR_L + GR^2}} \tag{10.11}$$

where G is the transistor power gain and the values of the resistors and the transistor gain β are such that the following relation is satisfied.

$$\beta = 23 + \frac{29R}{R_L} + \frac{4R_L}{R} \tag{10.12}$$

In general, for an RC phase–shift oscillator the frequency of oscillation (resonant frequency) can be approximated from the relation

$$\omega_0 = \frac{1}{RC\sqrt{n}} \tag{10.13}$$

where n is the number of RC sections.

10.4 LC Oscillators

The *LC type of oscillators* use resonant circuits. As we know, a resonant circuit stores energy alternately in the inductor and capacitor. However, every circuit contains some resistance and this resistance causes reduction in the amplitude of the oscillations. To sustain oscillations with constant amplitude it is necessary to use regenerative feedback. Figure 10.3 shows a block diagram of a typical LC oscillator. In an LC oscillator the sinusoidal signal is generated by the action of an inductor and a capacitor. The feedback signal is coupled from the LC tank of the oscillator circuit by using a coil (tickler* or a coil pair) as shown in Figure 10.4(a) and 10.4(b) or by using a capacitor pair in the tank circuit and tap the feedback signal between them as shown in Figure 10.4(c).

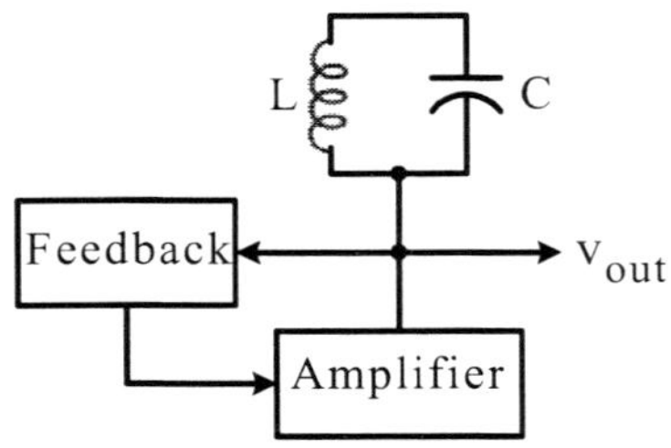

Figure 10.3. Block diagram of a typical LC oscillator

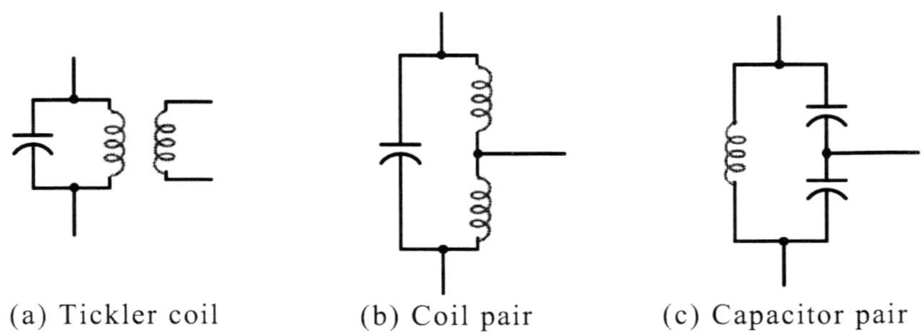

Figure 10.4. Feedback signal coupling for LC type oscillators

10.5 The Armstrong Oscillator

Figure 10.5 shows two circuits known as *Armstrong oscillators.*

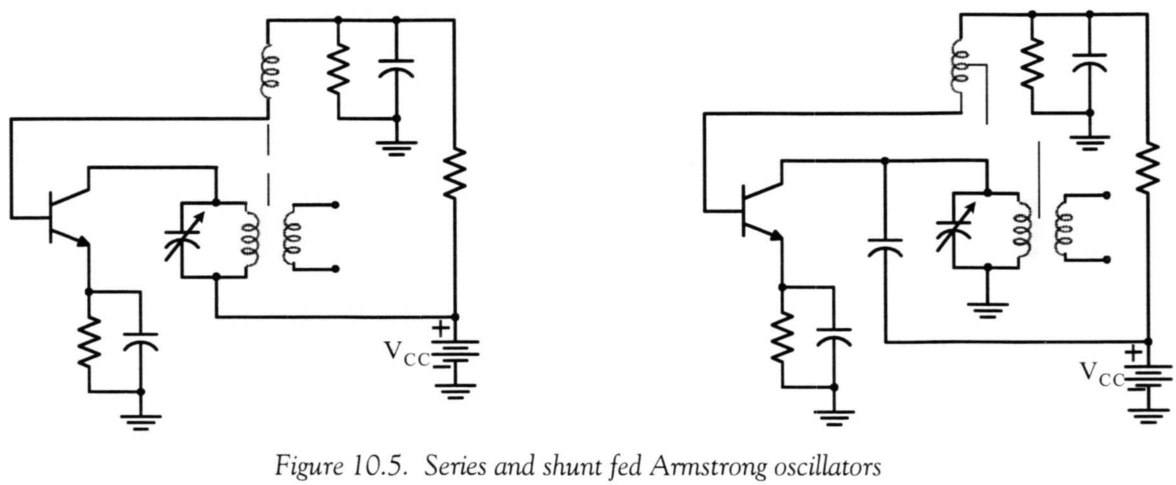

Figure 10.5. Series and shunt fed Armstrong oscillators

* *A tickler coil is an inductor that is inductively coupled to the inductor of the LC tank circuit.*

In an Armstrong oscillator, the feedback is provided through a tickler coil as shown in Figure 10.4(a). Figure 10.5(a) is known as *series–fed tuned–collector Armstrong oscillator* and it is so–called because the power supply voltage V_{CC} supplied to the transistor is through the tank circuit. Figure 10.5(b) is referred to as *shunt fed tuned collector Armstrong oscillator* and it is so–called because the power supply voltage V_{CC} supplied to the transistor is through a path parallel to the tank circuit. In either of the oscillator circuits of Figure 10.5, power through V_{CC} is supplied to the transistor and the tank circuit begins to oscillate. The transistor is operating as Class C amplifier, that is, the transistor conducts for a short period of time and returns sufficient energy to the tank circuit to ensure a constant amplitude output signal. As we've learned in Chapter 3, Class C operation provides the highest efficiency among all amplifier operations.

10.6 The Hartley Oscillator

Figure 10.6 shows a simplified version of the *Hartley oscillator*.

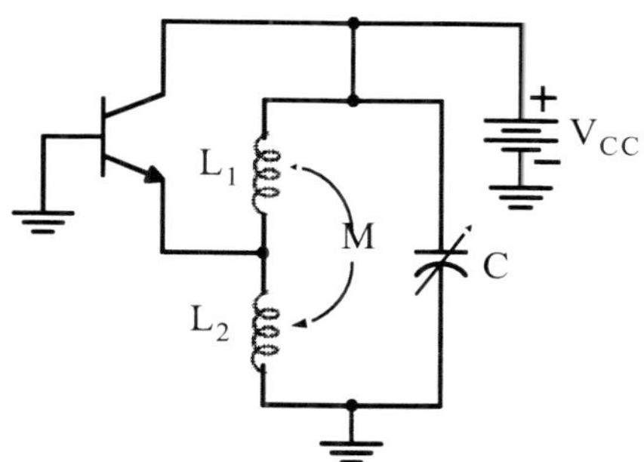

Figure 10.6. Simplified circuit for the Hartley oscillator

In a Hartley oscillator the feedback is provided through a coil pair as shown in Figure 10.4(b). For the oscillator circuit of Figure 10.6 the frequency of oscillation is

$$\omega_0 = \frac{1}{\sqrt{C(L_1 + L_2 + 2M) - (L_1 L_2 - M^2)(h_{ob}/h_{ib})}} \tag{10.14}$$

where M is the *mutual inductance** and h_{ob} and h_{ib} are the h–parameters representing the output admittance with open–circuit input and input impedance with short circuit output respectively, as discussed in Chapter 3.

10.7 The Colpitts Oscillator

Figure 10.7 shows a simplified version of the *Colpitts oscillator*. In a Colpitts oscillator the feedback is provided through a capacitor pair as shown in Figure 10.4(c).

* *For a detailed discussion on mutual inductance, please refer to Circuit Analysis II with MATLAB Applications, ISBN 978–0–9709511–5–1.*

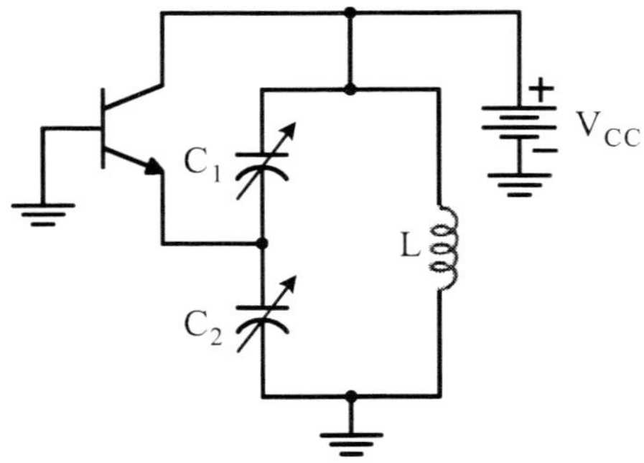

Figure 10.7. Simplified circuit for the Colpitts oscillator

The Colpitts oscillator provides better frequency stability than the Armstrong and Hartley oscillators. Moreover, the Colpitts oscillator is easier to tune and thus can be used for a wide range of frequencies.

For the oscillator circuit of Figure 10.7 the frequency of oscillation is

$$\omega_0 = \sqrt{\frac{C_1 + C_2}{LC_1C_2} + \frac{1}{C_1C_2}\frac{h_{ob}}{h_{ib}}} \tag{10.15}$$

where h_{ob} and h_{ib} are the h–parameters representing the output admittance with open–circuit input and input impedance with short circuit output respectively, as discussed in Chapter 3.

10.8 Crystal Oscillators

Crystal oscillators are oscillators where the primary frequency determining element is a quartz crystal. Because of the inherent characteristics of the quartz crystal the crystal oscillator may be held to extreme accuracy of frequency stability. The frequency depends almost entirely on the thickness where the thinner the thickness, the higher the frequency of oscillation. However, the power obtainable is limited by the heat the crystal will withstand without fracturing. The amount of heating is dependent upon the amount of current that can safely pass through a crystal and this current may be in the order of 50 to 200 milliamperes. Accordingly, temperature compensation must be applied to crystal oscillators to improve thermal stability of the crystal oscillator.

Crystal oscillators are used in applications where frequency accuracy and stability ar of utmost importance such as broadcast transmitters and radar. The frequency stability of crystal–controlled oscillators depends on the quality factor Q.* The Q of a crystal may vary from 10,000 to 100,000. Besides the quartz crystal oscillators, other precision oscillators are constructed with *cesium* or *rubidium.*

* *The quality factor Q is a very important parameter in resonant circuits and it is discussed in detail in Circuit Analysis II with MATLAB Applications, ISBN 978–0–9709511–5–1.*

The symbol and the equivalent circuit of a quartz crystal are shown in Figure 10.8 where capacitor C_1 represents the electrostatic capacitance between the electrodes of the crystal and in general $C_1 \gg C_2$.

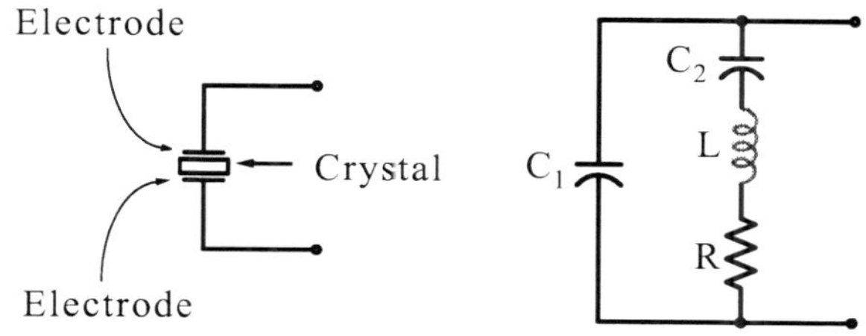

Figure 10.8. Symbol and equivalent circuit for a crystal

The impedance of the equivalent circuit of Figure 10.8 in the s – *domain* is

$$Z(s) = 1/sC_1 \parallel (R + sL + 1/sC_2) \tag{10.16}$$

We now recall that in a series resonant circuit the quality factor Q at resonance is

$$Q_{0S} = \frac{\omega_{0S}L}{R} \tag{10.17}$$

and since the Q of a crystal oscillator is very high, the value of the resistance R in (10.17) must be very low and thus it can be omitted in relation (10.16) which can now be expressed as

$$Z(s) = 1/sC_1 \parallel (sL + 1/sC_2) = (1/sC_1) \cdot \frac{(sL + 1/sC_2)}{1/sC_1 + sL + 1/sC_2}$$

or

$$Z(s) = (1/sC_1) \cdot \frac{s^2 + 1/LC_2}{s^2 + [(C_1 + C_2)/(LC_1C_2)]} \tag{10.18}$$

The denominator of (10.18) is a quadratic and it implies the presence of two resonant frequencies which can be found by inspection of the equivalent circuit of Figure 10.10. The resonance of the series branch occurs when the imaginary part of the impedance is equal to zero. Thus, letting $s = j\omega$ we obtain $Z(j\omega) = j\omega L + 1/j\omega C_2 = 0$ and denoting this frequency as ω_{0S} we obtain

$$\omega_{0S} = \frac{1}{\sqrt{LC_2}} \tag{10.19}$$

We also can prove* that the resonance of the parallel combination occurs when

$$\omega_{0P} = \sqrt{\frac{C_1 + C_2}{LC_1C_2}} \tag{10.20}$$

* *The proof is left as an exercise for the reader at the end of this chapter.*

As stated above, $C_1 \gg C_2$ and under this condition relation (10.20) reduces to that of relation (10.19).

Example 10.2

A crystal oscillator has a nominal frequency of oscillation at $5\ \text{MHz}$ with $Q = 85,000$ and with reference to the equivalent circuit of Figure 10.8, it is known that $L = 50\ \text{mH}$, $C_1 = 2\ \text{pF}$, and $C_2 = 0.02\ \text{pF}$. Find:

a. The series resonance ω_{0S}

b. The parallel resonance ω_{0P}

c. The value of the resistance R

Solution:

a. From (10.19)

$$\frac{1}{2\pi\sqrt{LC_2}} = \frac{1}{2\pi\sqrt{50 \times 10^{-3} \times 0.02 \times 10^{-12}}} = 5.033\ \text{MHz}$$

b. From (10.20)

$$f_{0P} = \frac{1}{2\pi}\sqrt{\frac{C_1 + C_2}{LC_1C_2}} = \frac{1}{2\pi}\sqrt{\frac{2.02 \times 10^{-12}}{50 \times 10^{-3} \times 2 \times 10^{-12} \times 0.02 \times 10^{-12}}} = 5.058\ \text{MHz}$$

c. From (10.17)

$$Q_{0S} = \frac{\omega_{0S}L}{R} = \frac{2\pi f_{0S}L}{R}$$

or

$$R = \frac{2\pi f_{0S}L}{Q_{0S}} = \frac{2\pi \times 5.033 \times 10^6}{85 \times 10^3} = 372\ \Omega$$

10.9 The Pierce Oscillator

The *Pierce oscillator* is a modified Colpitts oscillator that uses a crystal as a parallel–resonant circuit, and for this reason is often referred to as *crystal–controlled Pierce oscillator*. Figure 10.9 shows a Pierce oscillator with a PNP transistor as an amplifier in the common–base configuration.

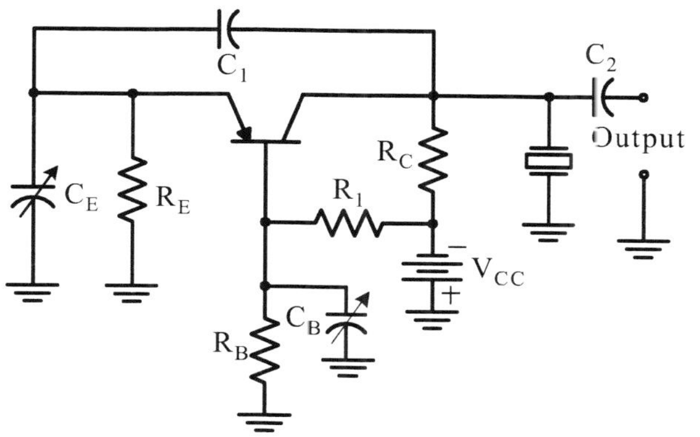

Figure 10.9. Pierce oscillator with PNP transistor in the common–base configuration

In the oscillator circuit of Figure 10.9, feedback is provided from the collector to the emitter of the transistor through capacitor C_1 and resistors R_1, R_B, and R_C are used to establish the proper bias conditions. Besides the crystal, the frequency of oscillation is also determined by the settings of the variable capacitors C_E and C_B.

10.10 Summary

- An oscillator may be defined as a class of wave generators that produce sinusoidal, square, triangular, and sawtooth waveforms. An oscillator is essentially an amplifier circuit that provides its own input signal. In this chapter we discussed sinusoidal oscillators. Non–sinusoidal oscillators are known as relaxation oscillators.
- Many low–frequency oscillators use resistors and capacitors to form their frequency–determining networks and are referred to as RC oscillators. These are used in the audio–frequency range. Other types of employ inductors and capacitors for its frequency–determining network. These are known as the LC oscillators and use tank circuits, are commonly used for the higher radio frequencies.
- A third type of sinusoidal oscillator is the crystal–controlled oscillator. The crystal–controlled oscillator provides excellent frequency stability and is used from the middle of the audio range through the radio frequency range.
- An oscillator must provide amplification where the amplification of signal power occurs from the input to the output of the oscillator and a portion of the output is fed back to the input to sustain a constant input.
- In an oscillator circuit the positive feedback must be large enough to compensate for circuit losses so that oscillations will be sustained. Moreover, a practical oscillator must oscillate at a predetermined frequency and thus the oscillator must include a frequency–determining device which essentially is a band–pass filter allowing only the desired frequency to pass.
- The output of the oscillator contains only a single sinusoidal frequency.
- In an RC oscillator the frequency of oscillations is determined by the values of resistance and capacitance in the three sections. Variable resistors and capacitors are usually employed to provide tuning in the feedback network for variations in phase shift. For an RC phase–shift oscillator, the amplifier is biased for Class A operation to minimize distortion of the signal.
- In an RC phase–shift oscillator the frequency of oscillation (resonant frequency) can be approximated from the relation

  $$\omega_0 = \frac{1}{RC\sqrt{n}}$$

  where n is the number of RC sections.
- In an LC oscillator the sinusoidal signal is generated by the action of an inductor and a capacitor. The feedback signal is coupled from the LC tank of the oscillator circuit by using a tickler coil or a coil pair.
- In an Armstrong oscillator, the feedback is provided through a tickler coil.
- In a Hartley oscillator the feedback is provided through a coil pair.

- In a Colpitts oscillator the feedback is provided through a capacitor pair.
- The Colpitts oscillator provides better frequency stability than the Armstrong and Hartley oscillators. Moreover, the Colpitts oscillator is easier to tune and thus can be used for a wide range of frequencies.
- In crystal oscillators the frequency determining element is a quartz crystal. Because of the inherent characteristics of the quartz crystal the crystal oscillator may be held to extreme accuracy of frequency stability. The frequency depends almost entirely on the thickness where the thinner the thickness, the higher the frequency of oscillation. However, the power obtainable is limited by the heat the crystal will withstand without fracturing.
- Crystal oscillators are used in applications where frequency accuracy and stability ar of utmost importance such as broadcast transmitters and radar. The frequency stability of crystal–controlled oscillators depends on the quality factor Q. Besides the quartz crystal oscillators, other precision oscillators are constructed with cesium or rubidium.
- The Pierce oscillator is a modified Colpitts oscillator that uses a crystal as a parallel–resonant circuit, and for this reason is often referred to as crystal–controlled Pierce oscillator.

10.11 Exercises

1. For the op amp circuit below, the open–loop gain is $A = 10,000$, and $v_S = 1\ V$.

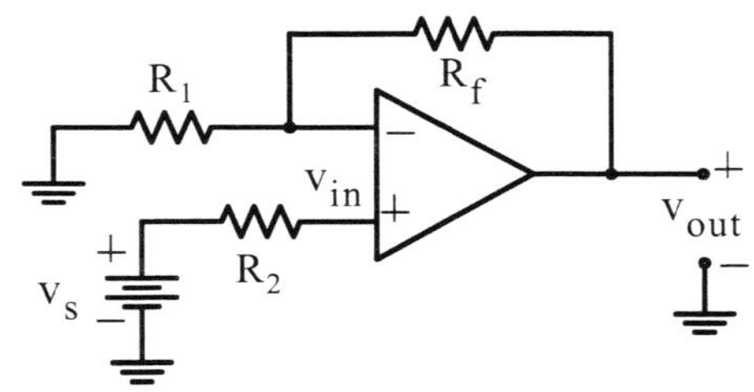

a. Derive an expression for the feedback factor β

b. Find an appropriate ratio of R_f/R_1 so that the closed loop gain will be $A_f = 10$

c. Express the amount of feedback in dB

d. Find the output voltage v_{out}

e. Find the feedback voltage v_f

f. Find the input voltage v_{in} at the non–inverting input.

g. Determine the decrease in the closed–loop gain A_f if the open–loop gain A decreases by 20%.

2. A sinusoidal oscillator consists of an amplifier with gain $A = 10$ and a bandpass filter whose center frequency is $f_0 = 20\ KHz$. Determine

a. The frequency ω_0 and the gain of the filter to produce sustained oscillations

b. The gain of the filter at the frequency ω_0

3. The figure below shows a crystal oscillator and its equivalent circuit.

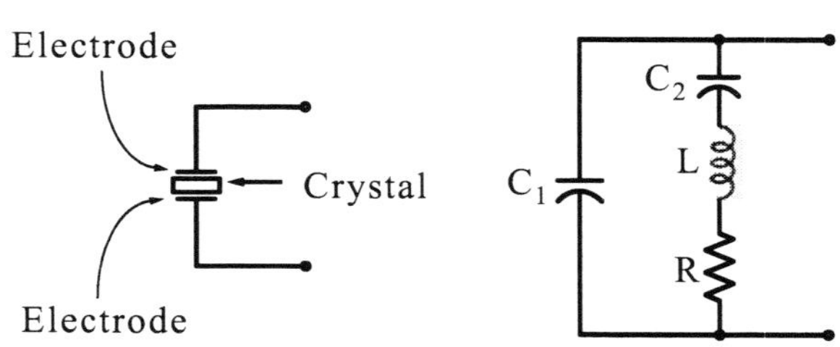

Prove that

$$\omega_{OP} = \sqrt{\frac{C_1 + C_2}{LC_1C_2}}$$

10.12 Solutions to End–of–Chapter Exercises

1.

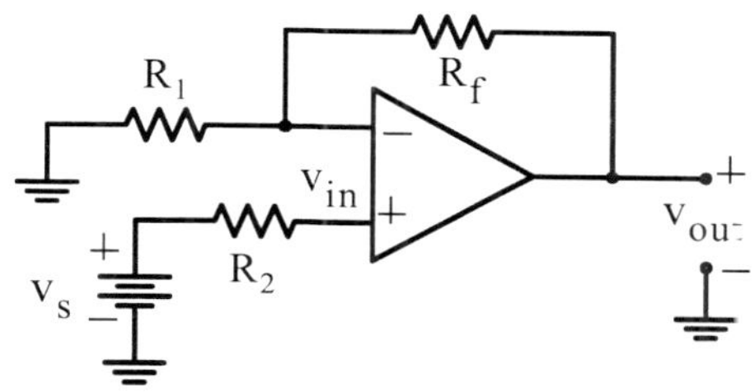

a. The feedback factor β is the portion of the output voltage v_{out} which appears at the inverting input, and by the voltage division expression

$$v_{(-)} = \left(\frac{R_1}{R_f + R_1}\right) v_{out}$$

and thus

$$\beta = \frac{R_1}{R_f + R_1} \qquad (1)$$

b. From relation (10.4) and with $A = 10^4$ and $A_f = 10$ we find that

$$A_f = \frac{A}{1 + \beta A} = 10 = \frac{10^4}{1 + 10^4 \beta}$$

from which

$$\beta = \frac{10^3 - 1}{10^4} = 0.0999 \quad (2)$$

With (1) and (2) above

$$\frac{R_F/R_1 + R_1/R_1}{R_1/R_1} = \frac{1}{\beta} = \frac{1}{0.0999} = 10.01$$

$$1 + R_f/R_1 = 10.01$$

$$\frac{R_f}{R_1} = 9.01$$

c. The amount of feedback is $1 + \beta A$ and in dB

$$Feedback_{dB} = 20\log(1 + \beta A) = 20\log(1 + 0.0999 \times 10^4) = 60 \text{ dB}$$

d.

$$v_{out} = A_f v_s = 10 \times 1 = 10 \text{ V}$$

e. The feedback voltage v_f is

$$v_f = \beta v_{out} = 0.0999 \times 10 = 0.999 \text{ V}$$

f. The voltage at the non–inverting input is

$$v_{in} = v_s - v_f = 1.000 - 0.999 = 0.001 \text{ V}$$

g. Let the reduced open loop gain be denoted as A' and the corresponding closed loop gain as A'_f. Then,

$$A'_f = \frac{A'}{1 + \beta A'} = \frac{0.8 \times 10,000}{1 + 0.0999 \times 0.8 \times 10,000} = 9.975$$

and the percent change is

$$\frac{9.975 - 10}{9.975} \times 100 = -0.025\%$$

where the minus (–) sign implies decrease is percentage.

2.

a. The frequency ω_0 at which sustained oscillations will be produced is the center frequency of the bandpass filter and thus

$$\omega_0 = 2\pi f_0 = 2\pi \times 20 \times 10^3 = 1.26 \times 10^5 \text{ r/s}$$

b. To sustain constant amplitude oscillations we must have $|\beta(j\omega_0)A(j\omega_0)| = 1$ and with $A = 10$,

$$\text{Gain}_{\omega = \omega_0} = 1/10 = 0.1$$

3.

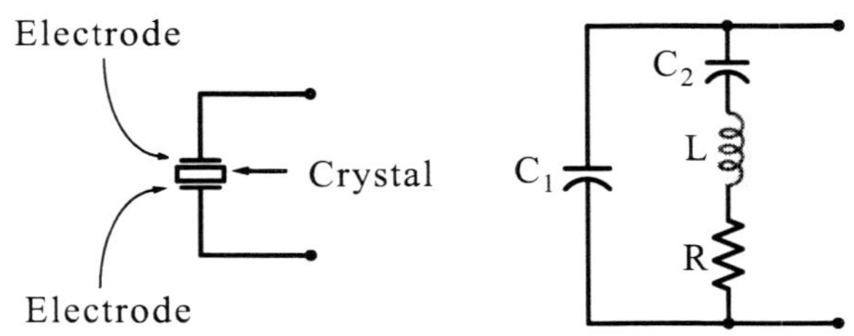

This is a parallel circuit and thus the resonant frequency occurs when the imaginary part of the admittance Y is equal to zero. Therefore,

$$Y = j\omega_0 C_1 + \frac{1}{j\omega_0 L + 1/j\omega_0 C_2} = 0$$

$$j\omega_0 C_1 + \frac{j\omega_0 C_2}{-\omega_0^2 L C_2 + 1} = 0$$

$$j\omega_0\left(C_1 + \frac{C_2}{-\omega_0^2 L C_2 + 1}\right) = 0$$

$$C_1 + \frac{C_2}{-\omega_0^2 L C_2 + 1} = 0$$

$$C_1 = \frac{C_2}{\omega_0^2 L C_2 - 1}$$

$$\omega_0^2 L C_1 C_2 - C_1 = C_2$$

$$\omega_{0P} = \sqrt{\frac{C_1 + C_2}{L C_1 C_2}}$$

NOTES:

Appendix A

Introduction to MATLAB®

This appendix serves as an introduction to the basic MATLAB commands and functions, procedures for naming and saving the user generated files, comment lines, access to MATLAB's Editor / Debugger, finding the roots of a polynomial, and making plots. Several examples are provided with detailed explanations.

A.1 MATLAB® and Simulink®

MATLAB and Simulink are products of The MathWorks,™ Inc. These are two outstanding software packages for scientific and engineering computations and are used in educational institutions and in industries including automotive, aerospace, electronics, telecommunications, and environmental applications. MATLAB enables us to solve many advanced numerical problems rapidly and efficiently.

A.2 Command Window

To distinguish the screen displays from the user commands, important terms, and MATLAB functions, we will use the following conventions:

Click: Click the left button of the mouse
`Courier Font:` Screen displays
Helvetica Font: User inputs at MATLAB's command window prompt >> or `EDU>>`*
Helvetica Bold: MATLAB functions

Times Bold Italic: Important terms and facts, notes and file names
When we first start MATLAB, we see various help topics and other information. Initially, we are interested in the *command screen* which can be selected from the Window drop menu. When the command screen, we see the prompt >> or `EDU>>`. This prompt is displayed also after execution of a command; MATLAB now waits for a new command from the user. It is highly recommended that we use the *Editor/Debugger* to write our program, save it, and return to the command screen to execute the program as explained below.

To use the Editor/Debugger:

1. From the *File* menu on the toolbar, we choose *New* and click on *M–File*. This takes us to the *Editor Window* where we can type our *script* (list of statements) for a new file, or open a previously saved file. We must save our program with a file name which starts with a letter. ***Impor-***

* EDU>> *is the MATLAB prompt in the Student Version*

tant! MATLAB is *case sensitive*, that is, it distinguishes between upper– and lower–case letters. Thus, *t* and *T* are two different letters in MATLAB language. The files that we create are saved with the file name we use and the extension ***.m***; for example, ***myfile01.m***. It is a good practice to save the script in a file name that is descriptive of our script content. For instance, if the script performs some matrix operations, we ought to name and save that file as ***matrices01.m*** or any other similar name. We should also use a floppy disk or an external drive to backup our files.

2. Once the script is written and saved as an *m–file*, we may exit the *Editor/Debugger* window by clicking on *Exit Editor/Debugger* of the *File* menu. MATLAB then returns to the command window.

3. To execute a program, we type the file name ***without*** the ***.m*** extension at the >> prompt; then, we press <enter> and observe the execution and the values obtained from it. If we have saved our file in drive ***a*** or any other drive, we must make sure that it is added it to the desired directory in MATLAB's search path. The MATLAB User's Guide provides more information on this topic.

Henceforth, it will be understood that each input command is typed after the >> prompt and followed by the <enter> key.

The command **help matlab\iofun** will display input/output information. To get help with other MATLAB topics, we can type help followed by any topic from the displayed menu. For example, to get information on graphics, we type help matlab\graphics. The MATLAB User's Guide contains numerous help topics.

To appreciate MATLAB's capabilities, we type demo and we see the `MATLAB Demos` menu. We can do this periodically to become familiar with them. Whenever we want to return to the command window, we click on the `Close` button.

When we are done and want to leave MATLAB, we type **quit** or **exit**. But if we want to clear all previous values, variables, and equations without exiting, we should use the command **clear**. This command erases everything; it is like exiting MATLAB and starting it again. The command **clc** clears the screen but MATLAB still remembers all values, variables and equations that we have already used. In other words, if we want to clear all previously entered commands, leaving only the >> prompt on the upper left of the screen, we use the **clc** command.

All text after the **%** (percent) symbol is interpreted as a *comment line* by MATLAB, and thus it is ignored during the execution of a program. A comment can be typed on the same line as the function or command or as a separate line. For instance,

```
conv(p,q)    % performs multiplication of polynomials p and q
% The next statement performs partial fraction expansion of p(x) / q(x)
```

are both correct.

One of the most powerful features of MATLAB is the ability to do computations involving *complex numbers*. We can use either i, or j to denote the imaginary part of a complex number, such as 3-4i or 3-4j. For example, the statement

z=3–4j

displays

```
z = 3.0000-4.0000i
```

In the above example, a multiplication (*) sign between 4 and j was not necessary because the complex number consists of numerical constants. However, if the imaginary part is a function, or variable such as cos(x), we must use the multiplication sign, that is, we must type cos(x)*j or j*cos(x) for the imaginary part of the complex number.

A.3 Roots of Polynomials

In MATLAB, a polynomial is expressed as a *row vector* of the form $[a_n \ a_{n-1} \ \cdots \ a_2 \ a_1 \ a_0]$. These are the coefficients of the polynomial in descending order. ***We must include terms whose coefficients are zero.***

We find the roots of any polynomial with the **roots(p)** function; **p** is a row vector containing the polynomial coefficients in descending order.

Example A.1

Find the roots of the polynomial

$$p_1(x) = x^4 - 10x^3 + 35x^2 - 50x + 24$$

Solution:

The roots are found with the following two statements where we have denoted the polynomial as p1, and the roots as roots_ p1.

```
p1=[1 -10 35 -50 24]     % Specify and display the coefficients of p1(x)

p1 =
     1    -10     35    -50     24

roots_ p1=roots(p1)      % Find the roots of p1(x)

roots_p1 =

   4.0000
   3.0000
   2.0000
   1.0000
```

We observe that MATLAB displays the polynomial coefficients as a row vector, and the roots as a column vector.

Example A.2

Find the roots of the polynomial

$$p_2(x) = x^5 - 7x^4 + 16x^2 + 25x + 52$$

Solution:

There is no cube term; therefore, we must enter zero as its coefficient. The roots are found with the statements below, where we have defined the polynomial as **p2**, and the roots of this polynomial as **roots_ p2**. The result indicates that this polynomial has three real roots, and two complex roots. Of course, complex roots always occur in *complex conjugate** pairs.

```
p2=[1 -7  0 16 25 52]
```

```
p2 =
     1    -7     0    16    25    52
```

```
roots_ p2=roots(p2)
```

```
roots_p2 =
   6.5014
   2.7428
  -1.5711
  -0.3366 + 1.3202i
  -0.3366 - 1.3202i
```

A.4 Polynomial Construction from Known Roots

We can compute the coefficients of a polynomial, from a given set of roots, with the **poly(r)** function where **r** is a row vector containing the roots.

Example A.3

It is known that the roots of a polynomial are $1, 2, 3,$ and 4. Compute the coefficients of this polynomial.

* *By definition, the conjugate of a complex number* $A = a + jb$ *is* $A^* = a - jb$

Solution:

We first define a row vector, say r3, with the given roots as elements of this vector; then, we find the coefficients with the **poly(r)** function as shown below.

```
r3=[1  2  3  4]          % Specify the roots of the polynomial
r3 =
     1     2     3     4
poly_r3=poly(r3)         % Find the polynomial coefficients
poly_r3 =
     1   -10    35   -50    24
```

We observe that these are the coefficients of the polynomial $p_1(x)$ of Example A.1.

Example A.4

It is known that the roots of a polynomial are -1, -2, -3, $4+j5$, and $4-j5$. Find the coefficients of this polynomial.

Solution:

We form a row vector, say r4, with the given roots, and we find the polynomial coefficients with the **poly(r)** function as shown below.

```
r4=[ -1  -2  -3  4+5j  4-5j ]
r4 =
  Columns 1 through 4
  -1.0000    -2.0000    -3.0000    -4.0000+ 5.0000i
  Column 5
  -4.0000- 5.0000i
poly_r4=poly(r4)
poly_r4 =
     1    14    100    340    499    246
```

Therefore, the polynomial is

$$p_4(x) = x^5 + 14x^4 + 100x^3 + 340x^2 + 499x + 246$$

A.5 Evaluation of a Polynomial at Specified Values

The **polyval(p,x)** function evaluates a polynomial $p(x)$ at some specified value of the independent variable x.

Example A.5

Evaluate the polynomial

$$p_5(x) = x^6 - 3x^5 + 5x^3 - 4x^2 + 3x + 2 \tag{A.1}$$

at $x = -3$.

Solution:

```
p5=[1 −3  0  5 −4  3  2];   % These are the coefficients of the given polynomial
                            % The semicolon (;) after the right bracket suppresses the
                            %  display of the row vector that contains the coefficients of p5.
%
val_minus3=polyval(p5, −3)  % Evaluate p5 at x=−3; no semicolon is used here
                            % because we want the answer to be displayed

val_minus3 =
        1280
```

Other MATLAB functions used with polynomials are the following:

conv(a,b) – multiplies two polynomials **a** and **b**

[q,r]=deconv(c,d) –divides polynomial **c** by polynomial **d** and displays the quotient **q** and remainder **r**.

polyder(p) – produces the coefficients of the derivative of a polynomial **p**.

Example A.6

Let

$$p_1 = x^5 - 3x^4 + 5x^2 + 7x + 9$$

and

$$p_2 = 2x^6 - 8x^4 + 4x^2 + 10x + 12$$

Compute the product $p_1 \cdot p_2$ using the **conv(a,b)** function.

Solution:

```
p1=[1 -3  0 5  7  9];            % The coefficients of p1
p2=[2  0 -8 0 4  10  12];        % The coefficients of p2
p1p2=conv(p1,p2)                 % Multiply p1 by p2 to compute coefficients of the product p1p2

p1p2 =
   2   -6   -8   34   18   -24   -74   -88   78   166   174   108
```

Therefore,

$$p_1 \cdot p_2 = 2x^{11} - 6x^{10} - 8x^9 + 34x^8 + 18x^7 - 24x^6$$
$$-74x^5 - 88x^4 + 78x^3 + 166x^2 + 174x + 108$$

Example A.7

Let

$$p_3 = x^7 - 3x^5 + 5x^3 + 7x + 9$$

and

$$p_4 = 2x^6 - 8x^5 + 4x^2 + 10x + 12$$

Compute the quotient p_3/p_4 using the **[q,r]=deconv(c,d)** function.

Solution:

```
% It is permissible to write two or more statements in one line separated by semicolons
p3=[1  0 -3   0  5  7   9]; p4=[2 -8  0   0  4  10  12]; [q,r]=deconv(p3,p4)

q =
      0.5000
r =
       0        4       -3        0        3        2        3
```

Therefore,

$$q = 0.5 \qquad r = 4x^5 - 3x^4 + 3x^2 + 2x + 3$$

Example A.8

Let

$$p_5 = 2x^6 - 8x^4 + 4x^2 + 10x + 12$$

Compute the derivative $\frac{d}{dx}p_5$ using the **polyder(p)** function.

Solution:

```
p5=[2 0 -8 0 4 10 12];       % The coefficients of p5
der_p5=polyder(p5)           % Compute the coefficients of the derivative of p5

der_p5 =
    12     0   -32     0     8    10
```

Therefore,

$$\frac{d}{dx}p_5 = 12x^5 - 32x^3 + 4x^2 + 8x + 10$$

A.6 Rational Polynomials

Rational Polynomials are those which can be expressed in ratio form, that is, as

$$R(x) = \frac{Num(x)}{Den(x)} = \frac{b_n x^n + b_{n-1}x^{n-1} + b_{n-2}x^{n-2} + \ldots + b_1 x + b_0}{a_m x^m + a_{m-1}x^{m-1} + a_{m-2}x^{m-2} + \ldots + a_1 x + a_0} \quad \text{(A.2)}$$

where some of the terms in the numerator and/or denominator may be zero. We can find the roots of the numerator and denominator with the **roots(p)** function as before.

As noted in the comment line of Example A.7, we can write MATLAB statements in one line, if we separate them by commas or semicolons. ***Commas will display the results whereas semicolons will suppress the display.***

Example A.9

Let

$$R(x) = \frac{p_{num}}{p_{den}} = \frac{x^5 - 3x^4 + 5x^2 + 7x + 9}{x^6 - 4x^4 + 2x^2 + 5x + 6}$$

Express the numerator and denominator in factored form, using the **roots(p)** function.

Solution:

```
num=[1 -3 0 5 7 9]; den=[1 0 -4 0 2 5 6];   % Do not display num and den coefficients
roots_num=roots(num), roots_den=roots(den)   % Display num and den roots

roots_num =
   2.4186 + 1.0712i    2.4186 - 1.0712i   -1.1633
  -0.3370 + 0.9961i   -0.3370 - 0.9961i
```

```
roots_den =
   1.6760 + 0.4922i      1.6760 - 0.4922i   -1.9304
  -0.2108 + 0.9870i     -0.2108 - 0.9870i   -1.0000
```

As expected, the complex roots occur in complex conjugate pairs.

For the numerator, we have the factored form

$$p_{num} = (x-2.4186-j1.0712)(x-2.4186+j1.0712)(x+1.1633)$$
$$(x+0.3370-j0.9961)(x+0.3370+j0.9961)$$

and for the denominator, we have

$$p_{den} = (x-1.6760-j0.4922)(x-1.6760+j0.4922)(x+1.9304)$$
$$(x+0.2108-j0.9870)(x+0.2108+j0.9870)(x+1.0000)$$

We can also express the numerator and denominator of this rational function as a combination of *linear* and *quadratic* factors. We recall that, in a quadratic equation of the form $x^2+bx+c = 0$ whose roots are x_1 and x_2, the negative sum of the roots is equal to the coefficient b of the x term, that is, $-(x_1+x_2) = b$, while the product of the roots is equal to the constant term c, that is, $x_1 \cdot x_2 = c$. Accordingly, we form the coefficient b by addition of the complex conjugate roots and this is done by inspection; then we multiply the complex conjugate roots to obtain the constant term c using MATLAB as follows:

```
(2.4186 + 1.0712i)*(2.4186 -1.0712i)
```

```
ans = 6.9971
```

```
(-0.3370+ 0.9961i)*(-0.3370-0.9961i)
```

```
ans = 1.1058
```

```
(1.6760+ 0.4922i)*(1.6760-0.4922i)
```

```
ans = 3.0512
```

```
(-0.2108+ 0.9870i)*(-0.2108-0.9870i)
```

```
ans = 1.0186
```

Thus,

$$R(x) = \frac{p_{num}}{p_{den}} = \frac{(x^2-4.8372x+6.9971)(x^2+0.6740x+1.1058)(x+1.1633)}{(x^2-3.3520x+3.0512)(x^2+0.4216x+1.0186)(x+1.0000)(x+1.9304)}$$

We can check this result of Example A.9 above with MATLAB's *Symbolic Math Toolbox* which is a collection of tools (functions) used in solving symbolic expressions. They are discussed in detail in MATLAB's Users Manual. For the present, our interest is in using the **collect(s)** function that is used to multiply two or more symbolic expressions to obtain the result in polynomial form. We must remember that the **conv(p,q)** function is used with numeric expressions only, that is, polynomial coefficients.

Before using a symbolic expression, we must create one or more symbolic variables such as x, y, t, and so on. For our example, we use the following script:

```
syms x % Define a symbolic variable and use collect(s) to express numerator in polynomial form
collect((x^2–4.8372*x+6.9971)*(x^2+0.6740*x+1.1058)*(x+1.1633))
```

```
ans =
  x^5-29999/10000*x^4-1323/3125000*x^3+7813277909/
  1562500000*x^2+1750276323053/2500000000000*x+4500454743147/
  5000000000000
```

and if we simplify this, we find that is the same as the numerator of the given rational expression in polynomial form. We can use the same procedure to verify the denominator.

A.7 Using MATLAB to Make Plots

Quite often, we want to plot a set of ordered pairs. This is a very easy task with the MATLAB **plot(x,y)** command that plots *y* versus *x*, where *x* is the horizontal axis (abscissa) and *y* is the vertical axis (ordinate).

Example A.10

Consider the electric circuit of Figure A.1, where the radian frequency ω (radians/second) of the applied voltage was varied from 300 to 3000 in steps of 100 radians/second, while the amplitude was held constant.

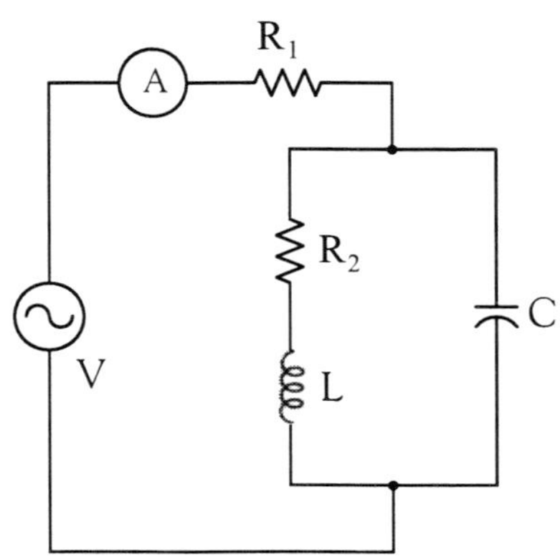

Figure A.1. Electric circuit for Example A.10

The ammeter readings were then recorded for each frequency. The magnitude of the impedance $|Z|$ was computed as $|Z| = |V/A|$ and the data were tabulated on Table A.1.

TABLE A.1 Table for Example A.10

ω (rads/s)	|Z| Ohms	ω (rads/s)	|Z| Ohms
300	39.339	1700	90.603
400	52.589	1800	81.088
500	71.184	1900	73.588
600	97.665	2000	67.513
700	140.437	2100	62.481
800	222.182	2200	58.240
900	436.056	2300	54.611
1000	1014.938	2400	51.428
1100	469.83	2500	48.717
1200	266.032	2600	46.286
1300	187.052	2700	44.122
1400	145.751	2800	42.182
1500	120.353	2900	40.432
1600	103.111	3000	38.845

Plot the magnitude of the impedance, that is, $|Z|$ versus radian frequency ω.

Solution:

We cannot type ω (omega) in the MATLAB Command prompt, so we will use the English letter w instead.

If a statement, or a row vector is too long to fit in one line, it can be continued to the next line by typing three or more periods, then pressing <*enter*> to start a new line, and continue to enter data. This is illustrated below for the data of **w** and **z**. Also, as mentioned before, we use the semicolon (;) to suppress the display of numbers that we do not care to see on the screen.

The data are entered as follows:

```
w=[300 400 500 600 700 800 900 1000 1100 1200 1300 1400 1500 1600 1700 1800 1900....
2000 2100 2200 2300 2400 2500 2600 2700 2800 2900 3000];
%
z=[39.339 52.789 71.104 97.665 140.437 222.182 436.056....
1014.938 469.830 266.032 187.052 145.751 120.353 103.111....
90.603 81.088 73.588 67.513 62.481 58.240 54.611 51.468....
48.717 46.286 44.122 42.182 40.432 38.845];
```

Of course, if we want to see the values of *w* or *z* or both, we simply type w or z, and we press

<*enter*>. To plot z (y–axis) versus w (x–axis), we use the **plot(x,y)** command. For this example, we use **plot(w,z)**. When this command is executed, MATLAB displays the plot on MATLAB's *graph screen* and MATLAB denotes this plot as Figure 1. This plot is shown in Figure A.2.

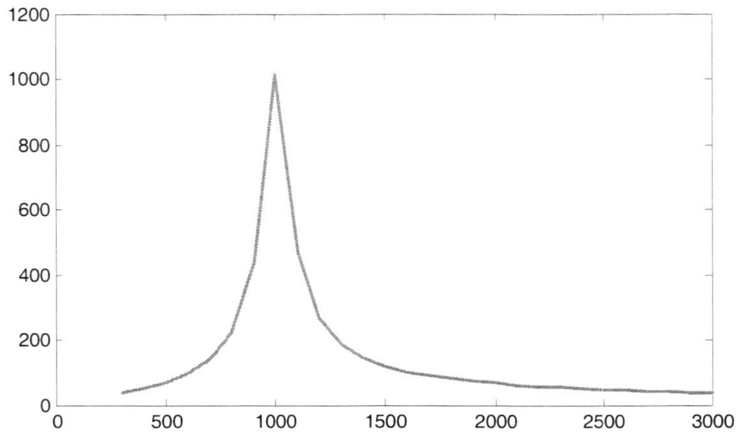

Figure A.2. Plot of impedance |z| versus frequency ω for Example A.10

This plot is referred to as the *magnitude frequency response* of the circuit.

To return to the command window, we press any key, or from the *Window* pull–down menu, we select *MATLAB Command Window*. To see the graph again, we click on the Window pull–down menu, and we choose *Figure 1*.

We can make the above, or any plot, more presentable with the following commands:

grid on: This command adds grid lines to the plot. The **grid off** command removes the grid. The command **grid** toggles them, that is, changes from off to on or vice versa. The default* is off.

box off: This command removes the box (the solid lines which enclose the plot), and **box on** restores the box. The command **box** toggles them. The default is on.

title('string'): This command adds a line of the text **string** (label) at the top of the plot.

xlabel('string') and **ylabel('string')** are used to label the x– and y–axis respectively.

The magnitude frequency response is usually represented with the x–axis in a logarithmic scale. We can use the **semilogx(x,y)** command which is similar to the **plot(x,y)** command, except that the x–axis is represented as a log scale, and the y–axis as a linear scale. Likewise, the **semilogy(x,y)** command is similar to the **plot(x,y)** command, except that the y–axis is represented as a

* *A default is a particular value for a variable that is assigned automatically by an operating system and remains in effect unless canceled or overridden by the operator.*

log scale, and the x–axis as a linear scale. The **loglog(x,y)** command uses logarithmic scales for both axes.

Throughout this text it will be understood that *log* is the common (base 10) logarithm, and *ln* is the natural (base e) logarithm. We must remember, however, the function **log(x)** in MATLAB is the natural logarithm, whereas the common logarithm is expressed as **log10(x)**, and the logarithm to the base 2 as **log2(x).**

Let us now redraw the plot with the above options by adding the following statements:

```
semilogx(w,z); grid;              % Replaces the plot(w,z) command
title('Magnitude of Impedance vs. Radian Frequency');
xlabel('w in rads/sec'); ylabel('|Z| in Ohms')
```

After execution of these commands, the plot is as shown in Figure A.3.

If the y–axis represents power, voltage or current, the x–axis of the frequency response is more often shown in a logarithmic scale, and the y–axis in dB (decibels).

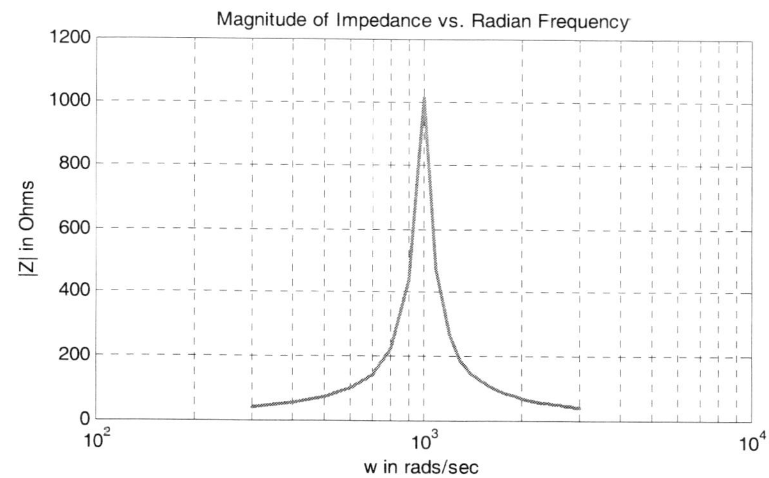

Figure A.3. Modified frequency response plot of Figure A.2.

To display the voltage v in a dB scale on the y–axis, we add the relation dB=20*log10(v), and we replace the **semilogx(w,z)** command with semilogx(w,dB).

The command **gtext('string')*** switches to the current *Figure Window*, and displays a cross–hair that can be moved around with the mouse. For instance, we can use the command gtext('Impedance |Z| versus Frequency'), and this will place a cross–hair in the *Figure* window. Then, using

* *With the latest MATLAB Versions 6 and 7 (Student Editions 13 and 14), we can add text, lines and arrows directly into the graph using the tools provided on the Figure Window. For advanced MATLAB graphics, please refer to The MathWorks* ***Using MATLAB Graphics*** *documentation.*

the mouse, we can move the cross–hair to the position where we want our label to begin, and we press <enter>.

The command **text(x,y,'string')** is similar to **gtext('string')**. It places a label on a plot in some specific location specified by **x** and **y**, and **string** is the label which we want to place at that location. We will illustrate its use with the following example which plots a *3–phase* sinusoidal waveform.

The first line of the script below has the form

linspace(first_value, last_value, number_of_values)

This function specifies ***the number of data points*** but not the increments between data points. An alternate function is

x=first: increment: last

and this specifies ***the increments between points*** but not the number of data points.

The script for the 3–phase plot is as follows:

```
x=linspace(0, 2*pi, 60);          % pi is a built–in function in MATLAB;
% we could have used x=0:0.02*pi:2*pi or x = (0: 0.02: 2)*pi instead;
y=sin(x); u=sin(x+2*pi/3); v=sin(x+4*pi/3);
plot(x,y,x,u,x,v);                % The x–axis must be specified for each function
grid on, box on,                  % turn grid and axes box on
text(0.75, 0.65, 'sin(x)'); text(2.85, 0.65, 'sin(x+2*pi/3)'); text(4.95, 0.65, 'sin(x+4*pi/3)')
```

These three waveforms are shown on the same plot of Figure A.4.

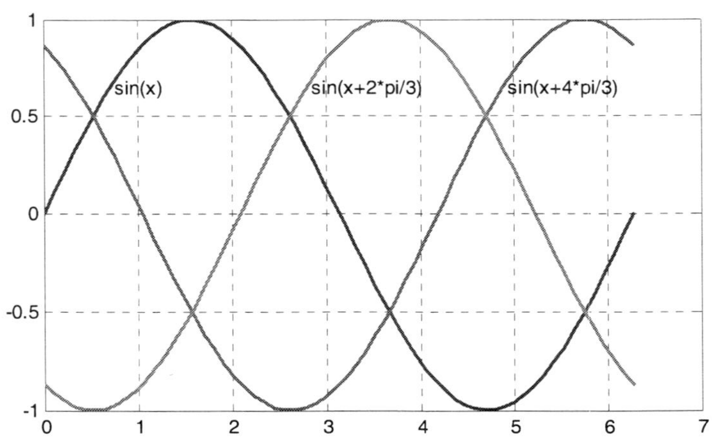

Figure A.4. Three–phase waveforms

In our previous examples, we did not specify line styles, markers, and colors for our plots. However, MATLAB allows us to specify various line types, plot symbols, and colors. These, or a combination of these, can be added with the **plot(x,y,s)** command, where **s** is a character string containing one or more characters shown on the three columns of Table A.2. MATLAB has no default color; it starts with blue and cycles through the first seven colors listed in Table A.2 for each additional line in the plot. Also, there is no default marker; no markers are drawn unless they are selected. The default line is the solid line. But with the latest MATLAB versions, we can select the line color, line width, and other options directly from the *Figure Window.*

TABLE A.2 Styles, colors, and markets used in MATLAB

Symbol	*Color*	*Symbol*	*Marker*	*Symbol*	*Line Style*
b	blue	.	point	–	solid line
g	green	o	circle	:	dotted line
r	red	x	x–mark	–.	dash–dot line
c	cyan	+	plus	––	dashed line
m	magenta	*	star		
y	yellow	s	square		
k	black	d	diamond		
w	white	v	triangle down		
		^	triangle up		
		<	triangle left		
		>	triangle right		
		p	pentagram		
		h	hexagram		

For example, **plot(x,y,'m*:')** plots a magenta dotted line with a star at each data point, and **plot(x,y,'rs')** plots a red square at each data point, but does not draw any line because no line was selected. If we want to connect the data points with a solid line, we must type **plot(x,y,'rs–')**. For additional information we can type **help plot** in MATLAB's command screen.

The plots we have discussed thus far are two–dimensional, that is, they are drawn on two axes. MATLAB has also a three–dimensional (three–axes) capability and this is discussed next.

The **plot3(x,y,z)** command plots a line in *3–space* through the points whose coordinates are the elements of x, y and z, where x, y and z are three vectors of the same length.

The general format is **plot3($x_1,y_1,z_1,s_1,x_2,y_2,z_2,s_2,x_3,y_3,z_3,s_3,...$)** where $\mathbf{x_n}$, $\mathbf{y_n}$ and $\mathbf{z_n}$ are vectors or matrices, and $\mathbf{s_n}$ are strings specifying color, marker symbol, or line style. These strings are the same as those of the two–dimensional plots.

Example A.11

Plot the function

$$z = -2x^3 + x + 3y^2 - 1 \tag{A.3}$$

Solution:

We arbitrarily choose the interval (length) shown on the script below.

```
x= −10: 0.5: 10;              % Length of vector x
y= x;                         % Length of vector y must be same as x
z= −2.*x.^3+x+3.*y.^2−1;      % Vector z is function of both x and y*
plot3(x,y,z); grid
```

The three−dimensional plot is shown in Figure A.5.

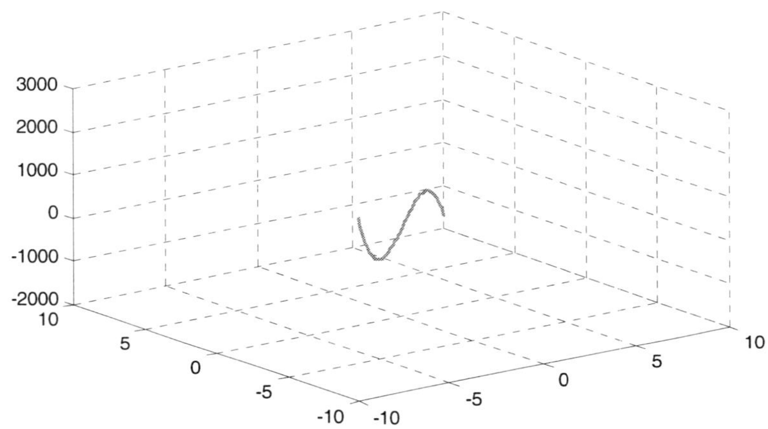

Figure A.5. Three dimensional plot for Example A.11

In a two−dimensional plot, we can set the limits of the *x*− and *y−axes* with the **axis([xmin xmax ymin ymax])** command. Likewise, in a three−dimensional plot we can set the limits of all three axes with the **axis([xmin xmax ymin ymax zmin zmax])** command. It must be placed after the **plot(x,y)** or **plot3(x,y,z)** commands, or on the same line without first executing the **plot** command. This must be done for each plot. The three−dimensional **text(x,y,z,'string')** command will place **string** beginning at the co−ordinate (*x,y,z*) on the plot.

For three−dimensional plots, **grid on** and **box off** are the default states.

* *This statement uses the so called dot multiplication, dot division, and dot exponentiation where the multiplication, division, and exponential operators are preceded by a dot. These important operations will be explained in Section A.9.*

We can also use the **mesh(x,y,z)** command with two vector arguments. These must be defined as length(x) = n and length(y) = m where [m, n] = size(Z). In this case, the vertices of the mesh lines are the triples {x(j), y(i), Z(i, j)}. We observe that **x** corresponds to the columns of Z, and **y** corresponds to the rows.

To produce a mesh plot of a function of two variables, say $z = f(x, y)$, we must first generate the X and Y matrices that consist of repeated rows and columns over the range of the variables x and y. We can generate the matrices X and Y with the **[X,Y]=meshgrid(x,y)** function that creates the matrix X whose rows are copies of the vector **x**, and the matrix Y whose columns are copies of the vector **y**.

Example A.12

The volume V of a right circular cone of radius r and height h is given by

$$V = \frac{1}{3}\pi r^2 h \tag{A.4}$$

Plot the volume of the cone as r and h vary on the intervals $0 \le r \le 4$ and $0 \le h \le 6$ meters.

Solution:

The volume of the cone is a function of both the radius r and the height h, that is,

$$V = f(r, h)$$

The three–dimensional plot is created with the following MATLAB script where, as in the previous example, in the second line we have used the dot multiplication, dot division, and dot exponentiation. This will be explained in Section A.9.

```
[R,H]=meshgrid(0: 4, 0: 6);    % Creates R and H matrices from vectors r and h;...
V=(pi .* R .^ 2 .* H) ./ 3;  mesh(R, H, V);...
xlabel('x–axis, radius r (meters)'); ylabel('y–axis, altitude h (meters)');...
zlabel('z–axis, volume (cubic meters)'); title('Volume of Right Circular Cone'); box on
```

The three–dimensional plot of Figure A.6 shows how the volume of the cone increases as the radius and height are increased.

The plots of Figure A.5 and A.6 are rudimentary; MATLAB can generate very sophisticated three–dimensional plots. The MATLAB User's Manual and the Using MATLAB Graphics Manual contain numerous examples.

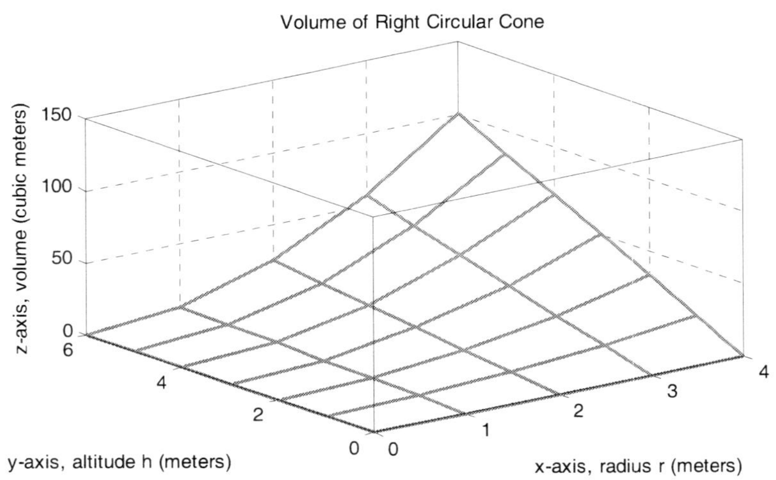

Figure A.6. Volume of a right circular cone.

A.8 Subplots

MATLAB can display up to four windows of different plots on the *Figure* window using the command **subplot(m,n,p)**. This command divides the window into an $m \times n$ matrix of plotting areas and chooses the *pth* area to be active. No spaces or commas are required between the three integers *m*, *n* and *p*. The possible combinations are shown in Figure A.7.

We will illustrate the use of the **subplot(m,n,p)** command following the discussion on multiplication, division and exponentiation that follows.

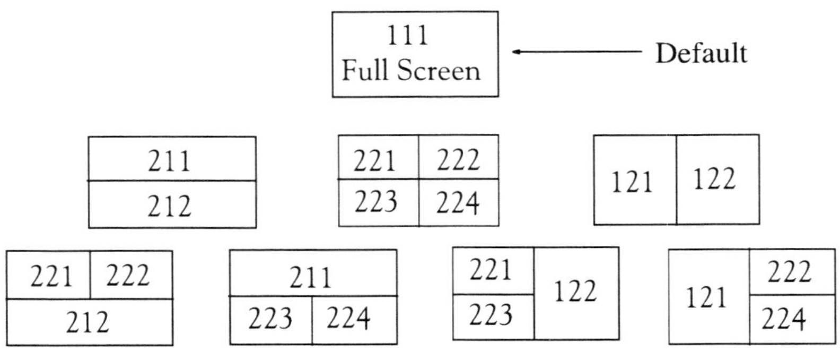

Figure A.7. Possible subplot arrangements in MATLAB

A.9 Multiplication, Division, and Exponentiation

MATLAB recognizes two types of multiplication, division, and exponentiation. These are the ***matrix*** multiplication, division, and exponentiation, and the ***element–by–element*** multiplication, division, and exponentiation. They are explained in the following paragraphs.

In Section A.2, the arrays [a b c ...], such a those that contained the coefficients of polynomials, consisted of one row and multiple columns, and thus are called ***row vectors***. If an array has one column and multiple rows, it is called a ***column vector***. We recall that the elements of a row vector are separated by spaces. To distinguish between row and column vectors, the elements of a column vector must be separated by semicolons. An easier way to construct a column vector, is to write it first as a row vector, and then transpose it into a column vector. MATLAB uses the single quotation character (') to transpose a vector. Thus, a column vector can be written either as

```
b=[-1; 3; 6; 11]
```

or as

```
b=[-1 3 6 11]'
```

As shown below, MATLAB produces the same display with either format.

```
b=[-1; 3; 6; 11]

b =
    -1
     3
     6
    11

b=[-1 3 6 11]'        % Observe the single quotation character (')

b =
    -1
     3
     6
    11
```

We will now define Matrix Multiplication and Element–by–Element multiplication.

1. **Matrix Multiplication** (multiplication of row by column vectors)

   Let

$$\mathbf{A} = [a_1 \;\; a_2 \;\; a_3 \;\; \ldots \;\; a_n]$$

   and

$$\mathbf{B} = [b_1 \;\; b_2 \;\; b_3 \;\; \ldots \;\; b_n]'$$

   be two vectors. We observe that $\mathbf{A}$ is defined as a row vector whereas $\mathbf{B}$ is defined as a column vector, as indicated by the transpose operator ('). Here, multiplication of the row vector $\mathbf{A}$ by the column vector $\mathbf{B}$, is performed with the matrix multiplication operator (*). Then,

$$\mathbf{A*B} = [a_1b_1 + a_2b_2 + a_3b_3 + \ldots + a_nb_n] = \text{single value} \tag{A.5}$$

For example, if

$$\mathbf{A} = [1 \quad 2 \quad 3 \quad 4 \quad 5]$$

and

$$\mathbf{B} = [-2 \quad 6 \quad -3 \quad 8 \quad 7]'$$

the matrix multiplication **A*B** produces the single value 68, that is,

$$\mathbf{A*B} = 1 \times (-2) + 2 \times 6 + 3 \times (-3) + 4 \times 8 + 5 \times 7 = 68$$

and this is verified with the MATLAB script

```
A=[1  2   3  4  5]; B=[ -2  6 -3  8  7]'; A*B   % Observe transpose operator (') in B
ans =
   68
```

Now, let us suppose that both **A** and **B** are row vectors, and we attempt to perform a row–by–row multiplication with the following MATLAB statements.

```
A=[1  2  3  4  5]; B=[-2  6  -3  8  7]; A*B          % No transpose operator (') here
```

When these statements are executed, MATLAB displays the following message:

```
??? Error using ==> *
Inner matrix dimensions must agree.
```

Here, because we have used the matrix multiplication operator (*) in A*B, MATLAB expects vector **B** to be a column vector, not a row vector. It recognizes that **B** is a row vector, and warns us that we cannot perform this multiplication using the matrix multiplication operator (*). Accordingly, we must perform this type of multiplication with a different operator. This operator is defined below.

2. **Element–by–Element Multiplication** (multiplication of a row vector by another row vector)

Let

$$\mathbf{C} = [c_1 \quad c_2 \quad c_3 \quad \ldots \quad c_n]$$

and

$$\mathbf{D} = [d_1 \quad d_2 \quad d_3 \quad \ldots \quad d_n]$$

be two row vectors. Here, multiplication of the row vector **C** by the row vector **D** is performed with the ***dot multiplication operator*** (.*). There is no space between the dot and the multiplication symbol. Thus,

$$\mathbf{C.*D} = [c_1d_1 \quad c_2d_2 \quad c_3d_3 \quad \ldots \quad c_nd_n] \qquad (A.6)$$

This product is another row vector with the same number of elements, as the elements of **C** and **D**.

As an example, let

$$\mathbf{C} = [1 \quad 2 \quad 3 \quad 4 \quad 5]$$

and

$$\mathbf{D} = [-2 \quad 6 \quad -3 \quad 8 \quad 7]$$

Dot multiplication of these two row vectors produce the following result.

$$\mathbf{C.*D} = 1\times(-2) \quad 2\times6 \quad 3\times(-3) \quad 4\times8 \quad 5\times7 = -2 \quad 12 \quad -9 \quad 32 \quad 35$$

Check with MATLAB:

```
C=[1 2 3 4 5];      % Vectors C and D must have
D=[-2 6 -3 8 7];    % same number of elements
C.*D                % We observe that this is a dot multiplication
```

```
ans =
    -2    12    -9    32    35
```

Similarly, the division (/) and exponentiation (^) operators, are used for matrix division and exponentiation, whereas dot division (./) and dot exponentiation (.^) are used for element–by–element division and exponentiation, as illustrated in Examples A.11 and A.12 above.

We must remember that ***no space is allowed between the dot (.) and the multiplication, division, and exponentiation operators.***

Note: A dot (.) is never required with the plus (+) and minus (–) operators.

Example A.13

Write the MATLAB script that produces a simple plot for the waveform defined as

$$y = f(t) = 3e^{-4t}\cos 5t - 2e^{-3t}\sin 2t + \frac{t^2}{t+1} \tag{A.7}$$

in the $0 \le t \le 5$ seconds interval.

Solution:

The MATLAB script for this example is as follows:

```
t=0: 0.01: 5;                % Define t-axis in 0.01 increments
y=3 .* exp(-4 .* t) .* cos(5 .* t)-2 .* exp(-3 .* t) .* sin(2 .* t) + t .^2 ./ (t+1);
plot(t,y); grid; xlabel('t'); ylabel('y=f(t)'); title('Plot for Example A.13')
```

The plot for this example is shown in Figure A.8.

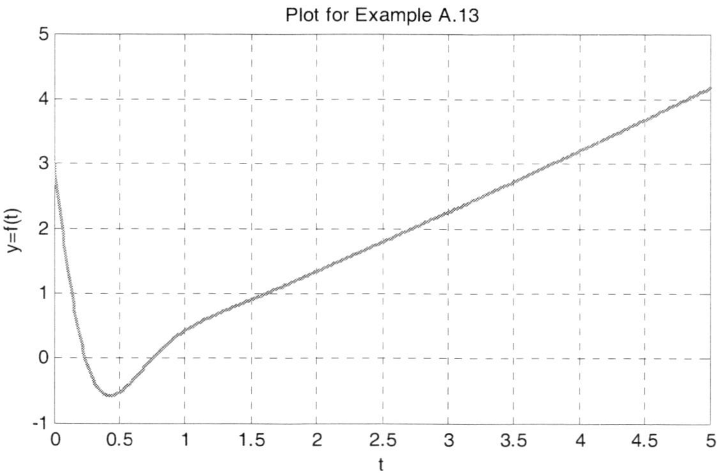

Figure A.8. Plot for Example A.13

Had we, in this example, defined the time interval starting with a negative value equal to or less than -1, say as $-3 \le t \le 3$, MATLAB would have displayed the following message:

```
Warning: Divide by zero.
```

This is because the last term (the rational fraction) of the given expression, is divided by zero when $t = -1$. To avoid division by zero, we use the special MATLAB function **eps,** which is a number approximately equal to 2.2×10^{-16}. It will be used with the next example.

The command **axis([xmin xmax ymin ymax])** scales the current plot to the values specified by the arguments **xmin, xmax, ymin and ymax.** There are no commas between these four arguments. This command must be placed ***after*** the plot command and must be repeated for each plot. The following example illustrates the use of the dot multiplication, division, and exponentiation, the **eps** number, the **axis([xmin xmax ymin ymax])** command, and also MATLAB's capability of displaying up to four windows of different plots.

Example A.14

Plot the functions

$$y = \sin^2 x, \quad z = \cos^2 x, \quad w = \sin^2 x \cdot \cos^2 x, \quad v = \sin^2 x / \cos^2 x$$

in the interval $0 \le x \le 2\pi$ using 100 data points. Use the **subplot** command to display these functions on four windows on the same graph.

Solution:

The MATLAB script to produce the four subplots is as follows:

```
x=linspace(0,2*pi,100);              % Interval with 100 data points
y=(sin(x).^ 2); z=(cos(x).^ 2);
w=y.* z;
v=y./ (z+eps);% add eps to avoid division by zero
subplot(221);% upper left of four subplots
plot(x,y); axis([0 2*pi 0 1]);
title('y= (sinx) ^ 2');
subplot(222);                        % upper right of four subplots
plot(x,z); axis([0 2*pi 0 1]);
title('z= (cosx) ^ 2');
subplot(223);                        % lower left of four subplots
plot(x,w); axis([0 2*pi 0 0.3]);
title('w= (sinx) ^ 2*(cosx) ^ 2');
subplot(224);                        % lower right of four subplots
plot(x,v); axis([0 2*pi 0 400]);
title('v= (sinx) ^ 2/(cosx) ^ 2');
```

These subplots are shown in Figure A.9.

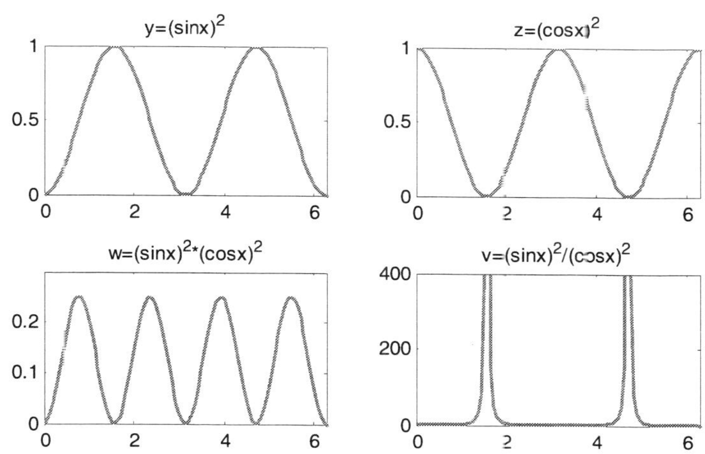

Figure A.9. Subplots for the functions of Example A.14

The next example illustrates MATLAB's capabilities with imaginary numbers. We will introduce the **real(z)** and **imag(z)** functions that display the real and imaginary parts of the complex quantity $z = x + iy$, the **abs(z)**, and the **angle(z)** functions that compute the absolute value (magnitude) and phase angle of the complex quantity $z = x + iy = r\angle\theta$. We will also use the **polar(theta,r)** function that produces a plot in polar coordinates, where **r** is the magnitude, **theta** is the angle in radians, and the **round(n)** function that rounds a number to its nearest integer.

Example A.15

Consider the electric circuit of Figure A.10.

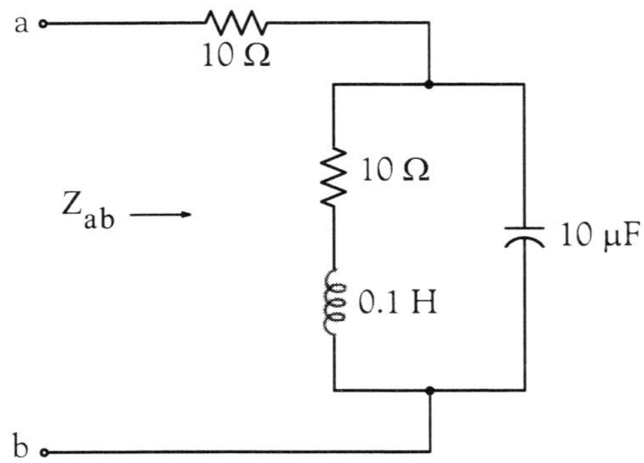

Figure A.10. Electric circuit for Example A.15

With the given values of resistance, inductance, and capacitance, the impedance Z_{ab} as a function of the radian frequency ω can be computed from the following expression:

$$Z_{ab} = Z = 10 + \frac{10^4 - j(10^6/\omega)}{10 + j(0.1\omega - 10^5/\omega)} \tag{A.8}$$

a. Plot $Re\{Z\}$ (the real part of the impedance Z) versus frequency ω.

b. Plot $Im\{Z\}$ (the imaginary part of the impedance Z) versus frequency ω.

c. Plot the impedance Z versus frequency ω in polar coordinates.

Solution:

The MATLAB script below computes the real and imaginary parts of Z_{ab} which, for simplicity, are denoted as **z**, and plots these as two separate graphs (parts a & b). It also produces a polar plot (part c).

```
w=0: 1: 2000;                    % Define interval with one radian interval;...
z=(10+(10 .^ 4 –j .* 10 .^ 6 ./ (w+eps)) ./ (10 + j .* (0.1 .* w –10.^5./ (w+eps))));...
%
% The first five statements (next two lines) compute and plot Re{z}
real_part=real(z); plot(w,real_part);...
xlabel('radian frequency w'); ylabel('Real part of Z'); grid
```

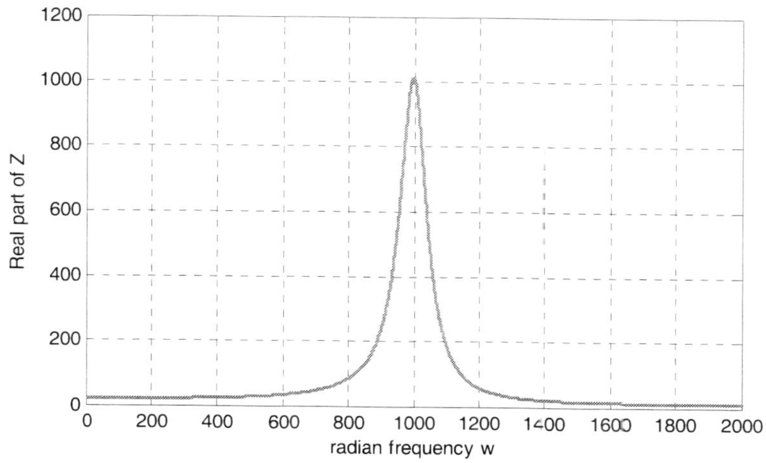

Figure A.11. Plot for the real part of the impedance in Example A.15

```
% The next five statements (next two lines) compute and plot Im{z}
imag_part=imag(z); plot(w,imag_part);...
xlabel('radian frequency w'); ylabel('Imaginary part of Z'); grid
```

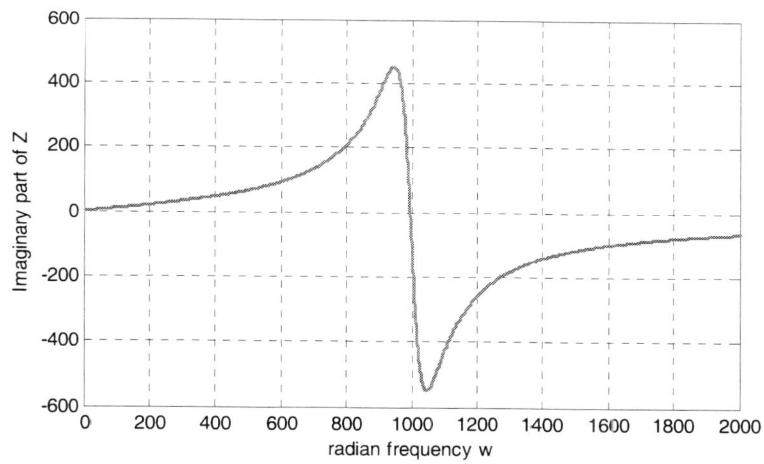

Figure A.12. Plot for the imaginary part of the impedance in Example A.15

```
% The last six statements (next five lines) below produce the polar plot of z
mag=abs(z);                    % Computes |Z|;...
rndz=round(abs(z));            % Rounds |Z| to read polar plot easier;...
theta=angle(z);                % Computes the phase angle of impedance Z;...
polar(theta,rndz);             % Angle is the first argument
ylabel('Polar Plot of Z'); grid
```

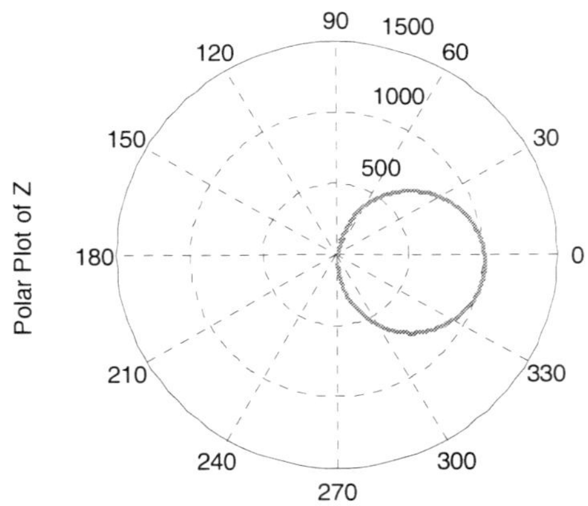

Figure A.13. Polar plot of the impedance in Example A.15

Example A.15 clearly illustrates how powerful, fast, accurate, and flexible MATLAB is.

A.10 Script and Function Files

MATLAB recognizes two types of files: ***script files*** and ***function files***. Both types are referred to as ***m–files*** since both require the *.m* extension.

A ***script file*** consists of two or more built–in functions such as those we have discussed thus far. Thus, the script for each of the examples we discussed earlier, make up a script file. Generally, a script file is one which was generated and saved as an m–file with an editor such as the MATLAB's Editor/Debugger.

A ***function file*** is a user–defined function using MATLAB. We use function files for repetitive tasks. The first line of a function file must contain the word *function*, followed by the output argument, the equal sign (=), and the input argument enclosed in parentheses. The function name and file name must be the same, but the file name must have the extension ***.m***. For example, the function file consisting of the two lines below

```
function y = myfunction(x)
y=x.^ 3 + cos(3.* x)
```

is a function file and must be saved as ***myfunction.m***

For the next example, we will use the following MATLAB functions:

fzero(f,x) – attempts to find a zero of a function of one variable, where **f** is a string containing the name of a real–valued function of a single real variable. MATLAB searches for a value near a point where the function **f** changes sign, and returns that value, or returns NaN if the search fails.

Important: We must remember that we use **roots(p)** to find the roots of polynomials only, such as those in Examples A.1 and A.2.

fplot(fcn,lims) – plots the function specified by the string **fcn** between the x–axis limits specified by **lims = [xmin xmax]**. Using **lims = [xmin xmax ymin ymax]** also controls the y–axis limits. The string **fcn** must be the name of an *m–file* function or a string with variable x .

NaN (Not–a–Number) is not a function; it is MATLAB's response to an undefined expression such as $0/0$, ∞/∞, or inability to produce a result as described on the next paragraph. We can avoid division by zero using the **eps** number, which we mentioned earlier.

Example A.16

Find the zeros, the minimum, and the maximum values of the function

$$f(x) = \frac{1}{(x-0.1)^2 + 0.01} - \frac{1}{(x-1.2)^2 + 0.04} - 10 \qquad (A.9)$$

in the interval $-1.5 \le x \le 1.5$

Solution:

We first plot this function to observe the approximate zeros, maxima, and minima using the following script.

```
x=−1.5: 0.01: 1.5;
y=1./ ((x−0.1).^ 2 + 0.01) −1./ ((x−1.2).^ 2 + 0.04) −10;
plot(x,y); grid
```

The plot is shown in Figure A.14.

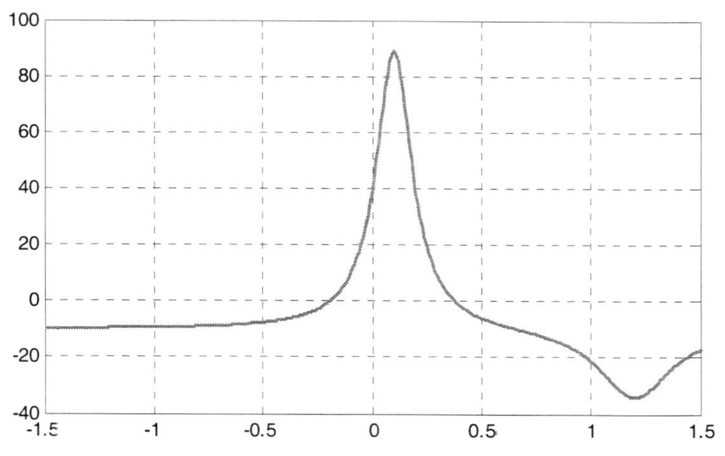

Figure A.14. Plot for Example A.16 using the plot command

The roots (zeros) of this function appear to be in the neighborhood of $x = -0.2$ and $x = 0.3$. The maximum occurs at approximately $x = 0.1$ where, approximately, $y_{max} = 90$, and the minimum occurs at approximately $x = 1.2$ where, approximately, $y_{min} = -34$.

Next, we define and save *f(x)* as the **funczero01.m** function m–file with the following script:

```
function y=funczero01(x)
% Finding the zeros of the function shown below
y=1/((x-0.1)^2+0.01)-1/((x-1.2)^2+0.04)-10;
```

To save this file, from the File drop menu on the Command Window, we choose New, and when the Editor Window appears, we type the script above and we save it as **funczero01.** MATLAB appends the extension **.m** to it.

Now, we can use the **fplot(fcn,lims)** command to plot $f(x)$ as follows:

```
fplot('funczero01', [-1.5  1.5]); grid
```

This plot is shown in Figure A.15. As expected, this plot is identical to the plot of Figure A.14 which was obtained with the **plot(x,y)** command as shown in Figure A.14.

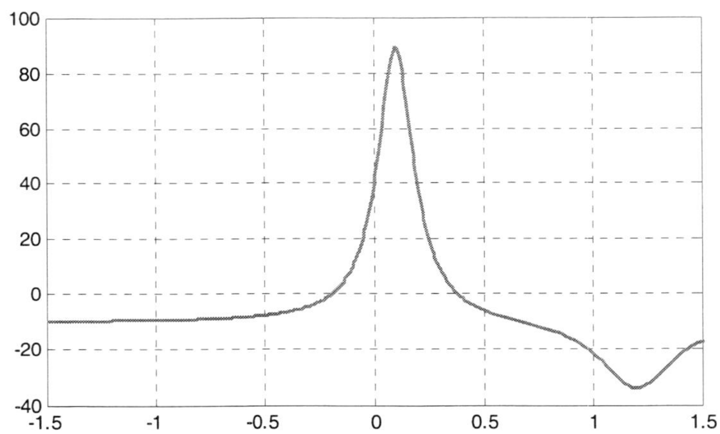

Figure A.15. Plot for Example A.16 using the fplot command

We will use the **fzero(f,x)** function to compute the roots of $f(x)$ in Equation (A.9) more precisely. The MATLAB script below will accomplish this.

```
x1= fzero('funczero01', -0.2);
x2= fzero('funczero01', 0.3);
fprintf('The roots (zeros) of this function are r1= %3.4f', x1);
fprintf(' and r2= %3.4f \n', x2)
```

MATLAB displays the following:

```
The roots (zeros) of this function are r1= -0.1919 and r2= 0.3788
```

The earlier MATLAB versions included the function **fmin(f,x1,x2)** and with this function we could compute both a minimum of some function $f(x)$ or a maximum of $f(x)$ since a maximum of $f(x)$ is equal to a minimum of $-f(x)$. This can be visualized by flipping the plot of a function $f(x)$ upside–down. This function is no longer used in MATLAB and thus we will compute the maxima and minima from the derivative of the given function.

From elementary calculus, we recall that the maxima or minima of a function $y = f(x)$ can be found by setting the first derivative of a function equal to zero and solving for the independent variable x. For this example we use the **diff(x)** function which produces the approximate derivative of a function. Thus, we use the following MATLAB script:

```
syms x ymin zmin; ymin=1/((x–0.1)^2+0.01)–1/((x–1.2)^2+0.04)–10;...
zmin=diff(ymin)
```

```
zmin =
-1/((x-1/10)^2+1/100)^2*(2*x-1/5)+1/((x-6/5)^2+1/25)^2*(2*x-12/5)
```

When the command

```
solve(zmin)
```

is executed, MATLAB displays a very long expression which when copied at the command prompt and executed, produces the following:

```
ans =
   0.6585 + 0.3437i
ans =
   0.6585 - 0.3437i
ans =
    1.2012
```

The real value 1.2012 above is the value of x at which the function y has its minimum value as we observe also in the plot of Figure A.15.

To find the value of y corresponding to this value of x, we substitute it into $f(x)$, that is,

```
x=1.2012; ymin=1 / ((x–0.1) ^ 2 + 0.01) –1 / ((x–1.2) ^ 2 + 0.04) –10
```

```
ymin = -34.1812
```

We can find the maximum value from $-f(x)$ whose plot is produced with the script

```
x=–1.5:0.01:1.5; ymax=–1./((x–0.1).^2+0.01)+1./((x–1.2).^2+0.04)+10; plot(x,ymax); grid
```

and the plot is shown in Figure A.16.

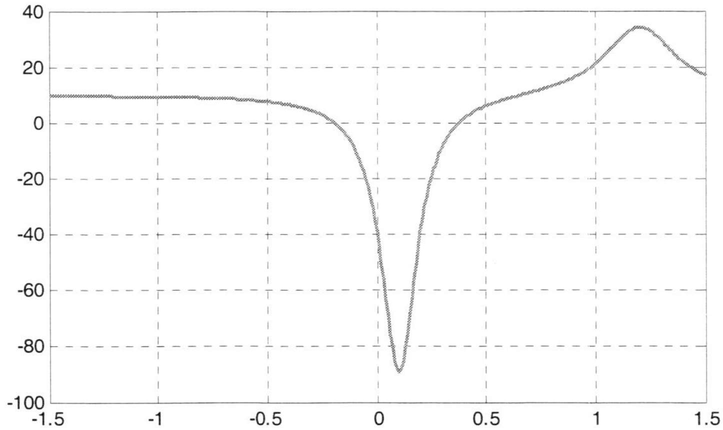

Figure A.16. Plot of −f(x) for Example A.16

Next we compute the first derivative of −f(x) and we solve for x to find the value where the maximum of ymax occurs. This is accomplished with the MATLAB script below.

```
syms x ymax zmax; ymax=-(1/((x-0.1)^2+0.01)-1/((x-1.2)^2+0.04)-10); zmax=diff(ymax)
```

```
zmax =
 1/((x-1/10)^2+1/100)^2*(2*x-1/5)-1/((x-6/5)^2+1/25)^2*(2*x-12/5)
```

```
solve(zmax)
```

When the command

```
solve(zmax)
```

is executed, MATLAB displays a very long expression which when copied at the command prompt and executed, produces the following:

```
ans =
   0.6585 + 0.3437i

ans =
   0.6585 - 0.3437i

ans =
    1.2012
ans =
    0.0999
```

From the values above we choose $x = 0.0999$ which is consistent with the plots of Figures A.15 and A.16. Accordingly, we execute the following script to obtain the value of ymin.

```
x=0.0999;                      % Using this value find the corresponding value of ymax
ymax=1 / ((x−0.1) ^ 2 + 0.01) −1 / ((x−1.2) ^ 2 + 0.04) −10

ymax = 89.2000
```

A.11 Display Formats

MATLAB displays the results on the screen in integer format without decimals if the result is an integer number, or in short floating point format with four decimals if it a fractional number. The format displayed has nothing to do with the accuracy in the computations. MATLAB performs all computations with accuracy up to 16 decimal places.

The output format can changed with the **format** command. The available MATLAB formats can be displayed with the **help format** command as follows:

```
help format

FORMAT Set output format.
All computations in MATLAB are done in double precision.
FORMAT may be used to switch between different output display formats
as follows:

FORMAT  Default. Same as SHORT.
FORMAT SHORT Scaled fixed point format with 5 digits.
FORMAT LONG Scaled fixed point format with 15 digits.
FORMAT SHORT E Floating point format with 5 digits.
FORMAT LONG E  Floating point format with 15 digits.
FORMAT SHORT G Best of fixed or floating point format with 5 digits.
FORMAT LONG G Best of fixed or floating point format with 15 digits.
FORMAT HEX Hexadecimal format.
FORMAT + The symbols +, - and blank are printed for positive, negative,
        and zero elements.Imaginary parts are ignored.
FORMAT BANK Fixed format for dollars and cents.
FORMAT RAT Approximation by ratio of small integers.

Spacing:

FORMAT COMPACT Suppress extra line-feeds.
FORMAT LOOSE  Puts the extra line-feeds back in.

Some examples with different format displays age given below.

format short  33.3335  Four decimal digits (default)
format long  33.33333333333334 16 digits
format short e  3.3333e+01  Four decimal digits plus exponent
format short g  33.333  Better of format short or format short e
format bank  33.33 two decimal digits
format +  only + or - or zero are printed
```

```
format rat 100/3 rational approximation
```

The **disp(X)** command displays the array **X** without printing the array name. If **X** is a string, the text is displayed.

The **fprintf(format,array)** command displays and prints both text and arrays. It uses specifiers to indicate where and in which format the values would be displayed and printed. Thus, if **%f** is used, the values will be displayed and printed in fixed decimal format, and if **%e** is used, the values will be displayed and printed in scientific notation format. With this command only the real part of each parameter is processed.

This appendix is just an introduction to MATLAB.* This outstanding software package consists of many applications known as *Toolboxes*. The MATLAB Student Version contains just a few of these Toolboxes. Others can be bought directly from The MathWorks,™ Inc., as add–ons.

* *For more MATLAB applications, please refer to Numerical Analysis Using MATLAB and Spreadsheets, ISBN 0–9709511–1–6.*

Appendix B

Introduction to Simulink®

This appendix is a brief introduction to Simulink. This author feels that we can best introduce Simulink with a few examples. Some familiarity with MATLAB is essential in understanding Simulink, and for this purpose, Appendix A is included as an introduction to MATLAB.

B.1 Simulink and its Relation to MATLAB

The MATLAB® and Simulink® environments are integrated into one entity, and thus we can analyze, simulate, and revise our models in either environment at any point. We invoke Simulink from within MATLAB. We will introduce Simulink with a few illustrated examples.

Example B.1

For the circuit of Figure B.1, the initial conditions are $i_L(0^-) = 0$, and $v_c(0^-) = 0.5\ V$. We will compute $v_c(t)$.

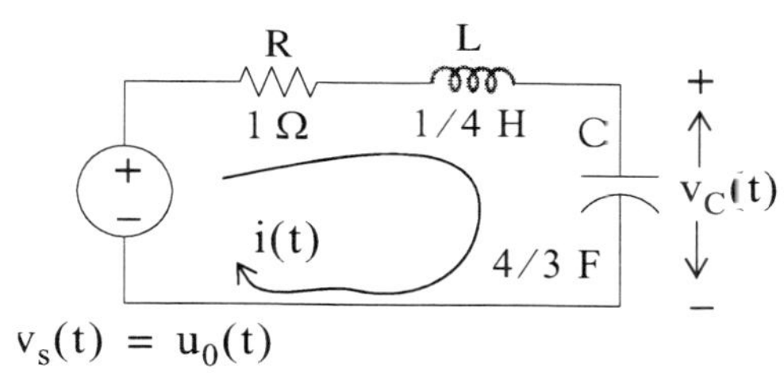

Figure B.1. Circuit for Example B.1

For this example,

$$i = i_L = i_C = C\frac{dv_C}{dt} \tag{B.1}$$

and by Kirchoff's voltage law (KVL),

$$Ri_L + L\frac{di_L}{dt} + v_C = u_0(t) \tag{B.2}$$

Substitution of (B.1) into (B.2) yields

$$RC\frac{dv_C}{dt} + LC\frac{d^2v_C}{dt^2} + v_C = u_0(t) \qquad (B.3)$$

Substituting the values of the circuit constants and rearranging we obtain:

$$\frac{1}{3}\frac{d^2v_C}{dt^2} + \frac{4}{3}\frac{dv_C}{dt} + v_C = u_0(t)$$

$$\frac{d^2v_C}{dt^2} + 4\frac{dv_C}{dt} + 3v_C = 3u_0(t) \qquad (B.4)$$

$$\frac{d^2v_C}{dt^2} + 4\frac{dv_C}{dt} + 3v_C = 3 \qquad t > 0 \qquad (B.5)$$

To appreciate Simulink's capabilities, for comparison, three different methods of obtaining the solution are presented, and the solution using Simulink follows.

First Method – Assumed Solution

Equation (B.5) is a second–order, non–homogeneous differential equation with constant coefficients, and thus the complete solution will consist of the sum of the forced response and the natural response. It is obvious that the solution of this equation cannot be a constant since the derivatives of a constant are zero and thus the equation is not satisfied. Also, the solution cannot contain sinusoidal functions (sine and cosine) since the derivatives of these are also sinusoids. However, decaying exponentials of the form ke^{-at} where *k* and *a* are constants, are possible candidates since their derivatives have the same form but alternate in sign.

It can be shown* that if $k_1e^{-s_1t}$ and $k_2e^{-s_2t}$ where k_1 and k_2 are constants and s_1 and s_2 are the roots of the characteristic equation of the homogeneous part of the given differential equation, the natural response is the sum of the terms $k_1e^{-s_1t}$ and $k_2e^{-s_2t}$. Therefore, the total solution will be

$$v_c(t) = \text{natural response} + \text{forced response} = v_{cn}(t) + v_{cf}(t) = k_1e^{-s_1t} + k_2e^{-s_2t} + v_{cf}(t) \qquad (B.6)$$

The values of s_1 and s_2 are the roots of the characteristic equation

* *Please refer to Circuit Analysis II with MATLAB Applications, ISBN 978–0–9709511–5–1, Appendix B for a thorough discussion.*

$$s^2 + 4s + 3 = 0 \tag{B.7}$$

Solution of (B.7) yields of $s_1 = -1$ and $s_2 = -3$ and with these values (B.6) is written as

$$v_c(t) = k_1 e^{-t} + k_2 e^{-3t} + v_{cf}(t) \tag{B.8}$$

The forced component $v_{cf}(t)$ is found from (B.5), i.e.,

$$\frac{d^2 v_C}{dt^2} + 4\frac{dv_C}{dt} + 3v_C = 3 \qquad t > 0 \tag{B.9}$$

Since the right side of (B.9) is a constant, the forced response will also be a constant and we denote it as $v_{Cf} = k_3$. By substitution into (B.9) we obtain

$$0 + 0 + 3k_3 = 3$$

or

$$v_{Cf} = k_3 = 1 \tag{B.10}$$

Substitution of this value into (B.8), yields the total solution as

$$v_C(t) = v_{Cn}(t) + v_{Cf} = k_1 e^{-t} + k_2 e^{-3t} + 1 \tag{B.11}$$

The constants k_1 and k_2 will be evaluated from the initial conditions. First, using $v_C(0) = 0.5\ V$ and evaluating (B.11) at $t = 0$, we obtain

$$v_C(0) = k_1 e^0 + k_2 e^0 + 1 = 0.5$$

$$k_1 + k_2 = -0.5 \tag{B.12}$$

Also,

$$i_L = i_C = C\frac{dv_C}{dt}, \quad \frac{dv_C}{dt} = \frac{i_L}{C}$$

and

$$\left.\frac{dv_C}{dt}\right|_{t=0} = \frac{i_L(0)}{C} = \frac{0}{C} = 0 \tag{B.13}$$

Next, we differentiate (B.11), we evaluate it at $t = 0$, and equate it with (B.13). Thus,

$$\left.\frac{dv_C}{dt}\right|_{t=0} = -k_1 - 3k_2 \tag{B.14}$$

By equating the right sides of (B.13) and (B.14) we obtain

$$-k_1 - 3k_2 = 0 \tag{B.15}$$

Simultaneous solution of (B.12) and (B.15), gives $k_1 = -0.75$ and $k_2 = 0.25$. By substitution into (B.8), we obtain the total solution as

$$\mathbf{v_C(t) = (-0.75e^{-t} + 0.25e^{-3t} + 1)u_0(t)} \tag{B.16}$$

Check with MATLAB:

```
syms t                                      % Define symbolic variable t
y0=-0.75*exp(-t)+0.25*exp(-3*t)+1;          % The total solution y(t), for our example, vc(t)
y1=diff(y0)                                 % The first derivative of y(t)

y1 =
3/4*exp(-t)-3/4*exp(-3*t)

y2=diff(y0,2)                               % The second derivative of y(t)

y2 =
-3/4*exp(-t)+9/4*exp(-3*t)

y=y2+4*y1+3*y0                              % Summation of y and its derivatives

y =
3
```

Thus, the solution has been verified by MATLAB. Using the expression for $v_C(t)$ in (B.16), we find the expression for the current as

$$\mathbf{i = i_L = i_C = C\frac{dv_C}{dt} = \frac{4}{3}\left(\frac{3}{4}e^{-t} - \frac{3}{4}e^{-3t}\right) = e^{-t} - e^{-3t}\ A} \tag{B.17}$$

Second Method – Using the Laplace Transformation

The transformed circuit is shown in Figure B.2.

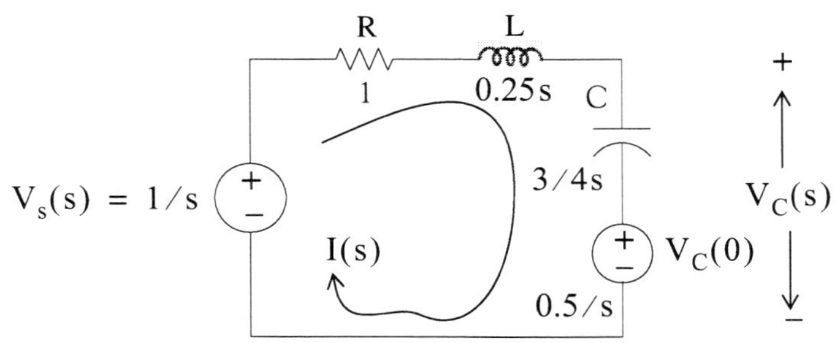

Figure B.2. Transformed Circuit for Example B.1

By the voltage division* expression,

$$V_C(s) = \frac{3/4s}{(1+0.25s+3/4s)} \cdot \left(\frac{1}{s} - \frac{0.5}{s}\right) + \frac{0.5}{s} = \frac{1.5}{s(s^2+4s+3)} + \frac{0.5}{s} = \frac{0.5s^2+2s+3}{s(s+1)(s+3)}$$

Using partial fraction expansion,† we let

$$\frac{0.5s^2+2s+3}{s(s+1)(s+3)} = \frac{r_1}{s} + \frac{r_2}{(s+1)} + \frac{r_3}{(s+3)} \tag{B.18}$$

$$r_1 = \left.\frac{0.5s^2+2s+3}{(s+1)(s+3)}\right|_{s=0} = 1$$

$$r_2 = \left.\frac{0.5s^2+2s+3}{s(s+3)}\right|_{s=-1} = -0.75$$

$$r_3 = \left.\frac{0.5s^2+2s+3}{s(s+1)}\right|_{s=-3} = 0.25$$

and by substitution into (B.18)

$$V_C(s) = \frac{0.5s^2+2s+3}{s(s+1)(s+3)} = \frac{1}{s} + \frac{-0.75}{(s+1)} + \frac{0.25}{(s+3)}$$

Taking the Inverse Laplace transform‡ we find that

$$\mathbf{v_C(t) = 1 - 0.75e^{-t} + 0.25e^{-3t}}$$

Third Method – Using State Variables

$$Ri_L + L\frac{di_L}{dt} + v_C = u_0(t)^{**}$$

* *For derivation of the voltage division and current division expressions, please refer to Circuit Analysis I with MATLAB Applications, ISBN 978-0-9709511-2-0.*

† *Partial fraction expansion is discussed in Chapter 3, Signals and Systems with MATLAB Computing and Simulink Modeling, ISBN 978-1-934404-11-9*

‡ *For an introduction to Laplace Transform and Inverse Laplace Transform, please refer Chapters 2 and 3, Signals and Systems with MATLAB Computing and Simulink Modeling, ISBN 978-1-934404-11-9*

** *Usually, in State-Space and State Variables Analysis,* $u(t)$ *denotes any input. For distinction, we will denote the Unit Step Function as* $u_0(t)$*. Please refer to Chapter 5, Signals and Systems with MATLAB Computing and Simulink Modeling, ISBN 978-1-934404-11-9*

By substitution of given values and rearranging, we obtain

$$\frac{1}{4}\frac{di_L}{dt} = (-1)i_L - v_C + 1$$

or

$$\frac{di_L}{dt} = -4i_L - 4v_C + 4 \tag{B.19}$$

Next, we define the state variables $x_1 = i_L$ and $x_2 = v_C$. Then,

$$\dot{x}_1 = \frac{di_L}{dt}^{*} \tag{B.20}$$

and

$$\dot{x}_2 = \frac{dv_C}{dt} \tag{B.21}$$

Also,

$$i_L = C\frac{dv_C}{dt}$$

and thus,

$$x_1 = i_L = C\frac{dv_C}{dt} = C\dot{x}_2 = \frac{4}{3}\dot{x}_2$$

or

$$\dot{x}_2 = \frac{3}{4}x_1 \tag{B.22}$$

Therefore, from (B.19), (B.20), and (B.22), we obtain the state equations

$$\dot{x}_1 = -4x_1 - 4x_2 + 4$$

$$\dot{x}_2 = \frac{3}{4}x_1$$

and in matrix form,

$$\begin{bmatrix} \dot{x}_1 \\ \dot{x}_2 \end{bmatrix} = \begin{bmatrix} -4 & -4 \\ 3/4 & 0 \end{bmatrix}\begin{bmatrix} x_1 \\ x_2 \end{bmatrix} + \begin{bmatrix} 4 \\ 0 \end{bmatrix}u_0(t) \tag{B.23}$$

Solution[†] of (B.23) yields

* *The notation* $\dot{x}$ *(x dot) is often used to denote the first derivative of the function* x*, that is,* $\dot{x} = dx/dt$.

† *The detailed solution of (B.23) is given in Chapter 5, Example 5.10, Page 5–23, Signals and Systems with MATLAB Computing and Simulink Modeling, ISBN 978–1–934404–11–9*

$$\begin{bmatrix} x_1 \\ x_2 \end{bmatrix} = \begin{bmatrix} e^{-t} - e^{-3t} \\ 1 - 0.75e^{-t} + 0.25e^{-3t} \end{bmatrix}$$

Then,

$$x_1 = i_L = e^{-t} - e^{-3t} \tag{B.24}$$

and

$$x_2 = v_C = 1 - 0.75e^{-t} + 0.25e^{-3t} \tag{B.25}$$

Modeling the Differential Equation of Example B.1 with Simulink

To run Simulink, we must first invoke MATLAB. Make sure that Simulink is installed in your system. In the MATLAB Command prompt, we type:

```
simulink
```

Alternately, we can click on the Simulink icon shown in Figure B.3. It appears on the top bar on MATLAB's Command prompt.

Figure B.3. The Simulink icon

Upon execution of the Simulink command, the **Commonly Used Blocks** appear as shown in Figure B.4.

In Figure B.4, the left side is referred to as the **Tree Pane** and displays all Simulink libraries installed. The right side is referred to as the **Contents Pane** and displays the blocks that reside in the library currently selected in the Tree Pane.

Let us express the differential equation of Example B.1 as

$$\frac{d^2 v_C}{dt^2} = -4\frac{dv_C}{dt} - 3v_C + 3u_0(t) \tag{B.26}$$

A block diagram representing relation (B.26) above is shown in Figure B.5. We will use Simulink to draw a similar block diagram.*

* *Henceforth, all Simulink block diagrams will be referred to as models.*

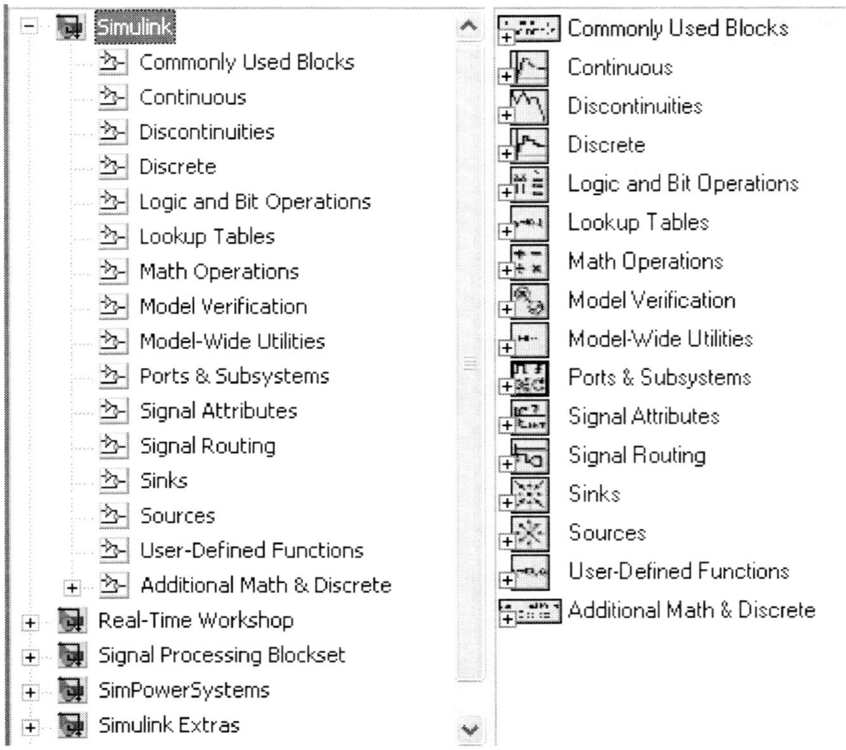

Figure B.4. The Simulink Library Browser

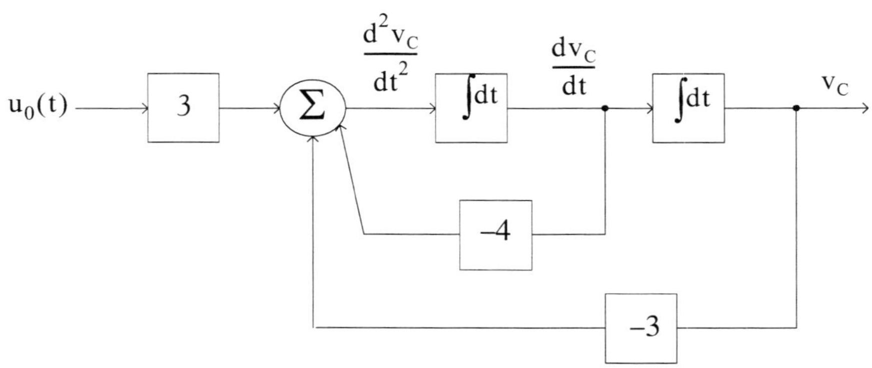

Figure B.5. Block diagram for equation (B.26)

To model the differential equation (B.26) using Simulink, we perform the following steps:

1. On the **Simulink Library Browser**, we click on the leftmost icon shown as a blank page on the top title bar. A new model window named **untitled** will appear as shown in Figure B.6.

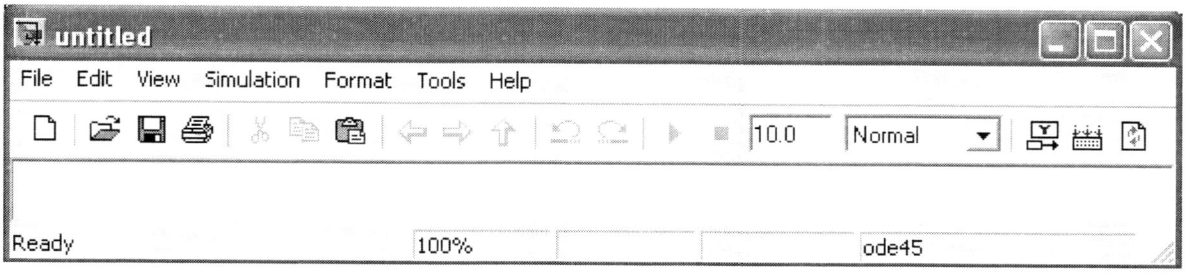

Figure B.6. The Untitled model window in Simulink.

The window of Figure B.6 is the model window where we enter our blocks to form a block diagram. We save this as model file name **Equation_1_26**. This is done from the File drop menu of Figure B.6 where we choose **Save as** and name the file as **Equation_1_26**. Simulink will add the extension **.mdl**. The new model window will now be shown as **Equation_1_26**, and all saved files will have this appearance. See Figure B.7.

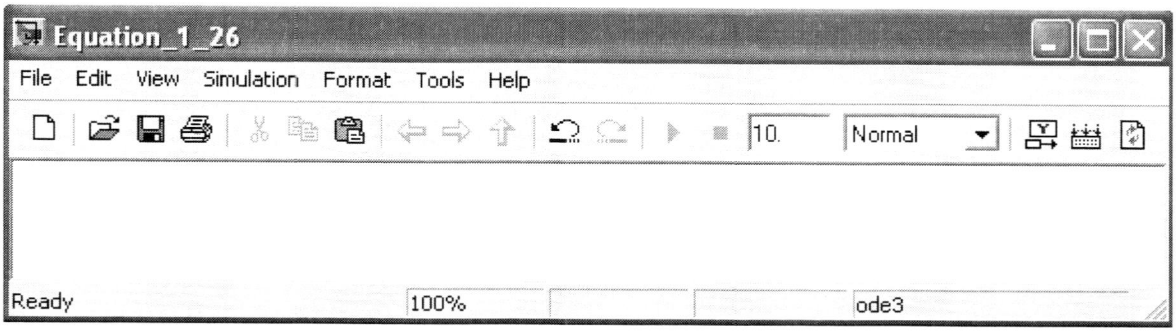

Figure B.7. Model window for Equation_1_26.mdl file

2. With the **Equation_1_26** model window and the **Simulink Library Browser** both visible, we click on the **Sources** appearing on the left side list, and on the right side we scroll down until we see the unit step function shown as **Step**. See Figure B.8. We select it, and we drag it into the **Equation_1_26** model window which now appears as shown in Figure B.8. We save file Equation_1_26 using the File drop menu on the **Equation_1_26** model window (right side of Figure B.8).

3. With reference to block diagram of Figure B.5, we observe that we need to connect an amplifier with Gain 3 to the unit step function block. The gain block in Simulink is under **Commonly Used Blocks** (first item under Simulink on the **Simulink Library Browser**). See Figure B.8. If the **Equation_1_26** model window is no longer visible, it can be recalled by clicking on the white page icon on the top bar of the **Simulink Library Browser**.

4. We choose the gain block and we drag it to the right of the unit step function. The triangle on the right side of the unit step function block and the > symbols on the left and right sides of the gain block are connection points. We point the mouse close to the connection point of the unit step function until is shows as a cross hair, and draw a straight line to connect the two

blocks.* We double–click on the gain block and on the **Function Block Parameters**, we change the gain from 1 to 3. See Figure B.9.

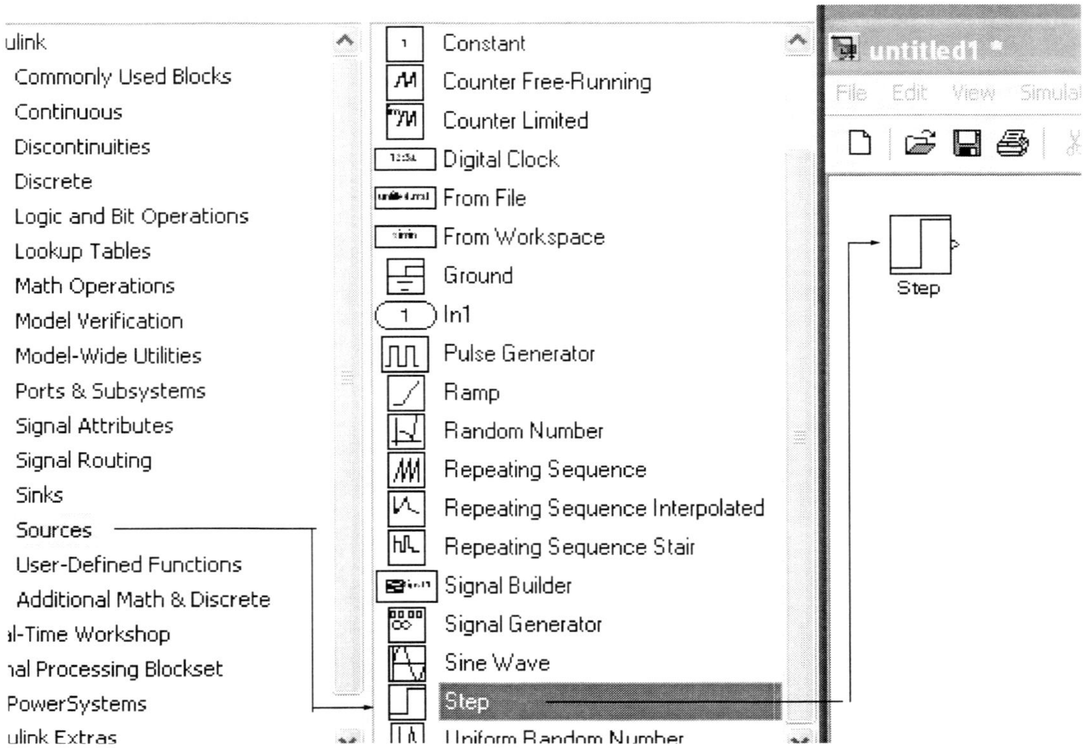

Figure B.8. Dragging the unit step function into File Equation_1_26

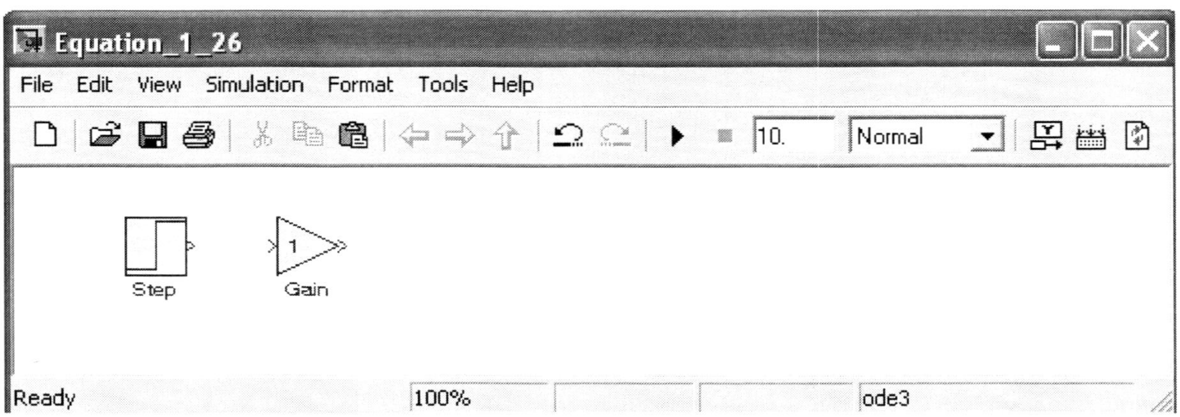

Figure B.9. File Equation_1_26 with added Step and Gain blocks

* *An easy method to interconnect two Simulink blocks by clicking on the source block to select it, then hold down the* ***Ctrl*** *key and left–click on the destination block.*

5. Next, we need to add a thee–input adder. The adder block appears on the right side of the **Simulink Library Browser** under **Math Operations**. We select it, and we drag it into the Equation_1_26 model window. We double click it, and on the **Function Block Parameters** window which appears, we specify 3 inputs. We then connect the output of the of the gain block to the first input of the adder block as shown in Figure B.10.

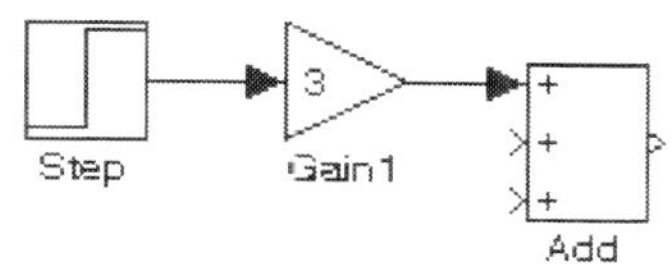

Figure B.10. File Equation_1_26 with added gain block

6. From the **Commonly Used Blocks** of the **Simulink Library Browser**, we choose the **Integrator** block, we drag it into the Equation_1_26 model window, and we connect it to the output of the **Add** block. We repeat this step and to add a second **Integrator** block. We click on the text "Integrator" under the first integrator block, and we change it to Integrator 1. Then, we change the text "Integrator 1" under the second Integrator to "Integrator 2" as shown in Figure B.11.

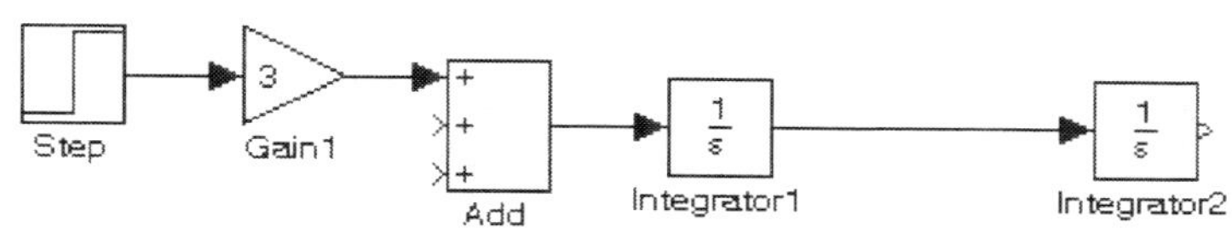

Figure B.11. File Equation_1_26 with the addition of two integrators

7. To complete the block diagram, we add the **Scope** block which is found in the **Commonly Used Blocks** on the **Simulink Library Browser**, we click on the Gain block, and we copy and paste it twice. We flip the pasted Gain blocks by using the **Flip Block** command from the Format drop menu, and we label these as Gain 2 and Gain 3. Finally, we double–click on these gain blocks and on the **Function Block Parameters** window, we change the gains from to –4 and –3 as shown in Figure B.12.

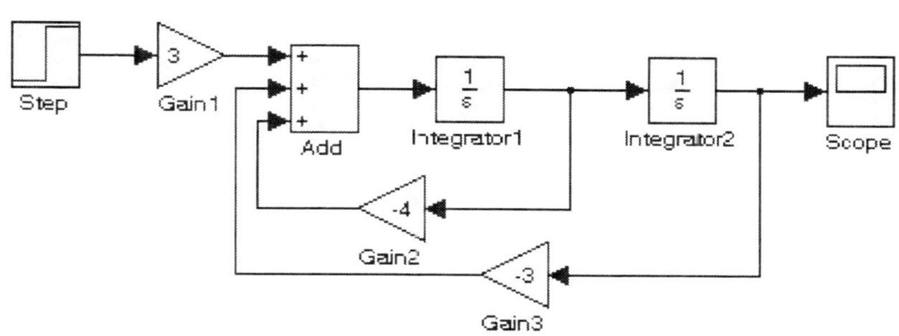

Figure B.12. File Equation_1_26 complete block diagram

8. The initial conditions $i_L(0^-) = C(dv_C/dt)|_{t=0} = 0$, and $v_c(0^-) = 0.5\ V$ are entered by double clicking the Integrator blocks and entering the values 0 for the first integrator, and 0.5 for the second integrator. We also need to specify the simulation time. This is done by specifying the simulation time to be 10 seconds on the **Configuration Parameters** from the **Simulation** drop menu. We can start the simulation on **Start** from the **Simulation** drop menu or by clicking on the  icon.
9. To see the output waveform, we double click on the **Scope** block, and then clicking on the Autoscale  icon, we obtain the waveform shown in Figure B.13.

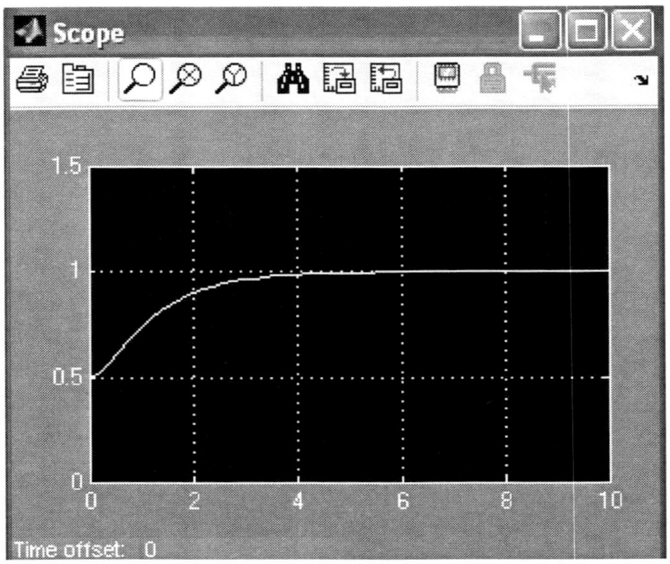

Figure B.13. The waveform for the function $v_C(t)$ for Example B.1

Another easier method to obtain and display the output $v_C(t)$ for Example B.1, is to use **State–Space** block from **Continuous** in the Simulink Library Browser, as shown in Figure B.14.

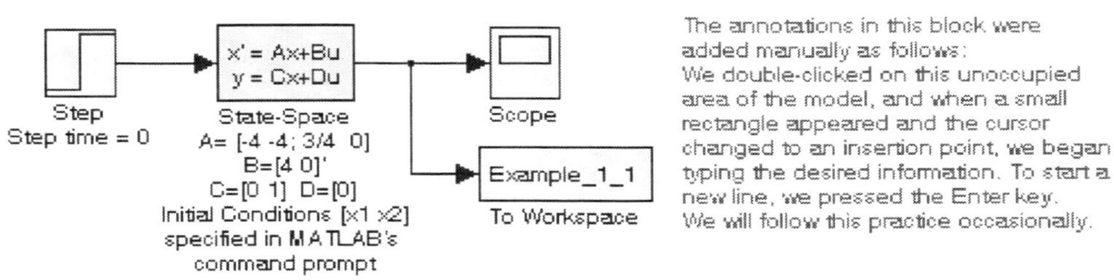

Figure B.14. Obtaining the function $v_C(t)$ for Example B.1 with the State–Space block.

The **simout To Workspace** block shown in Figure B.14 writes its input to the workspace. The data and variables created in the MATLAB Command window, reside in the MATLAB Workspace. This block writes its output to an array or structure that has the name specified by the block's Variable name parameter. This gives us the ability to delete or modify selected variables. We issue the command **who** to see those variables. From Equation B.23, Page B–6,

$$\begin{bmatrix} \dot{x}_1 \\ \dot{x}_2 \end{bmatrix} = \begin{bmatrix} -4 & -4 \\ 3/4 & 0 \end{bmatrix} \begin{bmatrix} x_1 \\ x_2 \end{bmatrix} + \begin{bmatrix} 4 \\ 0 \end{bmatrix} u_0(t)$$

The output equation is

$$y = Cx + du$$

or

$$y = [0 \quad 1] \begin{bmatrix} x_1 \\ x_2 \end{bmatrix} + [0]u$$

We double–click on the **State–Space** block, and in the **Functions Block Parameters** window we enter the constants shown in Figure B.15.

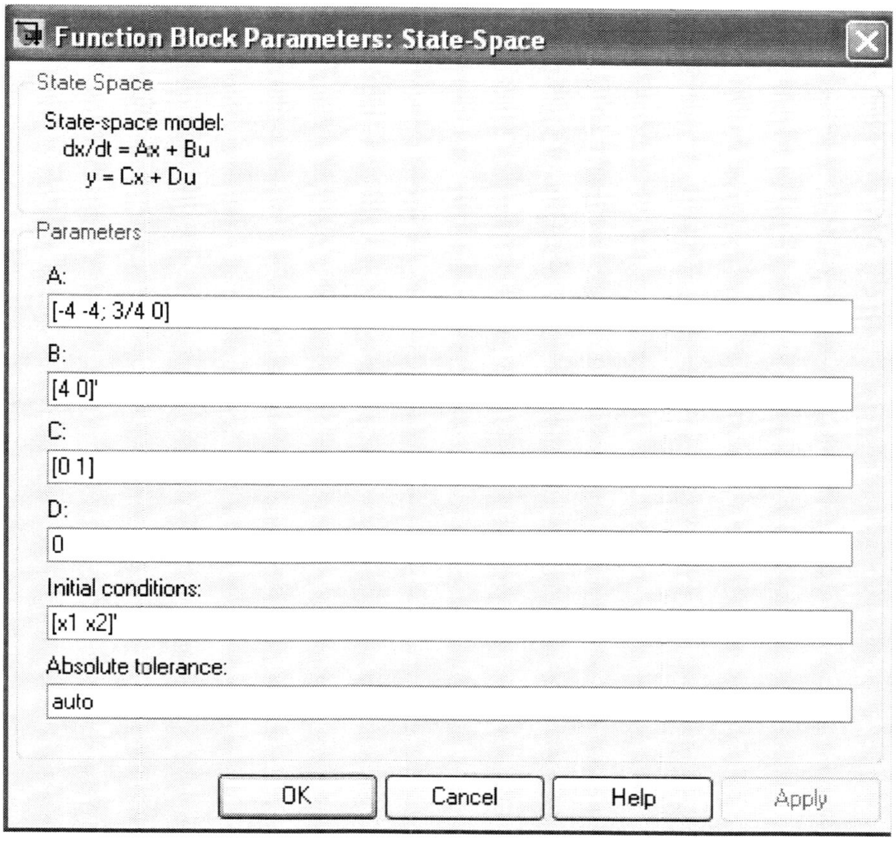

Figure B.15. The Function block parameters for the State–Space block.

The initials conditions [x1 x2]' are specified in MATLAB's Command prompt as

```
x1=0; x2=0.5;
```

As before, to start the simulation we click clicking on the ▸ icon, and to see the output waveform, we double click on the **Scope** block, and then clicking on the Autoscale icon, we obtain the waveform shown in Figure B.16.

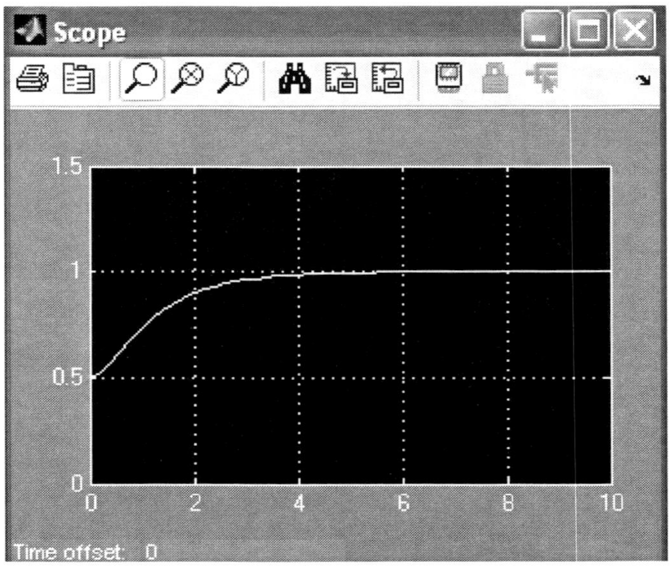

Figure B.16. The waveform for the function $v_C(t)$ for Example B.1 with the State–Space block.

The state–space block is the best choice when we need to display the output waveform of three or more variables as illustrated by the following example.

Example B.2

A fourth–order network is described by the differential equation

$$\frac{d^4y}{dt^4} + a_3\frac{d^3y}{dt^3} + a_2\frac{d^2y}{dt^2} + a_1\frac{dy}{dt} + a_0\, y(t) = u(t) \qquad (B.27)$$

where $y(t)$ is the output representing the voltage or current of the network, and $u(t)$ is any input, and the initial conditions are $y(0) = y'(0) = y''(0) = y'''(0) = 0$.

a. We will express (B.27) as a set of state equations

b. It is known that the solution of the differential equation

$$\frac{d^4y}{dt^4} + 2\frac{d^2y}{dt^2} + y(t) = \sin t \tag{B.28}$$

subject to the initial conditions $y(0) = y'(0) = y''(0) = y'''(0) = 0$, has the solution

$$y(t) = 0.125[(3 - t^2) - 3t\cos t] \tag{B.29}$$

In our set of state equations, we will select appropriate values for the coefficients a_3, a_2, a_1, and a_0 so that the new set of the state equations will represent the differential equation of (B.28), and using Simulink, we will display the waveform of the output $y(t)$.

1. The differential equation of (B.28) is of fourth–order; therefore we must define four state variables that will be used with the four first–order state equations.

   We denote the state variables as x_1, x_2, x_3, and x_4, and we relate them to the terms of the given differential equation as

$$x_1 = y(t) \qquad x_2 = \frac{dy}{dt} \qquad x_3 = \frac{d^2y}{dt^2} \qquad x_4 = \frac{d^3y}{dt^3} \tag{B.30}$$

   We observe that

$$\begin{aligned} \dot{x}_1 &= x_2 \\ \dot{x}_2 &= x_3 \\ \dot{x}_3 &= x_4 \\ \frac{d^4y}{dt^4} &= \dot{x}_4 = -a_0x_1 - a_1x_2 - a_2x_3 - a_3x_4 + u(t) \end{aligned} \tag{B.31}$$

   and in matrix form

$$\begin{bmatrix} \dot{x}_1 \\ \dot{x}_2 \\ \dot{x}_3 \\ \dot{x}_4 \end{bmatrix} = \begin{bmatrix} 0 & 1 & 0 & 0 \\ 0 & 0 & 1 & 0 \\ 0 & 0 & 0 & 1 \\ -a_0 & -a_1 & -a_2 & -a_3 \end{bmatrix} \begin{bmatrix} x_1 \\ x_2 \\ x_3 \\ x_4 \end{bmatrix} + \begin{bmatrix} 0 \\ 0 \\ 0 \\ 1 \end{bmatrix} u(t) \tag{B.32}$$

   In compact form, (B.32) is written as

$$\dot{x} = Ax + bu \tag{B.33}$$

   Also, the output is

$$y = Cx + du \tag{B.34}$$

   where

$$\dot{x} = \begin{bmatrix} \dot{x}_1 \\ \dot{x}_2 \\ \dot{x}_3 \\ \dot{x}_4 \end{bmatrix}, \quad A = \begin{bmatrix} 0 & 1 & 0 & 0 \\ 0 & 0 & 1 & 0 \\ 0 & 0 & 0 & 1 \\ -a_0 & -a_1 & -a_2 & -a_3 \end{bmatrix}, \quad x = \begin{bmatrix} x_1 \\ x_2 \\ x_3 \\ x_4 \end{bmatrix}, \quad b = \begin{bmatrix} 0 \\ 0 \\ 0 \\ 1 \end{bmatrix}, \quad \text{and } u = u(t) \tag{B.35}$$

and since the output is defined as

$$y(t) = x_1$$

relation (B.34) is expressed as

$$y = [1 \ 0 \ 0 \ 0] \cdot \begin{bmatrix} x_1 \\ x_2 \\ x_3 \\ x_4 \end{bmatrix} + [0]u(t) \tag{B.36}$$

2. By inspection, the differential equation of (B.27) will be reduced to the differential equation of (B.28) if we let

$$a_3 = 0 \qquad a_2 = 2 \qquad a_1 = 0 \qquad a_0 = 1 \qquad u(t) = \sin t$$

and thus the differential equation of (B.28) can be expressed in state–space form as

$$\begin{bmatrix} \dot{x}_1 \\ \dot{x}_2 \\ \dot{x}_3 \\ \dot{x}_4 \end{bmatrix} = \begin{bmatrix} 0 & 1 & 0 & 0 \\ 0 & 0 & 1 & 0 \\ 0 & 0 & 0 & 1 \\ -a_0 & 0 & -2 & 0 \end{bmatrix} \begin{bmatrix} x_1 \\ x_2 \\ x_3 \\ x_4 \end{bmatrix} + \begin{bmatrix} 0 \\ 0 \\ 0 \\ 1 \end{bmatrix} \sin t \tag{B.37}$$

where

$$\dot{x} = \begin{bmatrix} \dot{x}_1 \\ \dot{x}_2 \\ \dot{x}_3 \\ \dot{x}_4 \end{bmatrix}, \quad A = \begin{bmatrix} 0 & 1 & 0 & 0 \\ 0 & 0 & 1 & 0 \\ 0 & 0 & 0 & 1 \\ -a_0 & 0 & -2 & 0 \end{bmatrix}, \quad x = \begin{bmatrix} x_1 \\ x_2 \\ x_3 \\ x_4 \end{bmatrix}, \quad b = \begin{bmatrix} 0 \\ 0 \\ 0 \\ 1 \end{bmatrix}, \quad \text{and } u = \sin t \tag{B.38}$$

Since the output is defined as

$$y(t) = x_1$$

in matrix form it is expressed as

$$y = [1 \;\; 0 \;\; 0 \;\; 0] \cdot \begin{bmatrix} x_1 \\ x_2 \\ x_3 \\ x_4 \end{bmatrix} + [0]\sin t \tag{B.39}$$

We invoke MATLAB, we start Simulink by clicking on the Simulink icon, on the **Simulink Library Browser** we click on the **Create a new model** (blank page icon on the left of the top bar), and we save this model as Example_1_2. On the **Simulink Library Browser** we select **Sources**, we drag the **Signal Generator** block on the Example_1_2 model window, we click and drag the **State–Space** block from the **Continuous** on **Simulink Library Browser**, and we click and drag the **Scope** block from the **Commonly Used Blocks** on the **Simulink Library Browser**. We also add the **Display** block found under **Sinks** on the **Simulink Library Browser**. We connect these four blocks and the complete block diagram is as shown in Figure B.17.

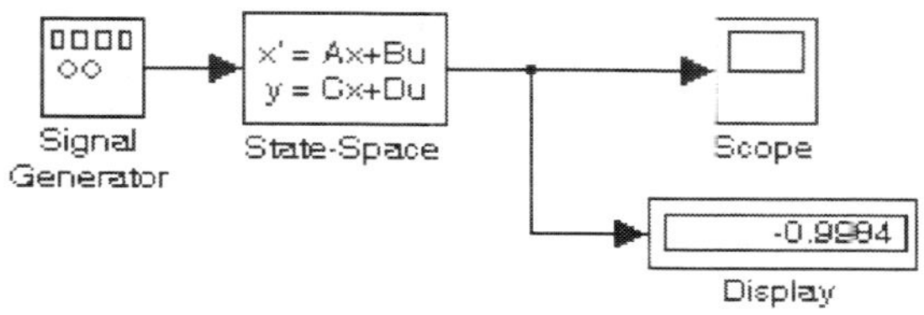

Figure B.17. Block diagram for Example B.2

We now double–click on the **Signal Generator** block and we enter the following in the **Function Block Parameters:**

Wave form: sine

Time (t): Use simulation time

Amplitude: 1

Frequency: 2

Units: Hertz

Next, we double–click on the **state–space** block and we enter the following parameter values in the **Function Block Parameters**:

A: [0 1 0 0; 0 0 1 0; 0 0 0 1; –a0 –a1 –a2 –a3]

B: [0 0 0 1]'

C: [1 0 0 0]

D: [0]

Initial conditions: x0

Absolute tolerance: auto

Now, we switch to the MATLAB Command prompt and we type the following:

```
>> a0=1; a1=0; a2=2; a3=0; x0=[0 0 0 0]';
```

We change the **Simulation Stop time** to 25, and we start the simulation by clicking on the ▶ icon. To see the output waveform, we double click on the **Scope** block, then clicking on the Autoscale icon, we obtain the waveform shown in Figure B.18.

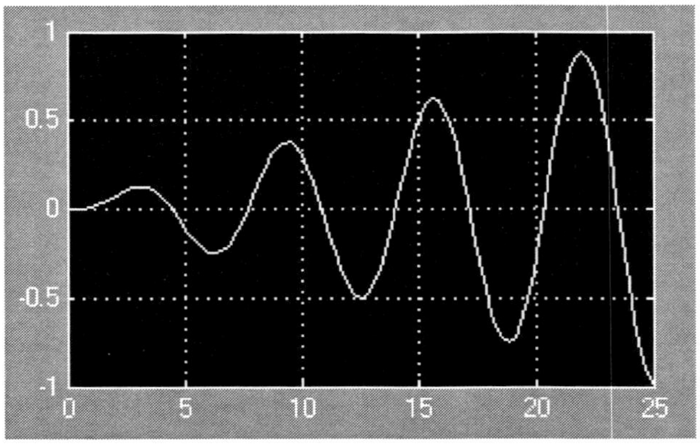

Figure B.18. Waveform for Example B.2

The **Display** block in Figure B.17 shows the value at the end of the simulation stop time.

Examples B.1 and B.2 have clearly illustrated that the State–Space is indeed a powerful block. We could have obtained the solution of Example B.2 using four Integrator blocks by this approach would have been more time consuming.

Example B.3

Using **Algebraic Constraint** blocks found in the **Math Operations** library, **Display** blocks found in the **Sinks** library, and **Gain** blocks found in the **Commonly Used Blocks** library, we will create a model that will produce the simultaneous solution of three equations with three unknowns.

The model will display the values for the unknowns z_1, z_2, and z_3 in the system of the equations

$$\begin{aligned} a_1 z_1 + a_2 z_2 + a_3 z_3 + k_1 &= 0 \\ a_4 z_1 + a_5 z_2 + a_6 z_3 + k_2 &= 0 \\ a_7 z_1 + a_8 z_2 + a_9 z_3 + k_3 &= 0 \end{aligned} \qquad (B.40)$$

The model is shown in Figure B.19.

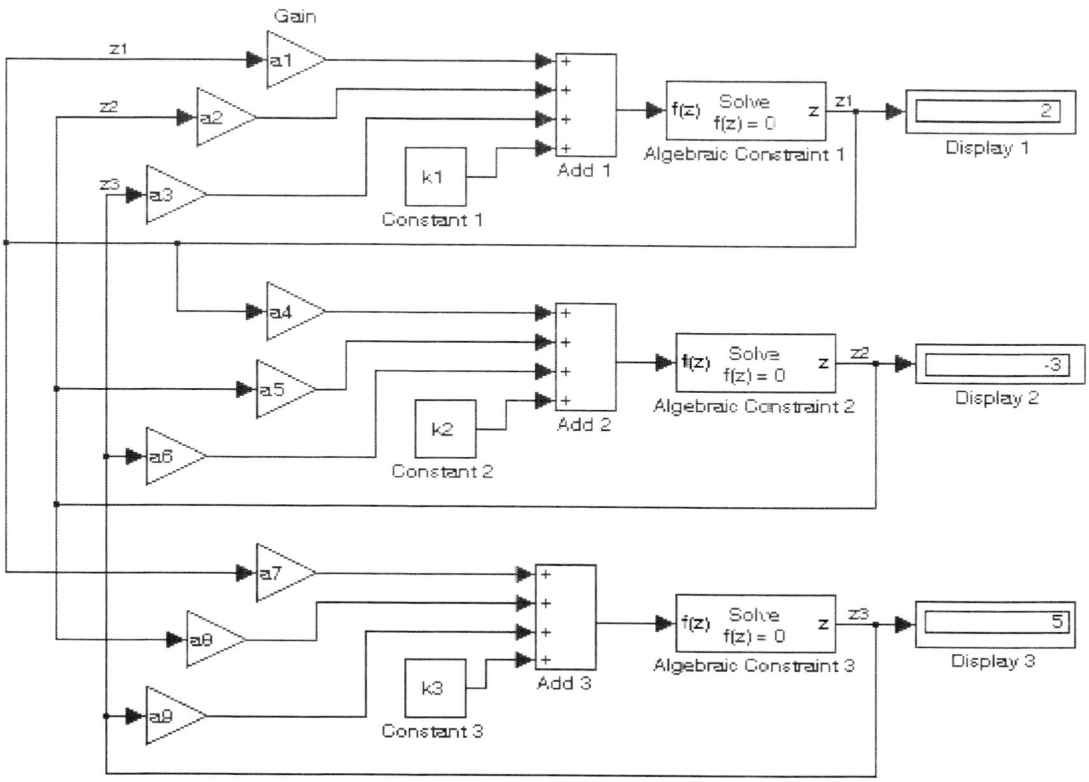

Figure B.19. Model for Example B.3

Next, we go to MATLAB's Command prompt and we enter the following values:

```
a1=2; a2=-3; a3=-1; a4=1; a5=5; a6=4; a7=-6; a8=1; a9=2;...
k1=-8; k2=-7; k3=5;
```

After clicking on the simulation icon, we observe the values of the unknowns as $z_1 = 2$, $z_2 = -3$, and $z_3 = 5$. These values are shown in the Display blocks of Figure B.19.

The **Algebraic Constraint** block constrains the input signal $f(z)$ to zero and outputs an algebraic state z. The block outputs the value necessary to produce a zero at the input. The output must affect the input through some feedback path. This enables us to specify algebraic equations for index 1 differential/algebraic systems (DAEs). By default, the Initial guess parameter is zero. We can improve the efficiency of the algebraic loop solver by providing an Initial guess for the algebraic state z that is close to the solution value.

An outstanding feature in Simulink is the representation of a large model consisting of many blocks and lines, to be shown as a single Subsystem block.* For instance, we can group all blocks and lines in the model of Figure B.19 except the display blocks, we choose **Create Subsystem** from the **Edit** menu, and this model will be shown as in Figure B.20† where in MATLAB's Command prompt we have entered:

```
a1=5; a2=–1; a3=4; a4=11; a5=6; a6=9; a7=–8; a8=4; a9=15;...
k1=14; k2=–6; k3=9;
```

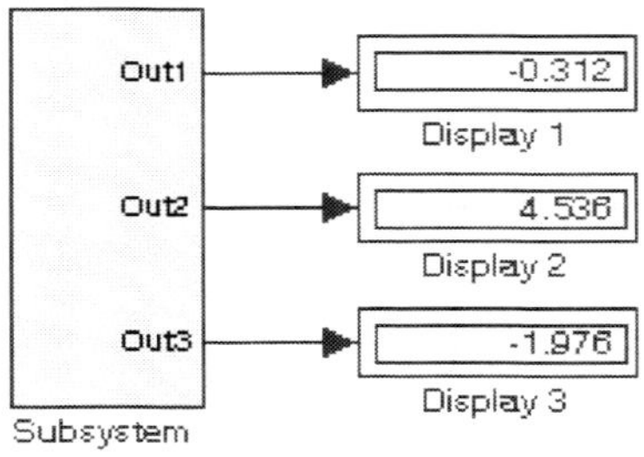

Figure B.20. The model of Figure B.19 represented as a subsystem

The Display blocks in Figure B.20 show the values of z_1, z_2, and z_3 for the values specified in MATLAB's Command prompt.

B.2 Simulink Demos

At this time, the reader with no prior knowledge of Simulink, should be ready to learn Simulink's additional capabilities. It is highly recommended that the reader becomes familiar with the block libraries found in the Simulink Library Browser. Then, the reader can follow the steps delineated in The MathWorks Simulink User's Manual to run the Demo Models beginning with the thermo model. This model can be seen by typing

```
thermo
```

in the MATLAB Command prompt.

* *The Subsystem block is described in detail in Chapter 2, Section 2.1, Page 2–2, Introduction to Simulink with Engineering Applications, ISBN 978–1–934404–09–6.*

† *The contents of the Subsystem block are not lost. We can double–click on the Subsystem block to see its contents. The Subsystem block replaces the inputs and outputs of the model with Inport and Outport blocks. These blocks are described in Section 2.1, Chapter 2, Page 2–2, Introduction to Simulink with Engineering Applications, 978–1–934404–09–6.*

Appendix C

Proportional-Integral-Derivative (PID) Controller

This appendix is a brief introduction to the Proportional–Integral–Derivative Controller, briefly known as PID. The components and the functions of a typical PID are described, and two examples are presented using the Simulink PID Controller Block and the Simulink PID Controller with Approximate Derivative Block.

C.1 Description and Components of a Typical PID

A **proportional–integral–derivative (PID) controller** is a control loop feedback system used in industrial control systems. A PID controller attempts to correct the error between a measured process variable and a desired setpoint by calculating and then outputting a corrective action that can adjust the process accordingly.

The PID controller calculation contains three separate parameters; the **Proportional**, the **Integral**, and **Derivative** values as shown in Figure C.1.

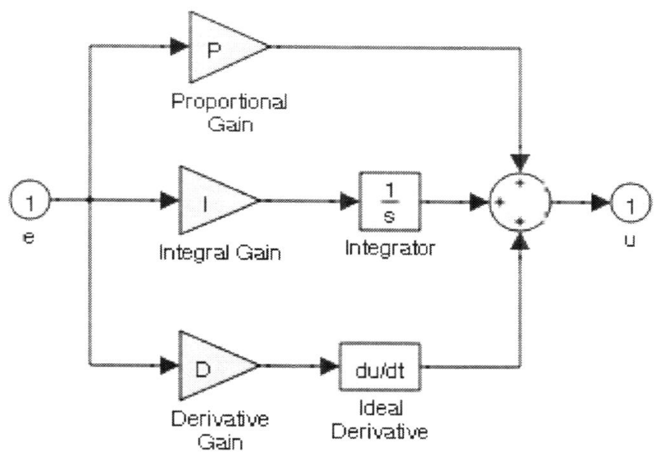

Figure C.1. The components of a typical PID

The Proportional value determines the reaction to the current error, the Integral determines the reaction based on the sum of recent errors, and the Derivative determines the reaction to the rate at which the error has been changing. The weighted sum of these three actions is used to adjust the process via a control element such as the position of a control valve or the power supply of a heating element.

By adjusting the three constants in the PID controller algorithm the PID can provide control action designed for specific process requirements. The response of the controller can be described in terms of the responsiveness of the controller to an error, the degree to which the controller overshoots the setpoint and the degree of system oscillation. The use of the PID algorithm for control does not necessarily result in an optimal configuration for the system.

Some applications may require using only one or two modes to provide the appropriate system control. This is achieved by setting the gain of undesired control outputs to zero. A PID controller will be called a PI, PD, P or I controller in the absence of the respective control actions. PI controllers are particularly common, since derivative action is very sensitive to measurement noise, and the absence of an integral value prevents the system from reaching its target value due to the control action.

A PID controller can be formed with four op amps as shown in Figure C.2.

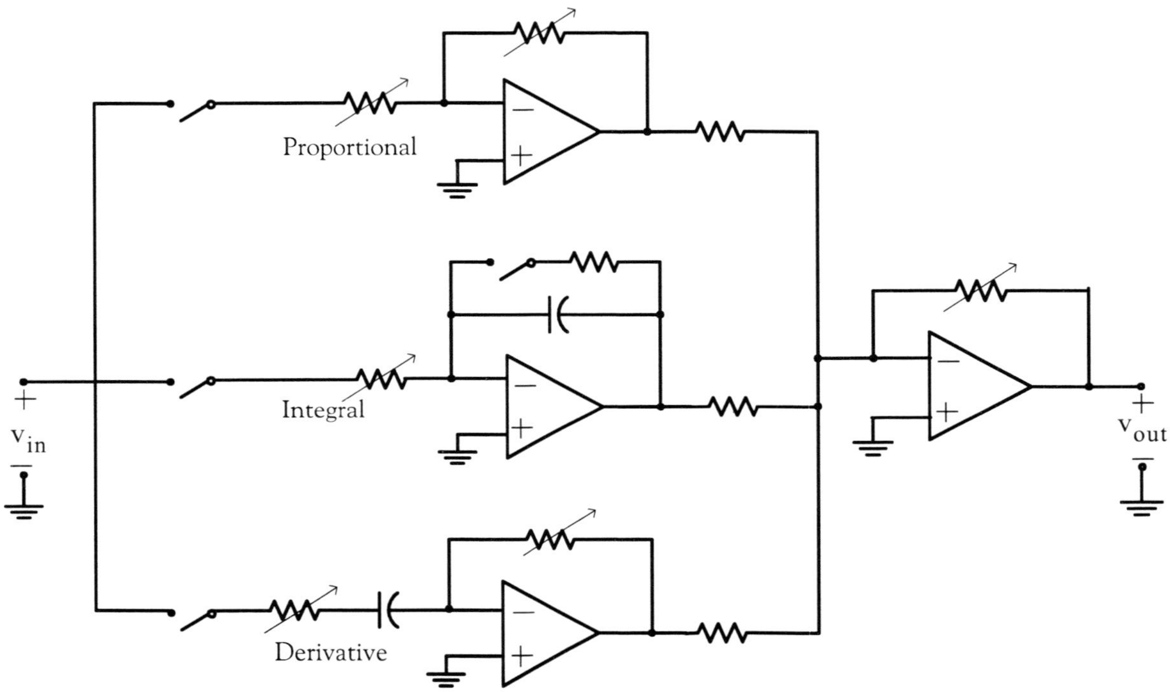

Figure C.2. PID with op amps

C.2 The Simulink PID Blocks

The Simulink PID block implements a PID controller where parameters are entered for the Proportional, Integral and Derivative terms, as illustrated by the example below.

Example C.1

The model in Figure C.3 produces the outputs of a transfer function without and with PID control. The output waveforms are shown in Figure C.4.

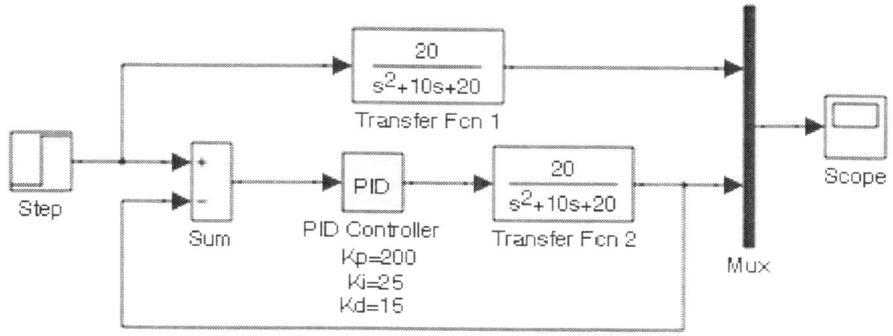

Figure C.3. Model for Example C.1

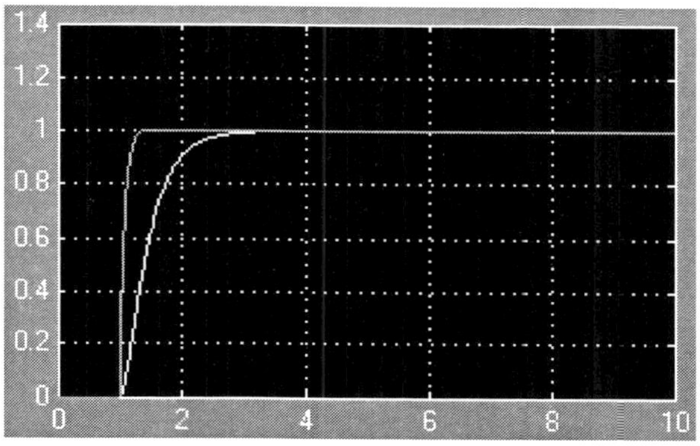

Figure C.4. Output waveforms for the model in Figure C.2 without and with PID control

The Simulink PID controller with approximate derivative calculation contains the functional blocks shown in Figure C.5.

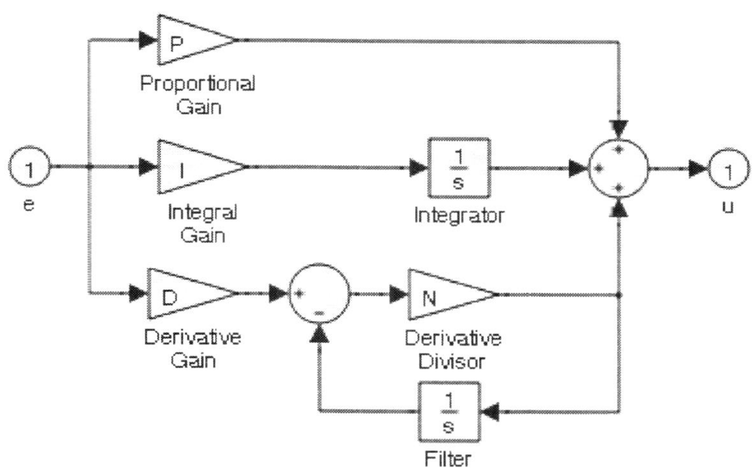

Figure C.5. The components of a typical PID with approximate derivative

Example C.2

The model in Figure C.6 produces the outputs of a transfer function without and with PID control with approximate derivative. The output waveforms are shown in Figure C.7.

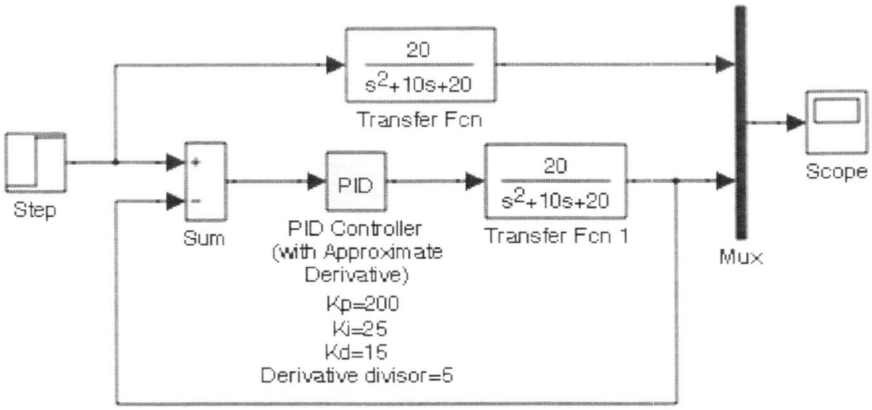

Figure C.6. Model for Example C.2

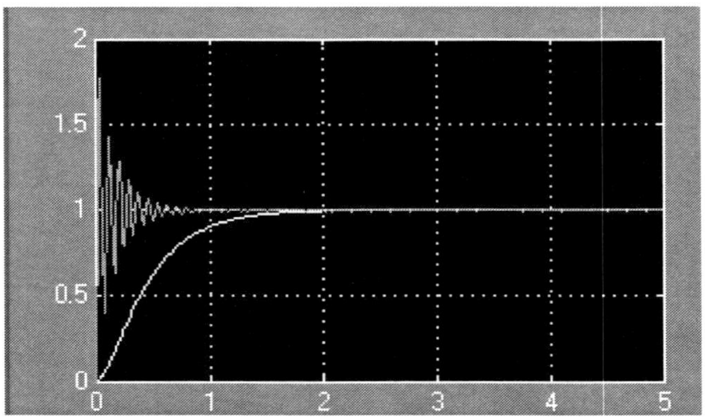

Figure C.7. Output waveforms for the model in Figure C.6 without and with PID control

The reader should try to minimize the oscillations shown in Figure C.7 by assigning other values for the Derivative divisor.

The examples presented in this appendix indicate that the Proportional Gain Kp, Integral Gain Ki, and Derivative Gain Kd are dependent on each another. Generally, Kp and Ki will decrease the rise time and increase the overshoot, and Kd will cause a small change in rise time and decrease the overshoot.

Appendix D

This appendix begins with a simple resistive attenuator which may be used to reduce the amplitude of a signal waveform. We will examine the conditions under which it is possible to ensure no distortion even if shunt capacitances are taken into consideration. Also, we will investigate the types of responses which are obtained with a step voltage input if the circuit is improperly adjusted.

D.1 Uncompensated Attenuator

Let us consider the simple resistor combination of Figure D.1(a).

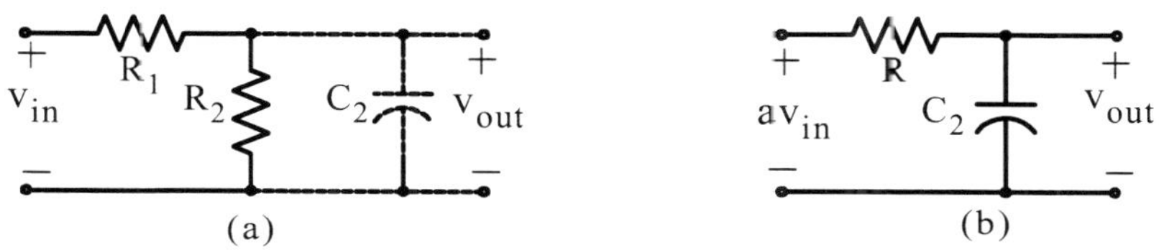

Figure D.1. Actual and equivalent circuit for an uncompensated attenuator

The presence of the stray capacitance C_2 in Figure D.1(a) represents an unavoidable condition since the output of the resistive attenuator in most cases is followed by the input capacitance of a stage of amplification. Using Thevenin's theorem, we can replace the circuit of (a) with that of (b) where R represents the parallel combination of R_1 and R_2 and a in av_{in} is the attenuation factor. We could make both R_1 and R_2 very large so that the input impedance of the attenuator would be large enough to prevent loading down the input signal but his may produce a large rise time which would, in most cases, be unacceptable. This is explained below.

The *rise time*, denoted as t_r, is defined as the time it takes the voltage to rise from 10% to 90% of its final value. It is an indication of how fast a circuit can respond to a discontinuity in voltage. In an RC network the time required for the output voltage v_{out} to reach 10% of its final value is $0.1RC$, and the time required for the output voltage v_{out} to reach 90% of its final value is $2.3RC$. The rise time t_r is the difference between these two values and in terms of the RC time constant τ it is given by

$$t_r = 2.2\tau = 2.2RC = \frac{2.2}{2\pi f_c} = \frac{0.35}{f_c} \qquad \text{(D.1)}$$

where f_c is the 3-dB frequency given by $f_c = 1/2\pi RC$.

It was stated above that we could make the input impedance of the attenuator large enough to prevent loading of the input signal. If, for example, we form an attenuator with $R_1 = R_2 = 1\ M\Omega$ and the input capacitance of a stage amplification is $C_2 = 15\ pF$. With these values the rise time for the circuit of Figure D.1(b) will be

$$t_r = 2.2\tau = 2.2RC = 2.2 \times 0.5 \times 10^6 \times 15 \times 10^{-12} = 16.5\ \mu sec$$

This is a very large rise time for a practical application and thus it is unacceptable.

D.2 Compensated Attenuator

We can make an uncompensated attenuator a *compensated attenuator*, meaning that its attenuation will once again become independent of frequency by placing a capacitor C_1 in parallel with resistor R_1 as shown in Figure D.2(a).

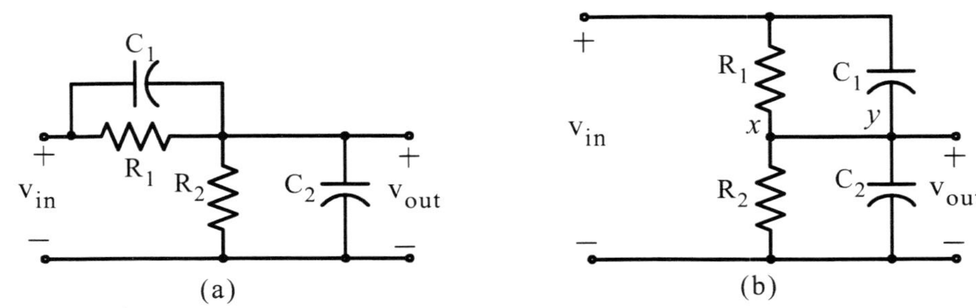

Figure D.2. A compensated attenuator and its equivalent drawn as a bridge

The circuit of Figure D.2(a) is shown as in Figure D.2(b) to form the four arms of a bridge. The bridge will be balanced if

$$R_1C_1 = R_2C_2 \tag{D.2}$$

and under this condition there will be no current flow in the branch connecting point x to point y, so when computing the output voltage v_{out} this branch may be omitted and the output will be equal to av_{in} and independent of frequency. Because relation (D.2) must hold precisely, it is practical to make capacitor C_1 variable and the final adjustment for perfect compensation may be made experimentally with the testing of a square-wave.

Figure D.3 shows the appearances of the output signal when the input is a step voltage of amplitude V if the compensation is not precise. In Figure D.3(a) the circuit is overcompensated and in Figure D.3(b) is undercompensated.

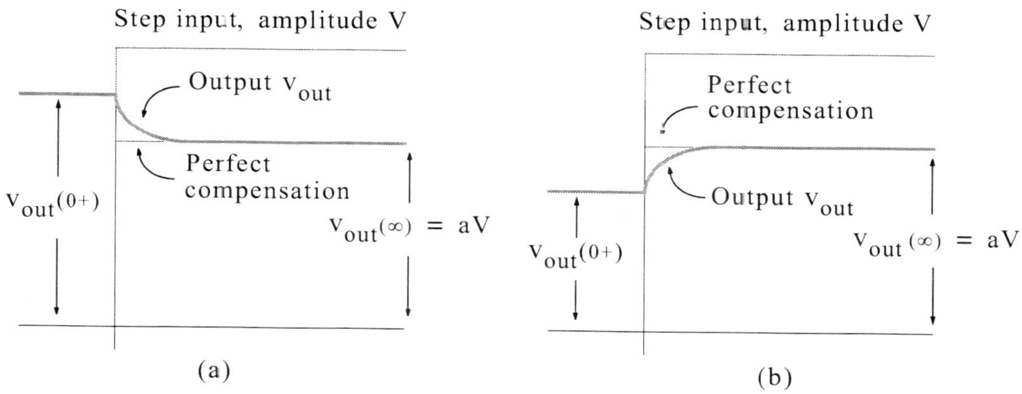

Figure D.3. Overcompensation and undercompensation responses of an attenuator to a step input

NOTES:

Appendix E

Substitution, Reduction, and Miller's Theorems

This appendix discusses three additional theorems that are especially useful in the simplification of circuits containing dependent sources. In our previous studies* we discussed the superposition principle and Thevenin's and Norton's theorems.

E.1 The Substitution Theorem

The *substitution theorem* states that if the voltage across a branch with nodes x and y of a network is v_{xy} and the current through this branch is i_{xy}, a different branch may be substituted in its place in the network provided that the voltage across the substitute branch is also v_{xy} and the current through it is also i_{xy}. The most common use of this theorem is to replace an impedance by a voltage or current source, or vice versa. The substitution theorem can best illustrated with the simple circuit and the substitute branches shown in Figure E.1.

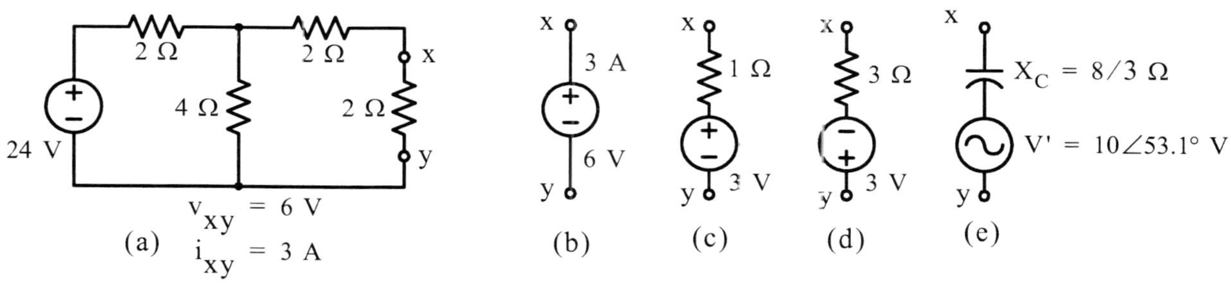

Figure E.1. Illustration of the substitution theorem

For the simple resistor circuit of Figure E.1(a) we find by series-parallel resistance combinations that $v_{xy} = 6\ V$ and $i_{xy} = 3\ A$. According to the substitution theorem, the $2\ \Omega$ resistor across terminals x and y can be replaced with a source with a 6 V source as shown in Figure E.1(b) and the rest of the network will be unaffected. The current in the branch will be 3 A as before.

Other substitutions are possible also. For instance, the substitute branch may consist of a resistance of $1\ \Omega$ and a voltage source of 3 V as shown in Figure E.1(c), or it may consist of a resistance of $3\ \Omega$ and a voltage source of 3 V of opposite polarity as shown in Figure E.1(d).

* *For a detailed discussion of Thevenin's and Norton's theorems and the superposition principle please refer to Circuit Analysis I with MATLAB Applications, ISBN 978–0–9709511–2–0.*

The voltage source in Figure E.1(a) could be DC or AC. If it is an AC source, we can substitute the $2\ \Omega$ resistor across terminals x and y with a branch that has capacitive reactance in series with a source V' as shown in Figure E.1(e). For this branch,

$$V_{xy} = -jX_C I_{xy} + V' \tag{E.1}$$

With the value $X_C = 8/3\ \Omega$, the value of the AC voltage source V' necessary to satisfy the substitution theorem must be such $V_{xy} = 6\ V$ and $I_{xy} = 3\ A$ and by substitution of these values into relation (E.1) we obtain

$$6 = -j(8/3)3 + V'$$

from which

$$V' = 6 + j8 = 10\angle 53.1°$$

as shown in Figure E.1(e).

Proof of the substitution theorem is based on the branch equations of the network. Operation of the whole network of B branches is defined by 2B simultaneous equations. Consider, for example, the bridge network shown in Figure E.2.

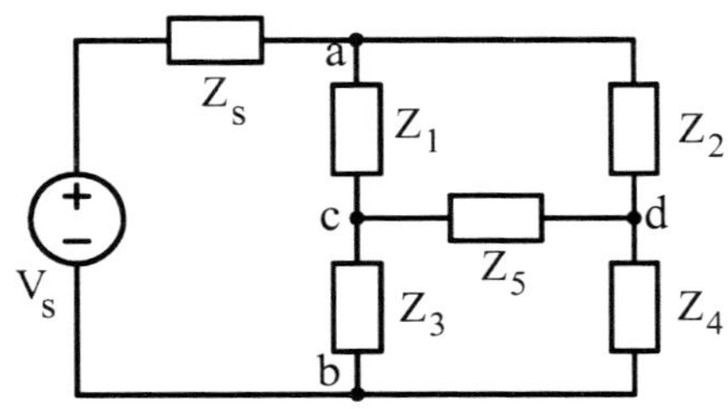

Figure E.2. Network for proof of the substitution theorem

The network of Figure E.2 has six branches and by application of Ohm's law,

$$\begin{array}{lll} V_{ab} = Z_{ab}I_{ab} + V_S & V_{ac} = Z_{ac}I_{ac} & V_{ad} = Z_{ad}I_{ad} \\ V_{cb} = Z_{cb}I_{cb} & V_{cd} = Z_{cd}I_{cd} & V_{db} = Z_{db}I_{db} \end{array} \tag{E.2}$$

If we assume that the source voltage V_S and six impedances are known constants, it becomes apparent that we have 6 equations with 12 unknowns, and thus we need 6 more equations to solve for all voltages and currents in each branch. We can obtain 3 additional equations by application of KCL, and the last three from KVL. These are shown below.

$$\begin{array}{lll} I_{ab} + I_{ac} + I_{ad} = 0 & I_{ca} + I_{cd} + I_{cb} = 0 & I_{da} + I_{dc} + I_{db} = 0 \\ V_{ab} + V_{ac} + V_{cb} = 0 & V_{ac} + V_{cd} + V_{da} = 0 & V_{cb} + V_{bd} + V_{dc} = 0 \end{array} \tag{E.3}$$

The 12 equations of (E.2) and (E.3) can now be solved simultaneously for the 12 unknowns.

From the foregoing discussion we conclude that it is possible to write as many equations as there are unknowns. Thus, if the number of branches is B, the number of unknowns is 2B. Let us now suppose that the branch for which substitution is to be made contains an impedance and a source and the equation for this branch is

$$V_{ab} = Z_{ab}I_{ab} + V_S \tag{E.4}$$

that is, the voltage across this branch is V_{ab} and the current is I_{ab}. Other branch voltages in the network are V_{bc}, V_{cd}, and so on, and currents are I_{bc}, I_{cd}, and so on, and all these voltages and currents simultaneously satisfy the 2B equations of the network, including equation (E.4).

Next, let us assume a substitute branch that has some different value of impedance Z'_{ab}, and some different voltage source V'_S, but with Z'_{ab} and V'_S so related that the same values of V_{ab} and I_{ab} that satisfied equation (E.4) will also satisfy the new branch equation

$$V_{ab} = Z'_{ab}I_{ab} + V'_S \tag{E.5}$$

When this substitution is made, operation of the altered network will be defined by 2B equations, as before; (2B – 1) of the equations will remain the same but equation (E.5) will have replaced equation (E.4). However, since equation (E.5) is satisfied by the same values of voltage and current that satisfied equation (E.4), it follows that the whole group of simultaneous equations will be satisfied by the same voltages and currents as before. Voltages and currents of all branches of the network will therefore be the same as before substitution, which was to be proved.

The substitution theorem is especially useful in the simplification of circuits containing dependent sources. Consider for example the current–controlled voltage source and the voltage–controlled current source shown in Figure E.3.

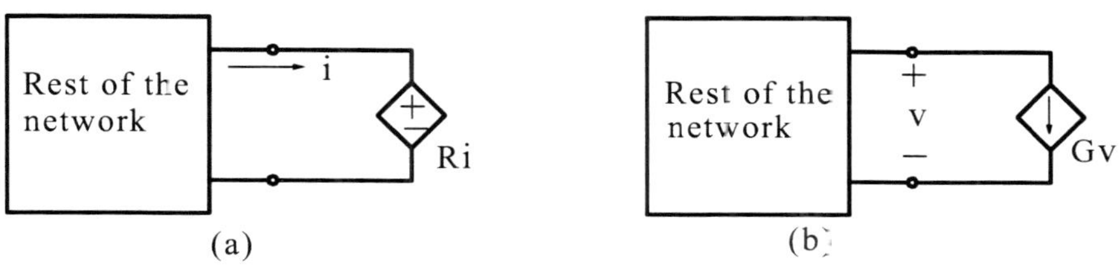

Figure E.3. Application of the substitution theorem in current– and voltage–controlled dependent sources.

The network of Figure E.3(a) can be any network having the terminal current i, and Ri is the voltage of a current–controlled voltage source. Figure E.3(b) shows the conditions under which the dual form can be applied. In this case the voltage–controlled current source Gv, which is connected across the voltage v. As another example, let us consider the equivalent circuit of some amplifier shown in Figure E.4(a).

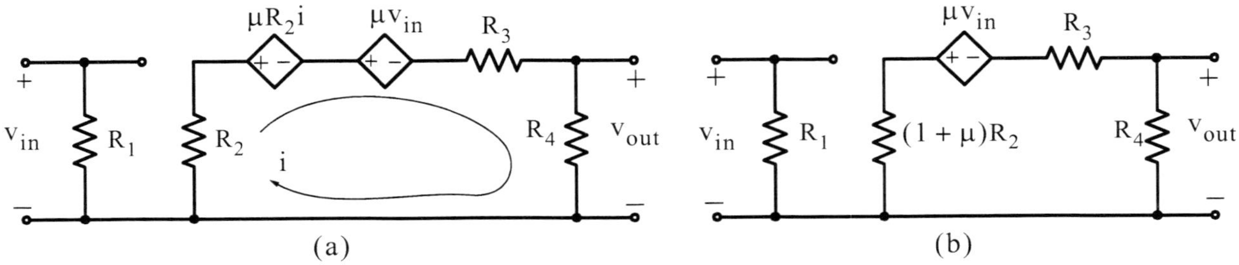

Figure E.4. Application of the substitution theorem in an amplifier equivalent circuit.

In Figure E.4(a), the currents and voltages are not changed if the dependent source $\mu R_2 i$ is replaced by a resistance μR_2, and the simplified equivalent circuit is as shown in Figure E.4(b).

A special case of the substitution theorem is the *source absorption theorem** and has two dual forms, the *voltage source absorption* and the *current source absorption* theorems.

The voltage source absorption theorem is illustrated in Figure E.5.

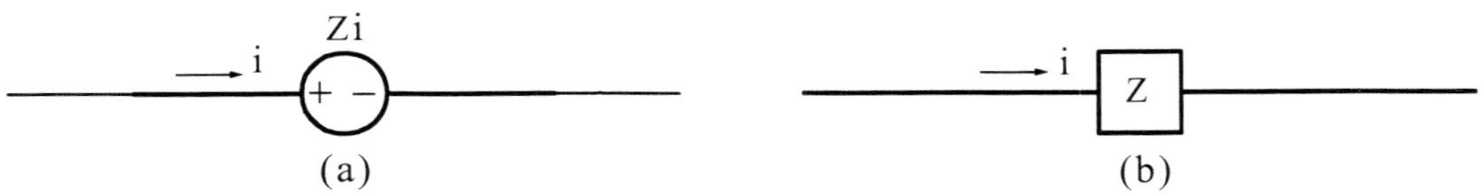

Figure E.5. Illustration of the voltage source absorption theorem

As shown in Figure E.5, the voltage source absorption states that if in a branch of the network in which the current through it is i, the voltage source Zi can be replaced by an impedance Z.

Example E.1

Use the voltage source absorption theorem to simplify the branch AB shown in Figure E.6.

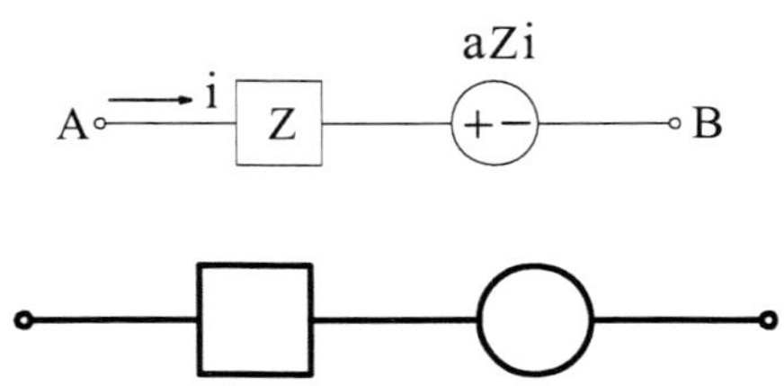

Figure E.6. Branch for Example E.1 to be simplified

* *This theorem is analogous to the absorption theorem in Boolean algebra which states that* $A = (\overline{A} + B) = AB$

Solution:

Application of the voltage source absorption theorem yields the simplified branch shown in Figure E.7(c) below.

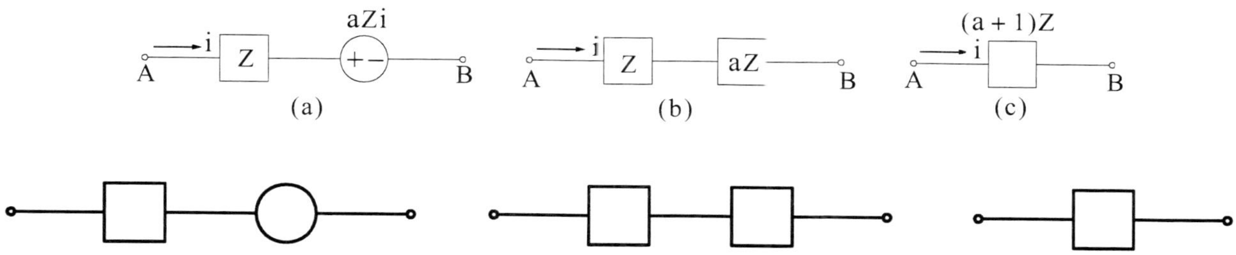

Figure E.7. Steps for simplification of the branch for Example E.1

The current source absorption theorem is illustrated in Figure E.8.

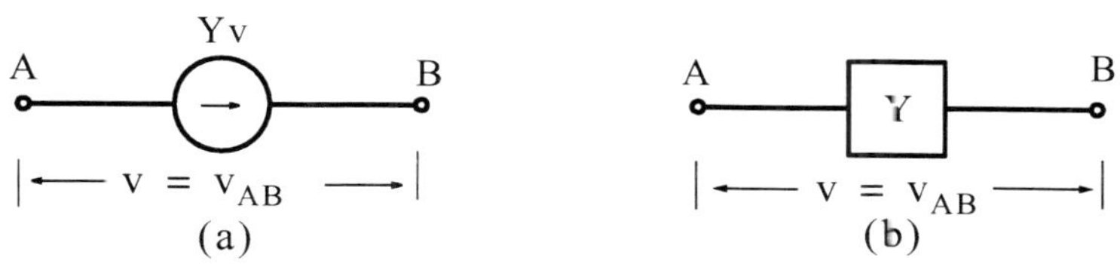

Figure E.8. Illustration of the current source absorption theorem

As shown in Figure E.8, the current source absorption states that if in a branch of the network in which the voltage across it is v, the current source Yv can be replaced by an admittance Y.

Example E.2

Use the current source absorption theorem to simplify the parallel combination shown in Figure E.9.

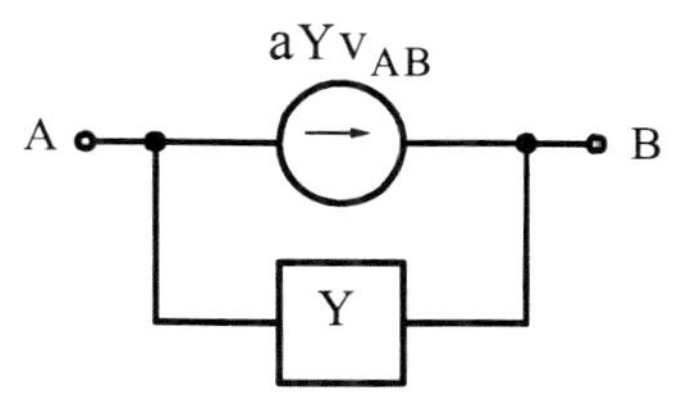

Figure E.9. Branch for Example E.2 to be simplified

Solution:

Application of the current source absorption theorem yields the simplified branch shown in Figure E.10(c) below.

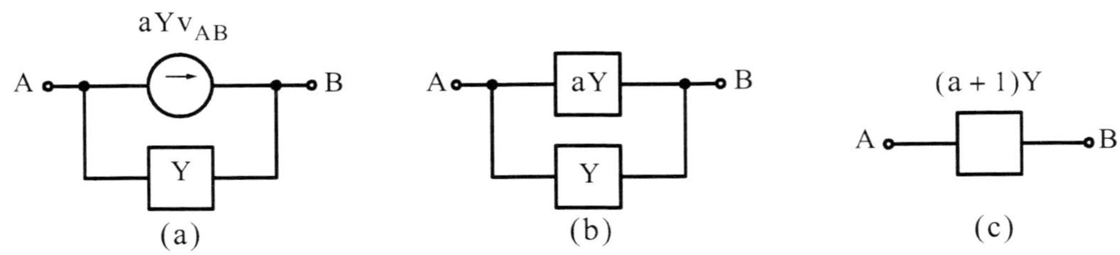

Figure E.10. Steps for simplification of the parallel combination of Example E.2

E.2 The Reduction Theorem

The *reduction theorem* is essentially an extension of the substitution theorem. This theorem applies to network configurations such as those shown in Figure E.11.

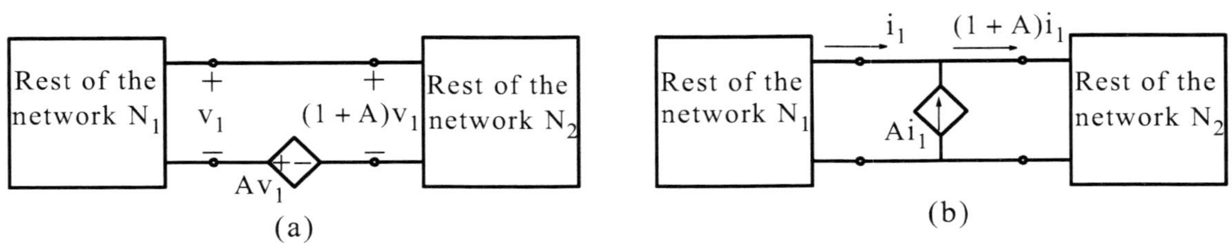

Figure E.11. Network configurations that can be simplified by the reduction theorem

The reduction theorem applies to the network configurations shown in Figure E.11. The network N_1 in Figure E.11(a) is any linear two–terminal network having the terminal voltage v_1, and N_2 is any other linear two–terminal network. The constant A of the voltage–controlled voltage source Av_1 is any real number, positive or negative. The two networks and the voltage source are connected in series so that the voltages v_1 and Av_1 are additive with respect to the loop that they form. The theorem states that all currents in N_1 and N_2 remain unchanged if the source Av_1 is replaced with a short circuit and if:

1. Each resistance, inductance, reciprocal capacitance, and voltage source in Network N_1 is multiplied by $1 + A$

   or

2. Each resistance, inductance, reciprocal capacitance, and voltage source in Network N_2 is divided by $1 + A$.

Proof of the reduction theorem is based on the form of the loop equations. Obviously, if all resistances, inductances, reciprocal capacitances, and voltage sources are multiplied by the same arbitrary factor, then every term in the set of loop equations is multiplied by the same factor, and all the loop currents remain unchanged. Thus, if all the parameters and voltage sources in N_1 are multiplied by $1 + A$ as specified in Part 1 of the theorem, and if Av_1 is replaced with a short circuit, then the voltage applied to N_2 is still $(1 + A)v_1$, and all currents in N_1 and N_2 remain unchanged. All voltages in N_1 are multiplied by the factor $1 + A$, but the voltages in N_2 are not changed. Any current sources in N_1 and N_2 are left unchanged in the reduction process.

By the same reasoning process, after the transformation described above is completed, all resistances, inductances, reciprocal capacitances, and voltage sources in the composite network $N_1 - N_2$ can be divided by the factor $1 + A$ without changing any of the currents in $N_1 - N_2$. Thus, Part 2 of the theorem is proved.

When all currents and voltages in the network are sinusoidal and of the same frequency, the constant of the controlled source may be expressed as a complex number; the reduction theorem holds in this case also.

The dual form of the reduction theorem is applicable to the network configuration shown in Figure E.11(b). It states that all voltages in the network of Figure E.11(b) remain unchanged if the current source Ai_1 is replaced with an open circuit and if:

1. All conductances, capacitances, reciprocal inductances, and current sources in N_1 are multiplied by $1 + A$

   or

2. All conductances, capacitances, reciprocal inductances, and current sources in N_2 are divided by $1 + A$

Proof of the dual form of the theorem follows from nodal equations and is dual to the proof outlined above. In applying the reduction theorem it is essential that the two networks N_1 and N_2 be properly identified before application of this theorem. The two necessary requirements are:

a. The terminal voltage (or current) of the network identified as Network N_1 must be the controlling quantity for the source to be eliminated; and
b. No current may enter or leave either network through any terminal other than the two terminals by which the networks and controlled source are joined in series (or parallel). It should be noted that connections joining separate parts of a network carry no current and can therefore be ignored in respect to item b.

A useful application of the reduction theorem is provided by the *cascode amplifier* shown in Figure E.12.

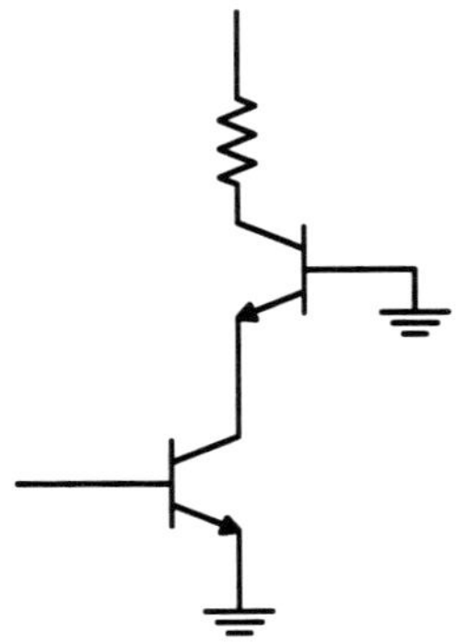

Figure E.12. Typical cascode amplifier with BJTs

Figure E.12 shows a partial circuit of a cascode amplifier where a common–base bipolar transistor sits on top of a common–emitter bipolar transistor. Of course, a cascode amplifier may use MOSFETs of CMOS devices. To illustrate an application of the reduction theorem, let us consider the the incremental model of some cascode amplifier shown in Figure E.13.

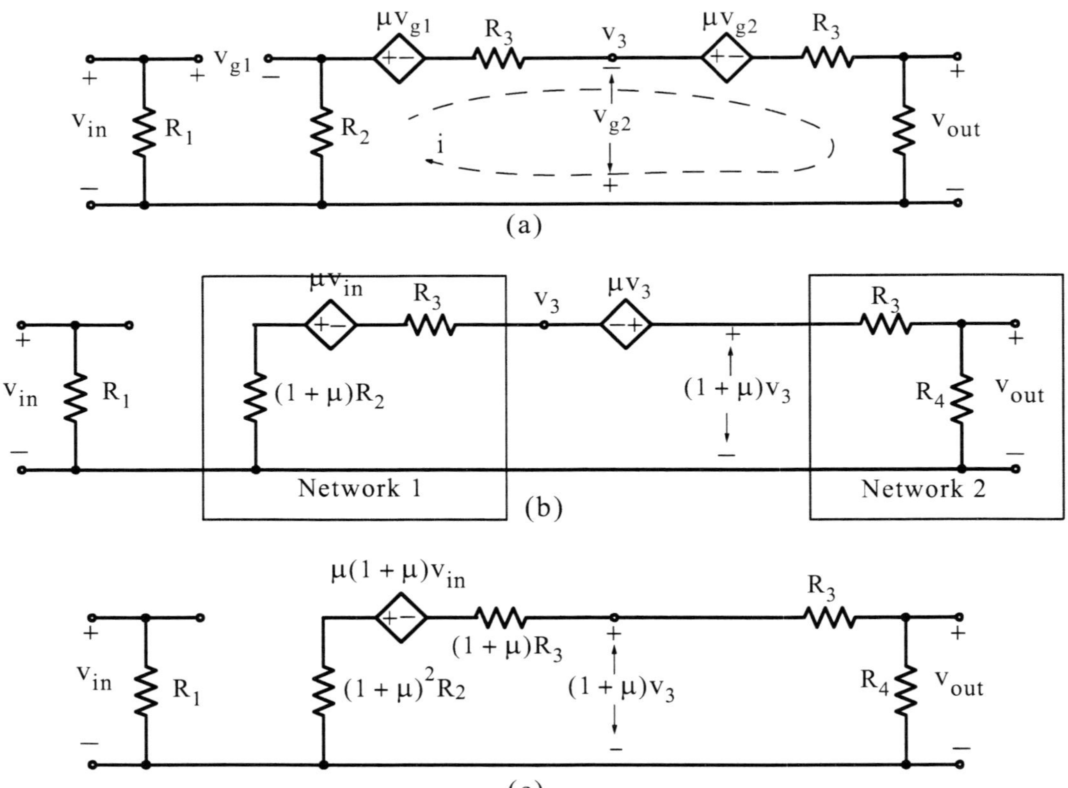

Figure E.13. Incremental model of a cascode amplifier and simplification by application of the reduction theorem

From the circuit of Figure E.13(a) we observe that $v_{g1} = v_{in} + R_2 i$; hence the circuit can be simplified by applying the substitution theorem to obtain the equivalent circuit shown in Figure E.13(b). This circuit has been separated into two networks, in N_1 and N_2, connected in series with the controlled source μv_3, in preparation for the reduction theorem. All the conditions of the reduction theorem are satisfied by this circuit; hence applying Part 1 of the voltage–source form of the theorem yields the reduced network shown in Figure E.13(c).

The emitter follower shown in Figure E.14(a) provides another illustration of the use of the reduction theorem in circuit analysis.

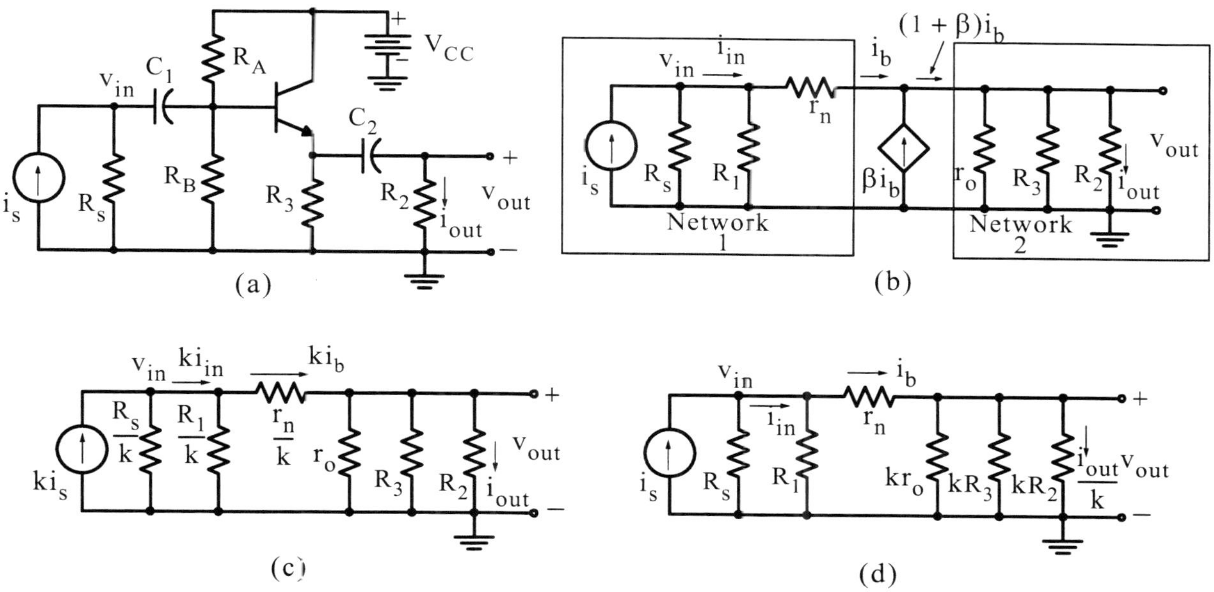

Figure E.14. Simplification of the emitter follower by application of the reduction theorem

As we know from Chapter 3, the principal properties of the emitter follower are that it provides a relatively low output resistance, it is capable of providing a relatively high input impedance, and it provides current and power amplification but no voltage amplification. Figure E.14(b) shows an incremental model for the circuit that is valid in the range of frequencies where the coupling capacitors act as short circuits and where the performance of the transistor is independent of frequency. The hybrid representation for the transistor is used with $u = 0$. The resistance R_1 represents the parallel combination of resistors R_A and R_B, and the circuit is separated into two networks in preparation for the reduction theorem.

Application of Part 1 of the current–source form of the theorem yields the reduced circuit shown in Figure E.14(c). For simplicity, the factor $1 + \beta$ is denoted as k in Figure E.14(c); in practice k is essentially equal β. Dividing the resistances R_S, R_1, and r_n, by k is equivalent to multiplying the corresponding conductances by this factor. Thus, application of the reduction theorem yields the simplified circuit of Figure E.14(c).

The current amplification of the circuit is accounted for by the fact that the input signal current is multiplied by the factor k in the reduced circuit. It is clear from Figure E.14(c) that the forward voltage transmittance of the emitter follower is less than unity, and for large values of β the output resistance of the emitter follower is very small.

If Part 2 of the theorem is applied instead of Part 1, the circuit of Figure E.14(d) is obtained; this circuit is convenient for studying relations at the input terminals since the input voltage and current are unmodified. For large values of β the output resistance of the emitter follower is very large.

E.3 Miller's Theorem

Miller's theorem states that under certain conditions, to be established below, the networks of Figures E.15(a) and E.15(b) are equivalent.

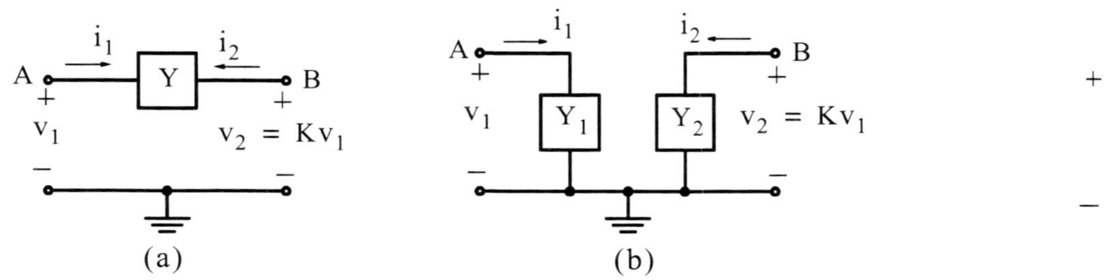

Figure E.15. Illustration for Miller's theorem

In the networks of Figure E.15, the constant K denotes the voltage gain from Node A to Node B, that is,

$$K = v_2/v_1 \tag{E.6}$$

or

$$v_2 = Kv_1 \tag{E.7}$$

At Node A of Figure E.15(a), we observe that

$$i_1 = Y(v_A - v_B) = Y(v_1 - v_2) = Y(v_1 - Kv_1) = Yv_1(1-K) \tag{E.8}$$

and at Node A of Figure E.15(b), we obtain,

$$i_1 = Y_1 v_1 \tag{E.9}$$

For equivalence, relations (E.8) and (E.9) must be equal. Thus,

$$Y_1 = Y(1-K) \tag{E.10}$$

At Node B of Figure E.15(a), we observe that

$$i_2 = -i_1 = -Y(v_1 - v_2) = Y(v_2 - v_1) = Y(v_2 - v_2/K) = Yv_2(1 - 1/K) \quad \text{(E.11)}$$

and at Node B of Figure E.15(b), we obtain,

$$i_2 = Y_2 v_2 \quad \text{(E.12)}$$

For equivalence, relations (E.11) and (E.12) must be equal. Thus,

$$Y_2 = Y(1 - 1/K) \quad \text{(E.13)}$$

We should remember that for the equivalent circuit of Figure E.15(b) the value of the gain K, once established, it cannot be changed. Thus, while Miller's theorem can be applied to find the input impedance, it should not be used to determine the output impedance of an amplifier since when determining the output impedance the input signal voltage v_S is normally set to zero volts.

Miller's dual theorem states that under certain conditions, to be established below, the networks of Figures E.16(a) and E.16(b) are equivalent.

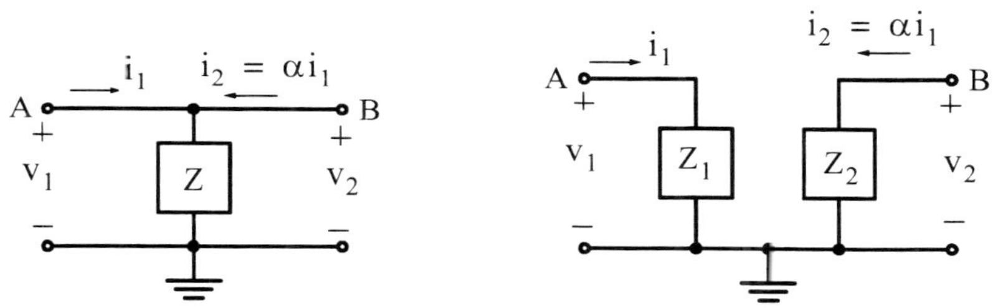

Figure E.16. Illustration for Miller's dual theorem

In the networks of Figure E.16, the constant α denotes the current gain from Node A to Node B, that is,

$$\alpha = i_2/i_1 \quad \text{(E.14)}$$

or

$$i_2 = \alpha i_1 \quad \text{(E.15)}$$

At Node A of Figure E.16(a), we observe that

$$v_1 = Z(i_1 + i_2) = Z(i_1 + \alpha i_1) = Zi_1(1 + \alpha) \quad \text{(E.16)}$$

and at Node A of Figure E.16(b), we obtain,

$$v_1 = Z_1 i_1 \quad \text{(E.17)}$$

For equivalence, relations (E.16) and (E.17) must be equal. Thus,

$$Z_1 = Z(1+\alpha) \tag{E.18}$$

At Node B of Figure E.16(a), we observe that

$$v_2 = v_1 = Zi_1(1+\alpha) \tag{E.19}$$

and at Node B of Figure E.16(b), we obtain,

$$v_2 = Z_2 i_2 = Z_2 \alpha i_1 \tag{E.20}$$

For equivalence, relations (E.19) and (E.20) must be equal. Thus,

$$Z_2 = Z(1+1/\alpha) \tag{E.21}$$

Example E.3

For the circuit of Figure E.17, it is given that $R_1 = 100R_2$. Use Miller's theorem to replace this circuit with an equivalent circuit where resistor R_1 is being replaced with another resistor in parallel with resistor R_S. Make any reasonable assumptions.

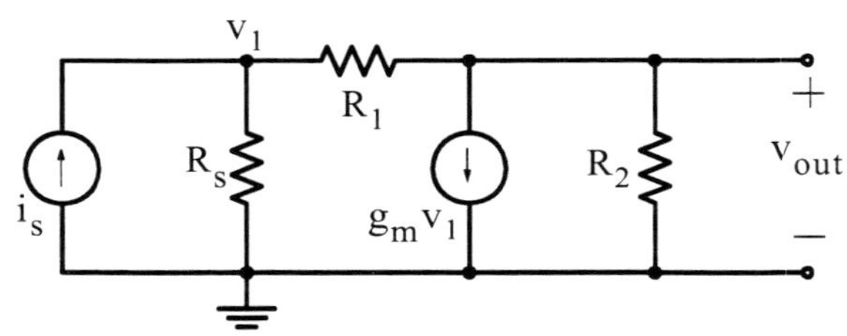

Figure E.17. Circuit for Example E.3

Solution:

Since $R_1 \gg R_2$ it is reasonable to assume that practically all the current of the current source flows through resistor R_2 and thus the output voltage is

$$v_{out} = -g_m v_1 R_2 = K v_1$$

from which

$$K = -g_m R_2$$

Figure E.18 shows the equivalent circuit after application of Miller's theorem where using relation (E.10) we obtain

$$R'_1 = R_1/(1-K) = R_1/(1+g_m R_2)$$

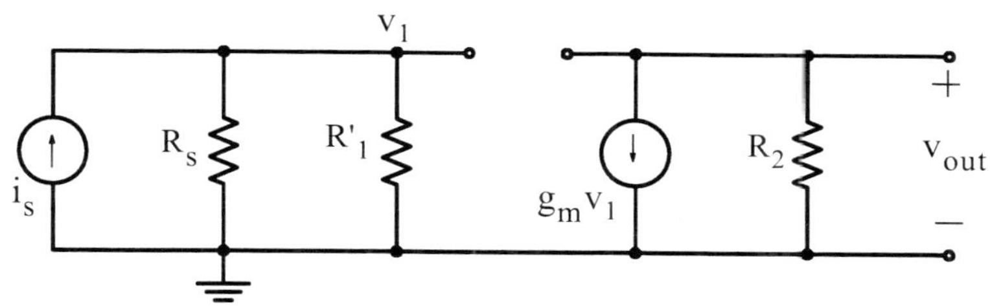

Figure E.18. Miller equivalent circuit for the circuit of Figure E.17

Being familiar with Miller's theorem helps us in better understanding the so–called *Miller effect*. This effect is very pronounced in amplifier circuits where the input capacitance caused by negative feedback from output to the input causes a significant increase in capacitance. As an illustration, let us consider the amplifier of the circuit in Figure E.19.

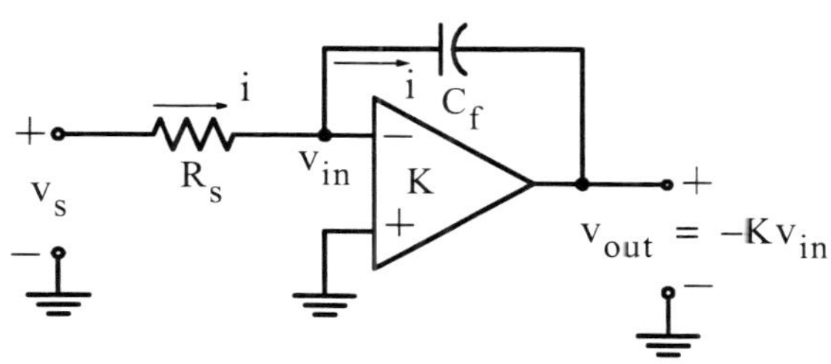

Figure E.19. Amplifier to illustrate the Miller effect

Assuming a sinusoidal input, the current i is

$$i = \frac{v_{in}-v_{out}}{1/j\omega C_f} = \frac{v_{in}+Kv_{in}}{1/j\omega C_f} = \frac{v_{in}(1+K)}{1/j\omega C_f}$$

and the input admittance is

$$Y_{in} = \frac{i}{v_{in}} = j\omega C_f(1+K) \tag{E.22}$$

Relation (E.22) indicates that the input voltage v_{in} sees an increase in capacitance by a factor of $C_f(1+K)$.

NOTES:

References and Suggestions for Further Study

A. Manufacturers Data Books and Sheets

1. *Fairchild TTL Data Book*, Fairchild Semiconductor, http://www.fairchildsemi.com/
2. *Texas Instruments TTL Data Book*, Texas Instruments, http://www.ti.com/
3. *MECL Device Data*, Motorola, http://www.motorola.com/home/
4. *Transistor Manual*, General Electric, http://www.ge.com/en/
5. *Rectifiers and Zener Diodes Data Book*, Motorola, ISBN 9–9991969–7–7
6. *SCR Manual*, General Electric, ISBN 0–1379676–3–2
7. *LM 555 Timer*, National Semiconductor, http://www.national.com/ds/LM/LM555.pdf

B. Texts and Handbooks

1. *Reference Data for Engineers Radio, Electronics, Computer & Communications*, ISBN 0–7506729–1–9
2. *Electronic Circuits*, E.J. Angelo, Jr. Mc-Graw-Hill Book Co.
3. *Microelectronic Circuits*, Oxford University Press, ISBN 0–19-511663–1
4. *Standard Handbook of Electronic Engineering*, ISBN 0–0713842–1–9
5. *Mathematics for Business, Science, and Technology*, ISBN 978–1–934404–01–0
6. *Numerical Analysis Using* MATLAB® *and Excel*, ISBN 978–1–934404–03–4
7. *Circuit Analysis I with* MATLAB® *Applications*, ISBN 978–0–9709511–2–0
8. *Circuit Analysis II with* MATLAB® *Applications*, ISBN 978–0–9709511–5–1
9. *Signals and systems with* MATLAB® Computing and *Simulink*® Modeling, ISBN 978–1–934404–11–9
10. *Digital Circuit Analysis and Design with Simulink Applications and Introduction to CPLDs and FPGAs*, ISBN 978–1–934404–05–8
11. *Introduction to Simulink® with Engineering Applications*, ISBN 978–1–934404–09–6
12. *Introduction to Stateflow® with Applications*, ISBN 978–1–934404–07–2

C. The following publications by The MathWorks, are highly recommended for further study. They are available from The MathWorks, 3 Apple Hill Drive, Natick, MA, 01760, www.mathworks.com.

1. *Getting Started with MATLAB®*
2. *Using MATLAB®*
3. *Using MATLAB® Graphics*
4. *Using Simulink®*
5. *Sim Power Systems for Use with Simulink®*
6. *Fixed–Point Toolbox*
7. *Simulink® Fixed–Point*
8. *Real–Time Workshop*
9. *Signal Processing Toolbox*
10. *Getting Started with Signal Processing Blockset*
11. *Signal Processing Blockset*
12. *Control System Toolbox*
13. *Stateflow®*

Index